Handbook of Research on Human Social Interaction in the Age of Mobile Devices

Xiaoge Xu
Botswana International University of Science and Technology, Botswana

A volume in the Advances in Human and Social
Aspects of Technology (AHSAT) Book Series

Information Science REFERENCE
An Imprint of IGI Global

Published in the United States of America by
    Information Science Reference (an imprint of IGI Global)
    701 E. Chocolate Avenue
    Hershey PA, USA 17033
    Tel: 717-533-8845
    Fax: 717-533-8661
    E mail: cust@igi global.com
    Web site: http://www.igi-global.com

                            Library of Congress Cataloging-in-Publication Data

Names: Xu, Xiaoge, editor.
Title: Handbook of research on human social interaction in the age of mobile
   devices / Xiaoge Xu, editor.
Description: Hershey : Information Science Reference, 2016. | Includes
   bibliographical references and index.
Identifiers: LCCN 2016010970| ISBN 9781522504696 (hardcover) | ISBN
   9781522504702 (ebook)
Subjects: LCSH: Social interaction. | Cell phone systems--Social aspects. |
   Information technology--Social aspects.
Classification: LCC HM1111 .H363 2016 | DDC 302--dc23 LC record available at https://lccn.loc.gov/2016010970

This book is published in the IGI Global book series Advances in Human and Social Aspects of Technology (AHSAT) (ISSN: 2328-1316; eISSN: 2328-1324)

British Cataloguing in Publication Data
A Cataloguing in Publication record for this book is available from the British Library.

For electronic access to this publication, please contact: eresources@igi-global.com.

Advances in Human and Social Aspects of Technology (AHSAT) Book Series

Ashish Dwivedi
The University of Hull, UK

ISSN: 2328-1316
EISSN: 2328-1324

MISSION

In recent years, the societal impact of technology has been noted as we become increasingly more connected and are presented with more digital tools and devices. With the popularity of digital devices such as cell phones and tablets, it is crucial to consider the implications of our digital dependence and the presence of technology in our everyday lives.

The **Advances in Human and Social Aspects of Technology (AHSAT) Book Series** seeks to explore the ways in which society and human beings have been affected by technology and how the technological revolution has changed the way we conduct our lives as well as our behavior. The AHSAT book series aims to publish the most cutting-edge research on human behavior and interaction with technology and the ways in which the digital age is changing society.

COVERAGE

- Gender and Technology
- Human Development and Technology
- Human Rights and Digitization
- Technology Adoption
- Cultural Influence of ICTs
- Technology Dependence
- Cyber Behavior
- Computer-Mediated Communication
- End-User Computing
- Technoself

IGI Global is currently accepting manuscripts for publication within this series. To submit a proposal for a volume in this series, please contact our Acquisition Editors at Acquisitions@igi-global.com or visit: http://www.igi-global.com/publish/.

Titles in this Series

For a list of additional titles in this series, please visit: www.igi-global.com

Defining Identity and the Changing Scope of Culture in the Digital Age
Alison Novak (Rowan University, USA) and Imaani Jamillah El-Burki (Lehigh University, USA)
Information Science Reference • copyright 2016 • 316pp • H/C (ISBN: 9781522502128) • US $185.00 (our price)

Gender Considerations in Online Consumption Behavior and Internet Use
Rebecca English (Queensland University of Technology, Australia) and Raechel Johns (University of Canberra, Australia)
Information Science Reference • copyright 2016 • 297pp • H/C (ISBN: 9781522500100) • US $165.00 (our price)

Analyzing Digital Discourse and Human Behavior in Modern Virtual Environments
Bobbe Gaines Baggio (American University, USA)
Information Science Reference • copyright 2016 • 320pp • H/C (ISBN: 9781466698994) • US $175.00 (our price)

Overcoming Gender Inequalities through Technology Integration
Joseph Wilson (University of Maiduguri, Nigeria) and Nuhu Diraso Gapsiso (University of Maiduguri, Nigeria)
Information Science Reference • copyright 2016 • 324pp • H/C (ISBN: 9781466697737) • US $185.00 (our price)

Cultural, Behavioral, and Social Considerations in Electronic Collaboration
Ayse Kok (Bogazici University, Turkey) and Hyunkyung Lee (Yonsei University, South Korea)
Business Science Reference • copyright 2016 • 374pp • H/C (ISBN: 9781466695566) • US $205.00 (our price)

Handbook of Research on Cultural and Economic Impacts of the Information Society
P.E. Thomas (Bharathiar University, India) M. Srihari (Bharathiar University, India) and Sandeep Kaur (Bharathiar University, India)
Information Science Reference • copyright 2015 • 618pp • H/C (ISBN: 9781466685987) • US $325.00 (our price)

Human Behavior, Psychology, and Social Interaction in the Digital Era
Anabela Mesquita (CICE – ISCAP/Polytechnic of Porto, Portugal & Algoritmi Centre, Minho University, Portugal) and Chia-Wen Tsai (Ming Chuan University, Taiwan)
Information Science Reference • copyright 2015 • 372pp • H/C (ISBN: 9781466684508) • US $200.00 (our price)

Rethinking Machine Ethics in the Age of Ubiquitous Technology
Jeffrey White (Korean Advanced Institute of Science and Technology, KAIST, South Korea) and Rick Searle (IEET, USA)
Information Science Reference • copyright 2015 • 331pp • H/C (ISBN: 9781466685925) • US $205.00 (our price)

www.igi-global.com

701 E. Chocolate Ave., Hershey, PA 17033
Order online at www.igi-global.com or call 717-533-8845 x100
To place a standing order for titles released in this series, contact: cust@igi-global.com
Mon-Fri 8:00 am - 5:00 pm (est) or fax 24 hours a day 717-533-8661

List of Reviewers

Seyedali Ahrari, *IPSAS, Malaysia*

Saleh Al-Shehri, *King Khalid University, Saudi Arabia*

Yulia An, *The Ilmenau University of Technology, Germany*

Philip J. Auter, *University of Louisiana – Lafayette, USA*

Florie Brizel, *BrizelMedia, USA*

Maria Alexandra Cunha, *Fundação Getulio Vargas Escola de Administração de Empresas, Brazil*

Yuchan Gao, *University of Nottingham – Ningbo, China*

Kenneth E. Harvey, *KIMEP University, Kazakhstan*

Nygmet Ibadildin, *KIMEP University, Kazakhstan*

Beatriz Barreto Brasileiro Lanza, *Universidade Federal do Paraná, Brazil & Center for Technology in Government at SUNY, USA*

Hans Karl Meyer, *Ohio University, USA*

Zeinab Zare Mohzzabieh, *UPM, Malaysia*

Jamilah Bt. Othman, *IPSAS, Malaysia*

Fernando de la Cruz Paragas, *University of the Philippines – Diliman, Philippines*

Bahaman Abu Samah, *IPSAS, Malaysia*

Burton Speakman, *Ohio University, USA*

Samantha Stevens, *University of Louisiana – Lafayette, USA*

Mei Wu, *University of Macau, Macau*

Janice Hua Xu, *Holy Family University, USA*

Qi Yao, *University of Macau, Macau*

Table of Contents

Preface xvii

Chapter 1
Learners and Mobile: A Reflexivity 1
    Fernando de la Cruz Paragas, University of the Philippines Diliman, The Philippines

Chapter 2
Language Learners and Mobile Technology: How They Interact? 22
    Saleh Al-Shehri, King Khalid University, Saudi Arabia

Chapter 3
Lifestyle Diglossia and Mobile: Ethnography of Multilingual Interaction 49
    Mukul Saxena, The University of Nottingham – Ningbo, China

Chapter 4
Educators and Mobile: Challenges and Trends 61
    Kenneth E. Harvey, KIMEP University, Kazakhstan
    Philip J. Auter, University of Louisiana – Lafayette, USA
    Samantha Stevens, University of Louisiana – Lafayette, USA

Chapter 5
Learners-Mobile Interaction: African Substance and Style 96
    Kushatha Kelebeng, Botswana International University of Science and Technology,
        Botswana
    Rebaone Mlalazi, Botswana International University of Science and Technology, Botswana
    Keorapetse Gosekwang, Botswana International University of Science and Technology,
        Botswana
    Pendukeni Phalaagae, Botswana International University of Science and Technology,
        Botswana
    Tebogo Mangwa, Botswana International University of Science and Technology, Botswana
    Tebogo Kebonang, Botswana International University of Science and Technology, Botswana
    Thotobolo Morapedi, Botswana International University of Science and Technology,
        Botswana

Chapter 6
Government and Mobile: Examining the Role of SMS 111
  *Beatriz Barreto Brasileiro Lanza, Companhia de Tecnologia da Informação e Comunicação
    do Paraná, Brazil & Center for Technology in Government at SUNY, USA*
  *Maria Alexandra Cunha, Fundação Getulio Vargas Escola de Administração de Empresas,
    Brazil*

Chapter 7
Government and Mobile: A Gear Change? 133
  Wendy Li, The University of Melbourne, Australia

Chapter 8
Voters and Mobile: Impact on Democratic Revolution 150
  Oarabile Sebubi, Botswana International University of Science and Technology, Botswana

Chapter 9
Local News and Mobile: Major Tipping Points 171
  Kenneth E. Harvey, KIMEP University, Kazakhstan

Chapter 10
Journalists and Mobile: Melding Social Media and Social Capital 200
  Hans Karl Meyer, Ohio University, USA
  Burton Speakman, Ohio University, USA

Chapter 11
Marketing and Mobile: Increasing Integration 220
  Kenneth E. Harvey, KIMEP University, Kazakhstan
  Yulia An, The Ilmenau University of Technology, Germany

Chapter 12
Branding and Mobile: Revolutionizing Strategies 248
  Yulia An, The Ilmenau University of Technology, Germany
  Kenneth E. Harvey, KIMEP University, Kazakhstan

Chapter 13
Public Relations and Mobile: Becoming Dialogic 284
  Yulia An, The Ilmenau University of Technology, Germany
  Kenneth E. Harvey, KIMEP University, Kazakhstan

Chapter 14
Business and Mobile: Rapid Restructure Required 312
  Nygmet Ibadildin, KIMEP University, Kazakhstan
  Kenneth E. Harvey, KIMEP University, Kazakhstan

Chapter 15
Advertising and Mobile: More than a Platform Shift 351
    Kenneth E. Harvey, KIMEP University, Kazakhstan
    Philip J. Auter, University of Louisiana – Lafayette, USA

Chapter 16
Examining the Role of WeChat in Advertising.. 386
    Qi Yao, University of Macau, China
    Mei Wu, University of Macau, China

Chapter 17
Reviewing Gratification Effects in Mobile Gaming... 406
    Yuchan Gao, The University of Nottingham, China

Chapter 18
Youth and Mobile: An Investigation of Socialization.. 429
    Zeinab Zaremohzzabieh, UPM, Malaysia
    Seyedali Ahrari, IPSAS, Malaysia
    Bahaman Abu Samah, IPSAS, Malaysia
    Jamilah Bt. Othman, IPSAS, Malaysia

Chapter 19
Left-Behind Children and Mobile: A Critical Discourse Analysis... 452
    Janice Hua Xu, Holy Family University, USA

Compilation of References 472

About the Contributors 543

Index 546

Detailed Table of Contents

Preface..xvii

Chapter 1
Learners and Mobile: A Reflexivity 1
Fernando de la Cruz Paragas, University of the Philippines Diliman, The Philippines

Studies about mobile phones, the learning process, and educational institutions have grown in recent years though research has mostly focused in the United States and in specific groups. This research contributes to the literature by looking at three educational levels in Singapore and by taking a two-pronged approach to the relationship among these three variables. It answers the following reflexivity: How do students learn to use mobile phone functions? How do they use mobile phones for learning functions? This chapter considers learning as the reflexive process where attitude and aptitude are acquired and shared for curricular and extra-curricular activities. Findings indicate two themes: how students 1) develop the skills to use the expanding array of mobile phone technologies, and integrate these in their daily life and 2) use mobile phones in school and in their schoolwork. Data for the study came from focus interviews with 36 informants who were selected through maximum variation sampling according to their age, educational level, and household income.

Chapter 2
Language Learners and Mobile Technology: How They Interact?....................................... 22
Saleh Al-Shehri, King Khalid University, Saudi Arabia

In order to understand the influence of mobile social media vs. formal learning platforms on creating effective student-student and teacher-student communication channels and linguistic outputs, this study was conducted. Using a qualitative approach, a number of 30 English language university teachers were interviewed. Evidence from their mobile and non-mobile interaction with their students was analyzed to support data from the interviews. The study evaluates the potential of both formal and informal communication mediums to maintain student-centered language learning experience. The study concludes that teachers still need to be aware of the potential of mobile technology and social media for language learning, and that there was a tendency among some teachers to implement formal technology tools for their teaching.

Chapter 3

Lifestyle Diglossia and Mobile: Ethnography of Multilingual Interaction 49

Mukul Saxena, The University of Nottingham – Ningbo, China

This chapter proposes to chart the development of understanding of literacy as a practice, which is now digital in nature and globally distributed, therefore digital literacy practices require new lenses and ways of research explorations. The notion of literacy has shifted from being autonomous to ideological during the last three decades. It is not seen anymore as a single unified competence, but as changing from place to place and varying in different social-cultural contexts. Despite the fact that there are different writing systems that are used in different ways in different contexts, the differences between them are no longer seen as primarily technical. The differences that do exist between literacies are seen as being due to differences in cultural practices, values and ideologies. As a consequence, the methodological shift towards ethnographic research on literacy has arisen from a fundamental change in thinking about the nature of literacy and the development of the "new literacy studies". Ethnographic approaches to literacy, such as those developed by Heath, Street, Barton, and others are based on the everyday uses of written language(s) by specific groups and subgroups in a specific locality. According to Graff, these approaches to literacy provide "both new and better cases for study, opportunity for explanations, and approaches to literacy's variable historical meaning and contribution". The ethnographic research on literacies in multilingual contexts further contributed to the development of 'new literacy studies' (NLS). However, with the development of mobile devices during the last decade and the allied software industries, digital literacy communication has become synonymous with globalisation and a divergence between the literacy practices in 'regulated spaces' and 'unregulated spaces' particularly among the youth is becoming a marked feature of inter-personal communication. Sebba defines 'unregulated spaces' as places where the prescriptiveness of standardisation and monolingualism has not yet reached, or where it holds no power and practices may deviate from the prescribed norms. Such spaces open up opportunities for identity construction and group definition.

Chapter 4

Educators and Mobile: Challenges and Trends 61

Kenneth E. Harvey, KIMEP University, Kazakhstan
Philip J. Auter, University of Louisiana – Lafayette, USA
Samantha Stevens, University of Louisiana – Lafayette, USA

More practical and experiential education was the demand of executives recently surveyed about how universities could better meet employers' needs. As an alternative, with enhanced Web and mobile technologies, executives are seeing the opportunity to provide employees access to essential education in the workplace. Global e-learning is expected to top $107 billion in 2015, and U.S. corporations each year are now spending $1,169 per employee on training. Bersin says organizations are facing not a lack of employees but a lack of key skills among employees, and that is driving the trend. Harvard's Clayton Christensen, famous for his theories on disruptive technologies, suggests that even Harvard could be in jeopardy if it does not respond to these trends. This chapter explores different strategies and technologies that can help meet these demands, and includes a case study of a university plan that makes distance learning more faculty-friendly, student-accessible, and cost-effective.

Chapter 5

Learners-Mobile Interaction: African Substance and Style 96

Kushatha Kelebeng, Botswana International University of Science and Technology, Botswana

Rebaone Mlalazi, Botswana International University of Science and Technology, Botswana

Keorapetse Gosekwang, Botswana International University of Science and Technology, Botswana

Pendukeni Phalaagae, Botswana International University of Science and Technology, Botswana

Tebogo Mangwa, Botswana International University of Science and Technology, Botswana

Tebogo Kebonang, Botswana International University of Science and Technology, Botswana

Thotobolo Morapedi, Botswana International University of Science and Technology, Botswana

The increasing demand for education in African continent has motivated more research in mobile learning to describe, explain and predict changes and chances in leveraging mobile technology to learn anytime and anywhere. Mobile technologies provide learning beyond the classroom walls where learners can have access to resources according to their needs without involving teachers. In developing countries like Africa, mobile learning is still gaining its ground and remains under-studied. With technological advancements and high usage of mobile phones in Africa, the educational horizon has broadened. Mobile learning plays a major role in delivering education in Africa. The rapid use of mobile devices in mobile learning projects in Africa confirm that mobile learning is transforming the traditional way of learning, teaching and delivery of education by integrating social media and advanced mobile technologies in education. After a critical look at the current status as well as the latest development in mobile learning, this chapter presents a road map of m-learning to guide future action and thinking of learners, teachers and institutions in Africa. It also tracks trends and their impact, key theories and key findings of mobile learning. The current chapter ends with a critical analysis of technologies and key theories that can contribute to the sustenance of mobile learning in the African continent.

Chapter 6

Government and Mobile: Examining the Role of SMS 111

Beatriz Barreto Brasileiro Lanza, Companhia de Tecnologia da Informação e Comunicação do Paraná, Brazil & Center for Technology in Government at SUNY, USA

Maria Alexandra Cunha, Fundação Getulio Vargas Escola de Administração de Empresas, Brazil

This chapter aims to study a fifteen-year project of Mobile Government (mGov) in the state of Paraná, Brazil. The precocious experience in Brazil in mGov and the number of cell phones used in the country are not arguments for these to be used in providing mass public services. Mason's Historical Method was used to get to know the project, raising available information and retrieving happenings and results. To identify events, relevant actors, their roles, aspects in collaboration, setting and density of the relations among these actors, Social Network Analysis was used. We conclude it is important to formalize the project, but that will not guarantee its continuity in the long run. Characterizing phases minimizes the importance given to the change of governors after new elections in large corporative projects. The networks forming favors the dissemination of IT knowledge among actors. This research shows little evidence that SMS still have long life.

Chapter 7
Government and Mobile: A Gear Change?.. 133
    Wendy Li, The University of Melbourne, Australia

This chapter presents a comprehensive review of what has been examined in the past and what needs to be explored in the future concerning mobile government (m-government). As an emerging branch of mobile services, m-government intends to help governments to better serve for the public, the business and non-government organizations with the assistance of mobile technologies. Although m-government originated from electronic government (e-government), it is not just a simple extension in respect to technological developments but a transformation from e-government to m-government. What matters most and is worthy of being further enhanced in this revolutionary process is the improvement of government's mobility instead of up-to-date technologies. That is to say, the shift from e-government to m-government is a game change instead of a gear change.

Chapter 8
Voters and Mobile: Impact on Democratic Revolution 150
    Oarabile Sebubi, Botswana International University of Science and Technology, Botswana

The objective of democracy is to allow people the freedom to vote at ease and according to their individual choices. Mobile Election has high potentials of transforming and improving efficiency of the current electoral process, thereby enabling convenient and ubiquitous elections, hence, revolutionizing the institution of democracy. With so much potential though, its adoption is extremely low worldwide because the barriers to adoption are extremely high as mobile election is still lacking in addressing the critical and sensitive requirements of the electoral process worldwide. This chapter explores the potential impact of mobile elections to democratic establishments such as politics and voter participation. It then adopts a comparative analysis approach in exploring the barriers to adoption and possible solutions. Lastly, it provides recommendations for future research in the areas of voter trafficking, network and data security, inter-operability issue, digital/democratic divide issue, and voter satisfaction.

Chapter 9
Local News and Mobile: Major Tipping Points.. 171
    Kenneth E. Harvey, KIMEP University, Kazakhstan

The first economic tipping for traditional media hit in 2006 when newspaper revenues began to collapse. Leonard Downie Jr., vice president at large for the Washington Post, notes how most newspaper executives knew that the time would come when the Internet would begin stealing newspaper readers and revenues, but they didn't expect it to happen so quickly. That is the nature of a tipping point. In one year, according to Downie, they were at an economic high and adding personnel to their news staff, and the next year they were laying people off. How quickly the economic downturn occurred took them and many other newspapers by surprise. In 2005 newspapers achieved $49.4 billion in advertising revenues. By 2014 revenues had fallen about 60% to $19.9 billion. There is evidence that such a tipping point is about to hit local TV stations. These two tipping points could lead to the end of quality local news coverage. This chapter explores how newspapers and TV stations can survive and flourish in the Digital Age.

Chapter 10

Journalists and Mobile: Melding Social Media and Social Capital 200

Hans Karl Meyer, Ohio University, USA
Burton Speakman, Ohio University, USA

Despite the high profile of many social media activist efforts, such as the Arab Spring, researchers are still questioning whether these mostly online campaigns can lead to real world impact. Journalists are also asking themselves what role they fill as they watch, comment on, and cover the deluge of activism on Twitter, Facebook, and other sites. Through a comprehensive review of the literature on social capital and its intersection with the Internet and social media, this chapter suggests that social media can lead to social capital, but journalists provide the key ingredient to lead to lasting social change. Literature on the goals and aims of journalism, coupled with a review of its vital role in a community, point to the need for the context and verification that journalism provides in order for social media to transform into social capital.

Chapter 11

Marketing and Mobile: Increasing Integration 220

Kenneth E. Harvey, KIMEP University, Kazakhstan
Yulia An, The Ilmenau University of Technology, Germany

Integrated Marketing Communications (IMC) was gaining popularity even before the World Wide Web. But with the explosion of online marketing, it is nearly impossible to avoid integration. This chapter will explore how digital and mobile marketing are dictating such changes in marketing. An overview of modern mobile marketing practices and trends builds upon a comparison of the traditional marketing mix (4 Ps) with a more consumer-oriented concept of 4 Cs. As products are being replaced with consumers, promotion with communication, place with convenience, and price with cost, the rise of mobile devices and the mobile Internet brings up a multitude of new concepts into traditional marketing theory and practice. Supplemented with best industry examples, the review of emerging mobile marketing concepts moves from broad online strategies, like inbound marketing, to very mobile-specific trends, like the Internet of Things.

Chapter 12

Branding and Mobile: Revolutionizing Strategies 248

Yulia An, The Ilmenau University of Technology, Germany
Kenneth E. Harvey, KIMEP University, Kazakhstan

An overview of mobile's impact on branding strategy expands from discussion of chronological ages of branding to mobile's effects on the brand management function. The latter are structured around four steps of the brand management process, including planning and brand positioning, implementing branding strategies, measuring and analyzing performance of a brand, and sustaining brand equity over time and across geographic markets. With a focus on cross-platform branding and establishment of seamless mobile experience, the chapter explores various aspects of mobile's influence on branding. A strategic view on the mobile-first approach that permeates all aspects of the branding process from logo design to brand equity measurement is accompanied with real-life applications. Best practices illustrate changing perspectives in branding and include examples of how new technologies like iBeacon and augmented reality are used by leading brands. Examples are drawn from recent academic and industry studies in business and management with links to classic brand management literature.

Chapter 13
Public Relations and Mobile: Becoming Dialogic 284
Yulia An, The Ilmenau University of Technology, Germany
Kenneth E. Harvey, KIMEP University, Kazakhstan

The chapter discusses impact of mobile technologies on public relations practice and scholarship by tracking the historical development of public relations as a distinct field, thus mapping a structural framework for the further discussion of new trends and areas of academic and industry research. Moving from functional to co-creational perspective, public relations enters the conversation age in which the nature of mobile technologies forces practitioners to adopt the dialogic approach to build trust and nurture relationships. Existing theoretical frameworks are supplemented with industry examples within the field of crisis communication, corporate social responsibility and customer and employee relations. An overview of some of the latest trends in social media research, public segmentation and video marketing applied to co-creational perspective of public relations organizes new trends around a more fundamental paradigm shift. This structure places industry practices within broader academic research, providing both tactical and strategic views on public relations in the mobile age.

Chapter 14
Business and Mobile: Rapid Restructure Required 312
Nygmet Ibadildin, KIMEP University, Kazakhstan
Kenneth E. Harvey, KIMEP University, Kazakhstan

This chapter will explore the peculiarities of business applications of mobile technologies, including a short history and a review of the current state of affairs, major trends likely to cause further change over the coming years, key theories and models to help understand and predict these changes, and future directions of research that may provide deeper scientific insight. M-commerce has many aspects from design and usability of the devices to monetization issues of mobile applications. M-enterprise is about drastic changes in internal and external communications and efficiency in the work of each business unit. M-industry reviews the impact of mobile technologies on traditional industries and the development of entirely new industries. M-style is how our everyday lives are changing in behavior, choices and preferences. After reading this chapter you will be able to differentiate m-business in many important areas: why is it important, where it is going, what is the value to consumers.

Chapter 15
Advertising and Mobile: More than a Platform Shift 351
Kenneth E. Harvey, KIMEP University, Kazakhstan
Philip J. Auter, University of Louisiana – Lafayette, USA

There is no field that has experienced a more positive financial impact from mobile technology than advertising. This is evident by billions of dollars in traditional media fleeing to online media, and increasingly to mobile. Yet, it is difficult to distinguish mobile totally from other online advertising approaches. Mobile is certainly not diverging from the other platforms, but rather driving some of the strongest advertising trends. Because the trends of all online channels overlap with mobile, it will be difficult to address mobile without addressing all – then clarifying and exploring how mobile is driving and will continue to drive those trends.

Chapter 16
Examining the Role of WeChat in Advertising..386

Qi Yao, University of Macau, China
Mei Wu, University of Macau, China

WeChat, the most popular social networking service mobile app in China, enables users to contact friends with text, audio, video contents as well as get to know new people within a range of a certain distance. Its latest version has gone beyond the pure social function and opened a window into marketing. This study applies activity theory as a framework to explore new features of WeChat as a new platform for advertising. Then the authors qualitatively analyze the implications that marketers adapted and appropriated WeChat to engage with their customers and in turn how these implications modified the ways of advertising. The significance of this study is that it applied activity theory as an attempt to complement the theoretical pillars of communication study. Activity theory focuses on the activity itself rather than the interaction between WeChat and the users. Second, activity theory reveals a 2-way process which emphasizes both on how advertisers adapt WeChat based on everyday practice; and how WeChat modifies the activities which the advertisers engage in.

Chapter 17
Reviewing Gratification Effects in Mobile Gaming..406

Yuchan Gao, The University of Nottingham, China

This chapter aims to critically review previous studies on the gratification effects mobile gaming provides in a variety of aspects including play experience, education and social networking. Previous studies reveal that mobile gaming gratifies a new play experience in casual, flexible and ubiquitous nature. Additionally, mobile gaming generates widely accepted gratifications in technology-based education, improving learning and teaching performance in education in general and specific courses. Meanwhile, mobile gaming demonstrates great gratification effect in quality, ability and physical training. Furthermore, scholars have widely researched on the gratification effect of mobile gaming to promote social networking with the applications of socially interactive technologies. Finally, this chapter delivers a forecast about the future development of mobile gaming into HR recruitment and management.

Chapter 18
Youth and Mobile: An Investigation of Socialization...429

Zeinab Zaremohzzabieh, UPM, Malaysia
Seyedali Ahrari, IPSAS, Malaysia
Bahaman Abu Samah, IPSAS, Malaysia
Jamilah Bt. Othman, IPSAS, Malaysia

While the rapid growth in studies on the effects of mobile phones has deepened our understanding of the role mobile phones play in the socialization process of youth, further work is required in reviewing the growing influence of mobile phones for continuing socialization. The objective of this paper therefore is to assess literature from a range of selected studies and in doing so, highlight the role of mobile phones in contributing to youth socialization. This state-of-the-art review demonstrates that mobile phones are a powerful socializing tool that can lead to plentiful consequences. It will show that the influence of mobile phones can be beneficial. It explores the harmful effects of mobile phones. Finally, this chapter will incorporate previous advancements in research to inform forthcoming research and identify new concepts, themes and theories to support or improve the role of mobile phones in increasing the socialization skills of youth.

Chapter 19
Left-Behind Children and Mobile: A Critical Discourse Analysis..452
    Janice Hua Xu, Holy Family University, USA

Through critical analysis of selected news stories from sina.com from 2010 to 2015 about "left-behind children" in China, the chapter examines media discourse on relationships between migrant families and communication technology. The author finds that the role of cell phones in their lives are portrayed in the following narratives: 1) Cell phones are highly valuable for connecting family members living apart; 2) Cell phones are used as a problem-solver in charity giving and rural development projects; 3) Cell phones can bring unexpected risks; 4) Cell phones could harbor or unleash evil—associated with increasing cases of crimes victimizing left-behind children and juvenile delinquency. The author discusses how institutional goals of social agencies, corporations, educators and law enforcement contribute to the polarity of cell-phone-related discourses, which reflect the societal anxieties over unsupervised access to technology by adolescents, as well as the cultural and political implications of empowering the "have-nots" of digital divide.

Compilation of References 472

About the Contributors 543

Index...546

Preface

Simply put, human-mobile interaction refers to a process in which human beings interact with mobile devices for different purposes, resulting in different experiences. Differing from human-computer interaction, human-mobile interaction is unique in many ways. For instance, it is ubiquitous since we can interact with mobile devices anywhere regardless of where we happen to be. It is wireless as we do not have to be tied by the cable connection like what we do with desktop or even lap computers. It is personal because figuratively it has become part of our bodies. And it is highly contextual since our interaction with mobile devices can be influenced by contextual factors, such as the mobile environment in which we find ourselves in, the activities in which we interact with mobile devices, and the purposes for which we interact with mobile devices. Similar to human-computer interaction, however, human-mobile interaction has been investigated largely on its technical side, mostly focusing on its areas, ease and enhancement.

The examined areas include mobile banking, mobile learning, and mobile content sharing. For instance, in the case of mobile payment, human-mobile interaction could be influenced by content, ease of use, promotion, made-for-the-medium and emotion (Liu, Wang & Wang, 2011). It may also happen in "ensuring ubiquity and mobility in learning without time, place and technical limitations" and "efficiency, effectiveness and usability of mobile learning applications" were also examined (Fetaji, 2008).

On how to ease human-mobile interaction, for instance, a proposed NeuroPhone project suggested that human-mobile interaction would be hands-free, silent and effortless if neural signals could be leveraged to control mobile phones as suggested by a group of researchers (see Campbell, Choudhury, Hu, Lu, Mukerjee, Rabbi, & Raizada, 2010). And it would also be hands-free in another proposed EyePhone, where "an interfacing system capable of driving mobile applications/functions using only the user's eyes movement and actions" (Miluzzo, Wang, & Campbell, 2010). Academic attention has also been paid to the fact that distractions of all kinds may affect user attention allocation during human-mobile interaction due to the scarcity of individual visual and cognitive resources available (Tsiaousis & Giaglis, 2008).

On how to enhance human-mobile interaction, earlier studies suggest that human-mobile interaction can be increased by "matching the object visibility to fulfill the users need" and simple interactivity (Fetaji & Dika, 2008). It can also be enhanced by improving personal, social, public, and hybrid participatory sensing (users involved) systems as well as personal, social and public opportunistic sensing (users not involved) systems (Khan, Xiang, Aalsalem & Arshad, 2013), by 3D augmented reality technology despite the small screen and limited input capabilities mobile phones (Koceski, & Koceska, 2011), through the installation of an emotional recognition system (Alepis & Virvou, 2012), or through integrating pervasive gaming elements into mobile content sharing" and annotate[ing] real world locations with multimedia content, and concurrently, provid[ing] opportunities for play through creating and engaging interactive game elements, earning currency, and socializing (Chua, Goh, Lee, & Tan, 2010).

The largely technologically-dominated studies of human-mobile interaction have left the human side unfortunately marginalized if not totally neglected. To fill the gap, as the editor of this volume, I invited 26 scholars from 10 countries where they are based to investigate 19 topics, focusing on the human side of human-mobile interaction in mobile education, mobile learning, mobile government, mobile journalism, mobile elections, mobile business, mobile advertising, mobile marketing, mobile branding, mobile public relations, and mobilc hcalth.

After more than one year of collaboration, the results have been put together in the form of this volume, which consists of 19 chapters. Among those chapters, 11 are research chapters while the rest are review chapters. Although research chapters have slightly different structures, review chapters follow largely the following structure: a. introduction, b. state of the art, c. to summarize previous studies, d. to map the future, and e. conclusions. The state of the art section is where major changes and trends of the chosen topic are elaborated. In summarizing previous studies, the focus lies in reviewing key concepts and themes, key theories and methods, key findings and achievements, and key weaknesses and problems. In mapping the future, major attention is paid to new concepts and themes, new research areas, new theories and methods, and recommendations.

Specifically, the topics investigated in this volume can be broadly grouped into the following six categories: a. mobile learning (Chapters 1-5), b. mobile government (Chapters 6-8), c. mobile journalism (Chapters 9-10), d. mobile persuasion (Chapters 11-16), and e. mobile youth (Chapters 17-19).

In Chapter 1, Fernando de la Cruz Paragas shared his findings regarding the interaction between learners and mobile in the context of learning with and about the mobile phone. The examination of human-mobile interactions took another turn in Chapter 2, where Saleh Al-Shehri investigated the situation in the case of mobile language learning. In his ethnographic examination of multilingual communication on mobile devices in the case of Brunei Darussalam, Mukul Saxena shared his findings on how multilingual mobile users interact with mobile in Chapter 3. Kenneth E Harvey, Philip J. Auter and Samantha Stevens offered a comprehensive review of changes, challenges and trends in discussing the interaction between education and mobile in Chapter 4. A group of young mobile scholars from Africa, namely, Kushatha Kelebeng, Rebaone Mlalazi, Keorapetse Gosekwang, Pendukeni Phalaagae, Tebogo Mangwa, Tebogo Kebonang, and Thototobolo Morapedi offered in Chapter 5 their insightful observations regarding the African substance and style of how learners interact with mobile based on their critical and comprehensive review of major published studies of mobile learning in Africa.

In Chapter 6, Beatriz Barreto Brasileiro Lanza and Maria Alexandra Cunha investigated how government interacts with mobile in their investigation of how SMS was used as a tool for mobile government in Brazil. In Chapter 7, Wendy Li offered her answer to the question whether it is a gear or game change for governments to move from e-government to m-government based on her critical review of the existing literature. When elections can also occur in the mobile space, what will happen to the interaction between voters and mobile? The answer can be found in Chapter 8 by Oarabile Sebubi.

In Chapter 9, Kenneth E. Harvey offered his insights on major tipping points amidst changes and trends in the interaction between local news and mobile. And another major case offered by Hans Karl Meyer and Burton Speakman in Chapter 10 focused on how journalism interacted with mobile in the case of how journalists played a major role in melding social media and social capital.

From Chapters 11 through Chapter 15, Kenneth E. Harvey and his research team (Philip J. Auter, Yulia An and Nygmet Ibadildin) provided insightful, critical and comprehensive review of major changes, challenges and trends in the interactions between mobile on the one hand and marketing, branding, public relations, business, and advertising, on the other hand. Qi Yao and Mei Wu, in Chapter 16, offered a case study of how WeChat, the most popular instant messenger in China, has been widely used in advertising activities in the world's second largest economy.

The last but not the least theme of this volume is the interaction between mobile and youth. This theme starts with Chapter 17, in which Yuchan Gao shared the results of her investigation of game players interacted with mobile in the case of mobile gaming in order to lead a gratifying mobile life. Chapter 18 contributed by Zeinab zare mohzzabieh, Seyedali Ahrari, Bahaman Abu Samah and Jamilah Bt. Othman investigated how socialization occurred in the interaction between youth and mobile. Janice Hua Xu, the last contributor, on the other hand, provided a very interesting and imperative critical discourse analysis of news coverage of how left-behind children interacted with mobile in China in Chapter 19.

Human-mobile interaction has been examined in this volume with a special focus on the human side by 26 scholars from 10 countries. Together, they have identified major opportunities and trends in the new age of human-mobile interaction locally or globally.

The main purpose of this volume is to serve as a stepping stone for further studies in examining how human beings have been interacting with mobile in different countries in different contexts in different activities for different purposes. It is the hope of the editor of this volume that further studies of human-mobile interaction will be conducted in a more comprehensive and comparative way so that findings from different countries can be triangulated to identify both universal and particular patterns of human-mobile interaction.

Xiaoge Xu
Botswana International University of Science and Technology, Botswana

REFERENCES

Alepis, E., & Virvou, M. (2012). Multimodal object oriented user interfaces in mobile affective interaction. *Multimedia Tools and Applications*, *59*(1), 41–63. doi:10.1007/s11042-011-0744-y

Campbell, A., Choudhury, T., Hu, S., Lu, H., Mukerjee, M. K., Rabbi, M., & Raizada, R. D. (2010, August). NeuroPhone: Brain-mobile phone interface using a wireless EEG headset. In *Proceedings of the Second ACM SIGCOMM Workshop on Networking, Systems, and Applications on Mobile Handhelds* (pp. 3-8). ACM. doi:10.1145/1851322.1851326

Chua, A. Y. K., Goh, D. H., Lee, C. S., & Tan, K. T. (2010, April). Mobile alternate reality gaming engine: A usability evaluation. In *Proceedings of Information Technology: New Generations (ITNG), 2010 Seventh International Conference*. Academic Press. doi:10.1109/ITNG.2010.47

Fetaji, M. (2008). Devising a strategy for usability testing of m-learning applications. In J. Luca & E. Weippl (Eds.), *Proceedings of EdMedia: World Conference on Educational Media and Technology 2008* (pp. 1393-1398). Association for the Advancement of Computing in Education (AACE). Retrieved from http://www.editlib.org/p/28565

Fetaji, M., & Dika, Z. (2008, June). Usability testing and evaluation of a mobile software solution: A case study. In *Information Technology Interfaces, 2008. ITI 2008. 30th International Conference on* (pp. 501-506). IEEE. doi:10.1109/ITI.2008.4588461

Khan, W. Z., Xiang, Y., Aalsalem, M. Y., & Arshad, Q. (2013). Mobile phone sensing systems: A survey. *IEEE Communications Surveys and Tutorials, 15*(1), 402–427. doi:10.1109/SURV.2012.031412.00077

Koceski, S., & Koceska, N. (2011, November). Interaction between players of mobile phone game with augmented reality (AR) interface. In *User Science and Engineering (iUSEr), 2011 International Conference on* (pp. 245-250). IEEE.

Liu, Y., Wang, S., & Wang, X. (2011). A usability-centred perspective on intention to use mobile payment. *International Journal of Mobile Communications, 9*(6), 541–562. doi:10.1504/IJMC.2011.042776

Miluzzo, E., Wang, T., & Campbell, A. T. (2010, August). EyePhone: Activating mobile phones with your eyes. In *Proceedings of the Second ACM SIGCOMM Workshop on Networking, Systems, and Applications on Mobile Handhelds* (pp. 15-20). ACM. doi:10.1145/1851322.1851328

Tsiaousis, A. S., & Giaglis, G. M. (2008, July). Evaluating the effects of the environmental context-of-use on mobile website usability. In *Mobile Business, 2008. ICMB'08. 7th International Conference on* (pp. 314-322). IEEE. doi:10.1109/ICMB.2008.25

Chapter 1
Learners and Mobile:
A Reflexivity

Fernando de la Cruz Paragas
University of the Philippines Diliman, The Philippines

ABSTRACT

Studies about mobile phones, the learning process, and educational institutions have grown in recent years though research has mostly focused in the United States and in specific groups. This research contributes to the literature by looking at three educational levels in Singapore and by taking a two-pronged approach to the relationship among these three variables. It answers the following reflexivity: How do students learn to use mobile phone functions? How do they use mobile phones for learning functions? This chapter considers learning as the reflexive process where attitude and aptitude are acquired and shared for curricular and extra-curricular activities. Findings indicate two themes: how students 1) develop the skills to use the expanding array of mobile phone technologies, and integrate these in their daily life and 2) use mobile phones in school and in their schoolwork. Data for the study came from focus interviews with 36 informants who were selected through maximum variation sampling according to their age, educational level, and household income.

INTRODUCTION

A mobile phone ringing in the classroom is one of the most oft-repeated breaches of netiquette, and research has found that people believe the classroom is among the places in which the mobile phone must not be used (Campbell, 2006). The reasons behind this disapproval of the use of the mobile phone in the classroom include the disruptive nature of ringing alerts for incoming calls and text messages, the tendency of students to fiddle or play with their mobile phones and be distracted from the lecture, and the use of mobile phones for cheating. Initial efforts to ban mobile phones from the classroom, however, have been tempered by the sense of security that the phones afford students and their parents. Moreover, mobile phones provide educational opportunities by facilitating access to informational resources in the Internet and by allowing students to micro-coordinate among classmates for school projects.

DOI: 10.4018/978-1-5225-0469-6.ch001

The increasing use of mobile communication technologies in education has expanded the domain of e-learning or learning that is supported by digital electronic tools and media (Milrad, 2003 in Cavus & Ibrahim, 2009). Called mobile learning, or m-learning for short, this relatively new practice is differentiated from traditional e-learning by its portable and wireless platform (Kang, Sung, Park, & Ahn, 2009). Moreover, it facilitates learning that is "self-paced, on-demand and real-time" (Chuang, 2009, p. 51) and enables ubiquitous or anytime, anywhere access to course contents (Al-Fahad, 2008; Kang, Sung, Park, & Ahn, 2009; Suki & Suki, 2007). The emergent attention on m-learning is a function of two developments. Firstly, there is increasing evidence about the viability and flexibility of mobile phones to enrich the learning experience (Al-Fahad, 2008). Secondly, mobile phone penetration rates are high in developed and developing countries (Chuang, 2009), which means there is less need to invest on the purchase of equipment and the training of students to use such equipment. According to Statistics Singapore 2008, Singapore, which is the context of this study, has a mobile phone density of 137 for every 100 people.

Definitions of m-learning, according to Peng, Su, Chou & Tsai (2009), can be grouped into three, depending upon the value of the mobile platform that is highlighted. The first definition focuses on the mobile phones' functional components, particularly on its ability to facilitate communication wirelessly. The second definition pertains on the portability of the mobile phone which then relates to values such as convenience, expediency, and immediacy which are important to learners and teachers. The third definition relates to the concept of ubiquitous computing. Regardless of their attribute of focus, these definitions remain skewed towards the medium and provide little attention to the pedagogical aspect of the m-learning equation (Peng, Su, Chou, & Tsai, 2009).

Indeed, the literature on m-learning has tended to underscore the technological platform. On the one hand, m-learning is discussed in terms of the choice of technology, particularly as regards the supposed effort by universities to pull students into online environments by investing heavily in their wired Internet infrastructure vis-à-vis the push among students for wireless devices such as laptops and mobile phones (Chuang, 2009; Al-Fahad, 2008) as well as smart phones and other hybrid devices (Cavus & Ibrahim, 2009).

On the other hand, the transition (Fallakhair, Pemberton, & Griffiths, 2007; Wang, Wu, & Wang, 2009) or complementation (Chao & Chen, 2009) between e-learning and m-learning is discussed by comparing the advantages and disadvantages of personal computers with mobile devices, the problems with configuring, reformatting and maintaining content that works across both platforms, and concerns about interoperability issues among the many variants of mobile devices. Collectively, these can then result in duplicated efforts and wasted resources (Goh & Kinshuk, 2006). Researchers have discussed how the effectiveness of m-learning can be weakened by the mobile devices' physical and functional limitations (Fallakhair, Pemberton, & Griffiths, 2007). These limitations include screen size and audio-visual quality, (Kang, Sung, Park, & Ahn, 2009; Wang, Wu, & Wang, 2009; Lu, 2008), keyboard interface (Kang, Sung, Park, & Ahn, 2009; Wang, Wu, & Wang, 2009), battery life, data security, and processing power (Wang, Wu, & Wang, 2009; Peng, Su, Chou, & Tsai, 2009). These limitations, in turn, can prevent students used to e-learning from transitioning to m-learning (Wang, Wu, & Wang, 2009) as these require them to adopt new learning strategies (Lu, 2008), or preclude the "the effective implementation of new learning paradigms" (Fallakhair, Pemberton, & Griffiths, 2007, p. 312).

REFLEXIVITY

The literature on m-learning is heavily grounded on technology. However, Fisher & Baird (2006, p. 8) said the focus on m-learning research should be on the relationships that transpire intra-personally and interpersonally within and between technological platform. They wrote,

To be clear: the mobile environment is merely another platform in which interaction, collaboration, and knowledge transfer can occur. The use of mobile technology defines only the parameters and building blocks on which the interaction can take place, providing opportunities for the social exchange of information, interaction, and instruction. Moreover, the ability for students to reconcile their authentic use of technology in a learning context can motivate and persuade users to actively engage in the course content.

The intra-personal and interpersonal relationships that inform m-learning can be studies through the concept of reflexivity. Reflexivity, as Lynch (2000) argues, has been a "central and yet confusing topic" (p. 26). To make sense of reflexivity, he identified six of its categories--mechanical, substantive, methodological, meta-theoretical, interpretative, and ethnomodological. Reflexivity, as conceptualized in this paper, falls in Lynch's first category of *mechanical reflexivity*. Within this category, Lynch (2000) identified *knee-jerk reflexivity* ("an habitual, thoughtless or instantaneous response" (p. 27)), *cybernetic loopiness* ("a circular, recursive process or pattern involving feedback loops" (p. 27)), and *reflections ad infinitum* ("the iteration of recursive patterns" (p. 28)). Of these three types of mechanical reflexivity, cybernetic loopiness best fits the conceptualization of reflexivity described in this paper which provides for a circular process between learning with and about the mobile (see Figure 1). It is important, to qualify, however, that mechanical in this case is not synonymous to automatic.

Reflexivity, in qualitative research, refers to the self-reflective process that researchers undertake to examine their subjectivities throughout the course of a study (Hansen, Cottle, Negrine, and Newbold, 1998; Jensen, 2002; Mason, 2002). Though this is the dominant use of the term *reflexivity* in research, Lynch (2000) argued that this *methodological reflexivity* had no particular ascendancy over the other conceptualizations of reflexivity. Indeed, Lynch argued against what he said was a self-conscious approach to reflexivity, which, he said, was no better than trying to be objective with regard to a particular phenomenon.

MOBILES IN SCHOOL

Using reflexivity helps de-center the discussion of m-learning from the mobile device itself and locate it to the venue and process of learning. Studies which locate the mobile device, particularly mobile phones, in the classroom have mostly focused on issues of etiquette and privacy. The extensive literature review by Campbell (2006) about mobile phone use in different locations indicates that classrooms and movie theaters are among the public places where mobile phone use is least accepted and most distracting. Campbell explains mobile phone use in the classroom is a private activity that goes against the collective experience and goal of learning that transpires there. During class breaks, however, Campbell notes significant changes in accepted behavior. Though still in the school but beyond the classroom, Campbell (2009) discovers that it was acceptable to make and accept private calls in the presence of classmates, a process he says has become "an identity construction device" (p. 99).

Sanctioned use of the mobile phone for learning opportunities, however, offers new avenues in education. Mobile devices can complement traditional learning materials (Chao & Chen, 2009) and practices (Wang, Wu, & Wang, 2009) or introduce new ways of learning such as through gaming simulations, videos, online discussions or interactive applications (Chuang, 2009). Moreover, patterns of mobile device use can be harnessed towards m-learning. Since these devices are constantly carried by their owners, they can be readily used to take notes, contact classmates, consult teachers (Chuang, 2009; Chao & Chen, 2009; Campbell, 2006) or pursue independent learning (Lu, 2008; Clough, Jones, McAndrew, & Scanlon, 2008). Because of their use for social networking, mobile devices can also facilitate community-based (Chuang, 2009; Fisher & Baird, 2006; Cavus & Ibrahim, 2009) and extra-curricular (Kang, Sung, Park, & Ahn, 2009) learning. In particular, mobile phones are always carried by students who can then use these to learn in their preferred time and place for quick lessons (Lu, 2008). Moreover, access to concise but ubiquitous information can increase awareness about available scholastic resources (Chao & Chen, 2009). Mobile phone services such as SMS can also help school performance by helping students with their time management (Cavus & Ibrahim, 2009).

According to literature, the direct application of m-learning, however, has been most successful in language learning. Mobile devices have significantly helped students learn second languages as they replace pocket dictionaries or vocabulary books (Fallakhair, Pemberton, & Griffiths, 2007) and can be used as part of everyday activities (Cavus & Ibrahim, 2009). Compared to e-learning, m-learning helped better improve the reading and speaking skills of intermediate and new students in as they learn a new language (Kang, Sung, Park, & Ahn, 2009). Conversely, e-learning helped writing skills better. Kang, Sung, Park & Ahn (2008) explained their findings through the inherent features and use of the platforms in either e-learning or m-learning modules. These findings, though currently confined to language learning, bear upon the integration of m-learning in the formal curriculum (Clough, Jones, McAndrew, & Scanlon, 2008, p. 369). Such integration has to consider potential problems in sanctioning mobile phone use in the classroom. The use of mobile phones and other mobile devices to play games and to cheat during exams pose serious challenges in the design of an m-learning curriculum (Campbell, 2006).

STATEMENT OF THE PROBLEM

Studies about the relationship between mobile phones and the learning process and educational institutions have grown in recent years though the literature remains scant. Most of the research has focused on the technological platform, the United States as context, and in specific disciplines. There is thus paucity in three areas of m-learning research which this research seeks to address. This study's focus on students' valuation and rituals as regards mobile phones serves as a counterpoint to the medium-centric focus of current m-learning research. Moreover, by conducting the study in Singapore and across three educational levels, the research enriches the discussion of how patterns of mobile phone use can bear upon the design and implementation of m-learning modules. Finally, across disciplinal applications, one of the strengths of e-learning is that it harnesses current patterns of mobile phone use. It is thus important to understand how, in the first place, students learn how to use mobile phone functions and whether this use translates to m-learning. The research is focused on mobile phones since they are the most popular type of mobile devices and their physical and functional features are a contrast to e-learning platforms. Accordingly, this exploratory research seeks to answer the following reflexivity:

RQ1: How do students in secondary school, junior colleges/polytechnics, and universities learn to use mobile phone functions?

RQ2: How do they use mobile phones for learning functions?

Answering these questions addresses the recommendations of Morgan & Kennewell (2005) in seeking to appreciate how students learn ICT within and beyond the school. They argue that it is important to "understand how our children best develop confidence and competence with new technologies if their potential is to be fully exploited. It is of particular interest to find out which previous experiences prove most valuable in allowing children of different ages to access new technologies" (p. 177-178).

STUDY FRAMEWORK

m-Learning research has focused on how mobile devices, particularly in terms of their technical design, can support formal curriculum-based education (Peng, Su, Chou, & Tsai, 2009; Clough, Jones, McAndrew, & Scanlon, 2008). In language learning, for instance, there is increasing evidence that m-learning can be an integral part of language learning pedagogy. At the same time, many researchers have argued about the benefits of m-learning towards informal learning, particularly in relation to intrapersonal and interpersonal learning. Clough, Jones, McAndrew & Scanlon (2008, p. 361) wrote, "Given the evidence that mobile devices have a role to play in formal learning scenarios, it seemed reasonable to expect that experienced mobile device users would include their mobile devices among the learning tools they used to support their informal learning". Despite the many arguments about the benefits and potential of m-learning, however, there remains a dearth in theorizing about it. According to Peng, Su, Chou, & Tsai (2009), this is a result of the limited attention to pedagogical concepts and the heavy focus on technology (which with its rate of development seems to preclude theorizing) in m-learning studies. They said that the focus on content delivery via mobile devices in these studies seems to equate m-learning with mechanical rather than mobile learning.

This study seeks to address this concern by focusing on the reflexivity between mobile phones and the learning process as a way to understand m-learning as a concept (see Figure). Reflexivity, in this regard, refers to a circular loop (Lynch, 2000) between learning about and with the mobile. In particular, this study looks into informal learning, or learning that transpires mostly interpersonally, in any locale or time, and outside of structured learning programs (Cavus & Ibrahim, 2009). There has been some research into informal learning as regards m-learning, but little has been done about it from an intrapersonal level. This oversight must be addressed because informal learning, even when it thrives in a community setting, is largely self-directed and self controlled (Clough, Jones, McAndrew, & Scanlon, 2008). Moreover, research has shown that self-efficacy (Morgan & Kennewell, 2005) and other intrapersonal variables such as performance expectancy, effort expectancy, social influence, perceived playfulness, and self-management (Wang, Wu, & Wang, 2009) relate to the use of technology. Thus, this research explores m-learning in intrapersonal and interpersonal informal learning situations. In so doing, the research helps identify grassroots learning of technology, which can be argued as the "desirable social practices of learning" (Roschelle, 2003 in Clough, Jones, McAndrew & Scanlon, 2008, p. 361), as a way to inform m-learning initiatives. Pettit & Kukulsa-Hume (2007 in Al Fahad, 2008, p. 102) call this approach the threading of "innovative uses of technology into the existing fabric of behavior".

The link between intrapersonal and interpersonal learning informs what is called constructive learning. Participants in constructive learning share their own knowledge and skills. In the process, exponents of constructive learning assert, participants find their own intellectual identity, develop their own knowledge base and articulate their own voice (Fisher & Baird, 2006). The pitfall of constructive learning in relation to m-learning is that the students might be technologically-savvy themselves but are unable to translate this proficiency towards learning experiences and outcomes (Al-Fahad, 2008). Researchers thus recommend integrating pedagogical techniques such as play-based learning sessions (Morgan & Kennewell, 2005) into constructive learning.

Informed by assertions about m-learning and informal learning, this research considers the mechanical reflexivity (Lynch, 2000) between learning about the mobile and learning with the mobile. Figure 1 depicts the indicators for each of these concepts. Learning about the mobile explores students' personal story about their mobile phone and traces how the unit has become integrated in their routines. Accordingly, the discussion revolves around how, when, and from whom they received their mobile phone, how they learned to use its functions, and how it is a part of their everyday life. Research has shown that histories and patterns of use relate to perceptions of mobile phones and self-efficacy (Campbell, 2006; Clough, Jones, McAndrew, & Scanlon, 2008). Learning with the mobile, meanwhile, focuses on the use of the mobile in school in general and for schoolwork in particular. It also discusses cheating which is one of the most popular concerns regarding the use of the mobile phone in the school setting. The two main concepts are engaged in a reflexive relationship such that how the students learned about the mobile phone can foreground how they may learn using the mobile phone. Conversely, learning with the mobile can enrich the students' use and evaluation of the mobile phone.

To qualify nuances across student profiles, income, gender, and educational level are considered in this research as possible variables that distinguish mobile phone uses and inform m-learning practices and potentials. Income is considered because it relates to the quality and quantity of mobile phones and

Figure 1.

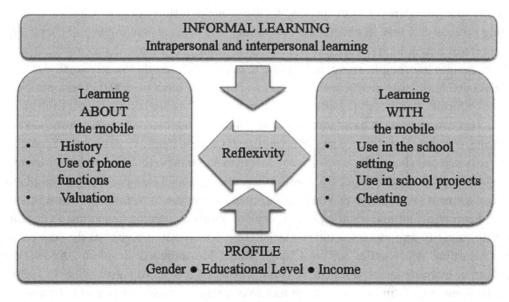

M-LEARNING

other mobile devices which the students maintain as well as to the software and hardware infrastructure that support these devices. Students from higher income families, for instance, will be able to afford higher and newer models of mobile devices and subscribe to faster and more expansive array of mobile network services. With less cost constraints, they can also use mobile services more heavily, which, in turn, can relate to perceptions of self-efficacy, an important variable in m-learning. Suki & Suki (2007) found that heavier mobile phone use is related to more frequent access to, and higher subscription and purchase of, m-learning content and mobile games.

Many m-learning studies or studies about the mobile phones in educational settings have focused on specific age groups, particularly those at the college level (Campbell, 2006; Fisher & Baird, 2006). However, there is a need to qualify informal m-learning across age groups because of the different educational requirements and social arrangements among them. For instance, even if students today have grown up with mobile phones and other relatively new communication technologies (Morgan & Kennewell, 2005; Fisher & Baird, 2006), studies have shown that younger students are more tolerant of mobile phones in the classroom (Campbell, 2006). Finally, gender can influence which mobile phone features people focus upon: men look at technical functions and women look at social aspects such as design, ring tone, and color (Campbell, 2006). Some studies also noted differences in adoption patterns across males and females though these diminished eventually (Campbell, 2006). Likewise, Wang, Wu, & Wang (2009) found that respondents with high performance expectancy and playfulness perception towards using m-learning, regardless of age and gender, had a higher intention to use m-learning than those with lower performance expectancy and playfulness perception.

METHODOLOGY

This qualitative research employs in-depth interviewing because, as Jensen (2002) writes, "in-depth interviewing, with its affinities to conversations, may well be suited to tap social agents' perspective on the media, since spoken language remains a primary and familiar mode of social interaction, and one that people habitually relate to the technological media" (p. 240). Thus, as this study looks at intrapersonal and interpersonal interactions regarding mobile phones, In-depth interviewing was the appropriate data-collection technique. Interviews were conducted by students in Singapore. Interviews began with a uniform message that introduced the topic, ensured informants' anonymity, and sought their informed consent. Accordingly, all names in this article are pseudonyms.

Interviewers used a semi-structured interview questionnaire which included guide questions (see Table 1). All interviewers were conducted face-to-face, which allowed the interviewers to probe informants' answers.

Based on the framework, 36 informants were purposefully-selected using maximum variation sampling. The sampling scheme considers gender, educational level (secondary school, junior college/ polytechnic, and university), and household monthly income (above or below SGD5460, which is the cut-off for the fifth decile in Singapore according to the Department of Statistics). Using these three attributes, the informants were identified using a selection tree. Three informants for each branch (for example, a male secondary school student whose family belongs to the upper income (UI) level) were interviewed (see Figure 2).

Table 1. Guide questions

Conceptual Domain	Questions
History	How long have you been using a mobile phone? How long did you learn how to use it? Who gave it to you and who taught you how to use it?
Phone functions	What functions do you use in your mobile phone? Do you use mobile internet? Why or why not?
Valuation	- What is the role of the mobile phone in your family and social life? How do you see the mobile phone as part of your everyday life? - What are its advantages and disadvantages? What are the things you like or dislike about it?
Use in the school setting	What is the role of the mobile phone in your school work? Do you use your mobile phone in the school? In the classroom? Why or why not?
Use in school projects	Do you use mobile Internet to surf for answers to projects or assignments? Why or why not?
Cheating	Are you aware of students who have cheated in their schoolwork using the mobile phone? Tell me about it

Figure 2.

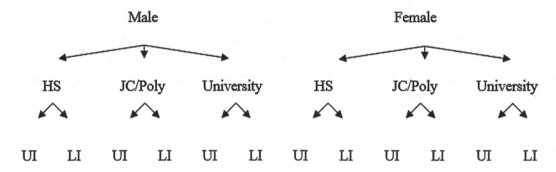

Interviews were recorded digitally and transcribed responses were arrayed in matrices for axial coding. Patterns were developed within the conceptual domains with due consideration of differences across informant groups according to gender and educational and income levels.

FINDINGS

Profile of Informants

Following the sampling scheme, the informants were distributed equally across gender, age, and income (see Table 2). All of the informants had a mobile phone, and two-fifths of them also had an Ipod. Half of the informants had a laptop, nine had a PC, and nine had both a laptop and a PC. A chi-square test revealed no significant differences in the ownership of technology across informants according to gender, age, and income. However, some differences across these groups were noticeable. For instance, only two of the 18 female informants had both a PC and a laptop compared to seven of the 18 male informants. Half of younger informants—those in secondary school and junior college/poly—had an Ipod compared to only a third of those in university. Conversely, more informants in the junior college/poly and the university had both a laptop and a PC. These differences require further investigation through a larger-scale quantitative study as they may impact initiatives on e-learning of which m-learning is a variant.

Table 2. Informants' profile

	Pseudonym	Gender	Educational Level	Income Level
1	Larry	Male	Secondary School	Upper
2	Jon	Male	Secondary School	Upper
3	Bert	Male	Secondary School	Upper
4	Alex	Male	Secondary School	Lower
5	Mark	Male	Secondary School	Lower
6	Wayne	Male	Secondary School	Lower
7	Lionel	Male	Polytechnic/Junior College	Upper
8	Danny	Male	Polytechnic/Junior College	Upper
9	Chris	Male	Polytechnic/Junior College	Upper
10	Steve	Male	Polytechnic/Junior College	Lower
11	Jim	Male	Polytechnic/Junior College	Lower
12	George	Male	Polytechnic/Junior College	Lower
13	Joe	Male	University	Upper
14	Henry	Male	University	Upper
15	Eman	Male	University	Upper
16	Omar	Male	University	Lower
17	Jake	Male	University	Lower
18	Ralph	Male	University	Lower
19	Michelle	Female	Secondary School	Upper
20	Angel	Female	Secondary School	Upper
21	Counille	Female	Secondary School	Upper
22	Chelo	Female	Secondary School	Lower
23	Luisa	Female	Secondary School	Lower
24	Mila	Female	Secondary School	Lower
25	Vicky	Female	Polytechnic/Junior College	Upper
26	Sandra	Female	Polytechnic/Junior College	Upper
27	Myra	Female	Polytechnic/Junior College	Upper
28	Vivien	Female	Polytechnic/Junior College	Lower
29	Ellen	Female	Polytechnic/Junior College	Lower
30	Karen	Female	Polytechnic/Junior College	Lower
31	Linda	Female	University	Upper
32	Sheila	Female	University	Upper
33	Emma	Female	University	Upper
34	Ceille	Female	University	Lower
35	Jenny	Female	University	Lower
36	Dianna	Female	University	Lower

HISTORY

All by Myself

Most of the informants got their first mobile phone when they were in late primary or early secondary school, around the time they became teenagers. Joe said his age then helped him learn how to use the phone quite fast. He said, "Easy to catch on and learn. For a young person, you can easily learn." Many of the informants could not remember the details of their first phone other than they learned how to use it quite easily. Cecille said, "I think from high school. I have used it for around 6 years already. I do not remember as it is quite easy to get to know the basic functions." Jim echoed this statement, "I can't remember how long I took to learn the functions but is should take a short while only as it is relatively easy to learn."

Alex said it took him a week to learn how to use his phone as it was, as Jim said, quite easy to learn. George took around a week, too. According to him, "Call was easy but SMS it took about one week to get used to the phone and the buttons are different, when I switched between Nokia and Sony Ericsson." Sheila, Vivien, and Emma cited a similar learning curve with the different functions. Sheila said, "I took about a month to learn everything. The essential ones I can learn it in a day, but to learn all the shortcuts and everything it takes about a month." Vivien, for her part, said she learned "Through trial and error and referring to the manual sometime. Didn't take very long, mainly I was using the basic functions, for the more advanced functions, it gradually build on, prolonged usage of the phone introduced you to the other functions of the phone." Emma said she "Just started to learn to use it, used it as a toy, play with it, and you can learn."

Informants cited the usability of the interface which made learning how to use the mobile phone quite easy. Danny referred to the simplicity of earlier models when he said, "The functions used to be quite simple at that time. It was easy to operate it." Omar, however, noted that even with more functions in newer models, they remain usable with some help from the manual. He said, "I find that phones these days are very easy to use and it doesn't take much time to figure out. However, I would usually start off with the user guide." Jake, a self-professed techie guy, also combined self-exploration with the phone manual. He said, "I think I have to figure out the functions and the way to use it through the reading of the manual myself. Because it was new technology back then. I think I am the more gadget guy in my family so I have to try everything by myself. If there's any function I don't know, I will approach the manual for instructions but normally would try and figure it out myself."

All in the Family

For some informants getting their first phone was not the first time they got to operate a mobile phone. Mila said it only "Took quite a short time to learn how to use it as I have been using other people's (parents') phones before." Similarly, Diana said, "I think before that my parents have a phone and sometimes I will use it, so when I got my phone I knew how to use it. It's quite easy to pick up." Almost all of the mobile phones were gifts from family members, mostly from parents and sometimes from siblings. Moreover, while 27 of the 36 informants said they learned the functions by themselves, others received some guidance from family members. Angel and Connille said their respective fathers gave them their phones and taught them how to use these. Jon, Mark, and Wayne received some help from their siblings. Jon said, "My older brother had a phone and showed me how to store numbers and call people," while Mark said, "My brother taught me how to use the functions: messaging, calling, and gaming."

PHONE FUNCTIONS

Mobile and Multi-Functional

Universally, mobile phones were used for voice calls and text messages, and less than half of the informants used other features of their mobile phone. Only 14 took pictures or listened to music, 12 played games or used the calendar, notepad and alarm clock features, and three watched video in their mobile phone. George said, "It's my organizer. I think that's the most important thing. And also the ability to connect to people. I can live without it but it will definitely come with a lot of inconvenience especially the organizer because I tend to forget a lot of things. Some people prefer diaries but I prefer the phone as it is convenient because it's just in my pocket and we don't need a pen so it's one whole thing together. I find it more essential than the internet." Omar, meanwhile, noted, "It helps me to keep track of my daily involvement as I use it to store my schedules in the calendar and also the reminder feature to track any important task ahead of time." Emma and Lionel noted the utility of the mobile phone's many functions. Lionel said, "I like it as it is multi-functional-it's a camera and very good communication device. I can't go without it," while Emma noted, "I think mobile phone is really convergent, I can try different functions only on mobile, I don't need to take any other things with me, the main function is still about communicating with other people."

Mobile and Online, But Not Quite

Of real interest, however, was the use of mobile internet, which would enable m-learning students to access the same online information that would be available to their counterparts in e-learning modules. Fifteen of the informants, of which ten are male, use mobile internet. Joe said he used it because it came as part of his phone service. He said, "obviously (I use it) because I have 12G of data plan." Others, meanwhile, used other people's phones. Larry said, "I use a simple phone that doesn't have the Internet app. I've tried accessing the net through a friend's phone." Saying that because she had to pay for mobile internet service, Angel said she used her father's phone instead.

Those who accessed the internet using their mobile phone said they only used it for specific types of information. Facebook, among other social networking sites, was among the popular sites among them. According to Joe, "Topics are mostly social networking such as Twitter, Facebook. There is nothing much to surf online because you can't download stuff. You can only check for information such as where to eat, bus arrival timing, maps and social networking such as Facebook updates." Jon, meanwhile, said, "Probably about the same topics that I would use on a normal computer. Facebook and You Tube." Other informants used mobile internet for other social activities, particularly for small bits of information. Larry said, "If we were having an argument and I want to prove a point for which I would have to check the Internet for facts." Jake said he used "it to check to check email at times using Wi-Fi and to check movie timings and listing." Danny also checked movie timings using his phone but he only did so occasionally because he did not have free data usage. Eman, meanwhile, said he has a smart phone which he uses "when there's Wi-Fi or GPRS access. I surf Wikipedia, Facebook on my mobile phone. And I book movie tickets through it as well."

One male and all of the five female informants who used mobile internet only did so as a back-up to online access through computers and a wired connection. Omar said, "I usually use my mobile phone to surf web only when it's emergency or when I do not have a computer beside me and I urgently needed to surf for information because it is quite inconvenient to surf from the phone. For instance, I may use it to check the movie timing, my school email, etc." Cecille and Chelo both used e-mail in their mobile phones only because there were no computers. Chelo said, "I have used my phone to surf the Internet. I wasn't able to access to internet at home, and needed to check some important mails, so had to use the phone." Cecille, meanwhile, said "I mostly check email or write email in some urgent situations where there are no computers on hand. Mila and Sheila also found themselves in similarly urgent situations during which they used mobile internet. Mila said "I used it once because computer broke down and needed information from the internet for school work," while Sheila explained "I did for a few times. I was out and I needed to find the address of a particular shop which has moved so I Google it. It's expensive, and I don't have an existing net plan on my phone, so I try not to use. I don't use it unless it's urgent." Myra said, "Yes, I did before but not often. I find that the screen is too small. I use my computer instead."

That mobile internet is only a back-up was partly explained by Emma who said, "I didn't try in Singapore at all, because I don't have the habit, and I can get Internet access from my laptop or computer very easily here. I tried using my father's phone, it was very slow." Diana, meanwhile, noted both the limitations of the phone unit and the mobile internet service. She said, "I think it's pretty slow and it's so small, like the screen is pretty small. I've tried using it on my phone and other phones, but I don't like it as it's pretty small and hard to scroll."

Mobile and Offline

As can be gleaned from the earlier section, the physical features of the mobile phone, the cost considerations, and the service speed as among the constraints in using mobile internet. These were the same reasons which those who did not use mobile internet cited. Jon, Vicky, Jim, and Jenny said their current unit is not internet ready and they were fine with this. Vicky said, "No, my phone can't access Internet. I just got it, and I don't think I will change it as soon" while Jim said, "My mobile phone is unable to connect to the internet and also I don't see the point in doing so. I can use my laptop to surf the internet."

Even those who could use the Internet in their mobile phones did not do so because of the physical features of their units. Wayne said, "I don't want. I use computers. It's hard to type the words (while using phone). Chris, Sheila, Myra, and Diana echoed this. Chris said, "No. The screen is too small, so usually I use my computer instead, while Sheila said "There's such function, but I don't use. And the screen is very small. I don't like to use it to surf." Karen noted both the small screen and the cost of service. She said, "I find it very troublesome as the screen is small and the price of using it is high."

Cost issues, particularly in relation to the information available and the speed of connection, precluded many of the informants from mobile internet. Bert said "it is too slow and takes too long to load a site and it's expensive." Chelo, Jim, Ellen and George agreed. George explained, "It's very expensive. My phone can connect to the internet but it's not the main function. It is a music phone so it's not used to connect to the internet. My phone doesn't have Wi-Fi so I have to pay and it's quite expensive." Mila, meanwhile, compared accessing the Internet in either the PC or the mobile. She said, "I would not really use mobile phone for surfing the internet because of the expensive data charges compared to using broadband on computer."

VALUATION

Mobile Connections

Almost all of the informants said their mobile phone is important, if not totally indispensable because, as Connille summarized, "it is slim, easy to carry and convenient to use." Jake said, "Like internet the phone is 100 percent of my life. I use it as an alarm clock. I keep it on it 24/7. I don't switch it off during sleep in case of emergency." Alex also compared it to the internet, but was more emphatic about his mobile phone. He said, "You can't live without it. Every day you get to know people through call and connect with people and can call people. I can live without internet but not without the mobile phone. Cecille also noted the idea of indispensability of mobile phones. She said, "It is of course a very indispensable tool for me as it combines functions such as communications, music listening, internet surfing, etc." Larry and Diana, meanwhile, were emotionally connected to their phone. Larry said "I feel very isolated without my phone," while Diana said, "I think it's important. If I don't have it, I feel kind of empty. And lost. I think I need my phone for communication."

The emotional investment on the mobile phone is perhaps a function of the emotional ties that the unit helps foster. Wayne, Jim, Angel, and Emma, for example, used their mobile to maintain friendships. Wayne said, "I can message friends, to ask what they are doing when I feel bored," while Jim said, "It is an essential part of my everyday life. I use it daily to chat with my friends and SMS them to keep up with our relationship." Meanwhile, Angela said, "I use it to communicate with parents and make plans with friends. I use my phone for 5 hours a day, after school."

This sense of connection is also related to concerns about safety and security. Mark said, "It's quite important. I can call my parents when I need to find them. They can contact me too when I've been outside for so long." Chelo said, "It is really useful when you go out. If I don't have a phone I don't feel safe because I don't have something to contact people with." Diana noted, "I think what's great about the phone is that you can communicate and keep up with people and if anything happens you have a source to turn to, like, for emergency you can call someone, call the police, call anything."

Mobile Distractions

Since mobiles operate on two-way networks, the blessing of contacting people can be tantamount to a curse of being unduly contacted by others. Omar said, "I don't find any disadvantage with mobile phones other than the part where it keeps me busy at times when people could just keep calling me for some reasons," while Ellen said, "People can keep on calling and calling me, at the time when I am busy." Sheila explained the situation well. She said, "I think what I like and dislike and the advantages and disadvantages are the same. It is useful when you want to find someone last minute or you have the urgent need to contact someone…. But at the same time since it is so easy to find people right, you feel that your life is being intruded… You can't avoid it as everyone knows you have a phone and if you ignore it you are obviously antisocial. Yet sometimes we need privacy and freedom so it sort of becomes an intrusion. "

Incessant connection can also result in other undue results. Jim said, "The disadvantages are that if it is overused, the phone bill can be quite expensive." Jon said he would get distracted by having to check his phone constantly, a circumstance which has had an unwarranted effect on Jake who said, "I feel that I've developed a symptom to be able to detect vibrations from other people's phone and thinks that it's from my phone. I think it's a big disadvantage. I can sense vibrations from other people's phone. Chelo, meanwhile, noted how "People tend to get distracted by their phones. When you talk to them face to face, when their phone ring they start to get distracted."

Mobile disconnections can also result from a failure of technology. Mark and Vicky said their phone would hang and be inoperable while Eman and Connille cited the short battery life of their units. George explained, "Some phones are quite retarded too because they keep hanging and all that. For example, my phone now gives this problem as it doesn't register my contacts."

Five informants aired health concerns about the use of mobile phones. Jim said, "Long period of use will also cause health problems like cancer, resulting from the radiation," while Eman said "it's reported that the radiation might cause health problem. But I don't really care." Myra explained, "I know about this medical thing from the news and I think it's true because I saw an egg corrode because if it."

USE IN SCHOOL

There were policy differences between secondary schools, which did not allow mobile phone use in their premises, and junior colleges, polytechnics, and university, which did so. Accordingly, the use of mobile phone across educational levels was markedly different. Mila said, "I can't bring phone to school therefore doesn't use it." Mark nevertheless used his, "Sometimes when there're emergency calls I use the phone. Not in the classroom though because we are not allowed to use the phone there. "

Learning Distraction

Regardless of whether phone use is allowed or not, informants generally did not take to the idea of using the phone in the classroom as it distracts from the educational goals of the class. Larry said "it would have to be for an educational purpose during class to surf the Internet using my phone, if I'm not carrying my laptop that day." Alex said, "I use in school but no classroom because you have to concentrate on your studies in class," while Luisa said, "I use it in school but not in the classroom. I focus on what the teacher is teaching." The distinction between the classroom and the school were noted by several other informants. Sandra said, "I use it in school but not in classroom. Normally I will put it to silent mode there." Danny, Diana, Angel and Linda would use their phone during breaks when they are still in school but outside their classroom."

Such a distinction, however, was not observed by other informants who also used their phone within the classroom, even if only discreetly using SMS. "I use SMS, but I would do it under the table when having class," according to Chris. Jake said, "If we do it during lectures, the lecturers won't see. If we don't converse too loud they wouldn't know. SMS with touch screen won't affect the people around you." Being caught by the lecturer was the biggest sanction against using the phone, as Myra noted, too, "Quite rarely. I don't really use SMS or anything because the teacher will catch. My school allows mobile phones but I cannot use in class. Even if it's not illegal I will still not use it unless it very urgent stuff or if the lesson is boring. I won't use it to SMS my friends for fun and all." Vivien and Sheila echoed

Myra's note about mobile phone use during boring lectures. Vivien said, "Because probably I feel that the SMS conversation at that time was more interesting than what my teacher was talking about," while Sheila said, "I seldom use it in the classroom unless it's urgent or if the lesson is very boring. Else it is rather distracting too."

USE FOR SCHOOLWORK

Curricular Coordination

Most mobile use for schoolwork was incidental. Diana said, "I will ask my friends questions sometimes, not like directly related to my school though. I think I text and call a lot to my family and friends though and it helps to maintain relationships." At most, students used the phone to ask friends about school work, as Alex, Steve, and Jim said. Alex said, "I will use it to call and message friends to ask for school work, while Jim said, "I use it to ask my friends about school stuff like what homework to submit and all that. They also use it to ask me." Coordination is another school-related use of mobile phones. George said, "For school work it's quite important especially in coordinating my friends especially during the Project Work period. I need to SMS my friends and all," while Sheila noted, "The mobile phone is the connection between me and my group mates. I use it mostly to find people. And I use it to ask for homework and all." Emma explained, "I mainly use it to connect with partners, to set up time when we can meet in school or some places else for school works, and also SMS or just call directly, because some of them may not online all the time."

Not related to school work but to school life is the use of the phone to coordinate with parents during school days, as Larry, Alex, Wayne, George, and Sheila noted. Larry said, "Like if I was getting home late I could text them. I check it fairly often, even in the morning before I leave for school." Alex said, "I will message my brother because my mum tells me to message to say what time I am coming home." "For family, my mum makes me call her when I reach home so I will just call. Now I have unlimited SMS so I will just SMS her," George said. Indeed, among the informants, there was but a cursory use of the mobile for curricular activities. Chelo said, "It does not affect my school work and family life but affect my social life as I have to contact my friends." Joe echoed this, "Hardly affect school work. Affect more my social life."

CHEATING

A notorious use of mobile phones is in cheating. Fourteen of the informants have heard of cheating instances in which mobile phones were used. The rest, however, said precautions against cheating are quite strong so as to prevent cheating in their school. Connille explained that her teachers are vigilant in observing them during exams.

Heard but not Seen

Six of those who knew about cheating using mobiles did not encounter it firsthand. Instead, they said heard about it from somewhere. Jim said, "I have heard of it but not around me," which Eman echoed,

"I heard of it about someone in my school. But it's not someone that I know." Two others, George and Myra, heard about such cheating but did not know the details. George said, "I heard before. I don't know but I heard that they will SMS during exam. I heard that someone got caught during O Levels and are barred for 5 years but I don't know how they do it. And I never witness a situation before," while Myra said, "I know of but I didn't see one face to face before. No idea how they cheat though, because I didn't ask. I know things like that happen but don't know how they do it." Vivien, meanwhile, placed this hearsay in perspective. She said, "I've heard stories but they're always stories. Probably these are cautionary tales to tell us not to cheat."

Multi-Media Cheating

Nine informants, however, directly knew of how their classmates cheated using mobile phones. Chelo and Mark said SMS was used. Chelo said, "They find answers during exams or class tests in their phone, or message their friends to find out the answers." "Mark said, "My primary school friends--they did it during the class test. They sent each other the answers through text message." Danny and Michelle, meanwhile, said their classmates recorded information in their phones which were then retrieved during the exam. Danny said, "They took pictures of work sheets and notes, and wrote down some answers or notes that they think were important in SMS. They would just put it in the pockets or under the table and took out when needed." Michelle said, "Before the class test or quizzes, they put answers in the phone and then cheat."

Calls were also made during exams in order to cheat. Jake said, "I think I heard of one case during a quiz. My friend went to the toilet to make a call seek for external help. Like in the game show who wants to be a millionaire," while Alex said, "I witnessed it before. For oral exams, when there was different timeslots, the first person left the exam hall SMS the topic to other classmates, who were in the exam hall and then they knew the topic for oral. It's hard to cheat in classroom during test using phone." Chris, however, said the mobile unit itself was used to cheat. He said his classmates, "kept a second phone with them and dropped down notes in the phone. Each of them did a portion and exchanged the phones."

DISCUSSION AND IMPLICATIONS

Lessons on using the Mobile that Inform How Students Can Learn with It

The sense of self-efficacy (Morgan & Kennewell, 2005) and the perceived ease with which the informants learned how to use the mobile phone indicates how this particular technology can be readily used to facilitate independent learning (Lu, 2008; Clough, Jones, McAndrew, & Scanlon, 2008). As the informants said, it was quite easy to figure out how to use the phone by them as the interface was usable and navigable. Accordingly, there is promise in introducing curricular applications in mobile phones and using mobile phones to implement class projects. Thus, in the same way that schools sought to pull students towards wired personal computers at the early stages of the internet, they can currently harness the students' self-developed skills in mobile phone use. Moreover, doing so will integrate school work within the medium which students believe is an important and indispensable part of their daily life. It will also be grounded on the students' informal learning of the mobile phone features and functions.

In pursuing the use of mobile phones for school work, however, schools must be conscious of several issues. Firstly, students' use of mobile phones is intensive but not extensive. There is considerable use of SMS and voice calls, but the use of the multiple functions of current models is still limited. Less than half of the informants took pictures, listened to music, played games, or used the memo and organizer features of their mobile phones—features which m-learning activities use. Thus, there is still some training to do in terms of making students use their mobile phones towards a holistic m-learning experience, though the learning curve in this case is not steep given the students' record of figuring out how to use their phones. Secondly, schools must note what students purport as limitations of mobile phone units in information exchange, particularly in relation to the Internet. These constraints can be categorized into two: the physical features of the mobile phones and the attendant data services (Kang, Sung, Park, & Ahn, 2009; Wang, Wu, & Wang, 2009; Lu, 2008; Fallakhair, Pemberton, & Griffiths, 2007; Peng, Su, Chou, & Tsai, 2009). Inasmuch as its compact size and portability are among the values of mobile phones, the suggestion is not to provide bigger models but to provide byte-sized information and lessons which can be easily and readily accessed and read in mobile phone screens. Such packages would feature light images and concise text. This strategy would also address the universal belief that data services are expensive, particularly in relation to the speed with which data are transferred through the mobile. Thirdly, though it is not a dominant concern, m-learning would involve additional, and in this instance required, use of the mobile phone which some informants believe is linked to health concerns. Thus, schools must be prepared to answer such concerns if they were to support officially the academic use of mobile phones.

One of the supposed strengths of m-learning is in collaborative learning since mobile phones are by nature a social medium. Interestingly, however, few of the informants learned how to use their mobile phone with the help of others, which indicates a more intrapersonal than interpersonal approach to learning about the mobile. Thus, even if m-learning strategies were to promote collaboration, they would still need to regard mobile use as inherently personal. For instance, m-learning activities can involve group work but independently use of mobile phone units. That the mobile phone units were gifts from, and introductions about the mobile first transpired, within the family can be an opportunity to involve the family unit in school affairs, if not in schoolwork. To do so would be to continue familial involvement in the use of the mobile phone.

Lessons on How Students Learn with the Mobile That Inform Their Use of It

The limited use of mobile phones within the school can be a function of school policy and social norms. Mobile phones are seen as inconsistent with the values of education. Whereas informants, for instance, see school as a place for focused attention, they regard mobile phones as a possible source of distraction. Thus, the greater challenge for schools which seek to pursue m-learning is not the training of students to use mobile phones but the changing of how mobile phones are regarded within the school. If schools would mainstream mobile phone use as part of school activities, however, they would have to contend with several extraneous issues. Students are quite concerned about respecting what they consider as etiquettes in mobile phone use in the school—limiting use to breaks outside of the classroom or using SMS so as not to distract lectures within the classroom—because they know this is a private activity (Campbell, 2006). Sanctioning use, however, would translate mobile phone use into a public activity which would then necessitate the development, if not the implementation, of new rules about it. This concern gains more credence especially if m-learning modules would include gaming and videos (Chuang, 2009).

Incidences of alleged cheating using the mobile phone (Campbell, 2006), even now when they are still considered as relatively taboo objects in the school, need to be considered in the introduction of m-learning. m-Learning exercises must be designed such that the exchange and the access of information among friends and from the internet is part of the learning experience and objectives. This way, cheating by way of retrieving supposedly not-to-be-accessed information becomes a moot issue. Instead, the processing, analysis, and synthesis of data using the mobile platform should be the focus rather than the rote memorization of school material. Testing students in a mobile platform but within the same set-up of a traditional closed-book exam would be against the ubiquity principle of mobile computing (Al-Fahad, 2008; Kang, Sung, Park, & Ahn, 2009; Suki & Suki, 2007; Fallakhair, Pemberton, & Griffiths, 2007).

Current patterns of mobile phone use for class work are limited to networking for school project, a practice which is aligned with the use of mobile phones for micro-coordination. Already, this indicates the potential of mobile phones for collaborative learning as schools can develop projects which require students to coordinate information using the mobile (Chuang, 2009; Chao & Chen, 2009; Campbell, 2006). However, juxtaposed with the dominant use of the mobile phone for social networking (Chuang, 2009; Fisher & Baird, 2006; Cavus & Ibrahim, 2009) and that this practice remains private in nature even if it is done for school activities, there remains the concern whether scholastic use of the mobile phone would be tantamount to a invading the private sphere of students. Informants complained about how their privacy seems to be intruded by undue calls from friends—imagine how they would feel if they were to be required using the phone for an official exercise.

LESSONS ACROSS DEMOGRAPHICS

An analysis of informants' experiences and insights across gender, educational level, and income surprisingly show little differences in how they learn about and with the mobile. Income is not a differentiating variable because all the informants had a mobile phone and the same sense of self-efficacy and experience in using it. Regardless of income, too, informants complained about the high cost of data services. In a country that is less affluent than Singapore, perhaps income would be a crucial variable since those in the lower income bracket would be less able to afford a mobile phone and this would then redound to issues of self-efficacy, patterns of use, and the valuation of the mobile phone. Educational level is important insofar as policy is concerned, particularly with secondary schools which do not allow students to use mobile phones in school. This poses a greater constrain among younger students than among older ones. Nevertheless, students, seemingly regardless of such policy, still maintain the same habits in the use of mobile phone within and outside the school and regard the mobile phone with as much importance. Having said that the practices and regard for the mobile phone is generally the same, one group distinguished itself in one measure. Male university students are far ahead in terms of using mobile internet, which is in line with literature on technological adoption that young, relatively well-educated men are more likely to avail of new technologies. However, literature also indicates that such a difference will eventually disappear as the new technology matures. Thus, basic m-learning can be applied with less regard for educational level and income but m-learning that uses the internet can favour male university students with their earlier adoption of this technology.

CONCLUSION

The mechanical, but not automatic, reflexivity (Lynch, 2000) between learning about and with the mobile surfaces lessons regarding m-learning. If m-learning is to be grounded on current practices, then it must harness independent learning that contributes towards collaborative work. Moreover, though m-learning strategies can readily build upon the self-developed skills of students in using the mobile phone, these initiatives need to consider instead the regard for the mobile phone as a device that contradicts the values of education and a device that serves a primarily social function. Since schoolwork is seen as requiring attention and mobile phones are considered as a course of distraction, such strategies need to highlight how mobile phone use can be seriously focused towards an educational goal without disrupting, on the contrary enriching, the learning experience. Moreover, appropriating the mobile phone for collaborative learning can be grounded on the current use of mobile phones for coordination and communication among friends and peers. However, this can also be seen as an intrusion of an official requirement in the primarily private domain that is the one's mobile phone. Finally, m-learning modules must be grounded on a pedagogy which values the access, processing and synthesis of ubiquitous material rather than the regurgitation of content that is considered off-limits during evaluation exercises. Such modules must also be concise both in terms of imagery and text so that they are tailored to currently small mobile phone screens and keyboards as well as slow data streams. Eventually, with improved network services, screen resolutions, and interfaces, m-learning modules can be expanded.

One of the major potentials of m-learning is towards informal learning. However, as this study shows, students' informal learning of the mobile phone can inform formal learning initiatives on m-learning. Similar to how m-learning necessitates pedagogy of ubiquitous computing, it also calls for a continuous stream of learning that is informed by formal and informal modes of learning.

As this is an exploratory study, future research can look at other variables that can possibly influence learning about and with the mobile. Among these variables are level of usage (which is not necessarily related to cost and income issues), school type (in terms of policy and practice towards information and communication devices), and student personality. Moreover, subsequent research can also determine quantitatively patterns of mobile phone use, mobile phone etiquette and evaluation so as to establish, if any, possible user-based digital divides among groups of young people. Current statistics indicate there is no ownership divide among them, but this does not necessarily mean all of them are harnessing the potential of technology similarly. As this study indicates, practices and perceptions of mobile phone use can redound to performance in future modes of learning.

REFERENCES

Al-Fahad, F. N. (2008). Student Perspectives About Using Mobile Devices in theirStudies in the King Saud University, Kingdom of Saudi Arabia. *Malaysian Journal of Distance Education, 10*(1), 97–110.

Campbell, S. W. (2006). College Classrooms: Ringing, Cheating, and Classroom Policies. *Communication Education, 55*(3), 280–294. doi:10.1080/03634520600748573

Caronia, L. (2005). Mobile Culture: An Ethnography of Cellular Phone Uses in Teenagers' Everyday Life. *Convergence (London), 11*(96), 96–103. doi:10.1177/135485650501100307

Cavus, N., & Ibrahim, D. (2009). m-Learning: An experiment in using SMS to support learning new English language words. *British Journal of Educational Technology, 40*(1), 78–91. doi:10.1111/j.1467-8535.2007.00801.x

Chao, P.-Y., & Chen, G.-D. (2009). Augmenting paper-based learning with mobile phones. *Interacting with Computers, 21*(3), 173–185. doi:10.1016/j.intcom.2009.01.001

Chuang, K.-W. (2009). Mobile Technologies Enhance The E-Learning Opportunity. *American Journal of Business Edueation, 2*(9), 49–54.

Clough, G., Jones, A., McAndrew, P., & Scanlon, E. (2008). Informal learning with PDAs and smart-phones. *Journal of Computer Assisted Learning, 24*(5), 359–371. doi:10.1111/j.1365-2729.2007.00268.x

Fallakhair, S., Pemberton, L., & Griffiths, R. (2007). Development of a cross-platform ubiquitous language learning service via mobile phone and interactive television. *Journal of Computer Assisted Learning, 23*(4), 312–325. doi:10.1111/j.1365-2729.2007.00236.x

Fisher, M., & Baird, D. (2006). Making mLearning work: Utilizing mobile technology for active exploration, collaboration, assessment, and reflection in higher education. *Journal of Educational Technology Systems, 35*(1), 3–30. doi:10.2190/4T10-RX04-113N-8858

Goh, T., & Kinshuk. (2006). Getting Ready for Mobile Learning—Adaptation Perspective. *Journal of Educational Multimedia and Hypermedia, 15*(2), 175–198.

Hansen, A., Cottle, S., Negrine, R., & Newbold, C. (1998). *Mass Communication Research Methods.* New York: New York University Press. doi:10.1007/978-1-349-26485-8

Jensen, K. B. (2002). *A Handbook of Media and Communication Research: Qualitative and Quantitative Methodologies.* London: Routledge. doi:10.4324/9780203465103

Kang, H. D., Sung, K. W., Park, M., & Ahn, B. K. (2009). The Effect of Blended English Learning Program of CDI on Students' Achievement. *English Teaching, 64*(4), 181–201. doi:10.15858/eng-tea.64.4.200912.181

Lai, C.-H., Yang, J., Chen, F., Ho, C., & Chan, T. (2007). Affordances of mobile technologies for experiential learning: The interplay of technology and pedagogical practices. *Journal of Computer Assisted Learning, 23*(4), 326–337. doi:10.1111/j.1365-2729.2007.00237.x

Lu, M. (2008). Effectiveness of vocabulary learning via mobile phone. *Journal of Computer Assisted Learning, 24*(6), 515–525. doi:10.1111/j.1365-2729.2008.00289.x

Lynch, M. (2000). Against Reflexivity as Academic Virtue and Source of Privileged Knowledge. *Theory, Culture & Society, 17*(3), 25–54. doi:10.1177/02632760022051202

Mason, J. (2002). *Qualitative Researching.* London: Sage Publications, Inc.

Morgan, A., & Kennewell, S. (2005). The Role of Play in the Pedagogy of ICT. *Education and Information Technologies, 10*(3), 177–188. doi:10.1007/s10639-005-2998-3

Peng, H., Su, Y.-J., Chou, C., & Tsai, C.-C. (2009). Ubiquitous knowledge construction: Mobile learning re-defined and a conceptual framework. *Innovations in Education and Teaching International*, *46*(2), 171–183. doi:10.1080/14703290902843828

Suki, N. M., & Suki, N. M. (2007). Mobile phone usage for m-learning: Comparing heavy and light mobile phone users. *Campus-Wide Information Systems*, *24*(5), 355–365. doi:10.1108/10650740710835779

Wang, Y.-S., Wu, M.-C., & Wang, H.-Y. (2009). Investigating the determinants and age and gender differences in the acceptance of mobile learning. *British Journal of Educational Technology*, *40*(1), 92–118. doi:10.1111/j.1467-8535.2007.00809.x

Chapter 2
Language Learners and Mobile Technology:
How They Interact?

Saleh Al-Shehri
King Khalid University, Saudi Arabia

ABSTRACT

In order to understand the influence of mobile social media vs. formal learning platforms on creating effective student-student and teacher-student communication channels and linguistic outputs, this study was conducted. Using a qualitative approach, a number of 30 English language university teachers were interviewed. Evidence from their mobile and non-mobile interaction with their students was analyzed to support data from the interviews. The study evaluates the potential of both formal and informal communication mediums to maintain student-centered language learning experience. The study concludes that teachers still need to be aware of the potential of mobile technology and social media for language learning, and that there was a tendency among some teachers to implement formal technology tools for their teaching.

INTRODUCTION

Mobile technology is being widely used by different users around the world for different purposes. Researchers started to study the behavior of mobile users, and how they are interacting via their handheld devices, and for what reasons. There have been different behavioral patterns that vary between users and contexts. In particular, researchers in education investigated the potential of mobile technology for learning, and attempted to establish specific theoretical approaches for mobile learning.

This study reviewed key pedagogical approaches that have been utilized in different mobile learning contexts. Using a constructivist framework, the study explored human-mobile interactional patterns, and how such patterns can inform mobile language learning. The study investigated prior and current practices of mobile users and their potential for language learning, and how mobile learning has influenced informal learning. Preferred mediums of interaction for both English language university teachers

DOI: 10.4018/978-1-5225-0469-6.ch002

and students were explored. The distinction between formal and informal technologies, and their affordances for mobile language learning have also been investigated. The study was conducted with 30 EFL (English as a Foreign English) teachers at a Saudi Arabian university. Qualitative data obtained from interviews with teachers illustrated that there were differences in terms of mobile literacy and conceptions of mobile learning between students and teachers. Teachers themselves viewed mobile learning differently based on their technological background, age, and teaching experience. The study concluded that teachers were hesitant to use innovative mobile and social media tools for several reasons. It also illustrated that students needed to be aware of mobile learning potential, and how their mobile devices could be powerful language learning tools.

LITERATURE REVIEW

One important aspect of the study of human-mobile interaction is investigating the relationship between users and mobile technology. According to Zimmermann, Henze, Righetti, Rukzio, and Enrico (2009), mobile devices, including mobile phones and mobile media players, have become the most pervasive ways that people interact with technology, and are used for a huge variety of services and applications. In de Sa and Carrico's (2010) words:

Our lives and society are rapidly gravitating towards a world in which mobile devices are taken for granted. These are no longer accessory but become natural extensions to our bodies. Their use has become quasi-permanent, following and supporting us in most of our activities and affecting the way we interact, share, and communicate with others. (p. 176)

This means that many people today are extremely emerged in mobile devices that have dramatically changed and are changing their way of life. Young people, often referred to as mobile or digital natives, are more frequent users of mobile devices since they usually tend to adopt more flexible lifestyles. Mobile devices have greatly changed the way they think, communicate, entertain, and above all, learn. In particular, students' life experiences and home culture can effectively be connected by mobile phone technologies, and students' own cultures were found to be apparent in their classroom learning activities thanks to mobile social media (Kolb, 2008). Traxler and Kukulska-Hulme (2016) note that the next generation of mobile learning is becoming 'context-aware', and that mobile learning takes advantage of students' rich environments inside and outside the classroom. Hence, students are encouraged to capture aspects of their environments, and integrate informal mobile media into their formal learning experiences. The divide between formal technologies, implemented by learning institutions, and informal technologies mobile tools needs to be narrowed.

The integration of technology into learning has considerably improved different pedagogical practices and teaching methodologies, and has helped to obtain better understanding of how mobile context-awareness derives informal learning activities (Clough, 2016). Both students and educators have benefited from the great potentials that technology offered for learning and teaching over the last 3-4 decades. As technology has shrunk into more powerful and portable forms, learning has effectively exploited the potential of these forms, i.e., mobile technologies, particularly when we realize that the vast majority of mobile users are our young students. However, systematic analysis of mobile practices is required before mobile technology is implemented in any educational context. Arrigo, Fulantelli, Gentile, and

Taibi (2016) suggest that analysis of mobile practices should be conducted in four areas: management, pedagogy, policymaking, and ethics. A synthesis of mobile pedagogies is reported in this chapter since it is not aimed at exploring the other areas. This should help to narrow the gap between formal and informal educational technologies, and develop our understanding of how informal mobile media can enrich language instruction in the Saudi context. In the context of the present study, it was noted that effective social mobile technologies, with which students were highly familiar, were not effectively employed for learning purposes. Students' lack of effective communication skills in English was also observed.

HUMAN-MOBILE INTERACTION: HISTORICAL/TECHNOLOGICAL OVERVIEW

Computer manufacturers started to design portable computers to enable users to benefit from certain computer functionalities at non-predetermined locations. Initial mobile computers, or laptops, did not contain all features that were available in desktop personal computers or PCs. Later on, laptops became more powerful devices with higher speed, larger memory, longer battery life, and overall semi-PC different functions. This improvement has led to the invention of other mobile devices such as mobile phones, media players, personal digital assistants or PDAs, digital cameras, and so on.

As modern mobile phones became equipped with most, if not all, of the functionalities of PDAs, mobile users started to rely more on mobile phones. Thus, PDA manufacturers started to focus on more effective devices that combined all the technological characteristics of both PDAs and mobile phones. Earlier mobile phones could receive (input) and send (output) data through different means including photography and audio-visual capture, synchronization to computers, and wireless connection (Infrared, Bluetooth, Wi-Fi, etc.). Current mobile phones can handle other computational (input) features such as typing (either through buttons or touch screens) and saving data on external memory (e.g. SD or mini/micro SD memory cards). Memory cards can be removed from a mobile phone and be directly connected to a computer in order to upload or download data. Other mobile phones are equipped with built-in memory.

Mobile devices, from a technological perspective, have been criticized for having a number of factors that may act as barriers to effective usage (see for example Mellow, 2005; Shuler, 2009). Screen size and the presence or absence of a keypad were essential issues that may have had a negative impact on mobile use in a way that made PCs or laptops more preferable. Small screen size has been considered by several mobile phone manufacturers. Many of the latest mobile phones have been equipped with larger screens with higher resolutions. For this purpose, the total shape of the mobile phone is usually dominated by the screen, and functions that were done by buttons are now being done through fully touch-screen control. Moreover, pop-up on-screen keyboards has have also been designed as compensation for the lack of physical traditional physical keypads.

However, the rapid development in mobile technology, mobile phones in particular, attracted more users (Shuler, 2009). Among the factors that increased the penetration of mobile phones were internet connectivity, more user-friendly interfaces and navigation, mobile entertainment, and mobile social media. According to Quinn (2012), access to mobile networks is available to about 90% of the world population.

Today's mobile phones, or smart phones, are meant to adopt most, if not all, of computers' applications and features, but with a smaller size and smarter appearance. In other words, functions that were previously limited to PDAs or the like are now within reach of mobile phone users including internet connectivity, data input and output, keyboards, and navigation (see Table 1). Due to this rapid history of development, it is hard to predict what the future of mobile phones, or mobile technology in general, will look like; most of the services and functionalities that are available now could not even be imagined by most mobile phone users 5 or 10 years ago. In the future we may experience a mobile phone that is as small as a pen, or as slim as a credit card, and future phones will no doubt have functions that are currently just as unpredictable. It is expected that mobile technology of the future will be more dynamic and functional, and that most, if not all, of the shortcomings of contemporary mobile devices will be dealt with sooner than we think (table 1).

HUMAN-MOBILE INTERACTION: EDUCATIONAL OVERVIEW

The wide use of mobile technology as well as the growing trends of mobility in the world have demonstrated the benefits of mobile devices as learning tools. This is true since mobile devices, mobile phones in particular, have become integral parts of students' social life. The potential of mobile phones for learning can be based on two attributes: the large rate of ownership of mobile devices among students,

Table 1. Summary of technological attributes of some current mobile phones

Type	Attributes			
	Communication	*Multimedia*	*Data Entry & Connectivity*	*Other Functions*
Nokia Mobile computer *(smart phone)*	• Phone & video calls • SMS & MMS • 3G Internet (push email, IM) • WAP Internet browser (HTML & RSS Feeds)	• MP3 Player • MPEG4 Player	• Bluetooth • Wi-Fi • 4G • Synchronization with PCs (micro USB) • Internal GPS • Full QWERTY keyboard • 2 cameras	• *Operating System*: Symbian • Pocket Office (Word, Excel, PowerPoint), PDF • *Memory*: Up to 64 GB (internal & external) • Voice command/dial • Digital compass • Stereo FM radio
Samsung Pocket PC *(smart phone)*	• Phone & video calls • SMS & MMS • 3G Internet (push email, Microsoft Outlook Mobile, & Microsoft Internet Explorer Mobile)	• Windows Media Player • MP3 Player • RealPlayer	• Bluetooth • Wi-Fi • 4G • Synchronization with PCs • Internal GPS • Touch screen with stylus • 2 cameras	• *Operating System*: Android (Google) • Microsoft Office Mobile (Word, Excel, & PowerPoint) • *Memory*: Up to 64 GB (internal & external) • Microsoft Voice Command (voice recognition)
iPhone Smart phone	• Phone calls FaceTime ® • SMS & MMS • 4G Internet (push email) • Safari web browser	• Digital media player • iTunes store (photos, videos, games, etc.)	• Bluetooth • Wi-Fi • 4G • Synchronization with PCs • Internal GPS • Multi-touch screen • 2 cameras	• *Operating System*: iPhone OS (Apple) • Compatible with Microsoft's Word, Excel, PowerPoint, & Outlook) • *Memory*: Up to 132 GB (internal only) • Text and voice recognition
HTC Smart phone	• Phone calls only • SMS & MMS • 4G Internet (push email) • WebKit-supported web browser	• MP3 Player • Media player	• Bluetooth • Wi-Fi • 4G • Synchronization with PCs (Mini USB) • Internal GPS • QWERTY keyboard • Capacitive touch screen • 2 cameras	• *Operating System*: Android (Google) • *Memory*: Up to 64 GB • Voice dialing • Digital compass • Motion sensor

on the one hand, and the educational need to engage students in informal learning environments without the constraints of typical classrooms or formal learning management systems (LMSs) on the other. Regarding the first attribute, students' familiarity with various mobile technologies can be attributed to the wide extent of ownership of mobile devices, which, in turn, means that there is no require need for additional training for either students or teachers. The other, as Kukulska-Hulme (2005) notes, requires nothing more than the motivation to learn wherever and whenever there is a learning opportunity. Thus, the awareness of the potential of mobile technologies as effective learning tools should be improved among both teachers and students.

The need for a theory of mobile learning is valuable in providing educators with a framework of learning with technology, when a learning approach or philosophy needs to be expanded to adopt new technologies (Levy, 1990). The recent implementation of mobile technology into learning has not yet led to a specific learning model or theoretical framework for mobile learning (Herrington, Herrington, & Mantei, 2009). According to Kolb (2008), even pre-service teachers, who grew up with mobile phones, have never experienced a learning model for mobile technology, and therefore, they have not realized the potential of mobile phones as learning tools. As a result, the lack of a mobile learning theory in different learning settings, Shuler (2009) argues, will affect the optimum potentiality of mobile technology. Consequently, unless practice is guided by theory, the effectiveness of mobile technology for learning will not be utilized to the fullest. In other words, theoretical foundations that can inform the instructional design of mobile learning should be made clear to educators and teachers.

For this reason, researchers like Naismith, Lonsdale, Vavoula, and Sharples (2004) insist on specifying a mobile learning theory that would enable an effective pedagogy and design of mobile learning applications. For example, based on the Activity Theory, Sharples, Taylor, and Vavoula (2005) analyze learning as an activity mediated by tools, and contend that tool-mediated activity can be classified into a semiotic layer (object-oriented learning through cultural tools) and a technological layer (technologically-mediated learning tools). Both layers can either be separately used to promote a semiotic framework (mobile education), or a technological framework (design of mobile applications), or collectively utilized to examine both learning and technology. In summary, Sharples et al. (2005) suggest that a theory of mobile learning must be tested against the following criteria:

- Is it significantly different from current theories of classroom, workplace or lifelong learning?
- Does it account for the mobility of learners?
- Does it cover both formal and informal learning?
- Does it theorize learning as a constructive and social process?
- Does it analyze learning as a personal and situated activity mediated by technology?

The answers to these fundamental questions not only help to establish solid theoretical foundation for mobile learning, but they also highlight more informal learning opportunities.

Despite the theoretical "pluralism", as described by Levy and Stockwell (2006), which has been a subject of debate in the literature, as well as the lack of specific ties to a learning theory, mobile technology is well placed to be educationally adaptive. For example, mobile learning has been implemented to support a variety of learning models and approaches (see discussion below). This indicates that mobile learning is flexible and adaptive, and can be utilized in different learning contexts with different populations, cultures, pedagogies, and technologies.

The lack of a theory for mobile learning, however, urges an 'integrated pedagogy' that is drawn from a number of theoretical areas (Naismith et al., 2004). Yet, given that mobile learning has not yet been adapted to a specific learning/teaching theory or approach, and does not currently even stand on solid theoretical ground at all, as Sharples et al. (2005) claim, researchers investigating its pedagogical potential are raising several research questions about the validity of mobile learning, which need to be operationalized in order to be answered. In other words, the soundness of mobile technology and recent practices are pushing the field of education towards the establishment of theoretical grounds for mobile learning, which is a matter of time and more practice.

In addition, the number of empirical studies dedicated to mobile learning are still limited and/or subject to the improvement of mobile technology. This makes the integration of new technology with pedagogy, as Mellow (2005) points out, a significant challenge. Accordingly, research has identified another pedagogical potential for mobile learning. For example, Armatas, Holt, and Rice (2005) argue that a proposal for the use of mobile phones in a learning context can be embedded within curriculum development, virtual online resources, and approaches of assessment. Stockwell (2008) agrees with the curriculum factor and asserts that mobile technology, in learning, has to have an "integral and integrated role in the curriculum rather than the more peripheral role that it is often placed in now" (p. 266). In other words, the use of mobile phones in learning is not just for the sake of technology, but rather, for improving pedagogy; i.e. technology-enhanced rather than technology-driven pedagogy.

In summary, mobile learning has been utilized in different contexts, using different learning approaches and theories and has been aimed at manipulating different pedagogical goals, in an effort to investigate the potential of mobile technology, mobile phones in particular, as effective learning tools. Furthermore, the application of different learning theories such as cognitive theory (individualized learning) or sociocultural theories (social aspects of learning) does not mean that cognitive approaches deny the impact of social learning, or that social approaches deny the role of individual learning, but rather, the choice of approach is a "matter of where the priorities are placed, and the territory over which the theory may be effectively applied" (Levy & Stockwell, 2006, p. 111). In this sense, mobile learning is, again, an adaptive form of instruction that can be based on different theoretical grounds in different contexts for different learners and different educational purposes. A range of mobile learning approaches and theories as well as learning practices are reviewed in the next section.

Mobile Learning: Key Theoretical Concepts

Learning theories and approaches that can provide a theoretical background for the practice of mobile learning can be categorized into three main paradigms:

1.   The behavioristic/structural and cognitive/psychological paradigm;
2.   The communicative, constructivist, or collaborative paradigm; and
3.   The supportive paradigm that comprises of different learning approaches including informal and lifelong learning.

This section is a brief discussion of these three theoretical paradigms, with a focus on the present study's constructivist framework. This section will also highlight how some elements in the constructivist framework can be translated in an informal language learning setting.

Behaviorist learning focuses on learning activities that promote observable changes in the behavior including the presentation of learning materials, students' responses and appropriate feedback. According to Naismith et al. (2004), the behaviorist paradigm is based on Skinner and Pavlov's work; learning is facilitated by the reinforcement between a stimulus (presenting a problem through the computer or the mobile device) and the student's response (a solution). This reinforcement-enabled association between students' stimuli and responses has also led O'Malley, Vavoula, Glew, Taylor, Sharples, and Lefrere (2003) to describe this approach as Associationism.

In second language teaching, the audio-lingual approach emerged in the late 1950s and emphasized exposure to the target language spoken forms through the process of habit-formation learning; i.e. imitation and repetition (Levy, 1997). Computer-based programmed instruction, Levy (1997) continues, was also a reflection of Skinner's behaviorist work which focused on learning individualism and the entailment of immediate feedback. Moreover, the "drill and practice" learning activity, highlighted by the audio-lingualist approach, is smoothly practiced due to the computer's programmability and systematic performance.

A mobile-based student response system involves a hybrid model of "drill and practice" (O'Malley et al., 2003) or "drill and feedback" (Naismith et al., 2004) that allows teachers to present questions and gather responses through students' mobile devices. A drill and feedback model can also maintain effective discussion, anonymity, and rapid collection of responses. However, although behaviorist mobile learning can cope with the routine daily use of mobile devices, designers of mobile applications may be challenged by providing entertaining and motivating learning applications (games or multimedia) that can maintain effective learning. This has led Göth and Schwabe (2008) to describe behaviorist mobile learning as a transmissive low-engagement learning, in which a learner is engaged with the learning environment by just observing it. Also, the behaviorist trial and error approach, known as generate-and-test in computer science, can entail practical learning steps and enhance knowledge. However, mobile applications, as Figueiredo, Almeida, and Machado (2004) point out, have to be tested in order to promote productivity and reuse. In other words, it is important to ensure, under a behaviorist mode, that mobile applications allow for frequent trial and error to happen in order to maintain learning. In addition, interactivity, rather than observation, which allows learners to perform tasks and effectively respond to the environment, also needs to be maintained.

Applications of mobile technology that support trial and error are well-characterized by mobile games. A 2006 study by the Federation of American Scientists, as cited by Pulman (2008), argues that "games could teach skills including strategic thinking, interpretive analysis, problem solving, forming and executing plans and adapting to rapid change" (p. 256). Specifically, mobile games can successfully train students' brains to think and solve problems in a creative way, and enhance students' activity-awareness (Liu, Tao, & Nee, 2008). Although they are useful and engaging, mobile games need to be carefully integrated into learning in order to ensure they are serving an educational purpose. On the other hand, mobile games are the best method for smoothly coping with the simplicity and representativeness of mobile technology. Simple games, as emphasized by Cheung (2004), have been always effective. Besides, mobile games are expected to play a partial and facilitative, not a fundamental, role in the learning process.

The other learning theory that resulted as an attempt to include other factors that were, to some extent, neglected by behaviorists is constructivism. The late 1960s and early 1970s, as Levy (1997) notes, witnessed this shift that expanded the scope of learning in order to include developmental, cognitive and psychological aspects such as attention, memory, language, reading and writing, problem-solving, etc. According to Naismith et al., (2004), constructivists believe that learning represents the construction of new concepts that are based on both prior and present knowledge. Constructivist learning was inspired by the cognitive theories of Bruner and Piaget. O'Malley et al. (2003) also state that Bruner's synthesis of the constructivist approach was inspired by the significance of the representational aspects of learning demonstrated by Piaget, and by the adaptation of the psychological Vygotskian theory in education. In other words, constructivism looks at learning from a panoramic position that takes into account how immediate knowledge affects previous knowledge, and how this mix of previous and existing knowledge can be affected by, or can even affect, different cognitive styles of learning.

In the field of second language acquisition or SLA, McLaughlin (1990) argues, as cited by Levy (1997), that cognitive psychologists investigate language learning with regard to language presentation, use, and communication, with emphasis on language automaticity and restructuring. Thus, it can be concluded that cognitive psychology cannot account for SLA from a holistic perspective due to linguistic universality. In other words, cognitive psychology resides deeply within linguistics, and therefore, it can only account for specific aspects of language learning. However, cognitive psychology has contributed to the development of sound instructional designs and cognitive approaches that are educationally effective. For example, cognitive load theory or CLT is a theoretical framework that accounts for the cognitive processing of memory and knowledge attainment. John Sweller's theory of CLT pays attention to multi-modal representations (e.g., audio-visual) as well as short term and long term working memory. CLT claims that optimal learning takes place when the cognitive load is kept to its minimum levels (Sweller, 1988). Notably, mobile learning, as claimed by Chen, Millard, and Wills (2008), has the potential to provide basic learning materials in order to avoid cognitive overload, and to enhance short-term working memory. In practice, although mobile technology can provide a variety of representations of learning, avoidance of oversimplification could be a challenge for both mobile designers and educators.

Cognitive-constructivists also believe that the learning process simply corresponds to the child's developmental stages (Naismith et al., 2004). Therefore, the computer is actively used by students to construct their own knowledge and learning. Hence, mobile devices are thought by these theorists to be superior in providing 'realistic contexts' to the learning environment. For example, mobile context-based learning allows learners to use their mobile devices in different locations. Moreover, mobile devices, as Cochrane (2005) notes, are meant to facilitate and construct a reflective student-student and teacher-student interaction in a convenient learning format. In other words, mobility reflects the learner's real-world in its actual forms, which may not be fully achieved within the boundaries of the classroom.

Naismith et al. (2004) also state that the implementation of constructivist concepts and mobile technology entails a 'participatory simulation' approach that allows students to be part of the simulation process of mobile learning. However, such a simulative model is best to maintain effective and dynamic feedback and guidance by the instructor that suit the mobile nature and constructivist scaffolding. Bruner's metaphor of scaffolding, as noted by Chen et al. (2003), involves the "interactive support that instructors, or more skillful peers, offer learners to bridge the gap between their current skill levels and a desired skill level" (p. 348). Most importantly, Bruner's recent extension of the simulative approach, as cited by O'Malley et al. (2003), sheds light on the social and cultural aspects of learning regardless of the age-related developmental changes basically assumed by constructivists. In other words, Bruner enabled the constructivist approach to be more applicable to mobile learning in terms of the learner's context (contextual) and the representational changes. However, instructors, in simulative mobile learning, should bridge the gap between present and intended skills including students' individual differences and mobile literacy. Furthermore, the simulative approach highlights the learning context and adopts a developmental procedure that could account for students' cognitive variability.

In practice, game-based mobile learning, for the trial and error behaviorist approach, is a representative of the participatory simulation approach, in which students are engaged in a learning activity in a simulative way. Game-enabled mobile vocabulary learning, for instance, was implemented by Fotouhi-Ghazvini, Earnshaw, Robison, and Excell (2008) who designed a Java-based mobile game called The MOBO City for 15 ESL Iranian university students. Their study investigated the influence of game-based learning on the hierarchical structuring of knowledge, incidental vocabulary learning, and how mobile games could promote the brain to store and organize information through verbal and visual materials in an interesting way. They concluded that educational mobile games were motivating and helpful in enhancing both incidental and deliberate vocabulary learning. It should be noted, however, that learning through entertainment has widely been utilized in many learning contexts and found facilitative. Accordingly, instructors should be cautious about the distinction between mobile phones as entertainment devices, and as learning tools. Matthee and Liebenberg (2007) implemented a learning application called M☺BI Maths, which adopted a joyful learning approach to teach mathematics. This is also supported by Suki and Suki (2007), who remark that mobile learning has the ability to educate and entertain through what can be called "edutainment".

O'Malley et al. (2003) add another dimensional theory that resides under the constructivist approach called cognitive flexibility theory. The major assumption of this theory is that multiple representations as well as the avoidance of oversimplification of the content have to be maintained. However, for the cognitive load theory, the current practices of mobile learning should take into consideration that mobile technology is rapidly evolving, and therefore, mobile technology can provide learning with comprehensive tasks and activities that are meant to be simple rather than oversimplified.

Information processing theory is another cognitive theory that focuses on mental representations and problem solving. According to O'Malley et al., (2003), information processing theory has two major paradigms. The first states that learning is a matter of problem-solving, memory operations, and chunking of information. The other paradigm focuses on memory processes and connects between existing and prior knowledge by making inferences.

A possible limitation of such learning theories is that they are basically teacher-driven. As a result, other student-centered theories that pay attention to the assessment of learners' development and individual learning have been introduced. One problem with a teacher-driven mobile practice is teachers' inability to manage and control every bit of a task due to their lack of knowledge about the potential of mobile learning. However, such mobile learning practices "are useful only once the learning goals and the knowledge domains have been well-articulated, in particular, for example, where a learner needs to learn a certain, well-specified procedure" (O'Malley et al., 2003, p. 14). Arguably, the openness of mobile learning makes it difficult for some learning activities to be well-specified. For example, a context-aware or location-based mobile learning activity relies on what a student is expected to encounter, rather than what has already been planned by the teacher. Thus, students are encouraged to take part in specifying their learning goals, procedures, and assessment.

Adult learning approaches, as categorized by O'Malley et al. (2003), including student-directed and autonomous learning, have been utilized under several frameworks such as experiential learning (transforming different experiences into learning opportunities). O'Malley et al. (2003) highlight the construction of meaning which takes place when a student reflects on his/her own experience and transforms it into meaningful schemes. Other theoretical frameworks such as the activity theory and conversation theory have paid attention to adult collaborative and social learning.

Although adult learning approaches have been focused on mobile learning by a number of researchers (see for example Ally, Schafer, Cheung, McGreal, & Tin, 2007; Nalder, Kendall, & Menzies, 2007), the learning domain is in need of developmental cognitive theories that pay attention to both younger and inexperienced students.

As far as cognitive theories are concerned, a number of research areas including cognitive psychology and educational psychology were focused in several contexts of computer-assisted language learning or CALL (Levy, 1997). Similarly, a number of mobile language learning studies were aimed at enhancing basic SLA approaches including communicative teaching/learning, negotiation of meaning, and motivation (Kiernan & Aizawa, 2004) For example, Nah, White, and Sussex (2008) conducted a mobile-based study which focused on a mix of SLA theories and approaches including comprehensible input of the target language or language intake, interaction or collaboration facilitated through the negotiation of meaning between students, and comprehensive output. Improvements in these theories and approaches of SLA are described by Bush (2008):

Attention turned from linguistic competence to communicative competence and the primacy of input was replaced by an emphasis on output. Now, instructed SLA is on the pedagogical scene, informing us that learning really can lead to acquisition. (p. 450)

As far as collaborative and communicative learning was concerned, several learning management systems or LMSs have been implemented at many academic institutions to improve learning outcomes, and to create better communications channels between students and teachers. Some of these learning systems are made available via mobile access as a result of the high mobile phone penetration rate among students. Synchronization between learning systems such as Blackboard and popular social media such as Facebook and Twitter has also been supported, in an attempt to invest students' time spent on social media for learning purposes.

MOBILE SOCIAL MEDIA IN THE ARABIC CONTEXT

In the Arab world, social media has found its way into education. Several Arabic institutions have implemented different social media for learning purposes after they realized the wide spread of social media among students. According to the Arab Social Media Report (2014), the country average for Facebook penetration in the Arab world was over 21.5 percent in 2014 compared to 15 percent in 2013, and 67 percent of Arab Facebook users were between 15 and 29 years as of May 2014. For Twitter, active users in the Arab world numbered over 5 million as of March 2014, and that Saudi Arabia had the highest number of active Twitter users in the Arab world with 2.4 million users, producing 40 percent of all tweets in the region. For other social media tools like YouTube, it was reported that Saudi Arabia leads the region with the most playbacks, with 50 percent of all views from mobile devices (Arab Social Media Report, 2013).

However, 2014 witnessed a drop among Arab Facebook users. This could be attributed to the fact that some users, particularly younger users, started to use other newer social media tools such as Snapchat. Khan (2015) argues that Arab youth found in this 'live stories' tool an effective mobile medium that supports geo-tagging for an entire city or live events that can be shared to other Snapchat users. This modern user-generated mobile tool "marks a change in the digital age – signaling the bolstering of bottom-up information transfer as a popular and effective model" (Khan, 2015, para. 4). Snapchat has dramatically globalized different events taking place in the Arab world. Arab Snapchat users usually create a unique hashtag that grabs users' attention from all over the world. For instance, millions of users from all over the world react to the live stream of rituals and prayers captured from Mecca. Many people appreciated Snapchat for "changing the perceptions others had of Islam, commenting that it made a change from the usual coverage" (Mandhai, 2015, para. 3). Other incidents that are taking place in the Arab world have also found their way onto social media and achieved global attention, including the demonstrations, protests and uprisings that have been dubbed the Arab Spring. Again, the high penetration of mobile social media among Arab users has great learning potential. Young students, in particular, will find in them effective learning tools with which they can not only share their experiences and cultural norms with the world, but also create their learning communities and interact in a meaningful way.

Formal online learning systems like Blackboard can be effectively used for learning purposes, and can to some extent create better student-student and student-teacher communication channels compared to traditional learning methods. However, informal communication tools, which incorporate students' own social spaces like mobile social media, are believed to have greater effect. Students can develop closer friendships with their classmates while communicating via social media and create sophisticated and reflective interaction. The teacher in such contexts can formulate a kind of student-teacher relationship that is informal and non-distant, in a way that complements the informal nature of mobile social media instruction. A less formal student-teacher relationship is also crucial in terms of predicting effective adjustment to the learning design and eliciting accurate feedback and reflection from the students on the whole learning task. Moreover, mobile social media can help each student to create a social community that includes family members and a number of close friends. Mobile social media can also promote extracurricular student-teacher interaction. However, further research is needed that explores teachers' willingness to implement such informal learning technologies with their students.

To sum up, the future will likely witness more applications of existing learning theories and approaches. New versions of such theories and approaches might be revealed as well. The application of such learning theories and approaches could reveal more evidence about the effectiveness of mobile language learning. In particular, informal learning opportunities, as noted by Clough (2016), can be seen in context-aware mobile learning settings. Clough also added that constructivist mobile learning allows students to create and participate in informal learning opportunities, and extends "our understanding of the mechanisms involved in informal learning with technology" (p. 65), which is the focus of this study. Further characteristics of constructivist informal learning need to be highlighted. Empirical studies on the distinction between formal and informal learning practices need to be conducted. We need to understand why our students are still reluctant to use 'expensive' learning technologies provided by their learning institutions. We also need to study the impact of other cheap or even free alternatives that hold rich and unique learning opportunities, and how our students can exploit them to the fullest. We still also need to know whether informal mobile learning is still in its infancy, or whether real and authentic learning opportunities are waiting. We also need to pay attention to different patterns of student-student and teacher-student interaction in specific contexts, and by which means these interactions occur.

METHODOLOGY

In order to understand the influence of informal mobile technologies vs. formal learning platforms in creating effective student-student and teacher-student communication channels and linguistic outputs, this study was conducted using a constructivist and a qualitative framework. Thirty English language teachers at King Khalid University were interviewed. Evidence from their mobile and non-mobile interaction with their students was analyzed to support data from the interviews, and to evaluate the learning potential of both formal and informal communication mediums. This study is an investigation of prior and current mobile learning practices, and an assessment of meaningful informal learning with mobile technology for both students and teachers. Since patterns of student-student and teacher-student communications were not investigated in this context, this study tries to deepen our understanding of how mobile technology drives informal learning activities that can inform the ways formal learning technologies are deployed. Possible consequences of the integration of informal mobile technology into this setting are also elaborated on by the study.

This study explored the following research questions:

1. What are participants' prior and current practices of mobile social media use and mobile communication?
2. What are the patterns of student-student and teacher-student communication that both mobile social media and LMSs can reveal? And to what extent might they differ?
3. To what extent can mobile social media provide authentic and real-world interaction that can be compared to human-human physical interaction? Does the use of informal tools help to make the interaction more real and reflective?
4. What kind of problems, if any, can be associated with the implementation of mobile social media for learning purposes?

Participants

This study was conducted with 30 university-level EFL (English as a Foreign Language) instructors. Participants were all male; segregated education is practiced in Saudi Arabia. Participants of the study were both native and non-native speakers of English. Native speakers were from the United Kingdom, the United States of America, and South Africa. Non-native speakers were of Asian origins: Pakistani, Indian, and Bangladeshi. Most participants held professional degrees in EFL education ranging from Diplomas to Doctorates. Participants had an average of 10 years of experience, and their ages ranged from 25 to 50. About 30 percent of the participants had taught at the university for more than 10 years, whereas the other 70 percent had only 1-2 years teaching experience at the university.

Students

In order to fully understand patterns of mobile and non-mobile interaction for teacher participants, a clear description of students and teaching programs in the context of this study is needed. Students at the university have to take compulsory English language courses that differ from one discipline to the other. Most English language courses can be considered as ESP (English for Specific Purposes) courses, and others are General English courses that focus more on teaching a range of language skills.

The students being taught by the teachers who participated in this study were taking both ESP and General English courses designed for the foundation year. ESP and General English courses are distributed over two semesters for students in medicine, pharmacy, nursery, and medical sciences. Engineering and computer sciences students take only General English courses over the first semester.

All students were believed to have excellent computer and mobile technology skills, and were frequent users of social media. Most of the courses they take at the university are considered blended in-person and online courses; students attend formal classes and usually complete quizzes and homework through Blackboard. Students are also expected to access the university's accounts on social media to get updates on courses, activities, and exam news. In a previous study conducted by Smala and Al-Shehri (2013), mobile phones were found to be the devices most frequently used to access the internet among the university's students.

Context

This study was conducted at the foundation year campus of King Khalid University in Saudi Arabia. The university started its e-learning initiative in 2005 to cope with global instructional technology trends. The Deanship of E-Learning is one of the most successful sectors of the university, and has won several excellence prizes locally and internationally. Blackboard is the official LMS of the university. It was launched in 2006 as an optional learning medium, and it became a compulsory learning tool two years later. The university was the first Saudi institution to launch Mobile Blackboard in 2011, which enabled most students of the university to access their learning materials and follow announcements via their mobile phones. Intensive e-learning training is available for all new students and teachers via both face-to-face and online courses. Wi-Fi networks are available for both students and teachers at all campuses of the university.

Procedure

A survey was distributed to the participants via their e-mail. The survey included 14 open-ended questions that asked about the participants' experiences as EFL teachers, technology literacy, communication preferences, patterns of student-teacher communications, patterns of mobile phone technology and mobile social media usage as well as their online interaction via Blackboard. All survey forms were completed within two weeks' time. Participants were informed that their participation was voluntary, and that their feedback would be used for research and professional development purposes.

Data Analysis

Data was initially analyzed in order to categorize and then describe common patterns among the data obtained from the surveys. Patterns were classified into common themes comprising (a) the use of mobile technology and social media vs. Blackboard, and (b) patterns of interaction on both mediums. Themes were then classified and analyzed into sub-themes that stemmed from the common themes. Major themes that were covered by the survey were:

1. Prior and current practices and experiences with mobile technologies and social media,
2. Preferred mediums of interactions and their impact on English language learning, and
3. The distinction between mobile social media and the LMS; i.e., Blackboard as a learning tool.

Qualitative data obtained from the surveys were coded into these thematic areas, and then analyzed. Analysis was aimed at justifying mobile and non-mobile constructive conversations, and identifying how teachers helped their students learn from online resources.

KEY FINDINGS

Participants were initially asked about their prior and current experiences with technology in general, and instructional technologies in particular. This included their mobile communication and mobile social media use. In general, most participants indicated that they have used some sort of technology in their previous education, and then started to implement different forms of technology with their own students. One participant, for example, stated that he used PowerPoint and Flash to access learning materials while he was a student. The same participant then indicated that he used other technologies with his current students including:

- Computer Based Learning – CBL and CBTs,
- PowerPoint and Flash,
- Audiovisuals,
- Moodls,
- Ed-line,
- CISCO Web Learning - CWL,
- Audible apps, and
- Multimedia. (Participant 2)

Other participants listed other tools that they had used while they were students including smart boards, Microsoft Office applications, and audio labs. Compared to younger participants, older participants indicated that more analog technologies were used such as cassettes, CD players, and overhead projectors. Overall, data showed that the majority of participants had adequate knowledge of technology and have used it in some way at their previous education.

Participants were also asked about tools or applications they were currently using with their students at the university. Mobile social applications such as WhatsApp were the most preferred way to connect with current students. As a response to the university's official guidelines, Blackboard was also used to connect with the students and inform them about learning-related announcements:

I am using Blackboard in my courses effectively to an extent. Sometimes, I download certain materials related to the text we are dealing in our course and forward them to the students through Blackboard. (Participant 4)

On top of Blackboard I made use of WhatsApp to communicate with my main classes. This is so because my students very regularly check WhatsApp, which made the flow and spread of information very easy and accessible. I preferred this over Blackboard when I needed immediate responses from my students because WhatsApp provided a faster platform of communication. (Participant 16)

Other participants indicated that Microsoft Office applications such as PowerPoint, Word, and Excel were frequently used to share learning materials with the students, and to inform the students about tests, classes, homework, attendance, etc. Online blogs, discussion boards, and Short Messaging Service or SMS messages were also used by other participants for similar needs.

Social media have also been a method of student-teacher interaction:

Apart from using most of Blackboard applications, I have opened a Facebook page named ELC KKU where I share EFL-related contents and updates for students' use. Further, I had been coordinating Listening and Language Labs for two-and-[a-]half years. During my tenure I used to take full care of both software and hardware components of all 8 language labs. I make full use of all the features of Sanako (the application for teaching Listening) in my class. (Participant 6)

More sophisticated learning processes, implemented via social media, were utilized by another participant:

I have used Facebook, Twitter, Google Plus, YouTube and various learning websites such as Khan Academy to help students. For example, I had students make movies, cooking videos, video logs, debates, documentaries, etc., in English and upload them onto YouTube. Students would complete [homework] semester projects and then review and analyze each other's work. Then they would criticize or correct each other's work in class/online. (Participant 12)

The previous quote indicates that teacher participants required their students to complete online activities and utilize them later in the classroom. For assessment, one participant indicated that he preferred using his mobile phone:

For marking listening exercises I downloaded the answers via the mobile app, so that it would be easier to mark students work. (Participant 7)

Whereas most participants were in favor of the use of mobile technology for learning purposes, other participants stated that they were more into using their laptops or computers. They attributed this to their familiarity with web-based Blackboard that made the use of mobile technology an additional function. However, some mobile phone technologies were helpful for some non-mobile advocates:

The mobile phone is not a frequent tool for me when carrying out classes. However, the only time I saw a mobile phone being useful was when I asked students to translate a word from English to Arabic. This helped with students understanding English vocabulary. (Participant 15)

Participants who were not in favor of mobile technology use stated that the university did not encourage them to use other applications than Blackboard. They also thought that Blackboard tools enabled them to communicate effectively with the students, so there was no actual need to use other mediums of communication.

Participants were also asked about the effectiveness of current available technologies and their potential for English language learning. It appeared from most responses that participants felt that technology should be implemented for the sake of language learning, not for the technology itself. One participant, for instance, indicated that:

We need a harmonized way of using technology for learning a language, which is not so easy to create considering the proper management of time.

In a language learning program, using technology should be for the sake of language learning and not for the sake of using technology itself or any other purpose. More importance should be given to the language and not the technology. (Participant 1)

The same participant argued that the proper use of technology, either mobile or not, can greatly improve students' critical thinking and problem solving skills. Thus, using technologies students are interested in, and with which students are already familiar, was much more effective. Another participant argued that current technologies provided by the university were effective, but the regular participation of the students in those technologies was not necessary. Options available via other informal technologies, including mobile technologies, were many:

Students can download e-books, read online, practice grammar, listen to English contents, watch educational movies, speak to online friends from English-speaking countries on Skype, etc. This proliferation of learning resources has left EFL teachers with the sole job of motivating, overseeing and guiding their students; [the] rest will be done by students themselves. (Participant 6)

The same participant added that:

Tools like YouTube and social media have made teaching easier and a lot more fun. They help to communicate and teach students in unconventional but effective ways, especially when it comes to skill teaching. Further, they offer options for students to choose the tool that suits their learning strategy. (Participant 6)

However, due to their effectiveness, heavy reliance on mobile technologies and social media can cause 'technology dependency', or students lacking their own learning independence and confidence, one participant claimed. Another participant indicated that the use of mobile technology for learning could imply some disadvantages such as cheating and copying work. The participant added that the lack of mobile internet for students from working class families could also be a barrier to the full integration of mobile technology into learning.

Participants observed that both teachers and students needed to be aware of the potential of mobile technology and social media for language learning. Otherwise, they will be distracting tools. This is in line with another participant's statement that such technologies were effective but most students did not take them seriously. Teachers, he added, should create a need for using technology for learning. Student-teacher relationships were of concern for some participants while using mobile social communication:

I did have concerns initially about breaching the student-teacher relationship by engaging with my students on WhatsApp, but our discussions and information sharing was kept very professional. (Participant 16)

Beside mobile social communication, the same participant suggested that the university should utilize 'smart classrooms' in a way that could minimize the religious and cultural divide between the students and the teachers:

Teachers have to prepare well and use such rooms carefully for the benefit of the students and the teaching process. This applies to any country, but more so to Saudi Arabia where cultural and religious [beliefs] are intertwined. (Participant 16)

The cultural divide between teachers and students was clear in some responses. As mentioned, some participants hold Western values, and indicated that they thought there was no adequate space, or freedom, to express or practice them. More research is needed to shed light on this crucial issue.

The distinction between formal and informal learning tools was the core subject of this study, especially identifying how mobile informal technology could enhance constructive communication. Participants were asked about the difference between formal learning mediums like Blackboard and informal tools that have educational potential such as mobile social media, YouTube, WhatsApp, SnapChat, etc. Formal learning tools were much easier to control and use to keep track of students' interactions compared to social media, one participant stated. With regard to mobile technology use, the participant claimed that it had more learning potential as it was user-friendly, and students felt free to talk and chat using mobile technology. Another participant advised that mobile social media had greater effect as:

Blackboard is used more as an academic method to assist students as well as teachers to manage their syllabus while social media is more of a pick and choose what interests you. Therefore, I believe social media most definitely would have greater effect. The more interested you are in a subject and the easier you have access to it the easier the learning process would be for you. (Participant 2)

Another participant indicated that, whereas Blackboard was an effective tool for delivering online classes, official announcements, evaluations, etc., the use of informal tools mentioned could be far more effective and varied than formal tools. Mobile social media could be a better means for rapid communication, learning vocabulary, and real-life interactions. He also added that by using such informal tools, teaching was extended to students' personal space without threatening their privacy, and that those tools were in most cases free of charge. In terms of student-student relationships, one participant indicated that:

I believe mobile-based communication can improve student-student relationship[s] and create social communit[ies] in a way that can improve EFL instruction. (Participant 8)

In another participant's words:

To put it succinctly, in the eyes of the students, Blackboard is mandatory, boring, dull, something you look at when you have to; mostly exams or attendance related. And mobile social media is fun, interesting, engaging, real, current, meaningful, relevant, and most of all something the students want to be engaged in. (Participant 12)

However, effectiveness, distraction, misuse, and irrelevance could all be associated with informal mobile tools like the social media and mobile-based communication, some participants said:

Great care needs to be placed on ensuring that quantifiable benefits are being gained by using these tools, otherwise they can simply become a distraction from learning rather than an addition to it. (Participant 3)

Other informal tools (YouTube, social media, Snapchat etc.) attract various students to indulge in irrelevant browsing. A lot of students are addicted to mobile use. They use mobile social media as the tool of their learning for pleasure. (Participant 9)

Notably, concerns over the use of informal tools, i.e., mobile social media and communication, were more readily apparent in responses from older participants. This obviously indicated that younger teachers were less conservative about implementing such tools in their teaching. The age factor also played an important role in visualizing students' needs and interests for the teachers.

Participants were finally asked about the role that mobile social media and communication can play in equalizing the members of a learning group, including the teacher. Additionally, participants responded regarding to what extent they thought mobile social media and communication could provide authentic and real-world interaction that could be compared to human-human physical interaction (face-to-face classroom instruction), and whether the use of informal tools helped to make the interaction more real and reflective.

Overall, most participants indicated that mobile social media and communication helped to make a balance between students' and teachers' interactions. One participant, for example, stated that:

While using social media and mobile chatting, students are not in front of teachers like in a physical classroom, which usually creates learning anxiety among students, and teachers try to dominate the class. Teacher-talk takes longer time than student-talk. While in mobile social media, students feel relaxed and are anxiety-free; they don't feel shy and they can use the language autonomously without being afraid of making errors. Teachers also learn something new while they take part in chatting. (Participant 13)

Another participant indicated that the use of mobile social media tools helped to follow modern theories of education that placed more emphasis on the new roles of the teacher as organizer, involver, enabler, and facilitator. Such technology tools, he added, enhanced the concept of student-centralism in a way that attention on students' needs and individual differences is now more focused than some decades ago. Thus, learning outcomes, he added, would be different in terms of critical thinking, multiple intelligences, and language outcomes. One participant argued that informal technology tools were more effective for English daily or casual use, whereas formal mediums like Blackboard were merely intended for academic purposes. Another participant noted that informal technologies including mobile social media and communication helped the teachers greatly to transmit information and deliver knowledge in a non-traditional way:

An effective teacher compels him or herself to [step] down to his or her students' level of understanding. (Participant 2)

The same participant indicated that, unlike face-to-face physical interaction, the use of the visual features of mobile devices helped in depicting about other people's lives and cultures in different parts of the world:

Students can travel to other parts of the world and interact with the others just as if they were actually there physically, to some extent. (Participant 2)

Another participant noted that social media and mobile communication were current and realistic as they draw their interactions from what is going on in different places. Therefore such tools, he added, were very real to nowadays students, and the interaction was more natural as opposed to out-of-date, laborious, and monotonous interactions available in formal tools.

However, regarding the effectiveness of mobile social media and communication, some participants were still cautious about the educational potential of these technologies. One participant, for instance, indicated that such technology tools could never take over the role of the teacher. Students in such contexts, he argued, tend to rely more on the teacher even with the help of technology. They feel sometimes that they were not on track, and that the teacher's presence and/or guidance are always vital, he added. This is in line with a statement from another participant who claimed that mobile technology and social media are just supplementary, and they would never be stand-alone learning tools. Another participant also claimed that:

Human-human interaction cannot be replaced. When human-human interaction takes place, the nuances, physical expressions and reactions, tone of voice, body language, etc. can be read and interpreted accordingly. The use of informal tools cannot replicate this. What it can be done though is to facilitate and aid human-human interaction/teaching by the easy spread and access to information. (Participant 16)

One participant pointed out that not all language learning skills could be promoted by mobile social media and communication. He felt that grammar skills, for instance, were best to be enhanced through classroom face-to-face practice under the guidance of the teacher. Another participant claimed that the use of informal tools might imply the lack of learning objectives and students' awareness of the potential of these tools for learning. In that participant's words:

Problems will arise if the objectives and parameters are not clearly defined from the inception of using any form of social media in the teaching/learning process. (Participant 16)

It should be noted that traditional grammar translation methods are still practiced by some EFL teachers in Saudi Arabia. The unwillingness among some of these teachers to utilize mobile technology tools for constructive communication is clear. Non-advocates of modern technology trends do not usually focus on the communicative elements of language learning. Hence, mobile technology tools that offer constructive and collaborative learning potential might not have a high priority for those teachers.

Other participants were skeptical about student-teacher relationships in terms of mobile communication and mobile social media. One participant argued that students may lose respect for teachers and assume that they can bypass the classroom boundaries when they have direct and informal access to the teacher. Another participant indicated that his students created a WhatsApp group for the class but removed him from the group after a while. He attributed this to the fact that it was still too early to interfere in their informal communication environments, and to utilize these informal tools for learning purposes. This is in line with another participant's statement that the parameters for the implementation of any new learning platform would be students' willingness, or rather their lack of willingness to participate and implement. Another participant indicated that the lack of solid learning objectives and/ or methods that quantify students' involvement in such informal tools would make the whole learning process meaningless.

Most participants were aware of the potential of mobile social media and communication for language learning, but either still unaware of their full potential or still cautious that such tools would take their respect and authority away:

Traditionalists might argue that the use of social media would degrade the profundity of pedagogy. However, for a teacher with a modern outlook (and I see myself as one of them), introducing social media in teaching will be a win-win situation for both teachers and students. (Participant 6)

However, when discussing the potential of mobile technology, participants made it clear that the teacher's role is still of great importance:

You need a blend of human interaction for facilitation and technology for support. (Participant 7)

Being unrealistic and mechanical was a concern for one participant:

The increased use of mobile social media and mobile chatting may enhance our students' mechanical mind set only, lacking humanistic feelings and approaches in the real world of human beings. (Participant 9)

The issue of teachers' being overloaded with additional duties was also a concern for one participant:

When students get accustomed to social tools, they tend to feel at ease in contacting teachers outside work hours. This may be a problem for academic institutes that limit/prohibit student-teacher interaction outside learning hours. (Participant 12)

In sum, it was shown in the data that both formal non-mobile and informal mobile learning media had great potential for language learning. In some cases, teachers were aware of the informal mobile learning potential but preferred not to take the risk. Based on their previous experiences at the university, teachers were not keen to implement technology in general into their language learning unless they had to; using technology was compulsory two years after its initial integration. Intensive training and compulsion by themselves cannot guarantee effective implementation of modern mobile technologies. Hence, awareness of young students' thinking strategies along with an analysis of their mobile technology behavior will be a necessary skill for professional teachers of the future. These issues are discussed in more detail in the next section.

Discussion and Limitations

In the previous section, an analysis of the prior and current mobile practices of EFL teachers was presented. Preferred mediums of interaction as well as participants' perceptions of the impact of mobile phone technologies and mobile social media on language learning were also analyzed and presented. The previous section also presented data on the distinction between formal and informal learning media and tools, as well as their potential for language learning. In this section, key findings presented earlier are discussed and examined in the light of the literature reviewed in section 2, particularly informal learning and constructivist approaches. This section also contains answers for the research questions laid out in section 3, and revolves around the main areas investigated by this study.

PRIOR AND CURRENT PRACTICES OF MOBILE SOCIAL MEDIA AND INTERACTION

This study revealed that there was a tendency among all participants to implement different technologies into their teaching experience. It was also shown that younger participants were more frequent users of mobile social media and communication compared to their elder counterparts. This is consistent with So (2008), and Ismail, Azizan and Azman (2013), who found that younger mobile users tend to use mobile phones for more sophisticated functions rather than elder mobile users. This is also in line with the work of Lin, Huang and Chen (2014), who found that younger teachers were more confident than older ones in terms of the integration of innovative technologies into their teaching. In the current study, it was obvious that most participants were aware of the potential of mobile technology and social media for learning. However, their awareness did not mean that they could integrate such technologies without caution. In other words, some participants believed that formal learning tools like Blackboard were more trustworthy and safer than informal tools like mobile social media and communication. In addition, these participants felt that informal tools were meant for informal communication, whereas formal learning mediums were better learning tools.

Teachers indicated that informal mobile learning occurred in different points where students participated and interacted via different mobile and social media tools. However, informal mobile learning, as Clough (2016) asserts, is difficult to assess since there are no official assessment criteria or even a curriculum. Thus, using a constructivist framework can be helpful in assessing meaningful learning in an informal setting (Clough, 2016). This is discussed in more detail in the next sections.

Patterns of Student-Student and Teacher-Student Mobile/Non-Mobile Interaction

Mobile technology has greatly enabled users to interact with other people and the surrounding context or environment (Kukulska-Hulme & Arcos, 2011). The convenience of communicating when and where desired is an exclusive functionality of mobile technology. Furthermore, an alternative to regular student-student and teacher-student interaction that is unique to mobile, as Quinn (2012) notes, is that "connection can happen in context" (p. 78). For the present study, the 'context' factor was not completely apparent in mobile interaction. Most responses indicated that mobile and non-mobile interaction was centered on learning-related issues. Cultural norms were not a major factor that students use to reflect on specific activities and articulate what they have learnt, and what language they use outside of the classroom. Thus, student-teacher mobile interaction was found to be merely pragmatic.

Participants demonstrated that the use of mobile social media and communication was focused more on assignments, announcements, and so on. Collaboration on non-learning-related issues was not mentioned as rich component of mobile interaction. It could be argued then, that participants were either not keen to interact with students on issues unrelated to their classes, or that they preferred not to give many details about their mobile interaction, which was deemed to be confidential and private. This might be true since the university does not encourage much student-teacher informal interaction in an attempt to maintain assessment reliability and objectivity. Therefore, it can be argued that the use of mobile technology in this context did not involve rich constructivist and informal learning opportunities.

Mobile Interaction vs. Human-Human Interaction

Although most participants admired the role that mobile social media and communication can play in enhancing interaction, some also indicated that human-human interaction could not be replaced by technology-based interaction. They attributed this to the fact that for language teachers, in particular, it is best to use face-to-face teaching techniques that can help the students to better learn the language. Grammar skills, body language, gestures, and handwriting exercises were among the language learning skills that participants thought could not effectively be enhanced by technology.

Whereas mobile technology can enhance language collaboration, collaborative learning, as Troussas, Virvou, and Alepis (2014) note, "is not a solution for every difficulty... in learning" (p. 290). Thus, there is logic behind some participants' thoughts in that the role of mobile social media and communication is supplementary, not essential. This is also in line with the work of Sims, Vidgen, and Powell (2008), who emphasize that engagement with technology that has learning potential requires more than just access; employment of appropriate pedagogies and fostering of cultural capital are also necessary.

Staff training plays another crucial role in enhancing their conceptualization of mobile social media and communication, and improving their awareness about the language learning potential of these technologies. However, mobile chatting could not only improve language use, but it could also be a method for involving shy students and engaging the learning community in casual and/or real-time conversations that might not find a space in the classroom, one participant claimed. However, the context is a central construct for understanding mobile learning in a particular setting. There is a need for mobile educators and practitioners to initially learn through context, in context, and about context (Sharples, 2016). Again, mobile learning in the setting of the study still needs ambitious steps to move forward, and integrate contextual elements that can enrich mobile learning activities.

Problems with Mobile Social Media and Interaction

Losing control of teachers' respect was the most apparent concern for some participants. However, none of the reviewed studies have demonstrated this as a concern. Thus, it could be argued that students in Middle Eastern learning contexts are required to pay higher respect to their teachers. Teachers, in turn, would therefore be conservative with any behavior or technique that may affect their image including the use of mobile social media or informal mobile communication.

Other problems that could be associated with the integration of mobile learning included the lack of internet access at students' own places of residence and the cost of mobile internet, some participants indicated. Although some earlier studies anticipated that there would be drop in mobile internet prices in the future (see Al-Fahad, 2008; Cochrane, 2005; Kolb, 2008), the current study indicated that cost of mobile internet is still a concern for some university students. The availability of Wi-Fi networks at the university campus did not seem to help to integrate contextual or cultural norms from students' own social lives.

Participants also listed other problems associated with the use of technology in general. The list included the lack of maintenance for the language lab computers, and the lack of awareness of the potential of technology for learning among elder teachers. They also thought that a cultural gap between local students and foreign teachers existed, although mobile social media helped to narrow it down. This is consistent with Rukzio, Broll, Leichtenstern, and Schmidt's (2007) findings, which show that mobile learning can enhance learning about different cultures in an entertaining manner. Again, teachers still need to integrate local cultural and context-aware elements into actual language learning activities in order to facilitate informal mobile learning.

Limitations

This study is not without limitations. Perceptions of students regarding the effectiveness of mobile social media and communication have not been explored. It was anticipated that a top-down look at this language learning setting would shape a clearer image of both student-student and teacher-student mobile interaction. This was true since students at the foundation year were still unaware of the potential of mobile technology for learning at tertiary education. Future research needs to be conducted with students and at advanced language classes.

It was not possible to obtain or to request screen captures of participants' own mobile social media interactions and mobile chatting. Although these tools would probably have had data that could have enriched data analysis and provided more evidence, such data was felt to contain too much private and personal information and therefore could not be used.

LOOKING AT THE FUTURE

Students are the most active motivators of mobile social media and mobile communication. Any future work should take into consideration students' own perceptions about mobile technology and social media, and how these shape their daily lives and learning interactions. We also still need to know what sorts of knowledge students can attain using their mobile devices, and which pedagogies are suitable for education that integrates innovative technologies. In their book '*Born digital: understanding the first generation of digital natives*', Palfrey and Gasser (2008) wrote about the concerns of both parents and educators over what kinds of information are seeping into young people's brains in the digital era. In their words, "the things that schools and teachers do best should not be scrapped in the rush to use technologies in the classroom. In every field, there are aspects of the curriculum that should be taught without screens or Net connection" (Palfrey & Gasser, 2008, p. 246).

This indicates that future research should decide what kind of knowledge best fits with mobile interaction, and what should be taught through face-to-face classroom instruction. Other future work should also demonstrate what aspects of learning interaction should remain within computers, and what other learning interaction can be done on the move.

CONCLUSION

The findings of this study illustrated that teachers' perceptions of mobile interaction, as opposed to face-to-face interaction, were positive overall. Teachers believed that mobile technology as well as social media had great potential for language learning, although some were reluctant to use them in a comprehensive way. The current study also illustrated the distinction between formal and informal technology tools, and how each paradigm can successfully be utilized for different constructivist mobile learning purposes. Some previous studies have investigated the role of mobile technology on students' achievement and attitudes as well student-centralism, and found that the impact was vital (see for example Hwang & Chang, 2011; Kukulska-Hulme, 2005).

Some of the reviewed studies also indicated that interaction via mobile devices has not only become a way of life or a daily human behavior, but rather a significant field that has its own theoretical and practical spaces. Many educational and behavioral theoretical approaches that have been reviewed so far have helped mobile researchers and educators to understand interactions in the mobile era, and how people behave and communicate using mobile devices. However, we still need to "find the optimal learning experience for the audience, taking into account their context, learning goals, preferences, and more and using all available channels" (Quinn, 2012, p. 93). Mobile learning also needs to move beyond specific contexts to contexts that are generic and abstract (Sharples, 2016).

The huge divide between present-day 'digital immigrants' (teachers) and 'digital natives' (students) needs to be narrowed. Intensive training and professional development courses may not be adequate at this stage, but rather, an awareness of students' needs and evidence-based analyses of mobile behavior for the next generation may be more useful. The potential of handheld devices for learning should also be highlighted among educators, policy makers, and parents. Students, on the other hand, should also be made aware of how their mobile devices can be turned into powerful learning tools. This necessitates looking at evolving factors and trends that may have an impact on future mobile learning.

REFERENCES

Al-Fahad, F. N. (2009). Students' attitudes and perceptions towards the effectiveness of mobile learning in King Saud University, Saudi Arabia. *The Turkish Online Journal of Educational Technology*, 8(2), 111–119.

Ally, M., Schafer, S., Cheung, B., McGreal, R., & Tin, T. (2007). Use of mobile learning technology to train ESL adults. In A. Norman & J. Pearce (Eds.), *Making the Connections:Proceedings of the mLearn Melbourne 2007 Conference*. University of Melbourne.

Arab, S. M. R. (2013). *Transforming education in the Arab world: Breaking barriers in the age of social learning*. Retrieved December 13, 2015, from http://www.arabsocialmediareport.com/home/index.aspx

Arab, S. M. R. (2014). *Citizen engagement and public services in the Arab world: The potential of social media*. Retrieved December 13, 2015, from http://www.arabsocialmediareport.com/home/index.aspx

Armatas, C., Holt, D., & Rice, M. (2005). Balancing the possibilities for mobile technologies in higher education. In *Proceedings of the 2005 ASCILITE conference*. Brisbane, Australia: ASCILITE.

Bush, M. D. (2008). Computer-assisted language learning: From vision to reality? *CALICO Journal, 25*(3), 443–470.

Chen, W. P., Millard, D. E., & Wills, G. B. (2008). Mobile VLE vs. mobile PLE: How informal is mobile learning? In J. Traxler, B. Riordan & C. Dennett (Eds.), *Proceedings of the mLearn2008 Conference*. University of Wolverhampton.

Cheung, S. (2004). Fun and games with mobile phones: SMS messaging in microeconomics experiments. In R. Atkinson, C. McBeath, D. Jonas-Dwyer & R. Phillips (Eds.), *Beyond the comfort zone:Proceedings of the 21st ASCILITE Conference*, (pp. 180-183). Perth, Australia: ASCILITE.

Clough, G. (2016). Mobile informal learning through geocatching. In J. Traxler & A. Kukulska-Hulme (Eds.), *Mobile learning: The next generation* (pp. 43–66). New York: Routledge.

Cochrane, T. (2005). Mobilising learning: A primer for utilising wireless palm devices to facilitate a collaborative learning environment. In *Proceedings of the 2005 ASCILITE conference*. Brisbane, Australia: ASCILITE.

de Sa, M., & Carrico, L. (2010). Designing and evaluating mobile interaction: Challenges and trends. *Foundations and Trends in Human-Computer Interaction, 4*(3), 175–243. doi:10.1561/1100000025

Figueiredo, A., Almeida, A., & Machado, P. (2004). Identifying and documenting test patterns from mobile agent design patterns. In A. Karmouch, L. Korba, & E. Madeira (Eds.), *Mobility Aware Technologies and Applications* (pp. 359–368). Berlin: Springer-Verlag. doi:10.1007/978-3-540-30178-3_35

Fotouhi-Ghazvini, F., Earnshaw, R. A., Robison, D. J., & Excell, P. S. (2008). The MOBO City: A mobile game package for technical language learning. In J. Traxler, B. Riordan & C. Dennett (Eds.), *Proceedings of the mLearn2008 Conference*. University of Wolverhampton.

Göth, C., & Schwabe, G. (2008). Designing tasks for engaging mobile learning, In J. Traxler, B. Riordan & C. Dennett (Eds.), *Proceedings of the mLearn2008 Conference*. University of Wolverhampton, UK.

Herrington, A., Herrington, J., & Mantei, J. (2009). Design principles for mobile learning. In J. Herrington, A. Herrington, J. Mantei, I. W. Olney, & B. Ferry (Eds.), *New technologies, new pedagogies: Mobile learning in higher education* (pp. 129–138). Australia: University of Wollongong.

Hwang, G., & Chang, H. (2011). A formative assessment-based mobile learning approach to improving the learning attitudes and achievements of students. *Computers & Education, 56*(4), 1023–1031. doi:10.1016/j.compedu.2010.12.002

Ismail, I., Azizan, S. N., & Azman, N. (2013). Mobile phone as pedagogical tools: Are teachers ready? *International Education Studies, 6*(3), 36–47. doi:10.5539/ies.v6n3p36

Khan, S. (2015). *Snapchat, the Arab world and global implications*. Retrieved December 13, 2015, from http://www.internationalpolicydigest.org/2015/07/27/snapchat-the-arab-world-and-global-implications/

Kiernan, P. J., & Aizawa, K. (2004). Cell phones in task based learning: Are cell phones useful language learning tools? *ReCALL, 16*(1), 71–84. doi:10.1017/S0958344004000618

Kolb, L. (2008). Toys to tools: Connecting student cell phones to education. Washington, DC: International Society for Technology in Education (ISTE).

Kukulska-Hulme, A. (2005). Introduction. In A. Kukulska-Hulme & J. Traxler (Eds.), *Mobile learning: A handbook for educators and trainers* (pp. 1–6). Wiltshire: The Cromwell Press.

Kukulska-Hulme, A., & Arcos, B. D. L. (2011). Researching emergent practice among mobile language learners. In *Proceedings of 10th World Conference on Mobile and Contextual Learning*.

Levy, M. (1990). Towards a theory of CALL. *CAELL Journal*, *1*(4), 5–8.

Levy, M. (1997). *Computer-assisted language learning: context and conceptualization*. Oxford, UK: Clarendon Press.

Levy, M., & Stockwell, G. (2006). *CALL dimensions: options and issues in computer-assisted language learning*. Lawrence Erlbaum Associate, Inc.

Lin, C., Huang, C., & Chen, C. (2014). Barriers to the adoption of ICT in teaching Chinese as a foreign language in US universities. *ReCALL*, *26*(1), 100–116. doi:10.1017/S0958344013000268

Liu, C. C., Tao, S. Y., & Nee, J. N. (2008). Bridging the gap between students and computers: Supporting activity awareness for network collaborative learning with GSM network. *Behaviour & Information Technology*, *27*(2), 127–137. doi:10.1080/01449290601054772

Mandhai, S. (2015). *Snapchat opens digital window on Mecca to millions*. Retrieved December 13, 2015, from http://www.aljazeera.com/news/2015/07/snapchat-opens-digital-window-mecca-millions-150714144609540.html

Matthee, M., & Liebenberg, J. (2007). Mathematics on the move: Supporting mathematics learning through mobile technology in South Africa. In *Making the Connections:Proceedings of the mLearn Melbourne 2007 conference*.

Mellow, P. (2005). The media generation: Maximise learning by getting mobile. In *Proceedings of the 2005 ASCILITE conference*. Brisbane, Australia: ASCILITE.

Nah, K. C., White, P., & Sussex, R. (2008). The potential of using a mobile phone to access the Internet for learning EFL listening skills within a Korean context. *ReCALL*, *20*(3), 331–347. doi:10.1017/S0958344008000633

Naismith, L., Lonsdale, P., Vavoula, G., & Sharples, M. (2004). *Literature review in mobile technologies and learning*. A Report for NESTA Futurelab.

Nalder, G., Kendall, E., & Menzies, V. (2007). Self-organising m-learning communities: A case study. In *Making the Connections:Proceedings of the mLearn Melbourne 2007 conference*.

O'Malley, C., Vavoula, G., Glew, J. P., Taylor, J., Sharples, M., & Lefrere, P. (2003). *Guidelines for learning/teaching/tutoring in a mobile environment*. Academic Press.

Palfrey, J., & Gasser, U. (2008). *Born digital: Understanding the first generation of digital natives*. Philadelphia: Basic Books.

Pulman, A. J. (2008). *The Nintendo DS as an assistive technology tool for health and social care students.* Paper presented at the mLearn 2008 Conference, Shropshire, UK.

Quinn, C. N. (2012). *The mobile academy: mLearning for higher education.* Jossey-Bass.

Rukzio, E., Broll, G., Leichtenstern, K., & Schmidt, A. (2007). Mobile interaction with the real world: An evaluation and comparison of physical mobile interaction techniques. In B. Schiele, A. K. Dey, H. Gellersen, B. de Ruyter, M. Tscheligi, R. Wichert, E. Aarts, & A. Buchmann (Eds.), *European Conference Proceedings, AmI 2007.*

Sharples, M., Taylor, J., & Vavoula, G. (2005). Towards a theory of mobile learning. In *Proceeding of the mLearn 2005 Conference.*

Shuler, C. (2009). *Pockets of potential: Using mobile technologies to promote children's learning.* New York: The Joan Ganz Cooney Center at Sesame Workshop.

Sims, J., Vidgen, R., & Powell, P. (2008). E-learning and the digital divide: Perpetuating cultural and socio-economic letisim in higher education. *Communications of the Association for Information Systems, 22*(1), 429–442.

Smala, S., & Al-Shehri, S. (2013). Privacy and identity management in social media: Driving factors for identity hiding. In J. Keengwe (Ed.), *Research perspectives and best practices in educational technology integration* (pp. 304–320). Hershey, PA: IGI Global.

So, S. (2008). A study on the acceptance of mobile phones for teaching and learning with a group of pre-service teachers in Hong Kong. *Journal of Educational Technology Development and Exchange, 1*(1), 81–92.

Stockwell, G. (2008). Investigating learner preparedness for and usage patterns of mobile learning. *ReCALL, 20*(3), 253–270. doi:10.1017/S0958344008000232

Suki, N. M., & Suki, N. M. (2007). Mobile phone usage for m-learning: Comparing heavy and light mobile phone users. *Campus-Wide Information Systems, 24*(5), 355–365. doi:10.1108/10650740710835779

Sweller, J. (1988). Cognitive load during problem solving: Effects on learning. *Cognitive Science, 12*(2), 257–285. doi:10.1207/s15516709cog1202_4

Traxler, J., & Kukulska-Hulme, A. (2016). Introduction to the next generation of mobile learning. In J. Traxler & A. Kukulska-Hulme (Eds.), *Mobile learning: The next generation* (pp. 1–10). New York: Routledge.

Troussas, C., Virvou, M., & Alepis, E. (2014). Collaborative learning: Group interaction in an intelligent mobile-assisted multiple language learning system. *Informatics in Education, 13*(2), 279–292. doi:10.15388/infedu.2014.08

Zimmermann, A., Henze, N., Righetti, X., & Rukzio, E. (2009). Workshop on mobile interaction with the real world. In A. Zimmermann, N. Henze, X. Righetti, & E. Rukzio (Eds.), *Mobile Interaction with the Real World* (pp. 9-14).

Chapter 3
Lifestyle Diglossia and Mobile:
Ethnography of Multilingual Interaction

Mukul Saxena
The University of Nottingham – Ningbo, China

ABSTRACT

This chapter proposes to chart the development of understanding of literacy as a practice, which is now digital in nature and globally distributed, therefore digital literacy practices require new lenses and ways of research explorations. The notion of literacy has shifted from being autonomous to ideological during the last three decades. It is not seen anymore as a single unified competence, but as changing from place to place and varying in different social-cultural contexts. Despite the fact that there are different writing systems that are used in different ways in different contexts, the differences between them are no longer seen as primarily technical (Graff, 1979; Heath, 1983; Street, 1993). The differences that do exist between literacies are seen as being due to differences in cultural practices, values and ideologies. As a consequence, the methodological shift towards ethnographic research on literacy has arisen from a fundamental change in thinking about the nature of literacy and the development of the "new literacy studies" (Street, 1993: 4). Ethnographic approaches to literacy, such as those developed by Heath (1983), Street (1984), Barton (1991, 1994) and others are based on the everyday uses of written language(s) by specific groups and subgroups in a specific locality. According to Graff, these approaches to literacy provide "both new and better cases for study, opportunity for explanations, and approaches to literacy's variable historical meaning and contribution" (1986: 127). The ethnographic research on literacies in multilingual contexts (e.g. Saxena, 1994; Hartley, 1994) further contributed to the development of 'new literacy studies' (NLS). However, with the development of mobile devices during the last decade and the allied software industries, digital literacy communication has become synonymous with globalisation and a divergence between the literacy practices in 'regulated spaces' and 'unregulated spaces' (Sebba, 2009) particularly among the youth is becoming a marked feature of inter-personal communication. Sebba defines 'unregulated spaces' as places where the prescriptiveness of standardisation and monolingualism has not yet reached, or where it holds no power and practices may deviate from the prescribed norms. Such spaces open up opportunities for identity construction and group definition.

DOI: 10.4018/978-1-5225-0469-6.ch003

INTRODUCTION

Studies on multilingual literacy practices in online communication are relatively new and largely unexplored. Yet, these practices are beginning to appear in a variety of digital genres: for example, multilingual emails (Hinrichs, 2005, 2006), blogging (Montes-Alcala, 2007) online discussion forums (Lewin and Donner, 2002; McLellan, 2005), instant messaging (Ling, 2005; Andoutsopoulos, 2006; Apriana, 2006; Sebba 2007), online chats (Warschauer et al., 2002; Lam, 2005) and Flickr (Barton, 2015). What these studies are showing is how through the transformative and hybrid practices observed in these digital genres the producers are constructing transcultural and global identities. For example, Lam (2000, 2005, 2009) has investigated the linguistically and culturally blended literacy practices of Chinese-American students in their online interactions with peers. She argues that such practices allow these young Chinese-Americans to engage in virtual crossings of physical and societal boundaries and to experience 'transculturation' rather than 'acculturation' (Lam, 2006). Warschauer et al. (2002) have also carried out a study of multilingual online communication among young Egyptian professionals in Cairo. They found that English was predominantly used in their formal e-mail correspondence and in web-browsing. However, communication in informal e-mail correspondence and in online chats was mostly in Egyptian Arabic, with frequent code-switching into English, and the Arabic was written in a Romanised script. Commenting on these findings, Warschauer (2009: 126) says:

The extensive use of English among this group reflected their elite background, their use of technology for professional purposes in a globalized economy, and the still undeveloped nature of text editors, e-mail clients, and other text software capable of handling Arabic script at the time of the study. Nevertheless, their use of Egyptian Arabic whenever possible allowed them...to express hybrid forms of communication and identity.

This hybrid form of communication, as Warschauer argues, expresses global identity. What we also need to recognise at the same time, as I will show in this paper, is that such identities may also come into conflict with local ideologies of nation-states, as in the case of Brunei Darussalam in South East Asia, which provides the context for my study.

The data for this paper is drawn from ethnographic research on multilingualism in Brunei Darussalam (Brunei hereafter) conducted as part of three funded projects[1]. The research was based on case studies of 23 students, aged 20-24 years, who were identified on the basis of their specific backgrounds: first, they were all of Malay background; second, they were pursuing degree programmes in English at the local national university (participants were selected from two programmes); third, they had their private primary education in English-medium schools; fourth, their parents had a university education and some of them had completed their higher education in English-speaking countries, mainly the UK; fifth, in their homes, English was used alongside Malay; and, finally, they were conversant with and used regularly mobile and web-based technologies (e.g. SMSs, emails, internet chat rooms and blogs). They were asked to keep two-week-long literacy diaries, along with samples of the texts they produced, and they were asked to write about their literacy histories. Following this, they were interviewed on the issues that emerged from the diaries and the narratives.

In the first section of the paper I will describe the sociolinguistic background of Brunei and particularly how the linguistic capital of Malay, the national language, and English are produced. Then I will show how young Bruneians are socialised into English in unregulated spaces and the role of global media. This helps to understand and explain the multilingual digital literacy practices of Bruneian youth presented in the following section. Then I look at how their global lifestyle comes into conflict with the reified notion of language, tradition and national ideology. The final section explains the overall findings drawing on the concept of 'lifestyle diglossia' introduced in an earlier paper (Saxena 2014). I will also present briefly an example of multilingual digital literacy practices among Chinese youth and argue that an ethnographic approach and the concept of 'lifestyle diglossia' will help in understanding the the interaction between the policies and micro-level practices and capturing the linguistic and cultural changes taking place in a rapidly globalising China.

LINGUISTIC CAPITAL OF MALAY AND ENGLISH IN BRUNEI

Brunei is located on the north-western coast of the island of Borneo (in Southeast Asia). This oil-rich Sultanate (with a total land area of 5,765 sq km and a population of around 348, 800) has transformed itself from a third world country to one with a GDP as high as in first world countries in the last four decades. Its economy is almost exclusively based on the production and export of oil and gas.

Malay and English derive their linguistic capital (Bourdieu, 1991) by being associated with tradition and modernity respectively. Historically, as a Muslim kingdom, all its policies, including language policies, are informed by its national ideology, *Melayu Islam Beraja* (Malay Islamic Monarchy) which invokes Islamic values in support of the Malay Sultanate and the absolute monarchy, *albeit* constitutional. Malay language (as the official/national language) is inextricably linked to Malay ethnic identity, Malay culture, Islam and the nation-state.

English was first introduced to Bruneian sociolinguistic ecology during the British Residency Period. During the 1950s and early 1960s, it experienced a brief setback due to high nationalistic feelings, particularly among the Malay-educated elites (Saxena, 2007). However, ever since the failed rebellion of 1962 and the decline in political and diplomatic relations with Malaysia during the mid-1970s (Jones, 1997), English has been gaining currency and prestige in the country. As in many other post-colonial situations, western-educated elites have played a central role in the preservation of English in Brunei. English is officially valued as an international language, and a language for gaining access to global trade and to scientific and technical knowledge. Therefore, it is used in offices, both in spoken and written form, alongside Malay (Wood & Swan et al., 2001). Malay and English have been compulsory school subjects since the independence of Brunei from British control and the introduction of the Bilingual Education system in 1985. They have become part of the communicative repertoire of many Bruneians, particularly the younger generation (e.g., McLellan & Noor Azam, 1998; Saxena & Sercombe, 2002; Saxena, 2007).

English is greatly emphasised in the education system for those subjects that can be pursued further in the core English-speaking countries. More recently, Brunei has introduced a new 'National Education System' for the 21st century called *Sistem Pendidikan Negara* (SPN21). The main change is that now English, rather than Malay, is the medium of instruction for Mathematics and Science right from the first year of schooling. Previously, while it was taught only as a subject from year 1, it was introduced as the medium from year 4 of the primary school. However, Civics (MIB), Brunei History and Religious Knowledge (Islam) continue to be taught in Malay. As in most postcolonial situations, code-switching (English-Malay) is commonplace in English subject classes and particularly, in English-medium content classes (Saxena, 2009, 2010). More recently, Islamic religious education and thus, learning of Arabic, has also been integrated into the state education system. Consequently, Malay, English and Arabic are progressively imbued with symbolic and communicative (particularly, Malay and English) currencies.

SOCIALISATION INTO ENGLISH: THE LOCAL VS. GLOBAL MEDIA

A growing institutional emphasis on English and its significance in the context of global connections for economic diversification has led to the 'symbolic capital' (Bourdieu, 1991) of English gaining grounds over that of the national language, Malay. This is seen in the increasing number of parents who have begun to send their children to English tuition schools or English-medium private schools to ensure that they have a better start to their education. These parents believe that a good command over English can only be acquired by sending the children to such fee-paying schools. However, the Bruneians with an average family size of 5-8 children who cannot easily afford to pay the fees in such schools have increasingly thrown their support behind an early introduction of English in the state schools which provide free education. This has influenced the most recent language education policy SPN21.

In addition, as the following quotes will show, the use of spoken and written English in the home environment has played an important role. The young people in my study had been introduced very early to English by their parents, through books, movies and TV Programmes. They were socialised early into the reading habit in English at home.

As a child, my first words, as my mother recalls, were in English. I learnt English rather quickly, through my parents, grandparents, cousins, aunts and uncles, and even television programmes like Sesame Street. By the time I started kindergarten, I was able to carry out whole conversations in English, with the expected grammatical errors, of course. (Nur)

When I was young, my mother made the effort of buying me English Practice Workbooks for me to practice my English language skills. My father never hesitated whenever I asked for him to buy me a book to read, such as ones written by Enid Blyton and R.L.Stine.... When I reached Secondary One, my interest changed to novels and also autobiographies. I was largely exposed to books and novels written by Stephen King, Dave Peltzer, Mitch Albom and also Shakespeare (due to my Literature classes). (Safiya)

The children of the English-educated elites in Brunei are therefore exposed to English canon both in regulated and unregulated spaces. However, many regulated spaces, like the national media and government affairs, require the knowledge of Malay.

If there was a choice, I would usually pick the English versions to read and write in for menus, brochures, subtitles and so on. Exceptions are watching movies on RTB or RTB2 (the local television stations) and previously TV2 and TV3 (Malaysian channels) which gives Malay subtitles for any movies not shown in Malay, road signs (the majority of which are in Malay), and also when filling in forms and basically everything to do with the government especially if you want to make a good impression in addition to ease of understanding. (Firdouse)

The broadcasts over the official national television channel are in Standard Malay (*Bahasa Melayu*), aimed at developing a distinctive Bruneian national identity. This reflects a policy of language imposed from above with the viewers having no choice in this matter. At the same time, Bruneian households have access to global media via the satellite/cable/online. It is not surprising then that the reading habits and script choices of many of the younger generation Bruneians are influenced by TV programmes for children made in the US or the UK. As Shahrin notes:

I recall learning the Roman script from watching children development shows like "Sesame Street" and from my parents, and this is what I have more knowledge on because I use it in everyday life, unlike the Arabic scripts which I only use and know well enough merely for religious or certain academic purposes... (Shahrin)

The Arabic script is called *Jawi* locally. It is taught and is required in some regulated spaces. However, as we shall see later, the response to the imposition of *Jawi* among the young people from English-educated elite background is generally negative.

MULTILINGUAL COMMUNICATION ON MOBILE DEVICES

Brunei's economy affords its population many of the comforts that are associated with developed countries. 21st century technological tools for communication, like mobile phones and the internet are ubiquitous. Affluence creates unregulated globalised spaces for communication in which young people can construct identities and build relationships. This is particularly true of the cohort selected for this study. The technoscapes (Appadurai, 1996) created through new media have opened up opportunities for them to communicate in styles of writing that are unfettered by the prescribed norms of reified language and literacy conventions promoted official institutions (e.g. schools, national television, etc.) and enter into the hybrid world of code-switching. The extract below is a SMS message:

Hope u [you] had a great smashing bday [birthday]. Sorry lmbt..[Sorry for the late (birthday wish)] Vry bzy ystrday [Very busy yesterday], **mnyiapkn** *[finishing] report. Coz [because]* **bru** *[just] finished my course.*

[Hope you had a great smashing birthday. Sorry for the late birthday wish. Very busy yesterday, finishing report. Because just finished my course.] (Roslan)

This kind of hybrid language is being produced in unregulated spaces where young people like Roslan negotiate different identities associated with reified Malay and English and with Brunei Malay and mixed codes. New technologies like the mobile phone, allow them to express globalised identities associated with global youth culture. This is reflected in the globalised features that are used by them, for instance, as can be seen in the SMS below sent by Shamdi to his friend.

Wil kol u l8er @ 7 pm ftr I came home 4rm d seminR. iatah tringat tia I hv sumtin 2tel u:)

[Will call you later after I come back from the seminar. So I remembered that I have something to tell you] (Shamdi)

Shamdi pointed out in my interview with him that the truncated forms (e.g. 'wil' and 'ftr'), icons ('U', 'l8ter', '@','d') and emoticon (:)) that he used are prevalent internationally. This obviously connects him and others like him to the global village in which he participates as a consumer of the 'global' lifestyles that satellite TV, magazines and the internet promote (see for details, Saxena, forthcoming)

At the same time, localisation of the applications of the mobile devices can be seen in the use of truncated Malay words, e.g. the use of **Pek** & **Ka2**, in the following SMS message:

Pek, *d u need* **ka2** *2 send u 2 tuition? Nur*

[Afiq, do you need sister to send you to tuition?][2]

Here **Pek** is *short for Afiq, a Malay Islamic name*, d is *short* for *'do'*, u for *'you'*, **ka** is short for *Malay 'kaka'* [*sister*]; 2 is the *convention for marking reduplication*, 2 for *'to'*. These local practices evoke the habitual patterns of spoken language use with the same interlocutors. This cohort of students and many other people, in similar age groups and with similar backgrounds, that I came to know in Brunei, pointed out that it had become a habit for them to code-switch between English and Malay while using the mobile devices, just like they did in their everyday conversations.

This hybrid language and literacy practice of mixing languages and of using local and global features is being documented among SMS users all around the world. In this particular age of high-tech communication and globalisation where our notions of time and space are being redefined, SMS and other forms of digitally-mediated media have acquired a marked significance in everyday life, particularly among young people around the world (for instance, see Hagood, et al. 2003; Sefton-Green, 2006; Coiro, et al. 2008; Lam, 2009). I have observed similar practices among Chinese students at an international university in China, as we will see below.

GLOBAL LIFESTYLE: TRADITION VS. MODERNITY

Speaking English in Brunei, as in most post-colonial and other contexts, is associated with being modern, educated and Western. Being Westernised is associated with, for example, "being cool", "having pre-marital sex" and, in case of women, "being able to voice their opinions" instead of being "traditional" (Saxena, 2007). Such Western influences are seen as associated with globalisation and the spread of a global lifestyle. These influences and modernisation are encouraging the spread of global consumer culture too. This is reflected in the restructuring of the landscape of Brunei's capital (e.g., high-rise buildings, modern malls, etc.). The landscape is adding consumer identities to the already multiple and overlapping identities found across different age groups, social and ethnic categories in Brunei.

Some of these modern landscapes are being claimed by youth subcultures which are, through their practices, trying to redefine them as postmodern landscapes. For example, certain shopping malls are frequented by Malay teenagers who project and claim their membership to western hip-hop culture. Their identification with this culture is symbolized by their talk, walk and dress copied from the global MTV culture and films beamed in via satellites. They wear, for instance, the baggy jeans, the over-sized t-shirts and the braided hair. Their English is laced with African-American slang which carries for them positive connotations and a marker of their global youth identity. Sometimes such youth subcultures come into conflict with other users of consumer cultural landscapes, who hold different *subject positions* (Fairclough, 1989). For instance, in one newspaper article (Azlan, 2005) a religious figure took an Islamist position in perceiving such subculture practices as a threat to MIB values. In another article (Disappointed Citizen, 2005), a "disappointed citizen" took a Malayist position in bemoaning the loss of cultural roots among such teenagers. The following excerpt from a local newspaper in Brunei reports the negative reaction that the hybrid language and literacy practice of mixing Malay and English, in speech and in informal writing, has evoked.

Yesterday, the Minister of Culture, Youth and Sports lamented over the usage of what he called "a rojak language" and called on authorities to implement and monitor the use of proper Malay language. He called for firm steps to be taken to stop the practice of such polluted language and save the dignity of the Malay language as the country's official language. "Nowadays, the usage of 'rojak' language among the generation of teenagers has become the norm. They do not speak proper Malay among their peers and even while speaking with their parents," he said. Defining 'rojak language' as a mix of languages involving the mother tongue with a second language, usually English, the minister said such spoken language only pollutes our mother tongue. "We know that language, community and culture are always evolving with time," said the minister. Conversing in 'rojak' language is now regarded by society as modern and advanced without realising that it is tearing apart the very fabric of our values and culture, he added. ('No to 'rojak' language', Borneo Bulletin, June 25, 2007).

Rojak, the Malay word which was used by the Minister as a metaphor for hybrid language is actually a traditional mixed fruit and vegetable salad – an everyday dish. By comparing the mixing of languages to this lowly salad, he has raised concerns about standards and correct usage of Malay. The Minister positions *Rojak* as symbolic of the erosion of Bruneian values and culture as well as the Malay language and points to English as the culprit in 'polluting' the mother-tongue by acting as the conduit for a 'modern' lifestyle.

The Malay-educated elites in Brunei defend Malay fiercely against the perceived onslaught of globalisation. Seminars organised by the National Malay Language and Literature Bureau and the Malay Department at the national university target the print media to further the case of Malay. Cited below is one newspaper report which warns about the challenges that Malay faces from globalisation forces:

Malay language and literature is definitely unique in facing new challenges of the new world order because of the current global interdependencies of one another.... This was stated by the Deputy Minister of Religious Affairs in his keynote address at a language and literature seminar held at the Orchid Garden Hotel in Berakas, yesterday morning. ('Deputy Minister calls to protect Malay language, literature', Borneo Bulletin, March 15, 2006).

As pointed out earlier, Malay derives its legitimacy and power from being associated with the national ideology of Brunei, called MIB which inextricably links the Malay language with the Malay culture and ethnicity, Islam, Monarchy and history. There is a growing body of research on the sociolinguistic history of Brunei (cf. for details, Saxena, 2007). This throws light on the ways in which Malay has taken on the symbolic value that it has today and how it has been reified as an emblem of the Islamic monarchy and of Bruneian citizenship, particularly when written in the *Jawi* script.

It is important to note here that, young elite Bruneians I interviewed see themselves as 'good' Muslims and 'true' Bruneians, but they do not like the imposition of *Jawi* because of its restricted function, *albeit* significant. *Jawi* is widely used in the multilingual literacy landscape of Brunei and it is represented as one of the central pillars of Bruneian identity. The script-based hierarchy on public signs reveals this iconic use of *Jawi*. As Sara pointed out in her narrative (below) all road signs, advertisement hoardings, and shop signs, in Malay, Chinese and/or English, are required to have *Jawi* script at the top. Those which do not comply with this regulation are torn down. This act of literacy policing is widely publicised in the media and many among the younger generation regard it as unnecessary. As Sara observes:

It can be seen on almost all sign boards and notices the roughly equal publicity of both the Roman and Jawi script, with the Jawi usually bigger and above the Roman. The government's way of thinking is that if there is more exposure of the Jawi script there would be more usage. The flaw in that argument is that it all depends on the choice of the reader and most would usually opt out of the attempt at reading the Jawi as it would defeat the purpose of sign boards; to provide information in the fastest way possible. (Sara)

Sara's observation that the majority of the young people do not bother reading the *Jawi* script and that its institutional imposition is a futile effort is brought home more emphatically by another student, Abdul. He directed my attention to the opinion he had posted on this topic on a public weblog, *Have Your Say* on 'Brudirect.com'[3], an unregulated space where the youngsters feel free to express their grievances:

The Jawi language law is not surprising, have you people realized Brunei is always desperate to promote the culture that it puts the road signs "Bahasa Melayu Bahasa Rasmi Negara" [Malay, the Official Language], "Utamakan Bahasa Melayu" [Give priority to Malay] followed by multiple "Gunakan Bahasa Melayu" [Use Malay] near airport, after Brunei/Miri border, and also at Brunei/Limbang border. This seems normal to us but the thing is they put it at borders and near airport. Let's imagine I travel to Russia, when I enter the country I see the road sign "Russian is the official language of the state", this is ok. After that come "Prioritize Russian language", I would be like, "err...", then after that I see many signs "Use Russian language! Use Russian language!" Surely I will say "what the... this is a desperate culture, putting that for tourists to see." There are proper ways to promote a culture, but putting road signs like that specifically for tourists to see is plain ridiculous.... Brunei, after numerous failures to promote its culture, uses forced regulations and put-signs-right-in-your-eyes method. Sword over brain is preferred. (Abdul)

Lifestyle Diglossia

21st Century globalisation has presented people with alternative lifestyles facilitated by the unprecedented nature of the flows of ideas, people, goods and language practices, particularly related to English. The concept 'lifestyle diglossia' (Saxena, 2014) contends that communication and other cultural behaviours in regulated and unregulated spaces reflects different lifestyles of individuals. The regulated spaces impose from above the dominant ideologies through various national policy initiatives, including language policies and language education policies. These lifestyles are meant to reproduce the traditional culture.

In contrast, although the communicative lifestyle in the unregulated spaces is condemned, that is where individuals exercise their choices in language use and lifestyles.

What I have shown in this paper is that the lifestyle chosen by the Malay youths in Brunei is indexical of their choices of global identities, as evident in their multilingual digital practices. As reflected in the socialisation practices of young people in unregulated spaces and their discourses, the global lifestyle commands higher prestige than the traditional cultural lifestyle, associated with the use of reified Malay. The lower prestige of Malay and the associated lifestyle is also evident in the fact, as I have shown elsewhere (Saxena 2014), that despite its 'symbolic capital' (Bourdieu, 1991) as the national language, Malay is experiencing semantic shift to English in the registers associated with health and food. This change is happening due to the shift from the traditional lifestyle to the modern lifestyle. It was found that Brunei's effort in setting up and providing access to modern infrastructure in health care, entertainment and food industry associated with 21st century globalisation, the occupational shift from nature to commodity based activities, and the change of identity from rural to urban are creating the 'lifestyle diglossia'. In the Bruneian context therefore modern ways of living are perceived as more prestigious than the traditional ones and the younger generations are leading the change.

My ongoing ethnographic research on multilingual digital practices in a British university in China is exploring similar questions: Are the Chinese students undergoing processes of lifestyle change? And, how can 'lifestyle diglossia' help explain the observed practices? I present here briefly some observations.

English has been gaining significance in China ever since 1978 when the Chinese government started implementing open-door and economic reform policies. It is closely associated with education and despite ups and downs in its fortune, it has increasingly acquired prestige since the internationalisation efforts in the 1990s (e.g. Lin and Block, 2011). This is also evident from the fact that in the last decade British and American universities have made their way into China. Drawing on data collected so far, I find the Chinese students in my study engaged in communication and following a global lifestyle similar to what I have shown above among the Bruneian youth. To give an example, the following is a SMS among a group of friends at the university:

我想吃芝士炒饭,go dutch呗
(I want to eat cheese fried rice and let's *go dutch*!)

This kind of code-switching is becoming quite common among the students on the campus. In addition, what is interesting to note is that it is reflecting also changes in cultural practices. As this group of students pointed out, traditionally in their culture, one person offers to pay the total bill and the others take their turn to pay on subsequent occasions. So, on this occasion, the practice of 'going dutch' reflects a change in their cultural values and their engagement with a global lifestyle of youth culture.

It would appear that ethnographic research on young people's micro-level practices of this nature in China may also be best explained by the concept of 'lifestyle diglossia'. However, what needs to be further explored how these kinds of cultural changes are viewed by the wider society. In the case of Bruneian youth, it is by capturing the micro-level practices and beliefs through 'lifestyle diglossia' which shows that people actively project their group affiliations and individuality by drawing on the global flows of lifestyles and local structural arrangements. It foregrounds the agentive dimension in the sense that individuals are not simply passive subjects of the structural arrangements of policies and ideologies, but actively and strategically deal with them or challenge them in their everyday practices.

REFERENCES

Andoutsopoulos, J. (2006). Multilingualism, Diaspora, and the Internet: Codes and Identities on German-based Diaspora Websites. *Journal of Sociolinguistics*, *10*(4), 429–450.

Appadurai, A. (1996). *Modernity at Large: Cultural Dimensions of Globalization*. Minneapolis: University of Minnesota Press.

Apriana, A. (2006). Mixing and Switching Languages in SMS Messages. *BAHASA DAN SENI*, *34*(1), 36–57.

Baetens Beardsmore, H. (1996). Reconciling Content Acquisition and Language Acquisition in Bilingual Classrooms. *Journal of Multilingual and Multicultural Development*, *17*(2-4), 114–127. doi:10.1080/01434639608666263

Barton, D. (1991). The social nature of writing. In D. Barton & R. Ivanic (Eds.), Writing in the community. Sage.

Barton, D. (1994). *Literacy: An introduction to the ecology of written language*. Oxford, UK: Basil Blackwell.

Barton, D. (2015). Tagging on Flicker as a Social Practice. In R. Jones, A. Chick, & C. Hafner (Eds.), *Discourse and Digital Practices: Doing discourse analysis in the digital age*. London: Routledge.

Baynham, M. (2004). Ethnographies of Literacy: Introduction. *Language and Education*, *18*(4), 285–290. doi:10.1080/09500780408666881

Baynham, M., & Prinsloo, M. (2009). *The Future of Literacy Studies*. Basingstoke, UK: Palgrave Macmillan. doi:10.1057/9780230245693

Bourdieu, P. (1991). *Language and Symbolic Power*. Cambridge, MA: Polity.

Brandt, D., & And Clinton, K. (2002). Limits of the Local: Expanding Perspectives on Literacy as a Social Practice. *Journal of Literacy Research*, *34*(3), 337–356. doi:10.1207/s15548430jlr3403_4

Coiro, J., Knobel, M., Lankshear, C., & Leu, D. (Eds.). (2008). *Handbook of research on new literacies*. Mahwah, NJ: Erlbaum.

Farrell, L. (2009). Texting the Future: Work, Literacies and Economies. In M. Baynham & M. Prinsloo (Eds.), *The Future of Literacy Studies*. Basingstoke, UK: Palgrave Macmillan.

Government of Brunei Darussalam. (1972). *Report of the Brunei Education Commission*. Brunei: Government of Brunei.

Government of Brunei Darussalam. (1985). Education System of Negara Brunei Darussalam. Bandar Seri Begawan: Jabatan Perkembangan Kurikulum, Jabatan Pelajaran, Kementerian Pelajaran dan Kesihatan.

Graff, H. J. (1979). *The literacy myth: Literacy and social structure in the nineteenth century city*. Academic Press.

Graff, H. J. (1986). The history of literacy: Toward a third generation. *Interchange*, *17*(2), 122–134. doi:10.1007/BF01807474

Hagood, M. C., Leander, K. M., Luke, C., Mackey, M., & Nixon, H. (2003). Media and online literacy studies. *Reading Research Quarterly*, *38*(3), 386–413. doi:10.1598/RRQ.38.3.4

Hartley, M. (1994). Generations of literacy among women in a bilingual community. In M. Hamilton, D. Barton, & R. Ivanic (Eds.), Worlds of literacies. Clevedon, UK: Multilingual Matters.

Heath, S. B. (1983). *Ways with words: Language, life and work in communities and classrooms*. Cambridge, UK: CUP.

Jones, G. M. (1996). Bilingual Education and Syllabus Design: Towards a Workable Blueprint. *Journal of Multilingual and Multicultural Development*, *17*(2-4), 280–293. doi:10.1080/01434639608666281

Lam, W. S. E. (2000). Second Language Literacy and the Design of the Self: A Case Study of a Teenager Writing on the Internet. *TESOL Quarterly*, *34*(3), 457–482. doi:10.2307/3587739

Lam, W. S. E. (2005). Second Language Socialization in a Bilingual Chat Room. *Language Learning & Technology*, *8*(3), 44–65.

Lam, W. S. E. (2006). Re-envisioning Language, Literacy and the Immigrant Subject in New Mediascapes. *Pedagogies*, *1*(3), 171–195. doi:10.1207/s15544818ped0103_2

Lam, W. S. E. (2009). Multiliteracies on Instant Messaging in Negotiating Local, Translocal, and Transnational Affiliations: A Case of an Adolescent Immigrant. *Reading Research Quarterly*, *44*(4), 377–397. doi:10.1598/RRQ.44.4.5

Lee, C. K. M. (2007) Linguistic Features of Email and ICQ Instant Messageing in Hong Kong. In B. Danet & S. Herring (Eds.), The Multilingual Internet: Language, Culture and Communication Online, (pp. 184-208). Oxford, UK: Oxford University Press.

Lewin, B. A., & Donner, Y. (2002). Communication in Internet Message Boards. *English Today*, *18*(3), 29–37. doi:10.1017/S026607840200305X

Lin, P., & Block, D. (2011). English as a "global language" in China: An investigation into learners' and teachers' language beliefs. *Science Direct,* *39*(3), 392-402. Available from: http://www.sciencedirect.com/science/article/pii/S0346251X11000972

Ling, R. (2005). The socio-linguistics of SMS: An analysis of SMS use by a random sample of Norwegians. In Mobile communications: Renegotiation of the social sphere (pp. 335–349). London: Springer.

Martin, P. W. (1999). Bilingual Unpacking of Monolingual Texts in Two Primary Classrooms in Brunei Darussalam. *Language and Education*, *13*(1), 38–58. doi:10.1080/09500789908666758

Mazawi, A. (2007). 'Knowledge Society' or Work as 'Spectacle'? Education for Work and the Prospects of the Social Transformation in Arab Societies. In L. Farrell & T. Fenwick (Eds.), *Educating the Global Work Force: Knowledge, Knowledge Work and Knowledge Workers*. London: Routledge.

McLellan, J. (2005). *Malay-English Language Alternation in Two Brunei Darussalam On-line Discussion Forum*. (Unpublished PhD Dissertation). Curtin University of Technology, Department of Language and Intercultural Education.

Saxena, M. (1994). Literacies among Panjabis in Southall. In M. Hamilton, D. Barton, & R. Ivanic (Eds.), Worlds of literacies. Clevedon: Multilingual Matters.

Saxena, M. (2007). Multilingual and multicultural identities in Brunei Darussalam. In A. B. M. Tsui & J. Tollefson (Eds.), *Language policy, culture and identity in Asian contexts*. Mahwah, NJ: Lawrence Erlbaum.

Saxena, M. (2009). Negotiating Conflicting Ideologies and Linguistic Otherness: Code-switching in English Classrooms. *English Teaching, 8*(2), 167–187.

Saxena, M. (2014). Critical Diglossia and Lifestyle Diglossia: National Development, English, Linguistic Diversity and Language Change. *International Journal of the Sociology of Language*. doi:10.1515/ijsl-2013-0067

Saxena, M., & Sercombe, P. (2002). Patterns and Variations in Language Choices and Language Attitudes among Bruneians. In D. W. C. So & G. M. Jones (Eds.), *Education and Society in Plurilingual Contexts* (pp. 248–265). Brussels: VUB Press.

Sebba, M. (2007). Identity and Language Construction in an Online Community: The Case of 'Ali G. In P. Auer (Ed.), *Style and Social Identities: Alternative Approaches to Linguistic Heterogeneity*. Berlin: Mouton/de Gruyter.

Sebba, M. (2009). Unregulated spaces. Plenary lecture at the conference on "Language Policy and Language Learning: New Paradigms and New Challenges". University of Limerick, UK.

Sefton-Green, J. (2006). Youth, technology, and media cultures. *Review of Research in Education, 30*(1), 279–306. doi:10.3102/0091732X030001279

Street, B. V. (1993). Cross-cultural perspectives on literacy. In J. Maybin (Ed.), *Language and literacy in social practice* (pp. 139–150). Clevedon: Multilingual Matters.

Swain, M. (1983). Bilingualism without Tears. In M. Clarke & J. Handscombe (Eds.), On TESOL '82: Pacific Perspectives on Language Learning and Teaching. Washington, DC: Teachers of English to Speakers of Other Languages (TESOL).

Warschauer, M. (2009). Digital Literacy Studies: Progress and Prospects. In M. Baynham & M. Prinsloo (Eds.), *The Future of Literacy Studies*. Basingstoke, UK: Palgrave Macmillan.

Warschauer, M., El Said, G. R., & Zohry, A. (2002). Language Choice Online: Globalization and Identity in Egypt. *Journal of Computer-Mediated Communication, 7*(4). Retrieved from http://jcmc.indiana.edu/vol7/issue4/warschauer.html

ENDNOTES

[1] University funded project (UBD/PNC2/RG/1(1)): Atlas of the Languages and Ethnic Communities of Brunei (2001-3); University funded project (UBD/PNC2/RG/4 (3)): Towards an Ethnography of Academic and non-Academic Literacies (2003-4); University funded project (UBD/PNC2/RG/1(45)): Globalisation & Language in Education Policy and Practice in Southeast Asian Societies (2006-7)

[2] Square brackets [ ] throughout the paper include translation of Malay or other comments by me.

[3] http://www.brudirect.com/index.php/200911039769/HYS-Topic-Of-the-Day/*Jawi*-language-law-for-cultural-preservation.html

Chapter 4
Educators and Mobile:
Challenges and Trends

Kenneth E. Harvey
KIMEP University, Kazakhstan

Philip J. Auter
University of Louisiana – Lafayette, USA

Samantha Stevens
University of Louisiana – Lafayette, USA

ABSTRACT

More practical and experiential education was the demand of executives recently surveyed about how universities could better meet employers' needs (Harvey & Manweller, 2015). As an alternative, with enhanced Web and mobile technologies, executives are seeing the opportunity to provide employees access to essential education in the workplace. Global e-learning is expected to top $107 billion in 2015 (Pappas, 2015), and U.S. corporations each year are now spending $1,169 per employee on training (Bersin, 2014b). Bersin says organizations are facing not a lack of employees but a lack of key skills among employees, and that is driving the trend (Bersin, 2014b). Harvard's Clayton Christensen, famous for his theories on disruptive technologies, suggests that even Harvard could be in jeopardy if it does not respond to these trends (Christensen, 2012). This chapter explores different strategies and technologies that can help meet these demands, and includes a case study of a university plan that makes distance learning more faculty-friendly, student-accessible, and cost-effective.

INTRODUCTION: STATE OF THE ART

The future of mobile education may depend as much on economics and lifestyle as on technology. It has been noted that technological application does not get interesting until technology itself gets "boring" (Shirky, 2009). The technology for exciting mobile-based distance-learning programs already exists and is getting better by the week. It is just waiting for the right people in the right situation to innovatively apply the technology.

DOI: 10.4018/978-1-5225-0469-6.ch004

Within educational institutions we already see mobile technology being used in conjunction with smart boards to create so-called smart classrooms. Teachers are able to interact with students within the classroom to share files of all kinds, to conduct assessments, etc. The same kind of interactions can be achieved over distance. Besides different asynchronous systems, such as the free Moodle learning management system, there are inexpensive synchronous systems such as Adobe Connect and HotConference (aka MeetCheap) that allow on-campus classes and presentations to be simulcast to students off-campus. These systems can share live video, panel discussions, Powerpoint slideshows, prerecorded videos, and other kinds of content. There is very little that can be done in a brick-and-mortar classroom that cannot be done in a virtual classroom via web-conferencing. And the systems allow the simulcast classes to be video-recorded for later asynchronous use. Such videos can easily be embedded into Moodle, for example, as part of a package of learning activities.

Like distance learning itself, research in the field has exploded in recent years – although both the services and research on them have been around for some time. As in traditional education, a number of key interactive elements must be considered when looking at the effectiveness of distance learning: appropriateness of the curriculum, availability (for both the student and the university) of the necessary technologies, teacher qualification, and student readiness.

CURRICULUM

In order for a university's curriculum to be successful, it has to address and fit the needs of its student body. In a recent study, Indonesian students cited concerns with their curriculum as an item of high priority (Budiman, 2015). At the completion of the study, the data suggested that lack of feedback, applying theoretical lessons, understanding different writing styles, forecasting the examination materials, ability to make translations, poor vocabulary, ability to understand grammar, the ability to write compositions based on specific instructions and ability to switch between writing styles were concerns shared overall by the students. While it is possible that this is the result of the appropriateness of the curriculum, it could also be a result of the instructor's readiness to teach online.

Joo, Andres, and Shearer (2014) found in a study of students in online courses at Costa Rican National University of Distance Education that after initially teaching a course, the outcomes should be assessed and the course retooled to better suit the needs of the students. After the course they had taken was reviewed by students, educators modified the types of assignments that would be assigned, altered the frequency of fact-to-face communication sessions, and suggested facilitation strategies that would result in a better experience for the students, both cognitively and in terms of their learning outcomes. It is important to point out that the researchers did not see a direct correlation between the changes in curriculum and student grades. However, if the curriculum were permanently altered to better suit student needs, it is not unreasonable to assume that over time, this change could result in greater student satisfaction and overall course success.

TECHNOLOGY

As more universities adapt their current curricula for mobile or online use, it has become apparent that a transitional issue between the two does exist. Rogerson-Revelle (2015) performed a study to determine what causes the disconnect between the two. The results of the study provided supportive evidence that online activities must be carefully aligned not only with desired learning outcomes, but also with the technological capabilities of the university and its students. Otherwise, it will be very difficult for faculty or students to see success in online courses.

Additionally, research has pointed to several technical issues that can hinder successful adaptation of technology in the classroom. Budiman (2015) surveyed students enrolled in distance learning (DL) courses working on English writing skills, and found that technical issues were among their key concerns. Seven themes were identified from the analysis: absence of communication with the lecturer and online tutor; absence of face-to-face tutorials; difficulty in purchasing course materials; limited opportunities to practice speaking and listening skills; limited time in the examination; being unfamiliar with modern technology; and poor quality of the course materials. Here, the research plainly shows that without proper adaptation for a course, technology will hinder more than help students succeed. Dr. John Orlando (2014) suggests "screencasting" as a way to provide real-time feedback to students during course instruction. First coined by journalist Jon Udell (2004), and defined by PC Magazine, "screencasting" is "screen recording software that turns screen output into a video to teach an application ... by demonstrating features" (2015). Screencasting includes several options for the instructor to customize the delivery of the material for specific class and student needs. For instance, screencasting allows instructors to turn on a live webcam video so that students can see the professors as they interact with the materials that they're teaching. Additionally, screencasting allows for on-the-spot editing of materials that are featured on the screen. "Screencast programs may allow narration during capture, and advanced versions allow editing and annotation after the capture" (PCmag.com, 2015). This feature offers a two-fold benefit. The first is that instructors can update or adapt materials as the class demands. The second benefit is that instructors can immediately address student questions on the materials as they arise over the class period. Finally, screencasting allows instructors to give immediate feedback, as opposed to written notes that will not be seen until after the materials have been reviewed. Screencasting is one of many advances in educational technology that allow both students and professors to interact as much like a traditional classroom setting as possible. As students become more dependent on mobile devices and technology, it is imperative that education continues to mobilize itself as well, if it has any hope of keeping student interest and involvement. While there is no total replacement for face-to-face interaction between students and teachers, adopting new technologies, such as recorded lectures, the option of practicing lessons online with other students, etc., could arguably improve the DL experience for all parties involved.

TEACHER QUALIFICATIONS

Just as it is necessary to evaluate and adapt curricula for integration with new technology, it is also necessary to evaluate who will conduct these DL courses. Stanisic Stojic, Dobrijevic, Stanisic, and Stanic (2014) compiled 71 student ratings of online teachers. The survey included questions on the teacher's gender, age, resources, overall involvement in online teaching, and overall teaching. After reviewing the results, researchers were able to determine that there was a correlation between all of the aforementioned factors and student assessments of teacher effectiveness.

The results indicated that overall scores seemed to be consistent for the same teacher when s/he taught online versus traditional classes. However, the results of the study indicated that it was difficult to compare multiple teachers of the same course. These results varied wildly and were generally lower than evaluations of teachers in courses where there were not multiple instructors. These results indicate that there is a need for universities to have a system that allows for direct comparison between faculty members teaching the same course. That way, it is possible to determine whether or not an instructor is a good fit for DL courses, which will ultimately improve both staff and student perceptions of online and mobile education.

Orlando (2010) suggests that in order for students to successfully use the technology necessary for completing a DL course, instructors must take part in the initial training process. Orlando points out that "a common fallacy is to believe that because students today are "digital natives" -- meaning that they grew up with technology -- they are good at using any technology" (2010). Therefore, at least some of the total course time should be spent introducing students to the technology they are expected to use throughout the duration of the course. Using a sort of virtual tour style, instructors can create a thorough explanation of the course needs and expectations for technology use, which can be recorded and then re-watched later should students need a reminder. While this does require an initial time investment up front from the professor, in the long run, both parties will save time and energy later on, when students can reference this tutorial on their own. It is the adaption of technology by both instructors and the student body that will allow the university itself to develop and manage a successful DL program to add to its complete curricula.

Instructors can employ several strategies in order to create a successful learning environment for their DL students. In a traditional classroom setting, the instructor interacts with the students and can gauge student engagement by "watching facial expressions and body language" and by "[monitoring] these, cause and [make] adjustments or [store] the information for later use" (Brady, 1998). Unfortunately, with a DL course, instructors often lack this feedback, especially if lessons are pre-recorded and viewed later by students.

It is possible to combat these issues by using an interactive video program (such as Skype or one of many web-conferencing programs), supplying students with a print component to accompany the lecture (Reed and Woodruff, 1996), or having assignments and/or quizzes that ask specific questions about the recorded material. Specifically, interactive video is a way for instructors to navigate the transition between a traditional classroom and a DL classroom: "[It] allows for 'real time' or synchronous visual contact between students and instructor or among students at different sites" (Brady, 1998). Video streaming of this nature allows for both the students and the instructor to familiarize themselves with DL technology without completely eliminating components of the brick-and-mortar instructional experience.

Instruction via DL does not mean that an instructor needs to forgo all traditional teaching tools, just that these elements should be incorporated in different ways. Instructors can require students to watch an informative video and, if not facilitating a live online audio or video class discussion via web-conferencing, can require students to engage in an online dialogue with their classmates in other ways (St. Francis Medical Center, College of Nursing, 2015). Instead of requiring students to pair up physically for an assignment, an instructor can randomly assign partners and require that they prove their collaboration using email exchange print outs, etc. Most modern learning management systems (LMSs) also include forum functions, or it is easy to create a Facebook page exclusively for use by a specific class. While fears may persist that DL courses reduce communication between students and instructors, it actually provides instructors with a unique opportunity to mandate exactly how often the parties interact with each other. With the popularity of numerous messaging programs, there is no reason why instructors cannot facilitate class conversation. Requiring students to communicate regularly with both the instructor or other students enrolled in the course can help create a dialogue between all parties and strengthen the chances for success in a DL course.

STUDENT READINESS

While appropriate course materials and leadership are among the necessary elements to see success in mobile education, student readiness for this type of course will also contribute to the overall success or failure of the course. Budiman (2015) found that some students studying English in Indonesia suffered from "personal issues" that had a direct effect on their ability to successfully complete their DL courses. The most prominent issues that emerged from the study were: ability to become an autonomous student; isolation; discipline; limited time to study; pace of learning; lack of motivation; lack of enthusiasm; and overall interest in writing. Among the issues, limited time to study and feelings of isolation received the most attention from study participants. In other words, the students that were the least prepared for the structure and requirements of an online course experienced the least success overall. Ozbek (2015) found that there are 58 skills and competencies that can benefit online learners. These skills were classified into meta-cognitive, cognitive, technological, and affective competencies and skills. It is arguable that a number of these skills are necessary for students to perform well in online classes. However, the question remains as to how to assess students for these skills.

Similarly, Cakiroglu (2014) looked at student learning styles and how those correlated with success in online education. Given that DL is a different type of educational experience, it is logical to assume that certain learning styles would be more successful in a distance learning environment than others. Using Kolb's learning style inventory (LSI) as well as a self-created study habits questionnaire, Cakiroglu surveyed 62 students enrolled in online courses. Findings suggested that there are significant relationships between learning styles, study habits, and learning performances. Regarding learning styles, accommodators and divergers both had a significantly higher average score than those of convergers. Not surprisingly as to study habits, students with "good" study habits typically did much better than those with weaker study habits. Techniques such as a required minimum GPA, a certain number of course hours successfully completed, or a recommendation from an academic advisor are a few of the many ways universities can determine eligibility for enrollment in DL courses.

Students need to be aware of the materials necessary to participate in a DL course. Internet access, basic computer proficiency, email access, and access to a computer (CCC&TI, 2016) are all things associated with modern DL courses. In order to improve computer skills, students can enroll in computer literacy courses and, with time and practice, acquire the skills necessary to complete a DL course. To ensure that only prepared students are enrolling in DL courses, universities can impose enrollment "qualifications" for student enrollment:

- Require a minimum GPA for enrollment (University of Alabama, 2015)
- Require a minimum on undergraduate test scores (University of Alabama, 2015)
- Impose course or degree prerequisites (Texas A&M University, 2015)
- Require that students complete a computer literacy course or pass a computer proficiency test (Blue Ridge Community College, 2015)
- Stipulate that students successfully complete a minimum number of hours prior to enrollment (Mississippi State University, 2015)

Perhaps the most direct method of ensuring that students are prepared for a DL course is to require that they first successfully complete a blended or "flipped" course that requires regular attendance in the brick-and-mortar classroom but also provides many lectures, assignments and assessments via the university's LMS. In this way students will have the face-to-face support of an instructor and maybe an assistant as they complete such tasks for the first time.

Employing measures such as these can help ensure that the students enrolled in DL courses or programs are qualified to do so, and will likely experience academic success while enrolled. Additionally, instructors can rest assured that their students will not only be able to complete the DL course but also will likely contribute positively to the course or program throughout the duration of enrollment.

Institutions should take note of implementation resources such as an information guide published by Caldwell Community College and Technical Institute (2016). It lists characteristics of students who succeed in DL courses, such as "know how to learn independently" or "can take notes well, whether from 'online lectures', textbooks, or television programs." While it is unwise for students with little or no technology experience to enroll in DL courses, there are some measures that students can take to improve their online proficiency and succeed in their coursework. If enrolled in a university with a physical location, resources such as "discussion groups, libraries, [and] writing guides" (CCC&TI, 2016) are available. Students can take advantage of these opportunities in order to improve their use of DL technologies. With a wealth of online tutorials available via popular video websites such as YouTube, students can practice taking notes and/or learning from tutorial videos. Ted Talks, lectures, and taped discussions (YouTube, 2016) are all available on YouTube and can serve as an excellent example of a DL lecture.

STUDENT BENEFITS

Students enrolled in DL courses can expect to gain more than simple convenience. Of course, the mobility that DL classes provide allows students to schedule classes when it best suits their needs, and provides a new way to engage with a traditional, accredited university. The US Journal of Academics (2015) suggests the following benefits, especially for international students:

- Access to courses at universities in the United States or other countries aside from one's own
- Completion of courses at a pace suitable for the students' needs
- Ability to continually review or re-study materials presented in previous lessons
- Opportunity to study and improve language skills without the pressure of a traditional classroom setting
- Increased scholastic opportunities for students with disabilities or with restricted mobility

Students who experience boredom in a traditional lecture course may experience greater course satisfaction with DL, since the rate of information consumption is frequently controlled by the individual students. DL courses often allow students to accommodate their studies to their personal schedules, allowing them to meet deadlines or submit assignments at more convenient times (mindful, of course, to an assignment due date). Students may be able to complete lessons when they wish, not when a course meeting schedule dictates that they must. This may particularly appeal to students who have full-time jobs or unconventional schedules. To keep its work force as well trained and competitive as possible, Korea has developed numerous "cyber universities" with very limited physical facilities, but with extensive and well developed virtual curricula. The Seoul (Korea) Cyber University, for example, was built at a cost of $54 million but has no regular classrooms. Instead, it has well-equipped video studios and teams with audio, video and graphic expertise (Uhm, 2015).

The degree to which DL courses provide such flexibility varies, however, depending to what degree their curriculum is synchronous (live) or asynchronous (on-demand). As discussed in the KIMEP case study later in this chapter, however, courses need not be totally one or the other. Indeed, they can be both. Regular classes can be held live with web-conferencing software and simultaneously recorded for on-demand viewing.

As with the university, students may also enjoy financial benefits from DL learning. Tuition decreases, little to no commuting costs (for that specific course), and lower costs of course supplies (OEDB.org, 2012) may entice students to consider DL courses in addition to or to completely replace a traditional classroom setting. Additionally, for students who do not have the means to purchase a computer or home internet access, courses can be completed using public computers – i.e. public libraries – to complete their school work, as long as they have access to an internet connection.

Smaller universities may have smaller course catalogues, which limits what courses students are exposed to. However, due to the ever-increasing number of universities offering DL courses – and even degrees – students have much greater control over what they are able to study (OEDB.org, 2012). Moreover, students can dictate how they complete their studies: "Online courses offer shy or more reticent students the opportunity to participate in class discussions or chats with more ease than face-to-face sessions. Some students even report that online courses are easier to concentrate in because they are not distracted by other students and classroom activity" (OEDB.org, 2012).

With these benefits and more, it makes sense for universities to incorporate distance learning into their traditional curricula. Many university faculty may remain unsure of distance learning, thus inhibiting the growth of DL. But there are clear arguments as to why faculty should support student learning via DL courses (Clay, 1999). DL can provide faculty with:

- Unique learning and teaching opportunities.
- The ability to teach an increased number of students.
- Increased possibility for creativity when developing lesson and course content.
- The opportunity to teach highly qualified and motivated students;
- A reduction in commuting to and from the university.
- Increased time to complete research.
- Greater schedule flexibility.

For students prepared to enroll in and devote time to DL courses, the benefits are clear. As traditional universities grow with and incorporate DL into the academic structure, education will be more and more accessible to prospective students, allowing many the opportunities that were previously only afforded to a few.

ACADEMIA AND BEYOND

There is no one who would argue that the mobile education industry is not growing rapidly and will not continue to grow. Mobile is revolutionizing society as a whole, as the number of mobile connections to the Internet will soon be triple that of desktop connections (Cisco, 2015). Truly, the only debate is how rapidly the industry will revolutionize itself, and to what degree mobile education will become integrated into and even replace some aspects of traditional education. Christopher Pappas (2015) identifies 10 key statistics on the state of mobile education:

- By the end of 2015, the mobile education industry will be a $107 billion industry, and revenue will reach nearly $49.9 billion.
- India (55%), China (52%), Malaysia (41%), Romania (38%), and Poland (28%) are experiencing the most growth in mobile education.
- The learning management system (LMS) market is expected to be worth $4 billion in 2015, and grow to over $7 billion by 2018, with the most significant contribution of revenue coming out of North America (hence its absence in the list of the top 10 countries experiencing growth in e-learning technologies).
- By the end of 2015, it is projected that the worth of the worldwide mobile learning market will reach $8.7 billion.
- About 8% of all companies currently utilize MOOCs (Massive Open Online Courses), another 7% are considering using MOOCS, and it is estimated that this technology will see a 28% increase in use over the next two years.
- Corporate use of online education and training is expected to grow up to 13% annually until 2017.
- Large companies constitute 30% of sales in the e-learning industry.
- In 2014, 28.5% of training hours were completed online or with computer-based technologies and 1.4% of trainings were completed via mobile devices.
- In 2014, 74% of companies used some sort of LMS for training, 48% used a rapid e-learning tool, 33% used an application simulation tool, 25% used a learning content management system, 21% used an online performance support system, 18% used mobile applications, and 11% utilized podcasts.
- Based on 2014 trends it is projected that 44% of companies intend to purchase online learning tools/systems; 41% intend to purchase LMSs; 29% intend to purchase content development products/services; 27% intend to purchase courseware design and presentation tools and software; and 18% intend to purchase audio/web-conferencing products/systems.

Clearly, universities and corporations alike are increasing their reliance on and use of mobile education. The question is not when mobile education will become a part of the education system, but how quickly and to what degree.

With the drop in prices for mobile education technologies, it won't take long before educational institutions begin questioning why they should spend millions of dollars on new brick-and-mortar classrooms when they can add virtual classrooms for less than the cost of just utilities and maintenance of the brick-and-mortar classrooms. And if the traditional institutions don't take advantage of such technology, as is the case in every other industry, their competitors will use the new technology to gradually put them out of business. Stiff competition in America by the for-profit University of Phoenix could be cited as one of the best examples of using technology to gain a competitive advantage. The university has hundreds of thousands of students, a large portion of whom work full-time and depend on the university's flexibility to learn while they earn.

The pressure to change, then, may come from clients as well as competitors. The "clients" for universities are both their students and their students' future employers. A recent survey conducted of media and public relations executives in North America resulted in some shocking results. Of 767 executives responding, a massive majority of 89% said that graduates of university communication programs need "more hands-on experience" (Harvey and Manweller, 2015). Only 1% disagreed with that statement. And 33% of the executives agreed that "mass communication education needs to be totally revamped." Most respondents were unsure or neutral in their response to that statement, but only 23% disagreed. The executives also responded very positively to a number of radical reforms presented. To the educational concept of "one year of general education, two years of intensive mass communication training, and a one-year professional apprenticeship," executives agreed 53%-11% overall that graduates would be better prepared for the real world (the remainder neutral or uncertain). Overall, executives also supported another statement 48%-20% that they "would be interested in hiring mass communication graduates as yearlong apprentices at lower pay while we help complete their professional training and consider whether to offer a permanent position." The strongest support came for perhaps the most radical idea of all, "a 4-year work-study apprenticeship program where students would work about 6 hours a day as news media apprentices and then study 6 hours a day using journalism as a general education learning method, as well as a set of professional skills." Overall, respondents interested in hiring graduates from such a program outpolled those opposed 63%-9%, including 65% approval by PR executives.

This survey was only of executives of companies or departments involved with communications. But it is important to note that PR executives who work with companies in all segments of the economy supported the most radical of proposals more strongly than did the media executives. If feelings are this strong in this field, it is likely that there are employers in many fields who would prefer an alternative to today's system of education. In short, there are a lot of people who hire the graduates of American universities who are unhappy with what they're getting and would prefer a more practical approach.

Harvard Professor Clayton Christensen addressed this particular dilemma at a conference concerning the future of higher education. He noted that the cost of hiring an Ivy League graduate, such as one from Harvard, limits the type of company that can afford to hire such individuals. However, big companies such as Johnson & Johnson, General Electric, and Intel have created their own sort of corporate education program within the company itself (2012). This set-up does away with the high price tag of the Harvard MBA graduate compensation, and also allows for individuals to learn and practice management on the job, working for the company that they are already a part of. Additionally, as these corporate colleges were built to address the exact needs of the respective company in real time, the internal design of the program can be easily adapted, changed, or expanded to meet the specific organization's demands. It can be hypothesizes that for MBA and other postgraduate programs to stay competitive in the market, adaptability will be the ultimate key to both success and survival.

A 2014 *Forbes* study (Bersin, 2014a) examined the then-current state of corporate learning management systems (LMS) and identified four key factors that were driving more companies to utilize in-house education for its employees:

- **Corporate Training:** As technology continues to create new opportunities for what is possible, companies may experience some difficulty in training their employees quickly enough through traditional education channels. In order to keep pace with industry change, it is becoming paramount for HR departments to be able to retrain employees at a faster rate than ever.

- **Learning Market:** The need for quick and reliable training has led to the rapid increase of online or mobile training and educational tools. With a number of major universities offering online courses as an alternative to traditional classroom settings, it has become easier than ever for individuals to receive the training that they need, nearly on-demand. Additionally, many of these LMS's have been optimized for corporate needs, allowing the mobile education market to constantly further its global reach.

- **Learning Platforms:** When technology grows and changes, so must the tools that accompany it. With 61% of companies planning to "replace their learning platforms in the next 18 months," it is crucial for LMSs to update at a parallel rate in order to stay competitive. Functionality, ease of access, comprehensiveness, and price are a few of the factors that will drive the further development of LMSs.

- **New Software:** With a constantly competitive market searching for the next best product, the LMS market is producing more options to choose from than ever before. "Today, the market has expanded by 100-fold, driven largely by mid-sized and now even small companies." The universal presence of the mobile world has driven down the cost of once very expensive software and technology, making the most up-to-date systems available to every member of the corporate world. In a study completed in Turkey, six main factors were identified when research was conducted that "address trends and gaps observed in... the integration of mobile learning" (Baran, 2014). Some of these factors included a lack of perspectives being reported, a variance in current "perceptions, attitudes and usage patterns," and a lack in reporting any challenges. However, it was noted overall that there was a positive relationship between mobile education and student happiness. In an additional study completed with participation from the Spanish National University of Distance Education (UNED), the research concluded that smart-device applications (apps) that are developed specifically for university coursework are preferred by students in terms of mobile education. Additionally, these customized apps can promote a higher rate of "collaborative work among students and professors" (Vazquez-Cano, 2014), making the decision to use mobile education at the university level even easier than in years past.

DL technology has and will continue to advance rapidly, especially as more top-tier universities, organizations, and companies adopt it as a means of learning and training. However, in order to best understand why DL has permeated both corporate and academic structures, a thorough understanding of DL techniques, technologies and history is necessary.

SUMMARIZING THE PAST

The term "distance learning" (DL) is by its nature similar to the definition of distance education (DE) or open and distance learning (ODL), as it is stated by Keegan (1996). Its key characteristics include some form of separation of teacher and student (either place or time, or both) during the process of education.

The usefulness and applicability of any education format including DL is usually accompanied by scrupulous research on how beneficial this format can be for students' success. Three major criteria for students' success include academic achievement, attitude, and persistence toward completion (or retention rate). As noticed by different authors (Bernard & Amundsen, 1989; Kember, 1995; Woodley, 2004), distance has been usually associated with a higher dropout rate. Indeed, as found by Frankola (2000),

DL students' dropout rates range from 20% to 50%, and these rates are often 10% to 20% higher than in on-campus education. However, Russell's review of more than 350 sources proves "no significant difference" in academic performance between DL and traditional learning (2001). Therefore, the next stage of DL research should focus on the exploration of DL's effectiveness in different formats.

Nowadays there are two popular forms of DL: asynchronous DL (ADL) and synchronous DL (SDL). ADL is defined by Hrastinski (2008) as the form of education when teacher and students can't be online simultaneously. ADL uses emails, discussions boards and other media that support asynchronous communication between participants. Since ADL provides more time for learners to engage in learning process, it is characterized by higher cognitive participation and seen as more thoughtful than SDL. SDL, in turn, is defined as the form of education that facilitates DL students' inclusion into a learning community. SDL uses videoconferencing, chats and live interactive media that, unlike ADL, prevent frustration due to inability to immediately participate in discussions and receive answers for questions. Thus, SDL includes more personal participation with higher sense of integration and motivation (Hrastinski, 2008).

There are also some findings that demonstrate that DL students can outperform on-campus students in some terms. As it has been acknowledged by Bernard et al (2004) that whereas students in classroom outperform SDL students by achievement outcomes, ADL students do better than on-campus students. At the same time, more ADL students tend to drop out of a course than their classroom peers (Bernard et al, 2004).

The results to date are inconsistent, but, more importantly, the questions or conditions being studied are inconsistent. For example, to say that SDL "uses videoconferencing, chats and live interactive media" does not specify how such techniques are employed. In addition, there is almost an infinite number of ways to intermix such approaches so that a university's plan may include several distinct approaches. When it comes to content delivery, these are a few such options:

- Text-based asynchronous curriculum on a learning management system
- Multimedia asynchronous curriculum on a learning management system
- One-way video-streaming for live viewing or recorded for use in an asynchronous multimedia learning management system
- Synchronous interactive web-conferencing with recording capability for use in an asynchronous multimedia learning management system

However, in reality, all of these approaches could be used not only within one university's broad DL program but even within any one course offered within that program. Then far more DL options open up that have never been fully researched. There are too many variables. And no previous study includes all of the variables that a university might employ.

Then, as a university begins creating its DL plan, new questions arise that may restrict or enhance these many options. Indeed, there are many challenges to launching an effective distance-learning program. They include:

- Finding or creating a platform, system and tools (NOT including content) that can handle the DL program.
- Achieving faculty buy-in, overcoming technophobia and other concerns, and incentivizing faculty to create and/or present the content.

- Achieving student buy-in, overcoming technophobia and their other concerns, and incentivizing students to participate in the DL program.
- Assessing student achievement of learning objectives.
- Funding the DL program.

In order to truly understand how students and instructors interact with DL programs, KIMEP University performed an extensive study and analysis of the effectiveness of DL courses in their specific academic setting. The case study in the following section provides a thorough summation of the findings. In addition to its detailed report on why and how the study was conducted, this case study also addresses its successes and failures, and gives anyone interested in distance learning much to consider and learn from.

PURSUING THE RIGHT FORMULA: KIMEP UNIVERSITY CASE STUDY

These are all major challenges, but KIMEP University in Almaty, Kazakhstan, concluded that the biggest of these challenges for them were creating a DL program in a very cost-effective way that would still allow them to achieve broad buy-in from a relatively small faculty base and to expand their student enrollment. This is a case study of their planning process and their testing of technologies in developing their DL plan.

The KIMEP DL chair had offered about 30 different workshops on DL skills and technologies, with an average attendance of under five faculty members. He also knew that convincing the administration of the financially challenged university to provide substantive financial incentives to professors would be a hard sale. The university has been struggling since the worldwide recession hit in 2008, but its economic challenges went beyond those global issues. The recession hit at the same time as unfavorable demographics. Students in Kazakhstan typically graduate from high school at age 17, and it was 17 years before the 2008 recession that Kazakhstan had gained its independence from the Soviet Union. With that independence came several years of economic chaos, rolling electrical blackouts, and dire living conditions that caused the birth rate to plummet 46% from 3.13 births per adult woman in 1988 to a low of 1.70 in 1999 (Google, 2015). While the birth rate has recovered to 2.59 as of 2012, all Central Asian universities are still faced with competing for far fewer university-age students until children being born today begin graduating from high school. And, while the private Western-style, English-language KIMEP University has long been considered the best in the region, the lower-cost, state-funded universities and the other private universities are becoming or at least claiming to become more competitive. And all of these factors have been further complicated by falling oil prices, causing Kazakhstan's GDP growth rate to drop from 6.0% in January 2014 to 1.7% in July 2015 (Trading Economics, 2015) and for its national currency, the tenge, to devalue by nearly two-third, from 120 per dollar in the fall of 2008 to 350 per dollar in December of 2015 (Bloomberg, 2015). Thus, while creating a high-quality distance-learning program was seen as a way for KIMEP to distinguish itself, it also needed to be very cost-effective.

Spurred by the possibility of partnering with an important university in the nation's capital of Astana, in January 2013 KIMEP President Chan Young Bang asked Dr. Ken Harvey, the university's most experienced communications professor and DL technologist, to head up a select committee to quickly review multimedia technologies and learning management systems and finalize a DL plan. Harvey had been experimenting with both LMSs and online multimedia since about the year 2000, including the use and testing of web-conferencing systems and the development of experimental multimedia Moodle sites, such as http://Virtual-University.us with over 3,000 instructional videos embedded. Other KIMEP had also used Moodle to a limited degree for about 10 years, had been further trained by Dr. Harvey, and had tested numerous other LMSs, trying to determine for their own purposes whether Moodle was the best option.

Wright, Lopes, Montgomerie, Reju and Schmol (2014) prepared in-depth recommendations on how an educational institution should go about choosing an LMS. Including the previous work by Harvey and KIMEP's EEC, many of these recommendations were followed. But probably the most important perspective the DL committee agreed on was also reflected by Wright, et al, that which LMS system is selected is not as important as buy-in by faculty and students. The proprietary Blackboard LMS dominated the U.S. higher education market until 2007 but saw its use drop about 30 percentage points to 41% by 2013 (Straumsheim, 2013). Open-source Moodle is now most popular among smaller American universities in the size range of KIMEP, whereas larger universities still tend to prefer Blackboard (Straumsheim, 2014). Desire2Learn (11%), Canvas (8%) and Sakai (5%) are also making significant inroads into the market (Straumsheim, 2013). When it comes to all LMS users worldwide, including other educational institutions and business organizations, Moodle far outdistances Blackboard, as does the Edmodo LMS, and a Capterra analysis of the top 20 systems rates Moodle now No. 1 in quality, too (Pappas, 2014). Regardless, successful DL programs have been set up with any of dozens of different open-source and proprietary systems, and it is not evident that any one LMS leads consistently to more success than another, as Wright, et al, (2014) pointed out:

The selection of an accessible, flexible, scalable, reliable/stable, robust, efficient, secure and cost-effective LMS for classroom, online and blended learning environments is vital for institutions. However, purchasing or leasing an LMS will yield good ROI only if instructors use it fully to engage and communicate effectively with students [and] receive support in developing and delivering courses.

In assessing these elements and possible obstacles, KIMEP decided to begin with a DL program emphasizing a simulcasting strategy by which courses taught on campus would be web-conferenced to students off campus in live synchronous fashion and recorded for later on-demand asynchronous learning. The committee determined there would be advantages in cost, faculty buy-in, and student buy-in to create a program involving:

- Live in-classroom learning.
- SDL simulcasting of the live classes to off-campus students via web-conferencing, with potentially all students able to be seen live via webcams, with additional interaction via live simultaneous text chat, and with the same kind of teaching tools as in the live classroom, such as Powerpoint slideshows, prerecorded videos and desktop viewing of literally anything available on the instructor's computer.
- ADL on-demand instruction, beginning with the video-recorded classroom lectures being embedded into the existing Moodle LMS, along with other asynchronous content the instructors choose to add, and with instructors incentivized to develop online quizzes and exams.
- The instructor's option to require students to come to campus or an established learning center to take exams.

This combination sets this case study apart, with potential for long-term comparisons among students enrolled in this program but selecting different learning platforms and between KIMEP students involved in this blended program and those in KIMEP's traditional classrooms. The advantages the committee saw in this strategy were that:

- Professors would simply do what they do best – teach, reducing several obstacles related to technophobia and work overload.
- Student technical assistants would handle cameras, recording, editing and uploading.
- Simulcasting would be more cost-effective since professors are teaching a classroom of students already paying his or her salary so that fewer distant students are needed for the program to produce a positive return on investment.
- Professors are typically more at ease teaching live students than speaking strictly to a camera.
- And interactive web-conferencing would allow multiple speakers to simultaneously converse on webcam, all students to simultaneously ask questions and make comments via text chat; and use of multimedia including PDFs, Powerpoint slideshows and pre-recorded videos.

This strategy also coincides with important changes in technology that will affect everything that we do. We are in the midst of a Web revolution that will dramatically change education in the coming years. The viewing of online video is exploding. The number of online videos viewed in June 2012 was about 550% greater than just one year earlier in June 2011 (comScore 2012), and that number is growing rapidly. The average power of broadband is also growing rapidly, greatly enhancing the quality of online video. Between 2011 and 2016, average broadband worldwide is expected to grow nearly 400% to 34 megabits per second (Cisco, May 30, 2012). Even television is expected now to quickly migrate from being broadcast through the airwaves and sent through separate cable networks to being accessed through the Internet. By 2017 half of all "cable type" paid TV may be received over the Worldwide Web (Xtreme 2013). Also driving this revolution is mobile technology. The easiest Internet medium to adapt to mobile is video, and the number of Internet devices is projected by Cisco to nearly double by 2016 to the point where there will be 2.5 devices per man, woman and child (Cisco, May 30, 2012). Already there are twice as many broadband mobile connections to the Internet than fixed-wire connections, and the trend is worldwide. More than half of all the mobile connections around the globe are in emerging countries (Pepper 2012).

Testing Technology

Once the KIMEP DL Committee decided in principle to explore interactive web-conferencing as the initial DL teaching platform, it contacted major technology firms to begin comparing the different technologies, from expensive high-end high-definition systems to low-end consumer systems. These included:

- High-end systems from Cisco, Microsoft and Samsung.
- Expensive TV-quality video-streaming through a Russian vendor.
- Expensive point-to-point virtual networking from China Telecom.
- Low-cost consumer web-conferencing systems: HotConference, TalkFusion and Adobe Connect.
- And free Google Hangout.

The DL Committee started with Cisco (Webex), Microsoft, and Samsung systems, along with the video-streaming option, using Adobe software and a leased server designed for video streaming. The main problems with all of the expensive, high-definition systems were: (1) bandwidth limitations of the Internet in Kazakhstan for which no one offered a viable alternative, and (2) their own poor demonstrations:

- **Cisco (Webex)** proposed hardware costing $135,575, but its demonstration was worse than that of the low-end consumer systems, using off-the-shelf webcams and mics on an inexpensive desktop computer. The high-definition video in the demonstration was technically high quality in detail but was stop-and-go jerky in delivery, and the site with which Cisco connected had such poor lighting that the committee was speaking, essentially, to a shadow.
- **Microsoft** vendors were not very definitive in price alternatives, and their demonstration was similar to Cisco's. They seemed intent on selling KIMEP a hardware/software package to cover the entire campus when the DL committee was looking primarily to set up a couple of studio classrooms. The video flowed a little better than with Cisco, but committee members were still speaking with a shadow in a closet. The vendors' poor presentation preparation aside, for a questionable improvement in video detail, they wanted to sell KIMEP an expensive Lync server hardware/software package (no precise price ever provided), high-definition cameras running between $27,505 to $58,362 apiece, and ongoing services.
- **Samsung** only showed the committee recorded videos of live conferencing in Korea, which has one of the most powerful Internet systems in the world. And they admitted they, too, would be limited by Kazakhstan's existing Internet system. Nevertheless, they wanted to sell KIMEP a site-to-site high-definition system and lower-grade home or office access by students for $500,000.
- **Direct video streaming** over a leased video server provided the highest quality video in terms of clarity and detail. The streaming video demo was created on a standard desktop computer and streamed simultaneously to a Moscow server, which then embedded the stream on a website in Astana, Kazakhstan. While it provided the highest quality of any system tested by the DL Committee, the video feed did often stop and start as other high-definition systems. The cost was $2,000-$12,000 per month, depending on the required bandwidth needed, and that was just for one video stream. To mix Powerpoints and other graphics would require an expensive switching board to move back and forth between camera video and graphics. And to achieve synchronous interaction rather than asynchronous streaming would essentially double the cost.

Consumer web-conferencing services tested were HotConference, TalkFusion, Adobe Connect and Google+ Hangout. All four services cost under $50 a month, with no cost to participating students, and were very competitive in quality with their high-price competitors – especially in an international environment of inconsistent broadband:

- **HotConference ($45/month with a dedicated URL for each virtual classroom 24/7):** This was found to be the easiest to use, fastest to set up, most intuitive, and easiest for technical assistants and faculty to learn. The quality of its audio and the streaming of Powerpoint slides and other learning aids was generally very good in 2013 and improved in 2014. Live webcam video is not the quality of high-definition video in clarity and detail but flowed more smoothly and still provided adequate quality. More important is that quality graphics are given emphasis, which is usually more important for online instruction. In other words, Powerpoint slideshows or other graphics carry more information than a talking head. Pre-recorded video can also be shown within the web-conferencing window. The HotConference window, illustrated below, provides a small live video feed of the speaker while a larger window is being used for slideshows, electronic white board, and other audio-visual content. There is also a live text chat box, as seen in the lower-left-hand corner of Figure 1, in which the presenter and audience can also interact. A video of the HotConference system in use can be seen at https://youtu.be/R76vJXEtzok. HotConference offers to versions of its software – one with the recorded output already shown and one with this recorded output: http://youtu.be/vo0Ecs1GNSc. The actual operation of both versions is essentially the same, however. This latter example is lower quality video after editing and uploading to YouTube. To see the quality of an original unedited video-recording like students would see, go to http://199.116.250.222/video/424/226M48kPnKNfGb.mp4.

- **TalkFusion ($200 setup + $35/month – new URL with each separate web-conference to set up):** This system provided better video but worse audio in the 2013 committee's last testing. In previous tests audio and video were good, and that company announced a number of upgrades to its system. It is problematic to have to schedule each session separately, however, and to then have to announce a different URL with each broadcast, whereas HotConference assigns permanent URLs. Even after the URL is scheduled and announced, the TalkFusion system involves more complicated setup. It takes about 10 minutes to set up, as opposed to about 3 minutes for HotConference. Its various functions are also not as simple and intuitive as HotConference, but somewhat simpler than Google Hangout.

Figure 1. Example screenshot from a HotConference webinar

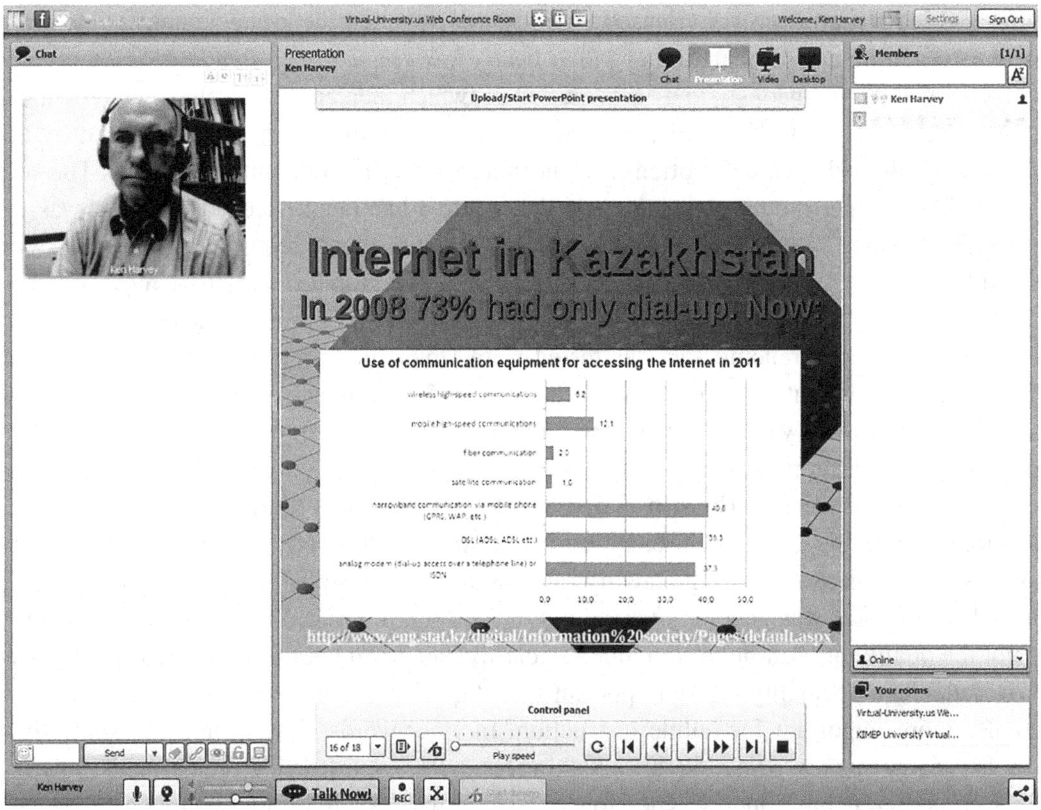

- **Google Hangout (free; invitation rather than URL based):** Hangout is not primarily a web-conferencing system. Web-conferencing is one of many tools within the Google+ toolbox. That has pros and cons. On the plus side, it offers many other tools, including a mini-learning management system – much less comprehensive than Moodle, but perhaps adequate for some faculty. Our recommendation, however, is that all faculty learn and support a single LMS; Moodle is the current preference. Google, of course, is a huge company that is likely to provide excellent support and expansion of its education-related tools long-term. And, while some participants participate only within Hangout, the video can also be streamed to YouTube, viewed live or recorded for later on-demand viewing. On the con side, without an assigned URL (neither temporarily nor permanent), it takes longer to set up and launch a class. Of these three free or inexpensive systems, this system is the hardest to learn and to operate but still not extremely difficult. It would still not take long to train students and student assistants (and others) in its use.
- **Adobe Connect (about $45 per month for 25 connections, depending on contract; more expensive with additional connections):** This system, chosen by Columbia University for its Global Classroom initiative, comes with many recommendations. It has many of the advantages of HotConference and more. The main disadvantage is that the cost goes up, depending on the number of participant connections. However, for the first 25 connections it is essentially the same price as HotConference. Adobe does not provide a single URL for all meetings, as does HotConference, but a series of classes or meetings can be set up with a single URL. The HotConference system is simpler in this regards, as well. Adobe's default conferencing set-up is almost identical to that of HotConference, but a major advantage is that each module within the default presentation window can be hidden or expanded, according to the needs and desires of the instructor. Thus, Figure 2, illustrates a customized set-up in which some modules have been hidden and the webcam and presentation modules have been made more equal in size. Also, while HotConference allows a handful of simultaneous webcam discussants, Adobe has no limit. In theory, all students could turn on their webcams during the entire class, but of course that would depend on available bandwidth. Allowing students to become presenters is simple for both systems. While Adobe's system is more complicated than HotConference, overall, it does offer a series of training videos, such as this 13-minute overview at http://www.youtube.com/watch?v=Oyx_hutZtzA.

The KIMEP DL Committee decided to continue working with both HotConference and Adobe Connect as very similar programs. Also, as will be discussed later, the committee saw some advantage in subscribing to both systems, using one as the preferred instructional tool and the second as a backup.

Upgrading Front-End Technology

It seemed clear to the KIMEP DL Committee that the expensive systems proposed by Samsung, Cisco and Microsoft were mostly proposing to enhance quality with high-end equipment at each end of a site-to-site DL system. High-quality HD cameras and large smart screens would enhance the video quality, but at a very high cost. Using off-the-shelf webcams, standard PCs, and large-screen projection systems could not match the quality of the higher-cost systems in video clarity and detail, but the committee's experimentation revealed that the quality of the consumer systems was still adequate. Nevertheless, the committee concluded that some limited enhancement of hardware could be justified. Whichever web-conferencing option was to be selected, higher quality will be achieved with superior equipment.

Figure 2. Example screenshot from an Adobe Connect webinar with a customized conference window with larger webcam module and smaller presentation module

- **Quality computers.** The DL Committee recommended a higher-quality computer with Core i7 processor, 8Gb RAM and increased storage for each "studio classroom," along with high quality, large-screen monitors, for a total cost of under $2000 per classroom or editing station.

- **Camcorders vs. webcams.** It was also recommended to replace the standard webcam feed in most cases with the feed of a medium-quality camcorder. Because of the limitations of the Internet to deliver the highest-quality video, the committee agreed that expensive TV cameras were not needed. Medium-quality camcorders at a cost under $1000 each are adequate for both quality and for their capability to zoom and pan. Current HD camcorders cannot be used in lieu of webcams without a conversion device. We are still testing conversion devices to see which is most cost-effective. One being tested is about $200 (http://www.blackmagicdesign.com/products/intensity/).

- **Wireless & parabolic/directional mics.** Quality audio is the most important element of a web-conference. To achieve that, the committee recommended a wireless mic for instructors and a parabolic/directional mic so that students could ask questions or make comments in the classroom without separate mics or without having to walk to a microphone.

- **Lighting.** The lighting demands for web-conferencing are not as great as for studio or film lighting. The webcam video feeds produced in the testing of the web-conference systems have generally been adequate even with standard office fluorescent lighting when light from outdoor windows is controlled. However, camcorders may be recording from farther away, where lighting and shadow control elements become more difficult. A TV producer recommended for more professional quality to purchase lighting such as the one shown above – one lighting bank with dimmer switch for direct lighting and one for fill lighting to remove shadows. Camera-top lighting should be available, and lighting in each studio classroom can be enhanced inexpensively if university maintenance personnel can mimic the professional direct and fill lighting recommended (see Figures 3 and 4).

Figure 3. Sample lighting

Figure 4. Studio classroom setup

A Need for Backup Technology

With reliability limitations of online video-based educational technologies, especially internationally, the "best practices" was deemed to develop a DL strategy with several backups. For example, Kazakhstan National University Al Farabi is part of Columbia University's Global Classroom. According to the Kazakhstan coordinator of the program, Rafiz Abazov (2013), they use Adobe Connect as their primary web-conferencing technology, backed up by Google or Skype video and by three audio options – Skype, regular telephone service, and/or Viber (a Skype competitor). The KIMEP DL Committee also proposed to enhance reliability by having a variety of backup options in case of poor Internet feed or technical failure:

Backup 1 – Webcast Video Recording: During a live web-conferencing feed produced with HotConference or Adobe Connect, a video recording of the online classes can and should be made. If a distant class or individual student cannot attend live online instruction, the class will be recorded and uploaded for later viewing. The video recording also has pedagogical value for all students – especially those of a different native language – who wish to re-watch a class.

Backup 2 – Second Web-Conferencing System: With the low cost of web-conferencing, the university can also subscribe to a second system. Committee members had conducted a simultaneous test of HotConference and TalkFusion during the 2013 KIMEP International Research Conference in accommodating a presentation by an American professor in Louisiana. They had started projecting the TalkFusion version to the audience, but the system's sound that day was not working well, so they switched to HotConference to complete the conference session. They knew from previous tests that this was not the typical audio quality of TalkFusion, but it did suggest an important future backup – a subscription to two different systems in case one is not working well.

Backup 3 - Camcorder Video: The highest-quality and most dependable video backup would be provided by a camcorder, which can record at the same time that it is functioning in lieu of a webcam. While HotConference and Adobe Connect allow simultaneous recording, they are not 100% reliable and will not record unless the instructor can himself connect with the selected web-conferencing URL. Camcorder video is much closer to 100% reliable and does not require a live connection, although as a separate recording it would not contain all graphics, interactive video, or text-chat elements of web-conferencing. It would be primarily a video of the instructor, perhaps a limited view of a white board or Powerpoint, and of students interacting within the university classroom. Distant students would not see this live but would be able to access it later when they have access to the university's network, to its YouTube channel where it could be uploaded, and/or to the university's LMS.

Backup 4 – Video Made with Audio Recording and Powerpoint Slides: If the instructor is using a comprehensive Powerpoint or similar graphic presentation with his lecture, a simple audio recording can allow the creation of what in most cases would be a pedagogically superior video in comparison with the camcorder video by itself. Powerpoint slides can be quickly exported as TIF files and mixed with the audio recording in video editing software. The results would be a video showing the slides very clearly and timed to match the audio recording.

Backup 5 – Mixing Backups 3 & 4: It requires additional editing, but Backups 3 & 4 can be fairly easily blended together to provide some direct video of the instructor, interweaved with his Powerpoint slides or other graphics. For long-term use to flip a classroom or to provide asynchronous distance-learning video, this could provide the best final product. However, more operating funds would be needed for video editing.

Backup 6 - Prepare & Post a Pre-Recorded Alternative Presentation in Advance: This cannot be done in all situations, but pre-recorded videos can be posted and designated for viewing by students if ever a live web-conference connection is not achieved. These videos may include content originally planned for later in the semester or they may be supplemental in nature and not seen as complete replacement for lost instructional content. In the latter case, students could be told to view the posted videos during class time and then expect to receive an email with the URL where the recorded lecture video could be viewed independently before the next class. This alternative is particularly valuable if students are traveling some distance to an off-campus learning center. For students to travel to the center and then have class canceled would not leave a favorable impression of the program.

Combining Webcast Video with Moodle and Other DL Tools

Besides the simulcast classes and their recordings, the KIMEP DL program would incorporate other DL tools, as well. The university already has a network drive that both students and faculty can access on campus or via the Internet from off-campus. For distributing documents to students, this is a convenient and familiar tool. However, in the long run a more robust learning management system is needed. Moodle is being used by millions of people around the world, and has been available at KIMEP for about 10 years. However, without the Administration providing any clear incentive to use the system, only a few instructors use it on a regular basis. To overcome instructor resistance to participating, the video-recorded lectures can be uploaded to Moodle automatically by technical assistants. In exchange for a course remission, instructors participating in the DL program would be asked in Year 1 to learn at least how to do one thing on Moodle – create and administer quizzes and exams. Moodle exams can be scheduled so that students can only access them between certain designated hours and for a limited period of time once initiated. Many types of objective questions can be asked and graded automatically by Moodle, including matching, fill in the blank, multiple choice, and true/false. Essay questions can also be asked, but they are saved for the instructor to download and grade separately. TurnItIn software can be linked to Moodle to check for plagiarism and to recommend a grade, based in the instructor's own scoring matrix. An instructor can create a bank of questions from which Moodle can specifically or randomly draw. The instructor, in using quizzes or tests for instructional purposes, can allow students to take a quiz more than once and can designate which to accept as the final score: the first, the last or the best. It is hoped that once instructors have their video-recorded lectures uploaded to Moodle and they have created quizzes and exams, they will be internally motivated to begin adding additional content to their course site. Additional workshops will be provided to teach instructors how to add text, graphics, audio, video and other content to their Moodle sites. Dr. Harvey created an example of a multimedia Moodle educational website that can be seen at http://Virtual-University.us. Dr. Harvey originally tried uploading video directly into the Moodle system, but the launch time for each video was unacceptably slow. Then he began embedding videos that had already been uploaded to YouTube. This experimental website now has over 3,000 videos embedded into it from YouTube, and he found that each embedded video launches quickly without any bandwidth drag. The International Education Institute, which he founded, allows guests to enter that site as a guest and access some content without a password. To experience the testing process, however, free registration is required so Moodle knows how to assign the score.

Further testing of Moodle is planned. KIMEP only recently upgraded from the 1.8 version to the 2.9 version of Moodle. As of its 2.0 version Moodle has become more user-friendly and mobile-accessible.

KIMEP has set up the current version of Moodle without taking down the old site, is transferring previous content to the new version, testing its mobile accessibility, and inviting faculty to experience and provide feedback on the new version before making a final decision to keep it. Mobile accessibility is of particular importance. The 2014 ECAR survey of university students, for example, showed that 99% of all American students now own at least one mobile Internet device, of which 92% own at least two and 59% own three or more. Mobile technology can be used in many ways, both in a brick-and-mortar or a virtual classroom (Dahlstrom & Bichsel, 2014, pp. 14-18). The vast majority of students say these mobile devices are important to academic success, but only 30% of instructors create assignments that incorporate the mobile devices, while 55% outright ban their use in a traditional classroom (pp. 18-19). Students, on the other hand, feel such technology makes them feel more connected to other students (51%), to instructors (54%) and to the institution as a whole (65%) (p. 10). Such devices during campus or online classes were considered important by over 50% of the surveyed students in looking up information, accessing the learning management system, capturing images of class activities and resources, participating in instructors' planned interactive activities, and recording lectures and activities for future review (p. 21). Over 80% of American students now prefer courses that have at least some online components (p. 23), which they are most likely to access via a mobile device. Nearly 70% of the students were very or extremely interested in their university's learning management system to include personal information, such as progress towards their degree, and 62% were very or extremely interested in having quizzes and practice questions included online (p. 27). The 2013 ECAR Study of Undergraduate Students and Information Technology (Dahlstrom, Walker & Dziuban, 2013) included such additional insights as:

- Students want "anytime, anywhere access to course materials" (p. 24).
- But they also want some face-to-face interaction with instructors and/or other ways to communicate with them directly (p. 6).
- Students feel they learn best with the combination of online and face-to-face components. (p.38).

The KIMEP DL Committee felt that its plan could achieve most if not all of these tasks. Instructors willing to participate would have their classes simulcast to students not in physical attendance. Their classes would be video-recorded with Powerpoint and other graphic illustrations by a technical assistant and the videos placed on the university network, the university's YouTube channel, and in the university's LMS, along with quizzes, exams and other learning content. Most of these DL tools would be accessible to mobile devices live and/or on demand, and, as students desire, would encourage instructors to use them for students involved in traditional, blended or distant instruction.

Recruiting and Training Faculty

The ultimate purpose of KIMEP's program is to set up attractive and effective education for distant learners. Beyond management systems and technology, the committee noted that serious consideration should be given to the skills and best practices of distance learning. Although many aspects of the DL educational process – especially as recommended in the KIMEP DL Plan -- are very similar to the traditional classroom, the differences are fundamental and significant enough to require training before and during implementation. In the end, the success of the DL project also depends on the capabilities and motivation of faculty and students.

Many professors who teach DL classes find it disconcerting to have little or no interpersonal interaction with students. Experienced professors find it difficult to teach to a camera rather than to individuals. Similarly, lack of student involvement is probably the central obstacle to effective DL, and must be addressed during preparation and teaching. While asynchronous teaching methods are more convenient for students, when students are distant from the professor and classmates geographically, interpersonally and in time, they frequently lose interest and motivation. Students are comfortable with online interactions, but are also used to content that is much more than simply "talking heads." Faculty must be able to produce interesting instructional material and build student interaction into the process. With the live web-conferencing tools, this is easy to do technically, but should still be built into the lesson plan. With asynchronous coursework, use of Facebook or other interactive tools can still be employed. Students can be required to make their own multimedia presentation online with Powerpoint and webcam and/or participate in panel discussions on camera. During their presentations, instructors can ask students to simultaneously and spontaneously respond to a question via the system's text chat or polling functions. Students can also be asked to respond individually to a question via webcam. In the HotConference system currently being used, 3-5 people can participate in a webcam panel discussion simultaneously. Adobe Connect has no limit. Instructors can also set up a Facebook page, and ask students to occasionally respond to and discuss a question posted there. Their interaction on both HotConference and Facebook can become part of their final grade.

Online classes and assignments should include such activities, and faculty members need to learn the variety of these that are available. Students also learn best in DL classes with "mini lectures," interspersed with activities. The DL technology allows virtual interaction among students, as well as the accountability of students being identified personally and involved in the educational process.

It has been difficult to get faculty to even attend workshops on Moodle, creating educational videos, and other aspects of distance learning. Some faculty members suffer from technophobia, some see no purpose in learning such skills, and some simply have other priorities, such as university service or research. The KIMEP DL plan has been developed in such a way as to remove as many of these obstacles as possible. KIMEP will train student technical assistants to run the web-conferencing equipment and software, the instructors will be teaching real students in a real classroom, and the assistants will worry about maintaining contact technically with those participating within the virtual classroom. But, as explained, the instructors should still adapt their classes to accommodate those distant students and should try to address them as directly as possible. Other related recommendations by the DL committee included that:

- Studio classrooms be set up where appropriate hardware and software are permanently installed and functional, and where the technical assistants are available to run the equipment.
- Instructors receive a one-course remission at least the first time they teach a new course online. They will need the time to improve their graphics, create new activities, develop online quizzes and exams, and in other ways accommodate the online environment.
- An instructor's newly developed DL course and videos be considered as partial fulfillment of the instructor's research requirements.
- Legal documents be signed making the new online course and its components the joint property of both parties. When instructors leave, they can take an electronic copy of their course material with them and use it at another university. But the current university – KIMEP in this case – can keep course content and continue to use it within their distance learning program until it can be replaced by a new instructor.

Distance-Learning Pilot Project

Testing of these technologies has continued since submission of the DL Plan, including a 2014 pilot project in which Dr. Harvey and Dr. Sholpan Kozhamkulova of KIMEP and Dr. Phil Auter of the University of Louisiana – Lafayette team-taught a 15-week Mass Communication & Society course to students gathered in a classroom at Kangwon National University (KNU) in Korea. The same course was taught again in 2015 by Dr. Harvey and Frederick Emrich, M.A., another KIMEP professor. When this pilot project was initiated, KNU was just launching its own inbound online instruction as a means to entice more native English instructors to provide courses to its students. Both courses taught as part of this pilot project had low enrollments because students were afraid their English skills would not be adequate, according to the program supervisor. As much of the KIMEP DL Plan as possible was included in this pilot project, but some elements could not be included. KIMEP, for example, had not purchased the higher-quality equipment requested, so older computers and laptops were used for the instructors' web-conferencing, which did cause occasional problems. In addition, no technical assistants were available. There were occasions when instructors forgot to turn on the system's recording function, which meant that instruction was not available later for asynchronous access. In addition, since the program was hosted by KNU, KIMEP had no control over certain classroom protocols. For example, KNU required that students attend in a traditional classroom on campus and did not give them the option of attending from home or other location, as some preferred.

At the end of the course both in 2014 and 2015 students were asked to participate in a survey and answer several questions regarding the course. As seen in Table 1, students had a good impression of this course, 100% of the 15 students referring to it as a good or very good experience (Question 7). To Questions 1 and 2, 60% of the students expressed specific appreciation for the opportunity to practice the English language and 40% the content of the course. To open-ended Question 3 students expressed how they enjoyed the opportunity to experience foreign education with its aspect of cultural diversity and English language practice without the need to travel abroad.

With faculty using older and/or lower-cost technical tools (contrary to DL plan recommendations), technical problems in course delivery did occasionally occur. But when students were asked in Question 4 what were the greatest challenges of being taught by foreign faculty from thousands of miles away via web-conferencing, 67% selected linguistic problems, 27% selected the understanding of course content, and only 1 student (7% of the students) identified technical problems. In contrast, in Question 1 20% of the students indicated that learning the technical tools was one of the greatest benefits of the course. And in ranking the various problems in the course in Question 5, technical problems were identified as the least of the problems. In the open-ended Question 6, two recommendations were made that should be considered, which related little to the technology:

1. Because of the linguistic challenges of international distance learning, the class should not meet only one day a week for three hours. Instructors had scheduled classes that way in case time had to be taken to solve technical problems, but since technical problems were minimal, shorter classes might be better.
2. More student discussion should be encouraged. That was the opinion of only one student, and most students, because of their English limitations, never volunteered to speak in class via webcam. They seemed to prefer to text their responses. Nevertheless, besides individual oral presentations, more student panel discussions (with webcams) or text chat questions could have been used to increase student interaction during the live course.

Table 1. KNU media and society student evaluations 2014-2015

QUESTIONS	RESPONSES 1	RESPONSES 2	RESPONSES 3	RESPONSES 4	RESPONSES 5
1) What do you feel were the greatest benefits of this course? Circle the letter for ALL that apply.	60% a) Linguistic: practice of English.	40% b) Content: learning more about media and communi-cations in society.	20% c) Technological: use of web-conferencing and related technical tools of distance learning.	13% d) Cultural: understand-ing of a foreign culture.	0% e) Other (explain):
2) Now rank these in order of importance, 1-5 (1 meaning most important, 5 least important). Mean for response options:	1.9 mean rating for a) Linguistic: practice of English.	2.1 mean rating for b) Content: learning more about media and communications in society.	3.4 mean rating for c) Technological: use of web-conferencing and related technical tools of distance learning.	2.9 mean rating for d) Cultural: understand-ing of a foreign culture.	N/A No responses for e) Other (explain):
3) Please explain the benefits of this course in your own words. Representative responses (edited to clarify student grammar):	*Speaking and writing English is not a common thing in my university. So this course is very impressive. And practice with real English is a great experience too.*	*It was precious time to me, because I could study with native professors. Especially, your class helped my listening skills.*	*I really like being in touch with distant places and people from there. It was about trying to understand different cultures and sharing our different opinions.*	*We had to write in English, so difficult but worthwhile.*	*Practice English and receive lectures by foreign professors.*
4) What were the greatest challenges in taking this course? Circle the letter for all that apply.	67% a) Linguistic: difficulty understanding instructors' English.	27% b) Content: difficulty understand-ing concepts taught about media and communications in society.	7% c) Technological: use of web-conferencing and related technical tools of distance learning.	0% d) Cultural: understand-ing concepts taught by instructors from a foreign culture.	7% e) Other (explain): Little opportunity to speak in English.
5) Now rank these in order of importance, 1-5 (1 meaning most important, 5 least important). (NOTE: There appears to have been some confusion among students that this question was directly related to Question 4, but most of students do seem to have understood, so we still share their responses.) Mean for response options:	1.7 mean rating for a) Linguistic: difficulty understanding instructors' English.	2.0 mean rating for b) Content: difficulty understand-ing concepts taught about media and communications in society.	3.4 mean rating for c) Technological: use of web-conferencing and related technical tools of distance learning.	3.1 mean rating for d) Cultural: understand-ing concepts taught by instructors from a foreign culture.	N/A No responses for e) Other (explain): Little opportunity to speak in English.
6) Please explain the challenges of this course in your own words. Representative responses (edited to clarify student grammar):	*It was hard because of linguistics… To understand was very difficult.*	*The lecture time was a challenge for me. I think we need more class time, not just 3 hours, because the lecture was presented in a foreign language.*	*Distance learning had some technical problems. If we could meet face-to-face, that would be better to concentrate.*	*Writing English is always challenge at me. (sic)*	*In Korea, debate is not used in class, so discussion does not go well in class, which is too bad. This was limited by technical problems and cultural differences.*

continued on following page

Table 1. Continued

QUESTIONS	RESPONSES 1	RESPONSES 2	RESPONSES 3	RESPONSES 4	RESPONSES 5
7) Considering both challenges and benefits of this course, was this course more of a good experience or bad experience?	60% a) Very good.	40% b) Good.	0% c) Neither particularly good nor bad.	0% d) Bad.	0% e) Very bad.
8) How did the technology and overall distance-learning approach used for this course work in your opinion. Select the letter for all of the responses that reflect your perspective.	73% a) There were a few problems, but nothing significant.	13% b) The benefits of accessing such internation-al professors far exceeded the challenges of distance learning.	13% c) Problems with the technology and approach of this course made it very difficult.	27% d) I would recommend courses like this to my friends.	0% e) I would NOT recommend courses like this to my friends.
9) Please compare this course to more traditional university courses you have taken. Select all of the responses that reflect your perspective.	40% a) I liked this course better than most traditional university courses.	40% b) I liked this course about the same as most other traditional courses.	0% c) I liked this course less than most other traditional courses.	33% d) I would like this course more if I could access it from my own computer.	20% e) I would like this course more if the lectures were video-recorded and included with other materials online to be accessed at students' own convenience.
10) Please explain your overall feelings toward the course in your own words. Representative responses (edited to clarify student grammar):	*It was very useful for brainstorming about mass media. It is good to get information from foreign professors. I mean I can now understand other views on mass commun-ication.*	*It is great to try to learn from other countries' professors. But my English skill is not enough to take classes this way for all topics.*	*Unbelievable!! Exciting!! I really liked it because it is the easiest way to get in touch with people from all over world and share different cultures and values.*	*I think class time is too long. Three straight hours listening to English makes it much harder than my other classes.*	*It would be better if a lecture was not just one-way teaching but two-way communi-cation.*

When asked specifically in Question 8 to what extent online technology was a problem, 73% said it represented no significant problem, 13% said benefits "far exceeded" the challenges, 27% said they would recommend "courses like this" to their friends, and only 13% (two students) indicated that "the technology and approach of this course made it very difficult," and one of the two students crossed out the word "very." No students indicated that they would "NOT recommend courses like this" to their friends. And in responding to Question 9, 40% of the students said they like this course better than most traditional courses, and another 40% said they liked it as much as traditional courses. No students said they liked it less than traditional courses. However, 33% did indicate that it would have been better if they could have attended class from another location of their choice, which in this case only KNU controlled, and 20% suggested that they would have preferred it if recorded lectures and all other content would have been accessible asynchronously at their own convenience. Of course, that is also part of the KIMEP Plan that was not always implemented in this pilot project. All content should have been provided for later review by students struggling to understand native English, but KNU policy would still have required them to attend the live classes. Responses to open-ended Question 10 were also mostly positive. The recommendations of a shorter class period and more student interaction were again expressed, however.

Unlike traditional expectations about the technical aspects of DL course delivery, these findings refute widespread conservative objections against DL as cumbersome to use. In fact, students' feedback explicitly demonstrated that more attention should be put into a completely different area of international DL experience -- intercultural communication and language barriers. These findings lead to two observations:

1. Language and cultural barriers in international distance learning are similar to studying abroad, where students, in addition to achieving learning objectives, face issues beyond those of their classmates. Here the conclusion can be made that when developing an international DL program a university might consult exchange and study abroad programs to find possible solutions that can be applied with DL tools. In fact, translation software and other learning tools may make international distance learning better for some students than a more expensive exchange program.

2. These linguistic circumstances can be seen both as challenges and advantages because students have the chance to experience new teaching-learning styles and broaden their understanding of the subject by including a multicultural aspect in their education while still studying in their native country. It can even ignite further interest in exploring the concepts taught in class not only in their native language and country but also abroad to gain yet broader perspective. This way DL provides a cheaper alternative for exposing students to a more cosmopolitan rather than local worldview and better prepares them for other international/intercultural experiences. And such global-thinking young professionals are much needed in the modern globalizing, interlinked world. Linguistic challenges faced in international DL prepare students to adapt better in a diversified work environment.

It should also be noted that almost all challenges described by students were people-related rather than technology-related. Students were primarily concerned about course content, linguistic/cultural barriers, and the need to adjust to several professors teaching the course. These are the questions that should not be dismissed in program discussion. The human component prevails both in DL program success and in general education. As this format is relying in part on technology and in part on human performance and international communication, one should not take DL as purely a technical innovation in education. In addition to admission benefits of higher recruitment rates and greater flexibility in education access, here lies another opportunity to integrate this international cultural experience for both professors and students. Indeed, further discussion about a DL program should take this into account and plan accordingly. Do professors manage to find common language with students? Are students proficient enough in foreign language knowledge (English in most cases) to communicate efficiently? This case study opens the door wide open to another step in DL study: the human execution of DL programs. It can be explored from three sides: 1) students with different levels of language capacity and thus ability to study all courses in DL; 2) universities with culturally aware professors who can teach DL classes for a diverse student body; and 3) society in the form of potential employers who might see this rich cultural experience as a valuable asset in future staff's competencies. Here DL programs can make this experience widely available for all students with no need to physically travel abroad.

As for technical aspect, students were mostly positive about technology and the DL approach, naming some problems as insignificant ones. Other respondents mentioned that enhanced technical convenience would actually boost their impression of the course (such as having an access from their own computer and having asynchronous learning opportunities to be accessed at student's own pace and convenience).

In other words, what few technology-related complaints existed could actually be solved by technology, not caused by it. In actual practice then, the students agreed that this kind of distance learning is as good or better than the on-campus experience.

MAPPING THE FUTURE

In the international, cross-cultural study with KNU students, we did not specifically address the use of mobile technologies, although students spontaneously expressed the desire to attend the virtual class without having to travel to the brick-and-mortar classroom. Clearly, the integration of mobile education into higher education, as shown by the annual ECAR studies, has boundless possibilities for growth and collaboration, both within universities and throughout the academic community. For instance, developing countries have different technological networks than those in developed countries. In order for mobile technology to succeed across the globe, it is necessary to customize electronic learning resources for the intended users (Keengwe and Bhargava, 2014). Cloud storage such as the Google Drive, Apple iCloud, or Microsoft's OneDrive makes it possible to create, edit, and store information virtually, and makes it available to anyone with an Internet connection. Developments such as cloud storage could help to bridge the gap between organizations and/or universities that have more access to technology and those that are limited to Internet access alone (Minjuan et al, 2014). Online databases are also available via university or independent subscriptions, and offer access to journals, news articles, books, etc., making research tasks easier for students, academics, and professionals alike to access and utilize. Moreover, it is clear that with the number of technological devices that emerge on the market each year, it is imperative that "a thorough analysis, from a pedagogical and technological perspective, is key to ensuring appropriate usage and implementation of mobile learning" (Ally and Prieto-Blazquez, 2014).

As noted previously, there are several approaches to online distance learning. For example, some universities almost exclusively use video of professors while they lecture and use the white board etc. This may be live or pre-recorded. Others are still providing old-style correspondence courses, with some aspects placed online rather than being sent through the mail. Most are currently asynchronous to accommodate varying schedules of students and to avoid the greater possibility of technical difficulties involved with live online courses, whether or not they include web video. Another version of this technology cited above is the Screencast-O-Matic program (Orlando, 2014) that allows an instructor to review students' work and other content so students see the material online while the instructor explains any problems with audio-video webcam feed and a marking tool that allows him or her to mark up the paper electronically. Video technology -- if implemented correctly -- can be an extremely helpful piece to the DL puzzle, to both students and teachers. The ease of access to online video via mobile devices as well as the level of interaction video lectures can provide to students suggest that this aspect of mobile education will only continue to develop and improve to meet student needs. Nonetheless, with the current "Second Internet Revolution," driven by online video and mobile broadband, the technical difficulties are declining and the opportunities are increasing.

However, there is still some lingering push-back from academics regarding the implementation of DL courses in the college setting. With asynchronous DL courses, instructors lose comforts of traditional classroom interactions such as taking visual cues from students, receiving immediate student feedback and checking student engagement by both asking and answering questions (Valentine, 2002). However, instructors can combat these misgivings by creating a community (Valentine, 2002) among their students and facilitating a dialogue between themselves and their students both collectively and independently.

Additionally, the further integration of DL technologies into classrooms will undoubtedly alter the role of an educator in higher education (Clark, 1993; Valentine, 2002). One might speculate that some of this resistance stems from instructors – especially those without tenure – fearing for job security. Additionally, those instructors that are less technologically savvy or experience difficulty in quickly adjusting to changes in technology will experience more difficulty with training. Frustration may get in the way of adaptation. Further issues may include concerns over the theft of materials posted online, a lack of buy-in or support from university administration, lack of professional recognition for DL courses, time required to prepare DL lessons, or lack of communication between administration and faculty. With proper support and training, however, educators, students and universities can see success with DL courses.

As the research suggests, mobile education allows for students to control nearly every aspect of when and where they learn, a concept not quite existent in traditional education. DL programs continue to simplify the structure of education and make it more affordable and accessible for students now than it ever has been. Christensen (2012) suggests that for universities to lower overhead cost and continue to adapt to the future of education, already established institutions have to let go of their outdated infrastructure. Teaching and researching can be handled in separate parts of the university, allowing for both to be done better, and for a lower cost. Technology can enable the traditional university to become more efficient and maximize both instructor efforts and student success. With each year, education becomes more geared towards the needs and specifications of the student, versus the needs of the university at large. Mobile education has the capability to take higher education learning and make it available to individuals globally, at any pace. New software, hardware, and cloud capability develop and improve each day, making now the best time for mobile education to solidify itself as a viable way of learning. However, it will require both student and university effort, training, and dedication to take full flight.

CONCLUSION

The blended program under development at KIMEP is a demonstration of why discussions comparing SDL to ADL is essentially irrelevant. KIMEP's plan attempts to incorporate the advantages of both synchronous and asynchronous education and respond to the recommendation of university students to include content access anytime from anyplace, plus face-to-face contact with the instructor in either the brick-and-mortar classroom or the virtual classroom (Dahlstrom, Walker & Dziuban, 2013). It also demonstrates that some of the most important DL issues only peripherally relate to technology. The KIMEP plan was developed to be faculty-friendly so more professors would agree to participate without compulsion. It was also designed to be cost-effective – indeed, financially profitable. And it was based on important technological trends, including the rapidly increased viewing of online video (550% just in 2012 and still surging). In essence, the KIMEP plan addresses all five of the biggest DL challenges:

- Developing a platform, system and tools that can handle the DL program.
- Achieving faculty buy-in, overcoming technophobia and other concerns, and incentivizing faculty to create and/or present the content.
- Achieving student buy-in, overcoming their other concerns, and incentivizing students to participate in the DL program.

- Assessing student achievement of learning objectives.
- Funding the DL program.

The KIMEP plan achieves these by simulcasting live on-campus courses to other sites around Central Asia, as well as to individual students who are able to dependably access from home or office, and making the video-recorded classes and other learning materials available asynchronously on demand when the live class is completed. The general rationales for using simulcasting as the initial and primary method of instruction in KIMEP's distance-learning program can apply to many other universities around the world. They are, in brief, that:

- Simulcasting can keep an institution's costs down while building the distance-learning program very quickly.
    - Web-conferencing systems such as HotConference or Adobe Connect are inexpensive but of rapidly increasing quality as broadband and other Internet technologies increasingly accommodate video.
    - Equipment needed for web-conferencing is not nearly as expensive as for professional TV studios. Studio classrooms can be set up for under $10,000 apiece with medium-quality equipment adequate for quality online video.
    - If a university already has 15 students physically sitting in the studio classroom, it has already paid for the instructor's time no matter how few students participate from a distance. Students who participate over the Internet therefore need only to cover the direct, fairly nominal costs related to broadcasting the live class.
- Web-conferencing (with the help of student assistants, as proposed in this plan) has relatively little impact on the function of instructors. Thus, an institution can require departments to participate without offering large financial incentives. While the KIMEP plan involves initially incentivizing voluntary participation by professors through course remissions, in the long run as most instructors become accustomed to simulcasting, such incentives may not be necessary.
    - Thus, the cost of simulcasting classes is so negligible, and the technology so simple to use, that a university could provide dozens of new live courses over the Internet every semester with only 2-3 studio classrooms.
- Recordings of simulcast classes can be available later as part of distance-learning courses, as supplements to traditional courses especially helpful for students with non-native English skills, or to help instructors "flip" their course content and require students to watch lectures as homework so they can come to class prepared to discuss, brainstorm and/or apply the material in project-based learning modules.
- The economic potential of online and hybrid education has been clearly demonstrated by such institutions as the University of Phoenix, which has achieved success by combining online education with on-site courses at 112 small satellite campuses around the world.
    - Most courses are taken online, but students have the option.
    - Enrollment peaked at 600,000 students in 2010, and the university is still one of the largest in 2015 with "only" 227,000 (McEvoy, 2015) as a result of major operational changes (WSJ, 2011).
    - Phoenix became the biggest beneficiary of student federal financial aid in America, reaping $2.5 billion in 2008 (USAspending.gov, 2008).

- A cost:benefit analysis in the KIMEP plan suggests that a system of 10 small satellite campuses could quickly and profitably increase overall enrollment at KIMEP by over 25%. The satellite campuses would only have to recruit about 20 new FTE bachelor students per year who otherwise would not enter the university in order to cover operational and setup costs.

This case study in some ways, however, is just a preview of a longer-term case study of KIMEP University's distance-learning program. It reviews the pre-testing and rationale behind the program, but KIMEP's blended program certainly merits longer-term study, comparing achievement, dropout rates and other data of students within the program who choose different learning approaches – on-campus classes, SDL at off-campus learning centers, SDL at sites of students' own choosing, and ADL.

Mobile technologies should be a primary consideration of any distance-learning program. Mobile education is a more narrowly defined segment of distance learning, being education that is more portable and, if you will, pocket-sized. Video in general and the newer versions of the Moodle LMS (and many of its competitors) make mobile education easy to achieve by any educational organization. With the rapid growth of mobile communications, every DL program should plan ongoing efforts to accommodate mobile. Like all distance learning, students must remain engaged in their mobile education or risk missing assignments and failing to absorb the education available to them. But, at the same time, mobile communication can make higher education available to an even greater portion of the world's populace. University faculty and administrators should proactively confront the challenges of ramping up their technology and instructor training so that their institutions are capable of delivering education effectively in ways that accommodate the needs of as many people as possible. Further assessments need to be made of the pros and cons of different methods of distance instruction. In so doing, early adopters should resist the temptation of adopting any specific educational approach only because it is new and exciting, while laggards should consider if they are resisting instructional innovation only because "that's not the way it has been done before."

REFERENCES

10 Advantages to Taking Online Classes. (2012, January 9). Retrieved from http://oedb.org/ilibrarian/10-advantages-to-taking-online-classes/

Ally, M. & Prieto-Blazquez, J. (2014). What is the future of mobile learning in education? *RUSC: Revista De Universidad Y Sociedad Del Conocimiento, 11*(1), 142-151.

Baran, E. (2014). A Review of Research on Mobile Learning in Teacher Education. *Journal of Educational Technology & Society, 17*(4), 17–32.

Bernard, R. M., Abrami, P. C., Lou, Y., Wade, A., & Borokhovski, E. (2004). The effects of synchronous and asynchronous distance education: A meta-analytical assessment of Simonson's "equivalency theory". In M. Simonson & M. Crawford (Eds.), *2004 annual proceedings of selected research and development papers presented at the national convention of the Association for Educational Communications and Technology* (pp.102-109). Chicago, IL: AECT.

Bernard, R. M., & Amundsen, C. L. (1989). Antecedents to dropout in distance education: Does one model fit all? *Journal of Distance Education, 4*(2), 25–46.

Bersin, J. (2014a). The Red Hot Market for Learning Management Systems. *Forbes*. Retrieved July 31, 2015, from http://www.forbes.com/sites/joshbersin/2014/08/28/the-red-hot-market-for-learning-technology-platforms/1

Bersin, J. (2014b). *The Corporate Learning Factbook 2014: Benchmarks, trends, and analysis of the U.S. training market*. Retrieved Sept. 7, 2015, from http://www.bersin.com/uploadedFiles/012714WWBCLF.pdf

Bloomberg. (2015). *USDKZT:CUR*. Retrieved 24 Dec. 2015 from http://www.bloomberg.com/quote/USDKZT:CUR

Brady, M. (1998). *Strategies for Effective Teaching: Using Interactive Video in the Distance Education Classroom*. Retrieved from http://www.designingforlearning.info/services/writing/interact.htm

Budiman, R. (2015). Distance language learning: Students' views of challenges and solutions. *International Journal on New Trends in Education, 6*(3), 137–147.

Cakiroglu, U. (2014). Analyzing the effect of learning styles and study habits of distance learners on learning performances: A case of an introductory programming course. *International Review of Research in Open and Distance Learning, 15*(4).

Caldwell Community College and Technical Institute. (2015). Hudson, NC: Distance Learning Student Information Guide.

Christensen, C. (2012). *Dr. Clayton Discusses Disruption in Higher Education*. [Video File]. Retrieved from https://www.youtube.com/watch?v=yUGn5ZdrDoU&feature=youtu.be

Cisco. (2015). *Forecast projects nearly 10-fold global mobile data traffic growth over next five years*. Retrieved from http://newsroom.cisco.com/press-release-content?type=webcontent&articleId=1578507

Clay, M. (1999). Development of Training and Support Programs for Distance Education Instructors. *Online Journal of Distance Learning Administration, 2*(3). Retrieved from http://www.westga.edu/~distance/clay23.html

Dahlstrom, E., & Bichsel, J. (2014). *ECAR Study of Undergraduate Students and Information Technology, 2014*. Louisville, CO: ECAR. Retrieved from http://www.educause.edu/library/resources/2014-student-and-faculty-technology-research-studies

Dahlstrom, E., Walker, J. D., & Dziuban, C. (2013). *ECAR Study of Undergraduate Students and Information Technology, 2013*. Louisville, CO: ECAR (EDUCAUSE Center for Analysis and Research). Retrieved from http://www.educause.edu/library/resources/ecar-study-undergraduate-students-and-information-technology-2013

Definition of Screencast. (2010). *PC Magazine: Encyclopedia*. Ziff Davis.

Distance Learning. (2015). Retrieved from http://www.sfmccon.edu/distance-learning/distance-learning.html

Distance Learning Requirements & Competencies. (2015). Retrieved from http://www.blueridge.edu/academics/distance-learning/distance-learning-requirements-competencies

DL Master of Engineering Graduate Program. (2015). Retrieved from https://engineering.tamu.edu/petroleum/academics/distance-learning/prospective-students/admission

Frankola. (2000). *Why online learners drop out.* Workforce.com. Retrieved August, 2012, from: http://www.workforce.com/archive/feature/22/26/22/index.php

Frequently Asked Questions. (2015). Retrieved from http://bamabydistance.ua.edu/faqs/

Google. (2015). *Public Data: Fertility rate.* Retrieved from https://www.google.kz/publicdata/explore?ds=d5bncppjof8f9_&met_y=sp_dyn_tfrt_in&idim=country:KAZ:UZB:KGZ&hl=en&dl=en

Harvey, K., & Manweller, M. (2015). *PR, media executives support radical education reform.* Retrieved 24 December 2015 from http://virtual-institute.us/Ken/CurrentResearch.html

Hrastinski, S. (2008). Asynchronous and synchronous e-learning. *EDUCAUSE Quarterly, 31*(4), 51–55.

Husbye, N., & Elsener, A. (2013). To Move Forward, We Must Be Mobile: Practical Uses of Mobile Technology in Literacy Education Courses. *Journal of Digital Learning in Teacher Education, 30*(2), 46–51. doi:10.1080/21532974.2013.10784726

Insook, H., & Seungyeon, H. (2014). Adoption of the Mobile Campus in a Cyber University. *International Review of Research in Open and Distance Learning, 15*(6), 237–256.

Joo, K. P., Andres, C., & Shearer, R. (2014). Promoting distance learners'; cognitive engagement and learning outcomes: Design-based research in the Costa Rican National University of Distance Education. *International Review of Research in Open and Distance Learning, 15*(8), 188–210.

Keegan, D. (1996). *Foundations of distance education* (3rd ed.). London: Routledge.

Keengwe, J., & Bhargava, M. (2014). Mobile learning and integration of mobile technologies in education. *Education and Information Technologies, 19*(4), 737–746. doi:10.1007/s10639-013-9250-3

Kember, D. (1995). *Open Learning Courses for Adults: A Model of Student Progress.* Englewood Cliffs, NJ: Educational Technology Publications.

Malinovski, T., Vasileva-Stojanovska, T., Jovevski, D., Vasileva, M., & Trajkovik, V. (2015). Adult Students' Perceptions in Distance Education Learning Environments Based on a Videoconferencing Platform -- QoE Analysis. *Journal of Information Technology Education, 14*, 2–19.

McEvoy, C. (2015). Enrollment Declines Equal Bad Fiscal Q1 for Apollo Education APOL – Investors. com. *Investor's Business Daily.* Retrieved 20 July 2015 from http://news.investors.com/010815-733803-apollo-education-student-enrollment-declines.htm?ven=yahoocp&src=aurlled&ven=yahoo

Minjuan, W., & Khan, M. (2014, April). Mobile Cloud Learning for Higher Education: A Case Study of Moodle in the Cloud. *International Review of Research in Open and Distance Learning, 15*(2), 254–267.

Mtebe, J. S., & Raisamo, R. (2014). Investigating students' behavioural intention to adopt and use mobile learning in higher education in East Africa. *International Journal of Education & Development Using Information & Communication Technology, 10*(3), 4–20.

Orlando, J. (2010, August 18). *Save Time and Teach Better with Screencasting - Faculty Focus*. Retrieved July 25, 2015, from http://www.facultyfocus.com/articles/effective-teaching-strategies/save-time-and-teach-better-with-screencasting/

Orlando, J. (2014, March 7). *Screencasting feedback with Screencast-O-Matic*. [Video File]. Retrieved July 25, 2015, from https://www.youtube.com/watch?v=CDcfX2Qj6k0&feature=youtu.be

Ozbek, E. A. (2015). A classification of student skills and compentencies in open and distance learning. *International Journal on New Trends in Education*, 6(3), 174–185.

Pappas, C. (2014). *The 20 Best Learning Management Systems*. Retrieved 20 July 2015 from http://elearningindustry.com/the-20-best-learning-management-systems

Pappas, C. (2015, January 25). The Top eLearning Statistics and Facts for 2015 You Need to Know. *eLearning Industry*. Retrieved August 5, 2015, from http://elearningindustry.com/elearning-statistics-and-facts-for-2015

Requirements. (2015). Retrieved from http://distance.msstate.edu/geosciences/bmp/requirements

Rius, A., Masip, D. & Clariso, R. (2014). Student projects empowering mobile learning in higher education. *RUSC: Revista De Universidad Y Sociedad Del Conocimiento, 11*(1), 192-207.

Rogerson-Revelle, P. (2015). Constructively aligning technologies with learning and assessment in a distance education master's programme. *Distance Education*, 36(1), 129–147. doi:10.1080/01587919.2015.1019972

Russell, T. L. (2001). *The No Significant Difference Phenomenon* (5th ed.). IDECC.

Shirky, C. (2009). *How social media can make history*. TED.com. Retrieved 25 March 2015 from http://www.ted.com/talks/clay_shirky_how_cellphones_twitter_facebook_can_make_history

Stanisic Stojic, S. M., Dobrijevic, G., Stanisic, N., & Stanic, N. (2014). Characteristics and activities of teachers on distance learning programs that affect their ratings. *International Review of Research in Open and Distance Learning*, 15(4), 248–262.

Straumsheim, C. (2013). Tech as a Service. *Inside Higher Ed*. Retrieved 20 July 2015 from https://www.insidehighered.com/news/2013/10/17/survey-shows-it-service-dominates-top-priorities-among-university-it-officials

Straumsheim, C. (2014). Moodle for the Masses. *Inside Higher Ed*. Retrieved 20 July 2015 from https://www.insidehighered.com/news/2014/02/13/moodle-tops-blackboard-among-small-colleges-analysis-sayshttps://www.insidehighered.com/news/2014/02/13/moodle-tops-blackboard-among-small-colleges-analysis-says

The Advantages of Distance Learning. (2015). Retrieved from http://www.usjournal.com/en/students/help/distancelearning.html

Torres Diaz, J., Moro, A. & Torres Carrion, P. (2015). Mobile learning: perspectives. *RUSC: Revista De Universidad Y Sociedad Del Conocimiento, 12*(1), 38-49.

Trading Economics. (2015). *Kazakhstan GDP Annual Growth Rate.* Retrieved 20 July 2015 from http://www.tradingeconomics.com/kazakhstan/gdp-growth-annual

Udell, J. (2004). Jon Udell: Name that genre: screencast. *InfoWorld.*

Uhm, J. H. (2015). *The goals and purposes of Seoul Cyber University.* Live presentation at KIMEP University.

USAspending.gov. (2008). *Top 100 Recipients of Federal Assistance for FY 2008.* Retrieved 20 July 2015 from usaspending.gov.

Valentine, D. (2002). Distance Learning: Promises, Problems, and Possibilities. *Online Journal of Distance Learning Administration*, *5*(3). Retrieved from http://www.westga.edu/~distance/ojdla/fall53/valentine53.html

Vasquez-Cano, E. (2014). Mobile Distance Learning with Smartphones and Apps in Higher Education. *Educational Sciences: Theory and Practice*, *14*(4), 1505–1520.

Wright, C. R., Lopes, V., Montgomerie, T. C., Reju, S. A., & Schmoller, S. (2014). Selecting a Learning Management System: Advice from an Academic Perspective. *EDUCAUSE Review Online.* Retrieved 20 July 2015 from http://www.educause.edu/ero/article/selecting-learning-management-system-advice-academic-perspective

WSJ. (2011, October 19). Apollo Group 4Q Net Soars on Fewer Charges: Enrollment Falls. *The Wall Street Journal.*

Chapter 5
Learners–Mobile Interaction:
African Substance and Style

Kushatha Kelebeng
Botswana International University of Science and Technology, Botswana

Pendukeni Phalaagae
Botswana International University of Science and Technology, Botswana

Rebaone Mlalazi
Botswana International University of Science and Technology, Botswana

Tebogo Mangwa
Botswana International University of Science and Technology, Botswana

Keorapetse Gosekwang
Botswana International University of Science and Technology, Botswana

Tebogo Kebonang
Botswana International University of Science and Technology, Botswana

Thotobolo Morapedi
Botswana International University of Science and Technology, Botswana

ABSTRACT

The increasing demand for education in African continent has motivated more research in mobile learning to describe, explain and predict changes and chances in leveraging mobile technology to learn anytime and anywhere. Mobile technologies provide learning beyond the classroom walls where learners can have access to resources according to their needs without involving teachers. In developing countries like Africa, mobile learning is still gaining its ground and remains under-studied. With technological advancements and high usage of mobile phones in Africa, the educational horizon has broadened. Mobile learning plays a major role in delivering education in Africa. The rapid use of mobile devices in mobile learning projects in Africa confirm that mobile learning is transforming the traditional way of learning, teaching and delivery of education by integrating social media and advanced mobile technologies in education. After a critical look at the current status as well as the latest development in mobile learning, this chapter presents a road map of m-learning to guide future action and thinking of learners, teachers and institutions in Africa. It also tracks trends and their impact, key theories and key findings of mobile learning. The current chapter ends with a critical analysis of technologies and key theories that can contribute to the sustenance of mobile learning in the African continent.

DOI: 10.4018/978-1-5225-0469-6.ch005

INTRODUCTION

M-learning is being applied to the use of portable, small, handheld and lightweight electronic devices used for educational activities in classrooms, fieldwork, at work, home and when travelling (John Traxler & Jenny Leach, 2004). The devices include mobile phones (in particular but not exclusively smartphones), Personal Digital Assistants (PDAs), audio players and new entrants include tablets (Regin Joy Conejar & Haeng-Kon Kim, 2014). Additionally John Traxler & Jenny Leach, (2004), argued that netbooks and laptops are technically not included as m-learning devices because they are not handheld, and they belong to the devices that employ the desktop WIMP (window, icon, menu, pointing device).

It is of great importance to firstly put forth difference between m-learning and electronic learning (e-learning). According Korucu & Alkan (2011) writers perceive m-learning as a naturally evolved form of e-learning. M–learning got its own distinguished technology and terminologies as opposed to e-learning as shown by the Table 1.

John Traxler (2007) states that the use of wireless, mobile, portable, and handheld devices are gradually increasing and diversifying across every sector of education and across both developed and developing countries. Technology-enabled needs a breakthrough every few years to reinvigorate it and drive thinking forward and the uptake and adoption of mobile technologies have opened a new learning platform, one that is ideally suited to increased informal learning and getting learning right into the personal space of learners.

Jiugen, Ruonan & Jianmin (2010) revealed that mobile learning (m-learning) is a new form of learning, using mobile network and tools, expanding digital learning channel, gaining educational information, educational resources and educational services anytime, anywhere as illustrated by. Asabere (2013) defined m-learning as a field that involves the use of mobile computing and wireless technology to enable learning occur anytime, anyplace and anywhere. Furthermore Asabere (2013) study shows that m-learning is the future of education in Africa and noted the benefits and disadvantages that come with it. Some of the benefits include:

- Mobile learning can occur at anyplace and anytime, and learning content can be accessed anywhere.
- Mobile learning process is not limited to one particular place.
- Mobile learning enhances the interaction between instructors and students.

Table 1. Terminology Comparison between M-Learning and E-Learning

E-Learning	M-Learning
Computer	Mobile
Bandwidth	GPRS,G3, Bluetooth
Multimedia	Objects
Interactive	Spontaneous
Hyperlinked	Connected
Collaborative	Networked
Media-rich	Lightweight
Distance learning	Situated learning
More formal	Informal
Simulated situation	Realistic situation
Hyperlearning	Constructivism, collaborative

- Mobile learning brings opportunity for students to learn while on the move.
- Mobile learning facilitates interactive and collaborative learning among students.

Barker, Krull & Mallinson (2005), found out in their research that indeed mobile technologies extends learning beyond the walls of lecture rooms as it provides portability, collaboration and motivation to learners, teachers and parents. These mobile technologies enable portable learning in the sense that learners can take their handheld mobile devices outside the confinement of the classrooms and acquire information by interacting with other colleagues. M-learning motivates and promote collaboration among students as they can use their handheld devices to form groups, aggregate and distribute information with ease as they can do that anywhere and anytime of the day. Barker et al. (2005) further highlighted the disadvantages of mobile learning and they include the following:

- Mobile learning can create isolation or a feeling of being left out-of-the-loop for learners who may not always have mobile connectivity.
- Mobile learning can also give technically savvy students an advantage in terms of system and device usage over non-technically inclined students, such as art students.
- Mobile learning cannot argument practical hands on lessons such as laboratory experiments.

John Traxler (2007) alluded that there are some emerging categories of mobile learning that include, "technology-driven mobile learning-this is where by technological innovation is deployed in an academic setting to demonstrate technical feasibility, remote/rural development mobile learning-these are the technologies that addresses environmental and infrastructural challenges to deliver and support education where 'conventional' e-learning would fail, miniature but portable e-learning-this category of mobile learning emphasize that mobile, wireless and handheld technologies are used to re-enact approaches that and solutions already used in e-learning. Last but not least category of mobile learning is informal, personalized, situated mobile learning whereby same technologies are enhanced with additional functionality e.g. location awareness and deployed to deliver educational experiences".

The advancement of technology and high penetration rates of mobile devices has widened the horizon of education in Africa. (Grimus, Ebner & Holzinger, 2011). M-learning has already started to play a very important role in e-learning environment and it should be noted that m-learning has brought e-learning to the rural communities of Africa, to learners that were never imagined as e-learning learners. Barker et al. (2005) highlighted that African continent has many developing countries which are technologically far behind the developed countries, and this lack of development has a detrimental effect on the education sector in Africa. Lack of infrastructure in Information Communication technology (ICT) in certain areas in Africa affect the optimal use of mobile technologies, in other areas the existing telecommunications infrastructure cannot reach the bulk of the population, with only around 50 percent of the telecommunication lines wholly directed to the cities, where only 10 percent of the people live.

In a nutshell implementing mobile learning program involves numerous considerations, multiple stakeholders and involves several steps namely, investigating: which answers the question 'why' certain district or school should be implemented with mobile learning, secondly scoping: where the stakeholders are determined and the scope of the implementation. The third aspect is planning whereby those responsible for the success of the m-learning program, followed by preparing for implementation which outlines the preparation needed for m-learning to be rolled out. Another step may involve rolling out where programs and procedures are outlined for a successful roll out of the mobile learning program followed by learning and teaching, which involves setting up and supporting professional learning communities. Last step for successfully implementing mobile learning program is evaluating and adjusting, whereby feedback is collected from learners and different stakeholders about a particular mobile learning program in order to make some refinements if need be.

THE STATE OF THE ART OF M-LEARNING

Current Status of Mobile Learning

Mobile devices today play a remarkable role in economies and society at large as they have strongly affected every field, from universities to banking and are currently used to increase efficiency and productivity across different sectors (Regin Joy Conejar & Haeng-Kon Kim, 2014). Mobile learning today encourages students to be independent learners and teachers being information guides rather than educational instructors at the forefront of delivering education. According to Brown (2003), m-learning currently contribute to the quality of education worldwide as it offers individuals with opportunity interacting and communicating among themselves anywhere, anytime without the restrictions of place and time. According to Francis Osang, Jey Ngole & Clive Tsuma (2013) mobile learning is considered as the future of education across the globe as mobile telecommunication devices provide learners with a learning platform that include:

- **Improved Feedback**: m-learning platform allows for immediate feedback to the learner, for example a learner may take a quiz via a mobile learning app and get feedback immediately after taking the quiz; they can view the answers to the questions they didn't get right.
- **Interest**: This mobile device excites and engage learners, young people who lost interest in education may prefer to play with these mobile devices and end up gaining educational material via learning games on their devices.
- **Automated Assessment**: the assessment of students learning is very vital to their education as it is a tool that measures the understanding of students. On the other hand it relieves teachers of the burden of marking each and every script of the learner.
- **Collaboration**: this platform makes it possible for students and educators to work collaboratively as a group and share vital information that can benefit each other.

According to Oller, R. (2012), the relevance of m-learning lies in the fact that majority of learners are without infrastructure for access, but interesting to note is that the adoption rate of mobile technologies in Africa is at an increasing rate globally. In 2005 there were nearly 100 million mobile users in Africa.

Latest Development

Today m-learning is undoubtedly integrated with mainstream education and has already begun to show a very strong potential to distort existing pedagogical infrastructure including that of online education and ignoring m-learning is not an option.(Grimus, M., Ebner, M., & Holzinger, A. 2012). In addition it was further emphasized that development in m-learning is likely to take the following directions:

- **Location-Based Learning**: Here the learner's location will be incorporated in GPS enabled devices, they can be location tracked in relation to other learners thus enabling them to interact with each other being aware of the geographical location of one another.
- **Ambient intelligence**: This is the presence of digital environment that is sensitive, adaptive, And responsive to the presence of people. It can also be understood as an environment where people are surrounded with networks of embedded intelligent devices that can sense their state, anticipate, and adapt to their needs.
- **Wearable Learning**: This is the kind of learning whereby individuals wear devices those are in the form of accessories such as jewellery, sunglasses, a backpack and even actual items of clothing such as jackets or shoes. For example the wearer's clothing can be studded with sensors that detect sensory level environmental data and gather inputs that are beyond human sensory capabilities, such as radiation levels, subsonic waves and electromagnetic fields.
- **Augmented Reality (AR)**: Augmented reality (AR) is a live, direct or indirect, view of a physical, real-world environment whose elements are augmented by computer-generated sensory input such as sound, video, graphics or GPS data. It is related to a more general concept called mediated reality, in which a view of reality is modified (possibly even diminished rather than augmented), by a computer. As a result, the technology functions by enhancing one's current perception of reality. By contrast, virtual reality replaces the real world with a simulated one. Augmentation is conventionally in real-time and in semantic context with environmental elements, such as sports scores on TV during a match. With the help of advanced AR technology (e.g. adding computer vision and object recognition) the information about the surrounding real world of the user becomes interactive and digitally manipulable.
- **Learning Implants**: Most research and experimentation with brain-computer interfaces focuses on neuroprosthesis-related applications, general –use brain implants intended as gaming interfaces are being developed.

MOBILE LEARNING IN AFRICA

This section presents an extensive review and a critical analysis of literature in mobile learning to determine the previous state of knowledge and research in Africa. The principal goal is to examine the occurring trends from year 2000 to year 2013, in order to present a road map of M-learning to guide future action and thinking of learners, teachers and institutions in Africa. This review has considered to over 150 articles with conference papers, reports, and research papers included. The preceding state of mobile learning in Africa is orderly presented with emphasis mainly oscillating around the changes in learning and limitations of mobile learning in Africa.

The Revolution and Impact of Mobile Learning in Africa

Mobile technologies have been accepted as learning tools by a significant number of teachers, learners and parents in Africa. Mobile learning programs have been implemented in other parts of Africa successfully and have received positive feedback (Network, 2014).Mobile technologies provides learning beyond the classroom walls and this section explores trends and benefits of mobile learning to learners, teachers and parents and other stakeholders engaged in the learning environment.

Desktops and SMS-enabled Mobile Phones

According to Shuler, C., Winters, N., & West, M. (2013) around year 2003 learners in Africa only had a periodic access to the internet via PC's at learning or community centres. The internet access at these centres were controlled, each learner was allocated a time slot. During these periods of access the focus of learning is normally on ICT literacy, downloading of content, access to study materials, e mail access. Nevertheless learners used mobile phones on daily basis for social interaction (peers and friends respectively) and communication with their educational institutions Via SMS. The educational institutions used the SMS service to communicate administrative information to their learners (exam reminders, notifications, urgent information).Again through SMS service learners were able communicate among themselves and in study groups.

Portable Devices

In 2005 portable mobile devices with more features that supported a significant number of functionalities such as MMS, WAP (web browsing technology) came into perspective. According to Prasertsilp (2013), mobile learning technologies can make a significant difference in a wide variety of settings, both outside and inside classrooms. The arrival of these portable devices brought about new ways of delivering educational content which brought about many opportunities and benefits. One of the benefits is that Learners and teachers took the learning experience outside the confinement of the classroom. The portability effect of these devices allowed learners to carry them from one class to the other exchanging, interacting and sharing information with other peers enabling them to learn on the go.

Collaborative Learning

According to Mayende, Divitini, and Haugaløkk, (2006) handheld devices provides a learner with the ability to disperse, aggregate and share information in a simple manner resulting in a successful collaboration. After the year 2005 mobile devices technologies became diminutively advanced; the innovative technology on mobile devices presented learners with the capability to communicate and interact with their educational institutions, browse e-learning course materials, download study guides, and receive tutorials at the palm of their hands using mobile phones. Collaborative learning allowed Institutions to provide course materials, course schedules, facilitate access to exam and test marks through a mobile portal (Brown, 2003).

Through mobile phone technologies learners were able to access a lot of information in order to meet their own learning objectives. This new pedagogy in learning enabled a learner centred education that is more contexts relevant, personalized and self-directed and very flexible; that is taking place anywhere.

In summary, mobile learning in Africa has evolved contributing to a paradigm shift in education bringing forth the twenty first century type of education that integrates the use of social media and advanced mobile technologies in education.

Key Theories and Methods

Learning theories are fundamental because they influence and inform the choice of learning strategies for mobile learning (Berking, Haag, Archibald, & Birtwhistle, 2012). Current mobile learning theories include Constructivism, Behaviourism; problem based learning; collaborative learning; conversational learning; Connectivism, Activity Theory; informal learning, lifelong learning, location based learning. This section discusses the application of collaborative learning in Africa.

Collaborative Learning in Africa

Students working together in teams can explore science and mathematics solutions better than students working alone. According to Nyarko and Ventura, (2010) collaboration learning can be used to connect students in different places. This type of learning theory can be extended to other areas of concern such as social media. Examples of social media collaborative platforms include Facebook and Twitter which allows users to share pictures, stories and other content. These social Medias have gained popularity and approval from millions of people around the continent including in Africa. Relative to mobile learning a study Nyarko and Ventura, (2010) that was conducted in South Africa, students were placed on a virtual platform through ICT to decrease the teacher to student ratios. In this study students were exposed to self-learning within groups in different locations and it was a success.

Key Findings and Achievements

The successful instances where mobile learning has shown to be effective in Africa provides a new path for understanding the perceived usefulness and contribution of mobile learning. In light to that this section presents key research findings with focus on developing African countries and further identifies the achievements of mobile learning programmes and also examines the potential of mobile learning to engage learners in a creative and collaborative learning environment.

Efficiency in Education

The rapid use of mobile devices in mobile learning projects in Africa confirm that mobile learning is transforming the traditional way of learning, teaching and delivery of education. According to Shuler, C., Winters, N., & West, M. (2013) it has been found that mobile learning is now opening up new opportunities for improving access to quality education. Mobile learning can be applied to African communities that have less access to books for example South Africa where 51 percent of households lack access to books and only 7 percent of public libraries are functional. A registered success of such implementation in other African regions is in Kenya where an SMS education management system was used to support teachers and learners for the purpose of efficiency (Sobrinho, 2014).

Motivation to Learners

Barker, Krull, & Mallinson (2005)'s findings indicate that learners using mobile phones demonstrate an increased autonomy in learning, as learners show amplified behaviour of self-conducted learning and take the initiative in finding ways to use the handheld devices for learning. Through mobile learning, learners can have access to resources according to their needs without involving teachers. Mobile learning gives students the opportunity to personalise learning according to their needs, helping them to engage more into the learning material. Students are motivated when they learn by themselves as this gives them the ability to focus on their interests, learning styles and abilities (Haji, Abdalla & Kombo, 2013).

Improvement of the Quality of Education

According to Network (2014)'s findings mobile technologies present an effective intervention that is acceptable to teachers, learners and parents to improve the learning outcomes. At least 5 key findings were identified, mobile devices can increase the learning outcome by 27 percent, mobile devices are effective in language learning, mobile devices are regarded as educational tools by teachers, learners, and parents; mobile devices create a teaching environment perceived as more professional and satisfying; mobile devices create a learning environment perceived as more interactive and informative.

Affordance of Mobile Devices

Handheld devices are much affordable as compared to laptop and desktop computers in Africa. In a study that was conducted in South Africa it was revealed that the majority of learners use handheld devices both at home and inside the classroom and while travelling (Traxler & Leach, 2006).This study found out that at least 56 percent of teachers and 75 percent of the school population have had no prior experience in using any form of computer but they willing to purchase handheld devices.

Key Weaknesses and Problems

A plethora of challenges confronts mobile learning when attempting to deliver and enhance learning through mobile devices in developing countries, this paper emphasis more in Africa.

Poor Infrastructure

James & Versteeg (2007) explained that mobile phones are a crucial mode of communication and welfare enhancement especially in countries with low economic status. One key weakness is that the physical infrastructure in Africa is poor when compared to developed countries. Traxler & Leach (2006) conducted a study in Kenya and South Africa respectively and the following findings were collected about the physical infrastructure of Africa. When generalizing these findings it can be established that they are true to almost majority of countries in Africa. It has been found out that the physical infrastructure in Africa can be characterized by sparsity, vast distance and low population densities, schools especially in rural areas, substandard buildings or none at all: poor roads, transport systems and postal services; poor landline phone networks, unreliable and often unprofitable; poor mains electricity, unreliable and concentrated in towns and cities; little or no Internet bandwidth outside major cities, often just internet cafes or hotel business centres in cities; very few modern PCs or peripherals in the public sectors, and little user expertise, especially in smaller towns and rural areas. However Africa has strong mobile phone networks that provide GSM and GPRS service and also there is potential for solar power.

Expensive Data Connections

Poor communities are faced with technical difficulties of mobile phones due to the fact that connectivity is restricted since data is expensive. To emphasise that the study conducted by Ngole & Jey (2013) in Nigeria exhibit that despite the ability of mobile devices to be accessed from any point of interest, outside and inside classroom environments data connection is relatively expensive.

Battery Power

In an effort to illustrate the difficulties of mobile learning a study was made in the Eastern Cape Province in South Africa. During the length of the study there were occasions where teachers suffered data loss due to loss of battery power, where data had not been synchronized with another machine. These instances were more common in rural schools than in urban contexts. Almost half of the schools in the study region are not connected to the electricity grid, to recharge the project equipment; teachers walk a few miles down an unmade track to the local hospital. Teachers who constantly used the handheld computers expressed distress and were displeased with the loss of data (Traxler & Leach, 2006).

THE FUTURE OF M-LEARNING

The adoption of mobile learning in educational environments and the pilot projects that comes with it show that m-learning is experiencing exponential growth. Mobile learning is undoubtedly the future of disseminating information or providing education to learners around the globe and in Africa in particular as it holds much promise. A study conducted by Jobe (2001) in Kenya where 30 Kenyan were provided with android smartphones and internet connection for a period of one year proved that indeed m-learning is the future when it comes to providing educational information. The participants of the study revealed that the smartphones provided a means to bridge the gap between informal and non-formal learning hence delivering educational opportunities to them anywhere and anytime of the day, and the statement from the Kenyan participant who were given smartphones was "it helps us a lot".

Parsons (2014) outlined top five mobile learning innovations that will benefit learners across the world and they include placing learning in a specific context-a learner can take a mobile device with him/her wherever he/she goes, augmenting reality with virtual information-with the use of mobile device something virtual can be overlaid onto something real as they may be used to give information about locations, artifacts in areas as diverse as history and geography by the help of augmented reality tools such as Wikitude, Google Goggles etc., contributing to shared learning resources-these devices allows learners to learn while communicating and contributing at a distance with other individuals, having an adaptive learning toolkit in the palm of a hand and taking ownership of learning-a mobile device can be looked at as a toolkit because it has tool-like functions that are built into the device(sound and video recorder, camera, multimedia messaging etc.) and it can be used as a guitar tuner, distance measuring device, a compass, a speedometer and a wide range of other things and last mobile learning innovation outlined by Parsons (2014) is that mobile devices enable one to take ownership of learning-mobile devices provide learners with a platform that empower them to acquire learning material that they want, whenever and wherever they wish to do so.

The potential of mobile learning in Education is reflected by gradually emerging technologies. M-learning innovations however pose opportunities and innovations that immensely contribute to the quality and adequacy of education.

The Potential of M-Learning in Africa

Mobile learning technologies can be used to enhance the learning process both inside and outside the classroom and the workplace. Mobile learning presents opportunities that transform the quality of formal and informal (outside-the classroom) education and makes the learning process to be more engaging. This section describes the future potential of mobile learning in Africa and it discusses future technologies in Mobile learning.

Bring your own Device (BYOD): Most learners are literate with popular devices such as e-readers and smartphones. Bring your own devices allows distant learners who live and work far from school to attend classes remotely instead of on campus facilities (Lennon, 2014). Also according to Lennon (2014) the 21st century learning characteristic uses the BYOD approach opens up digital learning opportunities for students in the classroom since the old constraints of centralized access to central resources is gradually fading away. This takes advantage of cloud-based services, as they are becoming more and more relevant and prominent enabling learners to use their own devices for learning hence leading to great efficiency and digital inclusion.

Augmented Reality with Virtual Learning: Augmented reality portrays the opportunity for using mobile devices to provide information about artefacts, locations etc. and is applied in diverse architectures such at architecture, history and geography (Ally & Tsinakos, 2014). There is a number on learning applications where virtual reality has been superimposed onto a physical location to provide students with a new learning experience such as Savannah (Facer et al, 2004) and Invisible Buildings (Winter & Pemberton, 2011). According to Brown & Mbati(2015) they also foresees the rise in usage of the simulations and augmented reality in m-learning because at first this technology has not become the mainstream due to high costs and lack of required skills for development. However the numbers of open source software and free augmented reality software to help create augmented reality in m-learning are increasing. Edugaming is also a new concept that will take the m-learning into victory and carry learning beyond the area of simulations and serious games.

Future Technology Advancement in Mobile Learning

During the next fifteen years, technology will continue to be dynamic in a number of ways that can be beneficial for education. It is important that educators understand these advances so as to inspire their development rather than simply react to it. Ideally technology and education will co-develop, with educational needs driving technological progress as well as adapting to it. In the next fifteen years, mobile learning will undoubtedly evolve and more incorporated with mainstream education (Shuler, C., Winters, N., & West, M. 2013). Just as computers are now viewed as a crucial component to learning in the twenty-first century, mobile technologies will soon become conventional in both formal and informal learning, and progressively even the term 'mobile learning' will fall into disuse as it is increasingly associated with learning in a more holistic rather than specialized or peripheral sense. As the links between technical and pedagogical innovations improve, mobile technology will take on a clearly defined but increasingly essential role within the overall education ecosystem. The report further explains the sections that summarize the anticipated focus areas for mobile learning concepts in the near future which are discussed fully below (Shuler, C., Winters, N., & West, M. 2013).

Distance Education and experiential learning: Technology-enhanced learning, particularly m-learning has been the expansion of MOOCs. It is predicted that this trends will continue to expand assuming the constant political support it has received across the world including Africa. In the next coming years, mobile technologies will allow MOOCs and other forms of distance learning to provide a more customized and user centric content (Walker, 2010). Distance learning will be able to extend into fields that typically require in situ learning, such as medicine and many forms of vocational training. Learners will be able to collect data on their practice and share and discuss information with professors, tutors, mentors and peers via mobile technologies. Gamification – the use of game mechanics in a non-game context to engage users – may also become more popular in distance education (Shuler, C., Winters, N., & West, M. 2013). The practice of rewarding points to people who share experiences and information on social networking sites like Facebook could be collaborated and/or incorporated with MOOCs to encourage experienced participants to mentor and support newcomers. All of these types of interactions can and will be facilitated by mobile technologies.

Furthermore, mobile learning will allow the expansion of experimental and location-based learning. In the future experiential learning as embedded and sophisticated location-aware technologies become more common in mobile devices (Walker, 2010).

Authentic and Personalized Learning: According to Leafsnap (2011) mobile technology will support learners and educators in exploring both the developing and the developed world while creating their own solutions to complex problems as well as working in collaboration among peers under the guidance experienced personnel or teachers. The author continues by saying already there are a number of applications that use image-capturing capabilities of mobile devices. The personalization of features in mobile technologies will allow learners with different capabilities to progress at their own pace. Artificial intelligence (AI) will also become widespread in the education domain and so its availability will increase on mobile devices. As an emerging field in mobile learning, AI will focus on simple and straightforward activities for educators in the coming years. They will have to ensure that it is a balanced phenomenon by personalized interventions that support more complex and multidimensional opportunities for learning (Shuler, C., Winters, N., & West, M. 2013).

Wearable Devices and the Future of Mobile Learning: According to (Mike Sharples, 2014), "the focus is not so much on the device but on the mobility of the learner, and this mobility of the learner is the key factor of mobile learning." On Mobile Learning and Wearable learning, (Manisha Reddy K, 2014) says wearable learning devices are gradually gaining significance the in both the education and business domains due to the fact that the devices in discussion are very much portable and comfortable. It is no longer only about smartphones and other mobile devices such as tablets. As wearables shift from becoming more of a need than a luxury (just as smartphones did, a couple of years ago) we can expect to see them being used actively for learning as well in the next 5 years or so (Manisha Reddy K, 2014). The wearable technologies are paving the way for inter-connectivity of learners to the content, co-learners and educators. It is the wearable technology which is creating shockwaves in the learning industry and changing the way we see mobile learning, indicating the new concepts in providing unprecedented mobility and learning freedom to the user.

Gamified mobile learning and assessments: According to Foreshew (2015), researchers are using gamification to explore m-learning tools that support learning. One of the projects that is being carried out in Australia's University of Newcastle the prototype app that improves the literacy of first year undergraduates. The project offers a fun and engaging way to improve literacy. This shows that gamification moving ahead in terms of delivering learning solutions through mobile technologies. In other cases, Christie Wrote (2014) reports that Gamification and mobile learning and two dynamic trends that are growing at a very high rate. Together they contribute to a learning platform that is powerful and combines engagement, collaboration together with convenience. Furthermore the report says that according to research M2 Research, the overall market for gamification tools, services and application is projected to be $5.5 Billion by 2018 this shows that mobile learning and gamification presents a huge potential in terms of growth in the next 4 years. When you combine gamification with mobile learning?

Gamification is engaging and mobile learning is convenient. The result when these two powerful trends meet is an awesome learning experience - that's both popular and effective.

Social constructivist learning environment: Social constructivism refers to the learning between the learner and the community where they co-construct new meanings and knowledge. One of the teaching strategies along the social constructivism is the class- sourcing which involves faculty members giving out assignment and learners making the assignment public to others via websites, blog, videos, podcast and others, allowing learning among themselves (Brown & Mbati, 2015). Future mobile learning environments will bring in an existence of personalized and contextual learning in a pervasive setting. What lies ahead on the further developments in social constructivist learning environment will be more example of lecturer reviewed resources and class-soured ready available to mobile learners around the world(Brown & Mbati, 2015).

Sharples, Taylor and Vavoula (2005) considers learning as a cultural-historical activity system, facilitated by tools which are categorized into two layers; that both constrain and support the learners in their goals to transform their knowledge and skills. The semiotic layer describes learning as a semiotic system in which the learner's actions promote objectives which are arbitrated by cultural tools and signs. Secondly it comprises of the technological layer which represents learning as an engagement with technology, in which tools such as computers and mobile phones function as interactive agents in the process of coming to know, creating a human-technology system to communicate, to mediate agreements between and to aid recall and reflection.

Socio Cultural Theory of Human Learning

According to Vygotsky (1978) social interaction plays a vital role in cognitive development. It categorizes interaction through peers which is later integrated into the individual's mental structure and secondly cognitive development influenced by a zone of proximal development. This "zone" is the area of exploration for which the student is cognitively prepared and needs to interact with others to gain full understanding. According to Rogers (2002) learning takes place in a social context and the structuring of knowledge concepts need not necessarily take place only at the level of the individual, but that collaborative group work and sharing with peers. Collaborative work can be a powerful way of confronting one's own conceptions, contributing to the need to restructure one's cognitive schemas.

CONCLUSION

Evidence from research exhibit that mobile learning in Africa is at its infant stage and seemingly enjoys a slow adoption rate despite its transformative effect and the advancement of wireless and mobile device technologies globally. Most studies show that a large population of people using mobile technologies in Africa have limited these digital devices to social interaction and a few have regarded mobile learning as a core pedagogical shift in learning. Even though m-learning has been used as a minor addition to e- learning, it is still not the primary mode of delivery in education.

M-Learning as a further expansion of e-learning it has a spanning potential to revolutionize the way education is delivered. The substance and style of African mobile learning is different from mobile learning in developed countries. African countries face a quite a significant number of challenges such as poor infrastructure, expensive data connections and this makes it difficult to implement mobile learning when compared to developed countries which have resources.

Moreover m-learning has shown to possess the potential to increase performance by presenting learning provisions anywhere and anytime, this gives the learner flexible time to engage in learning activities without the limitations of time and place. Collaboration, interaction and communication are key in the learning process. It is within this context that e-learning and m-learning can contribute to the quality of education. M- Learning offers a vast of opportunities for enhancing collaboration, interaction and communication between learners and lecturers and among themselves.

REFERENCES

Ally, M., & Tsinakos, A. (2014). *Increasing access through mobile learning*. Academic Press.

Asabere, N. Y. (2013). Benefits and Challenges of Mobile Learning Implementation: Story of Developing Nations. *International Journal of Computers and Applications*, *73*(1).

Barker, A., Krull, G., & Mallinson, B. (2005, October). A proposed theoretical model for m-learning adoption in developing countries. In Proceedings of mLearn.

Berking, P., Haag, J., Archibald, T., & Birtwhistle, M. (2012). Mobile learning: Not just another delivery method. In *Proceedings of the 2012 Interservice/Industry Training, Simulation, and Education Conference*.

Brown, T. (2003, June). The role of m-learning in the future of e-learning in Africa. In *21st ICDE World Conference*. Retrieved from http://www.tml.tkk.fi/Opinnot

Conejar, R. J., & Kim, H. K. (2014). The Effect of the Future Mobile Learning: Current State and Future Opportunities. *International Journal of Software Engineering & Its Applications*, *8*(8).

Edelson, D., & O'Neill, D. K. (1994). *The CoVis Collaboratory Notebook: Supporting Collaborative Scientific Enquiry*. Academic Press.

Grimus, M., Ebner, M., & Holzinger, A. (2012). Mobile Learning as a Chance to Enhance Education in Developing Countries-on the Example of Ghana. In mLearn (pp. 340-345).

Haji, H. A., Shaame, A. A., & Kombo, O. H. (2013, September). The opportunities and challenges in using mobile phones as learning tools for Higher Learning Students in the developing countries: Zanzibar context. In AFRICON, 2013 (pp. 1-5). IEEE.

Hull, R., Facer, K., Stanton, D., Kirk, D., Reid, J., & Joiner, R. (2004). Savannah: Mobile gaming and learning? *Journal of Computer Assisted Learning*, (6): 399–409.

INDEPTH Network. (2007). South Africa Country Report. Report by the Sonke Gender Justice Network, Johannesburg/Cape Town.

James, J., & Versteeg, M. (2007). Mobile phones in Africa: How much do we really know? *Social Indicators Research*, *84*(1), 117–126. doi:10.1007/s11205-006-9079-x

Jiugen, Y., Ruonan, X., & Jianmin, W. (2010, July). Applying research of mobile learning mode in teaching. In *Information Technology and Applications (IFITA), 2010 International Forum on, 3*, 417–420.

Korucu, A. T., & Alkan, A. (2011). Differences between m-learning (mobile learning) and e-learning, basic terminology and usage of m-learning in education. *Procedia: Social and Behavioral Sciences, 15*, 1925–1930. doi:10.1016/j.sbspro.2011.04.029

Lennon, R. G. (2012, October). Bring your own device (BYOD) with cloud 4 education. In *Proceedings of the 3rd annual conference on Systems, programming, and applications: software for humanity* (pp. 171-180). ACM. doi:10.1145/2384716.2384771

Mayende, G., & Divitini, M. (2006, July). MOTUS goes to Africa: mobile technologies to increase sustainability of collaborative models for teacher education. In null (pp. 53-54). IEEE. doi:10.1109/TEDC.2006.22

Mbati, L. S., & Brown, T. H. (2015). *Mobile Learning: Moving Past the Myths and Embracing the Opportunities.* Academic Press.

Murugesan, S. (2013). Mobile apps in Africa. *IT Professional, 15*(5), 8–11. doi:10.1109/MITP.2013.83

Oller, R. (2012). *The Future of Mobile Learning (Research Bulletin).* Retrieved from https://net.educause.edu/ir/library/pdf/ERB1204.pdf

Osang, F. B., Ngole, J., & Tsuma, C. (2013). Prospects and Challenges of Mobile Learning Implementation in Nigeria. Case Study National Open University of Nigeria NOUN. In *International Conference on ICT for Africa 2013* (pp. 20-23).

Osang, F. B., Ngole, J., & Tsuma, C. (2013). Prospects and Challenges of Mobile Learning Implementation in Nigeria. Case Study National Open University of Nigeria NOUN. In *International Conference on ICT for Africa 2013* (pp. 20-23).

Parsons, D. (2014). The future of mobile learning and implications for education and training. *Increasing Access*, 217.

Prasertsilp, P. (2013). Mobile Learning: Designing a Socio-Technical Model to Empower Learning in Higher Education. *LUX: A Journal of Transdisciplinary Writing and Research from Claremont Graduate University, 2*(1), 23.

Rogers, Y., Price, S., Harris, E., Phelps, T., Underwood, M., Wilde, D.,... Neale, H. (2002). *Learning through digitally-augmented physical experiences: Reflections on the Ambient Wood project.* Equator IRC. Available http://www.equator.ac.uk/papers/Ps//2002-rogers-1.pdf

Sharples, M., Taylor, J., & Vavoula, G. (2005). Towards a theory of mobile learning. *Proceedings of mLearn 2005, 1*(1), 1-9.

Tangney, B., Weber, S., Knowles, D., Munnelly, J., Watson, R., Salkham, A. A., & Jennings, K. (2010). *MobiMaths: an approach to utilising smartphones in teaching mathematics.* Academic Press.

Traxler, J. (2007). Defining, Discussing and Evaluating Mobile Learning: The moving finger writes and having writ..... *The International Review of Research in Open and Distributed Learning, 8*(2).

Traxler, J., & Leach, J. (2006, November). Innovative and sustainable mobile learning in Africa. In *Wireless, Mobile and Ubiquitous Technology in Education, 2006. WMUTE'06. Fourth IEEE International Workshop on* (pp. 98-102). IEEE. doi:10.1109/WMTE.2006.261354

Winter, M., & Pemberton, L. (2011). Unearthing invisible buildings: Device focus and device sharing in a collaborative mobile learning activity. *International Journal of Mobile and Blended Learning, 3*(4), 1–18. doi:10.4018/jmbl.2011100101

Winters, N., Sharples, M., Shuler, C., Vosloo, S., & West, M. (2013). *UNESCO/Nokia The Future of Mobile Learning Report: Implications for Policymakers and Planners*. UNESCO.

Chapter 6
Government and Mobile:
Examining the Role of SMS

Beatriz Barreto Brasileiro Lanza
Companhia de Tecnologia da Informação e Comunicação do Paraná, Brazil & Center for Technology in Government at SUNY, USA

Maria Alexandra Cunha
Fundação Getulio Vargas Escola de Administração de Empresas, Brazil

ABSTRACT

This chapter aims to study a fifteen-year project of Mobile Government (mGov) in the state of Paraná, Brazil. The precocious experience in Brazil in mGov and the number of cell phones used in the country are not arguments for these to be used in providing mass public services. Mason's Historical Method was used to get to know the project, raising available information and retrieving happenings and results. To identify events, relevant actors, their roles, aspects in collaboration, setting and density of the relations among these actors, Social Network Analysis was used. We conclude it is important to formalize the project, but that will not guarantee its continuity in the long run. Characterizing phases minimizes the importance given to the change of governors after new elections in large corporative projects. The networks forming favors the dissemination of IT knowledge among actors. This research shows little evidence that SMS still have long life.

INTRODUCTION: ELECTRONIC GOVERNMENT AND MOBILE GOVERNMENT

There have been many papers focusing on e-service within the field of e-government (Islam & Scupola, 2011), and the necessary transition from e-government to mobile government (Medeni et al., 2011). This paper dwells on the experience of Paraná State Government, Brazil, using cell phones to provide services to citizens. It reports on how the Mobile Government (mGov) Project was developed from 2000 to 2014. It reports especially on the relations among the actors in the project. Paraná State Government pioneered among the Brazilian governments (federal, state and local levels) in making cell phone services available to citizens and the government itself. In 2000, services already prospered in this platform, offering

DOI: 10.4018/978-1-5225-0469-6.ch006

traffic services, job opening listings, prices of agricultural products, frost warnings and giving access to cultural events. In the 1990s, in various countries, there was massive access to cell phone services. In Brazil, they greatly spread out after the regulation of pre-paid services. Brazil has the fourth largest number of cell phones in the world. In Latin America, it's the leader in cell phone lines. The potential of mobile coverage is affecting all aspects of society and economy, even Government services (Adams & Mouatt, 2010) and the numbers show the potential of this infrastructure for the government to offer m-services to citizens in Brazil. Mobile technologies can be expected to provide governments with significant opportunities to achieve greater cost optimization, and SMS has become a powerful land prevalent communication channel for citizens and government (OECD/ITU, 2011).

Despite over a decade of mGov projects in Brazil, these devices are not yet a channel for mass delivery of public services. In other developing countries, the situation is similar. One of the explanations is the difficulty to establish relations among multiple actors of mGov, necessary for a corporative model to sustain itself and be operative in the long run. Such actors include state government departments and autarchies, governmental computing organizations (computing companies), departmental computing structures within those departments and autarchies, telecommunication companies, brokers, Information Communication Technology (ICT) providers, and the main actors: citizens.

It's great the spread of mobile technology worldwide. Also great is the spread of so-called social technologies, in which the phenomena of networks are particularly visible. However, from an academic point of view, few studies have been identified in literature regarding the use of mGov, and such studies and researches do not show how and to what extent networks influence the adoption and implementation of projects using mobile technology.

The objective of this research is to document all steps faced by a pioneer mGov project in Brazil, located in the State of Paraná. The results of this work can be useful to public managers, including CIOs, who use this technology to provide new service channels to citizens, and for business partners, such as telecommunication companies. In addition, this study may bring insights to managers of corporative projects of governmental technology. In the theoretical context, this study establishes a relationship between mGov and the Social Network Analysis. The framework of social and organizational networks strength the theme Mobile Government in Social Sciences, especially in Brazilian Public Administration.

This chapter is comprised of five sections following this introduction. The first section provides background information about mGov in Brazil: Electronic Government (eGov) and Mobile Government. The second section introduces the methods used: Historical Method and Social Network Analysis. The third section describes Paraná Mobile Government. The fourth section is the discussion and interpretation. The last section provides the main conclusions from the research and finalizes with some questions for future researchs.

BACKGROUND

To present the context of the use of mobile telephony by the Government of Brazil, one must first understand how eGov and mGov take place.

In Brazil, the State reform is a historical process whose dimension is proportional to its crisis. It began in the 1970s, boomed in the 1980s, and brought about the reappearance of Liberalism with a profound critique of the forms of intervention or regulation of (or by?) the State. In the 1990s, this theme became very broad and complex, as it involved political, economical and administrative aspects. The intended result was a more efficient State, properly reaching the citizens. It became a state acting in partnership with society and in accordance with its desires, not exactly trying to protect the market, but rather, trying to render itself more able to compete (Bresser-Pereira, 1998).

In the scope of State Reform, actions of administrative reform gained force in the federal (national), state and local governments in Brazil. There were amplifications of the use of ICT in the implementation of public policies, in public management, in providing services for citizens' democratic practices and the relation between State and citizen. This new way in which the government relates with its citizens is the so-called Electronic Government, which Fang (2002) defines as how governments use new information and communication technologies. Particularly, it focuses on Internet-based applications aiming to provide citizens and companies with a more adequate access to governmental information and services, to improve the quality of such services and to provide better opportunities for participation. The objectives of the actions of eGov in the 1990s were, in the scope of administrative reform, the increase of efficiency and administrative transparency. The concept of eGov does not differ from the concept of government; only the manners and means to relate with the citizen change (Riecken & Lanza, 2007). But eGov assumed an important role in this new form of governing, with the participation of a broad network of actors (Cunha & Miranda, 2008). Providing governmental services is a complex and intertwined network, which needs to be coordinated. This can be done using all sorts of electronic services (eServices), including mobile devices: the so-called mGov.

The definition of mGov, according to Kushchu & Yu (2004), is a strategy of implementing public services made available through mobile platforms to provide citizens and society with the benefits and information anytime, anywhere. Diniz & Gregório (2007) consider mGov an essential communication platform between government and citizens. It is the migration of eGov services and applications to mobile platforms. The term 'mobile government' can be understood, according to Cunha *et al.* (2007), in three ways. The first regards the opportunity governments have to use mobile devices as a means to deliver services and information to the citizen. The second is about the use public agents make of these devices in their daily work. And the third concerns setting up mobile units to enable the government to provide electronic services to the population that lives in places with no access to public services. More recently, with social media, cell phones have been used as a powerful way to influence public decisions.

The needs and demands from citizens and society are growing, as well as the speed of technological advances and eGov infrastructure development. Thus, electronic service distribution channel strategies are fundamental for the advance of eGov services (Germanakos, 2009). Many governments, e.g., Dubai and Singapore, acknowledge the importance of mobile technology and have developed mGov initiatives. In the state of Virginia, USA, pioneer in the country to implement mGov applications, My Mobile Virginia was developed. It was the first portal to provide governmental services using mobile devices. SMS have been used, for instance, at London Police Department to inform citizens of matters concerning safety and emergency alerts (Trimi & Sheng, 2008). The Philippines government offers services using SMS and WAP (Wireless Application Protocol). In Kenya, the government uses cell phones as a tool to fight corruption, and SMS was indicated as the best platform to use in this kind of service for convenience, cost and simplicity (Salim & Wangusi, 2014). In Nigeria, the Ushahidi platform is used for citizens to monitor the electoral process and report incidents of electoral fraud (USHAHIDI, 2015). Information published by CONIP (2014) indicates that there are a few initiatives in Brazil, such as the SMS project in São Paulo city, sending text messages to alert patients about appointments scheduled at InCor Hospital. A few other initiatives were implemented, such as the one from Paraná State Government, where the Department for Labor and Jobs (SETS) sends text messages with job openings to workers registered with a profile compatible with the job description. In 2010, Porto Alegre City Hall (Rio Grande do Sul, Brazil) was recognized in the technological innovation category for the Mobile Porto Alegre project, an application to access different governmental services by means of mobile devices (CONIP, 2014).

The information made available by TELECO (2014) presents a series of indicators of communication in Brazil. For example, the number of cell phones is 278 million with a density of 137 phones for every 100 inhabitants. The most used function in these devices is SMS, representing 66% of functionalities used. This infrastructure can serve as a two-way communication channel between government and citizens, especially low-income citizens, as the number of pre-paid phones represents 76.61% of the total number of phones.

The research from the Brazilian Internet Steering Committee (CGI) pointed out that in 2013, among the cell phone-owning population, 31% have Internet access (CGI, 2014). These data point to an electronic channel already in existence and with great potential to deliver governmental services to citizens.

The data reveal that Brazil has not evolved in mGov proportionately to the exponential growth of mobile telephony. One of the explanations concerns the difficulty in establishing relations among multiple mGov actors. This relationship also involves the establishment of a business model to be sustained in the long run.

METHODOLOGICAL APPROACH

The methods used in this research were Mason et al.'s Historical Method (1997) and Social Network Analysis (SNA).

Mason's Historical Method

Historical research investigates the past, aiming to assess its influence in the present (Lakatos & Marconi, 2007; Neustad & May, 1996). It is also process-oriented, as it concerns understanding how and why projects evolve through time. Process research normally deals with retrospective data, based on narratives about what happened, who did what, and when. It examines events, activities or choices ordered in time (Langley, 1999).

Historical research methods can be of great use as they offer the business researcher the opportunity to obtain a good comprehension of situations and contexts where they take place (O'Brien, 2004). Vizeu (2010) highlights that research with these methods is interesting because it retrieves historical and intercultural aspects standing out in the dominant ideological reproduction, which tends to exclude the past or the context of organization theories and practices. Historical perspectives can contribute to advancement in the analysis of administrative phenomena, enriching and broadening the research, both by adopting a conceptual-theoretical framework built from the historical analysis, and by the application of historical research as method for analysis.

Organizations are rationally planned so as to resolve conflicts between collective needs and individual wishes, and their processes of action and creation are performed by individuals in a specific, historical context. Identifying and analyzing this context, by means of a theoretical-conceptual framework based on historical perspective, helps avoid attributing deterministic and historical character to the research. For example, using concepts and ideas from a given time to analyze factors from another historical time brings about methodological distortions by stepping further away and highlighting a universal timelessness (Reed, 1999).

This study was largely influenced by Costa et al. (2010), especially when they defend that the approach linking of Administration and History has not yet reached its ontological and epistemological potential. They believe that this discussion is transported to the administrative thought when appropriations of historical perspectives in administration begin. According to these authors, identifying the originating views of heated debate between the perspectives of traditional history and new history in the Administration field may contribute to:

- Better understanding of the administrative phenomena;
- Training of researchers more aware of their research methods;
- Strengthening of interdisciplinarity by creating stronger links between the areas.

They believe this incorporation cannot take place ingeniously or irresponsibly, as each ontological and epistemological assumed position will guide the researcher towards different sources, objects, problems and methods, hence the different attempts at organizational reality comprehension.

In order to retrieve the mGov history of Paraná, we came across three major challenges:

1. The project length was too long: fifteen years.
2. It was a complex project with many actors.
3. There was no formal registry of the project.

We found a rather large number of documents not in good order. Thus, to recover the necessary data, the method by Mason *et al.* (1997) was considered the most appropriate, especially as these authors proposed a historical method specifically concerning Information Technology research. It is of particular use when one wishes to know how these phenomena came to be or how they evolved through time, and their implications for the present. Mason *et al.* state that historical method research offers advantages in order to understand contemporary phenomena concerning the use of IT. Jayo (2010) also defends the use of this method, as opposed to a retrospective analysis, for the diversity of models practiced in operational management in Brazil since the historical analysis complements production of a given picture in management models, helping rebuild the diversity of processes.

Mason et al.'s method consists in the application of seven phases:

1. Formulating a research question;
2. Specifying a well-defined investigation domain;
3. Gathering evidences from documental sources and other registries;
4. Assessing critique of the gathered empirical material;
5. Spotting patterns, from the empirical phase to the inductive one;
6. Writing the report; and,
7. Situating contributions to the report regarding existing literature.

For expositional effects, the phases were presented in a sequence; however, when performing this research, they often suffered juxtapositions or repetitions. This included applying basic logic so as to check internal coherence of the data, and going back to the sources to remove inconsistencies. Using this method, we studied organizational processes of the mGov project of Paraná State, in addition to the official corporation history, interpreting existing organizational structures not as determined by law, but as a result of past decisions.

Social Network Analysis

The realization of innovative technological transformation in providing electronic services has been associated with the presence of a number of prime actors who perform their required functions in the organizations (Janssen, 2009). We also suspected the dynamics of the relationship among actors in the project could point at clues towards the identification of events, relevant actors, their roles, aspects in collaboration, setting and density of the relations among these actors, so Social Network Analysis (SNA) was used. SNA is the process of investigating social structures through the use of network and graph theories. This process uses nodes (individual actors, people, or things) and the ties or edges (interactions) that connect them (Otte & Rousseau, 2002).

Although it bears characteristics from a strict modeling sociological method, SNA is flexible as it proposes hybridism among different matrices to cause a series of conceptual elements to interact with new contributions coming from these interactions (Caballero, 2005). The networks are composed of aspects of trust and integrated information. There is a tangle of relations among individuals, requiring great efforts to maintain and support these relationships (Powell, 1990). Reciprocity is one of the characteristics of the relationship among actors, the determining factor for the formation of networks. It shows the degree to which the relationship is understood and agreed to by all parties. The theory defends that reciprocity is heightened when there is a long time perspective. Safety and stability encourage new ways to perform tasks, which lead to better learning, information exchange and confidence.

This method is used to set behaviors in context so as to reach systematic visualization of the relational dimension, sometimes invisible, of social connections within a society. It is a tool to serve collective action theory, which helps redefine disciplinary concepts and produces new knowledge from the perspectives set by the researcher's viewpoint (Pereira & Meirelles, 2009). In social networks, depending on the researcher's disciplinary background, a simplified interpretation can be done, reducing the complexity of the object of research. In order to neutralize this (possible) influence, the SNA combines a structural approach (visualization, graph generating) with other methods of network actions in context. Thus, the complex social interdependency of networks cannot be dealt with through/by mathematical formalization alone, but also with the elaboration of pertinent sociometric questions. Social networks are considered important partners to build knowledge about themselves and the individual parties within the relationships. This implies, as stated by Brandão (2008), a strong transdisciplinarity connecting experience with everyday life with knowledge and universities. Thus, it is possible to establish dialogues to deal with questions such as politics, technology management, world economy and the environment.

Steps and Products of the Empirical Analysis

To categorize the epistemological assumptions of a study, Orlikowski & Baroudi (1991) present three perspectives: positivistic, interpretational and critical. This classification has been useful to characterize researches in the area of IS (Information System) and IT (Chen & Hirscheim, 2004; Walsham, 1995; Diniz et al., 2006). This research is based on a group of positivistic epistemological assumptions, even if not explicit, regarding the nature of knowledge and how it is obtained (Myers, 1997).

Here the qualitative analysis (case study) allows to build the case, registering what changed throughout the lives of the organizations by means of interviews. The period for this research was from 2000 to 2014. In 2000, the first mGov movements in Paraná Government started and, in 2010, some state companies in Paraná still used mGov. In 2011 the government launched the portal for mobile technologies and in 2012 held a public tender to purchase SMS, thereby initiating a new phase of mGov Paraná. The case includes public and private institutions participating in mGov in Paraná State, totaling 22 institutions: AFPR[1], APPA[2], CASA CIVIL[3], CASA MILITAR[4], CCTG[5], CEASA[6], CELEPAR[7], DEPEN[8], DETRAN[9], e-PARANÁ[10], IAPAR[11], IIPR[12], JUCEPAR[13], POLÍCIA CIVIL[14], SEAP[15], SECS[16], SEFA[17], SEOG[18], SESA[19], SESP[20] and SETS[21].

The private company was Comunika, after the merger process turned Zenvia[22]. This broker includes all operators in the country: ALGAR, BRT, CLARO, GLOBAL, NEXTEL, OI, SERCOMTEL, and TIM e VIVO.

The tools used to collect data were documental research and interviews. For the documental research, written permission was obtained to observe administrative documents. The documents were analyzed using documental analysis technique which, according to Richardson (1989), is essentially thematic and aims towards a faithful identification of social phenomena. Richardson also defends that the observation of documents should be done carefully, not taking documents for literal registries of the events that took place. For Yin (2005), the most important use of documents in a case study is to corroborate and value evidences originated from other sources.

In order to obtain retrospective data to help get to know and acknowledge important happenings in the constitution of different events, 257 documents were examined. In the end, 55 of them were used, dating from 2000 to 2012, in addition to five undated documents, whose contents made it possible to situate them within the period. These documents are webpage registries based on databases from the researched organizations, digital files, printed documents such as press material, project descriptions, reports, meeting minutes, contracts and formal agreements, emails (digital and hard copies), and internal documents from the participating companies for Paraná State mGov project. The documental research data served as a source for consultation for the description of the case and the triangulation of the information. After documental analysis, eight in-depth interviews were performed in order to confirm and/or correct discrepancies found in the documental research. Technicians and managers from the organizations participating in Paraná State mGov project were interviewed.

The aims of, and reasons for, the research were made clear to the participants according to the Terms of Free and Clarified Consent. This document was signed by both researcher and respondents. Eight professionals were interviewed, identified in the documental research as actors relevant in the process of creating and building Paraná mGov Project. The interviews took place from July 29th to September 21st 2010, and November 02nd 2014. They were recorded in audio and video. The running time varied from 35 minutes to 140 minutes. In addition, contact was kept with interviewees, either by telephone, email or in person, to clear doubts and complement or correct information, all of which were added to the material gathered or transcribed. The interviews were transcribed and analyzed by means of content analysis technique.

Phase 4 of the method by Mason et al. (1997) consisted of the critical assessment of the gathered material, which involved a thorough reading of everything (documental sources, transcribed interviews and research notes) so as to spot information gaps or contradictions in the data. In some cases, interviewees were contacted again, either by phone or e-mail, in order to resolve doubts, update information and clarify contradictions found in the data.

Finally, the inductive analysis of the material – gathering and relating events, was conducted according to Phase 5 of the method.

To identify the aspects of collaboration, and collaboration density among actors, structured questionnaires were used and the 42 tabled answers were analyzed with the help of UCINET 6.289 for Windows. Square matrixes were used. To select the respondents, in addition to the names found in the documental research, the researchers also used the snowball technique recommended by Atkinson and Flint (2001). The research steps and products are summarized in Figure 1.

SIX PHASES OF PARANÁ mGOV

Paraná State mGov Project went through different phases throughout its existence, influenced by political happenings or choices the managers made. In order to describe it properly, it was necessary to characterize these phases and understand the influences and context of each of them. To mark the beginning of each phase, we used the criterion of event, i.e. a choice, activity, decision making or happening that strongly influenced how the project was conducted in the given period. The established phases have no standard duration. Their descriptions, main actors and relation among actors are detailed in Table 1.

Phase 1: mGov Exploration and Pre-Project

The first phase of the project — Phase one - mGov Exploration and Pre-project — was characterized by cooperative actions, outstanding partnerships for exploring, learning, discovering and experimenting. No concern was found regarding establishing commercial or contractual relations.

Figure 1. Steps and products of the empirical analysis

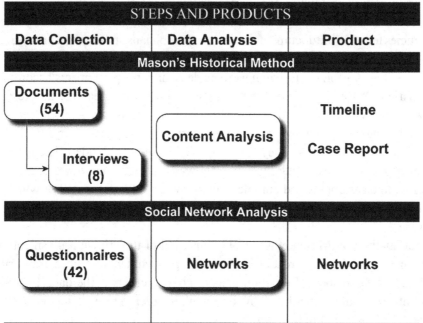

Table 1. mGov phases in Paraná State

Phase	Period
Phase 1: mGov Exploration and Pre-project From the Origin to the Formalization of the eGov	2000-2001
Phase 2: mGov Maturity From the Formalization of the eGov to the Government Change	2001-2003
Phase 3: mGov Formalization From the Government Change to the eGov Discontinuity	2003-2006
Phase 4: mGov Decline From eGov Discontinuity to the CELEPAR's Quitting the Project	2006-2009
Phase 5: mGov Survival From the CELEPAR's Quitting to the Government Change	2009-2010
Phase 6: mGov Return From the Government Change to the Return of the mGov Project	2011-2014

The initial phase of Paraná State mGov Project started in 2000, in CELEPAR. CELEPAR is responsible for the computer actions in the institutions of the whole State, and their coordination and integration. It is a company with approximately 1000 employees. At the time, cell phones were being introduced in Brazil. Then, mGov was initiated with no predetermined aim, deadline or budget — characteristics inherent to the project exploration phase. The initial phase was marked by partnerships. There was interest in mobile telephony, especially in WAP (Wireless Application Protocol) and SMS (Short Message Service) technologies, and partners noticed their potential. However, they did not know yet how to use them in governmental applications.

The actors' roles got defined as the project progressed. CELEPAR was in charge of keeping and providing for governmental database infrastructure. The purpose was to "learn how to do." The first service to be made available was experimental: checking traffic tickets from Paraná Traffic Department. The technology used was WAP, a protocol for wireless applications built in some cell phones. The service allowed to checking vehicle debts, including speed and parking tickets or vehicle tax debts.

It is important to highlight:

- The fact that the idea of using cell phones for work changed the dynamics of the adoption of this technology, both for governmental actors and private ones;
- Certain regions of the State were not covered by mobile telephony services, especially rural areas;
- A debate began among some of the technicians and managers from CELEPAR, as well as some representatives from the Traffic Department, about what today is called usability, and also about the high cost of the information traffic.

Here, CELEPAR was the central actor. It idealized and conducted the mGov project, especially considering its position in the government. It was responsible for the installations, equipment, software, maintenance and safety of State governmental data. In this phase, however, the Traffic Department, GLOBAL and TIM (Telecom service providers) were side actors. Analyzing the actors by their relationship nature, it is possible to notice at this point that there were no actors in contractual or commercial relation. All four actors had a partnership relation, with participation and mutual learning during the exploring phase. Concerning institutional and legal relations, the only interacting actors were the Traffic Department and CELEPAR, as shown in Figure 2.

Phase 2: mGov Maturing

In the second phase, the government showed interest in strengthening and disseminating the mGov project to State agencies in order to speed up the process and the quality of services to citizens. There was motivation to use the new technology. Many actors were involved together to discuss and discover possible uses for it. It was the phase with the highest number of actors and projects, with synergy among governmental communication agencies to be transparent about the project, headed by e-PARANÁ and CELEPAR.

The significant event marking the beginning of Phase 2 was the formalization of eGov in Paraná (e-PARANÁ). Its institutionalization took place on March 26th, 2001, with State Decree #3769, published in Diário Oficial (Official Gazette) no. 5954 on March 27th, 2001. The practice of eGov was then made official, although it actually started in 1995, with a team from CELEPAR. Once it gained official status, legal support was given for the involvement and integration of other governmental departments and agencies. This Decree also created the Executive Committee for Electronic Government in Paraná, aiming to formulate and establish policies, articulate actions of implementation, and operation of projects such as, for instance, mGov.

From 2001 to 2003, mGov was disseminated throughout the governmental structure, adding to the effort of institutionalizing it in a proactive way with the representatives of CASA CIVIL & SEOG. Nine projects or tests to offer services to the citizens or to the governmental organizations were identified. Discussions started on the commercial viability of the project, and the first proposal was presented for acquiring corporative SMS service. Governmental actors intensified partnerships with information providers and became concerned with the commercial viability of the project. There was no contractual or commercial relation among actors.

Figure 2. Nature of relation among actors in Phase 1 of mGov

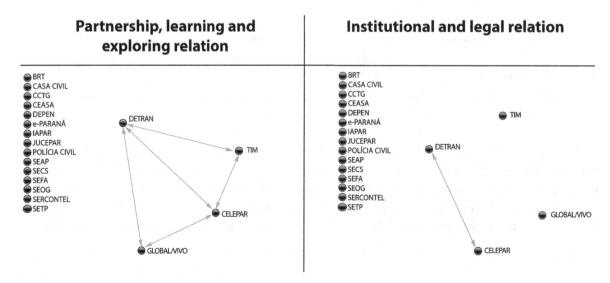

Partnership and learning were strong motives for relationship among actors. The institutional or legal relations among actors became stronger. SERCOMTEL was not involved in these relations, which can be explained by the geographical localization of this company and its partners: all based in Londrina, north of Paraná State. Figure 3 illustrates visually the nature of the relations of the actors in this phase.

Phase 3: mGov Formalization

Phase 3 saw mGov formalized by a contract to supply SMS to all government agencies. This contract later served as reference for the Federal Government and other States of the Federation. It was prized and presented abroad as a "business model. "

When the new governor took his position in January 2003, high-rank managers gave special attention to the governmental portal, which provided services to citizens, companies and the government itself. In order to deal with the high demands for services from citizens, business people and the governor himself, it was imperative that the technology used should not be very costly and should be developed and implemented in a short time.

On March 31st, 2004, the informal partnership with TIM finished. Paraná Government decided to use SMS services from all phone companies acting in the State. The idea was to sign a government-written, standard-defined contract with each one of the companies.

The eGov "Services" workgroup led the discussion. At the time, the estimated number of text messages per year was 1.8 million for the total lots, distributed according to the number of clients for each registered company in the systems of the Executive Branch. This project, called "SMS Corporative Use" was awarded with the 9th Prize of Excellency in Public Computing in 2006, presented in the Government Technology Event (GTEC) in Ottawa, Canada.

Four more new services were developed in the scope of the mGov project: e-Process, SEAP's Agenda, SECS News and eGov Executive Committee. The e-Torpedo system (SMS from the government) was developed in CELEPAR/eGov Division to support all the SMS project. It centralized, distributed and controlled all SMS applications in Paraná State.

A contractual relation was set up. With the adopted method, CELEPAR and SEAP signed a contract with private cell phone companies: BRT, GLOBAL/VIVO and TIM. CELEPAR provided governmental

Figure 3. Nature of relations among actors in Phase 2 of mGov

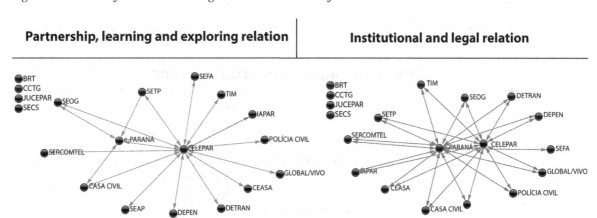

agencies with their SMS quota. As a result, SEAP now was, along with CELEPAR and e-PARANÁ, a central actor in institutional and legal relations. The partnership and learning relation remained strong in this phase, as displayed in Figure 4.

Phase 4: mGov Decline

With the discontinuation of e-PARANÁ, which gathered actions from the state electronic government, the mGov project started to fail. From the 16 actors in the previous phase, only six remained.

In 2006, there was a change in the coordination of the operations of eGov: the workgroup meetings of e-PARANÁ were interrupted; and, the project using mGov was discontinued. The CCTG was then added, with cultural programming and sales using the fidelity card from CCTG, an autarchy linked with Paraná Government and one of the largest theaters in Latin America. The outstanding fact in this period, more precisely in 2009, was the departure of CELEPAR, acknowledged as the central actor in the mGov project since its creation in 2000. TIM, considered another important actor/partner in the creation, development and maintenance of the mGov project, also left the project. The reason was because the contract with private cell phone companies was not renewed in 2009.

The relationship among actors continued, even though some had left the partnership. Regarding legal relation and roles, it was clear that CELEPAR had two different types of relationship: as the company related technically with the phone companies, and commercially with SEAP. DEPEN was isolated in the network, as it participated only as a test, as shown in Figure 5.

Figure 4. Nature of relation among actors in Phase 3 of mGov

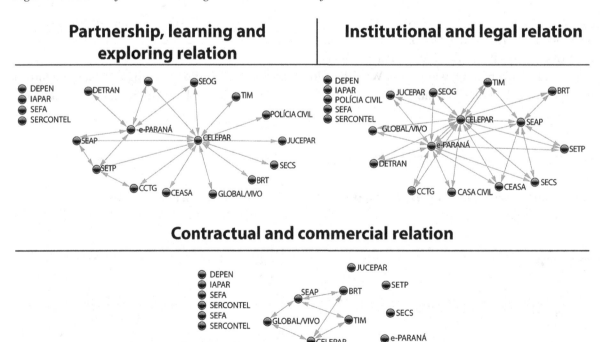

Figure 5. Nature of relation among actors in Phase 4 of mGov

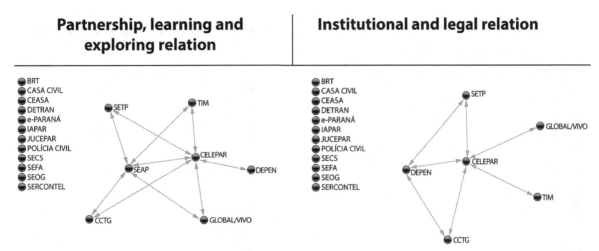

Phase 5: mGov Survival

By Phase 5, only three actors remained in the project – Global/Vivo, CCTG and SETP. The managers responsible for the service providing did not accept its deactivation. They insisted on continuing the project even without a contract for the supply of SMS and without the SMS system which was hosted on CELEPAR Data Center.

SETS and CCTG, not accepting the fact that the service they provided would be interrupted, adapted the technological infrastructure to support and maintain the project. Without the corporative contract renewal for the acquisitions of SMS, CCTG hired the mobile telephony company VIVO and the service was then made available only to those who owned a phone from that company. The services of Job Opportunity Warning and CCTG Cultural Program were still offered to the citizens.

Commercial relation took place between CTG and VIVO, since the corporative contract was not renewed and CCTG signed a contract directly with VIVO. In the institutional or legal relation, there was interaction with the three actors remaining in the project, as shown in Figure 6.

Phase 6: mGov Return

In Phase 6, the final phase, mGov Program was returned in 2011 with the launch of the portal for mobile technologies. In August 2012, a new contract to supply SMS to all government agencies was awarded. The previous contract served as reference for this new public tender. In March 2013, the government again started using SMS itself and delivered the same service to citizens.

Twelve new services were developed in the scope of the mGov project:

1. Confirmation of registered cargo in seaport;
2. Called for the truck that is waiting in the courtyard;
3. Notice of government policy staff meetings;
4. Community volunteers on notice for emergency situations;
5. Notice of unavailability of the central system;

Figure 6. Nature of relations among actors in Phase 5 of mGov

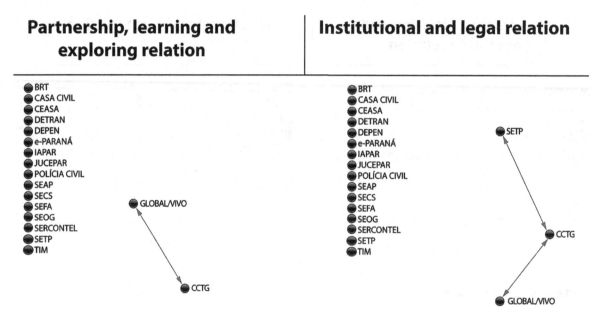

6.   Alert any unavailability of the mail server corporate;
7.   Consulting BI (?) information;
8.   Alert scheduling ID Cards;
9.   Alert medical appointments;
10. Alert medical exams;
11. Warn the population about theft; and,
12. Data queries by police vehicle (?)

In the messaging process, each government agency is responsible for the generation of its content. The system also allows receipt of SMS on the same architecture.

The following companies participated: AFPR, APPA, CASA CIVIL, CASA MILITAR, CELEPAR, DETRAN, IIPR, JUCEPAR, SEOG, SESA, SESP and SETS.

CELEPAR was the central actor. It idealized and conducted the mGov project, especially considering its position in the government. It was responsible for the centralization of communication with the broker, which effectively communicated with the mobile operators. Regarding contractual and commercial relation, Celepar continued to be responsible for software, maintenance and safety of State governmental data (Figure 7).

At its conclusion, a timeline of the mGov Project of Paraná Government was drawn (Figure 8), visually displaying the six phases and their nominations, events, decisions and projects in the 15 years of its existence.

Figure 7. Nature of relations among actors in Phase 6 of mGov

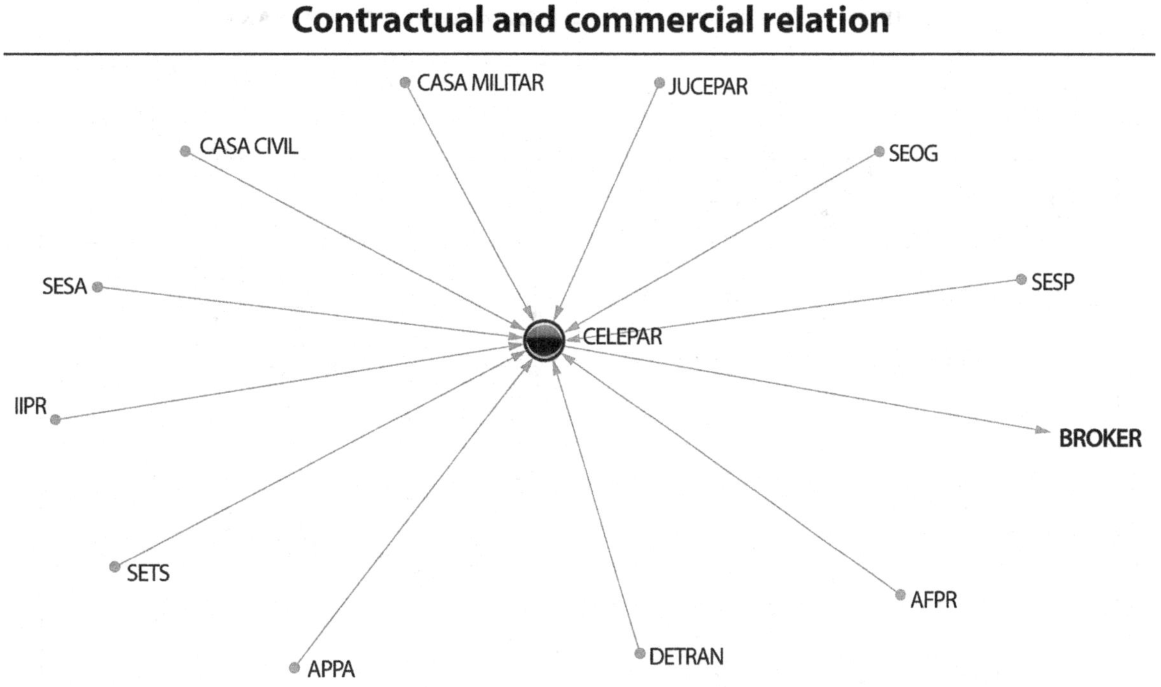

Contractual and commercial relation

CASA MILITAR   JUCEPAR

CASA CIVIL   SEOG

SESA   SESP

CELEPAR

IIPR   BROKER

SETS   AFPR

APPA   DETRAN

DISCUSSION AND INTERPRETATION

This paper outlines the history of the mGov project in Paraná Government from its creation, construction, survival and renaissance, with emphasis on the relationship among its actors. The approach used was Mason's Historical Method, as well as Social Network Analysis. Mason's Method was useful to retrieve, retain and create knowledge. Although it is a large project, with interaction among multiple actors, not all of the knowledge obtained was disclosed. It was restricted to those who took part in the project, and each of them had his/her own partial view of it. This method helped retrieve and disclose this knowledge, producing a complete view of the project, and making it possible to learn from it.

When characterizing the different phases of this fifteen-year project, the government changes influenced the large IT government projects, as expected. However, the phases characterized in this research do not always coincide with changes in the government. In this research, the mGov project was characterized by an event spawning it, and then by the relationship among the actors, and often differing from phase to phase. Power and politics in governmental practices were important, but it was interesting to observe that, in this case, the political power instituted in each election and the government changes are not the determining factors to characterize the path the project takes. It is also important to observe that the actors' roles are set differently in each phase. For instance: in initial phases, the tasks of dissemination, articulation among actors, and integration are more important than in later phases when the project is maturing. With a maturity of 15 years, the return on the mGov Paraná project in 2011 occurred quite quickly and naturally, since the learning of the previous phases have been incorporated by the actors involved.

Figure 8. Paraná mGov timeline

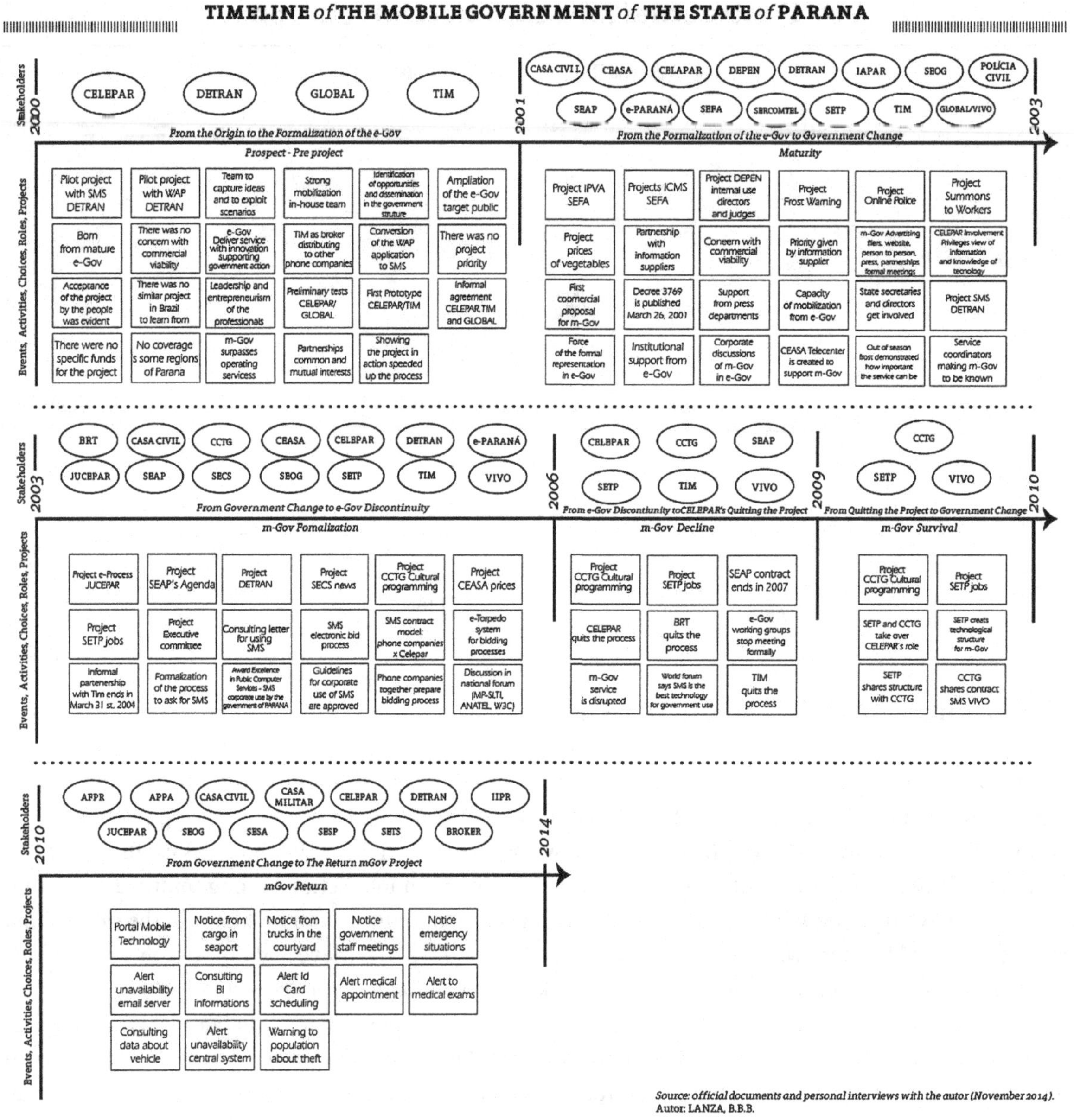

TIMELINE *of* THE MOBILE GOVERNMENT *of* THE STATE *of* PARANA

Stakeholders 2000: CELEPAR · DETRAN · GLOBAL · TIM

From the Origin to the Formalization of the e-Gov

Prospect - Pre project

Pilot project with SMS DETRAN	Pilot project with WAP DETRAN	Team to capture ideas and to exploit scenarios	Strong mobilization in-house team	Identification of opportunities and dissemination in the government structure	Ampliation of the e-Gov target public
Born from mature e-Gov	There was no concern with commercial viability	e-Gov Deliver service with innovation supporting government action	TIM as broker distributing to other phone companies	Conversion of the WAP application to SMS	There was no project priority
Acceptance of the project by the people was evident	There was no similar project in Brazil to learn from	Leadership and entrepreneurism of the professionals	Preliminary tests CELEPAR/GLOBAL	First Prototype CELEPAR/TIM	Informal agreement CELEPAR,TIM and GLOBAL
There were no specific funds for the project	No coverage s some regions of Parana	m-Gov surpasses operating services	Partnerships common and mutual interests	Showing the project in action speeded up the process	

Stakeholders 2001–2003: CASA CIVIL · CEASA · CELAPAR · DEPEN · DETRAN · IAPAR · SBOG · POLÍCIA CIVIL · SEAP · e-PARANÁ · SEFA · SERCOMTEL · SETP · TIM · GLOBAL/VIVO

From the Formalization of the e-Gov to Government Change

Maturity

Project IPVA SEFA	Projects ICMS SEFA	Project DEPEN internal use directors and judges	Project Frost Warning	Project Online Police	Project Summons to Workers
Project prices of vegetables	Partnership with information suppliers	Conern with commercial viability	Priority given by information supplier	m-Gov Advertising fliers, website, person to person, press, partnerships formal meetings	CELEPAR Involvement Privileges view of information and knowledge of tecnology
First coomercial proposal for m-Gov	Decree 3769 is published March 26, 2001	Support from press departaments	Capacity of mobilization from e-Gov	State secretaries and directors get involved	Project SMS DETRAN
Force of the formal representation in e-Gov	Institutional support from e-Gov	Corporate discussions of m-Gov in e-Gov	CEASA Telecenter is created to support m-Gov	Out of season frost demonstrated how important the service can be	Service coordinators making m-Gov to be known

Stakeholders 2003: BRT · CASA CIVIL · CCTG · CEASA · CELEPAR · DETRAN · e-PARANÁ · JUCEPAR · SEAP · SECS · SBOG · SETP · TIM · VIVO

From Government Change to e-Gov Discontinuity

m-Gov Fomalization

Project e-Process JUCEPAR	Project SEAP's Agenda	Project DETRAN	Project SECS news	Project CCTG Cultural programming	Project CEASA prices
Project SETP jobs	Project Executive committee	Consulting letter for using SMS	SMS electronic bid process	SMS contract model: phone companies x Celepar	e-Torpedo system for bidding processes
Informal partenership with Tim ends in March 31 st. 2004	Formalization of the process to ask for SMS	Award Excellence in Public Computer Services - SMS corporate use by the government of PARANA	Guidelines for corporate use of SMS are approved	Phone companies together prepare bidding process	Discussion in national forum (MP-SLTI, ANATEL, W3C)

Stakeholders 2006: CELEPAR · CCTG · SEAP · SETP · TIM · VIVO

From e-Gov Discontinuity to CELEPAR's Quitting the Project

m-Gov Decline

Project CCTG Cultural programming	Project SETP jobs	SEAP contract ends in 2007
CELEPAR quits the process	BRT quits the process	e-Gov working groups stop meeting formally
m-Gov service is disrupted	World forum says SMS is the best technology for government use	TIM quits the process

Stakeholders 2009–2010: CCTG · SETP · VIVO

From Quitting the Project to Government Change

m-Gov Survival

Project CCTG Cultural programming	Project SETP jobs
SETP and CCTG take over CELEPAR's role	SETP creats technological structure for m-Gov
SETP shares structure with CCTG	CCTG shares contract SMS VIVO

Stakeholders 2010–2014: AFPR · APPA · CASA CIVIL · CASA MILITAR · CELEPAR · DETRAN · IIPR · JUCEPAR · SEOG · SESA · SESP · SETS · BROKER

From Government Change to The Return mGov Project

mGov Return

Portal Mobile Technology	Notice from cargo in seaport	Notice from trucks in the courtyard	Notice government staff meetings	Notice emergency situations
Alert unavailability email server	Consulting BI informations	Alert Id Card scheduling	Alert medical appointment	Alert to medical exams
Consulting data about vehicle	Alert unavailability central system	Warning to population about theft		

Source: official documents and personal interviews with the autor (November 2014).
Autor: LANZA, B.B.B.

Another important observation is that in this corporative project, the relationship among the actors allowed the gathering of knowledge of ICT inside the organizations where ICT was not part of their core. In this case, the departments of Culture and Labor kept the project alive due to their personal relation and acquired ICT knowledge. In large corporative projects, knowledge exchange among the participating actors leads to the gathering of even more knowledge, and this can be a movement to be deliberately broadened and improved.

The formalization of eGov accelerated the appearance of the mGov project, which was not an initial objective. While the eGov project existed, mGov in Paraná clearly broadened (Phases 2 and 3). One possible interpretation is that in large corporative technology projects involving multiple actors, different spheres and interests, the formalization of a project in a greater government context gives potential and amplifies its running. However, formal contracts established among different parts and/or laws do not suffice to guarantee the commitment to keep a project in the long run, despite its importance. The governmental contract, with the actors in accord, did not suffice to ensure the sustainability of the project either. The structures formed in the corporative projects depend on formalization, be it by law or norm, but the history of this project was a result of the network relationships.

It is also important to observe in the last phase, all major stakeholders returned, except the e-Paraná, which in the third phase was one of those central actors in the legal and institutional relations center. This is a point that needs to be investigated because, despite its absence, the project returned strong, very similar third stage, despite the absence of a multi-institutional project bringing together all stakeholders. The interactions observed showed dynamism in the relation among actors, contributing towards the development of the idea of using a technology little explored at that time. In general, the actors who aimed towards providing eGov with information or service took part in the project to test the technology and were involved in the project for only one of the phases, a short period. Those with concrete relative proposals towards providing services aligned with the fulfillment of their organizational mission, who saw this technology as an opportunity to broaden their services in a long-term perspective, remained longer in the project.

The recovery of the history of the project identified and characterized six phases:

1. Exploration and pre-project
2. Maturing
3. Formalizing
4. Falling (failing?)
5. Surviving
6. Return

Throughout these phases, the participating organizations also played some specific roles:

- Infrastructure provider
- Content provider
- Disseminator
- Solution developer;
- Broker,
- Telecom service provider

In this study, we verified that in the chain of eGov, the role of the organization in charge of IT in government as a disseminator of the service and technology was as important as the role of the service providers.

CONCLUSION

This study aimed at reconstructing the start, development,survival and return of the mGov Project of Paraná Government, emphasizing the relations among actors.

Some findings may be used as insights for governmental technology corporate project managers. Firstly, in Paraná mGov, the formalization of eGov meetings with actors and group discussions caused the corporative project to grow – hence the importance of formalization. On the other hand, the formalization of eGov did not guarantee the continuation of the project. It "died" without being closed, hence the importance of central actors keeping the project "alive".

Secondly, despite the importance that government changes have in corporative projects, we identified phases in the project not coinciding with political orientation changes. In these phases, there were changes in the network setting and in the roles of the actors, demanding articulation, fulfillment of tasks, and the performance of roles typical of each phase. Understanding this fact and taking action can bring potential to the network results in each phase. Networks with a less defined central role are less vulnerable to political orientation changes.

Thirdly, the relationship the actors orchestrated in a corporative project favored knowledge exchange about ICT in organizations having little knowledge about the technology. The withdrawal of one of the central actors, the state company of informatics, caused the service providing to be nearly closed, but some applications survived despite that. Their managers did not allow the discontinuity and kept the project in the public interest. So new associations are formed in the absence of central actors, even when the actors' area is not technology – this is an aspect worthy of further investigation.

From a theoretical standpoint, this study contributes to strategy, organizational analysis and (understanding of) mobile information systems. Has stated by OCDE (2011) it is crucial for the governments to understanding the underlying concepts and motivational factors that explain the emergence of m-government to formulate adequate policies and set priorities.

Suggestions for future research are:

1. To replicate the study in different governments in Brazil and abroad;
2. To continue this longitudinal study, looking at how this project evolves in time;
3. To develop a study to assess and compare other enterprise technology projects;
4. To advance the study of the proposition of an mGov business model.

Albeit obvious, it is worth mentioning that corporative projects can be leveraged or discontinued by changes in government leadership. Actors who are technically responsible for these projects should be prepared for governmental changes.

It is not common, in academic research, to report on "the day after." However, it is interesting to note that the project of using mobile technology in Paraná State Government had a new thrust, "pushed" by the dissemination of knowledge and information acquired from and through this research over the past 15 years.

This article concludes with three questions. The first: Why is SMS, though an old and simple *technology, still used by governments (little used, actually), while ignored in academic research?* The second question is based on Molina-Castillo & Meroño-Cerdan's (2014) recently published research in IJESMA: *Are firms and industry aware of the drivers that empower consumers to use these kinds of applications?* Finally: *Is a lack of a business model the cause of the low use of SMS for delivery of government services?* (Lanza & Cunha, 2012).

REFERENCES

Adams, C., & Mouatt, S. (2010). Evolution of Electronic and Mobile Business and Services: Government Support for E/M-Payment Systems. *International Journal of E-Services and Mobile Applications*, 2(2), 58–0. doi:10.4018/jesma.2010040104

Atkinson, R., & Flint, J. (2001). Accessing hidden and hard-to-reach populations: snowball research strategies. *Social Research Update*, (33). Retrieved November 03, 2012, from http://sru.soc.surrey.ac.uk/SRU33.html

Brandão, C. A. L. (2008). A transdisciplinaridade. In *A transdisciplinaridade e os desafios contemporâneos* (pp. 17–39). Belo Horizonte, Minas Gerais: IEAT/UFMG.

Bresser-Pereira, L. C. (1998). A Reforma do Estado dos anos 90: lógica e mecanismos de controle. In CLAD (Ed.), *Centro Latinoamericano de Administración para el Desarrollo*. Retrieved December 17, 2015, from http://old.clad.org/congresos/congresos-anteriores/ii-isla-de-margarita-1997/a-reforma-do-estado-dos-anos-90-logica-e-mecanismos-decontrole/?searchterm=A%20Reforma%20do%20Estado%20dos%20anos%2090

Caballero, E. G. (2005). Pluralidad teórica, metodológica y técnica en el abordaje delas redes sociales: Hacia la "hibridación" disciplinaria. *Revista Hispana para el Análisis de Redes Sociales*, 1(9), 1–24.

Chen, W., & Hirscheim, R. (2004). A paradigmatic and methodological examination of information systems research from 1991 to 2001. *Information Systems Journal*, 14(3), 197–35. doi:10.1111/j.1365-2575.2004.00173.x

Comitê Gestor de Internet do Brasil. (2014). Retrieved November 7, 2014, from http://www.cgi.br

CONIP SP. (2014). *Congresso de Inovação e Informática na Gestão Pública*. Retrieved November 02, 2014, from http://www.conipsp.com

Costa, A. de S. M. da, Barros, D. F., & Martins, P. E. M. (2010). Perspectiva histórica em administração: novos objetos, novos problemas, novas abordagens. *RAE – Revista Administração de Empresas*, 50(3), 288-299. doi: 10.1590/S0034-75902010000300005

Cunha, M. A., Anneberg, R. M., & Agune, R. (2007). Prestação de serviços públicos eletrônicos ao cidadão. In P. T. Knight, C. C. C. Fernandes, & M. A. Cunha (Eds.), E-desenvolvimento no Brasil e no mundo: subsídios e Programa e-Brasil (pp. 559-584). São Caetano do Sul, São Paulo: Yendis.

Cunha, M. A., & Miranda, P. R. M. (2008). A pesquisa no uso e implicações sociais das tecnologias de informação e comunicação pelos governos no Brasil: uma proposta de agenda a partir da prática e da produção acadêmica nacional. In *Anais do XXXII ENANPAD*. Rio de Janeiro: ANPAD - Associação Nacional de Pós-Graduação e Pesquisa em Administração.

Diniz, E. H., Petrini, M., Barbosa, A. F., Monaco, H., & Christopoulos, T. (2006). Abordagens epistemológicas em pesquisas qualitativas: além do positivismo nas pesquisas na área de sistemas de informação. In *Anais do XXX ENANPAD*. Salvador, Bahia: ANPAD - Associação Nacional de Pós-Graduação e Pesquisa em Administração.

Diniz, V., & Gregório, A. (2007). Do e-gov governo eletrônico para o M-gov Cidadania Móvel. In E-desenvolvimento no Brasil e no mundo: subsídios e Programa e-Brasil (pp. 688-702). São Caetano do Sul, São Paulo: Yendis.

Fang, Z. (2002). E-Government in digital era: Concept, practice, and development. *International Journal of The Computer. The Internet and Management, 10*(2), 1–22.

Germanakos, P., Tsianos, N., Lekkas, Z., Belk, M., Mourlas, C., & Samaras, G. (2009). Human Factors as a Parameter for Improving Interface Usability and User Satisfaction. Academic Press.

Informações e Serviços de Telecomunicações Ltda. (2014). Retrieved November 1, 2014, from http://www.teleco.com.br

Islam, M. S., & Scupola, A. (2013). E-Service Research Trends in the Domain of E-Government: A Contemporary Study. In A. Scupola (Ed.), *Mobile Opportunities and Applications for E-Service Innovations* (pp. 152–169). Hershey, PA: Information Science Reference; doi:10.4018/978-1-4666-2654-6.ch009

Janssen, M. (2010). Electronic Intermediaries Managing and Orchestrating Organizational Networks Using E-Services. In Electronic Services: Concepts, Methodologies, Tools and Applications (pp. 1319-1333). Hershey, PA: Information Science Reference. doi:10.4018/978-1-61520-967-5.ch081

Jayo, M. (2010). *Correspondentes bancários como canal de distribuição de serviços financeiros: taxonomia, histórico, limites e potencialidades dos modelos de gestão de redes*. São Paulo: Tese de Doutorado em Administração, Escola de Administração de Empresas de São Paulo –Fundação Getúlio Vargas.

Kushchu, I., & Yu, B. (2004). *Evaluating mobility for citizens*. Niigata: mGovLab-Internacional University of Japan. Retrieved November 04, 2012, from http://www.mgovlab.org

Lakatos, E. M., & Marconi, M. A. (2007). *Metodologia científica*. São Paulo, São Paulo: Atlas.

Langley, A. (1999). Strategies for theorizing from process data. *Academy of Management Review, 24*(4), 691–710.

Lanza, B. B. B., & Cunha, M. A. (2012). Relations Among Governmental Project Actors: The case of Paraná mGov. *CONF-IRM 2012 Proceedings*. Retrieved from http://aisel.aisnet.org/confirm2012/64

Mason, R. O., McKenney, J. L., & Copeland, D. G. (1997). An historical method for MIS research: Steps and assumptions. *Management Information Systems Quarterly, 31*(3), 307–320. doi:10.2307/249499

Medeni, T. D., Medeni, I. T., & Balci, A. (2011). Proposing a Knowledge Amphora Model for Transition towards Mobile Government. *International Journal of E-Services and Mobile Applications, 3*(1), 17–38. doi:10.4018/jesma.2011010102

Molina-Castillo, F., & Meroño-Cerdan, A. (2014). Drivers of Mobile Application Acceptance by Consumers: A Meta Analytical Review. *International Journal of E-Services and Mobile Applications, 6*(3), 34–47. doi:10.4018/ijesma.2014070103

Myers, M. D. (1997). Qualitative research in information systems. *Management Information Systems Quarterly, 21*(2), 241–242. doi:10.2307/249422

Neustad, R. E., & May, E. R. (1996). *Thinking in time: the uses of history for decision makers.* New York: The Free Press.

O'Brien, J. A. (2004). *Sistemas de Informação e as decisões gerenciais na era da internet* (2nd ed.). São Paulo, São Paulo: Saraiva.

OECD/ITU. (2011). *M-Government: Mobile Technologies for Responsive Governments and Connected Societies.* Paris: OECD Publishing; doi:10.1787/9789264118706-en

Orlikowski, W., & Baroudi, J. J. (1991). Studying information technology in organizations: Research approaches and assumptions. *Information Systems Research, 2*(1), 1–28. doi:10.1287/isre.2.1.1

Otte, E. & Rousseau, R. (2002). Social network analysis: a powerful strategy, also for the information sciences. *Journal of Information Science, 28,* 441–453. doi: 10.1177/016555150202800601

Pereira, D. C., & Meirelles, M. R. G. (2009). Uma abordagem transdisciplinar do método Análise de Redes Sociais. *Informação & Informação, 14*(2), 84–99.

Powell, W. W. (1990). Neither market nor hierarchy: network forms of organization. In B. M. Staw & L. Cummings (Eds.), *Research in organizational behavior.* Greenwich, UK: JAI Press.

Richardson, R. J. (1989). *Pesquisa social: métodos e técnicas.* São Paulo, São Paulo: Atlas.

Riecken, R. & Lanza, B. B. B. (2007). E-Paraná: a rede de informações e serviços eletrônicos do governo do Estado do Paraná. *Informação & Informação, 12*(2).

Salim, A., & Wangusi, N. (2014). Mobile phone technology: an effective tool to fight corruption in Kenya. In *Proceedings of the 15th Annual International Conference on Digital Government Research* (pp.300-305). ACM. doi:10.1145/2612733.2612772

Trimi, S., & Sheng, H. (2008). Emerging Trends in M-Government. *Communications of the ACM, 51*(5), 53–58. doi:10.1145/1342327.1342338

Using the Ushahidi Platform to Monitor the Nigeria Elections. (2011). Retrieved November 29, 2015, from: https://www.ushahidi.com/blog/2011/03/30/using-the-ushahidi-platform-to-monitor-the-nigeria-elections-2011

Vizeu, F. (2010). Potencialidades da análise histórica nos estudos organizacionais brasileiros. *RAE - Revista de administração de Empresas, 50*(1), 37-47.

Walsham, G. (1995). The emergence of interpretivism in IS research. *Information Systems Research, 6*(4), 376–394. doi:10.1287/isre.6.4.376

Yin, R. K. (2005). Estudo de Caso: Planejamento e Métodos. 3a ed. Porto Alegre, Rio Grande do Sul. *The Bookman.*

ENDNOTES

[1]   Paraná State Sponsoring Agency

2   Association of sea ports of Paranaguá and Antonina

3   Paraná State Government Civil House

4   Paraná State Government Militar House

5   Guaíra Theater Cultural Center

6   Paraná Fueling Centers

7   Paraná Computing Company (a public computing company belonging to the State)

8   Paraná State Prison Department

9   Paraná Traffic Department

10   Paraná State Electronic Government

11   Paraná Agricultural Institute

12   Paraná State Identifying Institute

13   Paraná Board of Trade

14   State Police Department

15   Pension and Administration State Office

16   Social Communication State Office

17   Finance State Department

18   Special Department for Internal Affairs and Ombudsman

19   State office of Heath

20   State office of Public Safety

21   State office of Labor and Jobs

22   SENVIA Mobile Service broker (includes all operators in Brazil: ALGAR, BRT, CLARO, GLOBAL, NEXTEL, OI, SERCOMTEL, TIM and VIVO)

Chapter 7
Government and Mobile:
A Gear Change?

Wendy Li
The University of Melbourne, Australia

ABSTRACT

This chapter presents a comprehensive review of what has been examined in the past and what needs to be explored in the future concerning mobile government (m-government). As an emerging branch of mobile services, m-government intends to help governments to better serve for the public, the business and non-government organizations with the assistance of mobile technologies. Although m-government originated from electronic government (e-government), it is not just a simple extension in respect to technological developments but a transformation from e-government to m-government. What matters most and is worthy of being further enhanced in this revolutionary process is the improvement of government's mobility instead of up-to-date technologies. That is to say, the shift from e-government to m-government is a game change instead of a gear change.

INTRODUCTION

Mobile Government (m-government) refers to an emerging set of strategies conducted by public sectors using latest mobile and wireless communication technologies in order to achieve a more efficient and effective governance and serve better for the intended recipients (El-Kiki, & Lawrence, 2006). Therefore, government administrations equipped with mobile technological infrastructures such as mobile applications and portals on smart phones, portable PCs, and personal digital assistants(PDAs), together with 3G even 4G wireless networks are able to offer the public, the business and non-government organizations quick access to information and services through multi-channels (Kim, Yoon, Park, & Han, 2004; Chang & Kannan, 2002).

Mobile government arises not only as a major branch of mobile studies in the academia, but also as a more effective and efficient service model for governments. With the rapid adoption of mobile technologies, electronic government (e-government) evolves into mobile government (m-government) unsurprisingly. A higher penetration of mobile network and a wider application of Internet-based mo-

DOI: 10.4018/978-1-5225-0469-6.ch007

bile devices enable governments to improve their service delivery and increase the interaction between different objects, for instance, citizens, businesses and government employees. Hence, comparing to e-government, the future of m-government looks more promising, particularly in developing regions. A wide range of mobile services have been initiated in emergency management, election and voting, financial and banking businesses, police and agriculture services, etc. Yet, limitations of m-government still exist, such as the instability of mobile infrastructure and the questionable real capacity of improving two-way communication.

Most earlier studies on m-government focus on particular dimensions or m-government in a particular country. Few studies have offered a comprehensive review of what has been done and reveal what lies ahead for further studies. The current chapter fills the gap by summarizing earlier key studies and mapping new dimensions and directions for further studies.

The comprehensive and critical review is based on the search results generated by a combination of use of key words and the snowball method (Napoleon & Bhuiyan, 2010). A search using "mobile government" and "m-government" was conducted at Google Scholar search engine and the online database at the library of University of Melbourne. And the snowball method resulted in 53 most-cited articles.

STATE OF THE ART

At present, m-government is not only seen as the technical assistance to the governance (El-Kiki, & Lawrence, 2006), but also a new paradigm that can transcend its previous service delivery model (Song, 2005; Song & Cornford, 2006). It is believed that, as a competent method for reshaping governance, m-government is capable of bringing target-specific services to mobile audiences and enhancing the mobility of integrated governmental systems. Having transcending the mere potential of mobile technologies successfully, it is not just a supplement to e-government but an updated version with e-government as a basis (Song, 2005).

The current model of m-government contains four main types of relationship: G2C, G2B, G2G and G2E (Mengistu, Zo, & Rho, 2009). Here G refers to government, C refers to citizens, B refers to businesses, and E refers to employees in government organizations. M-government is being largely explored relying on this framework for it is basic yet inclusive. More details will be revealed below as a part of theoretical basis.

Characteristics and Advantages

Two distinctive characteristics of mobile technologies and thus of m-government are "mobility" and "wireless" (Trimi & Sheng, 2008). "Mobility" emancipates users from being tied to fixed locations and generates flexible channels of communication. "Wireless" renders data and information exchange to be freed from physical links even between digital devices. Therefore, instant and dependable access to m-government services stand out opposed to previous inflexible ways (Aloudat & Michael, 2011; Al-Hujran, 2012). Another essential characteristic of m-government is decentralization (Hellström, 2012). As m-government aims to improve democracy and become increasingly service-oriented, m-government is operated under the guidance of a decentralized approach.

Based on characteristics, several advantages of m-government services can be generalized as follow (Ntaliani, 2008). Ubiquity is generally acknowledged to be one as it is derived from mobility, which allows m-government services to be accessed anywhere any time. There is no doubt that previous lengthy office procedures will be accelerated with the help of mobile technologies so that time-saving becomes an obvious advantage. Instantaneity is another one because information is able to be delivered timely and efficiently under the framework of m-government compared to e-government.

Research Scope

M-government is involved in a wide range of research areas across multiple disciplines. Interactions between governments and citizens, businesses and government employees enable m-government to be established in various sectors in the society. The flexibility of m-government framework is required in order to meet demands of different business sectors (Ntaliani et al., 2008). M-government has been initiated in the field of health, education, public transport, agriculture, election and voting, banking and commerce, emergency and crisis management etc. (Maumbe, Owei & Taylor, 2007, p.212). Since each sector is beneficial to m-government with its unique strengths, the combination of m-government and services in a particular sector receive close attention. Intensive studies on m-transaction, m-police, m-democracy and so forth constantly emerge (Kuscu, Kushchu & Yu, 2007, p.5) with multidisciplinary research methods being put into practice. Based on the aforementioned operations, governments are expected to seek for cooperation with private sector partnership on m-government, and many more collaborations across sectors will be established in the future (Maumbe, Owei & Taylor, 2007, p.217).

KEY CONCEPTS AND THEMES

Since the 1990s, the increased use of Information and Communication Technologies (ICTs) has stimulated the public sector into providing latest information, delivering better services, engaging citizens actively and coordinating transactions seamlessly. This trend has developed into a common practice of governance currently known as Electronic Government (e-government) (Lee, Tan, & Trimi, 2005; Lee, Tan, & Trimi, 2006; Mengistu, Zo, & Rho, 2009; Trimi, & Sheng, 2008). Aiming at rendering more efficient and comprehensive services to citizens, businesses, government employees and governmental agencies, e-government has mainly adopted online portal based on wired network as the digital infrastructure.

As ICTs keeps advancing over times, mobile and wireless technologies appear on the stage with a wide use of mobile devices including cell phones, tablet PCs, laptops, PDAs etc. (Lee et al., 2006; Mengistu et al., 2009; Ntaliani, Costopoulou, & Karetsos, 2008; Trimi, & Sheng, 2008). Mobile Government (m-government) subsequently emerges as "the new frontier of service delivery" (Mengistu et al., 2009), mostly considered as the subset or the extension of e-government (Lee et al., 2006; Mengistu et al., 2009; Ntaliani et al., 2008). While e-government develops with its focus on wired Internet access (Trimi, & Sheng, 2008), m-government pays special attention to mobile-enabled services like M-notifications, M-votes and M-payments supported by mobile developments such as short message services (SMS), mobile applications and wireless unified portals (El-Kiki, & Lawrence, 2006).

Mobile Technologies and Infrastructures

Breakthroughs and innovations of ICTs is one of the main driving forces of the ever-increasing practices of m-government (Lee et al., 2006). In most developed countries, mobile technologies are widespread

(Kim et al., 2004), so it's natural for e-government to evolve into m-government. while in developing areas, although ICTs in developing countries are widely adopted to improve the governance, the lack of ICT infrastructures renders going mobile the only feasible choice (Hellström, 2012). Therefore, in developing regions, the adoption of mobile technologies is not as prevalent as in developed areas despite a high mobile penetration compared to the location-based network (Ghyasi & Kushchu, 2004; Al-Hujran, 2012). But there is no doubt that, the mobile infrastructures still need to be further developed and popularized in both regions.

Mobile devices, Internet, net applications and services all contribute to the constitution of mobile infrastructures (Kushchu & Kuscu, 2008). Mobile devices are mostly handled devices—internet-based mobile phones, laptops, PDAs etc.—that are portable and can be easy to use any where any time. In most cases, the mobile device is used as a platform for the spread of information, while its new function of being a communicative bridge between the government and citizens have been carried out. Naturally, short Message Service (SMS) is the most widely-used mobile technology worldwide (Amailef and Lu, 2008& 2010), together with Cell Broadcasting Service (CBS) (Aloudat & Michael, 2010; Aloudat, et al.,2014), in that the utilization of mobile phone is advantageous to mobile governance compared to limited web access. Therefore, text and voice messaging based on SMS technology is widely adopted even in East Africa (Hellström, 2012), in Arab countries (Al-Hujran, 2012), not less to say others. Multimedia Messaging Service (MMS) is an upgraded version of SMS, which is able to send message in multimedia forms.

Both SMS and MMS are functioning based on the second generation (2G) protocols known as Global System for Mobile Communication (GSM). And the third generation (3G) mobile internet, Universal Telecommunications Systems (UMTS) follows with Wideband Code Division Multiple Access (WCDMA) as the technological basis (Ntaliani et al., 2008). Not to mention 4G network is coming soon. Besides these general foundations, some other mechanisms have been applied in particular services of m-government. For instance, PIN (Personal Identification Number), infrared, RFID and Bluetooth are fully adopted to assist mobile commerce (Mallat et al., 2004).

M-Government Services and Practices

According to El-Kiki and Lawrence, m-government is applied to four main aspects in the public sector: M-communication, M-services, M-democracy and M-administration (2006; Amailef, & Lu, 2011). M-communication mainly provides information to citizens, while m-services supply the opportunity of two-way interaction for the government and citizens. M-democracy aims to promote citizens' democratic participation letting their voices to be heard, while m-administration intends to improve government's internal efficiency.

M-democracy in terms of elections have been closely examined in East Africa (Hellström, 2012). The political advertising, post-election monitoring and even riot-managing are successful applications in this field in Uganda, Kenya and Kampala. Hellström gives a summary of services that have been initiated by governments via cell phones *in East Africa*, including government information update on a daily basis, such as road safety and community news, law enforcement, citizens-to-government interaction, emergency management, health projects, employment and agricultural services etc. (Hellström, 2012).

The *M-service* as the most representative dimension of m-government covers a wide range of topics in various fields, which is going to be introduced next. The *emergency* service, including emergency management and crisis communication, is an important branch of m-services. Amailef and Lu establish a framework for mobile based emergency response system (MERS) which is able to deal with the traditional emergency response system's unsettled positioning problem (2008). MERS contains five application constituents and five databases are equipped with it in order to keep information updating (Amailef & Lu, 2008). They further develop this MERS into an upgraded intelligent system by aggregating an algorithm of text information retrieval (Amailef & Lu, 2011). What's more, case-based reasoning (CBR) combined with ontology approach are used to support MERS so as to facilitate interaction in emergency situations.

Another emergency system, a location-based national emergency warning system (NEWS), is constructed by Aloudat and Michael after the 2009 Victorian bushfires (2011). Even the idea of legislation on NEWS is introduced to meet the urgent demand of its deployment in Australia. User acceptance of location-based mobile emergency services have been further researched in order to put this system into wider practice (Aloudat, Michael, Chen & Al-Debei, 2014). While others focus on particular systems of mobile emergency, Moon reviews the historical development of emergency management and reveals several critical issues in m-government related to emergency services, such as security and interoperability, together with factors that will affect the implementation of these services (2010). The *m-police* service is proposed by Firoozy-Najafabadi and Pashazadeh as a new type of m-services to improve the efficiency of communication not only within police systems but also between police officers and citizens (2011). M-police services

Mobile *financial* services, among many other promising m-services, have been examined by Mallat, Rossi & Tuunainen as early as in 2004. Mobile payments and mobile banking are two main applications of m-commerce. While a broad range of mobile financial services have arisen, they haven't been fully adopted due to the immaturity of technological infrastructures. Therefore, a variety of services offered by different key players are expected to be available in the near future (Mallat et al., 2004).

M-government in other business cases also have received attention—*mobile agricultural* service as a subset has been explored by Ntaliani and his colleagues (2008). Deploying a set of cost-effective mobile services to prompt the communication between agricultural organizations and producers is proposed and practiced based on a case study of organic agriculture in Greece.

Despite the aforementioned literature specialized in particular services of m-government, a generic management framework for m-governance services has been proposed by El-Kiki, Lawrence and Steele so as to help m-government to get maximized profits and avoid risky challenges (2005). Later, a generic service platform designed to fit all m-government services based on a business management model is initiated by de Reuver, Stein & Hampe (2013). The latter one intend to promote mobile participation while the former attempt to improve the government management.

Furthermore, *strengths and weaknesses* of mobile services have been summarized by Ishmatova & Obi (2009). In general, mobile devices' portability and use of ease contribute to the widespread adoption of m-government services, but connectivity and security of m-government services are still left to be settled. Since different geographical areas have different social and economic contexts, the development of m-government services is unbalanced around the world (Lee et al., 2006). The industrialized countries are generally more innovative in generating new m-government services (Lee et al., 2006), while the deploying of m-government services is considered to be more beneficial to the developing countries compared to the developed one (Abdelghaffar & Magdy, 2012). Yet m-government is still immature no matter in leading countries or *in developing regions* (Mengistu, 2009) as it is still in its infancy and there is still a long way to go (Al-Hujran, 2012).

User-Centric Research

Since one of mobile government's ultimate goals is to provide better services for citizens, a suggestion, that the application of m-government should also be examined from users' viewpoint, has been put forward. Citizen acceptance is essential to m-government's development because user's response is an index to test whether m-government is on the way towards a service-centric as well as a user-centric paradigm.

Citizen Acceptance

As m-government develops over time, Rossel, Finger, & Misuraca suggest that there is a need to evaluate the performance of it in real-life situations so as to correct its focus regularly (2006). They mapped m-government with four types of initiatives based on different users: mobile e-government for mobile persons, mobile services for mobile organizations, mobile state, and mobile administration agents or agencies (Rossel et al., 2006). Then, each initiative will be examined regularly to see whether their goals have been achieved according to user's evaluations. And in the process of examination, priorities should be given to the characteristics of flexibility, easiness and efficiency that are vital to user experiences over others so as to see how well citizens have accepted (Rossel et al., 2006).

The concept of user needs reflected by customer desired value is put forwarded by Ishmatova & Obi (2009). According to their study, two dimensions of value—added value represented by mobility, as well as value in use represented by content feasibility and qualitative requirements are emphasized. It is believed that a better understanding on user value helps to reveal user needs and improve the applicability of m-government. Not only generalizations, but also empirical studies have been made.

How m-government is implemented in Malaysia has been researched by identifying user reactions to m-government services through focus group method (Al Thunibat et al, 2011a). It has been found that Malaysians are willing to adopt m-government services but several aspects of m-government services need to be improved, such as accelerating the access speed and decreasing the cost. Generally, users in Malaysia welcome the implementation of m-government and expect m-government services to be further expanded.

Hung, Chang and Kuo also mention that user acceptance will be affected by the effectiveness of m-government service delivery, which depends on both aspects of mobile service, like IT mechanisms, and non-mobile service, for instance, perceived usefulness and perceived use of ease (2013). These aspects noticed by different scholars gradually evolve into factors as a main topic received widespread research interest.

CSFs—Critical Success Factors

As mentioned above, factors affecting the implementation of m-government services is receiving increased attention. Al-Khamayseh and Lawrence first analyze and classify those factors from different perspectives and into three groups: technological, human and socio-economic, and organizational factors (2006). Then, Al-Khamayseh, Lawrence, & Zmijewska propose the initial model of m-government success factors including 18 items listed in their findings as a result of an online survey targeting experts in m-governance around the world (2006a). In Al-Hadidi and Rezgui's literature review on this topic, they name m-government success factors the "Critical Success Factors" (CSFs), and introduce this concept into the field based on a summary of five previous studies in different cultural contexts (2009).

Researchers mentioned above hold the global view, while others intend to perform region-specific analysis. Al Thunibat, Zin, & Sahari conduct their survey by issuing questionnaires to Malaysian communities with the objective of constructing a framework of m-government acceptance factors chiefly on the basis of technology acceptance model (TAM) (2011b). Unlike previous comprehensive studies, Al-Hujran simplifies those indices/ indicators and narrow down them to 5 main factors affecting m-government implementation based on interviews and qualitative data analysis in Jordan (2012). Later on, Sareen, Punia, & Chanana carry out an online research according to Al-Khamayseh and his colleagues' survey in the context of India to meet the need of exploring m-governance development at local (2013).

In order to have a clearer view on CSFs, a comparison has been made as listed below.

The authors and their literature's publication years, as representations of their work, are arranged in chronological order in the first row, while factors are ordered in two ways: in terms of the percentage of responses in Al Thunibat et al., Al Thunibat et al. and Sareen et al.'s cases, and in terms of the authors' original ranking of importance in other two cases.

As is shown in the table, there is no doubt that Al-Khamayseh et al.'s study, together with Sareen et al.'s one, have been quite thorough. Nevertheless, some slight modifications are still needed. From author's point of view, citizen's Trust and government's capability of protecting Privacy and providing Security are actually a coin's two sides, so these two factors should be classified into one item—User Considerations—as named by Al-Hadidi and Rezgui (2009). As the availability of e-government services has been considered as the prerequisite of m-government, "E-government" can be included in "Infrastructure". So can "Infrastructure Management" and "Mobile Penetration", for they are parts of the overall picture of mobile infrastructure. Transformation of Culture to some extent belongs to citizen's Acceptance, in that the former item refers to government's duty to guide users on mobile use, whereas the latter one refers to users' reaction to mobile use, including their reaction to this guidance.

What we can conclude directly from the comparison between Al-Khamayseh et al.'s survey in a worldwide view and Sareen et al.'s one in the context of India is that, some generic items of CSFs may be common across the world, but the importance of each differ in different areas. In other words, each CSF influences particular regions to varying degrees. Therefore, it's difficult to suggest an "one size fits all" ranking of CSFs by reviewing previous cases. Further researches in various geographical and cultural contexts should be empirically conducted (Table 1).

In order to achieve a better understanding of m-government, several models of m-government's development have been generalized from its practices. As m-government has long been considered as a subset of e-government, its framework is naturally originated from e-government's initial one. Layne and Lee initiates a four-stage growth model for e-government composed of cataloguing, transaction, vertical integration and horizontal integration, with specific transformative developments occurring in each stage when traditional governmental structure undergoes transition to e-government (Layne, Karen, & Lee, 2001; El-Kiki, & Lawrence, 2006).

According to the diagram, e-government is experiencing an evolutionary process, as the degree of complexity and integration goes up from cataloguing to horizontal integration. The authors provide a detailed explanation to each stage's definition, functions and challenges with technological and organizational issues. In the first two stages, from simply posting information on web portals to connecting government's intranet to online interfaces with online-transaction accomplished, the government becomes electronic technically. Driven by external forces such as citizens' demands and societal changes, the preliminary e-government aims at achieving the "one-stop-transaction" objective across not only

Table 1. Key theories and methods

CSFs	Al-Khamayseh, Lawrence, & Zmijewska, 2006a	Al-Hadidi & Rezgui, 2009	Al Thunibat, Zin, & Sahari, 2011b	Al-Hujran, 2012	Sareen, Punia, & Chanana, 2013
1	**Privacy & Security**	E-gov's Vision & Strategy	Social Influence	Public Awareness	E-gov
2	**Infrastructure**	Leadership & Support	Service Quality	**Trust**	Strategy
3	User Needs	ICT Infrastructure & Mobile Penetration	Perceived Usefulness	Cost	M-gov Awareness
4	Quality	Transformation of Culture	Perceived Risk	Infrastructure Constraints	User's Access
5	E-gov	Human Resource Management & Training/ ICT & Mobile Literacy	Cost of Service	Legal Framework	Quality
6	Acceptance	Inter- and Intra Organization Integration	Perceived Compatibility		Framework
7	Cost	E-laws	**Trust** in Government		Cost
8	Standards	**User Considerations** (Requirements, Trust, Privacy & Security)	**Trust** in Technology		User Needs
9	M-gov Framework	E-readiness & Marketing	Attitude towards Use		IT Literacy
10	**Mobile Penetration**	Funding			Privacy & Security
11	**Infrastructure Management**				Infrastructure
12	Awareness				Acceptance
13	Access				Mobile Penetration
14	Strategy				Standards
15	IT Literacy				Legal Issues
16	Portals & Gateways				Private Partnerships
17	Private Partnerships				Infrastructure Management
18	Legal Issues				Portals & Gateways

governmental levels vertically but also various functions and services horizontally in the next two stages (Layne et al., 2001; El-Kiki, & Lawrence, 2006).

Several issues—universal access, citizen's privacy and appropriate execution of e-governmental policy—that e-government needs to be considered are mentioned as a conclusion (Layne et al., 2001). The suggestion of offering "Internet access through public terminals" implicitly leads to the necessity of introducing mobile technologies into improving the current model of e-government in its near future.

Although the model is specifically based on the United States' governmental constitution with multiple levels, namely federal, state and local (Layne & Lee, 2001), the whole set of stages represents the developmental process of a particular level rather than all. Therefore, it is a model of universal applicability for the majority of e-governments in their preliminary stages.

Moon further develops this model into five stages based on the analysis of the 2000 E-government survey designed to examine e-government's implementation (2002; El-Kiki, & Lawrence, 2006). The fifth stage, political participation, is added and categorized as the political function while previous four stages are categorized into administrative functions. Meanwhile, e-government's practices are classified into internal and external ones according to the relationship between government and its service object, with examples given in each of five stages. Comparing to the four-stage model, this one exemplifies more specific and advanced technologies adopted in the developmental process—bulletin boards, electronic data interchange, digital signature, etc. (Moon, 2002). Moon also names the second stage as "two-way communication", laying stress on its interactive mode between government and its service objects. Thus, the interactive service delivery is highlighted thereafter, together with the demand of more sophisticated technologies for interactive communications (Moon, 2002).

In Lee, Tan and Trimi's study on the practices of leading e-government countries, they set down the fundamental types of e-government practices on the basis of an America's statute, the E-Government Act of 2002 (2005). According to their classification, the five categories are government to citizens (G2C), government to businesses (G2B), government to government (G2G), government internal efficiency and effectiveness (IEE), and overarching infrastructure (Cross-cutting). This categorization presents the relationship between government and its stakeholders both clearly and comprehensively.

Lee and his colleagues further narrow down and apply this model to m-government combined with three types of m-government's services—government's notification, user's access and transaction (Lee et al., 2006). Trimi and Sheng adopt and adapt it for their analysis of trends in m-government with mobile applications, SMS, Wi-Fi and mobile maps, exemplified (2008). Hence, m-government's interactive communication mode stands out as mobile technologies progress. Since then, a framework of m-government, categorized into G2C, G2B, G2G and IEE, have formed. In the following studies, Ntaliani et al. and Mengistu et al. continue to focus on the government's operations from these aspects, and change IEE into government to employees (G2E) (2008; 2009).

Another view put forward by Song on understanding m-government is that the overall mobility of the government should be taken into consideration besides mobile infrastructures (Song, 2005; Song and Cornford, 2006). As the mobile network enabled by mobile technologies is considered to have bridged the gap between people in the reality through the virtual connectivity, this prevailing trend is also to government's benefit. The increasing mobility brought by the burgeoning development of ubiquitous computing is able to result in a paradigm shift of government service delivery from manufacturing mentality to service mentality (Song and Cornford, 2006).

Being more action-oriented, responsive and individual-targeted in service delivery, the government's fluidity of interaction with its service objects is enhanced and the concept of the distinct fluid organization is introduced. Based on a case of mobile government in Beijing, "Social topology, ICT and Government Service Delivery Model" is proposed to enlighten the future development of mobile government. This model suggests that we should not only pay attention to the effective use of mobile technologies, but also be aware of the importance of the interaction paradigm shift of mobile government. Because in being mobile, the reshaping of government itself weighs more than the mere potential of mobile technologies (Song, 2005; Song and Cornford, 2006).

A wide range of research methods have been adopted including taking advantage of secondary data and statistics, carrying out online surveys and face-to-face interviews, hosting meetings and focus group discussions with key stakeholders and so on (Hellström, 2012). It's easy to see that both empirical evidences and theoretical foundations are put into practice in the studies on m-government.

Online web-based survey is the most common choice among all the above-mentioned methods, attracting close attention from scholars (Hung et al., 2013;). Case studies and fieldwork have been not only conducted but also reviewed in order to provide more detailed insights into the field of m-government (Rossel et al., 2006).

Sometimes, both quantitative and qualitative methods are closely interwoven with the former laying a statistical foundation and the latter offering specific analysis (Al-Khamayseh & Lawrence, 2006). Several notable studies can be cited as illustrations: online survey is combined with qualitative replies (Al-Khamayseh & Lawrence, 2006); data collected from the focus group method receive qualitative analysis (Al Thunibat et al., 2011a); fieldwork conducted to support the desk-based research (Hellström, 2012). To be specific, a quite comprehensive example recorded by Al-Hujran is the case study conducted through semi-structured interviews combined with qualitative research (2012). The interviews were done with 6 government officials who were in charge of m-government related issues in Jordan, and were recorded, decoded and interpreted by using software for qualitative data analysis accordingly. All in all, the research methods that have been adopted are all effective and innovative.

KEY FINDINGS AND ACHIEVEMENTS

As an evolutionary subset of e-government, m-government has been beneficial to people's daily life from several aspects.

The efficiency and effectiveness of governments' service delivery has been greatly enhanced. With the support of mobile technologies, citizens are able to get access to m-government services more conveniently through multi-channels instead of fixed portals, and governments are becoming advantageous to deal with crisis forecast and emergency management. The productivity of government employees also gets improved through simplified office procedures and accelerated logistics (Lee et al, 2006).

What's more, democracy has been promoted as a result of increased interaction between citizens and governments. Citizens take advantage of up-to-date mobile technologies to express their opinions freely, which can hardly be achieved in the past (Lee et al, 2006). As governments are becoming service-oriented and decentralized approach is adopted among the majority of m-government applications, citizens' opinions have been increasingly treasured (Hellström, 2012).

The digital divide resulted from unbalanced internet access of land-based networks can be partially reduced by adopting mobile devices for m-government services. The high penetration and overall stability of mobile technologies have rendered more information and services of m-government to citizens (Mallat et al., 2004; Lee et al, 2006; Shareef, Archer & Dwivedi, 2012).

The great proliferation of mobile technologies ensures further implementation and application of m-government, especially their use in special sectors, such as m-communication, m-services, m-democracy and m-administration (Shareef et al., 2012).

As mobile devices are mostly individually-possessed and more personal compared to office computer or other ICTs a generation ago, users tend to have positive experiences of adopting m-government services based on mobile technologies (Mallat et al., 2004).

User is definitely the necessary link in the whole chain of m-government operation, without which the development objectives of m-government are unable to be achieved. Therefore, user readiness and user willingness have to be taken into consideration (Roggenkamp, 2004). Users who first make use of m-government services should be exposed to certain guidance in order to avoid their misperceptions towards m-government. They also should be educated to gradually adapt themselves to the new conditions of m-government (Antovski & Gusev, 2005).

KEY WEAKNESSES AND PROBLEMS

Although m-government has already been established and extended around the globe, it still has limitations and the great potential is awaited to be fulfilled.

The m-government infrastructure should be improved and perfected further as the adoption of m-government services hasn't come up to expectation, which refers to several layers of meaning here. At present, part of m-government services fails to be provided for citizens in public sectors in that their technological basis, mobile architecture, hasn't been properly integrated to computation and networking (Al-Khamayseh et al., 2006a). The lack of standards of mobile systems resulted from diversified and therefore mismatched digital platforms has affected the the interoperability of m-government (Lee et al, 2006; Al-Khamayseh & Lawrence, 2006). And the usability of mobile devices is inherently limited by their own poor computational memory and short battery life (Trimi & Sheng, 2008; Al Thunibat et al., 2010)

Security and privacy are the biggest threats m-government faced, which badly influence the acceptance and diffusion of m-government services (Chang & Kannan, 2003; Al-Khamayseh & Lawrence, 2006; Trimi & Sheng, 2008). The wireless network is vulnerable to cyber attack and signal interception may happen because technological platforms are provided by private service suppliers (kim et al., 2004; Al-Khamayseh & Lawrence, 2006). The theft or lost of devices may result in data leakages or breaches, which concerns people who pay special attentions to personal information protection when they are about to adopt m-government services (Al-Khamayseh & Lawrence, 2006; Trimi & Sheng, 2008).

User readiness also concerns m-government initiatives in terms of two aspects: technological readiness and psychological readiness. The former refers to the capability of users getting access to mobile devices and networks as well as users' level of competence in managing mobile technologies (Lee et al, 2006). The latter means whether users are willing to interact with m-government through mobile platforms, in which uses' trust in mobile technologies and m-government is included (Lee et al, 2006; Al Thunibat et al., 2010).

Information overload and high cost are involved in users' concerns. Excessive or even useless information easily upsets users, not to mention increasing spamming and hoaxing (Carroll, 2005; Lee et al, 2006). Due to the limited adoption of wireless networks, the cost of accessing services is prohibitive, which may lead to users' rejections of m-government (Al-Khamayseh, & Lawrence, 2006; Al-Khamayseh et al., 2006a; Trimi & Sheng, 2008)

The future development of m-government in developing countries is even more challenging. Although there is a wide mobile penetration, the failure rate of m-government is high as well. In these regions, for example, in East Africa most citizens scratch a living through non-public sectors that are out of local government's control (Hellström, 2012). Therefore, how to empower people who live in grinding poverty in real ways and how to guide them to influence political decisions by using m-government are demanding.

NEW CONCEPTS AND THEMES

There is no doubt that the importance of user perspectives and the potential of citizens in the developmental process of m-government have become well aware. But previous researches on user-centric topics are far from comprehensive. Hence, in the next step, more detailed analyses of m-government from user's standpoint should be conducted. The case study carried out by Al Thunibat and his colleagues shows that there is a high awareness of m-government yet the actual rate of utilization is rather low, which means that users haven't been well-prepared to use m-government services (2010). It can not be resolved by simply assessing user requirements—other themes related to user also need to be evaluated.

User requirement is actually based on user readiness and user willingness in that only when users have prepared themselves for m-government adoption will they make requests. User readiness, further rely on positive user experiences and user perceptions (Roggenkamp, 2004). As mentioned before, users need to be guided so as to get accustomed to and feel comfortable with mobile technologies (Antovski & Gusev, 2005). Only then will users be willing to adopt m-government psychologically and intend to accept m-government services. Besides, all these are changing as m-government based on digital technologies is developing rapidly. Therefore, how to successfully assess user's changing need as a moving target as well as an open-ended question is worthy of researching.

In order to achieve this goal, the understanding of particular contexts of the application of mobile technologies, which refer to how, when and where mobile technologies are used, are needed (Carroll, 2005). Hence, how to identify appropriate contexts and inspect changes of "a portfolio of technology" promptly should be further studied (Carroll, 2005).

In addition to technology, the government complementary policy and program should render assistance in order to facilitate greater development of m-government (Ojo, Janowski & Awotwi, 2013).

Another trend of m-government we should concern beyond the initial implementation is sustainability (Hellström, 2012). How to push forward the continuance use of m-government (Wang, 2014) and how to bring m-government up to date level by level (Al Thunibat et al., 2011a; Al-Hujran, 2012) deserve careful consideration. Here, not only users but also workers within organizations, including employees of both governments and businesses, should be educated to cope with and catch up with changes (Borucki, Arat & Kushchu, 2005).

NEW RESEARCH AREAS

In terms of technological issues, more customized services with the use of the combination of personalization and location-based techniques has been proved to be a possible method to improve the efficiency of m-government (Al-Khamayseh & Lawrence, 2006b). As this is a promising research direction, what exact impacts of these techniques have on m-government services and how to apply them in practice should be investigated next. Despite advancing services by equipping with latest technologies, m-government should also take actions to protect users' privacy from being violated by these technologies (Al Thunibat et al., 2010). Therefore, there is a need to introduce specialized regulations on security of m-government or new mechanisms for helping users to deal with information security (Misuraca, 2009).

In terms of organizational structures, business management models can be popularized throughout the field of m-government (Misuraca, 2009). Management analysis from business perspectives, being able to reveal the relationship of m-government and its stakeholders, at first originates from e-commerce and has been largely adopted in m-commerce (Mallat et al., 2004; Jahanshahi, Khaksar, Yaghoobi & Nawaser, 2011). It has been found that not only external business relations but also the added value on organizations are easier to be captured, which is able to shed lights on how to improve m-government further (Carroll, 2006). Another direction of research within this realm is to seek how to promote public-private partnerships (Misuraca, 2009). The cooperation of private and public sectors in supplying m-government services is of critical importance particularly in developing regions. Because the majority of citizens in countries survive through informal sectors that are out of government's control, which means that they are out of reach of common m-government services tailored to users in developed areas (Hellström, 2012). Therefore, it is quite urgent to gap the bridge and then strengthen the cooperation between private and public sectors so as to advance m-government.

NEW THEORIES AND METHODS

Since m-government has already been implemented in wide areas for a while, it's the turn to upgrade its theoretical framework in order to help it to move to a higher level. Adaptive m-government framework has been put forward to further expand the concept of flexibility in m-government (Hassan, Jaber & Hamdan, 2009). It states that not only particular sectors in m-government need adapted service analyses but also individual users need adapted service contents. Content adaptation as a critical step is added to previous data transfer process so as to tailor m-government services according to each user's personal preferences and then get presented in different types. In this way, this adaptive model has the potential to improve the efficiency and effectiveness of m-government by delivering personalized services.

Another model of m-government built on an integration of value management theories provide us some illuminations as well (Wang, Feng, Fang & Lu, 2012). Based on value-focused theories associated with service science and transaction cost economics in the public management, a value-drivers model is integrated and constructed to explore how and what value is created in m-government. This research further inspires us to introduce theories in other fields to and integrate with m-government, unlike extending e-government framework into m-government in the past (Georgiadis, & Stiakakis, 2010).

SWOT analysis and other business management methods integrated to m-government should therefore be highlighted (Jahanshahi et al., 2011). Through SWOT analysis, strengths, weaknesses, opportunities and threats of particular services in m-government can be obviously seen. Other methods in business management also have this advantage of being intuitive compared to previous ones adopted in m-government researches. As m-government is an emerging area, Al Thunibat and his colleagues point out that research questions should be open-ended and unguided by a particular theory so as to have more space for innovation (2011a). At the same time, quantitative and qualitative approaches should be combined together in order to validate each research's result.

CONCLUSION

Mobile government is embarking on a period of exponential growth based on the rapid development of information and communication technologies. Despite depending on technological changes, m-government shouldn't be considered merely as the subset of electronic government any more. Rather, we should think beyond technology and reflect on the reshaping of government itself (Al Thunibat et al.,

2010; Song, 2005; Song & Cornford, 2006). In the next stage, the mobility of m-government instead of mobile technological infrastructures deserves more special attentions in order to bring m-government up to date. Therefore, how to optimize the mobility of m-government and how to increase the interaction between different stakeholders of m-government should be further studied.

Although a variety of m-government's services have been examined from the past decade to the present, still more detailed and indepth studies should be given to a broader scope with more cross-disciplinary and multi-disciplinary research methods adopted. As m-government continues to move ahead refreshingly, there is an ongoing need to keep track of where and how it goes. More up-to-date information should be recorded and added to m-government researches as new trends immediately. In this way, studies on m-government will be enriched and become increasingly diversified as m-government further expands. All in all, it is not a gear change but a game change from e-government to m-government. The mobility of the government deserves more attention than mobile technological developments in the future.

REFERENCES

Abdelghaffar, H., & Magdy, Y. (2012). The adoption of mobile government services in developing countries: The case of Egypt. *International Journal of Information*, 2(4), 333–341.

Al-Hadidi, A., & Rezgui, Y. (2009). Critical success factors for the adoption and diffusion of m-government services: A literature review. In *9th European Conference on e-Government* (pp. 21-28). London, UK: Academic Publishing Limited.

Al-Hujran, O. (2012). Toward the utilization of m-Government services in developing countries: A qualitative investigation. *International Journal of Business and Social Science*, 3(5), 155–160.

Al-Khamayseh, S., Hujran, O., Aloudat, A., & Lawrence, E. (2006b). Intelligent m-government: application of personalisation and location awareness techniques. In *Second European Conference on Mobile Government*.

Al-Khamayseh, S., & Lawrence, E. (2006). Towards citizen centric mobile government services: A roadmap. *CollECTeR Europe*, 2006, 129.

Al-Khamayseh, S., Lawrence, E., & Zmijewska, A. (2006a). Towards understanding success factors in interactive mobile government. In Proceedings of Euro mGov (pp. 3-5).

Al Thunibat, A., Zin, N. A. M., & Sahari, N. (2011a). Identifying user requirements of mobile government services in Malaysia using focus group method. *Journal of e-Government Studies and Best Practices,2011*, 1-14.

Al Thunibat, A., Zin, N. A. M., & Sahari, N. (2011b). Modelling the factors that influence mobile government services acceptance. *African Journal of Business Management*, 5(34), 13030.

Al Thunibat, A. A., Zin, N. A. M., & Ashaari, N. S. (2010, June). Mobile government services in Malaysia: Challenges and opportunities. In *Information Technology (ITSim), 2010 International Symposium in* (Vol. 3, pp. 1244-1249). IEEE.

Aloudat, A., & Michael, K. (2011). Toward the regulation of ubiquitous mobile government: A case study on location-based emergency services in Australia. *Electronic Commerce Research, 11*(1), 31–74. doi:10.1007/s10660-010-9070-0

Aloudat, A., Michael, K., Chen, X., & Al-Debei, M. M. (2014). Social acceptance of location-based mobile government services for emergency management. *Telematics and Informatics, 31*(1), 153–171. doi:10.1016/j.tele.2013.02.002

Amailef, K., & Lu, J. (2008, November). m-Government: A framework of mobile-based emergency response systems. In *Intelligent System and Knowledge Engineering, 2008. ISKE 2008. 3rd International Conference on* (Vol. 1, pp. 1398-1403). IEEE.

Amailef, K., & Lu, J. (2011). A mobile-based emergency response system for intelligent m-government services. *Journal of Enterprise Information Management, 24*(4), 338–359. doi:10.1108/17410391111148585

Antovski, L., & Gusev, M. (2005, July). M-government framework. In Euro mGov (Vol. 2005, pp. 36-44).

Borucki, C., Arat, S., & Kushchu, I. (2005, July). Mobile government and organizational effectiveness. In *Proceedings of the First European Conference on Mobile Government, Brighton, UK:Mobile Government Consortium* (pp. 56-66).

Carroll, J. (2005, July). Risky Business: Will Citizens Accept M-government in the Long Term?'. In Euro mGov (pp. 77-87).

Carroll, J. (2006). 'What's in It for Me?': Taking M-Government to the People. *BLED 2006 Proceedings*, 49.

Chang, A. M., & Kannan, P. K. (2002). Preparing for wireless and mobile technologies in government. *E-government*, 345-393.

de Reuver, M., Stein, S., & Hampe, J. F. (2013). From eParticipation to mobile participation: Designing a service platform and business model for mobile participation. *Information Polity, 18*(1), 57–73.

El Kiki, T., & Lawrence, E. (2006, April). Government as a mobile enterprise: real-time, ubiquitous government. In *Information Technology: New Generations, 2006. ITNG 2006. Third International Conference on* (pp. 320-327). IEEE.

El-Kiki, T., Lawrence, E., & Steele, R. (2005). A management framework for mobile government services. *Proceedings of CollECTeR, Sydney, Australia*, 2009-4.

Firoozy-Najafabadi, H. R., & Pashazadeh, S. (2011, October). Mobile police service in mobile government. In *Application of Information and Communication Technologies (AICT), 2011 5th International Conference on* (pp. 1-5). IEEE. doi:10.1109/ICAICT.2011.6110902

Georgiadis, C. K., & Stiakakis, E. (2010, June). Extending an e-Government Service Measurement Framework to m-Governement Services. In *Mobile Business and 2010 Ninth Global Mobility Roundtable (ICMB-GMR), 2010 Ninth International Conference on* (pp. 432-439). IEEE. doi:10.1109/ICMB-GMR.2010.31

Ghyasi, F., & Kushchu, I. (2004). *Uses of mobile government in developing countries*. Retrieved from http://www.movlab.org

Hassan, M., Jaber, T., & Hamdan, Z. (2009, November). Adaptive mobile-government framework. In *Proceedings of the International Conference on Administrative Development: Towards Excellence in Public Sector Performance*.

Hellström, J. (2012). Mobile Governance: Applications, Challenges and Scaling-up. In M. Poblet (Ed.), *Mobile Technologies for Conflict Management: Online Dispute Resolution, Governance*. Participation.

Hung, S. Y., Chang, C. M., & Kuo, S. R. (2013). User acceptance of mobile e-government services: An empirical study. *Government Information Quarterly*, *30*(1), 33–44. doi:10.1016/j.giq.2012.07.008

Ishmatova, D., & Obi, T. (2009). m-government services: User needs and value. *I-WAYS-The Journal of E-Government Policy and Regulation*, *32*(1), 39–46.

Jahanshahi, A. A., Khaksar, S. M. S., Yaghoobi, N. M., & Nawaser, K. (2011). Comprehensive model of mobile government in Iran. *Indian Journal of Science and Technology*, *4*(9), 1188–1197.

Kim, Y., Yoon, J., Park, S., & Han, J. (2004). Architecture for implementing the mobile government services in Korea. In *Conceptual modeling for advanced application domains* (pp. 601–612). Springer Berlin Heidelberg.

Kuscu, H., Kushchu, I., & Yu, B. (2007) Introducing Mobile Government. In I. Kushchu (Ed.), *Mobile Government: An Emerging Direction in E-government* (pp.1-11). IGI Global.

Kushchu, I., & Kuscu, H. (2003, July). From E-government to M-government: Facing the Inevitable. In *the 3rd European Conference on e-Government* (pp. 253-260). MCIL Trinity College Dublin Ireland.

Layne, K., & Lee, J. (2001). Developing fully functional E-government: A four stage model. *Government Information Quarterly*, *18*(2), 122–136. doi:10.1016/S0740-624X(01)00066-1

Lee, S. M., Tan, X., & Trimi, S. (2005). Current practices of leading e-government countries. *Communications of the ACM*, *48*(10), 99–104. doi:10.1145/1089107.1089112

Lee, S. M., Tan, X., & Trimi, S. (2006). M-government, from rhetoric to reality: Learning from leading countries. *Electronic Government. International Journal (Toronto, Ont.)*, *3*(2), 113–126.

Mallat, N., Rossi, M., & Tuunainen, V. K. (2004). Mobile banking services. *Communications of the ACM*, *47*(5), 42–46. doi:10.1145/986213.986236

Maumbe, B. M., Owei, V., & Taylor, W. (2007). Enabling M-Government in South Africa: An Emerging Direction for Africa. In I. Kushchu (Ed.), *Mobile Government: An Emerging Direction in E-government* (pp. 207–232). IGI Global. doi:10.4018/978-1-59140-884-0.ch011

Mengistu, D., Zo, H., & Rho, J. J. (2009, November). M-government: opportunities and challenges to deliver mobile government services in developing countries. In *Computer Sciences and Convergence Information Technology, 2009. ICCIT'09. Fourth International Conference on* (pp. 1445-1450). IEEE. doi:10.1109/ICCIT.2009.171

Misuraca, G. C. (2009). e-Government 2015: Exploring m-government scenarios, between ICT-driven experiments and citizen-centric implications. *Technology Analysis and Strategic Management, 21*(3), 407–424. doi:10.1080/09537320902750871

Moon, M. J. (2002). The evolution of e-government among municipalities: Rhetoric or reality? *Public Administration Review, 62*(4), 424–433. doi:10.1111/0033-3352.00196

Moon, M. J. (2010). Shaping M-Government for emergency management: Issues and challenges. *Journal of e-Governance, 33*(2), 100-107.

Napoleon, A. E., & Bhuiyan, M. S. H. (2010). Contemporary research on mobile government. In *Scandinavian workshop on e-Government (SVEG)* (p. 61).

Ntaliani, M., Costopoulou, C., & Karetsos, S. (2008). Mobile government: A challenge for agriculture. *Government Information Quarterly, 25*(4), 699–716. doi:10.1016/j.giq.2007.04.010

Ojo, A., Janowski, T., & Awotwi, J. (2013). Enabling development through governance and mobile technology. *Government Information Quarterly, 30*, S32–S45. doi:10.1016/j.giq.2012.10.004

Roggenkamp, K. (2004). Development modules to unleash the potential of Mobile Government. In *European Conference on E-government.*

Rossel, P., Finger, M., & Misuraca, G. (2006). Mobile" e-government options: Between technology-driven and usercentric. *The electronic. Journal of E-Government, 4*(2), 79–86.

Sareen, M., Punia, D. K., & Chanana, L. (2013). Exploring factors affecting use of mobile government services in India. *Problems and Perspectives in Management, 11*(4), 86–93.

Shareef, M. A., Archer, N., & Dwivedi, Y. K. (2012). Examining adoption behavior of mobile government. *Journal of Computer Information Systems, 53*(2), 39.

Song, G. (2005, July). Transcending e-government: A case of mobile government in Beijing. In *The First European Conference on Mobile Government.*

Song, G., & Cornford, T. (2006, October). Mobile government: Towards a service paradigm. In *Proceedings of the 2nd International Conference on e-Government* (pp. 208-218). University of Pittsburgh, USA.

Trimi, S., & Sheng, H. (2008). Emerging trends in M-government. *Communications of the ACM, 51*(5), 53–58. doi:10.1145/1342327.1342338

Wang, C. (2014). Antecedents and consequences of perceived value in Mobile Government continuance use: An empirical research in China. *Computers in Human Behavior, 34*, 140–147. doi:10.1016/j.chb.2014.01.034

Wang, C., Feng, Y., Fang, R., & Lu, Z. (2012). Model for Value Creation in Mobile Government: An Integrated Theory Perspective. *International Journal of Advancements in Computing Technology, 4*(2).

Chapter 8
Voters and Mobile:
Impact on Democratic Revolution

Oarabile Sebubi
Botswana International University of Science and Technology, Botswana

ABSTRACT

The objective of democracy is to allow people the freedom to vote at ease and according to their individual choices. Mobile Election has high potentials of transforming and improving efficiency of the current electoral process, thereby enabling convenient and ubiquitous elections, hence, revolutionizing the institution of democracy. With so much potential though, its adoption is extremely low worldwide because the barriers to adoption are extremely high as mobile election is still lacking in addressing the critical and sensitive requirements of the electoral process worldwide. This chapter explores the potential impact of mobile elections to democratic establishments such as politics and voter participation. It then adopts a comparative analysis approach in exploring the barriers to adoption and possible solutions. Lastly, it provides recommendations for future research in the areas of voter trafficking, network and data security, inter-operability issue, digital/democratic divide issue, and voter satisfaction.

INTRODUCTION

Mobile media is one of the contemporary research areas that have attracted a lot of attention from scholars interested in their emerging importance as a means of accessing, sharing, and disseminating news and information regarding politics, political events, and elections (Martin J., 2014; Martin J. A., 2015). One of such mobile media involves the use of mobile phone communication technologies to drive service delivery. Mobile communications have the potential to radically transform governments and to provide access to public information and services in areas where infrastructure required for Internet or wired phone service is not an option (OECD/International Telecommunication Union, 2011).

Mobile election is a special form of electronic service, commonly known as electronic voting (e-voting) driven by mobile phone based media. It is emerging as the future reformation of the current electoral process with a high likelihood of revolutionizing the democratic institution. Democratic election is a geographically dispersed activity which has to be organized and implemented within very tight time

DOI: 10.4018/978-1-5225-0469-6.ch008

frames, providing a cost-effective voting service for all eligible voters whilst maintaining high standards of integrity, security and professionalism and to achieve this is a major challenge to electoral management bodies (ACE Electoral Knowledge Network, 2015).

Therefore, mobile election, considered the core part of the democratic process, is the perfect solution to this challenge. It provides excellent democratic opportunities with the following benefits: increases citizen participation in the democratic process in both modern and developing societies. It allows voting to be done via mobile devices from home, work or while on the move, thereby, removing the inherited limitations of other voting systems that dictates physical interactions with the polling location and consequently, enhancing voter participation in elections. Above that, it makes voting easier for eligible voters and grant that their votes will be counted (Abdelghaffar, 2012).

Currently, most electoral management bodies around the world use new technologies with the aim of improving the electoral processes and one such is mobile election (ACE Electoral Knowledge Network, 2015). Mobile election, like other forms of mobile services offer *Enabled mobility* for more responsive public service delivery. The main reasons for the emergence of mobile services are the wider acceptance of these technologies by the public sector, high penetration of mobile devices, and ease of use for citizens, easier interoperability, and its ability to bring government closer to citizens and finally, it is less cost compared to computer-based services. Motivational factors for mobile services encompass better service accessibility, availability, responsiveness, quality, efficiency, and scalability. It also promotes better stakeholder participation, as well as improved communication (OECD/International Telecommunication Union, 2011).

The key objective of mobile phone voting is to eliminate going to polling booths, paper ballots, time and cost efficiency, tiredness and violence due to standing in line in pooling booths to cast their vote and to lessen the numbers of polling booths agents. Above that, it provides mobility feature which enhances turnout ratio in election (Ullah, 2013).

Mobile election has a wide range of application in areas such as shareholders' meetings, public policy initiatives, award nominations, opinion surveys, and school, club, and association elections. Each of these systems will have different requirements for security and auditability, depending upon their use (Mercuri, 2002). This chapter narrows the discussions only to multiparty/democratic elections.

METHODS

The chapter adopts an interdisciplinary literature review approach. In terms of topic, the review covers articles that discuss *mobile elections* and *democratic revolution*. The assumption here is that there is a strong correlation between mobile election and democratic revolution and the review is of the opinion that democratic revolution is possible with the adoption of mobile election. To perform such an analysis, the review devised different themes deemed to be relevant worldwide in the adoption of mobile election, as well as in ensuring that mobile election as an e-service yields tremendous change to the democratic institution. Therefore, discussions outlined below detail the process followed in selecting documents for the review. The review is restricted to documents that support chosen themes which are in turn categorized into broad analytical categories. There are two broad analytical categories for the review which reflects on the current issues pertaining to the topic of study from two angles, mobile election and democratic revolution dimensions.

The first analytical category, Mobile Election, encompasses themes related to the concept of Mobile Election. These themes are subdivided into two groups; General Review Themes entailing mobile phone voting systems and the world wide adoption rate of mobile election. The second subcategory of mobile election themes is a comparative study of three mobile election systems proposed in previous studies. The comparison is in terms of mobile phone voting system architectures, voter authentication, vote casting, verification, security, anonymity, auditability, and tally.

The second analytical category, democratic revolution, covers themes that explore the impact of mobile elections on the institution of democracy. The themes reflected on establish the relationship between mobile election and voter participation, political reformation, and lastly digital/democratic divide.

Discussions on themes are based on the following theoretical foundations: the calculus of voting, diffusion of innovations, technological determinism, and Uses and Gratification theory.

The review is a reflection of over 50 documents produced from 1968 to 2015. The period for old articles ranges from 1968 to 1994 and this group constitutes only 10% of the whole cited collection. All the documents in this group relate to theories that are still relevant today and tomorrow and thus form the review foundation. The period for the rest of the documents ranges from year 2000 up to 2015 and this gives a clear picture of the trend analysis of the evolution of mobile election that spans over a 15 year period.

MOBILE ELECTION GENERAL REVIEW

Mobile Election and Voting Evolution

Methods of voting have evolved through generations from manual to more innovative and efficient methods that embrace electronic and internet voting.

Popular manual voting systems currently in operation worldwide make use of paper ballot boxes; the open-response ballot, arrow ballot, bubble ballot and the optical scan ballot. Their major weakness is that they all require hand counting, except for the latter which can be tabulated by a machine (Everett, 2007). Another alternative manual voting system, postal voting, reviewed by Karp, employs mail interactions. Election officers mail out ballots to voters, who, on the other hand, fill them out and send them back within specified timeframe. Prior to voting, voters have to opt for absentee voter and the voting is done prior to national election date. This method decreases election administration costs, as well as, simplifies the voting process, thereby increasing election turnout. On the other hand, users cannot interact directly with the database to draw information that is for public consumption. This platform also does not allow them to communicate with each other and the service providers. Above that, it suffers delayed response as ballot casting is not processed in real time.

Over time manual voting systems evolved into various electronic media that makes use of punch cards, Direct Recording Electronic (DREs) and Radio-Frequency Identification (RFID) technologies. With punch card systems, voters punch holes in cards to indicate votes for their chosen candidates. The voter may feed the card directly into a computer vote tabulating device at the polling place, or the voter may place the card in a ballot box, which is later transported to a central location for tabulation (ACE Electoral Knowledge Network, 2015). DREs are paperless machines designed to allow a direct vote on the machine by the manual touch of a screen, monitor, wheel, or other device that records the individual votes and vote totals directly into computer memory (NCSL, 2015). With DREs, people with disabilities can vote unassisted. Ballots can be changed at the last minute and quickly personalized for local elections (Bruce Schneier, 2003). The major weakness of DREs is that all of the internal mechanics of voting are hidden from the voter. A computer can easily display one set of votes on the screen for confirmation by the voter while recording entirely different votes in electronic memory, either because of a programming error or a malicious design. Above that, election officers are powerless to prevent accidental or deliberate errors in the recording of votes (Bruce Schneier, 2003). Another election concept advanced by Hans Weghorn is a ticket system that makes use of Radio-Frequency Identification (RFID) technology. The technology makes use of the Identity (ID) card, a phone equipped with RFID communication hardware and the matriculation terminal. The ID card is a smart card which holds the matriculation number of the voter. It contains sectorised memory, and for each data sector there can be different access rights and keys defined, which are independent of the other data sectors. The phone is equipped with an additional RF hardware for communicating with the ID card (Weghorn, 2007). The RFID-based ticket system improves secrecy as it is not traced in any database, how often they vote, or whether they vote at all. Ensuring confidentiality is a very important issue for making the students to contribute their vote. The students do not have to come any more actively to an office, which is located in one single building of a widely distributed campus. The election office is made movable itself, and it can be placed to efficiently meet many students. The RFID concept has a lot of inbuilt security features that prevents abuse of voting (Weghorn, 2007).

Currently, electronic voting is evolving into various contemporary Internet voting methods that enable votes to be casted over the internet using either electronic equipment, web applications or mobile device technologies (Farik, 2015). It includes a wide range of possible implementations, which can either be voting at a supervised poll-site or unsupervised electronic kiosk using electronic equipment or remote voting from home or business using the voter's equipment, such as mobile phone (Rivest, 2004).

In overall, all electronic and internet voting methods are inexpensive, fastest, and most efficient ways to administer elections and count votes (Martin J. A., 2015). At the same time they are deficient of satisfying the electoral system requirements of a completely secure, trustworthy, verifiable, anonymous and usable system. All voting methods in exception of remote voting suffer the weakness of being constrained to a fix voting location, hence, hiking up the cost of voting.

In conclusion, of all contemporary voting methods, remote voting through mobile phone usage offers the best solution due to its ubiquitous and agile nature and its ability to revolutionize the democratic institution. It also offers mobility, flexibility and convenience as it enables votes to be cast remotely from any location at any time within the voting period. It also fosters increased participation in voter registration and turnout as it reduces voting costs, increase voter engagement and can be accessed by all. Accessibility and reach is increased to cater even for the old and disabled citizens who cast their votes from the comfort of their homes. It also lowers logistical and administrative costs with the reduction of materials required for printing and distributing ballots, as well as the personnel required to assist in voting stations. Above that, atomized ballot casting and counting votes is much faster and more accurate (Kogeda, 2013).

Mobile Election and Mobile Phone Voting Systems

Mobile election makes use of mobile phone voting systems. These are systems which allow users to engage in the electoral process through the use of cellular network service providers. That is, a voter accesses the Global System for Mobile Communications (GSM) network to cast their vote independent of the electoral office supervision. Previous studies record tremendous increase in the use of and access to mobile technology, in both developed and developing countries. Above that, mobile cellular is the most rapidly adopted technology in history and the most popular and widespread personal technology worldwide (OECD/International Telecommunication Union, 2011). In most countries, cellular phones are more feasible for human social interactions as they are the easiest, most convenient and least expensive mode of communication affordable to all despite economic status. Above that, they are more pervasive than Internet access (Kogeda, 2013).

Like most of the e-voting systems, mobile phone voting systems are required to ensure voter authenticity while ensuring vote privacy and vote-counting proof (Abdelghaffar, 2012). Authenticity dictates that only eligible voters can cast their votes (Kogeda, 2013). The privacy requirement means that analogous checks and balances cannot be employed to protect ballots and it is critical to a fair election as well as prevents voter coercion, intimidation, and ballot-selling. Vote counting proof instills auditability characteristic in which the system is expected to back track vote totals from actual ballots that come from legitimate voters voting no more than once and at the same time, and the ballot must not be traced back to the voter. The system must also alert the voter to mistakes such as flagging over-voting and under-voting (Mercuri, 2002). Other requirements include the following: vote integrity where once a voter cast a vote, no alternation to this vote is permitted. The system must also ensure accuracy as in capturing casted ballot, vote processing and counting. Democracy/ uniqueness requirement must be enforced to accept only one vote per voter. The verifiability requirement must also be met to allow voters to independently verify that their votes have been counted. Accessibility requirements must also be met to ensure that the system can be accessed by voters from any location. The availability requirement dictates that the system must have high-availability during the entire electoral process. Simplicity demands that the system be easy to use. The system must also allow multi-user access for simultaneous voting of multiple voters. Multi-lingual demands that it be translates to a diverse of citizens native and standard languages. Reliability provides that election system should work robustly, without loss of any votes, even in the face of numerous failures, including failures of voting machines and total loss of mobile communication (Kogeda, 2013). It is also important to create a multi-platform that caters for different mobile phone platforms.

Mobile Election and Application Platforms

The two forms of mobile phone voting systems are SMS applications and graphical user interface applications designed to run on various smart phone platforms i.e. Android devises, iPhone devises, etc. SMS applications, though limited in functionality, works well with all types of phones and this is good for reaching out to rural areas where most voters own simple phones with limited functionalities. On the other hand, graphical user interface applications make use of smart phones which are a common asset to urban populations. Therefore, the chapter maintains the view that to increase accessibility and use of mobile phone voting systems, a combination of SMS and graphical user interface applications must be availed to meet platform needs per voter device. Above that, interoperability feature must be taken into consideration in the development of graphical user interface applications to cater for devise differences in terms of platforms.

In the Ugandan 2011 general elections, a number of SMS enabled tools were deployed to increase political participation: Political campaigns using mass SMS broadcasts, SMS application to determine voter registration status, SMS news service subscription, voter education using bulk SMS, parallel voter tallying and crowdsourced election monitoring platforms that enabled citizens to share their observations via SMS on issues such as vote buying, registration hiccups, inappropriate campaign conduct, violence cases, complaints and feedback posts (Hellström J. K., 2012).

Mobile Election and Implementation

To ensure successful implementation of mobile phone voting systems, there must be control over the combination of technology and human labor. This will resolve the challenge of having the system operated by diverse groups of people each with diverse training, experience, motive and opportunity that might influence election results. This will also help in resolving the need for assuring the uniqueness of the vote and allowing for vote automation while guaranteeing avoidance of software or hardware problems (Abdelghaffar, 2012). The challenge to successful deployment and maintenance of mobile elections is costly to governments as it requires investment in the relevant technological skills (Farik, 2015).

Successful mobile projects are more aligned with existing practices, and more focused on intended outcomes. Therefore, approaches to success include: embedding the mobile element into an ongoing development effort, rather than creating the mobile service as the development effort itself; using mobile technology to reduce transaction costs and increase productivity of existing practices, rather than introducing entirely new behaviors; requiring only basic literacy or skills from users, rather than requiring additional technical knowledge or support (OECD/International Telecommunication Union, 2011). In order to meet user needs, the development of mobile phone voting systems requires a closer alignment to electoral processes and procedures.

In terms of usability of the end product, issues of concern encompass the diverse range of phones, frequent introduction and turnover of mobile technologies, small screen size, awkward data entry and slow or congested networks (Farik, 2015).

Mobile Election and Adoption Rate

The chapter also reviews the world wide adoption rate of mobile election by the two key stakeholders, citizens and government. This exploration is formulated within the context of the theory of 'Diffusion of Innovations'. This theory is concerned with the adoption and spread of inventions and major characteristics which determine an innovation's rate of adoption are relative advantage, compatibility, complexity, trialability and lastly observability to people within the social system (Rogers, 1983). The potential rate of adoption for mobile election is very high as it meets most of the adoption requirements. In terms of relative advantage, mobile election supersedes all predecessor election methods due to its agility and ubiquitous characteristics. Also, mobile election applications are fairly compatible as they may be developed for each phone type. Above that, prototypes of mobile election applications can be experimented on hence ensuring trialability and observability is attained as the results are visible, enabling stakeholder evaluation for adoption.

Despite all these strengths, the current electoral system operations dictate complex system requirements on mobile election and this has lowered its adoption rate worldwide by both key stakeholders. Farik et al. reports that although mobile phones are quite prevalent around the world and the amount of smart phones sold is increasing at a rapid rate, there have not been many elections which have capitalized on the use of mobile phones as a remote voting tool (Farik, 2015). On an international landscape, only a minute number of governments such as Estonia, Norway, Pakistan, Brazil, India and few others have shifted their democracies on a large scale towards the use of e-voting tools (Achieng M., 2013). On the other hand, despite the high adoption rate of mobile phones by individual worldwide, some societies have chosen not to consider them as democratic tools and this was depicted in a study carried out in Philippine to find out how mobile phones were used during the 2004 national elections. Study findings exposed that neither of the participating communities made significant use of the cell phone during the election period and people preferred face-to-face communication for local events, especially in rural communities (Pertierra, 2005).

A number of factors deter governments from adopting mobile elections. The current problems/challenges of adopting mobile elections range from security, trust, usability and technological skills requirements.

On the citizen side, though mobile election offsets most of the voting costs, the use of mobile phones for political participation comes with a financial cost of purchasing the device as well as cellular phone network service providers' subscriptions as in the case of Uganda where almost a third of the survey answers on why participants did not use SMS based platforms during elections expressed that 100 Uganda shillings was an obstacle (Hellström J. K., 2012).

MOBILE ELECTION COMPARATIVE REVIEW

Mobile Election and Voting System Architectures

Previous studies have categorized the electronic voting process into five stages. The first stage, setup stage, initializes voting parameters (candidates, voters, authorities' eligibility criteria, voting procedures, ballot validity and counting rules). Then registration and authentication of eligible voters is done and followed by the actual voting of candidates. The final stage, tally stage, encompasses vote collection, vote validity checks and final results production (Abdelghaffar, 2012).

Researchers have proposed different architectures to implement the voting process. Mpekoa proposes a system where the mobile voter connects to the mobile network using the 2G, 3G or 4G technology that allows the mobile voter to connect to the application server to download the application. Once the application is downloaded and installed, the mobile voter registers to vote using the application. During the registration, the application connects to the Staff database to verify the Identity Document (ID) number of the mobile voter. After a successful registration, the mobile voter can cast his/her vote (Mpekoa, 2014).

Farik et al. proposes a system architecture that involves two different servers – Voter Authentication Server (VAS) and Voter Processing Server (VPS), the use of Quick Response (QR) codes, checksums of QR codes and two different public key/private encryptions and the envelope method used in Estonia remote e-voting system (Farik, 2015).

The third system architecture proposed by Kumar et al. is a GSM mobile voting scheme. In this scheme, GSM is used for the voting system to introduce voter mobility and provide voter authentication. The system has four main components: Mobile Equipment/Voting Device dedicated for electronic vote casting i.e. GSM mobile equipment consisting of a GSM SIM card; Authentication Centre being an entity within the GSM network that generates the authentication parameters and authenticates the mobile equipment; Verification Server belonging to the voting authority, the organizers of the voting event who verifies the legitimacy of the voter and issues a voting token to the voter; and lastly, Collecting and Counting Server that collects and counts the votes to give the final result, whose actions need to be audited by all candidate parties. The Scheme is divided into three phases; voter authentication phase, voting phase and lastly counting phase (Kumar, 2008).

Mobile Election and Voter Authentication

One of the major features of mobile election mode is its ability to authenticate voters so as to give vote access to only eligible voters. Voter authentication is crucial for both the registration and voting processes. Voter Authentication for mobile services poses security problems (Jamnadas, 2015). There is controversy concerning voter authentication performed by the same system that records the ballots, thereby exacerbating the auditability and privacy problems. Above that, verifying a person's right to vote is difficult. Civil rights groups have objected to the use of bio-identification through fingerprints and retinal scans. Alternative log-in mechanisms, like personal identification numbers or smart cards can be easily transferred, sold, or faked (Mercuri, 2002).

A number of previous researches have advanced various alternatives that address voter authentication during the voter registration process. Mpekoa proposes a system that authenticates users at registration using their ID numbers. Voter inputs an ID number which is then validated against the Staff database. If the ID number is invalid, or the person is found to be inactive, the voter receives the proper message with another opportunity to enter a valid ID number. If the person is under 18 years of age, they also receive a proper message and given two option, to either continue with the registration knowing that they are not able to continue with voting otherwise, they terminate the registration process. After successful registration of voter details, the voter creates a pin which is purely a numeric value, a password which should be ten or more mixed characters of letters, numbers and symbols. The voter also selects a secret question and its relevant answer that is used when the voter recovers a pin or a password. If the password does not meet the above criteria, the voter receives the necessary message and prompted to enter it again. If the security question and its relevant answer are not selected then the system also sends the voter a message and given a chance to select these again. Once everything has been checked and is ok, the voter information is saved in the voter's roll. Finally, the registration flag automatically becomes *true* once the voter has registered (Mpekoa, 2014). In evaluating this scheme, the recommendation is that the system must not allow a person under voting age to continue any transaction.

The registration authentication of Kumar's scheme is left in the hands of mobile phone network service providers. If the voter is authentic then only he will be allowed to participate in the next steps of voting (Kumar, 2008). The challenge with this approach is that network service providers may not have up-to-date client information i.e. dead people may still be in their active client files, thereby compromising the integrity of the authentication output.

Researchers have also proposed various authentication techniques that could be used to authenticate users during the voting process. According to Mpekoa, the voter opens the application and login by using his/her ID number, pin and password and he claims that the combination of three keys strengthens security. If the combination is incorrect, the voter is given two more attempts then the system blocks them and in such a case the voter has to vote manually (Mpekoa, 2014). The researcher here needs to find better ways of handling blocked accounts instead of proposing manual voting. If a system is so relaxed that it permits voters to just easily switch voting methods, such a system would attract corrupt practices and the whole system would not be economical as government would have to print ballot papers that may either be under-utilized or in shortage. Therefore, it has to be clear well in advance; the method opted by voter to assist the government in planning for expenditure. The chapter recommends the establishment of a help center to resolve password issues.

On the other hand, the system that Farik et al. advances relies on the use of QR codes for authentication purposes. The QR code that is obtained after registration contains voter details such as voter id, voter name, phone number and password is used for voter authentication (Farik, 2015). The only challenge with this way of authentication is that it creates a digital/democratic divide as the technology is suitable only for smart phones.

According to the system proposed by Kumar et al. at voting time, voter is authenticated after submitting the ballot. The system allows the voter to fill in the ballot, encrypts the ballot, blinds it using the blinding technique of a blind signature scheme, and sends it to Authentication Centre (mobile phone network service provider) which in turn authenticates the voter, signs the encrypted ballot and forwards the encrypted ballot along with the signature to Verification Server which in turn checks the signature of Authorizing Centre and the eligibility of the voter, and signs the encrypted ballot with its private key (Kumar, 2008). The major weakness of this scheme is in collecting data from every person before verifying their eligibility to vote. This wastes transmission and processing resources that could have otherwise increased system efficiency and effectiveness.

Mobile Election and Vote Casting

Due to the sensitivity of elections, a robust system is required to allow one and only one vote session per registered voter. Here is a comparison of the voting process by different researchers.

According to Mpekoa's proposal, once authentication succeeds, the voter is given a menu where they can choose option *vote* to cast their vote (voting is not linked to a specific mobile phone or mobile phone number, thereby, enabling device sharing). The electronic ballot appears and the voter has to make a choice from a list of contestants. The ballot design takes form of the traditional paper-based ballot. To further enhance usability and friendliness, he proposes a multi-lingual application, to cater for different languages. The system allows voters to pause the voting process at any time before they commit the transaction. The voting flag automatically becomes *true* once the voter commits the transaction (Mpekoa, 2014).

Vote casting according to the system proposed by Farik et al. is as follows: voter cast the ballot through the mobile app residing in their phone. The app forwards the vote and voter details in an encrypted form to the Vote Authentication Server where voter authentication is performed. Once voter is successfully authenticated, vote is passed on to the Vote Processing Server for tally operations (Farik, 2015).

Mobile Elections and Vote Verification

Vote verification is employed to ensure that votes are correctly recorded as they may be intercepted and delayed or stopped in a denial of service attack or modified by malicious attackers. In contrast to manual voting methods, the level of transparency with mobile election systems is very low as all the processing is concealed from the public eye. Transparency in the voting process is essential, not only to provide auditability, but also to enhance voter confidence and it can be provided through the use of a voter-verified physical audit trail for use in recounts (Mercuri, 2002). In turn, lack of transparency raises the question of trust in the election process and results (Farik, 2015).

On the one hand, research suggests that using online or mobile channels to interact with citizens and engage them in decision making has a positive impact on trust (OECD/International Telecommunication Union, 2011). On the other hand, researchers claim that electronic systems cannot be fully trusted as code produced by someone else cannot be trusted. Above that, no manner of system self-reporting is sufficient to ensure that intentional tampering, equipment malfunction, or erroneous programming has not affected the election results and the proliferation of programmed e-voting systems invites opportunities for large-scale manipulation of elections (Mercuri, 2002).

A number of solutions have been proposed in previous studies to address the issue of vote verification and one of such is the Mercuri method of electronic voting which is also applicable to mobile application designs. It requires that the voting system prints a paper ballot containing the selections made on the computer. This ballot is then examined for correctness by the voter through a glass or screen, and deposited mechanically into a ballot box, eliminating the chance of accidental removal from the premises. If, for some reason, the paper does not match the intended choices on the computer, a poll worker can be shown the problem, the ballot can be voided, and another opportunity to vote provided (Mercuri, 2002).

The review maintains that Mercuri method can be adapted for m-voting systems. The paper printouts may be replaced with screen printouts for voter verification. The screen print outs will be identified with a unique code per voting session and will make no reference to user details. The transactional code verifies to voters that the ballots produced are a perfect representation of their vote, hence increasing voter confidence in the system. Once voter approves the transaction, then the system would print an electronic copy to user phone as well as a physical copy scanned into a ballot box. The physical copies are meant to preserve audit trails in case audit queries arise. Such a system ensures both privacy and auditability requirements of the electoral system.

A QR code and checksum of the QR code are used for vote verification. The checksum of the QR code on the mobile app, Voter Authentication and Processing Servers must be the same and point to one and only one voter for each voting instance. The checksum value on the Voter Authentication Server is forwarded to voter's mobile app for voter verification notification. Above that, only the voter can connect from their phone to the Voter Processing Server and verify the code (Farik, 2015).

Mpekoa's system ensures variability by producing a confirmation before the choice is saved into the database and then the voter receives a message that they have voted for their contestant (Mpekoa, 2014).

According to Kumar et al. once voter's account is verified by the Verification Server, it generates one ID for the ballot with one asymmetric key and then encrypts the ballot, ballot ID and public key of generated asymmetric key, generated for a particular ballot and sends the signed encrypted ballot with added attributes back to the voter (Kumar, 2008).

Mobile Election and Vote Security

Vote Security can only be assured if network and data are fairly protected. Network security poses controversy on the safety of elections run over the Internet. The internet is not safe for elections, due to its vast potential for disruption by viruses, denial-of-service attacks, spoofing, and other commonplace malicious interventions (Mercuri, 2002). The integrity of the mobile phone as a voting tool may be compromised by viruses and Trojan horses designed for mobile phones (Jamnadas, 2015). Other researchers have also reported on the potential threat of botnets based on mobile networks resulting from the current integrated of mobile networks with the Internet. Previously, the two networks operated separately, and this lessened the threat. Botnets of malware injected into mobile devices will affect the effectiveness and efficiency of mobile phone voting systems (Flo, 2009). Researchers have devised means to solve the security issues.

According to Mpekoa's proposed system, once the voter has voted, the vote menu is no longer accessible and he claims this offers uniqueness to the system (Mpekoa, 2014). Above that, it ensures data integrity in controlling access to meet the electoral requirements of "one voter, one vote".

On the other hand, the system proposed by Farik et al. employs a combination of Quick Response (QR) code which is a special barcode that can typically hold more information than other types of barcode and checksum of the QR code created using a hashing function to check the integrity of transmitted data, and ensuring that files have not been corrupted or modified. The process is as follows: first, the voter casts a vote using a mobile app. Second, the mobile app encrypts vote and voter details as a QR code. Third, the mobile app creates a checksum of the QR code. Fourth, the encrypted packet is then transmitted to the Voter Authentication Server (VAS) which decrypts the packet to obtain the voters details and the encrypted QR code. The VAS then authenticates the voter and once authenticated as a valid voter, the VAS calculates the checksum of the QR code and submits the checksum back to the voter's mobile app. It may also publish the checksum on a public website and in such a case then the voter can compare his checksum stored on his/her phone with the one published online. On the other hand, the mobile app compares the checksum it received from the VAS (step 4) to the one it calculated earlier (step 3). If it is the same, the mobile app will then display the message that the vote was successfully delivered and was not in any way modified (Farik, 2015).

This scheme has fairly handled the crucial electoral system requirements of a secure, verifiable, auditable, anonymous and transparent system. The only issue of concern is the loss of resources in collecting data from every person before verifying their eligibility to vote. For instance, vote is collected in step1 and voter authentication is processed at step4.

According to Kumar et al., the voter fills in the ballot, encrypts the ballot, blinds it using the blinding technique of a blind signature scheme, and sends it to Authorizing Centre that authenticates the voter, signs the encrypted ballot and forwards the encrypted ballot along with the signature to Verification Server. Verification Server checks the signature of Authorizing Centre and the eligibility of the voter, signs the encrypted ballot with its private key. It then generates one ID for this ballot with one asymmetric key and sends the signed encrypted ballot with added attributes back to the voter. It also sends the pair of ballot ID with the encrypted private key to the Counting Server. The voter checks the signature and retrieves the Verification Server-signed ballot, ballot ID and Public key from the message using the retrieving technique of the blind signature scheme. The voter sends the voting token along with ballot ID and the public key correspond to this particular ballot ID to Authorizing Centre. This three items ballot, ballot ID and Public key will be encrypted with Counting Server's public key to avoid Authorizing Centre decrypting the ballot and compromising the privacy of the vote (Kumar, 2008).

Mobile Election and Vote Anonymity

Some security problems result from an underlying fundamental conflict in the construction of e-voting systems: the simultaneous need for privacy and auditability. That is, anonymity of votes should be maintained to ensure secrecy of votes and at the same time the votes must be auditable. Currently, these two constraints cannot be mutually satisfied by any fully automated system (Mercuri, 2002).

According to Mpekoa' scheme, the vote choice is not linked with the voter but to the contestant to provide privacy to the voters (Mpekoa, 2014). The challenge is that, the proposal talks about the ballot which contains vote and voter details but is silent on how it separates vote from voter. This is crucial for stakeholder's trust of the system.

With Farik's scheme, once voter is successfully authenticated, the Voter Authentication Server will then strip all identifying voter information and submit only the QR code containing the encrypted vote to the Vote Processing Server (Farik, 2015).

Mobile Election and Vote Auditability

In general, all over the world, the main reason for the low adoption of e-voting systems by governments is due to their limited capabilities in double checking votes for fraud, especially in cases where voting systems are entirely paperless. Electronic balloting has yet to reconcile two conflicting needs in the polling place, anonymity and verifiable audit trail. The anonymity requirement of the voting system dictates that there be no link between the vote and voter entities, yet, if a recount is required, election officials are expected to reproduce a verifiable audit trail (Greenemeier, 2015). Above that, if a machine collects and records votes with an electronic ballot box, it is impossible to audit the transaction and establish the cause of the mistake and this also applies to mobile devices.

While the other two researchers remained silent on this issue, Farik et al. proposes an architecture in which the QR code stored on the mobile phone can also be used to verify if the vote was for the right candidate. All the voter has to do is scan or upload the QR code from the app or other designated devices to the Vote Processing Server which in turn reads the encrypted vote from the QR code, decrypt it and then transmit it back (Farik, 2015).

Mobile Election and Vote Tally

Mpekoa claims that accuracy is provided in mobile election as each vote is counted immediately and the automatic tallying capability of mobile voting systems means that the results are available instantly after closing time (Mpekoa, 2014).

With Farik's proposed scheme, once the QR code successfully reach the Verification Processing Server, it reads the encrypted vote from the QR code and decrypts the encrypted vote and tallies the vote accordingly (Farik, 2015).

According to Kumar, at the scheduled time of counting, Counting Server decrypts the ballot and checks whether the voting token is valid or not. If it is valid it will be counted else it will be rejected (Kumar, 2008). The weakness with this scheme is that the counting is not in real-time.

Mercuri proposes a dual tally method that reconciles manually counted figures to electronically generated ones. The ballots are in machine and human readable format, hence, can be optically scanned for a tally, or hand-tabulated for a recount. At the end of the election, electronic tallies produced by the machine can be used to provide preliminary results, but official certification of the election must come from the paper records. The dual tally method enables other entities to verify the ballots using their own scanning equipment (Mercuri, 2002).

DEMOCRATIC REVOLUTION THEMES

Mobile Elections and Voter Participation

Theories of political participation identify reward as being the main motivation for voter participation. Reward is expressed as a function of benefits over costs. Costs are important determinants of voting and vary greatly from voter to voter. Studies have shown that the lower the costs of participation, the greater the benefits and consequently, the greater the reward to be gained from participation. That is, if the individual's perceived benefit in casting the vote overpowers the costs of voting, then the greater likelihood of voter participation (Karp, 2000). Therefore, this chapter views intervention in voter participation as involving any attempts to decrease voter costs so as to increase voter reward. Voter participation encompasses the processes of voter registration and voter turnout. Voter turnout is one of the measures of citizen participation in politics and is usually expressed as the percentage of voters who cast a vote at an election (Pintor, 2012). Optimal voter participation describes the highest achievable level of voter response to the voting responsibilities.

Various studies have reported low voter participation due to various cost related factors, ranging from time, distance, health status, economic adversity, information acquisition and weather conditions. Therefore, the chapter explores the likelihood of offsetting low voter participation factors through mobile elections. Mobile Elections (m-elections) is synonymous to Mobile Voting (m-voting) and describes the use of Information, Communications and Technology (ICT) tools to foster remote voting (OECD/ International Telecommunication Union, 2011).

There is a strong correlation between mobile election and voter participation. The socio-economic models of voting directly relate the act of voting to cost of accomplishing so. The calculus theory of voting claims that it costs more to vote than one can expect to get in return, thus the low voter participation. The decision to vote is motivated by the expected rewards and is expressed as:

$$R = PB - C + D$$

where R is the reward that an individual voter receives from his act of voting; B is the differential benefit that an individual voter receives from the success of his more preferred candidate over his less preferred one; P signifies the probability that by voting, a citizen will bring about the benefit (B); C represents the cost to the individual of the act of voting. D is the individual benefit expressed as satisfaction from compliance with the ethics of voting, affirming allegiance and efficacy to the political system, affirming a partisan preference, and finally the satisfaction of deciding and going to the polls (Ricker, 1968). In summary, this theory claims that an individual will vote only if they perceive a positive reward from the voting exercise. On the other hand, a positive reward is only possible if the sum of the differential and individual benefits outweighs the voting costs which range from time, weather, distance, health status, economic adversity and information acquisition.

According to Sanders, time is a major component of the cost of voting and the costs vary according to voter differences. For instance, self-employed or non-working persons and adults without children can more easily accomplish the prerequisites of voting. In general, voters consider the time needed to acquire and assess information, to register, to get to the polls, as well as the waiting period at the polling stations (Sanders, 1980).

Economic adversity such as unemployment, poverty, and a decline in financial wellbeing suppress voter participation. Opportunity costs related to economic adversity affect an individual's decision to participate in politics. When a person votes, attends political meetings, or works for a candidate he foregoes spending scarce resources on more personal concerns. When the return from attending to an immediate stressful personal problem, such as unemployment, is greater than the return from participating in politics, the opportunity costs of participation are higher and this in turn lowers the probability the citizen will participate in politics (Rosenstone, 1982).

Being informed affects the propensity to vote. That is, people who are more knowledgeable in political affairs are more likely to participate in voting (Lassen, 2005). Also, the intellectual demands of obtaining and processing political information are greater for persons who possess only limited contextual knowledge relevant to the electoral process (Sanders, 1980). Above that, search costs incurred in accessing and using information also suppress participation (Aker, 2010).

Weather also is a determinant of voter participation. It reduces participation where election procedures are less convenient. For instance, rainfall or extreme cold on Election Day is expected to lower voter turnout (Eisinga, 2012).

A research was carried out to examine how distance factors into the costs associated with political participation with the hypothesis that the political geography of a voter's residence affects not only the likelihood that he or she will vote, but also the voting method. The findings of the study revealed that indeed travel costs are taken into account in both the decision to participate and just how one should cast one's vote (Dyck, 2005).

In one of the after elections studies on voter experience in USA, 21% of non-voter participants cited illness as the reason for their lack of participation in the democratic elections of 2008 (Stewart III C. A., 2008). This percentage rose up to 28% in the 2014 elections (Stewart III C., 2014).

Mobile Election has a high potential to reduce costs, thereby increasing the reward and consequently, participation. This phenomenon can best be explained from the psychological and communication perspective, by the *Use and Gratification* theory. The theory seeks to explain why and how people actively seek out specific media to satisfy specific needs (Hanjun, 2005). The core reason for using mobile election is to increase voter reward which in turn yields increased participation. Services optimized by smartphones results in better perception and higher participation (OECD/International Telecommunication Union, 2011).

A study of voter and non-voter satisfaction survey carried out after 2014 New Zealand democratic elections revealed the four major reasons for lack of participation from high to lower percentages as follows: lack of interest in political participation, followed by reason of self-stated personal barriers to voting, either due to personal access restrictions (e.g. health reasons, religious reasons) or other commitments (e.g. work). Next in prevalence is reason of not knowing who to vote for with the least as not knowing how, when or where to vote (TNS, 2014). The 2008 survey identified only three main reasons for non-voting with the reason of having other commitments being the highest, followed by work commitments and overseas trips at the same rate (Brunton, 2009). All these barriers of can be removed by mobile election.

Mobile Election and Political Reformation

The relationship between mobile elections and political reformation is best explained within the context of the *Technological Determinism* theory. This theory of large scale socio-economic transformation

(Heilbroner, 1994), claims that social structures evolve by adapting to technological changes and cultural and social change follows from technology (Bimber, 1990). Indeed, mobile election, enabled by the use of mobile phone voting systems, is highly likely to evolve political structures to adapt to mobile election systems in use. Mobile election systems make use of mobile phones which help to create an informative, connected, innovative, participative and converging society all over the world (Hellström J.). The unique qualities of mobile devices are contributing to new and different pathways to political engagement while also retaining significance in relation to traditional forms of offline political participation (Martin J. A., 2015). Due to its agility and ubiquitous nature, mobile election changes political collaborations at the different political levels – party, social localities, national and international levels.

At the party level, Mobile election is now a core element of modern political campaigns (Arulchelvan, 2014). Mobile cellular phones enhance interactions through the use of cellular network service providers (OECD/International Telecommunication Union, 2011) and allow party meetings and talks to be organized quickly and effectively. It also serves to mobilize volunteer workers as well as party supporters, enhances collaborations between party leaders and support workers, as well as fosters collaborations with relevant government wings. Senior leaders give guidance and reinforcement to junior workers through mobile phone communications. Apart from that it kept workers motivated by reminding them of their tasks and keeping them in regular touch with the party leadership (Robin, 2012).

In terms of social localities, it empowers the marginalized societal groups to exercise their democratic rights as in the case of India. The utilization of mobile phones in the Dalit politics in Uttar Pradesh in 2007, overcame physical barriers imposed by social superiors, distance and cost. It bypassed mainstream media controlled by caste-Hindus, unsympathetic to Dalit causes. It enabled the difficult message of Brahmin-Dalit alliance to be explained personally, relentlessly and widely. It fostered workers' sense of importance and purpose by enabling person-to-person conversations to instruct, inform, rally and praise. And it ensured a fair, free election by providing rapid access to responsive election officials. Incidences of fraud, misconduct and intimidation by election officials were reported with accompanying visual images (Robin, 2012).

At the national level, it influences the way citizens collaborate with government on election related matters. It has a high potential to increase political interactions and engagements as literature exposes that mobile cellular phones give governments the opportunity to interact with specific groups of users who otherwise may not be reached through conventional communication approaches. It also increases social engagements between different social groups (ITU Publications, 2011). During elections, it enables workers to report on the needs and requirements at sector and booth levels. It also fosters effective and efficient dissemination of political information and education. M-voting systems empower voters with basic information and education on electoral systems, election calendar and timetable, political parties, election officials, use of election sites, election management authority offices, voting sites education, and counting stations education (ACE Electoral Knowledge Network, 2013). This absorbs the voting cost of information acquisition and voter participation measures goes beyond quantity to embrace the dimension of quality. That is, mobile election is not only concerned with getting a large number to vote but also concerned with empowering voters to make informed decisions with their vote.

At the international level, it resolves one of modern democracy's greatest challenges: the overseas voter (Hall, 2010) by facilitating external voting in which the voter may use a mobile phone to cast his or her vote from abroad. While the constitutions of many countries guarantee the right to vote for all citizens, in reality voters who are outside their home country are often disenfranchised because of lack of procedures enabling them to exercise that right. There are four main groups of people residing abroad who are entitled to vote: workers; internally displaced persons (IDPs); certain professional groups, such as military personnel; public officials or diplomatic staff (and their families) and citizens living or staying abroad, temporarily or permanently (ACE Electoral Knowledge Network, 2015). The chapter claims that mobile elections will increase voter participation for external voters by eradicating barriers identified by the ACE project such as fear of immigrants being located at places where they are not expected to be either for work or political reasons; being distanced from the political issues in their own country; and the complex or costly logistical efforts that have to be gone through in order first to register and then to vote (ACE Electoral Knowledge Network, 2015).

Mobile Elections and the Digital/Democratic Divide

Digital divide identifies the first-level divide associated with socio-demographic factors (Seong-Jae, 2010). The concept of the digital divide refers to inequalities in the access to and the use of ICTs. It is about the existence or nonexistence of infrastructure, the provision of the physical access and most importantly, real access, which includes cognitive and cultural capital as well as technical resources. Real access goes beyond infrastructure and refers to people's actual possibilities to use technology to improve their lives. It is important to note that technology itself does not ensure its equal and efficient use but real access is ensured only when appropriate technologies are introduced into political, economic and social environments conducive to people's participation (Pertierra, 2005). Therefore, mobile election is highly likely to create digital divide as *real access* is determined by financial capability. Jamnadas has also echoed that mobile voting creates a digital divide between the rich and the poor (Farik, 2015). The second form of digital divide concerns the level of computer knowledge. Individuals familiar with computers found electronic voting systems easier to use than those with less computer experience (Mercuri, 2002). The other form of digital divide has to do with inequalities in ICT infrastructure. In the study carried out by Robin et al. the distribution of mobile infrastructure was highly skewed towards middle-class people in towns and cities. In rural areas, where most mobile-phone coverage was patchy, and phone penetration was estimated to have reached 10 phones per 100 people only in 2009 (Robin, 2012).

Democratic divide, on the other hand, identifies the second-level divide which concerns Internet usage inequalities and is associated with factors such as motivation and Internet skills. The democratic divide concerns the differences between those who actively use the Web for politics and those who do not. Analysis of General Social Survey data shows there is a democratic divide where political Internet users are individuals with high Internet skills and political interest (Seong-Jae, 2010). Taewoo has found out that with the exception of income, the subjects of digital divide are also subject to the democratic divide. He noted the existence of a demographic divide in participation and involvement in election campaigns in the form of social inequality in access, skills and participation in online politics. Demographic groups with respect to age, education, race and gender make a gap between participation and non-participation in various types of political activities online (Taewoo, 2010).

FUTURE REFLECTIONS

Ubiquitous Elections/Voting

The chapter sees the future of mobile elections/voting fully evolving into the concept of ubiquitous elections/voting through the integration of mobile and cloud computing. This would satisfy the World Wide Web Consortium (W3C) device independence principles of "anywhere, anytime, anyhow access to any service by anybody" (OECD/International Telecommunication Union, 2011). Therefore, the future infrastructure for ubiquitous voting would comprise of a complete wireless world of technology as in electronic services running on wireless communication devices, and deployed in wireless platforms.

Modified Vote Verification Method

The chapter proposes a modified vote verification method that is adapted from the mercuri scheme discussed above. The architecture will involve two servers, one hosting the voting system, and the other will host the vote processing system. The system models vote verification as follows:

1. A registered voter fills in the ballot through the mobile election application residing on their phone.
2. On commit, the ballot details update a temporary file in the voting system which will automatically allocate a unique identification code to it and in return the temporary record is printed out on the screen for voter verification.
3. Once voter verifies the record to be fine, then, vote processing will proceed in the following chronological order:
   a. Ballot record will be moved from temporary file in the voting system to the vote processing system.
   b. Once the record successfully updates the vote processing system, it is allocated vote processing code.
   c. Then, voter is updated with the vote status (i.e. vote successfully received) and the two codes (ballot and vote processing codes) on their phone.
   d. Once that is done, the corresponding record in the temporary file that is located in the voting system will be deleted, followed by the separation of voter from vote details in the vote processing system.
   e. Now the system will print out modified paper ballots into the ballot box and vote details will be published for public viewing.

The chapter argues that this method will ensure that the most critical election procedures and requirements are met:

1. **Vote Anonymity:** Destroys the relationship between vote and voter entities so that voter's democratic choice remains private to themselves. That is, the system does not hold the voter's secret but it lies entirely with the voter. The audit trail record of the vote transaction recorded in the system is in exception of the actual vote value.
2. **Auditability:** The vote transaction can always be back tracked with the two secret codes given out in an sms format to voter.

3. **Trust:** The relationship between vote and voter is destroyed only after assuring voter that their vote is part of the vote processing system input. This is necessary to maintain voter confidence on the system.
4. **Vote Counting Proof:** The system produces modified paper ballots for electronic counting by any interest group so as to counter check the vote processing system counts. Printing out paper ballots opens what is inside the system to public eyes, thereby resolving the black box controversy associated with all electronic elections.
5. **Transparency:** Publishing of votes per candidate fulfil the free and fair electoral requirement.

New Research Areas

For mobile election to be of relevance worldwide, researches have to be carried out to address the current pending issues, some of which involves devising strategies and ways to handle voter trafficking; draw a balance between anonymity of votes and ensuring verifiable audit trails at the same time; integrity of vote and voter, and ensuring accuracy in data entry and processing; digital divide; system usability in a rapidly fluctuating technologically innovative environment; and lastly how the system will instill stakeholder trust.

As mobile phone technology advances, security challenges also grow. Therefore, there are numerous security issues that also need to be addressed by future research initiatives such as robust authentication and authorization techniques, mobile phone device and data integrity. As mobile networks are integrating with the Internet, research studies must be pro-active in devising mitigation strategies to handle the flow of security threats from Internet to mobile devises.

The review recommends future studies to establish the relationships that exist between the wider acceptance of mobile phone technologies by the public sector and the service value derived from these devises. As previous studies on mobile phones have revealed that many see them as leisure tools more than tools for serious activities and politics being a serious business involving difficult choices, it is difficult to align these two mismatched worlds (OECD/International Telecommunication Union, 2011). This then necessitates that in the future studies be carried out to establish if in reality people are deriving optimal value from those devises and if so to what extent they make life easier for people. Such studies would also draw relationships between devise concentration and device type. Another dimension of looking at such an analysis is from the dimension of devise concentration versus ease of use for citizens.

Interoperability issues must also be addressed by future studies. This is important as mobile election apps are developed and deployed on various platforms. Studies must address the possibilities of establishing interoperable platforms that cut across all deployments.

As mobile elections is an emerging/developing concept, studies must determine its future evolution for optimal service accessibility, availability, responsiveness, quality, efficiency, scalability, stakeholder participation, and finally, optimal communication, integration and interactions.

Future studies must also address policy actions that address the digital/democratic divide issue associated with mobile elections so as to close the current digital/democratic gaps.

CONCLUSION

As the next generation of public service will thrive on societal participation and engagement, systems that foster human social interactions must be devised in every sphere of life. Mobile communication-based technologies are emerging as the future source of realistic, affordable, convenient and easily accessible human social interaction medium. Mobile phone technologies have the highest potential of transforming the electoral process so as to yield a democratic revolution in which we will see mobile devises acting as powerful democratic tools with increased levels of mobile election adoption, increased voter motivation and participation in politics, and increased political reformation. Mobile election will relieve governments from high costs of election administration in setting up voter registration and polling stations, securing and rewarding the labor force, printing ballot papers and providing all the necessary stationaries, and support materials for the labor force.

Therefore, researches must be carried out to unleash the value adding potential of mobile election as well as resolve the current barriers to its adoption.

REFERENCES

Abdelghaffar, H., & Galal, L. (2012). Assessing Citizens Acceptance of Mobile Voting System in Developing Countries: The case of Egypt. *International Journal of E-Adoption*, *4*(2), 15–27. doi:10.4018/jea.2012040102

ACE Electoral Knowledge Network. (2013). *The ACE Encyclopaedia: Civic and Voter Education*. ACE Electoral Knowledge Network. Retrieved from http://www.aceproject.org

ACE Electoral Knowledge Network. (2015, December 09). *Elections and Technology*. Retrieved from The Electoral Knowledge Network: https://aceproject.org/ace-en/topics/et/eth/eth02/eth02a

ACE Electoral Knowledge Network. (2015, December 09). *Voting From Abroad*. Retrieved from ACE Electoral Knowledge Network: http://aceproject.org/ace-en/topics/va/onePage

Aker, J. C., & Mbiti, I. M. (2010, June). Mobile Phones and Economic Development in Africa. *The Journal of Economic Perspectives*, *24*(3), 207–232. doi:10.1257/jep.24.3.207

Arulchelvan, S. (2014, July). New Media communication Strategies for Election Campaigns: Experiences of Indian Political Parties. *Online Journal of Communication and Media Technologies*, *4*(3), 124–142.

Bimber, B. (1990). *Karl Max and the Three faces of Technological Determinism*. Working Paper No.11. Massachusetts Institute of Technology.

Bruce Schneier, D. L. (2003, August). *Voting and Technology: Who Gets to Count Your Vote?* Retrieved from Schneier on Security: https://www.schneier.com/essays/archives/2003/08/voting_and_technolog.html

Brunton, C. (2009). *Final results: Voter and non-voter satisfaction survey 2008*. Colmar Brunton.

Cambridge Dictionaries Online. (2015, December 9). *Definition of interaction from the Cambridge Advanced Learner's Dictionary & Thesaurus*. Retrieved from Cambridge Dictionaries Online: http://dictionary.cambridge.org/dictionary/english/interaction

Dyck, J. J., & Gimpel, J. G. (2005, September). Distance, Turnout, and the Convenience of Voting. *Social Science Quarterly*, *86*(3), 531–548. doi:10.1111/j.0038-4941.2005.00316.x

Eisinga, R. G., Te Grotenhuis, M., & Pelzer, B. (2012). Weather conditions and voter turnout in Dutch national parliament elections, 1971–2010. *International Journal of Biometeorology*, *56*(4), 783–786. doi:10.1007/s00484-011-0477-7 PMID:21792567

EstonianE. (n.d.).

Everett, S. P. (2007). *The Usability of Electronic Voting Machines and How Votes Can Be Changed Without Detection.* Rice University.

Greenemeier, L. (2015, December 09). *Ballot Secrecy Keeps Voting Technology at Bay.* Retrieved from Technology for Scientific American: http://www.scientificamerican.com/article/2012-presidential-election-electronic-voting/#

Hanjun, K. C.-H. (2005). Internet uses and gratifications: A Structural Equation Model of Interactive Advertising. *Journal of Advertising*, *34*(2), 57–70. doi:10.1080/00913367.2005.10639191

Heilbroner, R. (1994). *Technological Determinism Revisited.* EBSCOHOST.

Hellström, J. (2012). Mobile Participation? Crowdsourcing during the 2011 Uganda General Elections. In Proc. M4D 2012 (pp. 411-424). New Delhi: Excel India Publishers.

Hellström, J. (n.d.). Mobile phones for good governance – challenges and way forward. *Stockholm University.*

Jamnadas, H. K. (2015, October). Challenges & Solutions Of Adoption In Regards To Phone-Based Remote E-Voting. *Int J Scientific & Technology Research, 4*(10).

Karp, J. A. (2000). Going postal: How all-mail elections influence turnout. *Political Behavior, 22*(3), 223–239. doi:10.1023/A:1026662130163

Kogeda, O. K. (2013). Model for a Mobile Phone Voting System for South Africa. *Proc. 15th Annual Conference on World Wide Web Applications (ZAWWW 2013).* Cape Town: Cape Peninsula University of Technology.

Kumar, M. K. (n.d.). *Secure Mobile Based Voting System.* Academic Press.

Lassen, D. (2005, January). The Effect of Information on Voter Turnout: Evidence from a Natural Experiment. *American Journal of Political Science, 49*(1), 103–118. doi:10.1111/j.0092-5853.2005.00113.x

Martin, J. (2014). Mobile Media and Political Participation: Defining and Developing an emerging Field. *Mobile Media and Communication, 2*(2), 173–195. doi:10.1177/2050157914520847

Martin, J. A. (2015). Mobile News Use and Participation in Elections: A bridge for the Democratic Divide? *Mobile Media @ Communication, 3*(2), 230-249. doi: 10.1177/2050157914550664

Mercuri, R. (2002, October). A better ballot box (S. Charry, Ed.). Academic Press.

Mpekoa, N. (2014). Designing, developing and testing a mobile phone voting system in the South African context. In J. V. Steyn (Ed.), *ICTs for inclusive communities in developing societies* (pp. 372-385). Port Elizabeth.

NCSL. (2015, July 7). *Voting Equipment: Paper Ballots and Direct-Recording Electronic Voting Machines*. Retrieved from National Conference of State Legislators: http://www.ncsl.org/research/elections-and-campaigns/voting-equipment.aspx OECD/International

Pertierra, R. (2005, April). Mobile phones, identity and discursive intimacy. *An Interdisciplinary Journal on Humans in ICT Environments, 1*(1), 23–44. doi:10.17011/ht/urn.2005124

Pintor, R. L. (n.d.). *Voter Turnout Rates from a Comparative Perspective. Academic Press.*

Ricker, W. H. (1968, March). A Theory of the Calculus of Voting. *The American Political Science Review, 62*(01), 25–42. doi:10.2307/1953324

Rivest, R. L. (n.d.). *Electronic Voting*. Massachusetts Institute of Technology.

Robin, J. (2012, February). Mobile-izing: Democracy, Organization and India's First "Mass Mobile Phone" Elections. *The Journal of Asian Studies, 71*(1), 63–80. doi:10.1017/S0021911811003007

Rogers, E. M. (1983). Elements of Diffussion. In E. Rogers (Ed.), *Diffusion Of Innovations* (pp. 1–37). London: Collier Macmillian Publishers.

Rosenstone, J. (1982, February). Economic Adversity and Voter Turnout. *American Journal of Political Science, 26*(1), 25–46. doi:10.2307/2110837

Sanders, E. (1980, August). On the Costs, Utilities and Simple Joys of Voting. *The Journal of Politics, 42*(3), 854–863. doi:10.2307/2130557

Seong-Jae, M. (2010, February03). From the Digital Divide to the Democratic Divide: Internet Skills, Political Interest, and the Second-Level Digital Divide in Political Internet Use. *Journal of Information Technology & Politics, 7*(1), 22–35. doi:10.1080/19331680903109402

Stewart, C. III. (2014). *2014 Survey of the Performance of American Elections*. The Massachusetts Institute of Technology.

Stewart, C. A. III. (2008). *2008 Survey of the Performance of American Elections*. The Massachusetts Institute of Technology.

Storer, T. L. (2006, June). *ResearchGate*. Retrieved from ResearchGate: http://www.researchgate.net/profile/Ishbel_Duncan/publication/250889738_An_Exploratory_Study_of_Voter_attitudes_towards_a_Pollsterless_Remote_Voting_System/links/5447aac90cf22b3c14e0f845.pdf

Taewoo, N. (2010). Whither Digital Equality?: An Empirical Study of the Democratic Divide. *Proc. 43rd Hawaii International Conference on System Sciences* (pp. 1-10). IEEE Computer Society.

Telecommunication Union. (2011). *M-Government: Mobile Technologies for Responsive Governments and Connected Societies*. Geneva, Switzerland: OECD Publishing; doi:10.1787/9789264118706-en

TNS. (2014). *Report into the 2014 General Election*. TNS New Zealand.

Weghorn, H. E. (2007, April 23). Mobile Ticket Control System with RFID Cards for Administering Annual Secret Elections of University Committees. *Informatica*, 161-166.

Chapter 9
Local News and Mobile:
Major Tipping Points

Kenneth E. Harvey
KIMEP University, Kazakhstan

ABSTRACT

The first economic tipping for traditional media hit in 2006 when newspaper revenues began to collapse. Leonard Downie Jr. (2009), vice president at large for the Washington Post, notes how most newspaper executives knew that the time would come when the Internet would begin stealing newspaper readers and revenues, but they didn't expect it to happen so quickly. That is the nature of a tipping point. In one year, according to Downie, they were at an economic high and adding personnel to their news staff, and the next year they were laying people off. How quickly the economic downturn occurred took them and many other newspapers by surprise. In 2005 newspapers achieved $49.4 billion in advertising revenues. By 2014 revenues had fallen about 60% to $19.9 billion (Pew, 2015). There is evidence that such a tipping point is about to hit local TV stations. These two tipping points could lead to the end of quality local news coverage. This chapter explores how newspapers and TV stations can survive and flourish in the Digital Age.

INTRODUCTION

The efforts of American newspapers to transition their business onto the Internet have been an abysmal failure. So much so that Chyi (2015) argues that it is time for newspaper publishers to realize that they are no good at online publishing and return to focusing their efforts on their printed products.

[T]he truth is, most newspapers are stuck between an unsuccessful experiment (for their digital product) and a shrinking market (for their print product). Even more embarrassing is the fact that the (supposedly dying) print edition still outperforms the (supposedly hopeful) digital product by almost every standard, be it readership, engagement, advertising revenue, or paying intent. According to Scarborough, a research firm collecting readership data for the industry, U.S. newspapers reached a total of 67% of American adults through multiple platforms including print, Web, and e-editions during a given week, but the print

DOI: 10.4018/978-1-5225-0469-6.ch009

product alone covered 61% of American adults within the same timeframe (Newspaper Association of America, 2012b), suggesting that the Web edition contributed only 6% of total readers who were non-print, online-only. Additionally, users are not engaged with the online edition — throughout the month of November 2012, an average online reader spent a total of 39 minutes on a newspaper site, which translates into 4.4 minutes per visit—the best in recent years (Newspaper Association of America, 2012c). It is therefore unsurprising that advertisers are less than enthusiastic about placing ads on newspaper sites — the Web edition generated 15% of total newspaper advertising revenue, while the print edition accounted for 85%, according to data released by the Newspaper Association of America. Note that this is after print advertising revenue dropped dramatically from $47.4 billion in 2005 to $18.9 billion in 2012 (Edmonds, Guskin, Mitchell, & Jurkowitz, 2013). To compensate for the substantial loss of print advertising revenue, many newspapers have instituted aggressive price increases on their print product (Edmonds et al., 2013) and erected paywalls around their online content. More than 450 newspapers today are charging users for online news access (News & Tech, 2014), but subscription rates for most local newspaper sites linger in the single digits (Mutter, 2013). Sooner than most had expected, some major metro papers (The San Francisco Chronicle and The Dallas Morning News) have put an end to their short-lived paywall experiment by the end of 2013. In contrast, more than 44 million Americans are still paying for the dead-tree edition during the week (and 48 million on Sundays) despite recent price spikes (Newspaper Association of America, 2012b). To sum up, after 20 years of trial and error, the performance of U.S. newspapers' digital products remains underwhelming. From a business perspective, despite all the efforts made, newspaper firms have been "exchanging analog dollars for digital dimes. (quoted in Dick, 2009, para. 1)

Alan Mutter (2015), a former newspaper and cable TV executive who now focuses his Reflections of a Newsosaur blog on the changing media landscape, refutes Chyi's conclusions and suggests that newspapers face a do-or-die scenario if they cannot transition to the Internet. He cites several related facts:

- **Print Circulation is Falling:** Readership of printed newspapers has fallen nearly in half since 2005, and the trend seems unlikely to change.
- **Current Newspaper Readers Are Literally a Dying Breed:** The median age of the New York Times is 60, compared with the median age of 37 for the general public. A large portion of their readers will die over the next 10-15 years, and only a small portion of the younger generations subscribe to any printed newspaper. Other newspapers face a similarly aging subscription base.
- **Ad Sales are Declining for the 10th Straight Year:** During that time newspapers have lost 2/3 of their print advertising revenue, and that, too, appears not to be a temporary trend.
- **Economy of Scale Will Ultimately Doom Newspapers:** While websites cost about the same to service one reader or one million, or to offer one page of news or one million pages, that is not true for printed newspapers. Press staff and equipment cost a lot to maintain no matter how few copies or how few pages are printed. The same is true for many other costs in producing a printed newspaper as opposed to an online newspaper. Current downward trends related to revenues and circulation could kill most printed newspapers within 10 years.

All traditional media are losing audience (and thus advertising revenues) as people turn more to computers, smart phones and tablets for their news and information. This is especially true among younger generations. Mutter (2012a) contrasted the use of different technology by Boomers and Millennials to access news:

- **Newspaper (print):** 53% Boomers, 22% Millennials.
- **TV**: 92% Boomers, 75% Millennials.
- **Radio:** 65% Boomers, 47% Millennials.
- **Magazine (print):** 13% Boomers, 9% Millennials.
- **Computer**: 56% Boomers, 69% Millennials.
- **Tablet:** 12% Boomers, 17% Millennials.
- **Smartphone:** 12% Boomers, 40% Millennials.

Whichever side one might support, the Chyi-Mutter debate highlights two grim realities for newspapers:

1.   Their efforts to date to go digital have failed badly; and
2.   If they don't find a solution, neither their printed newspaper nor their digital newspaper is likely to survive much longer.

This is further underscored by their failure to increase substantially their digital advertising sales, which is, ultimately, the key to survival. In 2014, newspapers' print advertising revenues fell 900% faster than their digital advertising grew (Pew, 2015, p. 27). In 2007 newspapers had already achieved $3.2 billion in digital ad revenues, but by 2014 that number had only grown to $3.5 billion. During that same time, while digital advertising gained $300 million, print ad revenues fell about $30 billion. Figure 1 illustrates even more discouraging news, contrasting newspapers' stagnant digital revenues to Google's skyrocketing revenues. In 2003 Google digital ad sales and newspaper digital ad sales were about equal. Since then, newspaper digital ad sales have grown to $3.5 billion while Google digital ad sales have grown to nearly $50 billion. And ad sales by Facebook and other online media have also been skyrocketing. Of all the digital advertising sold in America, newspapers' combined digital sales represent only 7%. Newspaper publishers need to realize that their main competition is NOT other traditional media, it is new media such as Google and Facebook. If they cannot compete effectively with major online media, that is also a sign that most newspapers will face a worsening struggle to survive.

The trend to new media, especially mobile media, has increased dramatically since 2012 and demographic-related data suggest the trend is certain to continue. The primary sources of news in the past – printed newspapers – are fading rapidly in their importance, and the collapse of advertising revenues suggests the slide has far from bottomed out (Mutter, 2012b). As of 2013, 50% of all American adults now report the Internet as a primary source of national and international news – second only to TV at 69%. Newspapers dropped all the way to 28%, only 5 percentage points higher than radio (Pew, 2013). Again, demographics indicate the trend will continue. Among 18- to 29-year-olds, the Internet is the top source of national and international news by a large margin – 71% Internet, 55% TV, 22% newspapers and 19% radio. Among 30- to 49-year-olds, the survey results were perhaps more shocking in the sense that newspapers even dropped below radio (27% radio vs. only 18% newspapers), while TV and the Internet were tied at the top at 63%. Among 50- to 64-year-olds, TV was No. 1 at 77%, but the Internet still beat newspapers for the second spot, 38%-29%, and radio was last at 25%. The only age group that

Figure 1. U.S. newspaper print and digital revenues vs. Google ad revenues (billions)
Sources: Newspaper Association of America and Google Inc.

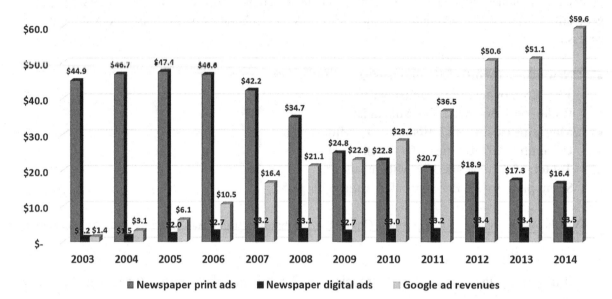

still prefers newspapers over the Internet are those 65 and older, where preferences are 84% TV, 54% newspapers, 18% radio and only 15% Internet. As with the New York Times, newspapers, in general, appear doomed as the Boomer generation diminishes.

But these data suggest TV is doing OK, right? Revenues for 2014 still looked good. But this chapter will review other data that suggests that 2015 for television will be like 2005 for newspapers – a tipping point that could spell the doom of local TV stations and difficult times for pay-TV (e.g., cable) and the networks if they fail to adapt quickly. TV audiences are now moving rapidly online for their entertainment, and, as with newspapers, advertising will almost certainly follow the audience. But the chapter will also discuss strategies that might save at least some of the local news media. This is important since a successful democracy depends on a well-informed public, and most online news is still initiated by or strongly tied to the traditional media.

STATE OF JOURNALISM

These trends are having a dramatic impact on the business of journalism, and both economics and new technologies are impacting the practice of journalism. In this section we will review more in depth how news media are fairing in maintaining audience and economic resources, as well as their efforts to follow the technological trends. Only then can we attempt to predict the future of journalism.

The Pew Research Center has been providing an in-depth annual report on the status of news media, based on several studies conducted each year, in order to track these changes. In many other parts of the developed world, similar changes are likely to parallel those in America. In the developing world, different patterns are emerging because the press there, in most cases, has never achieved the dominant role seen in the developed world, and the growth of new media in developing parts of the world is more gradual, since a smaller portion of its population has transitioned to the Internet as their primary news source, and changes thus produce less dramatic effects on the traditional media in those countries. But even this assumption is not as true as many may think. In fact, young professionals from countries like China, India and Korea rely on their mobile devices for news, information and entertainment more than those from the USA, Canada and Western Europe and, if given only one choice, would be far more likely to choose a smartphone over a TV (Cisco, 2014). Young professionals were asked how often they went through a day using only mobile apps and never a website. Among Generation Y professionals, 15-25% of those from the USA, Canada and Western Europe indicated multiple times weekly, the rest less often. But of Generation Y professionals from Mexico, India, China and Korea, 36-43% said that occurred multiple times weekly. Among Generation Y professionals, 18-28% from the USA, Canada and Western Europe said multiple times weekly, as opposed to 31-46% from India, China and Korea (p. 29). Ask which they would choose if they could have only a smartphone or a TV, but not both, 57-76% of Generation X professionals from the USA, Canada and Western Europe chose the smartphone, but from India, China and Korea, 87-97% chose the smartphone. Among Generation Y professionals, 41-62% of those from the USA, Canada and Western Europe chose the smartphone, but 85-97% from India, China and Korea (p. 36).

Because of its in-depth research, much of the following information about traditional and new media will be summarized from the Pew Research Center State of Journalism reports of 2014 and 2015. The reports note that advertising still accounts for about two-thirds of all news-related revenues, although print advertising is continuing its steep decline of about 60% since 2005. While TV ad revenues have remained fairly stable during this same time, the migration of TV audience to the Web is increasing and expected eventually to mirror that of newspaper readers, which naturally leads to revenue losses. Advertising follows the audience. So far, none of the traditional media have shown the ability to replace the loss of traditional ad revenues with those of digital (Pew, 2015, p. 3). According to eMarketer (2015a), digital advertising in all formats in the U.S. was almost $51 billion in 2014 – almost 30% of all U.S. ad spending of $178 billion. But the traditional media saw very little of the money going to digital, even though most do offer digital advertising services. While traditional media seem stuck in old paradigms, such digital giants as Google and Facebook continue to innovate new advertising approaches.

Newspapers

Newspaper print advertising is continuing to fall dramatically. In 2014 print advertising revenues for American newspapers fell about $900 million (4%) from 2013 revenues, but newspapers' digital advertising only increased about $100 million (Pew, 2015, p. 27). Overall newspaper advertising revenues have dropped from a high of $49.4 billion in 2005 to $19.9 billion in 2014 – a decline of almost 60%. With these revenue losses, U.S. newspaper employment has also dropped dramatically from about 55,000 in 2007 to 36,700 in 2014 (p. 28). More discouraging is to compare newspaper digital trends to digital advertising overall, as illustrated in Figure 2. Digital advertising in America almost doubled between 2010 and 2014, increasing from $26 billion to $51 billion. During that same time, newspapers' digital advertising increased less than 17%, from $3 billion to $3.5 billion. And the further you look back, the worse the trend line appears. Newspapers already had $3.2 billion in digital revenues in 2007, then fell below $3 billion before beginning their relatively slow upward trend again (p. 27). Instead of newspapers converting ad revenues from print to digital, most of the digital display advertising has gone to the young and aggressive digital giants. Of $22.2 billion spent on digital display ads in the U.S. in 2014,

Figure 2. Newspaper revenues from print and digital advertising: annual revenue in billions of dollars
Data Source: Newspaper Association of America and BIA (Kelsey).

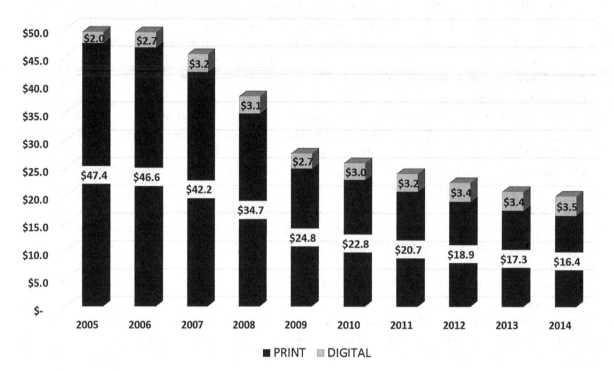

nearly half ($11.2 billion) went to Facebook, Google, Yahoo, AOL and Twitter, with Facebook leading the way with 24% of all digital display ad revenues and Google not far behind (p. 22). These digital giants, however, don't just sell display ads. According to eMarketer (2015a), the total revenue of digital advertising in the U.S. in 2014, including display, video, native advertising, search, etc., was more than double the display ad revenues cited by Pew. Newspaper digital display advertising was only about 7% of the total digital advertising sold in America.

Magazines

The magazine industry as a whole is also hurting. In 2014 average single sale circulation fell 14% and overall circulation fell for the seventh year in a row (-2% in 2014). Sixteen news magazines tracked by Pew, however, did much better, on average. Some experienced sizable reader increases during the year while others experienced substantial declines, but, overall, sales of printed and digital copies fell only 1%. The Nation was hit the hardest, with a decline in subscriptions of 18%, whereas Wired did the best with a 6% increase in overall circulation. The top news magazines by circulation were Time (3.3 million), Rolling Stone (1.5 million), Vanity Fair (1.2 million), New Yorker (1 million), Bloomberg BusinessWeek (1 million), Forbes (0.9 million) and Wired (0.9 million). Digital subscriptions and single sales are important to these magazines – but some a lot more so than to others. New Yorker has the most digital subscribers, at 80,153, followed by Vanity Fair, Bloomberg BusinessWeek, and Time. New York Magazine has the most digital single sales, which quadrupled in 2014 to 26,112, making about 70% of its single sales digital. Rolling Stone had the second-most digital single sales, with 23,506 – about 30%

of its total single sales. Forbes had the most monthly unique visitors to its website, about 50% more than the nearest rival, Time. About half of the 16 news magazines had at least 10% more mobile visitors than desktop visitors, while only four had significantly more desktop visitors (Pew, 2015, pp. 70-76). In spite online readership increasing, magazines nationally are losing advertising revenues as their total circulation declines. In 2013 ad revenues fell 3%; in 2014 they dropped 5%.

Radio

About 91% of all Americans 12 and older listen to broadcast radio, even though 53% also listen to on-line radio. About 73% of those listening to online radio use their smartphones, and an overlapping 61% use a desktop or laptop. Across America there were about 2,000 news/talk/information radio stations in 2014 – a number that had remained fairly steady over the previous five years. All-news radio stations, however, dropped from 37 in 2012 to 31 in 2014, of which CBS Corporation owned 10. The remaining 21 were distributed among 14 different owners.

Radio spot advertising represents about 76% of radio stations' revenue, overall. In 2014 that revenue source dropped about 3%. That was offset partially by lesser revenue sources – digital and off-air commercial activities – which experienced substantial increases of 9% and 16%, respectively. Still total revenues, on average, dropped 1%. Despite its increase of 9%, digital advertising still represents less than 6% of total radio revenues (Pew, 2015, pp. 57-62).

Because of commuting listeners, broadcast radio's audience has been more consistent than newspapers'. However, radio and other audio services are also moving online. The number of online radio listeners has doubled since 2010, with 53% of Americans over age 12 saying in 2014 that they had listened to online radio in the previous month. Most are, as with broadcast radio, commuters. They listen to it over their smartphones as they drive (Pew, 2015, p. 59). Similar is the growth of podcasting. In January 2015, 17% of Americans said they had listened to a podcast during the previous month – an increase from 15% in 2014 and from 9% in 2008. The biggest promoter of podcasts was LibSyn (Liberated Syndication), the website of which hosts 22,000 podcast sites from which 2.6 billion podcasts were downloaded in 2014. Of these, 63% were requested through mobile devices, again reflecting the powerful trend to mobile (p. 56).

Television

The most stable of the traditional news media is television. Viewership for the ABC, CBS and NBC evening news, combined, grew 5% in 2014. There is not a strong trend line over the past few years, but at least it is not a downward trend as with most traditional media. In 2008 there was an average of 22.9 million Americans watching network news each evening. The average from 2009-2013 was lower than in 2008 but never below 22 million. Since 2008, 2014 was the first year the average has surpassed 23 million, reaching 23.7 million in 2014. NBC achieved the greatest growth, at 6%, and continued to lead the competition with an average evening viewership of almost 9 million. ABC lost anchor Diane Sawyer but still experienced 5% growth to 8 million, and the average CBS audience grew by 5% to almost 7 million. (Pew, 2015, pp. 36-37).

Morning news viewership for the three networks is more than 30% lower than evening news, totaling about 14 million combined viewership, with ABC leading the competition with 5.4 million, trailed by NBC with 4.9 million and CBS with 3.3 million (p. 38). Besides their broadcast news programs, the three networks have three of the most successful online news sites. Yahoo/ABC leads the way with 65 million unique visitors in January 2015, followed by NBC with 56 million and CBS with 47 million. All three have far more mobile visitors than desktop visitors – 11 million more for ABC, 15 million more for NBC and 13 million more for CBS (p. 43), as seen in Figure 3.

Local news viewership has also been fairly stable over the past few years. In 2014 (Pew, 2015, p. 44) early morning news, starting at 4:30 a.m., on average, grew 6%, regular morning news 2% and evening news 3%, while late night news dropped 1%. Despite the growth in the experimental early morning news shows, their ratings were still not high enough to satisfy many local stations, causing many of the early morning shows to be canceled by year's end. The highest level of viewership in the past five years for evening and late night news broadcasts was in 2011, but there has been no strong trend line up or down since then.

Pew analyzed Fox affiliates' broadcast news separately, perhaps because Fox offers support for neither early morning nor early evening news programs – and affiliates' late prime time news programs are one hour long instead of 30 minutes. Despite being the clear leader in all-day cable news, Fox affiliates are struggling with their nightly news, with a drop of 4% in combined average viewership in 2014 and a decline of 17% since 2010. Still they have over 5 million nightly viewers. The affiliates' morning news, on the other hand, has steadily gained viewership, increasing 9% in 2013 and 5% in 2014, but it still trails the other network morning shows with about 2.5 million viewers (p. 45).

While advertising revenues for the local affiliates fell dramatically in 2009 with the recession, the trend line has been positive since then. In 2014 combined ad revenues reached $20 billion, up 7% from 2013 but down 3% from 2012. Pew attributes an up-and-down advertising trend to the Supreme Court's Citizens United ruling. Since it gave corporations and other organizations the same free speech rights as individual citizens, advertising during even-numbered election years is consistently up, and that of

Figure 3. Mobile key to TV/cable online traffic: total unique visitors for January 2015 (in thousands)
Source: comScore Media Metrix.

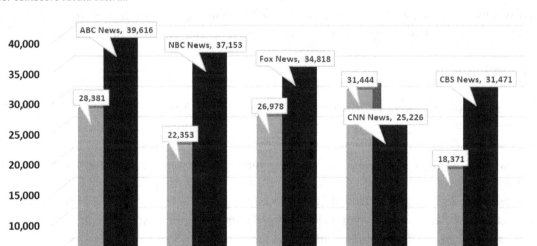

odd-numbered years down. During a presidential election year, such as 2012, stations can expect to get an additional boost. Only 68% of the local TV stations offer their own locally produced news programming but achieved 84% of the advertising revenues in 2013, the most recent data available. Local stations with news programs in 2013 attributed 50% of their entire local advertising revenues to commercials run during their news shows, which Pew noted was an all-time record. In 2002 the percentage of local advertising linked to stations' news shows was below 40%.

Among all local news media, TV stations appear to be the most stable, which has led to the sale of nearly 500 U.S. TV stations in two years – nearly 300 sales worth $9.7 billion in 2013 and 171 sales worth $5 billion in 2014. Over a period of 15 years, only the sale of stations in 2006 surpassed those of 2013. Station sales in 2014 were only surpassed by those in 2006, 2013 and 2000, in that order (p. 51).

Cable

Cable's 24-hour news channels have seen their median prime time viewership continue to fall, as illustrated in Figure 4, but Fox has faired far better than competitors MSNBC and CNN. Fox declined 1% in 2014 while MSNBC dropped 8% and CNN 9%. CNN and MSNBC combined for less than 1.1 million average prime time viewership, while Fox by itself averaged 2.8 million – substantially more than double its competitors'. Prime time viewership is particularly important for advertising, which actually increased in spite of the drop in audience. Comparing total median viewership throughout the 24-hour period, CNN viewership increased 1%, seemingly at the expense of MSNBC, which declined 14%. Fox declined slightly, as well, by 2%. Fox's viewership was still more than MSNBC and CNN combined (Pew, 2015, pp. 32-33).

As with the other media, cable news channels' ability to cover the news is dependent on their revenues – which for cable channels includes advertising and subscriber fees. The three channels combined for

Figure 4. Cable news audience shrinking: median total viewership (in thousands)
Source: Nielsen Media Research.

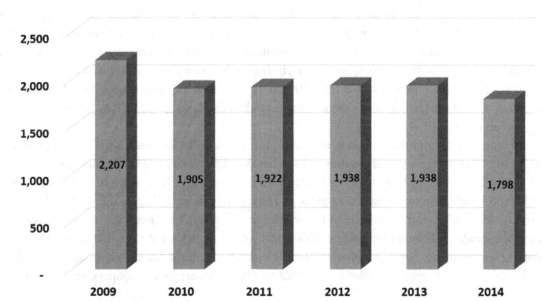

an increase in revenues of 4% to a total of $3.7 billion in 2014, but again Fox outdistanced its competitors, up 6% to $2.04 billion. CNN's revenues rose 3% to $1.13 billion and MSNBC revenues fell 1% to $501 million. Only Fox was projected to achieve an increase in profits in 2014 – up 10% to a healthy $1.2 billion. Nonetheless, all three were projected to experience healthy profits – Fox 61%, MSNBC 41% and CNN 29% (pp. 32-34).

Despite their drop in viewership, none of the channels was expected to reduce its news operation expenses. CNN and Fox both spend a little over $800 million on their news operations, while MSNBC spends $296 million but is able to piggyback on the NBC broadcast news operation. With their budgets, CNN operates 33 foreign bureaus and 11 domestic bureaus; NBC operates 11 foreign and nine domestic bureaus; and Fox operates 11 domestic bureaus but only four foreign bureaus (p. 34).

In competition for unique visitors to their websites, in January 2015 FoxNews.com had over 55 million and CNN.com had almost 50 million. MSNBC trailed far behind with under 7 million, but it is embedded within the NBCNews.com website, which overall had about 30 million unique visitors. CNN had about 6 million more desktop users than mobile users visiting their website, while Fox had about 8 million more mobile users than desktop, which bodes well for its future. MSNBC visitors were slightly more mobile (p. 35).

New Media Convergence

While traditional news media are generally losing audience, online media are gaining – and mobile is now leading the way. Citing comScore data, Pew's 2015 State of the News Media report (Pew, 2015, p. 4) noted that of the 50 most popular online news sites, 39 had more mobile visitors in 2014 than desktop visitors, as shown in Figure 5, four sites had about equal traffic, and only seven of the sites had more desktop traffic than mobile. That is in spite of the fact that many websites do not yet fully support mobile. While no analysis was found regarding these top news sites, of the Fortune 500 commercial websites, as of April 2015, 46% of their websites would not meet Google's mobile-friendly ranking criteria about to be imposed (Merkle/RKG, 2015, p. 21). Even of those companies that seemed to have set digital marketing as their highest priority, as indicated by their inclusion in the Internet Retailer 500, 29% of them still did not comply with Google's criteria. More related to the news industry was an analysis of the Web presence of 100 of the top magazines in the world (Ayres, 2014), which concluded that 93% of them were failing to fully accommodate mobile visitors. Perhaps for this reason, the Pew 2015 study reported that mobile visitors did not spend as much time on the news sites as did desktop visitors. On those top 50 sites, desktop users spent at least 10% more time than did mobile users on 25 of the sites. On only 10 of the sites did mobile users spend more time than did desktop users. Time spent on the other 15 sites was about even (Pew, 2015, p. 4).

It is no longer just about the transition of traditional media to the Internet, either. All-digital news organizations are growing in size and number. Together they had reached 3,000 employees in 2013, including a growing number of foreign bureaus. Vice Media, for example, had grown to 35 bureaus around the world, and Huffington Post planned to grow to 15 bureaus during 2014. BuzzFeed hired a foreign editor to oversee foreign correspondents in many foreign countries, and Quartz accumulated reporters who together speak 19 different languages. This growth came at the same time that traditional media were cutting back staffing. Network newscasts in 2013 included half the foreign coverage as they did in the 1980s. The number of reporters working internationally for American newspapers was 24% lower in

Figure 5. Traffic of world's top 10 digital newspapers: total number of unique visitors (in thousands)
Source: comScore Media Metrix, January 2015.

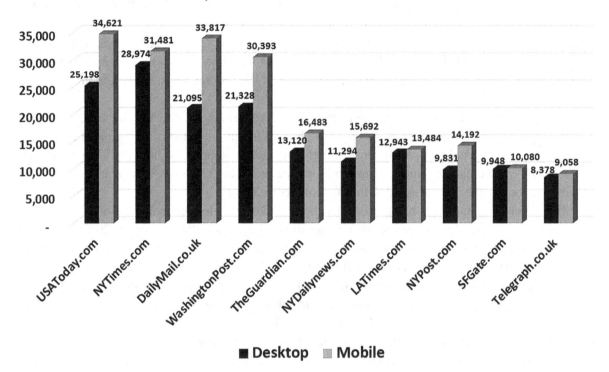

■ **Desktop** ■ **Mobile**

2010 than in 2003 (Pew, 2014, p. 3). But, counting dedicated journalists working for traditional media, Pew estimates there are now 5,000 online journalists in America (2014, p. 2).

But news sites are not the only online sites where people get news. Nearly half of American adults who use the Web say they get news about politics and government from Facebook. Some of that "news" may be generated by political organizations spreading information that supports their causes, but all major news media in America also have Facebook sites that people can follow, from which they will receive much of the same news as if they went to the media's own websites.

HOW TECHNOLOGY, ECONOMICS DRIVE JOURNALISM

These technological developments and socio-economic trends together will impact future journalistic trends. New technologies are changing the way reporting is done and how news is presented.

Powerful Twin Trends: Mobile, Video Intertwine

Trends reveal a new direction not just for traditional media but also within the new media. It could be argued that there have been two Internet revolutions occur over the past few years. The First Internet Revolution was the Social Media Revolution. It was obvious and dramatic, as demonstrated by Barrack Obama's presidential campaign. Social Media had been picking up popularity, but in 2009, the year after

Obama's Internet-driven victory, the number of Facebook users increased 350%, and 1,000% by 2012. In 2009 alone, the number of Twitter users skyrocketed 1,500%, and by 2012 10,000% (White, 2013).

The Second Internet Revolution was not so obvious and took even professionals a couple of years to appreciate. But it has been just as dramatic in its impact. This was a two-pronged revolution. The Video Revolution is so intertwined with the Mobile Revolution that it is almost impossible to separate the two. Note just a few statistics that reveal the dramatic changes of the Second Internet Revolution:

- The number of videos viewed online is exploding. The number viewed in June 2012 compared to June 2011 increased by 550%, meaning the hunger for online video is very great and growing (xStream 2013). And that phenomenal growth rate has not stopped. In June 2011 the number of videos viewed online in America was 6 billion; one year later 33 billion. Two years later in 2014 the number worldwide crossed 300 billion (comScore 2014).
- This revolution, in part, is being driven by the spread of mobile broadband. As of 2012, mobile connections were already double the number of fixed wire broadband (Pepper 2012). By 2019 Cisco projects that 97% of all mobile traffic will be via smart phones, and overall data traffic on mobile – driven by video demand – will increase 1,000% (Cisco 2015).
- Average broadband speed will nearly quadruple between 2011 and 2016 from 9 megabits per second to 34 mbps, meaning the quality of online video is getting much better (Cisco 2012). This, too, is in part because of the spread of mobile broadband.
- Internet connections, largely because of the rapid spread of mobile devices, will nearly double worldwide between 2011-2016, at which time there will be an estimated 2.5 connections for every man, woman, child and infant in the world (Cisco 2012).

Changing "Best Practices"

The "best practices" of journalism have changed dramatically over the past 10 years, and they will undoubtedly continue to change.

Video Journalism

The "2000-pound pencil" that made broadcast journalism so difficult, time-consuming and costly now weighs a lot less physically, economically and organizationally. Citizen journalists can now take reasonably good video with their mobile devices and transmit them to news media in moments (Kamdar, 2015). Even TV news organizations don't always need to send a camera crew if the reporter can carry a much smaller video camera. Newspapers can provide video-based news now on their websites without expensive broadcasting equipment and highly trained technicians, and because of economic pressures, any technology that can cut costs while enhancing or at least maintaining acceptable quality has to be considered.

Twitter Journalism

Twitter has emerged as an increasingly effective tool for journalists. Citizen journalists using Twitter first broke the news of U.S. Airways flight 1549 landing in the Hudson River bordering New York City in 2009. Witnesses used their mobile devices and began spreading the breaking news via social media just

4 minutes after the incident. The first tweets beat traditional media by some 15 minutes and were quickly accompanied by multiple photos via Twitpic and by video footage (Wilkinson, Miller, Burch & Halton, 2014, p. 9). The 2004 Indian Ocean tsunami and the 2008 Sichuan (China) earthquake were other early examples of citizen journalism using mobile media to report breaking news ahead of traditional media (Shirky, 2009). Of course, professional journalists can also use increasingly superior mobile devices for audio, video or photos, and news organizations can develop advanced strategies to encourage citizen journalism to take advantage of a billion potential citizen "reporters" around the world carrying such devices (The Economist, 2014).

Journalists, bloggers and citizen journalists can now use such tools not only to report on occasional catastrophic events but can provide live tweets from a restricted courtroom, minute-to-minute updates from council meetings, and play-by-play descriptions from sports events. Text, audio and video can go immediately back to the newspaper or broadcast station to be disseminated quickly to the public. Many news media now have Twitter feeds on their websites where news reporters and citizen journalists can post news as a live stream. Using such applications as Flipboard, these multimedia elements can be quickly combined for websites and for mobile-friendly distribution to tablets and smartphones. (Wilkinson, el al, 2014, p. 10).

Twitter now offers its own video app, with improving capabilities. It purchased Vine in 2012 for $30 million. In 2015 it announced improved video capabilities to "seamlessly capture, edit and share videos right from the Twitter app" on a person's mobile device. So now it's not just text that can be immediately inserted into a news organization's Twitter feed. Now videos of up to 30 seconds in length can be taken by journalists, quickly edited and uploaded to the news feed. Uploading videos to Twitter from an iPhone camera roll was also made possible, with the same capability promised for Android phones (Kamdar, 2015). These capabilities can help newspapers become more competitive with TV stations, which, if newspapers don't do, might be the media most likely to drive the final nail in the coffin of the growingly obsolete printed news.

News is no longer a one-way system of communication from professionals to the public. And the news ecosystem is also broader than the billion prospective citizen journalists carrying news production devices in their pockets. Readers also mediate, viralize and provide perspective on the news real-time. An Ofcom study cited by Wilkinson, et al (2013, p. 10) found that 25% of UK adults are talking on the phone, texting or using social media, sharing their thoughts on what they are simultaneously watching on TV. While watching the 2013 Wimbledon men's tennis final match, 1.1 million viewers shared 2.6 million tweets using hashtags related to the sports event, and 80% of those tweets came from mobile devices.

Making News More Social

Social and mobile developments are reshaping the very process of news, as we've known it. Half of all social network participants share and repost news on those sites, and nearly half (46%) discuss such news posts on social networks. Currently, however, most of this interaction with news appears more coincidental than purposeful. Only about a third of those who read and comment on news actually follow any particular news organization or journalist. Most of the other two-thirds are just reading news stories shared by friends and adding their own news and comments, as they choose. The percentage of participants who also act as citizen journalists and post news videos and content of their own is significant though hardly overwhelming. About 10% of all social media users say they have taken and posted their own news videos. Most major news media have a Facebook presence, and when clicking on news postings, readers are typically and strategically taken to the news organization's own website for details. But, according to Pew (2014), relatively few Facebook participants actually follow those links to the news sites themselves. This led Pew to conclude that to enhance their audience and their ad revenues, news providers need to develop a more comprehensive online strategy beyond the social media postings.

Nevertheless, since the audience is getting more and more of its news from the Internet, it is important to find them where they are, and Facebook is the top site on the Internet. Facebook provides some good advice on how news media can enhance their Facebook sites to attract more fans (Hershkowitz and Lavrusik, 2013), which can then be linked to the media's regular websites. This advice should also be helpful with other social media and on the news media's own sites. Indeed, to the degree the news media can create their own online interaction with their readers, the more they are likely to take that media time away from their new, most-dangerous competitors – the major social media sites, such as Facebook and Twitter. The suggestions for attracting and keeping readers interacting with journalists include:

- **Share Breaking News:** Posts announcing "breaking news" achieve 57% greater engagement. And updates in quick succession increases engagement 10% more.
- **Use Conversational Tone:** Posts with such a tone or with clever language can boost engagement by 120%.
- **Provide Analysis:** Five lines of text including analysis averages 60% greater engagement, and any post with analysis increases referrals by 20%.
- **Initiate Conversation with Questions and Responses:** Including a question and prompt for readers to respond increases engagement 70%. And if new staff respond to initial readers' comments, 14% more readers will likely participate.
- **Include Graphics, Photos, Cartoons and Videos with Stories:** Photos increase likes by 50%. Changing the site's cover photo frequently to promote a top story can boost engagement even more. Graphics and cartoons can also enhance engagement and promotion of stories. Another study not mentioned in this article says video has a 137% greater impact than photos (Ross, 2015).
- **Reward Your Social Media Audience with Exclusive Content:** Story advances, early content, and personal and behind-the-scenes content help keep fans coming back. Using the Pages Manager mobile app can allow staff to update Facebook on the fly.

This entire book could be filled with new "best practices" of journalism in the Digital Age. Indeed, many books and journals are already filled with them. But if the "best practices" were adequate in turning around the fate of the newspaper industry, the trends indicating their impending doom would have at least leveled off. They haven't. Revenues and readers continue to decline, suggesting that new journalism practices by themselves will not save newspapers. Local TV stations will almost certainly face the same grim reality. Journalistic best practices aren't enough to save traditional media from a technological revolution with such economic implications.

SECOND MAJOR TIPPING POINT WILL ENDANGER LOCAL TV NEWS

Local television still reaches 90% of all Americans and is enjoying relative financial success among the traditional media. TV advertising revenues fell as did that of all traditional media in 2009 but have recovered substantially since then. Consequently, local channels have suddenly become prime commodities on the market. In 2013 the sale of stations increased by 205% over 2012 and the value of sales increased 367%. Sinclair Broadcasting has become one of the biggest players in local TV, now with 167 stations in 77 markets, reaching almost 40% of America. Thus, all appears to be going well for local TV – just like it did for newspapers in 2005 when the first major tipping point occurred.

New data suggests a second major tipping point occurred in 2015, this time likely to do to television what the first tipping point did to newspapers. What has made local TV stations profitable has not been the growth of advertising but rather the growth of cable retransmission fees – what stations are paid for having their programming carried on their local cable networks (Pew, 2014, p. 6-7). As shown in Figure 6, industry projections are for these fees to continue climbing. But what happens as both the TV and cable audiences move online, their local advertising begins to diminish as it has for local newspapers, and the cable fees disappear? Then, as with newspapers, their online advertising revenues will become critical to survival. But for local stations online advertising represented only 4% of total revenues in 2014 (Pew, 2015, p. 46-47), which indicates that they are doing no better at making the digital transition than have newspapers. With more and more entertainment video moving online and the TV networks providing their own national and international news, local TV stations are likely now to find their audience and their ad revenues falling quickly. An eMarketer (2015d) study found that nearly 70% of all Internet users were already watching online TV-like entertainment – 94% watching YouTube, 63% Netflix, 36% Amazon and 33% Hulu. eMarketer projected that the so-called OTT (over the top) Internet-based TV viewing was already plateauing and would only expand to 72% by 2019. That projection could prove very conservative for several reasons:

- New projections by Cisco (2015) that the number of mobile Internet devices globally will increase more than 50% from 7.4 billion in 2014 to 11.5 billion in 2019.

Figure 6. Retransmission fee revenue for local TV stations in the U.S. (in billions of dollars)
Estimates and projections by SNL Kagan.

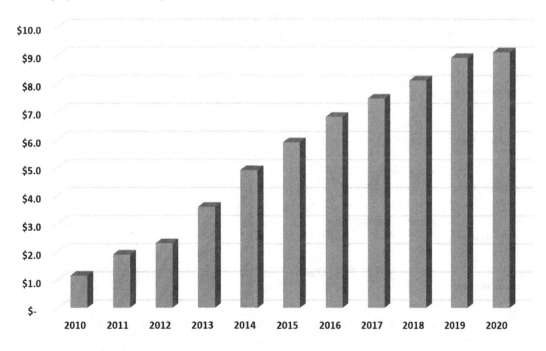

- Global mobile network speed will increase 2.4 times between 2014 and 2019 from 1.7 Mbps to 4.0 Mbps (Cisco, 2015).
- Mobile video will increase globally from 55% of all data traffic to 72% (Cisco, 2015).
- Mobile data traffic just in North America will increase nearly 7 fold between 2014 and 2019 (Cisco, 2015).
- More directly to the point, a study by Blueshift (2015, p. 2) found that the rapid growth of OTT (online TV) is also leading to more and more Americans deciding to drop their pay-TV services – 7.4 percentage points just in the first six months of 2015. The American use of pay-TV dropped from 69.3% to 61.9% during that time.
- The monthly rate of pay-TV cancellations increased to an annualized rate in the previous month of 15.6%. But 10.2% of the survey respondents said they were very or extremely likely to drop pay-TV within the next six months (Blueshift, 2015, pp. 2-3), as illustrated in Figure 7. Thus, by the end of 2015, cable may have lost nearly 20% of its customers, and it would not be unlikely that 30% more could abandon cable by the end of 2016.
- Among the younger generation 18-29 years old, 47% have never had pay-TV and, based on current trends, never will (Blueshift, 2015, p. 3).
- A study by DEFY Media focused on 13- to 24-year-olds. It found the younger generation more tuned into free online content such as YouTube than broadcast or online TV (Carufel, 2015). Online videos make them feel better than TV (62%-40%). They enjoy online videos more (69%-56%), relate to them more (67%-41%), prefer relaxing to them more (66%-47%) and find them more entertaining (76%-55%). Consequently, 96% of them watch online videos, only 57% watch free online TV, and only 56% watch recorded TV.
- Blueshift's study says more than 72% of all Americans were already viewing online streaming video in 2015. Just of their respondents 75% were already paying for streaming online video services – 9.9 percentage points more than there were six months earlier (Blueshift, 2015, p. 4).

Figure 7. How likely are you to cancel pat-TV within 6 months?
Survey by Blueshift Honest Research in conjunction with SurveyMonkey, June 2015.

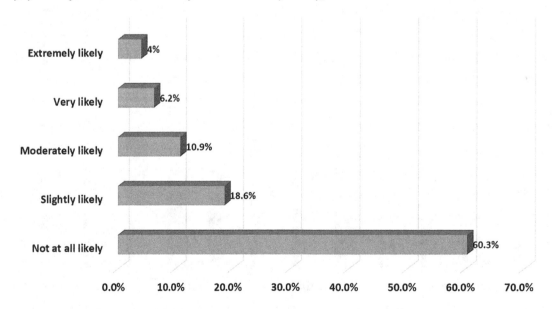

- More advertising-based "free" TV is also moving online, including more professional content on YouTube, which was not included in the Blueshift study.

In the same way that advertising follows the audience, the communications industry follows the advertising. Among advertising executives who have already begun advertising online with video, 77% say this has given them "superior targeting" and measurability, and 61% said they are achieving greater ROI than with TV advertising (eMarketer, Feb. 2015). If Blueshift's 1,129 respondents, selected to mirror census data in age and gender, accurately reflect reality, the negative tipping point for broadcast and cable TV may have already arrived, and their audience and advertising may go the way of newspapers' previously. Betting that local TV will continue to prosper in the long run is risky.

News media, overall, are still dependent on advertising for about two-thirds of their revenues. They are trying to develop new revenue streams. Such new strategies as event hosting and Web consulting accounted for 7% of the news media revenues in 2013. For large traditional media, especially newspapers that have lost 60% of their advertising revenue in the past 10 years, this is relatively miniscule. Another new revenue stream for some news media is so-called native advertising – paid news-like stories written for advertisers, frequently written by journalists. This trend was initiated by such early adopters as The Atlantic and Mashable, but spread rapidly in 2013 to such venerable newspapers as the New York Times, the Washington Post and the Wall Street Journal. eMarketer predicted that native ad revenue in America would reach $2.85 billion by 2014. Many news executives initially expressed concern about advertising that readers might confuse as news stories, including Wall Street Journal editor-in-chief Gerard Baker, who described it as a "Faustian pact." But ultimately even Baker could not resist the potential revenue stream that native advertising could create. After adopting the new form of advertising, Baker said he was "confident that our readers will appreciate what is sponsor-generated content and what is content from our global staff," but others, including Pew, expressed some concern (2014, pp. 3-4). The Pew report also ponders as to whether such new revenue plans may at least allow the digital-only media to survive (Pew, 2014, p. 4), but with that very suggestion implies that they will fall far short of saving most other media.

STRATEGIES FOR SAVING TRADITIONAL NEWS MEDIA

Chyi (2015) makes convincing argument that newspapers' two decades of efforts to move online have been an abysmal failure. While their digital advertising now makes up almost 18% of their revenues, that percentage is only increasing because overall ad revenues are declining so drastically – 900% faster than digital advertising is increasing. In contrast, local television revenues are currently more stable, but online advertising represents only 4% of their total revenue – not enough to pay their bills. Chyi is not very convincing in her argument for newspapers to abandon their online efforts and focus on their printed product. Mutter (2015) makes more convincing argument that newspapers have no choice but to move online if they want to survive another decade. And evidence of the second major tipping point and its probable impact on local television suggests TV stations are now on the same treacherous path as newspapers, just a little further behind.

If local media do not find a solution, neither most newspapers nor most local TV stations are likely to survive much longer. Some national newspapers, such as the New York Times, which now claims over 1 million online subscribers, may survive longer and perhaps make the transition to a fully online product. And at the other end, perhaps some local weekly newspapers will survive in print because of their hyperlocal content and community support, but even that is unlikely to last long without some major changes. As their readers turn more to their computers and, increasingly, to their mobile devices for news and shopping information, printed products will seem increasingly archaic and cumbersome. And the younger generations are already turning their backs on local media as an indicator of worse business conditions to come for even hyperlocal media. More importantly, hyperlocal newspapers will gradually lose their financial foundation of hyper-focused advertising. Their readers are almost all on-line, and local advertisers -- as regional and national advertisers before them -- will gradually realize the advantages of reaching out to their customer base through less expensive, more targeted and more measurable online advertising. Most local advertisers may not realize it yet, but already they can buy pay-per-click advertising by zip code at a cheaper price than printed advertising, be fairly confident in what they are getting for their money, and have those prospective customers hyperlinked directly to their specially designed website landing pages, where the advertisers can feature not only their best buys but every single item in inventory. Advertisers cannot afford to do that in their local newspaper. Once their consumers become more comfortable with online shopping, many advertisers may then offer home delivery and begin closing their storefronts -- or they will face competitors who will. If large regional newspapers survive much longer, they will probably cut back soon to just three profitable printed editions per week – typically Wednesday for grocery ads, Friday for entertainment, and Sunday because of its higher overall readership – then offer daily news only online. But the same forces that will gradually kill most hyperlocal media will also require all newspapers to either adapt or die.

Similar to hyperlocal newspapers, some local TV stations may survive another decade or maybe even two -- but three or four stations in one small city will almost certainly not survive much longer as their cable fees disappear and their advertisers also discover the advantages of online advertising. TV networks will be forced to increasingly compete with their own affiliates for online advertising and will probably follow the lead of Facebook and Google in using technology to allow advertisers to buy advertising focused on specific geographic regions and demographic profiles. Then they won't need affiliates at all. The time for traditional media to figure out if they have a future role in society is now.

Don't Think Outside Box – Destroy the Box!

Why won't more of the traditional media survive? Primarily for the reasons Mutter offered: rapidly declining circulation, demographics that will keep audience declining, rapidly falling advertising revenue, and an inability to trim their budgets adequately to match the declining revenues. If labor unions cooperated and media executives took radical action soon enough, more could probably successfully transition to digital-only media but would have to begin while their brand is strong, without a lot of debt and with a vastly reduced staff of perhaps 10-20% of what it used to have.

Christensen (American Press Institute, 2006), an expert on the impact of "disruptive technologies," tried to help the newspaper industry make the transition. However, he mostly taught them general strategies, such as looking for "jobs undone" – things not being done by anyone else, which their community would pay for with either their attention (for advertising) or their money. While Christensen's insights are valuable, he had no background in news media in order to make specific recommendations, and those trying to implement his strategies were stuck within their traditional paradigm.

Jarvis (2009), a longtime journalist, author and educator, on the other hand, declared soon after the first tipping point that newspapers should proceed as quickly as possible to close down their presses and welcome their digital future. He recognized that the expensive infrastructure and large organization required to put out a daily newspaper was previously a guarantee of limited competition, but with the arrival of the Digital Age, could be an "albatross around their necks." He realized that the transition would be traumatic, but concluded: "The longer we try to delay that inevitable change, the more we risk putting papers out of business. The sooner we try to experiment with new models, the sooner we can reinvent news for a new age."

Compete Against the Right Competition

There is another important reason why most of the traditional media will not survive. They are also trapped inside the traditional paradigm of competition. They seem to have no clue with whom they are competing or how to compete with their real competitors. They are still stuck in the mode of competing for declining local advertising dollars against their traditional local competitors when most of their revenue decline is caused by advertising moving to their online competitors. To compete with the new media giants, traditional media must be able to compete against the new strategies of Google (including YouTube), Facebook, et al. Those strategies that are destroying the traditional media include:

- **Pay-Per-Click (PPC) Advertising:** Advertisers can pay only if someone comes to their website landing page.
- **Greater Segmentation of Audience:** Only a carefully selected audience even sees the ad delivered over the Internet.
- **Social Media Marketing:** This is very enticing because it can begin with no budget whatsoever as so-called "inbound marketing" or "content marketing." Many companies are reporting that most of their sales leads now are coming from free social media posts and other inbound marketing, such as e-books, webinars, games, reports, instructional videos, and viral videos. As companies see high response rates from free social media postings, it is easier even for small enterprises to justify PPC advertising, native advertising (advertising that looks like the surrounding content but is, in reality, paid content), and other forms of social media advertising.
- **Video Advertising:** Television has been very successful to date with traditional video advertising, but so are online advertisers. As noted previously, marketers believe video is innately the most powerful form of advertising (Aberdeen, 2014). And the new tipping point relates to the Internet becoming the preferred medium for TV and other video entertainment. As video entertainment expands on the Internet, so will video advertising opportunities.
- **Programmatic Advertising:** Advertisers can use new computerized systems to find, negotiate, contract and deliver advertising. Programmatic advertising grew by about 250% in 2014 and is still growing. Programmatic ads can be video ads, social media ads, etc., so the distinction is in how they are contracted, produced and delivered, not in the content. This has facilitated advertising with more than just the biggest online media since the computer programs search throughout the Internet for the best advertising opportunities.
- **Mobile Advertising:** Like programmatic advertising, mobile advertising can be in the form of social media, PPC, video, etc. But it is important to see mobile advertising differently for several reasons. Mobile advertising can be seen anytime, anywhere – including while shopping, while traveling, on the job, when approaching a certain geographic location, etc. Geo-triggered advertising can be associated with special sales offers when consumers approach certain stores, for

example. Mobile marketing also involves creation of mobile-accessible websites, which Google now rewards with a higher search ranking. But traditional media must understand, as major online media already have, that mobile advertising is about to dominate all other venues of advertising. In the U.S., mobile advertising has been skyrocketing by over 170% in 2012 and 2013 and 78% in 2014, and becoming a larger portion of overall digital advertising – jumping from 25% in 2013 to 37% in 2014 (Pew, 2015, p.19). eMarketer (2015b) projects that mobile advertising in America and worldwide will continue to grow rapidly -- surpassing desktop and becoming the top digital platform by the end of 2016. Digital ad sales just in the U.S. are expected to grow from $19 billion in 2014 to $28 billion in 2015, $40 billion in 2016, $50 billion in 2017, and $58 billion in 2018. World mobile ad sales are expected to hit $69 billion in 2015 and jump to $167 billion by 2018.

Possible Strategies for Traditional Media vs. New Media

To compete against Google and Facebook is far more challenging than to compete against local media, but could be a challenge taken on by a major newspaper like the New York Times, a major chain like Gannett or McClatchy, or some other association of traditional media. Here are possible strategies for traditional media to become competitive.

- **Printed Media Should Sell Video Advertising:** Newspapers are already expanding their story-telling techniques, especially with online news videos. Since 90-95% of online marketers believe video to be the most powerful online marketing tool (Aberdeen, 2014), it makes sense to tie more online video-based news to online video advertising. As of 2013, however, no study had specifi-cally assessed how much of that is currently being done. Across the Internet, however, Pew notes that advertising tied to video increased about 44% in 2013 and represented about 10% of all online advertising (2014, p. 4). Whether part of an entirely new online TV channel or only within local news coverage, if print media are not increasing their video news and selling video advertising, they seem to be missing out on an important and increasing revenue opportunity. About one-third of Americans watch online news videos, according to Pew, but more would almost certainly do so as accessibility and quality increase. Besides TV and cable, relatively few news media have made a commitment to online advertising video. Of course, there are expenses involved for other news media, particularly print media, to substantially increase and improve their video content, but not as much as some might think. To stream high definition video can be quite expensive, but to stream standard definition video can be very cheap – as little as $45 a month for a 24/7 virtual newsroom (Harvey, 2015). As audience grows, expenses for live video would grow, too, but "almost live" might be good enough for the print media. For example, Twitter video can now be taken and edited on a good mobile phone in just a couple of minutes. It no longer requires a major investment in equipment or training to make newspaper reporters into multimedia reporters (Kamdar, 2015). And advertisers frequently provide their own commercials, and, if not, would pay for production.
- **Traditional Newspapers Could Launch Their Own Global Online TV Channel:** A major newspaper, newspaper chain, or association of local media could provide free, high quality global programming 24/7, associated or not associated with a major TV network. Local programming in either case can be part of the attraction to associated communities, but global programming must be available, too, in order to compete effectively with the online organizations currently grabbing

up so much advertising. Programming can include news-related shows, but almost certainly needs to include entertainment, as well. Entertainment could begin with old reruns, as have done some of the successful cable channels, then expand with new proprietary programming. Launching such a TV channel would seem rather easy for a strong brand like the New York Times. On the other hand, because it is so easy to duplicate websites, each local news medium within a chain or association could have its own customized version of a global channel.

- **Become More Mobile and More Social:** The strong trends related to mobile, to social media and to video are intertwined, and to compete effectively, traditional media need to consider how to tap into all three trends. More and more of the social media interaction is now mobile-driven. Facebook, according to Mixpo (2015), recently overtook YouTube as the site with the most video views via desktop computers. According to Mixpo's report, using data gathered by comScore, Facebook's desktop views skyrocketed from 4.9 billion views in January 2014 to 9.6 billion in February 2015, overtaking YouTube, which shrank in video views from 11.9 billion desktop views to 9.4 billion during that same period of time. More unique visitors watched videos on YouTube, but Facebook's views per person were nearly double those of YouTube – 107.4 vs. 69.5. But desktop view statistics provided by comScore were not an accurate measurement of total usage, or of the true trend line, according to Facebook. The social media leader claims that its video views climbed steeply from September 2014, when its total (desktop plus mobile) video views were about 1 billion per day (30 billion per month), to January 2015, when its total views reached 3 billion per day (90 billion per month). If that is true, then desktop views in January 2015 represented only about 10% of Facebook's total video views. According to comScore, the desktop video views on Facebook in September were about the same as in January, which means mobile users would have driven almost all of that meteoric increase during those four months. This has implications for both traditional and online news media. To mirror Facebook's success, traditional news media should consider incorporating all of those same elements – become ultra-mobile friendly, use a lot of video in both content and advertising, and find a way to incorporate more social interaction. An organization like Associated Press might be able to provide a news-based social network involving all of its members, helping to drive more audience to individual members' websites and to get the audience more involved with the news. Sharing news in existing social networks or creating their own, according to Pew (2014, p. 4), could also enhance citizen journalism. But using established social media in this effort seems rather self-defeating since the current social media and Google are the main beneficiaries of the traditional media's lost advertising. Nonetheless, Pew reports that about 10% of all social media users say that they have taken news videos of their own and posted them to social networks. About that same percentage reports to having posted some kind of news content (video, photos or articles) to regular news websites or blogs. With their own social site, news media could foster greater audience interaction and citizen journalism, as well as benefit financially, but it is unlikely that a local medium by itself could achieve the critical mass needed to create a viable social medium. It would require broader online cooperation between traditional news outlets.
- **Use GPS-Associated Services to One-Up Global Media:** Facebook and Google can provide local advertisers with a geographically restricted audience. But their local news and advertising services are still otherwise quite limited. A whole new kind of localized advertising can be developed with GPS-associated information and advertising. Consumers could point their mobile phone at a restaurant and call up its menu, point it at an apartment building to see if there are any

vacancies and at what price, point it at a grocery store and call up a list of items on sale, or point it at another store and call up its virtual coupons. Local media are in a better position to provide such promotional services if they are willing.

- **Think Beyond the News:** For a large newspaper or chain to create a full-service online TV channel is certainly going beyond the news, but there are other possible ways to utilize a traditional medium's brand, advertising capability, customer relationships, distribution network, etc. For example, as an incentive for a three-year subscription contract to the online version of a newspaper, a subscriber could be given a smartphone or a tablet with mobile phone capability, and could have a one-touch connection to its online TV channel. The home page of the device would be set to the newspaper's website. The newspaper could piggyback on an optional cell phone service, negotiated with an existing cell phone company but white-labeled with the newspaper's brand. For a larger newspaper, the smartphone's cost would be covered within one or two years just by the savings in printing and delivering a traditional newspaper. A revenue-sharing plan negotiated with an online university would be another non-news product. If a story about an upcoming election is run, for example, an accompanying link could lead the reader to a university political science course of possible interest. University professors might be invited to become bloggers for the newspaper, with links from their blogs to their courses. Book publishing is another possibility.

- **With Expanded Online Presence, Initiate Other Popular Advertising Options:** If a large newspaper or news media chain or association initiates an online TV channel and/or popular online social medium, it is also in position to begin offering the same advertising options as do Google and Facebook. PPC, native, video, programmatic and segmented advertising will all then be possible. It is only then that traditional media will become competitive with their primary competitors of the Digital Age.

MAPPING FUTURE RESEARCH

This chapter has reviewed the current and near-future reality of the Internet, and these revolutionary trends must be taken into account as we consider what impact technical innovations will have on the socio-political and journalistic future.

Applying Communications Theory to the New Information Ecosystem

Before 1940, the prevalent theory of mass communication attributed great power to the media. The magic bullet theory suggested the media had an almost universal and powerful impact on individuals. However, in 1944 Lazarsfield's People's Choice study showed that messages delivered during the 1944 presidential campaign did not have as much a direct effect as an indirect, moderated effect. Individual opinion leaders shared the news with others, intertwined with their own interpretations and perspectives. And these opinion leaders had a greater impact on voters than did the media. Thus, in the 1940s Katz and Lazarsfield rejected the magic bullet theory and developed the two-step flow theory of mass communications, later expanded into the multistep flow model. With the audience now enabled by mobile devices and social media, evidence of the multistep flow model has never been stronger, and opinion leaders have never had so many tools with which to spread their moderated information (Baran, 2006, pp. 421-22; Lule, 2013, pp. 68-73). This should not be seen as a threat to professional journalists but an opportunity create a virtual town hall-style discussion of the news, which will entice more citizens to become factual reporters, as well.

The new information ecosystem shines a whole new light on old communication theories, which could lead to verification of some theories and creation of new ones. It was not long ago when no one had ever studied a case of "shocking good news" that would spread through society quickly and mostly person to person rather than through the media, such as how news of President John Kennedy's assassination and later that of Robert Kennedy's viralized so that essentially everyone had heard the news within one day. The first case of shocking good news was documented by Haroldsen and Harvey in 1979, but now such viral action is seen every day on the Internet, and "shocking" continues to be one of the elements that professionals cite as a key to viralizing a message. Studies about viral messages might well lead to a new related theory that would be of value to industry and academia.

Dissonance theory suggests that people consciously and subconsciously resist new and conflicting ideas and information that cause them to feel mental dissonance or discomfort. This affects how people selectively expose themselves to such conflicting information, how they perceive or comprehend it when exposed, and even if and how they retain it after they are exposed. New experimentation can be conducted relating to the selective processes. Selective exposure, attention, perception and retention can all be examined within the new Web ecosystem Baran, 2006, pp. 423-426). And questions are being raised about the socio-political impact of the Web facilitating the selective processes in such a way that the gulf between political extremes seems to be widening and the ability of people to understand one another's position and to find compromise seems to be disappearing.

While the almost limitless informational, educational and entertainment outlets available online allow people to easily select or avoid dissonant situations, they also allow for expanding uses and gratifying opportunities anywhere and anytime. This should lead to new research related to the uses and gratifications theory of media communications, which maintains that media don't just do things to people but people purposefully select what they will do with media (Baran, p. 427). When media communication was limited in available channels and almost totally one-way in nature, it was easier to discount the importance of this theory. Now it is the quintessential nature of modern communication and demands further research.

Interactivity theory began as a subject related to interpersonal communications but quickly encompassed mediated communications. Through media, to what degree could true interactivity be achieved? Rafaeli (1986) sought a way to measure interactivity as an independent variable in research. The degree to which interactivity is achieved, he argued, depends on how much current communication depends on previous communications. One-way mediated communications is clearly not interactive. A quasi-interactivity can be achieved when information is dispensed on demand, such as an online library or some online news media. That is no more truly interactive, he would argue, than a vending machine that dispenses a candy bar once a person has inserted a coin and punched a button or two. True interactivity requires two-way communication and increases in degree and complexity as each message responds to and is influenced by not just one previous message but an entire series of bidirectional or multi-directional messages.

Later research by Rafaeli and associates (Kalman, Raban and Rafaeli, 2013) suggests that as interactivity increases, social cognition is facilitated. Individual cognition – the process of perceiving and acquiring knowledge and understanding – is greatly influenced by social environment, and social behavior is influenced by cognition (p. 23). Certainly the interactivity and social cognition theories can be applied to social media, but news media are also becoming increasing interactive themselves. Not only do most news media now offer a way for news consumers to comment on professionally prepared news stories and commentaries, which can lead to a varying degree of bidirectional and multi-directional conversation, but consumers can now become citizen journalists and commentators within the professional media or as independent bloggers, still influencing the overall ecosystem. Professional news media and citizen journalists also interact with other news consumers within the major social media, such as Facebook. This is the new normal, and, as previously suggested, opens new opportunities for professional media as they understand the importance, power and potential of enhanced interactivity.

The "netification of social cognition," according to Kalman, et al, also offers a wealth of research opportunities for academics. There is room within this realm for research by experts in behavioral sciences, sociology, communication, computer science and information systems, as well as those who would apply findings to such specific disciplines as education, political science and marketing. And netification changes the parameters and perimeters of these concepts, they conclude (p. 36-37). This invites an interdisciplinary and multidisciplinary approach, as sought with this publication. The new technology also facilitates such research by accumulating and storing immense data-sets. To the degree that industry is willing to share "big data," scholars can help both academics and professionals to better understand the functions and impacts of interactivity and social cognition as they are affected by new technologies, including the increasingly impactful mobile technologies (p.37).

Social cognition is already being explored within the realm of mass media. The social cognitive theory of mass communications suggests that people model behaviors they see via media (Baran, pp. 429-430). What if the images and behaviors observed on a widening array of media are increasingly fragmented, as is news and information? How will this change the impact? Also in the past, cultural theories focused mostly on how media sustain and affect local, regional or national cultures. As Web- and mobel-driven global communication exposes the world audience to numerous cultures, the focus of cultural research must change dramatically.

While these theories may become increasingly important with new media, other theories may diminish in their importance. For example, the agenda-setting theory suggests that news media don't always tell us effectively what to think but do tell us what to think about (Baran, pp. 427-428). That was almost certainly true when a community was limited primarily to one daily newspaper and a handful of TV stations affiliated with national news organizations. To what extent will this theory still have a place in modern communications research as the number of media grows and their messages fragment? Certainly this should be explored within the new ecosystem.

Researching the Practical Side of Journalism

Journalism research was initiated by academics in the fields of sociology and psychology and continues largely within that framework. With all of the technology-driven changes, industry's need for research in the fields of business and marketing increases. Their survival depends on it. The newspaper industry lost $29.5 billion in annual revenues between 2005 and 2014. Where did it go? Not to TV. TV is still the most successful of the traditional media, which is evidence that newspapers should adopt much more video into their repertoire. But their main competitors are Facebook and Google, and all of the traditional media must figure out how to compete with these new media to survive in the virtual world. What should the converging news media that are competing against Facebook look like? A lot can be learned in studying the current battle between the titans of the online world. What has occurred just in 2014-2015 with Facebook's rising stature suggests that a successful online strategy should be mobile and multimedia. Once Facebook incorporated those two elements into its social setting, it began overtaking Google/YouTube in video views and in advertising revenue. Where that battle is right now is open to debate because of the different definitions of a "video view" and the lack of outside verification of mobile views, which Facebook claims to dominate (Nasr, 2015). However, the trend line greatly favors Facebook, and its 2015 innovation of auto-run videos will likely put it over the top by any standard. Already in the first quarter of 2015, Facebook's advertising revenues were up 46% over the same time in 2014, and mobile accounted for 73% of its total ad sales (Heine, 2015). The Mixpo study (2015) claimed that Facebook had already overtaken Google/YouTube in video views by early 2015, and that more advertisers were planning to run video ad campaigns on Facebook than on YouTube during 2015 (87%-81.5%). This survey showed a great turnaround and great overall growth in video advertising from the previous year's survey, when advertisers planning video ad campaigns favored Google/YouTube over Facebook 77.8%-63%. Google and its YouTube subsidiary appear to have been too one-dimensional within their separate entities, and Facebook has been able to exploit that weakness. But there are other elements the converging media might adopt from Facebook. It may not be enough to offer news that involves in-depth text and compelling "as if you were there" video. Newspapers may need to become more dialogic, personalized and viralized. Citizens of the Web now expect to have their say. They want to be part of the conversation. And this can be the key to the converged media being able to capture the power of citizen journalists, as well. If the news is provided in a stream, like Facebook, the invitation at the top might not read, "What's on your mind?" but rather, "Share your news, your photos, your videos and your analysis." Each subscriber might have his own personalized page, like on Facebook, viewed only by his friends, fans and followers, but all subscribers could receive a constant stream of multimedia news from the converged medium throughout the day, and surrounding links could allow readers with one click to see all of the day's news related to specific topics such as breaking news, sports, international news, national news, politics, state news, local news, entertainment, education and human interest. New media theorist Jeff Jarvis (2009) foresees new ways that media and readers can work together, and a Facebook approach might facilitate that. Partnerships might also be explored that would allow multiple converging media to contribute to the same news feed, for example, or it can be an informal partnership developed by the subscribers themselves. Already on Facebook a subscriber can follow and share a mixture of stories from the Washington Times, the Washington Post, the Wall Street Journal, NBC and Fox News, for example, personalizing and sharing these within his own news feed. A social media news site could foster this kind of pluralistic interaction, formally or informally creating a subscriber-driven news partnership. Indeed, a local medium might include on its website RSS feeds from national and international media to facilitate this subscriber-driven news partnership. While allowing subscribers to personalize their news stream, there should also be a mechanism to allow them to viralize their contributions. That may be achieved automatically within the community served by a news medium, but it could be to the medium's benefit to draw more readers who are not formal subscribers. For most news media, subscriptions within the Web environment might have to be free, with perhaps some paid additional services, such as a social club for singles, online university courses, investigative services, entrepreneurial club for those wanting to start their own business, promotional consulting services, professional photography for special events, customized mobile devices and telephonic services, special entertainment services, sponsored "native messages" distributed to all subscribers beyond friends and fans, etc. With free subscriptions, a converged medium's audience could be expanded beyond its previously recognized community. Also, a way for subscribers to upload their non-subscribing friends' email addresses could be provided so the subscriber could automatically send their customized news feed to other friends via email.

In the end, financial and competitive forces within the converged environment may force successful news media to include such successful online elements into their own Web presence. Media convergence onto the Internet was anticipated for decades and is now nearing reality. A new theory might predict "convergence within convergence." This theory would suggest that competition will drive the evolving media into essentially all becoming multimedia, dialogic, mobile and personalized. To compete online, newspapers might also be forced to become TV stations, and TV stations might also have to be newspapers, so to speak. And both may be forced to become social media. In the real world each had distinct advantages. In the virtual world they may not be able to remain one-dimensional and survive. Part of this will be driven not by the effectiveness of multimedia news but by the need to compete for advertising. This will also be driven in part by the ultimate and necessary realization that the traditional media's main competitors are NOT each other but Facebook, Google/YouTube and the other online media.

Facing the current realities, the following questions may become research topics over the next few years:

- Have traditional news media reached a second tipping point? If so, will it play out the same as the first tipping point and its impact especially on print media? By experiencing the first tipping point, can the news industry mitigate the impact of the second tipping point?

- Can video advertising help newspapers to stop the rapid decline of advertising revenues? Could the demand for online and mobile video provide the opportunity to sell video ads and embed them within news videos?

- Could the Twitter video app lead to a new kind of "broadcast journalism," where journalists will produce video journalism reports and post them immediately to their Twitter stream? Print-based reports could then be filed later. As print media begin using more video, will their audience demand the quality achieved by the TV stations, or will they be able to compete with lower-quality but more timely news video, and perhaps higher-quality video for in-depth interviews, talk shows, debates, etc.?

- Might larger newspaper organizations, such as the New York Times, Gannett or McClatchy, benefit by taking even greater advantage of such trends by launching full online TV networks, including entertainment and educational video content, as well as news content?

- Likewise, is there a viable opportunity for traditional media to collaborate in creating a social medium with enough interesting, interactive content and critical mass to become competitive with other social media in both audience and advertising.

- While the convergence of media onto the Internet is clearly occurring before our eyes, will we witness that such pressures and opportunities as those discussed above drive all online news organizations to converge media within their own operation, becoming much more alike than they are now different?

CONCLUSION

No industry is changing faster than the communications industry, and caught in the middle of all these disruptive changes are the news media. Since an informed public is critical to the successful function of a democracy, there are few organizations more critical to society than those that gather and report the news. While most people are now getting their national and international news from the Internet, most of it is originating with the traditional media. Even more so for local and regional news. This makes their survival of vital interest to everyone. While the Internet is opening up all kinds of new research possibilities, none are more important than those that can help news media to continue fulfilling their function in a democratic society.

REFERENCES

Aberdeen Group. (2014). *Analyzing the ROI of video advertising.* Retrieved 16 Feb. 2015 from http://go.brightcove.com/bc-aberdeen-analyzing-roi-b

American Press Institute. (2006). *Newspaper Next: Blueprint for Transformation.* Retrieved 15 Jan. 2015 from http://www.americanpressinstitute.org/training-tools/newspaper-next-blueprint-transformation/

Ayres, N. (2014). Adventures in Publishing: The new dynamics of advertising. *Brand Perfect.* Retrieved 14 April 2015 from http://brandperfect.org/brand-perfect-dynamics-of-advertising-report.pdf

Baran, S. (2006). *Introduction to Mass Communication: Media Literacy and Culture* (4th ed.). New York City, NY: McGraw-Hill.

Blueshift Research. (2015). *June 2015 Trends Tracker Report.* Retrieved 2 July 2015 from https://smaudience.surveymonkey.com/download-trends-tracker-report-june-2015.html?utm_source=email&recent=email&program=Q2_15_June_Trends-Tracker_Wrap-Up&source=email&utm_campaign=trends-tracker&utm_content=trends-tracker-june-2015&mkt_tok=3RkMMJWWfF9wsRoiu6vJZKXonjHpfsX66e4vW6C1lMI%2F0ER3fOvrPUfGjI4FTcFiI%2BSLDwEYGJlv6SgFSrbFMaJy2LgJWBb0TD7slJfbfYRPf6Ba2Jwwrfg%3D

Carufel, R. (Ed.). (2015). *Millennials Moving Away from TV: Boob Tube Content "Doesn't Cut It"—Digital Delivers More Relatable and Entertaining Programming.* Retrieved 15 Oct. 2015 from https://www.bulldogreporter.com/millennials-moving-away-from-tv-boob-tube-content-doesn-t-cut-it-digi/

Chyi, H. I. (2015). *Trial and Error: U.S. Newspapers' Digital Struggles toward Inferiority.* The University of Navarra. Retrieved 15 Oct. 2015 from http://irischyi.com/

Cisco. (2012). *Cisco's VNI Forecast Projects the Internet Will Be Four Times as Large in Four Years.* Retrieved 15 May 2015 from http://newsroom.cisco.com/press-release-content?type=webcontent

Cisco. (2014). *2014 Connected World Technology Final Report.* Retrieved 15 Oct. 2015 from http://www.cisco.com/c/dam/en/us/solutions/collateral/enterprise/connected-world-technology-report/cisco-2014-connected-world-technology-report.pdf

Cisco. (2015). *Forecast projects nearly 10-fold global mobile data traffic growth over next five years.* Retrieved 16 Feb. 2015 from http://newsroom.cisco.com/press-release-content?type=webcontent

comScore. (2014). *Half of Millennial Netflix Viewers Stream Video on Mobile.* Retrieved 20 Feb. 2015 from http://www.comscore.com/Insights/Data-Mine/Half-of-Millennial-Netflix-Viewers-Stream-Video-on-Mobile

Downie, L. (2009). *The Future of Journalism: Where We've Been, Where We're Going.* Speech given at John S. Knight Fellowships Reunion 2009 given at Stanford University. Retrieved 15 Oct. 2015 from https://youtu.be/JYtOKnk5fiw?list=PLfSN69RCPDM-HID_MaJ7TxlfwMsnMAqPc

eMarketer. (2013). *Digital TV, Movie Streaming Reaches a Tipping Point*. Retrieved 20 Feb. 2015 from http://www.emarketer.com/Article.aspx?R=1009775

eMarketer. (2015a). *Worldwide ad spending*. Retrieved 30 May 2015 from http://www.emarketer.com/adspendtool

eMarketer. (2015b). *Mobile Ad Spend to Top $100 Billion Worldwide in 2016, 51% of Digital Market*. Retrieved 30 May 2015 from http://www.emarketer.com/Article/Mobile-Ad-Spend-Top-100-Billion-Worldwide-2016-51-of-Digital-Market/1012299/1

eMarketer. (2015c). *Cross-platform video trends roundup*. Retrieved 20 Feb. 2015 from https://www.emarketer.com/public_media/docs/eMarketer_Cross_Platform_Video_Trends_Roundup.pdf

eMarketer. (2015d). *Seven in 10 US Internet Users Watch OTT Video*. Retrieved 15 Oct. 2015 from http://www.emarketer.com/Article/Seven-10-US-Internet-Users-Watch-OTT-Video/1013061?ecid=NL1001

Gladwell, M. (2000). *The Tipping Point: How Little Things Can Make a Big Difference*. New York: Little, Brown and Company.

Haroldsen, E., & Harvey, K. (1979). The Diffusion of Shocking Good News. *The Journalism Quarterly*, *56*(4), 771–775. doi:10.1177/107769907905600409

Harvey, K. (2015). *How to create your own informational, promotional & educational videos*. Retrieved 15 June 2015 from http://iei-tv.net/0MakeVideos.htm

Heine, C. (2015). Facebook's Q1 Ad Revenue Increased 46 Percent to $3.3 Billion. *AdWeek*. Retrieved 30 July 2015 from http://www.adweek.com/news/technology/facebooks-q1-ad-revenue-increased-46-percent-33-billion-164234

Hershkowitz, S., & Lavrusik, V. (2013). *12 Best Practices for Media Companies Using Facebook Pages*. Retrieved 15 Oct. 2015 from https://www.facebook.com/notes/facebook-media/12-pages-best-practices-for-media-companies/518053828230111

Ingram, M. (2014). NYT Reporter Shows the Power of Twitter in Journalism. *GigaOm*. Retrieved from https://gigaom.com/2011/05/27/nyt-reporter-shows-the-power-of-twitter-as-journalism/

Jarvis, J. (2009). *Newspapers in Decline?* Video presented on the CBC (Canadian) TV network. Retrieved 15 Oct. 2015 from https://youtu.be/YjUeJH4mdF4

Kalman, Y. M., Raban, D. R., & Rafaeli, S. (2013). Netified: social cognition in crowds and clouds. In Y. Amichai-Hamburger (Ed.), *The Social Net: Human Behavior in Cyberspace* (2nd ed.). Oxford, UK: Oxford University Press. doi:10.1093/acprof:oso/9780199639540.003.0002

Kamdar, J. (2015). *Now on Twitter: group Direct Messages and mobile video camera*. Retrieved 15 Oct. 2015 from https://blog.twitter.com/2015/now-on-twitter-group-direct-messages-and-mobile-video-capture

King, C. (2015). *28 Social Media Marketing Predictions for 2015 from the Pros*. Social Media Examiner. Retrieved from http://www.socialmediaexaminer.com/social-media-marketing-predictions-for-2015/

Merkle/RKG. (2015). *Digital Marketing Report Q1 2015*. Retrieved 21 May 2015 from http://www.rimmkaufman.com/content/quarterly/Merkle-RKG-Q1-2015-DMR.pdf

Mixpo. (2015). *The state of video advertising.* Retrieved 1 May 2015 from http://marketing.mixpo.com/acton/fs/blocks/showLandingPage/a/2062/p/p-00b4/t/page/fm/0

Mutter, A. D. (2012a). *The incredible shrinking newspaper audience.* Academic Press.

Mutter, A. D. (2012b). Print ads fell 25x faster than digital grew. *Reflections of a Newsosaur.* Retrieved from http://newsosaur.blogspot.com/2012_09_01_archive.html

Mutter, A. D. (2015). Should newspapers abandon digital? *Editor & Publisher.* Retrieved 15 April 2015 from http://newsosaur.blogspot.com/2015/10/should-newspapers-abandon-digital.html

Nasr, R. (2015). Munster: Facebook is Quickly Catching Up to Facebook. *CNBC.* Retrieved 15 Oct. 2015 from http://www.cnbc.com/2015/07/27/munster-facebook-is-quickly-catching-up-to-google.html

Nguyen, J. (2014). *Online Video Consumption in APAC and Total Video.* Retrieved 16 Feb. 2015 from http://www.comscore.com/Insights/Presentations-and-Whitepapers/(offset)/10/?cs_edgescape_cc=KZ

Pepper, R. (2012). *The impact of mobile broadband on national economic growth.* Retrieved 16 Feb. 2015 from http://newsroom.cisco.com/press-release-content?type=webcontent

Pew Research Center. (2014). *State of the News Media 2014.* Retrieved 30 May 2015 from http://www.journalism.org/packages/state-of-the-news-media-2014/

Pew Research Center. (2015). *State of the News Media 2015.* Retrieved 30 May 2015 from http://www.journalism.org/files/2015/04/FINAL-STATE-OF-THE-NEWS-MEDIA1.pdf

Rafaeli, S. (1986). The electronic bulletin board: A computer-driven mass medium. *Computers and the Social Sciences, 2*(3), 123–137. doi:10.1177/089443938600200302

Reflections of a Newsosaur. (2012). Retrieved 15 April 2015 from http://newsosaur.blogspot.com/2012_10_01_archive.html

Ross, P. (2015). *Native Facebook Videos Get More Reach Than Any Other Type of Post.* Retrieved 1 May 2015 from http://www.socialbakers.com/blog/2367-native-facebook-videos-get-more-reach-than-any-other-type-of-post

Shirky, C. (2009). *How social media can make history.* TED.com presentation. Retrieved from http://www.ted.com/talks/clay_shirky_how_cellphones_twitter_facebook_can_make_history

The Economist. (2014). *The world in numbers: Industries.* Available at http://www.economist.com/news/21632039-telecoms

White, D. S. (2013). *Social Media Growth 2006-2012.* Retrieved from http://dstevenwhite.com/

Wilkinson, M. B., & Halton. (2014). *Mobile and Public Relations.* Retrieved 25 May 2015 from http://newsroom.cipr.co.uk/ciprsm-launch-new-guide-to-mobile-and-public-relations

Xstream. (2013). *Online video trends & predictions for 2013.* Retrieved 31 May 2013 from http://www.xstream.dk/content/online-video-trends-predictions-2013

Chapter 10
Journalists and Mobile:
Melding Social Media and Social Capital

Hans Karl Meyer
Ohio University, USA

Burton Speakman
Ohio University, USA

ABSTRACT

Despite the high profile of many social media activist efforts, such as the Arab Spring, researchers are still questioning whether these mostly online campaigns can lead to real world impact. Journalists are also asking themselves what role they fill as they watch, comment on, and cover the deluge of activism on Twitter, Facebook, and other sites. Through a comprehensive review of the literature on social capital and its intersection with the Internet and social media, this chapter suggests that social media can lead to social capital, but journalists provide the key ingredient to lead to lasting social change. Literature on the goals and aims of journalism, coupled with a review of its vital role in a community, point to the need for the context and verification that journalism provides in order for social media to transform into social capital.

INTRODUCTION

When Facebook become became the megaphone for Tunisian citizens to organize themselves and trumpet their causes to the world, many journalists, including Andy Carvin (2014) took notice. What he first saw in Tunisia and ended up chronicling closely throughout Egypt, Libya and other Middle Eastern and North African countries became what is now known as the Arab Spring, a movement that capitalized on social media in several parts of the world to enact real world government reforms. For Carvin, it also represented a "stunning revolution in the way breaking news is reported around the world – and who controls the news"

DOI: 10.4018/978-1-5225-0469-6.ch010

With countless revolutionaries using the Internet as part of their protests, anyone online could gain direct access to the news moment by moment – no filters, no spin, no delay. No longer did media outlets have a monopoly on international reporting; people on Twitter or YouTube could patch directly into the revolution of their choice. (p. xii)

For Carvin, social media director for National Public Ration in Washington D.C., the revolution couldn't be ignored because journalism had an important role to play. Through his book and his reporting on social media, Carvin, a journalist, told the stories that "would have otherwise fallen through the cracks of history if people on the ground weren't using social media to tell them" (p. xv). While he does not take credit for enabling or sustaining these movements, he does ask "What exactly does it take for one of these movements to go all the way and overthrow a regime?" (p. 7).

Within a less specific and less revolutionary frame, this chapter asks the same question. What does it take for an online movement such as those Carvin describes or other even less impactful movements such as the ALS Ice Bucket challenge, discussed later in this chapter, to gain real world traction after getting a start in online social media? In other words, can social media lead to real world social capital? And what is journalism's effect on the social capital that can be generated? Does a movement that begins with a niche audience on social media need the mass communication influence of mainstream journalism to really make a difference? Finally, how can academic research help denote the impacts of each and on each other?

We explore these questions by first defining social media in terms of real world and social media application. Next we explore the connection between the goals and mission of journalism as it relates to social capital and community building. Finally, we examine several examples, including what has been written of Carvin's work during the Arab Spring, to suggest that while social media has evolved to defy early social media theorists' predictions, it often needs the influence of journalists to shine a light on what's most important and help vet information to reveal what is true and what is unsubstantiated rumor or needless hyperbole.

DEFINING SOCIAL MEDIA

Social Capital

Social capital, in its simplest form, is the "ways our lives are made more productive by social ties" (Putnam, 2000, p. 19). In fact, society is most powerful when civic virtue is "embedded in a dense network of reciprocal social relations" (p. 19). In other words, social capital has transformative power when individuals can use their family, friends, and group affiliations to accomplish things for their lives, namely to get jobs, help government run efficiently, or just pass the time pleasurably with one another.

In the United States, social capital has always ebbed and flowed, Putnam (2000) says, even at some points collapsing and needing complete renewal (p. 23). However, the major premise of his book and the reason behind the title *Bowling Alone* is that Americans are becoming increasingly individualized and less likely to build social capital through leisure and political activism. Two of the main culprits Putnam blame are the media and technology. First, news and entertainment have become increasingly individualized (p. 216). People no longer need to attend a concert with scores of other community members to get their music fix or visit theaters to see plays or movies. Even the community building elements of news where people stand around the watercooler discussing the day's events are fading because electronic technology allows us to learn about the world entirely alone. Often modern media also blur the line between news and entertainment, further hindering society's ability to do something useful with what it learns. "At the very least, television and its electronic cousins are willing accomplices in the civic mystery we have been unraveling, and more likely than not, they are the ringleaders" (p. 246).

Social Media as Democratizing Force

The Internet, as one of television's electronic cousins, was not supposed to pull us away from civic engagement. In fact, Howard Dean, former chairman of the Democratic Party, called the Internet the "the most important democratizing invention since the printing press, 500 years ago" (quoted in Mayo and Newcomb, 2008). In fact, many people disagree that the Internet, in particular, has undermined social capital creation. Vergeer and Pelzer (2012) found that online activities that are supposed to be solitary, such as surfing for news or entertainment are increasingly becoming intertwined with socializing. Online and offline social capital are positively associated, they say, because people may simply be copying their offline social networks to the online realm.

This is particularly true of social media networks, such as Facebook and Twitter, where Valenzuela et. al (2009) said social media have developed, in part, as places to foster civic and political engagement, especially among young people. In fact, those who are most civic minded choose to join Facebook more often than those who are not. One reason, Ellison, Steinfeld and Lampe (2007) suggest is that Facebook makes it easier to communicate about civic issues and can even engage those who might otherwise be shy about engaging in discussion. In the social capital realm, social media such as Facebook might increase the potential to turn lists of online friends into social capital because it is easier to see from a list of connections which might be most useful in the real world (p. 1162). On the other hand, the authors found that social capital might be more difficult to sustain through social networks because how easy it is to make friends on social networks does not encourage the creation of strong ties. This effect cannot be explained by Internet use alone. Users need social media to make connections clear (Ellison, Steinfeld & Lampe, 2007).

Shah et al. (2001) also disputed the notion that all electronic activity blurs the lines between entertainment and news. The creation of social capital online relies mainly on the motives of the users. Those who actively choose to go online to exchange information are more likely across age groups and motivations to generate social capital. In fact, these active users are exposed to more mobilizing information and opportunities for recruitment than those who rely on traditional media, and they are able to more easily control the opportunities they select (p. 154).

Even Putnam backpedals a bit from his earlier assertion about the detriment to social capital that the Internet represents. In *Better Together*, co-written with Feldstein and Cohen (2004), he examines craigslist as a case for the creation of weak ties and social capital online. While not a perfect example of the kinds of offline communities that lead to real world social capital, they (2004) admit that craigslist has strong elements of community because it embodies many of the ideals that make real world communities vibrant: "localness, member participation in defining the norms of the group, aims and purposes beyond that of simply being together" (p. 240). While this one example does not suggest that everyone will abandon face-to-face relationships for online ones, craigslist does "suggest a role for the Internet in the mix of ways that people come to know, trust, and connect with one another" (p. 240).

The kinds of interactions Putnam and his colleagues describe on craigslist make other social networks effective as well. The key to creating bridging social capital through social media, Ellison et al. (2014) said, was making calls to action and receiving responses. This exposes users to not just their friends online, but friends of friends and enhances their network of weak ties. Nearly one-fourth of active Facebook users are sharing at least one request each month. "This represents a huge number of potential social capital conversions on and through the site" (p. 1118) and therefore a sizeable potential increase in the bridging social capital people can create through social networks.

Social Media Activism

Whether these connections through friends of friends are actually making a difference, however, remains up in the air. New media could create social capital, but are people actually taking advantage of their new power? The prevailing opinion, which has led to the creation of derisive terms such as hashtag activism, clicktivism, or slacktivism, is that they are not. The Internet has changed what were once primarily collective efforts for social action and turned them into mostly individual actions online (Shah, Friedland, Wells, Kim, & Rojas, 2012). In other words, people are largely content to use their new connections to advocate for a cause online without actually making a difference in the real world. This has become increasingly common, as society has moved many social activist campaigns online. The question that must be asked is if these online movements have the same influence as more traditional protest movements in enacting social change (Shah, Friedland, Wells, Kim, & Rojas, 2012).

Criticism of hashtag advocacy centers around one simple idea: It is easy for someone to "like" something on Facebook, or retweet something about a charitable effort. But is liking something on Facebook the equivalent of digital graffiti with no real benefit (Carr, 2012), or does social media activism go deeper and result in real engagement? Carr argued that only a few of the people who comment on social media about a social cause will do anything; the rest are simply burnishing their profiles. The ties of the digital world do not equal real world engagement (Carr, 2012). Furthermore there are those who claim the brief synopses offered on social media oversimplify issues (Dewey, 2014). In addition, social media has the ability to complicate things for the people campaigns are attempting to help (Welles, 2014). The international attention that comes with these efforts is powerful, but the added exposure can make things dangerous for those on the ground (Welles, 2014).

While social media allows these campaigns to gain traction quickly, it is difficult for those who started them to define objectives. Efforts quickly move beyond the intent of their creators (Herman, 2014). Organizers must understand that there is some element of narcissism involved in posting, and not everyone who likes something will take the cause seriously (Herman, 2014).

But there are positive signs using terms such as "slacktivism" may simply be an easy way to write off the influence of social media in an era where they have become the primary method for advocating social change (Ojigho, 2014). The Internet and social media are effective channels for learning about social issues and supporting causes or candidates (Wihbey, 2013). In the same vein, technology has created a new type of "media activist" who posts firsthand accounts from events and can provide coverage from the ground that might not be possible from traditional media (Simon, 2014). These social media movements can have sizable reach, and the most successful gain additional exposure through media coverage (Welles, 2014). Social media sources are gaining in popularity at a time of growing mistrust of the press (Kovach & Rosenstiel, 2001). A number of people even question if the media are invested in the public interest (Kovach & Rosenstiel, 2001). The challenge for social media campaigns is to capitalize on the potential for social capital to create genuine activism.

This challenge grows as some of the issues with technology that Putnam highlighted create difficulties in building engagement in the modern media era (Herman, 2014). The Web has increased connectivity and involvement, yet it can reduce their commitment to community (Wellman, Haase, Witte, & Hampton, 2001; Shah, Kwak, & Holbert, 2001). It is important to determine if people are working with others and building social capital or participating in asocial activities that turn them away from the community (Wellman, Haase, Witte, & Hampton, 2001) because connecting online and the real world can have lasting results. Online communities grow faster when they are created within traditional communities (Blanchard & Horan, 1998). Social media has been more effective than other sources in creating the intent to join a real-world cause (Jeong & Lee, 2013). Increased visibility from social media has also made individuals more likely to get involved in real-world causes (Jeong & Lee, 2013).

In addition, those who receive information from social networking sites like Facebook and Twitter are more likely to participate in social and political behavior, either online or offline (de Zuniga, 2012). Furthermore, social capital increases in virtual communities when opportunities exist for civic engagement (Blanchard & Horan, 1998) because people will actively choose the media sites that provide them with best gratification of their wants and needs (LaRose & Eastin, 2004). Online efforts for social causes can occur through social media without any sort of recognized leaders and allow for decentralized movements in a manner not previously possible (Howard & Hussain, 2011). This has led to more people willing to become involved in activism using social media if they believe the efforts will help to influence others (Lim & Golan, 2011). The online world provides the younger generations with a new outlet to become involved in traditional political behavior (Bode, Vraga, Borah, & Shah, 2014).

Social Media Advantages

The real measure of the efficacy of social media campaigns is the difference they have made, according to those who use social media to spur activism. They do not believe hashtag activism is a copout, or worse yet an act of vanity (Bertlatsky, 2015). Social media sites provide a vital way for protesters separated by geography to connect with one another (Bertlatsky, 2015; Gheytanchi & Moghadam, 2014). Further, social media allow activists to present their own message and create a community around an issue (Bertlatsky, 2015). This direct connection allows them to frame their stance without media interference. In addition, those who support social media campaigns compare liking something on Facebook to simply signing a petition (Dewey, 2014). Both are passive activities that require no further action but each holds the potential to become something greater. "Clearly, re-tweeting a message does not constitute activism the same way that donating time or money does, but re-tweeting a message could be part of a set of behaviors that lead to social change" (Welles, 2014). Social media have been a significant predictor an individuals' involvement in protest and traditional political behavior (Boyle & Schmierbach, 2009). Overall the benefits associated with hashtag activism outweigh possible problems based on the potential to quickly disseminate a message (Ojigho, 2014).

Hashtag activism has the possibility to be a successful tool for mobilization (Ojigho, 2014), but the media plays an important role in any potential success. Media exposure helps offer social capital to causes (Beaudoin, 2009; Shah, Kwak, & Holbert, 2001). However, the degree of success depends upon the type of media outlet and the group targeted (Beaudoin, 2009). All forms of Internet-based media supplement network capital by serving as an extension of face-to-face or telephone content (Wellman,

Haase, Witte, & Hampton, 2001). In addition, the same people who are involved in social efforts online are the same people likely to engage in the same behavior offline (Wellman, Haase, Witte, & Hampton, 2001; Shah, Kwak, & Holbert, 2001). This would seem to indicate that criticism of hashtag activism is misguided. "The efficacy of hashtag activism has been challenged and questioned, but critics are missing a crucial point: hashtag activism is only one tool in the advocacy circle and not a sole agent of change," (Ojigho, 2014).

JOURNALISM'S ROLE IN SOCIAL MEDIA ACTIVISM

All of these increased tweets and retweets, however, increase exposure and the potential for media coverage. In fact, online activists and the success of digital campaigns in generating media coverage have altered the traditional method of gaining exposure for social causes (Dewey, 2014; Bertlatsky, 2015). In the past, activists who wanted to raise awareness about their efforts would have to convince some level of media to write or broadcast about their issues. The Internet has allowed people to bring their issues directly to the people without any type of media gatekeeper (Bertlatsky, 2015). And the media are watching. Social media activism is easy for the media to follow and cover movements (Ojigho, 2014).

Beyond attracting attention for movements, journalism has a greater purpose. Kovach and Rosenstiel (2007) defined journalism as a "discipline of verification." The major goal of journalism in a democracy is to educate audiences so they can be free and self-governing. Often this goes beyond simply sharing information. It means helping audiences mobilize themselves so they can make the best decisions about their future as possible. This can seem to cross over into the realm of activism, which may appear contrary to the values of fairness and balance Kovach and Rosenstiel (2007) also espouse, but it also helps fulfill another important mission of journalism: to provide a forum for public criticism and compromise.

The journalism industry has a long and complicated history with activism, and the Internet has only added to it. It is difficult to understand how the Internet influenced advocacy journalism without exploring the past. Changes in technology have simplified advocacy journalism, but the genre has been part of the industry dating back to the early 19th century (Burns, 2014). Muckrakers, activist journalists in war-torn areas, and the alternative media all have a history of promoting causes in reporting. Journalism has struggled nearly since its inception with balancing objectivity and other goals. Journalist Finley Peter Dunne (1968) wrote that the media "comforts the afflicted and afflicts the comfortable." Even the Society of Professional code of ethics has stated for years that journalists should "serve as a voice for the voiceless," and do no harm (2014). This is necessary because if everyone in the media reports using the same frame it becomes difficult for readers to gain a full picture of any situation (Ruigrok, 2010) and technology, such as social media, have only made this easier if journalists take advantage.

Some already have. Vice News during the protests in Ferguson, MO chose to simply record the protests to provide full coverage as opposed to the snapshot that comes from photographs (Peterson, 2014). The outlet then provided streams of the video that were hours long so the audience could watch the incidents unfold unfiltered (Peterson, 2014). This is a type of consumer experience impossible with traditional media. The expectation is that technology and new forms of media will provide a space for activist reporting that is decreasing in the modern corporate media (Kovach & Rosenstiel, 2001) Technology has changed traditional reporting methods to such an extent that instead of simply informing the public, journalists are now pushing public agendas (Witschge, 2013).

Glenn Greenwald, one of the journalists who worked with Assange to publish information from Wikileaks, said the journalism that makes the biggest difference is the kind that pushes toward activism.

I find much to admire in America's history of the crusading journalist, from the pamphleteers to the muckrakers to the New Journalism of the '60s to the best of today's activist bloggers. At their best, their fortitude and passion have stimulated genuine reforms. (Hare, 2013)

Like Greenwald, a number of journalists are proud of their positions of advocacy on issues, especially those involving social justice (Sullivan, 2013; Ferdus, 2014). The main argument against journalistic objectivity is that reporters should write for the public good. Altschull (1996) argues that journalists should take action for what is best for the community. Objectivity simply supports the status quo (Altschull, 1996).

In addition, social responsibility theory has been used as justification for the public journalism movement in which some journalists and even a few publications have moved away from the role of being detached observers (Coleman, 1997). Public journalism, which placed the needs of the community before other news criteria, gained traction on the early 1990s and has been revived by the emergence of hyperlocal, non-profit news websites.

The second argument against objectivity is that people are no longer limited geographically to news sources. Objectivity is no longer necessary because there are so many online news sources, every opinion will be represented. The proliferation of news online has made objectivity "outdated and pointless," (Sullivan, 2013). There is no longer the expectation that journalists should have no opinion about issues (Hare, 2013; Berman, 2004). Transparency about their beliefs is what matters. It is debatable whether it is even possible for journalists to be objective, or if they should try.

The simple journalistic role of seeking truth and telling stories that are otherwise unknown is in some ways activism (Brewer, n.d.).

We need to be honest with ourselves about our motives and reasons for covering a story. The key is to ask searching questions to all sides, particularly those who hold public office, and, in doing so, provide the basis for a healthy and robust public debate. (Brewer, n.d.)

When people read about suffering, they could be motivated to act upon it, regardless of the reporter's intention. Journalists do not have to allow their ideology to seep into their writing for activism to result. Some in the industry consider themselves advocates for certain issues, such as Connie Schultz, a columnist at the Cleveland Plain Dealer. She made clear her activism for workers' rights (Schultz, 2011), and presented a unique news frame for her coverage. Frames such as the one Schultz chose partially dictate how readers perceive information (Olesen, 2008). Journalists who actively frame their stories as advocacy help keep the public sphere alive and shine light on wrongdoing (Olesen, 2008). Online news has strengthened the potential impact non-objective media have by extending their reach.

Reciprocity

Advocating for a community, whether geographic, like-minded, or connection socially online, represents opportunities for journalists to expand their reach and audience. Meyer and Carey (2014) found that journalists can have the greatest impact on online community formation by maintaining an active presence in online forums. An active presence on social media helps as well because many of these digital tools exist to bring journalists "closer to readers and readers closer to journalism by removing barriers to a more networked conversation" (Briggs, 2008). In fact, journalism as a conversation (Gillmor, 2006) should be the ultimate goal of a journalist in the digital age, both to engage the audience and advance journalism's democratic mission.

Social media enhances this mission immeasurably. Its very structure is designed to facilitate discussion and increase the relevance of news within a person's social network (de Zuniga, 2012). Increased discussion leads to additional elaboration and reflection, which in turn can lead to more action, de Zuniga (2012) adds; both are key components of the formation of social capital. Social networks also make users feel more connected to the community and more likely to reciprocate and trust other community members (de Zuniga, 2012). "After all, learning through social media may indeed contribute to not only the proliferation of a networked society, but also it may facilitate a healthier democracy. Or at least, a more participatory one" (p. 332).

Local news organizations are often either the only sources of information about the local community or the authority on what the community needs most, and Fleming and Thorson (2008) found that local media can make the biggest difference in building relationships and developing social norms among individuals. Journalists, by extension, need to encourage citizens to "exchange information, get connected for common causes, and effectively discuss issues that are important to their communities" (p. 415). The Internet and social media are vital cogs that, Fleming and Thorson emphasize, have the potential to help citizens become connected with civic life.

Encouraging participation cannot be a one-way street, Lewis, Holton and Coddington (2014) write, where journalists simply share information, advocate for causes, and expect the audience to act. They call for "reciprocal journalism" in which journalists engage with the audience and foster lasting conversations with them. This reciprocity is a key component of social capital, Putnam (2000) suggests, and a vital step toward fulfilling the public service mission of journalism.

Reciprocal journalism suggests seeing journalists in a new light: as community builders who can forge connections with and among community members by establishing patterns of reciprocal exchange. "By more readily acknowledging and reciprocating the input of audiences, and by fostering spaces for audiences to reciprocate with each other, journalists can begin to fulfill the normative purpose as stewards of the communities they serve" (p.236-237).

Social media plays a central role in reciprocity because it embodies the easy access to audiences and the two-way discussion spaces reciprocity needs. Hashtags, for example, allow users and journalists to actively seek topics that interest them and build conversations with others who share interests. Social media also involve other users within each other's networks and expand the shared experience (Lewis, Holton & Coddington, 2014).

Within three local markets practicing a form of reciprocal journalism, Richards (2012) found local news contributed to social capital primarily by providing information that spurs increased cooperation. Social media was especially adept because it is the most authentic. "Social media are providing a form of communication which is helping community members maintain personal relationships as well as encouraging them to participate in community dialogue which extends to local media" (p. 637).

What powers reciprocal journalism is the expectation of future interaction, not just a one-and-done interplay (Lewis, Holton & Coddington, 2014). In this area, journalists need improvement. Too often they approach social media as either a place to promote their stories to an audience or to find a quick source for a story they are working on. Rarely do they seek conversation in online forums (Meyer & Carey, 2014). Hermida (2013) found that even when journalists used social media for reporting, traditional voices and sources remain the loudest. Tensions also emerged between real-time publication on social media and verification, a key element of journalism. "Conceiving of the social media platform as a networked space where news is filtered, discussed, contested, and verified in the open in collaborations with the public appears to be inconceivable for the profession as a whole" (p. 304).

However, this needs to happen for journalism to become the bridge between social media and social capital formation. The kinds of open dialogs the research suggest can take place through social media that lead to reciprocity need a guide throughout the process (Meyer & Carey, 2014).

EXAMPLES FROM SOCIAL MEDIA CAMPAIGNS

The role of journalism as a facilitator of social media activism needs further research. However, a few examples exist that underscore its potential. Many of those focus only on how journalists use the tools to further their work. Journalists who used Twitter to cover the 2011 riots in the United Kingdon found it a rich source for leads (Vis, 2012). What made their social media use most effective is they went beyond the basic one-way relationship to help new journalism conventions emerge, ones that blurred the lines between journalist and audience member. In fact, it may be helpful to examine several other real world examples to understand how journalism can serve as the facilitator of social capital formation through social media.

Arab Spring

This chapter began with a discussion of the Arab Spring because it is one of the best-known examples of those outside the media using social media as a way to organize protests and civil disobedience and to create political change (Matyasik, 2014; Frangonikolopoulos & Chapsos, 2012). But Carvin's work also highlights the important magnification a journalist working to create discussion and reciprocity, even to the point of advocating a cause, can have on applying social capital to political action.

The Arab Spring was a series of political activities in the Middle East that ranged from political protests to Civil War. Social media accelerated a push for democratic change in the Middle East (Frangonikolopoulos & Chapsos, 2012). Furthermore, social media efforts allowed journalists to provide coverage that might not have otherwise been possible (Frangonikolopoulos & Chapsos, 2012; Galal & Spielhaus, 2012). "Social media filled the vacuum where a strong press should be... When traditional news coverage was blocked, Twitter and Facebook carried the news, though the standards were spotty," (Donnelly, 2012). During the Arab Spring, social media emerged as a vital tool for reporters attempting to cover the situation because government crackdowns made traditional on-the-ground reporting nearly impossible (Galal & Spielhaus, 2012).

Carefully analyzing Carvin's coverage, including the sources he cited most often, demonstrates the impact journalism had on fostering the movement (Hermida, Lewis & Zamith, 2014). Carvin's Twitter coverage cited more alternative voices than any other medium. In fact, much of his coverage focused on "messages from citizens who were expressing their demands for social change" (p. 493). Outside of the traditional role of journalist as "gatekeeper," Carvin operated more as "gatewatcher" (Bruns, 2005), a journalist who relies on and publicizes a more diverse and extensive set of sources from social media (Hermida, Lewis & Zamith, 2014).

Even after the Arab Spring, disagreement remains about whether social media alone is a tool for change or if others factors must be combined with social media efforts (Matyasik, 2014). Despite a proliferation of new media sources, affected governments responded by cracking down on traditional press freedom (Donnelly, 2012). This could indicate the government's acknowledgement of the vital role press coverage played, or it could suggest the government was merely attacking the sources it found easiest to control.

While journalists cannot control government's reaction to social media campaigns, they have the greatest effect when they connect their audiences. Carvin suggested they can even support movements from afar. This is especially true when the movements themselves are promoted through social media without the context necessary for the audience to fully understand the matter and make a difference beyond slacktivism as the next two examples suggest.

#Kony2012

The #Kony2012 campaign has generated the most criticism of any of the popular social media movements based on its simplicity and the deceptive nature of its content. However, simplicity is precisely why it was successful (Keating, 2014). Joseph Kony was the leader of the guerilla group the Lord's Resistance Army, which was estimated to have forced thousands of children to serve as soldiers and killed or abducted thousands more. The campaign made Kony an enemy and was able to generate action, at least initially, after becoming viral (Keating, 2014). The awareness was high among the young with 58% of young adults who said they heard a lot about the Kony video (Rainie, Hitlin, Jurkowitz, Dimock, & Neidorf, 2012). More than a quarter of young adults said they learned about the video through social media (Rainie, Hitlin, Jurkowitz, Dimock, & Neidorf, 2012). But the goal of this campaign was to prompt people to take action outside of social media by using social media to change the way people interacted (Briones, Madden, & Janoske, 2013).

However, the video that started the campaign was inaccurate (Muneer, 2012; Rainie, Hitlin, Jurkowitz, Dimock, & Neidorf, 2012). The quick nature of social media is a "double edged sword" in that it allows information to spread quickly, but also allowed questions about the Kony video to appear just as quickly (Muneer, 2012; Rainie, Hitlin, Jurkowitz, Dimock, & Neidorf, 2012). Several tweets, posts, and videos questioned how accurate and fair the video was. Without the efforts of journalists to clarify the Kony2012 video, audiences would have largely been duped by the simplistic and deceptive nature of the campaign (Curtis & McCarthy, 2014). The Guardian collected messages from social media and working journalists on the ground to clarify many of the myths of the campaign, including Kony's real power in Uganda. The newspaper also urged its audience to direct their political activity toward the real "villain" per se, not the U.S. but the Ugandan government. However, the follow up coverage did not have the same impact socially and materially as the initial video (Muneer, 2012). A lack of support in Uganda for grassroots efforts prevented any real effect on the ground (Cohen, 2012).

#BringBackOurGirls

A similar media campaign shed light on a well-meaning social movement that highlighted failings in Western coverage of African countries (John, 2014). The #BringBackOurGirls campaign began after a group of more than 100 schoolgirls were kidnapped by Boko Haram insurgents in Nigeria. Those behind the social media effort were upset with the way that the Nigerian government was handling the investigation (Ojigho, 2014). In terms of the raw figures of the effort's reach, #bringbackourgirls was retweeted more than 2 million times (Youngblood, 2014). The effort caught the attention of celebrities, global leaders, and the media (Ojigho, 2014). The exposure combined with media coverage resulted in protests around the world. In this instance the social media campaign was a response to a lack of initial coverage of the incident in the international press (Youngblood, 2014).

Initially, it had important effects. "The #BringBackOurGirls campaign has helped to proliferate worldwide awareness of Boko Haram's ideals, and in turn forced the UN to take action against the group to avoid losing face and appearing impotent," (Wren, 2014). Coverage of the campaign seems to have influenced the U.S. into sending troops into Nigeria (Ojigho, 2014; Wren, 2014). This engagement and action may not have occurred otherwise (Campbell, 2014).

The issue was staying power. People lost interest far before any crisis was solved (Anderson, 2014). People want stories that need action, but lose interest if a resolution does not come quickly (Anderson, 2014). It also highlighted the "white savior" motif that "Americans are slightly more likely to care about something is there is a Twitter hashtag associated with it" (John, 2014). A lot of the publicity surrounding #bringbackourgirls supported a for-profit documentary about Boko Haram, whose director took credit for creating a hashtag she didn't actually create.

#bringbackourgirls demonstrates both the power to create and to dethrone social media campaigns. At first Western media gleefully embraced the idea of a "Los Angeles Mother of Two who created a viral hashtag" but ended up having to correct their headlines when other media highlighted the real causes of the movement (John, 2014). Without news coverage, audiences would not have known the true motives of the movement, but then again, they might not have known about the movement in the first place.

Ice Bucket Challenge

Along the same lines, news media efforts to determine the true genesis of the Ice Bucket Challenge viral campaign and ensure that efforts and donations went to the right place made a significant difference. "The Ice Bucket Challenge may soon have run its course, but the lesson it provides about the nature of 21st Century charity will love on," Townsend (2014) wrote for the BBC. Despite criticism, it is hard to dispute the results of the campaign. Success can be measured through both fundraising and exposure for Amyotrophic Lateral Sclerosis. The worldwide effort that began on social media raised $98.2 million for A.L.S. between July 29 and Aug. 28 of 2014. The fundraising portion of the Ice Bucket Challenge was successful enough that those within the medical community are thinking about additional ways to use social media as a fundraising effort (Koohy, Koohy, & Watson, 2014). The lesson learned from the Ice Bucket Challenge is social media promotion should be interactive and engage directly with the public (Koohy, Koohy, & Watson, 2014).

There was widespread media criticism of the large number of people who participated online by dumping ice water on their heads without donating to A.L.S. (Herman, 2014). "I hope that most of the folks capturing their Ice Bucket Challenge on video and then posting it on Facebook aren't doing it for the publicity, but I'm disheartened by the way this has become donation-as-self-promotion," (Payne, 2014). This writer's concern was that people were overlooking local organizations that might be more deserving (Payne, 2014). Further she believed that while A.L.S. was a good organization the success of the challenge would bring similar efforts from less scrupulous groups (Payne, 2014). Dishonorable charitable groups are a legitimate concern, but one that has not occurred on a widespread level. Those involved with the A.L.S. Foundation considered the effort successful because of the newfound awareness of ALS, even online promotion from those who did not donate (Steel, 2014). Only half of Americans were aware of A.L.S. in the month prior to the challenge (Steel, 2014). Now it seems difficult to believe there are many adults in the United States who are not familiar with A.L.S.

CONCLUSION AND RECOMMENDATIONS

In the current media environment nearly every event has some type of hashtag activism component. While success varies, it is clear that social media activism can result in engagement beyond slacktivism. Activity will range from the individual who simply likes or retweets something as a way to bolster her self-image to those who become involved actively in the cause outside of cyberspace. However this varying level of engagement does not differ from what people have experienced with more traditional causes.

What sets social media activism apart is the vast amount of publicity it can generate thanks to the very nature of the media themselves. The role journalists can play in whether social media efforts lead to social capital, which in turns creates sustained activity is a vital research question to ask and consider. This chapter provides a start in defining social capital in the traditional sense and extending that definition to social media. It catalogs a wealth of research that suggests both the positive and negative effects social media, such as Twitter and Facebook, can have on social capital creation.

Most importantly, it applies the traditional roles and values of journalism to both social capital and social media and suggests a synthesis that helps define journalism's role in the 21st century. No longer does the news simply report on what happened. Journalists are becoming more connected to communities by necessity just to know what is going on. Through these community connections, journalists can re-establish themselves as trusted information providers, but they need to get their hands dirty in social media, sometimes even becoming advocates for causes their community needs. This new relationship asks a lot of journalists to apply the essence of verification to a vast network of tweets, posts, blogs and comments. But if done correctly, the research cataloged here suggests that journalism can have an even greater impact than ever before.

Future Research Direction

Questions remain about the connection between journalism, social media and social capital, which Hermida (2014) suggest researchers need new approaches to tackle. Several different areas could be researched within these questions. This chapter focuses on three particular areas that seem logical for future research: examining specifically social media, activism and journalism. First, longitudinal studies may be required to really explain the real world impact social media and journalistic campaigns can have. Several of the examples this chapter cited showed broad immediate impact for social media activism, but little lasting difference. For a charity such as ALS that might not mean a lot, but it could have great meaning for citizens of oppressive government regimes in Egypt or Nigeria. Each case helps researchers and professionals understand the others.

In fact, greater attention to case study research with a focus on triangulating methods could greatly benefit the academy's understanding of how social capital can be created online and what ultimately journalism's role can be. The next step is attempting to quantify activism. There are several potential research tools that could be used. The first would be to review the financial statements provided by charitable organizations such as A.L.S. that have experienced a social media campaign. It would be possible through these reports to compare donations prior to the campaign, after the campaign became popular on social media, and finally when it received coverage by mainstream media. This type of examination would show how much real world activism is experienced financially when an organization generates buzz through social media. It would also provide insight about hashtag activism's impact. Is press coverage necessary for offline success? The goal is to test the role of, and importance of journalism, as an industry in making hashtag activism into something more tangible.

Activism could also be quantified by reviewing how many people liked a cause or made a comment during a social media campaign. Then it would be necessary to evaluate later posts to determine how many of those who posted later made a statement about giving a donation, or taking some other action to further the cause. In turn, researchers could ask if they noticed media coverage of the event and what impact it had on their decisions. Of course it is expected that not everyone who acted in the offline world would post about it; however, it is expected a large enough sample would enable researchers to determine how online activism translates to offline activism.

There is also potential for content analysis of social media. It is possible to track campaigns by looking at tweets or posts to determine how frequently those running the campaign interact with their audience. Interactivity theory would argue that the most successful campaigns would help not only to start the conversation in social media, but interact with the audience to keep it moving forward (Ha & James, 1998; Rafaeli & Sudweeks, 1998). The content analysis of social media could then be compared to media coverage in the same mold as agenda setting research.

Furthermore social media could be examined in a qualitative manner by conducting interviews with those who comment on something like the Ice Bucket Challenge about why they commented online, what made them want to participate, and if they frequently are involved with causes either online or offline. This would provide perspective on what makes a social media campaign successful and if those who become involved online are those who will engage in activism offline.

There is also the possibility to examine how the journalism industry covers social media. Where and how do members of the media learn about social media campaigns? Are there certain types of social media efforts that are more likely than others to receive coverage in the mainstream media? For example it would be expected members of the media would find Twitter campaigns more easily because of the inherently open nature of the medium. A content analysis could look at stories about social media campaigns and look at what type of sources do the journalists select. For efforts that start on social media, do journalists use online sources for their coverage or do they the same type of offline sources they would use when covering local charity event?

Despite changes in the industry and challenges adapting to technology, journalism retains its central role in a democracy. It remains the fifth estate tasked with holding those in power accountable and giving a voice to the voiceless (Kovach & Rosenstiel, 2007). That will not change. Much needs to change in how journalists approach their jobs. They need to get past the cynicism about whether the ephemeral nature of the Internet leads to lasting important change and learn to embrace the responsiveness the Internet gives them to their audiences. Through reciprocity (Lewis, Holton & Coddington, 2014) journalism can regain some of its audience and educate everyone on how to be free and self-governing. Whether it is called social capital, activism, advocacy, or slacktism, action is required, and journalism has the power to spur those actions forward.

REFERENCES

Altschull, J. H. (1996). A crisis of conscience: Is community journalism the answer? *Journal of Mass Media Ethics, 11*(3), 166–172. doi:10.1207/s15327728jmme1103_5

Anderson, L. (2014, August 6). *From #BringBackOurGirls to Syria, why do we forget about a crisis long before it's over?* Retrieved from Deseret News: http://national.deseretnews.com/article/2065/from-bringbackourgirls-to-syria-why-do-we-forget-about-a-crisis-long-before-its-over.html

Arit, J. (2014). What's Wrong With Our Well-Intentioned Boko Haram Coverage. The Wire –. *Atlantic (Boston, Mass.), 9*(May). Retrieved from http://www.thewire.com/politics/2014/05/whats-wrong-with-our-well-intentioned-boko-haram-coverage/361993/

Beaudoin, C. (2009). Exploring the association between news use and social capital: Evidence of variance by ethnicity and medium. *Communication Research, 36*(5), 611–636. doi:10.1177/0093650209338905

Beaudoin, C. (2011). News Effects on Bonding and Bridging Social Capital: An Empirical Study Relevant to Ethnicity in the United States. *Communication Research, 38*(2), 155–178. doi:10.1177/0093650210381598

Berman, D. (2004, June 29). *Advocacy Journalism, The Least You Can Do, and The No Confidence Movement.* Retrieved from Independent Media Center: https://www.indymedia.org/en/2004/06/854953.shtml

Bertlatsky, N. (2015, January 7). *Hashtag activism isn't a cop-out.* Retrieved from The Atlantic: http://www.theatlantic.com/politics/archive/2015/01/not-just-hashtag-activism-why-social-media-matters-to-protestors/384215/

Blanchard, A., & Horan, T. (1998). Virtual communities and social capital. *Social Science Computer Review, 16*(3), 293–307. doi:10.1177/089443939801600306

Bode, L., Vraga, E. K., Borah, P., & Shah, D. V. (2014). A new space for political behavior: Political social networking and its democratic consequences. *Journal of Computer-Mediated Communication, 19*(3), 414–429. doi:10.1111/jcc4.12048

Boyle, M., & Schmierbach, M. (2009). Media use and protest: The role of mainstream and alternative media use in predicting traditional and protest participation. *Communication Quarterly, 57*, 1-17.

Brewer, D. (n.d.). *Are journalism and activism compatible?* Retrieved from Media Helping Media: http://www.mediahelpingmedia.org/43-news-archive/global/341-are-journalism-and-activism-compatible

Briggs, M. (2008, Winter). The End of Journalism as Usual. *Nieman Reports*, 40-41.

Briones, R., Madden, S., & Janoske, M. (2013). Kony 2012: Invisible children and the challenges of social media campaigning and digital activism. *Journal Of Current Issues In Media & Telecommunications, 5*(3), 205–234.

Burns, S. (2014, October 16). *'Advocacy' Is Not a Dirty Word in Journalism.* Retrieved from MediaShift: http://www.pbs.org/mediashift/2014/10/advocacy-is-not-a-dirty-word-in-journalism/

Campbell, A. (2014, December 10). *#TBT: #BringBackOurGirls—Exploring The Potential (and Perils) of Hashtag Activism.* Retrieved from Medium: https://medium.com/@internetweek/tbt-bringbackourgirls-exploring-the-potential-and-perils-of-hashtag-activism-ab862425f2a0

Carr, D. (2012, March 25). *Hashtag Activism, and Its Limits.* Retrieved from New York Times: http://www.nytimes.com/2012/03/26/business/media/hashtag-activism-and-its-limits.html?_r=0

Carvin, A. (2012). *Distant witness: Social media, the Arab Spring and a journalism revolution.* CUNY Journalism Press.

Chan, M. (2015). Mobile phones and the good life: Examining the relationships among mobile use, social capital and subjective well-being. *New Media & Society, 17*(1), 96–113. doi:10.1177/1461444813516836

Cohen, R. (2012, April 26). *Why did "Kony 2012" fizzle out?* Retrieved from NonProfit Quarterly: https://nonprofitquarterly.org/policysocial-context/20216-why-did-kony-2012-fizzle-out.html

Coleman, R. (1997). The intellectual antecedents of public journalism. *The Journal of Communication Inquiry, 21*(1), 60–76. doi:10.1177/019685999702100103

Cook, F. L., Tyler, T. R., Goetz, E. G., Gordon, M. T., Protess, D., Leff, D. R., & Molotch, H. L. (1983). Media and agenda setting: Effects on the public, interest group leaders, policy makers, and policy. *Public Opinion Quarterly, 47*(1), 16–35. doi:10.1086/268764 PMID:10261275

De Tocqueville, A., Mansfield, H. C., & Winthrop, D. (2002). *Democracy in America.* Folio Society.

de Zuniga, H. (2012). Social media use for news and individuals' social capital, civic engagement and political participation. *Journal of Computer-Mediated Communication, 17*(3), 319–336. doi:10.1111/j.1083-6101.2012.01574.x

Deuze, M. (2005). What is journalism? Professional identity and ideology of journalists reconsidered. *Journalism, 6*(4), 442–464. doi:10.1177/1464884905056815

Dewey, C. (2014, May 8). *#Bringbackourgirls, #Kony2012, and the complete, divisive history of 'hashtag activism'.* Retrieved from Washington Post: http://www.washingtonpost.com/news/the-intersect/wp/2014/05/08/bringbackourgirls-kony2012-and-the-complete-divisive-history-of-hashtag-activism/

Donnelly, J. (2012, February 15). *Freedom of the Press panel explores 'Arab Spring' aftermath.* Retrieved from The National Press Club: http://www.press.org/news-multimedia/news/freedom-press-panel-explores-arab-spring-aftermath

Dunne, F. P. (1968). *Observations.* Grosse Pointe, MI: Scholarly Press.

Ellison, N., Gray, R., Lampe, C., & Fiore, A. (2014). Social capital and resource requests on Facebook. *New Media & Society, 16*(7), 1104–1121. doi:10.1177/1461444814543998

Ellison, N., Steinfield, C., & Lampe, C. (2007). The Benefits of Facebook ''Friends:'' Social Capital and College Students' Use of Online Social Network Sites. *Journal of Computer-Mediated Communication, 12*(4), 1143–1168. doi:10.1111/j.1083-6101.2007.00367.x

Ellison, N. B. (2007). Social network sites: Definition, history, and scholarship. *Journal of Computer-Mediated Communication, 13*(1), 210–230. doi:10.1111/j.1083-6101.2007.00393.x

Ettema, J. S., & Glasser, T. L. (1984). *On the epistemology of investigative journalism.* Gainesville: Educational Resources Information Center.

Ferdus, I. (2014). Photography as activism: The role of visual media in humanitarian crisis. *Harvard International Review*, (Summer), 22–25.

Fleming, K., & Thorson, E. (2008). Assessing the Role of Information-Processing Strategies in Learning From Local News Media About Sources of Social Capital. *Mass Communication & Society*, *11*(4), 398–419. doi:10.1080/15205430801950643

Frangonikolopoulos, C. A., & Chapsos, I. (2012). xplaining the role and the impact of the social media in the Arab Spring. *Global Media Journal: Mediterranean Edition*, *7*(2), 10–20.

Galal, E., & Spielhaus, R. (2012). Covering the Arab Spring: Middle East in the Media – the Media in the Middle East. *The Editorial.* Retrieved from Academia.edu: http://www.academia.edu/2279607/Covering_the_Arab_Spring_Middle_East_in_the_Media_the_Media_in_the_Middle_East._The_Editorial

Gheytanchi, E., & Moghadam, V. (2014). Women, social protests, and the new media activism in the Middle East and North Africa. *International Review of Modern Sociology*, *40*(1), 1–26.

Gillmor, D. (2006). *We the media: Grassroots journalism by the people, for the people.* O'Reilly Media, Inc.

Gross, D. (2014). *Social Web tackles the #IceBucketChallenge.* CNN.com. Retrieved from http://www.cnn.com/2014/08/13/tech/ice-bucket-challenge/

Ha, L., & James, E. L. (1998). Interactivity Reexamined: A Baseline Analysis of Early Business Web Sites. *Journal of Broadcasting & Electronic Media*, *42*(4), 457–474. doi:10.1080/08838159809364462

Hare, K. (2013, October 28). *Keller, Greenwald debate whether journalists can be impartial.* Retrieved from Poynter: http://www.poynter.org/news/mediawire/227386/keller-greenwald-debate-whether-journalists-can-be-impartial/

Herman, J. (2014). Hashtags and human rights: Activism in the age of Twitter. *Carnegie Ethics Online*, 1–6.

Hermida, A. (2013). #Journalism. *Digital Journalism*, *1*(3), 295–313. doi:10.1080/21670811.2013.808456

Hermida, A., Lewis, S. C., & Zamith, R. (2014). Sourcing the Arab Spring: A Case Study of Andy Carvin's Sources on Twitter During the Tunisian and Egyptian Revolutions. *Journal of Computer-Mediated Communication*, *19*(3), 479–499. doi:10.1111/jcc4.12074

Howard, P., & Hussain, M. (2011). The role of digital media: The upheavals in Egypt and Tunisia. *Journal of Democracy*, *22*(3), 35–48. doi:10.1353/jod.2011.0041

Jeong, H., & Lee, M. (2013). The effect of online media platforms on joining causes: The impression management perspective. *Journal of Broadcasting & Electronic Media*, *57*(4), 439–455. doi:10.1080/08838151.2013.845824

Kaplan, A. M., & Haenlein, M. (2010). Users of the world, unite! The challenges and opportunities of Social Media. *Business Horizons*, *53*(1), 59–68. doi:10.1016/j.bushor.2009.09.003

Keating, J. (2014, May 20). *The less you know*. Retrieved from Slate: http://www.slate.com/blogs/ the_world_/2014/05/20/the_depressing_reason_why_hashtag_campaigns_like_stopkony_and_bring-backourgirls.html

KONY2012: What's the Real Story? (2014). *Reality Check: Uganda*. The Guardian.com. Retrieved from http://www.theguardian.com/politics/reality-check-with-polly-curtis/2012/mar/08/kony-2012-what-s-the-story

Koohy, H., Koohy, B., & Watson, M. (2014). A lesson from the ice bucket challenge: Using social networks to publicize science. *Frontiers in Genetics*, *5*, 1–3. doi:10.3389/fgene.2014.00430 PMID:25566317

Kovach, B., & Rosenstiel, T. (2001). Are watchdogs an endangered species? *Columbia Journalism Review*, (May/June), 50–53.

Kovach, B., & Rosenstiel, T. (2007). *The elements of journalism: What newspeople should know and the public should expect*. Three Rivers Press.

Kristofferson, K., White, K., & Peloza, J. (2014). The nature of slacktivism: How the social observability of an initial act of token support affects subsequent prosocial action. *The Journal of Consumer Research*, *40*(6), 1149–1166. doi:10.1086/674137

LaRose, R., & Eastin, M. S. (2004). A social cognitive theory of Internet uses and gratifications: Toward a new model of media attendance. *Journal of Broadcasting & Electronic Media*, *48*(3), 358–377. doi:10.1207/s15506878jobem4803_2

Lewis, S. C., Holton, A. E., & Coddington, M. (2014). Reciprocal Journalism. *Journalism Practice*, *8*(2), 229–241. doi:10.1080/17512786.2013.859840

Lim, J., & Golan, G. (2011). Social media activism in response to the influence of political parody videos on YouTube. *Communication Research*, *38*(5), 710–727. doi:10.1177/0093650211405649

Matyasik, M. (2014). Secure sustainable development: Impact of social media on political and social crisis. *Journal Of Security & Sustainability Issues*, *4*(1), 5–16. doi:10.9770/jssi.2014.4.1(1)

Mayo, K., & Newcomb, P. (2008). How the Web was Won: An Oral History of the Internet. *Vanity Fair*. Retrieved from http://www.vanityfair.com/news/2008/07/internet200807

Meyer, H. K., & Carey, M. C. (2014). In Moderation: Examining how journalists' attitudes toward online comments affect the creation of community. *Journalism Practice*, *8*(2), 213–228. doi:10.1080/175127 86.2013.859838

Muneer, F. (2012, July 10). *The Kony 2012 controversy: A look at its coverage in American media vs. Ugandan media*. Retrieved from Huffington Post: http://www.huffingtonpost.com/fatima-muneer/the-kony-2012-controversy_b_1503990.html

Nisbet, M. (2007). The future of public engagement. *Scientist (Philadelphia, Pa.)*, *21*(10), 38–44.

Ojigho, O. (2014, November 16). *Hashtag activism makes the invisible visible*. Retrieved from The Mantle: http://www.mantlethought.org/other/hashtag-activism-makes-invisible-visible

Olesen, T. (2008). Activist journalism? The Danish Cheminova debates, 1997 and 2006. *Journalism Practice*, 2(2), 245–263. doi:10.1080/17512780801999394

Payne, M. (2014, August 20). *Tell Mel:Throwing cold water on Ice Bucket Challenge*. Retrieved from News Press: http://www.news-press.com/story/news/investigations/melanie-payne/2014/08/19/throwing-cold-water-ice-bucket-challenge/14309687/

Peterson, T. (2014, August 20). *Vice News keeps the lens open during Ferguson protests*. Retrieved from Ad Age: http://adage.com/article/media/vice-news-lens-open-ferguson-protests/294641/

Poell, T., & Borra, E. (2011). Twitter, YouTube, and Flickr as platforms of alternative journalism: The social media account of the 2010 Toronto G20 protests. *Journalism*, *13*(6), 695–713. doi:10.1177/1464884911431533

Protess, D. L. (1992). *The journalism of outrage: Investigative reporting and agenda building in America*. Guilford Press.

Protess, D. L., Cook, F. L., Curtin, T. R., Gordon, M. T., Leff, D. R., McCombs, M. E., & Miller, P. (1987). The impact of investigative reporting on public opinion and policymaking targeting toxic waste. *Public Opinion Quarterly*, *51*(2), 166–185. doi:10.1086/269027

Putnam, R. D. (2001). *Bowling alone: The collapse and revival of American community*. Simon and Schuster.

Putnam, R. D., Feldstein, L., & Cohen, D. J. (2004). *Better together: Restoring the American community*. Simon and Schuster.

Rafaeli, S., & Sudweeks, F. (1998). Interactivity on the Nets. In *Network Netplay: Virtual Groups on the Internet* (pp. 173–189). Menlo Park, CA: AAAI Press/MIT Press.

Rainie, L., Hitlin, P., Jurkowitz, M., Dimock, M., & Neidorf, S. (2012, March 15). *The viral Kony 2012 video*. Retrieved from Poynter: http://www.pewinternet.org/2012/03/15/the-viral-kony-2012-video/

Richards, I. (2012). Beyond city limits: Regional journalism and social capital. *Journalism*, *14*(5), 627–642. doi:10.1177/1464884912453280

Roberts, G., & Klibanoff, H. (2006). *The Race Beat*. New York: Vintage Books.

Ruigrok, N. (2010). From Journalism of Activism to Journalism of Accountability. *The International Communication Gazette*, *72*(1), 85–90. doi:10.1177/1748048509350340

Schultz, C. (2011, March 4). *Why Connie Schultz Won't Give up on the Fight for Good Journalism*. Retrieved from Poynter: http://www.poynter.org/news/mediawire/106009/why-connie-schultz-wont-give-up-on-the-fight-for-good-journalism/

Shah, D., Friedland, L., Wells, C., Kim, Y., & Rojas, H. (2012). Communication, consumers, and citizens: Revisiting the politics of consumption. *The Annals of the American Academy of Political and Social Science*, *644*(1), 6–18. doi:10.1177/0002716212456349

Shah, D. V., Kwak, N., & Holbert, R. L. (2001). Connecting' and 'disconnecting' with civic life: Patterns of Internet use and the production of social capital. *Political Communication, 18*, 141–162. doi:10.1080/105846001750322952

Simon, J. (2014, December 11). *What's the Difference Between Activism and Journalism?* Retrieved from Nieman Reports: http://niemanreports.org/articles/whats-the-difference-between-activism-and-journalism/

Society of Professional Journalists. (2014, September 6). *SPJ Code of Ethics*. Retrieved from Society of Professional Journalists: http://www.spj.org/ethicscode.asp

Steel, E. (2014, August 21). *'Ice Bucket Challenge' donations for A.L.S. research top $41 million.* Retrieved from New York Times: http://www.nytimes.com/2014/08/22/business/media/ice-bucket-challenge-donations-for-als-top-41-million.html?_r=0

Sullivan, M. (2013, October 26). *As Media Change, Fairness Stays Same.* Retrieved from New York Times: http://www.nytimes.com/2013/10/27/public-editor/as-media-change-fairness-stays-same.html?_r=0

Thurman, N. (2011). Making 'The Daily Me': Technology, economics and habit in the mainstream assimilation of personalized news. *Journalism, 12*(4), 395–415. doi:10.1177/1464884910388228

Townsend, L. (2014). How much has the ice bucket challenge achieved? *BBC News Magazine*. Retrieved from http://www.bbc.com/news/magazine-29013707

Tsentas, T. (2011, November 12). *Challenging the information landscape: WikiLeaks' effect on the media, activism and politics.* Retrieved from Media @ McGill: http://media.mcgill.ca/en/content/challenging-information-landscape-wikileaks%E2%80%99-effect-media-activism-and-politics

Valenzuela, S., Park, N., & Kee, K. (2012). Is There Social Capital in a Social Network Site? Facebook Use and College Students' Life Satisfaction, Trust, and Participation. *Journal of Computer-Mediated Communication, 14*(2009), 875–901.

Vergeer, M. (2015). Peers and Sources as Social Capital in the Production of News: Online Social Networks as Communities of Journalists. *Social Science Computer Review, 33*(3), 277–297. doi:10.1177/0894439314539128

Vergeer, M., & Pelzer, B. (2009). Consequences of media and Internet use for offline and online network capital and well-being. A causal model approach. *Journal of Computer-Mediated Communication, 15*(1), 189–210. doi:10.1111/j.1083-6101.2009.01499.x

Vis, F. (2013). Twitter as a Reporting Tool for Breaking News. *Digital Journalism, 1*(1), 27–47. doi:10.1080/21670811.2012.741316

Wallsten, K. (2015). Non-elite Twitter sources rarely cited in coverage. *Newspaper Research Journal, 36*(1), 24–41.

Welles, B. (2014, June 17). *3Qs: A closer look at hashtag activism.* Retrieved from News @ Northeastern: http://www.northeastern.edu/news/2014/06/3qs-hashtag-activism/

Wellman, B., Haase, A., Witte, J., & Hampton, K. (2001). Does the Internet increase, decrease, or supplement social capital? Social networks, participation, and community commitment. *The American Behavioral Scientist, 35*(3), 436–455. doi:10.1177/00027640121957286

Wihbey, J. (2013, August 1). *Digital activism and organizing: Research review and reading list.* Retrieved from Journalist's Resource: http://journalistsresource.org/studies/society/internet/digital-activism-organizing-theory-research-review-reading-list#

Witschge, T. (2013). Transforming journalistic practice: A profession caught between change and tradition. In C. Peters & M. Broersma (Eds.), *Rethinking journalism: Trust and participation in a transformed news landscape* (pp. 160–172). New York: Routledge.

Wren, A. (2014, May 25). *Why #BringBackOurGirls is NOT another slacktivism campaign.* Retrieved from Asher Wren: http://asherwren.svbtle.com/why-is-bringbackourgirls-not-another-slacktivism-campaign

Youngblood, S. (2014, May 12). *#BringBackOurGirls fills void left by sluggish press.* Retrieved from Peace & Collaborative Development Network: http://www.internationalpeaceandconflict.org/profiles/blogs/bringbackourgirls-fills-void-left-by-sluggish-press?xg_source=activity#.VUu4s5TF9rg

Chapter 11
Marketing and Mobile:
Increasing Integration

Kenneth E. Harvey
KIMEP University, Kazakhstan

Yulia An
The Ilmenau University of Technology, Germany

ABSTRACT

Integrated Marketing Communications (IMC) was gaining popularity even before the World Wide Web. But with the explosion of online marketing, it is nearly impossible to avoid integration. This chapter will explore how digital and mobile marketing are dictating such changes in marketing. An overview of modern mobile marketing practices and trends builds upon a comparison of the traditional marketing mix (4 Ps) with a more consumer-oriented concept of 4 Cs. As products are being replaced with consumers, promotion with communication, place with convenience, and price with cost, the rise of mobile devices and the mobile Internet brings up a multitude of new concepts into traditional marketing theory and practice. Supplemented with best industry examples, the review of emerging mobile marketing concepts moves from broad online strategies, like inbound marketing, to very mobile-specific trends, like the Internet of Things.

INTRODUCTION

Over the past two decades, marketing, advertising, branding, public relations, direct sales, packaging, point of purchase, sales promotions, special events, direct response and social media have all become intertwined by the increasingly popular strategic approach of "integrated marketing communications" (IMC). In many organizations now one unit encompasses all of these activities. The theoretical and pragmatic basis for this approach is that all of these activities in an organization can and should help overcome common challenges, influence the same target audiences, pursue the same goals and objectives, and coordinate their strategies and tactics.

DOI: 10.4018/978-1-5225-0469-6.ch011

One influence in popularizing IMC was the development of technologies that could facilitate such integration and coordination of marketing activities. Inexpensive personal computers and communication software, as well as, powerful databases to help segment and track prospective consumers' purchasing habits, media preferences, demographics, psychographics, etc., all provided impetus for the adoption of the IMC approach.

The Internet and digital communications have provided further impetus for this integration. Consider, for example, one of the strongest trends in digital marketing – those activities we call content marketing or inbound marketing. These activities include social media interaction; webinars and other virtual activities, viral videos, blogs, white papers and ebooks. In the old days these might have been assigned to the PR department primarily based on the fact that they are free promotional activities – free in the sense that these are not paid advertising. No one outside the organization itself need be paid a significant amount for the organization to offer these free products or services to their target audience. The organization's own staff can develop the website where free offers will be hosted and where social media contacts will be referred. If any advertising is purchased, it is primarily to help build the database of interested individuals, who, by the very nature of the free offers, also represent potential customers and social opinion leaders.

Such free offers, however, clearly have other marketing purposes. Webinars, blogs and ebooks can help the organization achieve thought leadership and credibility. In other words, they help build the brand. Participants are typically required to provide the organization with their name, company and contact information in order to receive such free offers. This information goes into a database to support other forms of marketing. At first it may just be to alert the prospect to more free offers. Then if the participant returns for future freebies, the landing page on the website is likely to ask if a representative from the organization can call them to discuss how else the organization can assist. This may provide the bulk of the contacts for the direct sales group. And within the webinar or ebook, there should be included some advertising content. While the organization may not have to pay for the content, by its very nature it is advertising. During the last 10-15 minutes of a webinar, for example, the host is typically presenting a "commercial" for the organization's paid products and services, and within the ebook there will be printed "advertisements" to entice the reader to purchase paid products and services. So, who should be in charge of content marketing? It almost demands an integrated approach to be successful.

Nonetheless, marketing includes far more activities than advertising, branding and PR. These are strong elements of marketing, but certainly not all. Of the so-called 4 Ps of traditional marketing, only one is promotion. The 4 Ps are:

1.   **Product:** Determining the goods and services to be offered.
2.   **Promotion:** Carrying out marketing communication to attract prospective customers.
3.   **Place:** Establishing where and how to make the product available for purchase.
4.   **Price:** Setting the price point at which the products can bring the organization the greatest profits.

Digital communications in general and mobile technologies specifically have led to new products, new methods of promotion, new virtual "places" where customers can make their purchases, and new pricing considerations. However, the impact of digital communications and mobile technologies on marketing has gone even beyond that and dramatically reinforced the stature of newer forms of marketing strategy that already started appearing in mid-1980s, yet often had been met as just other managerial fads (Eagle & Kitchen, 2000; Kitchen et al., 2004). While the concept of a classic marketing mix hasn't been completely abandoned by the marketing scholarship, it is difficult to reject the influence of a new communication-based paradigm, like relationship marketing and IMC, that have come on stage during that period. In further sections we will thus discuss the evolution of marketing practices across the four marketing mix elements within both product and communication paradigms.

STATE OF THE ART

Some concerns about the dominating paradigm of the traditional marketing mix as conceived of by Mc-Carthy (1964) were raised by other practitioners and scholars of the marketing theory. For instance, one of the reasons why Duncan and Moriarty (1998) questioned the 4Ps was that it couldn't properly categorize some existing marketing activities like direct marketing or personal selling. Should we categorize them as place or promotion? With the rapid emergence of information technologies the early concerns of Duncan and Moriarty and a number of other marketing scholars (Grönroos, 1997; Constantinides, 2006) have become even more pronounced. For instance, commercial websites can be used to promote goods and services, as well as serve as distribution channels. At the same time a newly emerged freemium pricing strategy converges the price and promotion elements together.

A rampant growth of mobile technologies made the boundaries between separate elements even more blurred. Increasing mobility, instant connectivity, and a personal touch of mobile communication give way to instruments like mobile branded wallets. A consumer brand Starbucks, for instance, has left banks far behind in the sector of mobile payments that is expected to be worth $721 billion (here and later in USD) by 2017 (Gartner, 2013a). With its weekly multimillion mobile transactions in the United States within a Starbucks branded mobile wallet, the brand has already secured a position in this young, highly competitive and prospective market (Pope et al., 2011; Valentine, 2013). The Starbucks' mobile payment app enables the coffee chain conduct loyalty programs and grant bonus points and personalized store discounts for loyal customers. The company even gives away physical and electronic freebies to grab from stores or download inside the mobile wallet and unleashes the power of prepaid cards. Only these mentioned features of a branded mobile wallet already integrate all four elements of the traditional marketing mix and still don't grab a more strategic concept of an overall branding message.

In the wake of these technological changes, the need for integration of all marketing activities (on a tactical level) and business processes (on a more strategic level) is strong as never before. Moreover, even the concept of IMC itself has gone through dramatic transformations in the recent years. If in its infancy IMC was more of a tactical perspective that promoted a unified and integrated message in diverse marketing communications channel, it now shapes a strategic view on marketing theory in general with a special focus placed on communications and relationship building (Duncan & Mariarty, 1998; Kitchen et al., 2004; Schultz & Patti, 2009).

One of the ways to approach this new paradigm is to take a consumer-oriented perspective and to replace the traditional "inside-out" 4Ps with a new "outside-in" approach of 4Cs with *consumer* instead of *product*, *communication* instead of *promotion*, *cost* instead of *price*, and *convenience* instead of *place*. A short but influential piece by Lauterborn (1990) has attracted a lot of attention as one of those mapping a new consumer-oriented approach. Even though the 4Cs is more used among marketing practitioners, rather than scholars, it provides a straightforward framework for comparing the new consumer-oriented paradigm with the traditional production-oriented one. Hence, the further discussion will compare the McCarty's 4Ps with the Lauterborn's consumer-oriented 4Cs, and the impact of mobile on marketing within these elements.

Products/Consumers

The proposition to focus on customer needs instead of a product is not new. One of the first powerful critiques appeared as early as when the expansion of railroads had significantly dropped, with an influential article of Theodore Levitt for Harvard Business Review (Levitt, 1960). In this seminal piece Levitt first introduced the concept "marketing myopia" that described marketers' obsession with their own product that they sell to consumers. This preoccupation with the product makes them blind to the wants and the needs of consumers and leads to the failure of capturing new growing markets and losing their existing consumers whose needs were not fulfilled. Levitt thus explains the failure of the railroads by arguing that

The railroads did not stop growing because the need for passenger and freight transportation declined. That grew. The railroads are in trouble today not because the need was filled by others (cars, trucks, airplanes, even telephones), but because it was not filled by the railroads themselves. They let others take customers away from them because they assumed themselves to be in the railroad business rather than in the transportation business. The reason they defined their industry wrong was because they were railroad-oriented instead of transportation-oriented; they were product-oriented instead of customer-oriented. (p.45)

This is what happened to typewriters when personal computers completely replaced them, and this is what is now happening to such offline businesses as video rental stores whose customers shifted to Internet-based services. While Netflix has already surpassed 62.3 million members this year (Sharf, 2015) by adding almost 5 million new users in just the first three months of 2015, video stores are struggling to survive by migrating into something like "a physical space for people to talk and learn about movies, knowledgeable staffs, and vast catalogues of rare titles on VHS tapes and DVD" (Moskowitz, 2013).

Meanwhile, the rapid intrusion of mobile businesses into our everyday routines shattered existing paradigms about business in general and introduced a vast diversity of new business models and markets to serve. The outbreak of violence when taxi drivers in Paris protested against Uber, a rapidly growing international transportation network facilitating so-called gypsy taxi services, is an alarming signal of the expanding power of mobile businesses (Verbergt & Schechner, 2015). Now valued at more than $50 billion (Macmillan & Demos, 2015), the company facilitates its ridesharing service via a mobile app and manages the whole interaction process between drivers and smartphone users requesting a ride, including online bill payment, all done over a smartphone. The company went beyond the product-oriented thinking and served the customers' need to find a ride quickly, get flexible pricing, and be able to pay with a smartphone – the one device that is almost always on hand.

One of the many reasons for such disruptive changes in how business is done and in how products and services are marketed lies in the changing nature of the products and services themselves. Digital products and services, as seen from the economics standpoint, possess a number of specific characteristics. With the marginal replication and distribution cost of digital products, many businesses now employ "long tail" product strategies (Anderson, 2006). Digital products can be "inventoried" now profitably even with relatively few sales per item. A traditional music store might carry 1,000 of the most popular recordings, but an online music store can carry more than 100,000 digital music products and make much greater profit even if the extra 99,000 products only sell a few copies per year. Online digital product sales shown on a chart begin with a few hundred items with a lot of sales, followed by a long, long tail of items with only a few sales.

But the long tail is no longer just for digital products. Manufacturers and major wholesalers have also recognized a huge opportunity by offering drop-ship products directly to the customers of online marketers. TeeSpring specializes in producing customized T-shirts and hoodies for online marketers, processing payments, shipping the products and sending the profits to the marketers. The service has been so successful that TeeSpring made millions of dollars from 6 million sales in 2014, while hundreds of entrepreneurial marketers made over $100,000 each, and at least 10 associated marketers have become millionaires (Konrad, 2014). Some companies like eBay essentially broker the sales of items between others for a small commission, and they allow online entrepreneurs to receive a small portion of the commission for recruiting customers. Other companies like Amazon do carry large inventories to sell and ship, and they also now are recruiting affiliate marketers. Because of its high growth rate and profit margin, Amazon surpassed Walmart as the world's highest valued retailer in July 2015 despite generating far less in total sales revenue (Krantz, 2015). And there are now many "regular" stores that encourage customers to buy online and then drop by their brick-and-mortar store to pick up the purchased items and quickly be on their way. All of these are offering long tails of products and online marketing opportunities.

Yet, the argument that digitalization of products has driven many firms to employ long-tail strategies, in which niche products are supposed to be profitable too, doesn't necessarily hold true. The father of the long-tail rationale, Chris Anderson, argued that in the contemporary economic and technological environment, businesses that could successfully market the niche (and not blockbusters) will prosper, but another side of the story told by Anita Elberse (2008), a professor of business administration at Harvard Business School, suggests that this is not the case. In her analysis of sales data from the music and home-video industries, Elberse shows that abandoning blockbusters is not the best strategy. While the long tail does, indeed, become longer, no evidence supports the argument for this tail being "thicker," i.e. higher sales of niche products. Instead, the opposite is true. Blockbusters become even more popular, while the sales of obscure music and video titles remain flat. For online and mobile marketing it means that niche marketing is not necessarily a must-have differentiation strategy in the new age. As a result, a clever balance between keeping a few popular titles and promoting bestselling units, as well as expanding the diversity of niche content should make the product assortment work by serving both customer groups: light consumers, who only purchase popular content, and heavy users, who purchase both popular and obscure titles.

Apart from new product portfolio strategies that the digitalization and mobile technology has brought, an array of new kinds of products and services comes into play. And one of the most obvious ones is the market of applications for smartphones. Mobile apps is a huge market for both developers and app stores. According to Gartner (2013b), the market of mobile apps will surpass 268 billion downloads and be worth more than $76.5 billion by 2017, up from just $8.3 billion in 2011. With this in mind, marketing mobile apps has become an important issue for many businesses. As a result, the importance of an app as a *product* has become prominent, too. According to a qualitative study of Chiem et al. (2010), "companies can use apps to create customized software solutions that not only promote the brand but also add sustainable utility to the mobile handset" (p.43).

In an article for Kissmetrics Robi Ganguly (2013), the CEO of mobile CRM platform Apptentive pointed out the "five biggest mistakes in mobile app marketing." One of the strongest messages of the article was that *mobile* is different from just *online*, and this is what marketers of both online and offline products and services should consider while using apps in their marketing strategies, regardless of whether they are using a mobile app as a distribution platform or as a product/service itself. A mobile app is a new product category that requires a completely different approach. Some of the most frequent mistakes that app developers do with their *product* are "treating mobile experience like it's the desktop" and "passing off a mobile website for a mobile app" (ibid.).

Yet, from the consumer-oriented perspective, it is not only about the apps, but rather about the value that they bring. This is one of the key perspectives of the "disruptive innovation" guru, Clayton Christensen, PhD, one of Harvard's most-published researchers. He cites a number of case studies where companies have successfully entered and eventually taken over a market against monopolistic or oligopolistic competition by starting their competition against "nonconsumption," such as Sony entering the market with cheap transistor radios while RCA kept producing the larger, higher-quality but more expensive vacuum tube radios. Sony was selling to teenagers and others who were not yet RCA customers. "The incumbent leaders feel no pain and little threat until the disruption is in its final stages" (Christensen, 2014). In looking for nonconsumers Christensen has challenged newspaper executives and others to look for the "jobs not done," in other words, products or services that would fill an unmet need of consumers (Tran, 2014). This consumer orientation gives rise to innovative out-of-the-box thinking and creation of new products to solve existing consumers' problems. Mobile technology dramatically impacted the features of a *product* element and gave birth to a number of new product and service categories that have never existed before. The rise of customer-to-customer (C2C) markets is one of those new markets. In C2C traditional business relationships, like B2C and B2B, are replaced by networks of interconnected consumers. A previous example of the ridesharing service Uber and a mobile C2C marketplace SimplyListed, an app that enables users to take pictures of their items from smartphones and list them for sale right away, are only a couple from a growing list of C2C players. This gives rise to an incredible power of the "sharing economy," sometimes even coined as a "third mode of production" (Bauwens, 2005).

Promotion/Communication

Most companies have now integrated digital communication tools into their overall promotion strategies, and expectations are that mobile will eventually dominate digital communications. This chapter cannot ignore this prominent part of marketing, but subsequent chapters on advertising, branding and public relations will go into more depth.

As discussed earlier, mobile apps have become important venues for communication between businesses and consumers. With their rising importance and frequency of use, marketers' efforts in in-app advertising are soaring, too. A recent analysis of mobile marketing costs by Fiksu (2015), a data-fueled mobile marketing tech company, suggested that overall marketing costs spent on in-app advertising have increased over years. This change is linked to changing app usage patterns. Despite an increase in total time spent on using mobile apps, in the previous two years the number of apps in use hadn't increased. This makes it more difficult to target loyal customers and break through the advertising clutter within the most-used apps. Rising costs of in-app advertising are leading to new methods of attracting consumers' attention. One of the solutions might be attracting more organic inbound traffic.

"Inbound marketing" and "content marketing" have become some of the most important marketing tools of the digital age. Mallikarjunan (2014, p. 7) says inbound marketing has been the most effective marketing method for online business since 2006 and is "a holistic, data-driven strategy that involves attracting and converting visitors into customers through personalized, relevant information and content … and following them through the sales experience with ongoing engagement." As suggested earlier, inbound and content marketing don't fall exactly into any of the other promotional categories, such as advertising, branding and PR. For that reason we will cover this type of marketing more in-depth within this chapter.

First, let's clarify these two terms. "Inbound marketing" was a term coined by one of the top online marketing consulting agencies, HubSpot, in 2006 (Chernov, 2014, p. 12). A related term is "content marketing." Most professional marketers are familiar with both terms, but there is disagreement on whether they are the same or related. HubSpot surveyed over 6,000 marketing professionals in June 2014 to see how they considered the terms to relate. The majority (59%) considered "content marketing" to be a subset of "inbound marketing." Of the remaining respondents 33% were divided fairly equally between the other three options, that they were synonymous terms, that they were totally distinct concepts, or that inbound marketing was a subset of content marketing (Chernov, p. 13). So, for our purposes we will proceed as the majority of marketers believe the terms to relate and suggest that they are correct in that "content marketing" suggests strategies that provide content (usually free educational content) that attracts people toward the company and its paid products and services. "Inbound marketing," as the superset of content marketing, would include additional strategies that provide the same effect. For example, "freemiums" (free trial offers) could be considered part of inbound marketing but not content marketing. In reality the vast majority of inbound marketing efforts also fall into the subset category of content marketing.

The most popular forms of inbound marketing, according to the HubSpot survey, are, in order (Chernov, p. 11):

1. Blogging
2. Organically growing SEO (influenced strongly by all inbound marketing techniques)
3. Content distribution, in general
4. Webinars
5. Visual content
6. Interactive content (games, etc.)
7. How-to videos
8. Online tools
9. Freemium trials

Other key takeaways from HubSpot's survey include (Chernov, p. 6-10):

- Inbound marketers who routinely track ROI in their inbound marketing efforts are 1,200% more likely to report a year-to-year increase in ROI. Chernov says a major reason for this is that tracking ROI results in an organization increasing inbound budget and utilization.
- Inbound marketing helps a company to be found on the Internet. Marketers who blog are 1,300% more likely to drive positive ROI than those who don't.
- Organization leaders and communication practitioners prioritized the purposes of inbound marketing exactly the same and with almost identical scores. These are, in prioritized order:
  ○ Increasing the number of contacts/leads
  ○ Converting contacts/leads to customers
  ○ Reaching the relevant audience
  ○ Increasing revenue derived from existing customers

- ○ Proving the ROI of our marketing activities
- ○ Reducing the cost of contacts/leads/customer acquisition
- Inbound marketing is becoming more important to vendors and marketing agencies. "More companies are running inbound than ever before," writes Chernov (p. 9), but marketing agencies are more committed to inbound than vendors. About 47% of the agency respondents reported that inbound marketing was their primary lead generator, as opposed to about 27% of the vendors. In both cases, about 40% of their leads were attributed to neither inbound nor outbound marketing and probably included such sources as referrals.
- Inbound is becoming a vital part of sales. More than 25% of HubSpot's respondents reported that their sales teams are themselves employing inbound tactics. But more important was the response by sales staff themselves in answering the question, "Which lead sources have become more important over the last 6 months?" The top four in order were email marketing, social media, SEO and blogs. Ranked last was traditional advertising, followed by direct mail, pay per click (PPC), trade shows and telemarketing.
- While organization staff members provide content for most inbound marketing worldwide, guests and freelancers show the greatest ROI.
- For North American B2B companies with 1-200 employees, the average cost per lead is about 66% less for inbound marketing than outbound marketing. For larger companies inbound marketing leads are about 40% less expensive.
- Respondents were from around the world, and the results showed that these are global trends, with little variance between locations.

In authoring an earlier 2014 survey study involving 1,099 ecommerce marketing and business professionals from 45 countries in six continents, Mallikarjunan (pp. 10-19) wrote that 60% of the B2B respondents reported that their companies use inbound marketing, whereas 54% of B2C respondents so reported. Overall, 22% of the respondents were not sure whether their organizations were using inbound strategies. Apparently some respondents did not understand that blogging is an inbound strategy because 67% of the respondents reported that their organizations do invest in blogging. Of those organizations that reported a positive ROI from inbound marketing, 72% were blogging at least once a week. Those that blog were 155% more likely to report a positive ROI. Blogging also helps generate more indexable pages that lead to more and higher listings in organic searches. Of companies that reported that most of their customers come from organic or direct traffic, 69% blog at least weekly. Other findings from the study included:

- Once inbound marketing attracts a visitor to an organization's website, 25% of the marketers follow up primarily with email and 24% primarily via social media (p. 29).
- Businesses that emphasize educational content in their inbound marketing are 527% more likely to report a positive ROI than those that emphasize coupons (p. 33).
- Twice as many ecommerce marketers preferred social media as a means to provide educational content over any other method. HubSpot principal ecommerce inbound marketing expert Morgan Jacobson at first glance found the results surprising, but noted that each medium has its distinct benefits and drawbacks. "Social media posts tend to be more effective when used in a communicative manner rather than educational due to the limited amount of time that consumers are exposed to them. Email's popularity (20%) makes sense since that's still the primary way that people digest

new information. Ebooks and webinars and podcasts are all harder to produce, so their lack of popularity is also understandable" (p. 34).

- Organic searches are credited with driving the most traffic to a website (22%), although such inbound techniques as blogging, videos, podcasts, webinars and ebooks all help achieve higher SEO. Social media, at 17.4%, are credited with providing the second-most impact on traffic, followed by PPC and email.

While GPS geolocation and mobile location-based services like Foursquare and Google Maps have been in place for a while, newer technologies are entering the marketing reality too. Beacons and geofencing are some of the new trends in mobile marketing. With the help of these, advertising and promotions can be tied to individual customers' physical proximity to a specific store location. However, even simpler technology of GPS geolocation can provide incredible results for location-tied ads. A beauty brand Sephora, for example, provides an excellent case of using *geolocation* to turn consumer smartphones into their local store magnets (Google, b). The brand integrates all its digital marketing communications and simultaneously leverages channels like social media, mobile marketing and in-store digital. Sephora's mobile strategy used an insight that their consumers search for products even before they visit a store. So, the company used geotargeting with Google's local inventory ads.

Yet, the focus of all these new mobile-empowered promotional instruments is clear, and it's communication with consumers. As you will see from our chapter on public relations, two-way communication has grown into an important paradigm in the age of mobile. The early comparison between marketing and communication frameworks by Duncan and Moriarty (1998) that intersect at three key points – messages, stakeholders, and interactivity – showed how important the two-way communication is. And the mobile technology has intensified this tendency by increasing the number of possible touchpoints, blurring the borders between consumers and employees, and adding a new dimension of instant connectivity.

Place/Convenience

Mobile has obviously transformed the perceived physical space, too. While the *place* element focuses on a certain distribution channel, *convenience* focuses on consumer perceptions about space and on the need to get things done easier and more efficiently. Even before an outburst of the Internet and, later, mobile technologies, Lauterborn wrote,

Forget place. Think convenience to buy. People don't have to go anyplace any more, in this era of catalogs, credit cards and phones in every room. On the other hand, when they do decide to go somewhere, it's no longer only to Kroger's. What's a poor marketer to do? Think beyond those nice, neat distribution channels you've set up over the years. Know how each subsegment of the market prefers to buy, and be ubiquitous.

In the age of growing m-commerce and ubiquitous Internet, *convenience* is something that is a top priority, and being *ubiquitous* is now not just a metaphor, but rather a mobile-driven reality. From a corporate website to m-commerce and context marketing the *place* element is not static anymore. It is flexible and follows a consumer whenever she wants to make a purchase or learn about a product or service.

The place to explore products and to make purchases in the digital world is typically a company website. The website sales influences and is influenced by the other 3 Ps. Web pages to display and demonstrate products and services must be created, and because of the long-tail aspect, there may be many product pages. Then how do you get people to come to the virtual store? As noted previously, inbound marketing tools provide more searchable content for an organization's website. A HubSpot study of its own 7,000+ customers revealed that those with over 1,000 pages of content can achieve 3,500% more traffic than those with fewer than 50 pages of content. Inbound marketing techniques, such as blogging and free ebooks, provide those 1,000 pages of searchable content (HubSpot, 2014a, p. 9). Other inbound marketing requires customized landing pages to become part of the website. A business might place a video on YouTube promoting a new product or buy PPC advertising on Facebook to promote a free webinar, but the website landing page provides the details, requests the visitors' contact information, etc.

The greater power of video also impacts the design and content of a website. Almost 95% of online marketers believe video is the most powerful tool for digital marketing, and website video can increase conversion rates by 60% (Aberdeen, 2014). A study now somewhat dated showed that videos or web pages that embed videos are 53 times more likely to get on Google's front search page (Elliott, 2009). Other studies have shown that very few people go past the first search page, so if you're not on Page 1, your SEO efforts have essentially failed. Elliott reprinted the study in 2012 with the warning that the results in a replicated study now would almost certainly be different, and a Google software engineer and blogger noted that even Elliott's study showed that video pages still only had 11,000-1 odds of making the front search page (as opposed to 500,000-1 odds for text pages), which makes videos hardly an SEO "magic bullet" (Gannes, 2009). However, online entrepreneur Jon Penberthy decided to run a series of marketing experiments to test definitive strategies of using video to enhance SEO and marketing, overall. In the process of producing and strategically applying 85 short and highly targeted marketing videos, he was able to achieve front-page rankings on Google and YouTube searches numerous times, and over a few years of refining his strategies has made nearly $1 million in profits without any products of his own – just collecting commissions on products available on Clickbank (Penberthy, 2015). In his experimentation, some of the keys Penberthy found to employing the video SEO strategy were:

- Make the video highly targeted. His example is NOT to do a video about weight loss in general, but rather how to get rid of fat on a specific part of the body – the upper arms, the butt or the thighs. There is too much competition to get to Page 1 with a very general video. Use the Google AdWords Keyword Planner to find popular key words and phrases, but don't be satisfied with the results of a single search. A first search for "weight loss" will provide a number of options to choose from. Again, a marketer should not looking for the most popular but one that matches his product most precisely but still indicates numerous monthly searches for those specific keywords. In this example it might be "weight loss programs," which gets 18,100 searches per month. Then replace "weight loss" as your product with "weight loss programs" and see what Google suggests. If appropriate, you might then choose "free weight loss programs" with 2,900 searches per month. Continue this process several more times until you feel you have the right combination of precise description with high demand. With his array of videos and products, Penberthy averages a little over 1,000 new leads per video and makes an average of about $9 per lead. He also found that he could convert leads from more highly targeted videos at a rate of about 50% – far above industry average.
- Make sure the first 7 seconds of the video are both catchy and refer directly to the point of the video. Make sure that the video title, content and even the channel include and focus on the key

phrase. (Another strategist – not Penberthy – suggests placing the entire text of the video script in the video's upload Description section.)

- Make short videos. Part of a favorable video ranking, he discovered (and Google later confirmed to him), is the percentage of the video viewed on average. A short video that entices viewers to watch almost all of the video typically ranks higher than longer videos that most viewers quit watching midway – even if significantly more people start watching the longer video.

- The video does NOT have to be highly professional. It can be quite simple, but highly credible and compelling enough in 1-2 minutes to capture the viewer's interest and entice them to take the next step. To be compelling, the presenter does not need to be a professional model or actor, but should look the part of someone who might have used the product or have some knowledge of the product. Such a person, Penberthy suggests, can frequently be found at http://Fiverr.com at a very reasonable cost to small businesses and individual entrepreneurs. (Of course, the bigger the company, the greater the expectations.)

- Make a clear call to action to get viewers to go to the website for more detailed information. Viewers would typically be invited in the first video to go to the website for another more-detailed video or a free ebook that addresses the message suggested by the key phrase. The call to action should be included in the video, in the video description, and in other associated promotions.

- Create a powerful landing page on the website. Always use content marketing to drive people to the website and to a very focused landing page. The landing page itself may also have a short video. Then to access another highly targeted but a more detailed video, webinar or ebook, the prospect would provide his name, email address and additional information for strategic follow-up communication, leading ideally to conversion of the prospect to a customer. Videos in the conversion process do nearly double the conversion rate, as noted in other research (Aberdeen Group, 2014, p. 2).

Another failure for many organizations is to fully prepare their websites for easy mobile access. This is now vitally important because mobile accessibility has been added to Google's and to other search engine algorithms. The number of mobile searches in some countries is now equal or greater than desktop searches, and overall daily use of mobile devices has now surpassed that of desktop (Gibbs, 2015). Use of mobile is expected to increase since there are now three times as many mobile devices accessing the Internet as fixed-wire connections, according to the International Telecommunications Union (2015). In a live webcast, eMarketer CEO Geoff Ramsey (2015) noted that 86% of so-called Millennials still complained that most commercial websites were still not mobile-friendly, and 71% of them found that most online businesses fail to offer a mobile app as an alternative. Ramsey noted that of the billions of dollars in purchases made on mobile devices, 72% are via mobile apps and 28% on websites. Mobile customers who arrive to a website that does not accommodate mobile almost always go to a competitor's website thereafter.

The design and functionality of an online organization's "place" has a powerful effect on prospective customers. According to Adobe, of 2,146 executives in early 2015 who responded to the question, "Over the next five years, what is the primary way your organization will seek to differentiate itself from competitors?" the No. 1 priority for 44% of the respondents was "Customer service/customer experience – making it easy, fun, valuable and/or pleasurable to shop from us" (Adobe, 2015, p. 15). How does an organization enhance its customers' online experience? The Figure 1 below shows how executives responded to the Adobe survey (pp. 17-18).

Figure 1. Responses for directions where organizations place the highest emphasis on terms of improving the customer experience
Adobe (2015)

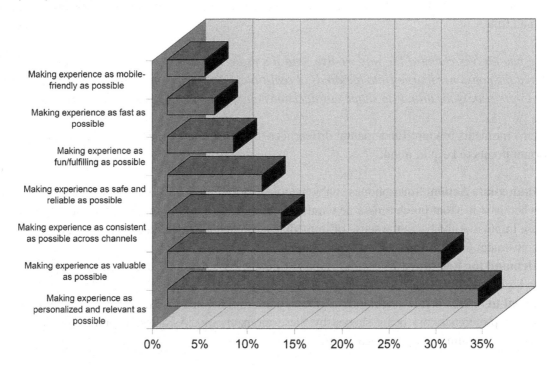

Some new forms of distribution channels include mobile retail commerce and in-app purchases. It is estimated that by 2019 the share of sales generated via mobile devices will reach almost a third of total e-commerce sales (eMarketer, 2015). Despite a growing share of m-commerce in total e-commerce sales, retailers are not so fast in keeping up with consumers' expectations about their shopping experience, at least the ones in the U.S. As findings of a recent Accenture Seamless Retail Research study (Accenture, 2015) suggest, there is a big gap between consumer expectations and actual mobile shopping experience, and retailers need to close this gap as soon as possible in order to win customer loyalty. Despite almost a third of respondents reporting more shopping activity via smartphones (35%) and tablets (31%) in the past year, less than half found it easy to complete a purchase via a mobile device. An additional emphasis on in-store experience is closely related to the mobile. The results of the Accenture Consumer Survey (ibid.) highlighted five new improvements to existing in-store experience:

- Ability to order out-of-stock products via mobile devices
- Free Wi-Fi
- Geo-prompted promotions via mobile devices while in store
- Ability to scan products and ship them home
- Interactive store displays with product information

These changes lead marketers to understanding consumer *micro-moments* and the importance of just-in-time communication. The concept of *micro-moments* introduced by the Google research team (Google, a) is a spur of the mobile revolution in its purest sense that transforms how consumers perceive space and time.

Mobile has forever changed the way we live, and it's forever changed what we expect of brands. It's fractured the consumer journey into hundreds of real-time, intent-driven micro-moments. Each one is a critical opportunity for brands to shape our decisions and preferences (ibid.).

Micro-moments brought fundamental differences to mobile marketing. The concept leads to three important points to keep in mind:

- **Immediate Action**: Smartphones enable people to act immediately whenever they feel prompted, whether it is about purchasing a new music album or just looking for information. These actions are impulsive and bring the sense of immediacy. Marketers should keep this in mind and make it convenient for consumers to act immediately.
- **Demand for Relevance**: The value of marketing quality, relevance, and usefulness is growing due to high expectations and low patience of consumers.
- **Loyal to Needs**: While existing in fragmented micro-moments characterized by immediacy, consumer preferences are being shaped by these moments too. As a result, brands should address specific consumer needs in each micro-moment.

As an example of how these micro-moments fragment and shape the consumer journey, the Google research argues that "people pursue big goals in small moments." The results of their March 2015 survey of 5,398 Internet users (ibid.) say that "90% of smartphone users have used their phone to make progress toward a long-term goal or multi-step process while 'out and about.'" So, as big goals of consumers can be broken into multiple micro-moments, the relevance of *when* and *where* to reach consumers is essential.

Price/Cost

Pricing is certainly affected by online promotional efforts, virtual storefronts and the digitization of products. Even the pricing of products that are essentially free to produce is a science. The lowest price does not always achieve the most unit sales, much less the greatest profits. A friend laughs about how she tried futilely to give away a litter of new puppies on Craigs List until she decided instead to charge $50 apiece. Apparently a free puppy is not perceived to be a quality puppy. The same has been found to be true for digital books. A $1 book does not necessarily sell more copies than the same book priced at $3 or $5 or $10. Pricing research must frequently be done on specific product lines to be most effective, but other pricing research can give organizations some idea of the most likely considerations. However, according to executives who responded to Adobe's question about their organization's "primary way … to differentiate itself from competitors" over the next five years, only 5% made price the top priority. (Adobe, 2015, p. 15)

As a result, similarly to the *place/convenience* dichotomy, Lauterborn recommends to focus on consumer perceptions, not a marketable product, and to:

Forget price. Understand the consumer's cost to satisfy that want or need. Price is almost irrelevant; dollars are only one part of cost. What you're selling against if you're selling hamburgers is not just another burger for a few cents more or less. It's the cost of time to drive to your place, the cost of conscience to eat meat at all, versus perhaps the cost of guilt for not treating the kids. Value is no longer the biggest burger for the cheapest price; it's a complex equation with as many different correct solutions as there are subsets of customers.

This new cost orientation may include considerations other than monetary costs, that is, the price of a product/service. Another example of such a non-monetary cost would be the memory storage that could bar a consumer from downloading a new app.

SUMMARIZING PAST RESEARCH

Despite an obvious hype around mobile technology, the academic research on mobile marketing is still slow to keep up with the mapped business opportunities and industry practices. The field is quite fragmented, and a great part of the research is focused on technology adoption and consumer attitudes (Lamarre et al., 2012; Ismail & Razak, 2011). Much of the early research was quite speculative, discussing opportunities that mobile technology might provide, while most of the first empirical studies focused generally on SMS marketing and mobile advertising (Figure 2) (Drossos & Giaglis, 2010).

Figure 2. U.S. retail m-commerce sales, 2011-2016
eMarketer (January 2013)

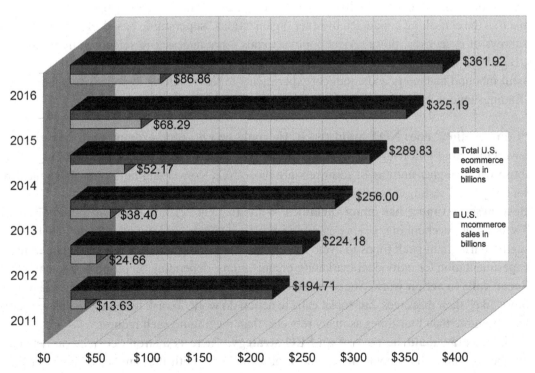

Hence, because financial opportunities available through online and mobile media are so great, the industry itself is conducting much of the research. Mobile sales are expected to approach $70 billion in 2015 – and that's only 21% of all ecommerce sales. During this dynamic decade of commercial development on the Internet, every organization is anxious to find the formula for success. Industry research is not always the best research, but even when it is done poorly it at least focuses light on an issue requiring more scientifically stringent research. Much of the industry research, for example, is done by survey. One study asked B2B marketers where they planned to increase funding the most during the next year, and the three top items for increased funding all related to "inbound marketing" techniques, such as webinars, search engine optimization, blogging, e-books and white papers, etc. (MarketingSherpa, 2010). However, just because that was the latest trend in marketing does not mean these were necessarily the most effective tools. That would require more intensive research. Another study (Content Marketing Institute, 2013) showed that the most effective web marketers had a documented strategy, a person dedicated to managing online marketing, and used a greater number of online tools than did the least effective online marketers. Those corollaries were used to suggest that all online marketers should have that level of commitment. However, the top online marketers may also have more effective traditional advertising campaigns to supplement their online campaigns. Correlation, as most academic researchers understand, is not the same as causation.

Of all the industry researchers, none has gained greater acclaim nor addressed online PR/marketing from a more scientific perspective than Dan Zarrella, who became the "social media scientist" for HubSpot online marketing company. The use of inbound strategies, especially free ebooks and webinars, helped HubSpot soar to the top of the online marketing consulting industry, and Zarrella became their superstar. He spoke at Harvard and has now published four books on the subject, including, "The Science of Marketing: When to Tweet, What to Post, How to Blog, and Other Proven Strategies" (2013), "The Facebook Marketing Book" (2012), "Zarrella's Heirarchy of Contagiousness: The Science, Design and Engineering of Contagious Ideas" (2011), and "The Social Media Marketing Book" (2009). Zarrella calls much of the consulting conducted by other organizations "superstition based." Even when it comes from surveys of executives, those reflect popular trends – not necessarily scientifically founded principles. Besides working with databases with thousands or even millions of files, he analyzes content of successful inbound marketers and conducts experiments to test resulting hypotheses. Here are several of his findings among many that should be given more attention in future research.

- **"Conversation" does NOT build reach.** He could find no good examples of organizations that have built large followings with social media "conversation." In analyzing millions of tweets, those with a higher number of followers are those NOT conversing but rather those that are essentially broadcasting interesting content.
- **Best message timing has many variables.** While there are clearly peaks of traffic for emailing, Tweeting or Facebooking, getting an organization's own message out effectively may be by messaging during non-peak times to avoid the clutter. The timing of messages should be a matter of experimentation for individual marketing organizations, depending on purpose, content, etc.
- **Social calls to action work.** Social marketers have been hesitant to ask their audience to retweet or to "like" their messages, but social calls to action do work. To ask followers to "please retweet" leads to more than four times as many retweets than making no such request.
- **Social media should be part of an SEO strategy.** There is a strong correlation between the number of tweets, Facebook shares and LinkedIn shares with the number of inbound links an

organization's website has from other websites. In other words, the more exposure a site has on social media, the more frequently those who control other websites create links to the target site.

- **PPC and emailing programs also build SEO.** Analyzing organic search traffic to the HubSpot website, Zarrella found a positive correlation of higher organic search rates with paid search tactics, such as pay-per-click ads, and also with higher emailing rates, such as times when they are sending out information about new ebooks or webinars to people in their database.
- **Be interesting, not self-promoting.** In analyzing the content of tweets, Zarrella found that the more an organization refers to itself, the fewer followers it has. In other words, talk about interesting things, but don't talk about yourself too much.

The smartphone has now provided a direct pipeline into the consumers' pocket. In 2014 the Burberry Clothing Brand made their 2014 fashion show available via the Weixin network in China, giving hundreds of thousands of interested viewers an opportunity to join in on the exclusive experience (Stevenson & Wang, 2014). This direct access to the consumer will undoubtedly create a larger pool of potential customers that may otherwise not have been reached. Smart use of mobile devices makes it possible for the marketer to follow all interactions the consumer has with the brand. This invaluable data allows the marketer to track customer reactions to advertisements and marketing strategy in real time, and create a very fluid dialogue between the two parties.

This leads to new "personalization" efforts by online organizations. HubSpot's survey of 1,099 ecommerce marketing and business professionals found that 65% of businesses are investing in marketing personalization (Mallikarjunan, 2014, p. 39-41), and those surveyed responded that personalized emailing and personalized websites are about equally important. "Personalization is a harder inbound tactic to deploy, usually requiring specialized software and a centralized contacts management database," said HubSpot ecommerce expert Ted Ammon. "Although some shopping carts have the functionality built in natively for product recommendations, the percentage of marketers using it in email indicates that it's rapidly expanding into other channels." Of businesses that get most of their customers from email, 74% felt that personalizing email was at least "somewhat important." Of businesses that get most of their customers from websites (direct or organic search), 82% felt that personalizing website functions was at least somewhat important. The 2015 Adobe survey (p. 21) found that 38% of respondents planned to achieve "omnichannel personalization" by the end of 2015.

Real-time data access is one key to enhanced personalization and contextual communications. "A wave of marketing that has been growing for years – contextual marketing – will reach new heights in 2015," concludes marketing consultants Strongview (2014). "The increased availability of real-time data and innovative approaches for harnessing it will enable marketers to engage in more contextually relevant messaging than ever before (figure 3). The next generation of competitive advantage in digital marketing will be driven by organizations that can execute messaging relevant to their customers' current contexts at the moment of engagement (Strongview, 2014, p. 23).

An Evergage survey of marketers identified eight major purposes to utilizing real-time data. The top two purposes suggested received far more support than the rest. The benefit identified by 81% of the respondents was to "increase customer engagement." Second, identified by 73% of the marketers, was to "improve customer experience" (Evergage, 2014).

By using geotargeting, customers are receiving advertisements that are more relevant to their buying experiences and are also within a reasonable distance to their location. About 45% of consumers noted that they would be "more likely to click on a mobile ad that is applicable to their location" (Eddy 2014).

Figure 3. Benefits of real-time marketing according to marketers worldwide
Evergage (March 2014)

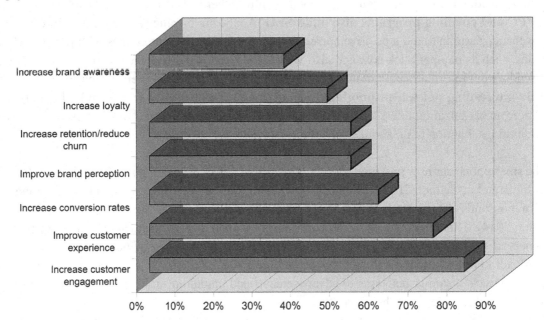

Eddy argues that mobile marketing should become part of the "traditional marketing mix" as "relevancy is the key to success." While some business owners – and in particular small business owners – worry that the mobile experience will discourage customers from frequenting the store's physical location, this simply isn't the case. When asked, 63% of consumers said they would be very likely to use a coupon if it were delivered in mobile form. The 2015 Adobe survey (p. 25) found that 24% of all respondents plan to be using geo-targeting by the end of 2015, and 25% are actively exploring the technology for possible implementation.

Mobile marketing can be used to manage the before, during, and after of the shopping experience. Customers can receive coupons or advertisements while in a brick-and-mortar store or a virtual store, prompting increased spending and, potentially, an increase in brand loyalty. Moreover, mobile marketing can be used to help manage the relationship between the consumer and retailer; mobile surveys, communication by automated text, and store emails are just a few examples of the tools that can be used to reach the consumer.

With the increased availability of mobile devices and access to the Internet, groups such as the banking industry have been able to increase their contact with the customer, as well as make this contact easier than ever before. There is even an opportunity for crossover between companies looking to more accurately market products to their consumer base. For instance, banks could market specific items and coupons to their customers. These recommendations would align with the spending habits and typical price range of items that mobile bank users generally purchase in (Brister 2010).

What are the top three online marketing priorities for executives? Adobe's 2015 survey (p. 20) provides a valuable list of options the 2,748 respondents are considering. A top priority to 30% of the respondents is "targeting and personalization," followed by "content optimization" to 29%, "social media engagement" to 27%, "brand building/viral marketing" to 24%, multichannel campaign management to 22%, and "conversion rate optimization" to 20% (Figure 4).

Figure 4. Which three digital-related areas are the top priorities for your organization in 2015
Adobe (2015)

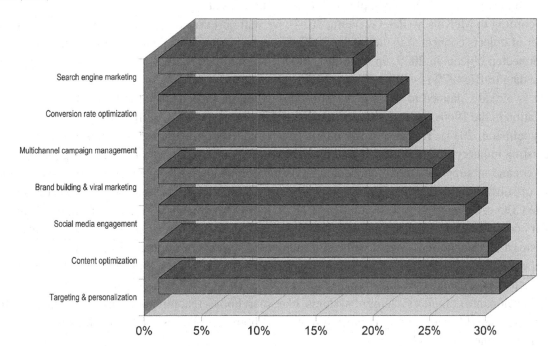

While it is clear that mobile marketing appeals to both the consumer and the marketer, there is a lack of academic research to parallel industry practice (Okazaki, 2012). Perhaps this is due to the speed at which the mobile market changes and evolves, leaving academia in the wake of its progress. However, it is reasonable to assert that unless further research is done to more clearly define the theories that drive mobile marketing, the mass communication community at large will be unable to harness the full power of mobile marketing.

MAPPING THE FUTURE

However, even though the bulk of research is done by the industry, there are some new trends in marketing with mobile coming on stage in academic research too. In this section we will discuss some of new methods and theories that emerged in the age of mobile.

Internet of Things

Mobile is not only about smartphones and tablets anymore. The Internet of Things (IoT) is a new trend that is already here and will not leave the scene for the next decades. Smartphones and tablets represent only a small portion of what this trend can potentially grow into. Neither the term of *smart wearables*, like smartwatches, smart glasses or fitness bands, can embrace the IoT. In fact, the IoT is a complete ecosystem of interconnected smart devices with sophisticated sensors and unique device identification to seamlessly communicate with each other and create a continuous flow of data. In general terms, this means complete integration of all our smart devices and other smart objects, like home-automation systems and smart grids, into a complex universal system of data.

According to McKinsey (Patel & Veira, 2014), we will observe an almost three-fold increase in the number of objects connected to the IoT by 2020. The agency's analysis predicts around 26 to 30 billion IoT-connected objects in 2020, up from just 7 to 10 billion in 2013, that will contribute to almost all mobile data traffic (97%) by 2019 (Cisco, 2015).

IoT is already claimed to become the third IT revolution, preceded by the first wave of industrial automation in the 1960s and 1970s and a more recent Internet revolution of the 1980s and 1990s (Porter & Heppelmann, 2014). This new smart mobile interconnectivity revolution will significantly reshape the existing industry structure affecting all dimensions of the classic Porter's Five: the bargaining power of buyers and of suppliers, the rivalry among existing competitors, the threat of new entrants, and the threat of substitutes to firm's products or services. According to an article of Michael Porter himself and the CEO of computer-software company PTC, James Heppelmann (ibid.), apart from reshaping the industry competition, even the boundaries between industries will be redefined. The growing functionality of smart interconnected devices will result in bundled systems of products, like a whole package of related equipment and services. For example, in farming this might be a family of farming equipment with tractors and planters all interconnected and optimized to work as a system. And further on this might develop into complex systems of systems, like smart homes that integrate various product families within a household, like lighting control systems and home appliances, or smart cities that integrate smart homes and smart grids.

For marketing this means huge structural changes in many instances from new customer relationship building mechanisms to sophisticated segmentation techniques and enormous opportunities for product differentiation. Stuart Leung from Salesforce (2014) in an article for Forbes listed several implications of IoT for marketing. Firstly, equipped with numerous sensors, instantly interconnected smart objects enable marketers to easily collect and exchange sales data with information about purchased products and services. Along with more sophisticated customer feedback systems, these data can contribute dramatically to improvement of ROI on future sales and smarter CRM systems. Coupled with social media, IoT can also lead to predictive social media and possibly "new online communities to develop centered around users of particular devices."

Predictive social media will be a step forward from one of the current themes – participative marketing. In participative marketing marketers involve consumers in social interaction with the brand by letting them decide, for example, what promotions they would like to be offered or even which price they want to pay (Santana & Morwitz, 2013), and thus generating a social buzz and a dedicated community around a product or service. The IoT as a network of smart objects will thus facilitate the transition of an online social network of Web 2.0 into a smart social network, Web 3.0. The latter was coined by John Markoff in an article for the New York Times (2006) as "the intelligent Web." With a horde of real-time data about human behaviors, preferences, and activities on hand, "the potential of the artificial intelligence, data mining and the so-called Big Data to build solutions based on the understanding of human behaviours" is increasingly big. This makes possible research into the branches of complex network theory like that of Human Dynamics (Jara et al., 2014a). In a similar fashion, research on Embedded Intelligence (EI) attempts to reveal the "individual behaviors, spatial contexts, as well as social patterns and urban dynamics by mining the digital traces left by people while interacting with Internet of Smart Things" (Guo et al., 2011).

Moreover, in transiting from the social communication-oriented Web 2.0 we can now observe a network of people voluntarily willing to share their knowledge and user experience with others (Kotler et al., 2010). With this in mind, a mobile-enabled social platform integrating such identification technologies as barcodes, QR codes and RFID/NFC tags was envisioned. However, according to Solima et al. (2015), a turning point in predictive marketing will take place when not only buyers can generate data but even the smart objects themselves:

This transition will be achieved with the shift from smart devices, currently at the core of the data-gathering stage of the process of purchasing a product, to smart objects, capable of automatically activating the flow of information about itself and its surrounding environment. This condition will have an impact on a person's decision-making process, on the methods employed to access sufficient elements to make well-informed decisions and, more in general, on each person's learning process. (p.12)

Overall, IoT transforms the traditional marketing mix completely with new product features, tailor-cut promotional possibilities, smart pricing algorithms, and ubiquitous connectivity. On the other hand, communication is in its heart too: providing accurate information about *consumers* and their wants to build solutions that they can *conveniently* access anytime, wherever they are with minimum monetary and non-monetary *costs* and the possibility of interactive *communication*.

Mobile-Driven Big Data

This instant interconnectivity of smart devices means that mobile-driven Big Data is now another big thing. Big Data accumulated from mobile users can provide invaluable insights into understanding consumers. From health-related to geolocation data, personal mobile devices open up a breadth of opportunities for consumer research. Even such generic information like battery life attributes can be used to track web browsing habits of smartphone users. The findings of a group of French and Belgian researchers (Olejnik et al., 2015) suggested that "the capacity of the battery, as well as its level, expose a fingerprintable surface that can be used to track web users in short time intervals." The work highlighted existing privacy risks, as the researchers could track smartphone users' visits across multiple websites with the help of a third-party script that used battery life information, like charging and discharging time, to link users' concurrent visits to those sites without their permission.

But even legal use of data that doesn't affect user privacy can make a huge difference. The impact of mobile-driven Big Data on marketing is thus enormous. MIT's Andrew McAfee and Erik Brynjolfsson (2012), authors of the two books *Race against the machine* (2011) and *The second machine age* (2014), discussed in an article for Harvard Business Review implications of Big Data in management decision-making. Technology has already reached the point at which it can enable businesses to accurately track hidden causes of acute drops and increases in sales or to enhance customer service by scanning health care practices in a multimillion-entry database. What is not yet in place is the management readiness to use this data smartly in important decision-making processes. And the payoff for the smart use of Big Data is huge too: on average, data-driven companies were 6% more profitable and 5% more productive than comparable companies from the industry that were not data-driven.

The concept of data analytics is not new. But the rapid growth in speed and accessibility of online and especially mobile technology has radically transformed the simple concept of analytics in three important instances. The most obvious one is the *volume* of data generated. According to Alon Even (2015), VP

Marketing for Appsee, mobile traffic will largely contribute to this growth: "By 2016, 61% of web traffic will come from wireless devices as opposed to desktops, contributing heavily to the growth of Big Data." According to Cisco's predictions (2015), by 2019 monthly global mobile data traffic will reach 24.3 exabytes (EB), this will be a tenfold increase in data volume generated since 2014. For comparison, trying to store only 1 EB of data on "32GB iPhone 5 devices stacked one on top of the other" (Data-Pop Alliance) will result in a pile that would be 566 times higher than the Empire State Building.

The speed of data creation and analysis has changed, too. One of the practical examples of the Big Data *velocity* is analysis of real-time data. McAffee and Brynjolfsson (2012) reported the study of their MIT colleague Alex Pentland, the Toshiba professor of media arts and sciences and the director of Media Lab Entrepreneurship Program, who used geolocation data gathered from mobile phones of people who parked their cars at Macy's on Black Friday. This real-time data enabled the research group to estimate the sales volume of the retailer the same day, even before official records.

New mobile handsets and mobile Internet substantially contributed to increased *variety* of data. In the single year of 2014 almost half a billion new mobile devices were introduced to the market with 88% of them being smartphones (Cisco, 2015). The data generated by all those devices includes (but is not limited to) GPS, VoIP, web data, video, audio streaming, and file sharing. Moreover, even the nature of analytics is changing from conventional statistical tests to artificial intelligence and machine learning algorithms.

McAffee and Brynjolfsson (2012) provided an example of Sears' personalized promotions that made use of huge amounts of consumer data generated by its Sears, Craftsman, and Lands' End brands. The data included information about customers, products and historical promotion data and was dispersed across numerous data warehouses maintained separately by each brand. This made the data difficult to be analyzed. However, the use of the software framework Handoop dramatically cut the time needed to generate personalized promotions.

According to the company's CTO, Phil Shelley, the time needed to generate a comprehensive set of promotions dropped from eight weeks to one, and is still dropping. And these promotions are of higher quality, because they're more timely, more granular, and more personalized. (p.65)

Results of a large-scale study by researchers from Norwegian Telenor Research and the MIT's Media Laboratory (Sundsoy et al., 2014) showed that machine learning techniques made possible with the analysis of Big Data collected from a mobile network operator (MNO) can indeed outperform the best conventional marketing practices in terms of conversion rates. With the help of metadata and social network analysis the researchers could conduct customer segmentation for text-based marketing and identify groups of users who would be most likely to start using mobile Internet. Resulting customer clusters were tested on two groups of customers: an experimental group segmented on a basis of Big Data-driven marketing and a control group selected using traditional marketing practices. The study results were quite impressive:

Experimental results with 250 000 customers show a 13 times better conversion-rate compared to the control group. [...] The model also shows very good properties in the longer term, as 98% of the converted customers in the treatment group renew their mobile Internet packages after the campaign, compared to 37% in the control group. These results show that data-driven marketing can significantly improve conversion rates over current best-practice marketing strategies. (p.367)

However, the use of Big Data from mobile devices is not limited to consumer research, though. McKinsey's seven traits of effective digital enterprises include being quick and data driven. An example of an American transportation company U.S. Xpress demonstrated how the company in the first year alone could save over $20 million USD by smarter fuel consumption. U.S. Xpress uses real-time data from a huge number of in-vehicle sensors and geospatial systems to "extract game-changing insights about its fleet operations" (Olanrewaju et al., 2014).

DIRECTIONS FOR FUTURE RESEARCH

There are many subjects to explore concerning mobile marketing. Here are a few more:

1. Explore why smartphone conversion rates are low and how to increase them. According to eMarketer CEO Ramsey (2015), shoppers use smartphones (82%), smartpads (89%), and desktops and laptops (90%) at about the same rate, but when it comes to actual purchases, smartphone conversion rates are much lower – 43% compared to 72% for tablets and 76% for desktops/laptops. Why is there such a disparity and now can merchants overcome related obstacles?

2. Explore virtual "point of purchase" advertising. Another condition that can enhance mobile advertising, in particular, is the rapid growth in mobile commerce. The more money people are spending with their mobile devices, the more that advertisers will naturally want to reach out to them. It's comparable to point-of-purchase promotion, which companies utilize regularly in brick-and-mortar outlets. They want to reach out to consumers at the time they are ready to make a purchase. One of the preferred ways to do this online is to make customized purchase recommendations. This function is built into some shopping cart systems and simply recommends items that are frequently purchased by those buying the customer's original cart items. Other systems can be linked real-time to the customer's previous purchasing or shopping habits. Even eMarketer (2013a) was surprised at the rapid growth of mcommerce, which shot up 81% in 2012 and continues its rapid ascent. Now the eMarketer CEO says mobile will reach 60% of all online purchases in 2016 and 70% in 2017 (Ramsey, 2015). Enhanced ecommerce, in general, and mcommerce specifically will attract an increasing share of advertising dollars. How can virtual point-of-purchase advertising enhance the marketing mix?

3. Explore the nature and opportunities in enhancing "customer experience." At the start of 2015, the 6,000 business executives surveyed by Adobe (2015) ranked "customer experience" as the most exciting marketing opportunity. By far the executives (44% compared to next highest of 28%) want to make their customers' experience easier, more pleasurable or more valuable in their effort to distinguish themselves from their competitors. What can be or should be organizations' efforts to achieve this goal? The survey suggests some, but this represents opinions and trends. Research could reveal the truly best practices.

4. Explore best ways to "experiment" with digital. Because digital advertising is so new, 69% of the executives responding to the Adobe (2015) survey agreed that they "are going to be experimenting heavily with digital next year." What will be the nature of this experimentation, and how can organizations best conduct their marketing experimentation? How can scholars facilitate these efforts?

5.   Explore mobile ROI. Executives' rapid move to digital reflects in part their continued distrust in the ROI of traditional media in comparison to digital. More than half (51%) of Adobe's (2015) survey respondents gave only 1-2 points on a 5-point scale (5 being maximum trust) to their ROI confidence in offline ad expenditures and 45% for offline marketing programs. Only 27% of them had the same concerns for digital ad expenditures and 26% for digital marketing programs, overall. ROI concerns for mobile advertising and marketing fell in between. About 36% of the executives rated mobile ad expenditures as 1 or 2 in ROI confidence, and 35% for mobile marketing programs. Why do executives distrust mobile more than digital in general, and how can mobile media raise that trust level to mirror that of other digital media?

6.   Explore how to measure mobile ROI. Related to the previous item is how to measure ROI for mobile. In an eMarketer survey (Ramsey, 2015) executives who were asked about their confidence in measuring mobile ROI expressed a lot of uncertainty. Only 9% indicated "extreme confidence" in measuring mobile ROI while, at the other extreme, 13% express no confidence. Investigating and developing new methods of measuring mobile ROI would be of extreme value to the industry, as well as to academic scholars.

7.   Explore video-based SEO. As cited earlier (Elliott, 2009), a 2009 study showed that videos and mixed-content web pages had a 5,300% greater chance than text- and graphic-based based web pages of achieving a first-page position in a Google search. There appears to be no replication of this study since then, nor any such research by academics, but an entrepreneur's experimentation based on this study has proven very lucrative (Penberthy, 2015). The power of online video, overall, is being supported by more and more industry research, and this is certainly worthy of additional academic research, as well.

REFERENCES

Aberdeen Group. (2014). *Analyzing the ROI of video advertising*. Retrieved from http://go.brightcove.com/bc-aberdeen-analyzing-roi-b

Accenture. (2015). *Seamless Retail Research Report 2015: Maximizing mobile to increase revenue*. Retrieved from https://www.accenture.com/_acnmedia/Accenture/Conversion-Assets/Microsites/Documents15/Accenture-Seamless-Retail-Research-2015-Maximizing-Mobile.pdf

Adobe. (2015). *Quarterly Digital Intelligence Briefing: Digital Trends 2015*. Retrieved 20 Feb. 2015 from http://offers.adobe.com/content/dam/offer-manager/en/na/marketing/Target/Adobe%20Digital%20Trends%20Report%202015.pdf

Anderson, C. (2006). *The long tail: Why the future of business is selling less of more*. Hyperion.

Bauwens, M. (2005). The political economy of peer production. *CTheory, 1*. Retrieved from http://www.ctheory.net/articles.aspx?id=499

Brister, K. (2010, March 1). Making the Most of Mobile; Banks can capitalize on marketing strategies for handheld devices, even if mobile banking is still short of being mainstream. *US Banker,* 17-17.

Brynjolfsson, E., & McAfee, A. (2011). *Race against the machine*. Lexington, MA: Digital Frontier.

Brynjolfsson, E., & McAfee, A. (2014). *The second machine age: Work, progress, and prosperity in a time of brilliant technologies.* New York, NY: WW Norton & Company.

Chernov, J. (2014). *State of Inbound 2014.* Cambridge, MA: Hubspot.

Chiem, R., Arriola, J., Browers, D., Gross, J., Limman, E., Nguyen, P. V., & Seal, K. C. et al. (2010). The critical success factors for marketing with downloadable applications: Lessons learned from selected European countries. *International Journal of Mobile Marketing, 5*(2), 43–56.

Christensen, C. M. (2014). Disruptive Innovation. In *The Encyclopedia of Human-Computer Interaction* (2nd ed.). Retrieved 26 June 2015 from https://www.interaction-design.org/encyclopedia/disruptive_innovation.html

Cisco. (2012). *Cisco's VNI Forecast Projects the Internet Will Be Four Times as Large in Four Years.* Retrieved from http://newsroom.cisco.com/press-release-content?type=webcontent&articleId=888280

Cisco. (2015). *Forecast projects nearly 10-fold global mobile data traffic growth over next five years.* Retrieved from http://newsroom.cisco.com/press-release-content?type=webcontent&articleId=1578507

Constantinides, E. (2006). The marketing mix revisited: towards the 21st century marketing. *Journal of Marketing Management, 22*(3-4), 407-438.

Content Marketing Institute. (2013). *B2B Content Marketing 2014 Benchmarks, Budgets, and Trends – North America.* Retrieved 20 Feb. 2015 from http://contentmarketinginstitute.com/wp-content/uploads/2013/10/B2B_Research_2014_CMI.pdf/

Data-Pop Alliance. (n.d.). *Data "inflation" table.* Retrieved from http://www.datapopalliance.org/resources#data-inflation-table

Drossos, D., & Giaglis, G. M. (2010). Reviewing mobile marketing research to date: Towards ubiquitous marketing. In K. Pousttchi (Ed.), *Handbook of Research on Mobile Marketing Management* (pp. 10–35). Hershey, PA: IGI Global. doi:10.4018/978-1-60566-074-5.ch002

Duncan, T., & Moriarty, S. E. (1998). A communication-based marketing model for managing relationships. *Journal of Marketing, 62*(2), 1–13. doi:10.2307/1252157

Eagle, L., & Kitchen, P. J. (2000). IMC, brand communications, and corporate cultures: Client/advertising agency co-ordination and cohesion. *European Journal of Marketing, 34*(5/6), 667–686. doi:10.1108/03090560010321983

Eddy, N. (2014, May 22). Mobile Advertising, Push Notifications Beneficial for Retailers. *EWeek*, 7-7.

Elliott, N. (2009). *The Easiest Way to a First-Page Ranking on Google.* Retrieved 20 August 2015 from http://blogs.forrester.com/interactive_marketing/2009/01/the-easiest-way.html

eMarketer. (2013a). *eMarketer: Tablets, Smartphones Drive Mobile Commerce to Record Heights.* Retrieved 20 Feb. 2015 from http://www.emarketer.com/newsroom/index.php/emarketer-tablets-smartphones-drive-mobile-commerce-record-heights/

eMarketer. (2013b). *US Total Media Ad Spend Inches Up, Pushed by Digital.* Retrieved 20 Feb. 2015 from http://www.emarketer.com/Article/US-Total-Media-Ad-Spend-Inches-Up-Pushed-by-Digital/1010154 eMarketer. (2014). *Despite Time Spent, Mobile Still Lacks for Ad Spend in the US*. Retrieved 20 Feb. 2015 from http://www.emarketer.com/Article/Despite-Time-Spent-Mobile-Still-Lacks-Ad-Spend-US/1010788

eMarketer. (2015). *Cross-platform video trends roundup*. Retrieved 20 Feb. 2015 from https://www.emarketer.com/public_media/docs/eMarketer_Cross_Platform_Video_Trends_Roundup.pdf

Even, A. (2015, January 22). Big Data and mobile analytics: Ready to rule 2015. *Venture Beat*. Retrieved from http://venturebeat.com/2015/01/22/big-data-and-mobile-analytics-ready-to-rule-2015/

Evergage. (2014). *Real-time for the rest of us: Perceptions of real-time marketing and how it's achieved*. Retrieved 25 March 2015 from http://info.evergage.com/perceptions_of_realtime_marketing_survey_results

Fiksu. (2015, June). *Fiksu's analysis: June 2015* [Press release]. Retrieved from https://www.fiksu.com/resources/fiksu-indexes#analysis

Ganguly, R. (2013). The 5 biggest mistakes in mobile app marketing. *Kissmetrics*. Retrieved from https://blog.kissmetrics.com/mistakes-in-app-marketing/

Gannes, L. (2009). *The Great Video SEO Frontier*. Retrieved 20 August 2015 from https://gigaom.com/2009/02/12/the-great-video-seo-frontier/

Gartner. (2013a, June 4). *Gartner says worldwide mobile payment transaction value to surpass $235 billion in 2013* [Press release]. Retrieved from http://www.gartner.com/newsroom/id/2504915

Gartner. (2013b, September 19). *Gartner says mobile app stores will see annual downloads reach 102 billion in 2013* [Press release]. Retrieved from http://www.gartner.com/newsroom/id/2504915

Gibbs, S. (2015, April 20). Google's 'mobilegeddon' will shake up search results. *The Guardian*.

Google. (n.d.a). Micro-moments. *Think with Google*. Retrieved 1 August 2015 from https://www.thinkwithgoogle.com/micromoments/

Google. (n.d.b). Sephora turns smartphones into local store magnets. *Think with Google*. Retrieved 1 August, 2015 from https://www.thinkwithgoogle.com/interviews/sephora-turns-smartphones-into-local-store-magnets.html

Grönroos, C. (1997). From marketing mix to relationship marketing-towards a paradigm shift in marketing. *Management Decision*, *35*(4), 322–339. doi:10.1108/00251749710169729

Guo, B., Zhang, D., & Wang, Z. (2011, October). Living with internet of things: The emergence of embedded intelligence. In *Internet of Things (iThings/CPSCom), 2011 International Conference on and 4th International Conference on Cyber, Physical and Social Computing* (pp. 297-304). IEEE.

Hershkowitz, S., & Lavrusik, V. (2013). 12 Best Practices for Media Companies Using Facebook Pages. *Facebook Media*. Retrieved from https://www.facebook.com/notes/facebook-media/12-pages-best-practices-for-media-companies/518053828230111

HubSpot. (2014a). *Marketing benchmarks from 7000+ businesses*. Cambridge, MA: Hubspot.

HubSpot. (2014b). *State of Inbound 2014*. Retrieved 4 July 2014 from http://offers.hubspot.com/2014-state-of-inbound

International Telecommunications Union. (2015). *Global ICT developments*. Retrieved 20 March 2015 from http://www.itu.int/en/ITU-D/Statistics/Pages/stat/default.aspx

Ismail, M., & Razak, R. C. (2011). A Short Review on the Trend of Mobile Marketing Studies. *International Journal of Interactive Mobile Technologies, 5*(3), 38–42.

Jara, A. J., Bocchi, Y., & Genoud, D. (2014a, September). Social Internet of Things: The potential of the Internet of Things for defining human behaviours. In *2014 International Conference on Intelligent Networking and Collaborative Systems (INCoS)* (pp. 581-585). IEEE. doi:10.1109/INCoS.2014.113

Jara, A. J., Parra, M. C., & Skarmeta, A. F. (2014b). Participative marketing: Extending social media marketing through the identification and interaction capabilities from the Internet of things. *Personal and Ubiquitous Computing, 18*(4), 997–1011. doi:10.1007/s00779-013-0714-7

King, C. (2015). 28 Social Media Marketing Predictions for 2015 from the Pros. *Social Media Examiner*. Published 1 Jan. 2015, downloaded 13 Feb. 2015 from http://www.socialmediaexaminer.com/social-media-marketing-predictions-for-2015/

Kitchen, P. J., Brignell, J., Li, T., & Jones, G. S. (2004). The emergence of IMC: A theoretical perspective. *Journal of Advertising Research, 44*(01), 19–30. doi:10.1017/S0021849904040048

Konrad, A. (2014, November 18). Teespring says it's minting new millionaires selling its T-shirts, raises $35 million of its own. *Forbes*. Retrieved 25 March 2015 from http://www.forbes.com/sites/alexkonrad/2014/11/18/teespring-says-it-making-millionaires-raises-millions/

Krantz, M. (2015, July 23). Amazon just surpassed Walmart in market cap. *USA Today*. Retrieved 12 August 2015 from http://www.usatoday.com/story/money/markets/2015/07/23/amazon-worth-more-walmart/30588783/

Lamarre, A., Galarneau, S., & Boeck, H. (2012). Mobile Marketing and Consumer Behaviour Current Research Trend. *International Journal of Latest Trends in Computing, 3*(1), 1–9.

Larivière, B., Joosten, H., Malthouse, E. C., van Birgelen, M., Aksoy, P., Kunz, W. H., & Huang, M. H. (2013). Value fusion: The blending of consumer and firm value in the distinct context of mobile technologies and social media. *Journal of Service Management, 24*(3), 268–293. doi:10.1108/09564231311326996

Lauterborn, B. (1990). New Marketing Litany: Four Ps Passé: C-Words Take Over. *Advertising Age, 61*(41), 26.

Leung, S. (2014, August 30). 5 ways the internet of things will make marketing smarter. *Forbes*. Retrieved from http://www.forbes.com/sites/salesforce/2014/08/30/5-ways-iot-marketing-smarter/

Levitt, T. (1960). Marketing myopia. *Harvard Business Review, 38*(4), 24–47. PMID:15252891

Mallikarjunan, S. (2014). *The State of Ecommerce Marketing*. Cambridge, MA: Hubspot. Retrieved 15 August 2015 from http://offers.hubspot.com/state-of-ecommerce-marketing-2014

MarketingSherpa. (2010). *2010 Social Media Marketing Benchmark Report*. SherpaStore.com.

Markoff, J. (2006, November 12). Entrepreneurs see a web guided by common sense. *The New York Times*. Retrieved from http://www.nytimes.com/2006/11/12/business/12web.html?ei=&_r=0

McAfee, A., & Brynjolfsson, E. (2012). Big Data: The management revolution. *Harvard Business Review*, *90*(10), 60–68. PMID:23074865

McCarthy, E. J. (1964). *Basic Marketing. A managerial approach*. Homewood, IL: Richard D. Irwin.

Moskowitz, G. (2013, November 15). Comedy nights and party supplies: How local video stores are scrambling to survive. *Time*. Retrieved from http://entertainment.time.com/2013/11/15/film-camp-and-party-supplies-how-local-video-stores-are-scrambling-to-survive/

Mutter, A. D. (10 Sept. 2012). Print ads fell 25x faster than digital grew. *Reflections of a Newsosaur*. Retrieved from http://newsosaur.blogspot.com/

Mutter, A. D. (2014). The plight of newspapers in a single chart. *Reflections of a Newsosaur*. Retrieved 20 Feb. 2015 from http://newsosaur.blogspot.com/2014/04/the-plight-of-newspapers-in-single-chart.html

Mutter, A. D. (2012). The incredible shrinking newspaper audience. *Reflections of a Newsosaur*.

Nguyen, J. (2014). *Online Video Consumption in APAC and Total Video*. http://www.comscore.com/Insights/Presentations-and-Whitepapers/(offset)/10/?cs_edgescape_cc=KZ retrieved 16 Feb. 2015.

Okazaki, S. (2012). Teaching mobile Advertising. In *Advertising Theory* (pp. 373–387). New York: Routledge.

Olanrewaju, T., Smaje, K., & Willmott, P. (2014, May). The seven traits of effective digital enterprises. *McKinsey Insights*. Retrieved from http://www.mckinsey.com/insights/organization/the_seven_traits_of_effective_digital_enterprises

Olejnik, L., Acar, G., Castelluccia, C., & Diaz, C. (n.d.). *The leaking battery A privacy analysis of the HTML5 Battery Status API*. Retrieved from http://eprint.iacr.org/2015/616.pdf

Patel, M., & Veira, J. (2014, December). Making connections: An industry perspective on the Internet of Things. *McKinsey Insights*. Retrieved from http://www.mckinsey.com/insights/high_tech_telecoms_internet/making_connections_an_industry_perspective_on_the_internet_of_things

Penberthy, J. (2015). *Weird YouTube Method Pulls in $816,481.53*. Live webinar presented 30 July 2015.

Pepper, R. (2012). *The impact of mobile broadband on national economic growth*. Retrieved from http://newsroom.cisco.com/press-release-content?type=webcontent&articleId=1110575

Pew Research Center. (2014). *The revenue picture for American journalism, and how it is changing*. Retrieved from http://www.journalism.org/files/2014/03/Revnue-Picture-for-American-Journalism.pdf

Pope, M., Pantages, R., Enachescu, N., Dinshaw, R., Joshlin, C., Stone, R., & Seal, K. (2011). Mobile payments: The reality on the ground in selected Asian countries and the United States. *International Journal of Mobile Marketing*, *6*(2), 88–104.

Porter, M. E., & Heppelmann, J. E. (2014). How smart, connected products are transforming competition. *Harvard Business Review*, *92*(11), 11–64.

Ramsey, G. (2015). *The State of Mobile 2015*. eMarketer live webcast 25 March 2015.

Santana, S., & Morwitz, V.G. (2013). *We're in this together: How sellers, social values, and relationship norms influence consumer payments in pay-what-you-want contexts*. Manuscript submitted for publication.

Schultz, D. E., & Patti, C. H. (2009). The evolution of IMC: IMC in a customer-driven marketplace. *Journal of Marketing Communications*, *15*(2-3), 75–84. doi:10.1080/13527260902757480

Sharf, S. (2015, April 15). 62 million: The only number Netflix shareholders care about. *Forbes*. Retrieved from http://www.forbes.com/sites/samanthasharf/2015/04/15/netflix-subscriber-count-crosses-62-million-sending-stock-above-500/

Smutkupt, P., Krairit, D., & Esichaikul, V. (2010). Mobile marketing: Implications for marketing strategies. *International Journal of Mobile Marketing*, *5*(2), 126–139.

Solima, L., Della Peruta, M. R., & Del Giudice, M. (2015). Object-Generated Content and Knowledge Sharing: The Forthcoming Impact of the Internet of Things. *Journal of the Knowledge Economy*, 1-15.

Stevenson, A., & Wang, S. (2014, December 20). A Jump to Mobile Ads in China. *The New York Times*, p. 2.

Sundsøy, P., Bjelland, J., Iqbal, A. M., & de Montjoye, Y. A. (2014). Big Data-Driven Marketing: How machine learning outperforms marketers' gut-feeling. In Social Computing, Behavioral-Cultural Modeling and Prediction (pp. 367-374). Springer International Publishing.

Swain, H. (2007, January26). Have You Got The Message Yet*? Times Higher Education Supplement*, 54–54.

Tran, M. (2014). *Revisiting disruption: 8 good questions with Clayton Christensen*. Retrieved 15 August 2015 from http://www.americanpressinstitute.org/publications/good-questions/revisiting-disruption-8-good-questions-clayton-christensen/

Valentine, L. (2013). Payments Landscape: Still crazy after all these years: Few clear winners stand out, leaving banks mostly still watching. Herewith an informal "new payments scorecard.". *ABA Banking Journal*, *105*(7), 24.

Verbergt, M., & Schechner, S. (2015, June 25). Taxi drivers block Paris roads in Uber protest. *The Wall Street Journal*. Retrieved from http://www.wsj.com/articles/taxi-drivers-block-paris-roads-in-uber-protest-1435225659

White, D. S. (2013). *Social Media Growth 2006-2012*. Retrieved from http://dstevenwhite.com/

Xstream. (2013). *Online video trends & predictions for 2013*. Retrieved from http://www.xstream.dk/content/online-video-trends-predictions-2013

Zarrella, D. (2011). *Zarrella's Heirarchy of Contagiousness: The Science, Design and Engineering of Contagious Ideas*. Dobbs Ferry, NY: Do You Zoom Inc.

Zarrella, D. (2013). *The Science of Marketing: When to Tweet, What to Post, How to Blog, and Other Proven Strategies*. Hoboken, NJ: John Wiley & Sons.

Chapter 12
Branding and Mobile:
Revolutionizing Strategies

Yulia An
The Ilmenau University of Technology, Germany

Kenneth E. Harvey
KIMEP University, Kazakhstan

ABSTRACT

An overview of mobile's impact on branding strategy expands from discussion of chronological ages of branding to mobile's effects on the brand management function. The latter are structured around four steps of the brand management process, including planning and brand positioning, implementing branding strategies, measuring and analyzing performance of a brand, and sustaining brand equity over time and across geographic markets. With a focus on cross-platform branding and establishment of seamless mobile experience, the chapter explores various aspects of mobile's influence on branding. A strategic view on the mobile-first approach that permeates all aspects of the branding process from logo design to brand equity measurement is accompanied with real-life applications. Best practices illustrate changing perspectives in branding and include examples of how new technologies like iBeacon and augmented reality are used by leading brands. Examples are drawn from recent academic and industry studies in business and management with links to classic brand management literature.

INTRODUCTION

Never before has media been as fragmented as now. Internet users around the globe have on average 3.6 devices that they are constantly switching (TNS Global, 2014), and 36% of total time using all those devices is spent on mobile (Sinton, 2014). With almost two-thirds of mobile users worldwide using their second screens (smartphones or tablets) while watching TV (Statista, 2015), it is more difficult to repeat Apple's early success with its only-once-aired-on-national-television "1984" commercial that immediately received unprecedented publicity and skyrocketed the Macintosh sales.

DOI: 10.4018/978-1-5225-0469-6.ch012

Brand managers need to integrate another essential dimension into their current branding strategies, and this dimension is *mobile*. From Burberry's virtual kisses to the augmented reality of the movie Ice Age, global brands are actively embracing mobile opportunities. They know that if they are unable to integrate mobile into the flowing cross-platform experience, they risk being thrashed by faster and more mobile-oriented competitors.

In order to avoid overlapping with the chapter on mobile marketing and advertising, this chapter will take a more strategic stance and will specifically focus on implications of the mobile revolution on strategic brand management, with brand equity as its central concept.

STATE OF THE ART

In today's world, brands are omnipresent. What started as trademarks on pottery to identify their hand-crafter, brands are now irreplaceable, intangible assets that are worth billions of dollars. Interbrand estimated the monetary value of Apple and Google each in its 2014 annual report as more than $100 billion (Interbrand, 2014). Brands today have gone far beyond simple trademarks and reside in the minds of consumers rather than on product labels. With the new digital mobile revolution taking place and shifting media consumption habits of consumers, building strong brands has become a challenging task for many companies.

Mobile triggered two fundamental changes. First, it created new markets and companies that couldn't be imagined even 10 years before. Swiping pictures of potential dates left and right (Tinder) or sending photos to friends that automatically disappear after 10 seconds (Snapchat) might have been unimaginable in the end of 1990s but are parts of our daily routine today. Second, mobile introduced new battlefields to existing firms. With mobile Internet surpassing that of desktop computers, companies are pressed to take the mobile-first strategy and m-branding seriously. If the mid-1980s were about the economic value of brands as business assets (Keller, 2013), and the 1990s and 2000s were about the social value of brands and trust building (Clifton & Simmons, 2003), now it goes even deeper to perceptions and seamless consumer experience across different media platforms (Welsbeck & Berney, 2014). In championing this trend, the individualization of the consumer experience plays an increasingly important role. The Adobe Marketer's Guide to Personalization (2013, pp. 2-3) summarizes the essence of this rising trend:

Who doesn't enjoy that feeling of walking into a shop and being recognized, welcomed and treated as an individual? … Increasingly, customers have come to expect a consistent, personalized service across all touch points: in-store, online, on mobile – wherever and whenever they choose to engage. It's up to you to ensure you are providing your customers with a service that makes them feel welcome. A good personalized web experience is smooth, non-intrusive and relevant to the customer. The information and content they are presented with should meet their needs and encourage them to interact with your site, deepening the engagement. As the customer starts to interact, they generate data that can be used to enhance their experience in future situations and validate the impact of the programs that you put in place. Personalization is the use of data to deliver a relevant and engaging experience to a consumer across channels and devices.

In Adobe's Digital Trends 2015, based on its survey of 6,000 organization executives, the strong plurality of respondents proclaimed "customer experience" as the "single most exciting opportunity" of

2015 (p. 11). About 78% of the respondents agreed or strongly agreed that they are trying to differentiate themselves from their competitors through enhanced customer experience (p. 13), and a strong plurality of 44% (in a selection between six options) proclaimed that customer experience would be their primary means of differentiating themselves from their competitors over the next five years (p. 15). And how will they do this? In choosing among seven options, No. 1 with 33% was "making experience as personalized and relevant as possible" (p. 18). From among 18 options, what was their top digital priority for 2015? "Targeting and personalization" (p. 20). And 67% agreed or strong agreed that in 2015 "omnichannel personalization will become a reality" (p. 21). In an open-ended response, one of the executives summarized, "As mobile is gaining scale, it becomes more important to offer a multichannel customer experience to users through personalization and targeting on different platforms/channels" (p. 23).

SUMMARIZING THE PAST

Despite a long history of brands as trademarks, branding as an indispensable part of the business strategy was accepted only in the 1980s with inception of the brand equity concept. Yet the concept itself is sometimes open to dispute. The two main paradigms on brand equity focus on either customer- or asset-based definitions. The former one, or the consumer brand equity, delves into consumers' minds and deals with the perceived value that a brand creates. In Feldwick's (1996) three-tier classification, the consumer brand equity can be expressed as both a degree of consumers' attachment to a brand (also referred to as *brand strength*) and the associations and beliefs that consumers hold about the brand (or *brand assets*). Both concepts belong to the consumer brand equity and are a preamble to the financial value of the brand and the asset-based definition of brand equity (Kapferer, 2008; Wood, 2000).

The asset-based brand equity can be also referred to as the *brand value*. It describes the total financial value of a brand as an intangible and conditional asset. Even though the financial perspective on brand equity doesn't tap into minds and perceptions of consumers, it plays a major role in wide acceptance of brand equity by senior management and influences how companies see branding today. One of the reasons is that the concept of brand equity along with an insightful psycho-sociological perspective also allows for "hard" financial measures. Thus, strategic brand management, as conceived by some of its early scholars, takes in consideration both perspectives.

The 1980s brought an important change to the understanding of brand management as a financial asset. The economic value of brands was accepted by senior management, and the importance of an integrated branding strategy was finally acknowledged. Branding became an essential part of business strategy (Keller, 2013). This global view on strategic brand management hasn't changed. However, the processes incorporated in building brands have undergone considerable transformations. For example, one of the most dominant ideas in branding – the famous funnel metaphor – that marketing have been applying to touch points with consumers seems to have fallen into decay. The funnel metaphor describes the way that consumers go from getting to know about the brand and considering it among a vast array of other brands to narrowing down the choices and finally making an actual purchase. With this metaphor in mind, marketers use traditional push strategies with a wall of paid media to lead the group. Though proven to be effective in older times, what this metaphor fails to grasp in the age of mobile is the changing nature of consumer behavior and engagement (Edelman, 2010). Similarly, other dominant strategies like "push" and "pull" are no longer effective, as they totally ignore the crucial mobile concepts of the "right time" and the "right place" (Ramsey, 2015).

With the recent rise of such digital and mobile giants as Google, Facebook and Apple, branding has become all about customer experience and communication. Initially triggered by the grassroots power of social media and the web 2.0, advocacy and engagement have become central to brand building (Parganas et al., 2015). And even though the relationship between social media and branding still has not been fully grasped by academic research, current technological tendencies make this view gradually give way to a new thinking about brands – as integrated cultural and technologically bound systems. The ecosystem view on branding is the future that hasn't yet fully developed but is gradually taking shape. The Internet of things, smart wearables, ubiquitous computing, personalization and integration of this personalized data into the interactive ecosystem – all of this can be coined into one concept of the Mecosystem, recently presented as a new model for brands by the Interbrand agency at SXSW Interactive 2015 (SXSW, 2015).

Results of the latest Annual Global CEO Survey conducted by PwC show mobile technologies is the key to customer engagement as the top strategically important category of digital technologies, according to an astounding 81%. These conclusion is not groundless. A study by Wang and Li (2012) shows that three attributes of mobile service – personalization, identifiability, and perceived enjoyment – can have a substantial effect on a company's brand equity, especially its key factors like brand loyalty, perceived quality of the branded products, and other brand assets, like brand awareness and brand associations.

Thus, two factors come to the fore. First, that mobile is a strategic matter, and companies should be ready to integrate it into their existing business strategy. And second, that mobile is not just a fad – it will stay around (think about the Internet of things and smart wearables) and will definitely bring changes to the existing brand management process. For this reason, the following section is structured to show the effects of the changing digital landscape on four main stages of the brand management process:

1. Planning and deciding on brand positioning,
2. Connecting it to consumers,
3. Measuring, and
4. Sustaining brand equity.

BRAND PLANNING: ESTABLISHING BRAND VALUES AND POSITIONING

Like any strategic process, brand management starts with planning. This includes development of core values, main positioning anchors and the brand's business model. The mobile-first approach starts from these branding cores. And it is first and foremost about changes in the vision and perspective.

According to recent studies of McKinsey, the "mobile tidal wave" requires changes in the entire value chain of a business. Among the six "digi-shifting" trends that the consulting agency named, five were directly related to the mobile transformation (Duncan, Hazan, & Roche, 2014). The first one, the device shift, describes consumers' transition from PCs to mobile and touch devices, the change that has been already mentioned. The second shift in communications touches upon consumers' transition from traditional voice-based services to the ones that are data- and video-driven. The agency stated that only about 20% of total time spent on phones is allocated to voice-calling (that is, 40% less in only five years), while the rest belongs to web surfing, music streaming, and playing games. The content shift is also best illustrated in the example of mobile phones. A clear transition from bundled app services to the more fragmented app ecology with a number of task-specific applications establishes serious challenges to businesses that want to engage with their customers on a deeper and more personal level.

In contrast, social networking, which was rapidly growing in the last few years, has now reached maturity. Over 75% of all Internet users belong to some social networking platform, and with ever-present and more powerful mobile devices, users are increasing the amount and nature of their social interaction. The social shift is now following a common pattern from rapid expansion to monetization. According to eMarketer's research (eMarketer, 2015), Facebook has already left Google far behind in its mobile display ad revenues in the U.S. market (Table 1). The second-most-popular social medium, Twitter, is now in third place but appears positioned to overtake Google, as well. The market itself (in terms of total ad revenues) is predicted to grow fivefold over the five years between 2013 and 2017, approaching $26 billion (Table 1).

With a growing number of location-based and augmented-reality technologies on hand that are intangible to the mobile nature, retailers are also pushed to go through the transformation from channel to experience. Already now, half of smartphone users look for retail-related information on their smartphones, and soon they will be completing their transactions via mobile too, McKinsey principals predict. "The combination of mobile retail and true multichannel integration will have a transformative effect on the retail experience and ring in the era of Retail 3.0," Duncan, Hazan, and Roche write in a publication on telecommunications, media, and technology.

Finally, the video shift from programmed to user-driven, though not explicitly articulated in the report, still has a strong relationship to mobile. Beyond watch-on-demand video services like Netflix and Amazon Instant Video, mobile's share of online video shows a rampant growth in the last few years. For instance, the mobile traffic of YouTube has skyrocketed from mere 6% in 2012 to some 40% in just two years (Danova, 2014). And an analysis by Mixpo suggests that about 90% of Facebook's video views are now on mobile devices, which has allowed it to surpass YouTube in total videos viewed (Mixpo, 2015, p.3).

Table 1. Net US mobile display ad revenues, by company, 2013-2017, millions

	2013	2014	2015	2016	2017
Facebook	$1,532.0	$3,541.8	$4,911.1	$6,288.4	$7,525.5
Google	$629.7	$1,133.5	$1,473.5	$1,886.1	$2,376.5
Twitter	$320.1	$694.1	$1,192.3	$1,732.8	$2,290.2
Apple (iAd)	$260.2	$487.1	$795.0	$1,166.8	$1,464.1
Pandora	$369.2	$559.7	$737.0	$926.0	$1,125.1
Yahoo	–	$189.7	$425.9	$756.1	$990.7
Amazon	$12.4	$88.1	$163.0	$265.1	$395.1
LinkedIn	$5.4	$54.4	$108.6	$148.6	$196.0
Millennial Media	$79.4	$83.9	$96.8	$111.0	$125.8
Yelp	$2.1	$6.6	$14.9	$28.3	$43.3
Other	$2,098.4	$2,806.9	$4,750.1	$7,502.4	$9,153.7
Total	**$5,308.9**	**$9,645.8**	**$14,668.1**	**$20,802.6**	**$25,685.9**

Note: net ad revenues after companies pay traffic acquisition costs (TAC) to partner sites; includes display (banners and other, rich media and video); ad spending on tablets is included; excludes SMS, MMS and P2P messaging-based advertising; numbers may not add up to 100% due to rounding.
Source: company reports; eMarketer, March 2015

These shifts necessitate changes in the entire value chain. On an example of European telecom players, McKinsey provided a step-by-step guide on how mobile providers must migrate to fit the dynamic mobile environment (Banfi et al., 2014). And this migration encompasses all the organizational levels from both front- and back-end offices to big data analytics.

Visual Identity in Web 3.0

Even such an unambiguous and straight-forward brand element as a corporate logo should be designed with the changing nature of mobile in mind. The balance between consistency and change, like yin and yang, might speak to the hearts and minds of the new mobile audiences. With 90% of all media interactions happening now on a screen, according to Google's study, "New multi-screen world" (2012), the visual identity of brands is gradually becoming digital, too. Fluid brands are not uncommon. It is what Absolut Vodka does with its subtle but ever present bottle silhouettes. It is what Google does on its main page with new Doodles every day. What has changed is the degree of consumer participation in shaping the contours of fluid identities and the overarching digital immersion.

The music channel MTV has been communicating its brand to the generation of digital natives for ages. MTV's logo variations reflect the channel's bold identity and are recognizable despite the changing context and filling. The brand was one of the first to use its logo in such a wide array of contexts already in the distant 1980s. Following an example of MTV, big brands like Saks Fifth Avenue, AOL, and even the city of Melbourne are now embracing the concept of "fluid brand identity" (Salmeron, 2013).

And who knows better about design than art schools? New logos of several art schools, including the famous London-based Croydon School of Art, the Dutch Design Academy Eindhoven (DAE), and the Canadian OCAD University, brightly display the tendency of consistency and change in their corporate identity. Their logos themselves become a part of the consumer experience. For example, a strikingly simple and minimalist DAE logo (three bold stripes) lets all students and faculty communicate their own associations with the school and to express what the brand means for them individually. This triggers a literal transition of brands from "consumers' minds" to the external world. From simple "Design Academy Eindhoven" to inspiring "Dare To Dream" and "Inspirational People Required," DAE's redesigned brand identity kills two birds with one stone – it provides a highly interactive consumer experience and taps into content personalization with ease (Figure 1).

These new tendencies in brand identity design in the dynamic digital media landscape are not groundless. Advances in current brand management research show that fluid identities, indeed, might have positive impact on how consumers perceive a brand. Sääksjärvi and colleagues (2015) recently reported that slight changes in logo design can add 'freshness' to the brand perception and enable the brand to cut through the media clutter. In two experimental settings they found results that contrast previous findings on consumer resistance to logo changes (Pimentel and Heckler, 2003; Walsh et al., 2010, 2011):

We attribute this to the fact that we used very slight logo variations. ... By using such logo varieties, consumers can subconsciously recognize the logo as belonging to the brand and thereby build brand prominence. Nevertheless, slight variations of the logo also make it distinctive, which can foster brand freshness.

A recent concept of User-Generated Branding (UGB) goes even further and engages consumers into the process of the brand value co-creation. In the study on two user-generated brands, Couchsurfing

Figure 1. New logo of Design Academy Eindhoven with gaps filled in by its students
(© 2010 The Stone Twins. Used with permission.)

and AirBnb, researchers Yannopoulou, Moufahim, and Bian (2013) found that UGBs add the human dimension to the companies' brand equity, making it authentic and personal. The only challenge of such flexibility is to stay focused on and consistent with "the source-identifying pillars of a brand" (Pearson, 2013, 2014). Consistency is central to a strong brand identity.

Personalization of Consumer Experience

As defined by Adobe, personalization can refer to "the use of data to deliver a relevant and engaging experience to a consumer across channels and devices" Adobe (2013, p. 3). Citing other recent studies, Adobe (p. 4) says company websites that create a personalized experience see an increase of 19% in sales. Personalized email is also much more effective, with 14% more clicks on embedded links, leading to 10% more sales. Lenovo reported a 14% increase in sales after establishing cross-channel personalization.

While most online organizations are still collecting little more than basic contact information, a technology-enhanced personalization strategy allows the gathering of much more information to enhance every individual's online interaction. Adobe suggests (2013, p. 5):

- Where they are based, using geo-targeting, either through a lookup or by asking them for their location through the browser.
- Which social network they subscribe to, based on the use of social plugins to share the product information they researched.
- Whether they are a first-time or returning visitor.
- The search terms they have used to reach your site.
- Which device and browser they are using.

And from the individual's exploration on a company's website, certain offers can be made. Even after they leave the site, data can help a company present to them individualized online advertisements. And as the consumer relationship and data grow, the company can cross-sell and up-sell customers in such a pleasant, individualized way that the customer can be turned into an active advocate of the brand. At a more advanced stage a company can expect to use data to (Adobe 2013, p. 6-9):

- Serve content and offers that are targeted to the user's needs.
- Use automated recommendations and test to see where they are most effective.
- Use behavioral data to trigger targeted emails at the right time.
- Look at automated behavioral targeting software to predict user behavior and match to content.

Shawn Burns, global vice president of digital marketing for SAP, says using such omnichannel personalization strategies to deliver the right messages at the right time on the right device has enhanced their conversion rates by 300% (Adobe 2013, p. 12).

Personalization can help companies enhance their brand, but most consumers say companies are failing miserably (Econsultancy, 2015). A U.S. study completed by Econsultancy on behalf of IBM found that while 90% of all customer personalization is critical to success, 78% of consumers do not believe that the average brand knows them as individuals. Econsultancy conducted two surveys, one of 276 marketing professionals from consumer companies with revenues exceeding $1 billion, and a second of 1,135 consumers from across America.

The Bulldog PR consulting company summarized why this study is important. "The consumer/brand relationship has evolved into a two-way partnership where consumers are willing to share their most personal details with trusted businesses in exchange for experiences that are unique to them—and the onus is on brands to deliver" (Bulldog Reporter, 2015).

Most marketers agreed or strongly agreed that their companies "deliver a superior … customer experience" offline (75%), online (69%) and via mobile (57%) (Econsultancy, 2015, p. 3). But their consumers disagree. Only 22% feel the average company does a good job of personalizing its relationship with customers (p. 8), and 49% of respondents had changed service providers in such a critical and time-consuming area as banking, mobile, Internet or cable within the past 12 months. Of those who had recently switched brands, 30% said their previous provider had failed them – most commonly with poor customer experience (pp. 6-7). Consumers' feelings of alienation may explain why the percentage of online shopping carts being abandoned is increasing, 75.8% on Thanksgiving 2014 and 72.4% on Black Friday 2014, for example (IBM, 2015, p. 5).

According to a study by Yesmail Interactive and analyst Gleanster (Carufel, 2013), there is a disconnect between executives' perception and reality. The report, Customer Lifecycle Engagement: Customer engagement imperatives for mid-to-large companies, reveals that 88% of the senior-level marketers surveyed say their companies are using data well, but 80% of their companies are really using little more than basic customer profile and purchasing data to personalize marketing messages. Only 20% of the respondents have and use data on the level of customers' social media participation, 21% data on their channel preference, 21% life triggers such as birthdays, 27% data on household composition, 41% data on web browser history/online behavior, 38% social data, 36% attitudinal data, 56% point-of-sale data, and 67% CRM (customer relations management) data.

While the marketers felt they were doing a good job personalizing their messages, 86% admitted they could do better. Their limitations were blamed on inadequate marketing tools (42%), fragmented marketing systems (34%), and poor data quality (34%) (Carufel, 2013).

Deepak Advani, general manager of IBM Commerce (Bulldog Reporter, 2015), says these limitations are not necessary. "The customer is in control, but this is not the threat many marketers perceive it to be. It's an opportunity to engage and serve the customer's needs like never before. By increasing investments in marketing innovations, teams can examine consumers at unimaginable depths, including specific behavior patterns from one channel to the next. With this level of insight, brands can become customers' trusted partner rather than an unwanted intrusion," he said.

A survey of about 6,000 business executives by Adobe (2015, p. 22) found that the top three "most exciting opportunities" over the next five years are all related. No. 1 according to 20% of the executives is "customer experience," No. 2 according to 16% is "personalization," and No. 3 according to 14% is "big data," on which personalization activities can be based. And 78% of respondents agreed or strongly agreed that the most important way to distinguish their company from competitors was through customer experience (p. 13). The most important way to improve customer experience, according to a plurality of 33% of the respondents, is by making experience as personalized and relevant as possible (p. 18). And of 18 possible responses, a plurality of 30% concluded that their top digital priority of 2015 is "targeting and personalization" (p. 20).

All of the marketers surveyed by Econsultancy (2015, p. 9) felt their companies were attempting to personalize their customers' experience. The chart below shows different methods being used, and every method listed achieved over 40% penetration rate among surveyed marketers (Figure 2).

Figure 2. Components of a holistic customer view
(Adapted from Adobe Quarterly Intelligence Briefing)

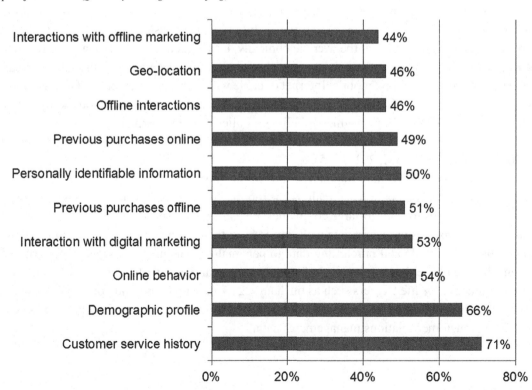

Attempting and achieving, according to Econsultancy, do not always equate. Adobe (2013, p. 13) encourages companies to measure the results of their personalization efforts to justify their investment. When it asked how they were measuring results at that time, the chart below reveals the various metrics they employed (Figure 3):

Cross-Platform Branding

Another area in which organization executives think they are doing better than they really are, according to consumers, is in their cross-platform branding. Many organizations have failed to fully prepare their websites for easy mobile access. That automatically undermines efforts to achieve positive cross-platform branding. CEO Geoff Ramsey of eMarketer (2015) says that 86% of millennials complain that most commercial websites are not mobile-friendly, and 71% of them find that most online businesses also fail to provide mobile apps as an alternative. This is at least part of the reason that far more consumers shop on mobile than buy on mobile. According to Ramsey, 82% of smartphone users shop on their smartphones, but only 43% make purchases. In contrast, 89% of smartpad users shop, and 72% actually

Figure 3. Measuring impact of personalization on ROI and engagement
(Adapted from Adobe Quarterly Intelligence Briefing)

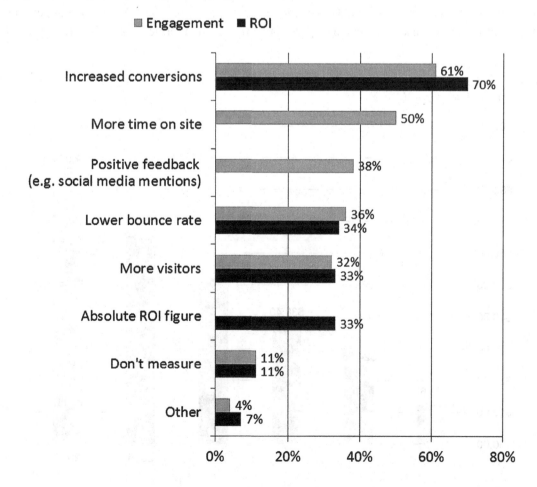

make purchases on smartpads, and 90% of desktop/laptop users shop on their computers, and 76% make purchases. So, 84% of the PC shoppers actually purchase, but only 52% of the smartphone shoppers.

With the world switching from brick-and-mortar to virtual storefronts, as demanded by consumers, and with online consumers transitioning to mobile, $5.9 trillion dollars are up for grabs (Accenture, 2013). Companies that do not make the technological and marketing changes required will not share fully in this unprecedented bonanza.

eMarketer (2013) predicts that mobile commerce ("mcommerce") will grow to nearly $87 billion by 2016, representing 24% of all ecommerce retail sales, but it could be much more. Worldwide, there are now three times as many mobile devices accessing the Internet as fixed-wire connections, according to the International Telecommunications Union (2015) (Figure 4).

Despite the billions of dollars already at stake, most online retailers do not satisfactorily accommodate online shoppers. Of the billions of dollars in purchases made on mobile devices, Ramsey (2015) said mobile customers who arrive to a website that does not accommodate mobile almost always go a competitor's website thereafter. A study conducted by Forrester Consulting (2013b) on behalf of KANA Software confirmed eMarketer's findings. Only 39% of all companies received an "excellent" or "good" rating by consumers, and if they run into problems, 75% of the prospective customers move to other communication channels. Forrester concluded that recontacting prospective buyers can cost millions of dollars.

Besides the shopping experience itself, companies are recognizing the need to communicate with prospective customers across all channels. An Adobe (2015, p. 21) survey of 6,000 business executives found that 38% plan to make omnichannel personalization of marketing messages a reality in 2015,

Figure 4. U.S. Retail mcommerce sales 2011-2016 in billions
(Adapted from eMarketer, Jan. 2013)

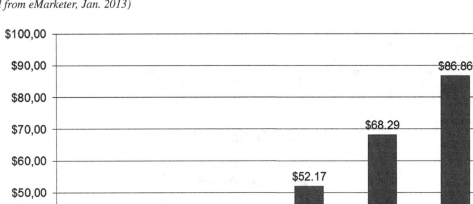

but that leaves 62% lagging behind. Only 12% judge multichannel campaign management as the "most exciting opportunity" of the next five years. One of their respondents, however, noted "As mobile is gaining scale, it becomes more important to offer a multichannel customer experience to users through personalization and targeting on different platforms/channels" (Adobe, 2015, p. 22-23).

To achieve ideal cross-channel communications depends on enough real-time data to track consumers in the various channels and to personalize those messages. An Evergage (2014) survey of marketers identified major purposes to utilizing real-time data in cross-channel communications. The seven top purposes were to:

- Increase customer engagement 81%
- Improve customer experience 73%
- Increase conversion rates 59%
- Improve brand perception 52%
- Increase retention/reduce churn 52%
- Increase loyalty 46%
- Increase brand awareness 35%

Perhaps of all organizations, magazines are the ones whose future is most dependent on achieving high-quality communications across all platforms. They need to achieve this in order to achieve maximum readership, which is the basis on which advertising is priced and sold. And that includes content readership as well as advertising readership. But a 2014 study of 100 leading consumer magazines by Brand Perfect revealed that even magazines are struggling with this. The study (Ayres, 2014) found that:

- 93% of leading magazines do not offer complete cross-platform digital support (p. 3).
- While HTML5 can potentially provide seamless access across the web from desktop, tablet, smartphone and television, relatively few magazines have yet made a full commitment to the new programming language.
- While website readers are more plentiful, more targetable and easier to analyze, many publishers aren't providing positive website experiences for mobile users. Instead, they have invested heavily in apps, yet many of their apps have limited audiences.
- Of the 78 magazines examined in the U.S. and U.K., 83% had at least one app available, all of which published to iPads. But only 65% published to iPhones and 40% to Android. And content was often obscured by the reduced width of the smartphone display (p. 5).
- 55% of the German magazines had mobile-accessible websites, mostly aimed at smartphones (p. 16).
- Tablets in Germany and elsewhere are primarily served by simply scaling down desktop pages, which provides a poor experience for small tablets. Only 10% of Germany's magazine websites are optimized for tablets. As with their British and American counterparts, most German magazines have opted to invest in apps rather than tablet-optimized websites (p. 16).

The researchers found that the content of many of the online magazines accessed by smartphones and small tablets was very difficult to read (p. 20). The report concluded (p. 21): "Realistically, when it comes to tablet computing and mobile content, these are still early days for consumer publishers, who are only just beginning to explore how the brand presence of their individual magazines might be expressed through mobile content beyond re-formatted print material."

It is now possible for magazines and all online organizations to overcome these problems. The Brand Perfect report noted (p. 24):

Thanks to the principles of Responsive Design, websites can now look and feel as if they were designed specifically for the device on which we are viewing them. Put simply, a responsive website uses "media queries" to determine the resolution of the desktop, laptop, tablet or smartphone it is being served to, before a flexible layout then sizes the page to fit the screen. For publishers, it's an elegant solution to the problem of presenting content to readers across multiple devices. For users, it ensures a great experience on every screen. It makes for perfect publishing.

The report also quotes Ed Barnes, VP of Rich Media at ADTECH, who said that Using HTML5 enables brands greater control of what reaches readers, and lets them adapt their campaigns on the fly. "Brands could ... produce the creative content once and use web standards such as HTML5 to ensure their advertising is serving in how they want it regardless of the device it's being encountered on" (p. 25).

Josh Clark, one of the designer/developers behind People magazine, was also quoted in the report. "It's not just a technical issue. It's also a sales issue. 'Separate creative for separate devices' is a reflection of the way these ads are sold. Trouble is, it's people who form market segments, not devices. Segmenting by device — whether that's for content or for advertising — just doesn't reflect the way we consume information today. Mobile, tablet, and desktop versions of websites are presented as completely separate properties instead of simply 'the website'" (p. 25).

Cameron Connors, managing director of The Studio at Condé Nast, leads a creative team that pushes the boundaries of technology on behalf of the entire portfolio of Condé Nast publications so that the individual brands can focus on "storytelling and innovation across existing and emerging digital platforms." When a new project begins, Connors' team is brought on board as quickly as possible so technical challenges can be addressed up front (pp. 31-32). Technology is advancing so rapidly, teams of this kind may be more and more common – and not just in the publishing industry. Brand executives frequently don't even know what they don't know. If they don't understand the technology, they can't understand how it can best be applied.

Brand as a Narrative

With a sharp increase in the number of media channels and considerable fragmentation within the mobile environment, engaging audiences across different platforms and bringing a consistent message has become a big challenge for brand managers. In addition to that the growing mobility and interactivity of new communication forms have made control over the disseminated messages far more difficult.

Research has shown that brand stories engage consumers by reinforcing positive brand associations, shape brand experiences, and can even decrease consumers' price sensitivity (Benjamin, 2006; Chang, 2009; Lundqvist et al., 2012; Mossberg, 2008; Woodside et al., 2008). But while in the previous decade brand stories were anchors to drive brand loyalty and had to be preceded by commercial advertising (Lundqvist et al., 2012), now, due to enormous fragmentation of the media channels, stories gained another functional role as "glue" to keep dispersed communication tactics together. Convergent storytelling employs the maximum number of media platforms to communicate the same story and brand identity to various audiences across varying contexts. Such an approach creates synergies between dispersed communication channels and maximizes "touch-points" at which brands can reach their consumers (Granitz & Forman, 2015; Martin & Todorov, 2010; Robideaux & Robideaux, 2012). Above all, convergent storytelling adopts both top-down and bottom-up approaches (Edwards, 2012) that are essential for strategic communication in the age of mobile. First of all, because it helps organizations to retain strategic focus and control over their meta-stories and brand identities, but also because it gives the brand stories enough space to be flexible and emerge naturally though the process of co-creation. Stories generate "an immersive brand experience and by doing so, create a higher level of engagement" (Robideaux & Robideaux, 2012).

The best way to craft stories like this is to employ brand personas, which are central to good brand narratives. They build personal connections as "human brands" and persist through time. They can be expressed in different forms and shapes without losing integrity. The effect that brand personas have on consumers is attributed to stories speaking to both their rational and emotional needs (Herskovitz & Crystal, 2010). Certain brands are used by consumers "as protagonists to enact roles that give them the feelings of achievement, well-being, and/or emotional excitement" (Muniz, Woodside, & Sood, 2015). The personification approach has been found to be effective in social media, as it enhances the human side of the brand by constructing a holistic corporate character, and it also triggers natural brand co-creation processes that paradoxically go in line with a bigger corporate branding strategy (Fournier & Avery, 2011; Men & Tsai, 2015; Singh & Sonnenburg, 2012).

Apart from that, brands are now transforming into content publishers on their own. Given a wide array of non-paid media channels (both earned and owned), they can communicate their own content to engage consumers and build personal relationships. Hence, the corporate content strategy becomes an essential element of branding across platforms, mobile branding included. Both scholars and industry experts claim that the content strategy in the age of mobile and transmedia storytelling should be well suited for an interactive and dynamic context of the digital age (McGrane, 2012; Pulizzi, 2012, 2013; Sabatier, 2012; Lieb, Silva, & Tran, 2013). This means that the content strategy for mobile should have a holistic approach, and yet be flexible and open for conversation with consumers. There are some general lessons for mobile branding to take away, and they all focus on the kinds of content that will best fit the mobile context. According to Pulizzi (2012, 2013), storytelling in the modern age spins around social media, search engine optimization (SEO), and lead generation. The content should be thus consistent, useful and human. For Sabatier (2012) it is also dynamic, searchable, convergent, user-oriented, and user-generated. In other words, what we can expect is the holistic, adaptive, social, and intuitive mobile content strategy.

But besides the content itself, mobile as a new channel for communication also demands new approaches in bringing messages to the screen. Even though the principles of storytelling haven't changed in centuries, the ways to communicate stories to the audience have undergone drastic transformation. The developments in information interfaces and image processing offer a wide array of tools and techniques for adaptive content presentation on mobile screens that taps into visual and interactive brand experience. Augmented graphics, interactive animations and touch interfaces help to design immersive and truly interactive narratives (Kim et al., 2012; Marchesi & Riccò, 2013). Various techniques for visual storytelling on mobile devices allow users to explore narrative elements pertaining to the central story, while also viewing story-related visual context (Elmieh et al., 2014). In-built sensors and touchscreen opportunities open up a myriad of opportunities for engaging storytelling. And even though the mobile dimension is challenging, the social and technological complexity of this new dimension can be also empowering for both consumers and brands. At the end of the day, it is all about consistency, flexibility and trust.

Personal Branding

One of the forms of storytelling to build trust is personal branding. Studies show that reputation of a company's CEO affects corporate capital investments and predictions of the company's future performance (Jian & Lee, 2011; Cianci & Kaplan, 2010). Also, it is the credibility of the CEO that is associated with engagement of employees and perceived organizational reputation (Men, 2012). The latter, according to Graffin, Pfarrer and Will (2012), over time converges with executive reputation, making it difficult to separate the two. Thus, besides brands for companies, subsidiaries and individual products, it is also valuable for individuals to have brands. Brand management techniques are valuable for all of those wanting to sell themselves, as well as organizations wanting to build their brand(s) on the individual strengths of their employees.

For some organizations, this is obvious. Consulting firms routinely provide the CVs of their team members in order to land a major contract, and if they are smart, they take more care in boosting the brand of these employees than simply compiling the CVs that employees provide. Similarly, universities frequently build their brand on the accomplishments of their professors. Most other organizations could also build their own brand by building their employees' individual brands.

In "Approaching PR Like a Journalist" (Harvey, 2011), it's noted how a web search for "executive biography" brings up biographies for executives from some of the largest companies in the world with one thing in common. All the executive biographies are boring. Some of the best-paid PR professionals in the world seem to think that writing a journalistically enticing story about organization executives is of no value. The video suggests other ways to help build the individual brand of executives, as well, such as more regular use of executive attribution in press releases and other PR products.

Steve Jobs may have been the epitome of CEOs in the digital age. A survey of 1,700 senior executives from 19 countries around the world found that 81% now believe that visibility and engagement by CEOs has become critical to a company's brand (Weber Shandwick, 2015).

"Years ago, CEOs and those around them confused CEO visibility with CEO celebrity. Today, it is not about CEO celebrity, but CEO credibility that can be built through multiple channels that add value inside and outside the organization. Today, CEO visibility means having a greater presence with greater purpose and in more ways than one," says Leslie Gaines-Ross, chief reputation strategist for Weber Shandwick, named the 2015 PR Agency of the Year by PR Week (Figure 5).

The study by Weber Shandwick, in conjunction with KRC Research, found that a strong CEO reputation:

- Attracts investors, according to 87% of senior executives.
- Generates positive media attention, according to 83% of executives.
- Affords crisis protection, according to 83% of executives.
- Attracts new employees, according to 77% of executives.
- Retains current employees, according to 70% of executives.
- Will become even more valuable in the coming years, according to 50% of executives.
- Represents 45% of the entire company's reputation, according to the average executive.
- Represents 44% of the entire company's market value, according to the average executive.

Figure 5. Why CEO reputation matters
(Adapted from Weber Shandwick, 2015)

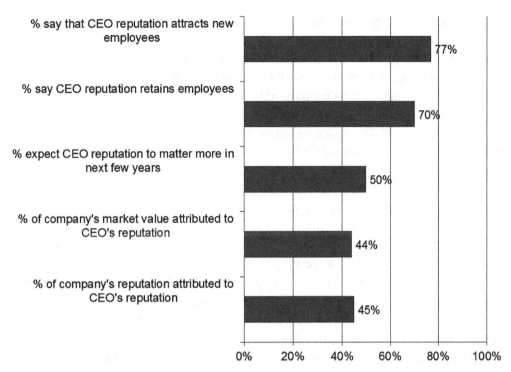

According to a Weber Shandwick press release:

There is a close tie between reputation and external relations. Admired CEOs are four times more likely to be seen as being good at engaging the public than those with less admired status (50 percent vs. 13 percent, respectively). The question is: Which of the many available platforms are mission critical for CEOs when their time is so limited and they are understandably risk-averse? The majority of global executives (82 percent) consider speaking engagements job No. 1 for engaging with external stakeholders, but there are many other important external CEO responsibilities as well:

- *Speak at public events. 82%*
- *Be accessible to the news media. 71%*
- *Be visible on the company website. 68%*
- *Share new insights and trends with the public. 67%*
- *Be active in the local community. 64%*
- *Be visible on the corporate video channel. 63%*
- *Hold positions of leadership outside the company. 53%*
- *Publicly take positions on issues that affect society at large. 52%*
- *Participate in social media. 43%*
- *Publicly take positions on policy and political issues. 36%*

Based on their study's findings, Weber Shandwick makes the follow 12 recommendations:

1. Assess the CEO's reputation premium.
2. Develop the CEO's "equity" statement.
3. Identify and develop the CEO's story on behalf of the company.
4. Be an industry champion by having a visible and involved industry presence.
5. Leverage the senior management team, in addition to the CEO.
6. Bulk up on media training.
7. Carefully evaluate the CEO's stance on public policy.
8. Decide which venue is right for the CEO.
9. Develop a solid social media strategy.
10. Keep reputation drivers at the top of the to-do list.
11. Bolster CEO reputation among employees.
12. Don't view CEO humility as a weakness.

To summarize, organizations need to develop executives' individual brands in order to enhance the entire organization's brand. Today's consumers see the organization's brand inexorably intertwined with the brand of the CEO and other senior management (Weber Shandwick, 2015).

If a personal brand is valuable to a CEO, it is also valuable to an employee trying to advance in an organization. All of these recommendations that can be applied to an individual employee or prospective employee would benefit an individual's brand and, to some degree, the organization's brand. For the individual, Entrepreneur.com (Lee, 2015) recommends beginning by Googling yourself on an "incognito browser" unrelated to you. Search for your name, including variations and likely misspellings. With a common name, try it with such other associated details as business and education affiliations. People might then contrast their search with that of other people they would like to emulate. See what they have that you don't, and then begin building your brand.

Communicating Brand to Consumers

Building bridges with consumers is one of the most demanding and important stages of the brand management process. Hence, understanding how consumers engage with brands is crucial. While the earlier discussed funnel metaphor is no more applicable to the way we think about consumer touch points, a new perspective is needed. David Edelman (2010) in his article for Harvard Business Review suggested we think about consumer brand engagement in the digital age as of the consumer decision journey (CDJ). Instead of a downward narrowing spiral of the funnel metaphor, CDJ portrays a looped behavior model that flows from a consideration stage to evaluation, and to a final purchase, from which the so-called "loyalty loop" follows (Figure 6).

The loyalty loop incorporates such experience-based stages as "enjoy, advocate, and bond." If consumers like the product quality, they will be more likely to spread positive word-of-mouth about the brand and build a stronger emotional connection (bond). Indeed, the deeper connection begins after the purchase, and this is what marketers often lose sight of. Mobile phones might be one of those rarely considered touch points, as more and more mobile users actively engage with brands on various stages of CDJ via their mobile devices. This is not limited to product-related research and transaction completion (with

Figure 6. The consumer decision journey
(Adapted from Edelman, 2010)

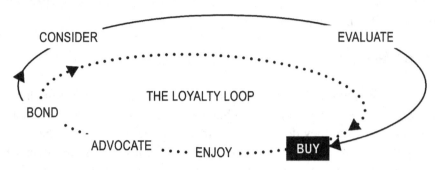

growing m-commerce) only. With at least 66% of social networking activities starting from a smartphone (Google, 2012; Welsbeck & Berney, 2014), stages of the loyalty loop are central to mobile publics. In fact, loyalty is exactly what many brand managers might find missing with the start of the mobile age.

Loyalty Is Dead, It's Experience Instead

As young mobile savvy consumers are growing up, brand managers need to start building relationships from scratch. The recent Mobile People's study (Welsbeck & Berney, 2014) shows that 91% of the surveyed mobile web users have decided to switch to a competitor's website after a failed attempt to access a badly mobile-optimized website of their first choice. The average among the Chinese was even more striking – 99% of surveyed mobile users.

"For mobile applications, branding is experience," claimed David Rondeau, the design chair at In-Context Enterprises, in 2005, even before the world witnessed the birth of touchscreen smartphones as we know them today. Indeed, in the world of mobile data, it's all or nothing. The same Mobile People's study reported that 80% of mobile users said that they gave brand recommendations based only on their mobile web experience. Again, China along with India leads – with 95% of those advocating the brand because of the positive mobile experience. The study results of McRae et al (2013) might shed some light on this behavior. Researchers reported that mobile drives more trust and initiates a stronger emotional response to mobile websites and branded apps than any other personal communication device. Furthermore, customer experience was found to play a mediating role in the relationship between product attributes and mobile brand equity in the study of Sheng and Teo (2012).

M-Branding Techniques and Tactics

From branded mobile games and apps to geo-location services, mobile opens up a wide array of branding techniques to engage with consumers. Yet, with so many tools on hand, creativity and appropriateness are still the drivers of success. This is what the Gold Lion winners of the Cannes Lions 2013, Ogilvy Deutschland working for the German record label Kontor Records, teach us. In their "Back to Vinyl – The Office Turntable" campaign, the agency used mobile technology to create seamless analogue-digital experience with analogue vinyl records played via digital smartphones. Kontor's vinyl records were sent to creative directors packed into a carton "turntable" that could be activated with a smartphone. According to the agency, out of 900 Turntable QR codes sent, 71% were activated – a 64% increase from the average response across the industry.

Mobile Branded Wallets

While the struggle for the reigns in the future mobile payment market is now in full play, and everyone, big and small, tries to get his piece of a pie in mobile payments, consumer brands can also potentially win from joining the race. And they do. Starbucks' mobile wallet has become the most used mobile payment app in the USA (Bertoni, 2014). The coffeehouse's 10 million customers contribute to over 5 million mobile transactions per week. And their secret is the enhanced customer experience. Starbucks uses an array of tools: morning coffee gamification, special offers, freebies, and a great loyalty program. Branded mobile wallets are changing traditional business models of payment systems by bringing onboard revenues from advertising and licensing fees (Mercator Advisory Group, 2013). Forrester Research (Husson, 2015) predicts branded mobile payment systems to transform into full-fledged marketing platforms, accumulating promotional offers, digital coupons, loyalty rewards and other non-payment functionality.

Mobile Geo-Location Services

Location-based mobile apps like Foursquare or Yelp are no longer a novelty. So, when Apple demonstrated its iBeacon at work in December 2013, it sent an alarming signal to marketers, a promise that brick-and-mortar retail stores will never be the same. But even before, geo-fencing had already prepared the soil for the geo-location-related media buzz. Geo-fencing uses GPS coordinates to set geographic boundaries on a virtual map and send notifications when a smartphone enters or leaves the designated area. Such a technology can have many applications, from location-based reminder apps – when a user is reminded about an activity once he enters a specified area, e.g. buy bread when near a bakery – to marketing indoor offers of retail stores based on user's proximity to the store location. Mobile geo-fencing today, however, is even more sophisticated and leverages on technological advancements of big data by integrating social networking dynamics, text data mining, and dynamic behavioral profiling (Brown & Harmon, 2014).

Beacons follow the same goal as geo-fencing, yet they approach this goal from the opposite side. Small, low-power, and inexpensive, beacons possess unique identification numbers and thus transmit their location to a Bluetooth-enabled device. Unlike geo-fencing, beacons cannot map the smartphone user's positioning. What they can do is triggering specific actions when in the beacon's range. The device can track the proximity of a smartphone user, and, for example, also send push-notifications. It is best illustrated by an example of another Cannes Lion Winner. The "Nivea Sun Kids" campaign in Brazil elegantly deployed beacons as part of its beach-season m-branding. The brand helped young mothers track their kids on a beach using a pre-downloaded branded mobile app and Bluetooth beacons integrated into Nivea's magazine advertising page. The beacons, attached to moisture-resistant paper bracelets, could be adjusted to children's hands and sent notifications to the mother when a child left the beacon range.

Augmented and Virtual Reality

Technologies of augmented reality (AR) have already been in place for quite a while. As soon as the first AR-enabled smartphones appeared in 2009 (Liao, 2014), mobile augmented reality was almost immediately appropriated by marketers and advertisers. The technology that places virtual objects into the real-world environment has become accessible to any smartphone user and enables consumers to interact with brands, directly mediated by their smartphone only. Augmented Reality Experiential Marketing (AREM) aims at reaching deeper connection than what one-time "shiny object" syndrome can offer. This approach focuses on the long-term experiential value of the AR technology in sustaining a lasting customer-brand relationship and brand satisfaction (Bulearca & Tamarjan, 2010). Some examples of mobile AR by the leading brands show how dynamically this tool develops in terms of technical sophistication. With its AR catalogue, IKEA made it possible for shoppers to see how this or that furniture piece can look in their own apartments even before making a purchase and bringing the piece home. The outdoor apparel brand Moosejaw took the opposite approach and enabled its customers to "undress" models on the pages of the brand's catalogue with the help of its X-Ray app (Evans, 2014).

Branded Mobile Games

With the appearance of app stores, users got a chance to download games on their own, which has driven enormous demand for mobile applications and created a new technological and mobile service ecosystem. App stores as multisided platforms boosted a huge number of applications by bringing together mobile users and app developers (Basole, & Karla, 2012). And mobile games were the biggest hits among downloaded apps. According to Gizmodo (Wagner, 2012), among 10 top paid apps for iPhone, only one was a non-game application.

Though mobile games have been in place since the early inception of mobile phones and were pre-installed on virtually all feature phones, they rapidly took off in late 2007 after the first smartphones with touchscreens appeared. In 2009 the first mobile social games emerged in Japan and are now extremely popular in East Asian countries (Yamakami, 2012; Seo, 2014). Since then mobile gaming has evolved into a serious industry (Feijoo et al., 2012).

Branded mobile games, or advergames, caught on almost immediately. Already in early 2006, NBC Sports launched a series of branded mobile games (Ivanov, 2005) featuring sports like alpine racing and speed and figure skating. A more recent example, Red Bull Air Race, got into the list of top mobile racing games, according to the tech portal Tom's Guide (Corpuz, 2015), along with traditional racing games like Need for Speed and Drift Mania. Branded mobile apps in general are reported to make personal connections with brands and trigger more loyalty and subsequent word-of-mouth (Bellman et al., 2011; Sprosen, 2014), as mobile games are engaging users by providing enjoyment, social interaction, and time flexibility to play games at any time and for any longevity (Wei, & Lu, 2014).

However, today the appeal of advergames is slowing down. Despite a positive impact of branded apps on brand attitude and purchase intention, game-like apps were reported to be less successful than apps with an informational and user-centered focus (Bellman et al., 2011). Moreover, experts consider them less financially viable too. According to Grossberg (2014), native games are expensive to produce and require considerable efforts to maintain. So, instead, marketers are shifting towards in-game advertising, branded entertainment, and gamification, applying psychological mechanics and design of games to non-game contexts (McGonigal, 2011). In-game advertising, however, was also found to be less effective for advertising placement, as compared to entertainment apps (Valvi & West, 2015). This, yet, doesn't imply that advergames and in-game advertising are inefficient at all. Current state of research suggests that multiple factors are involved in the effectiveness of the branded message, like placement prominence and congruity between the game and the brand's segment among others (Terlutter & Capella, 2013). Taken these results, branding in game-like apps should be taken with care and be well thought out.

Measuring Brand Equity

The brand equity measurement system includes a number of procedures that collect data about and evaluate the brand performance. As mobile users generate a lot of data to study, a strategic approach to data analytics and mobile data mining provide valuable consumer insights on brand engagement. The changes brought in by the mobile revolution can be observed on both ends of the brand equity measurement process.

On the one hand, the nature of mobile communication is deeply personal. Existing systems of brand equity measurement often fail to capture this relationship, as well as manifold consequences that follow this higher degree of personality in mobile-mediated communication. For brand mangers this means a fundamental change in the approach to measuring brand equity – from the mass communication perspective to the one of a continuous dialogue. Christodoulides et al. (2006) developed a scale for measuring online retail/service (ORS) brand equity that incorporated five distinct dimensions:

1. Emotional connection,
2. Online experience,
3. Responsive service nature,
4. Trust, and
5. Fulfillment.

This scale takes into consideration the essence of the online consumer experience – the co-creational character of brand construction. The highly interactive nature of the mobile environment accelerates this tendency. Traditional measurements usually track formal behavioral expressions, like willingness to pay or brand awareness. Though still important, these measurements might not take into account changing priorities of consumers, for instance, mobile experience or the sense of security. In other words, they often miss personal connectedness to brands that arises through a dialogic process of co-creation.

On the other hand, the amount of personal data generated by mobile (taken its very personal character) leads to an overwhelming myriad of opportunities. Inspired by these new chances, brand managers try to embrace them all, yet are often driven to frustration. According to executives from leading firms, including AIG, American Express, Samsung Mobile, Siemens Healthcare, TD Bank, and Wal-Mart Stores, the buzz created around big data has oversimplified the companies' approach towards data and its analysis (Brown, Court, & McGuire, 2014). The focus on how data analytics can be applied in practice may help companies to depart from this too global view on data and dramatically improve their performance. Companies, thus, can focus on one of the three applications:

1. Analyzing mobile consumer-generated data to fix marketing budgets and decide on pricing strategies;
2. Optimize back-end performance, like transportation services; or
3. Use data analytics for both front- and back-end activities.

Another question raised is how to approach enormous volumes of data generated? Alex Pentland from the Massachusetts Institute of Technology (MIT) labeled reality mining as "one of ten technologies that will change the world," referencing an article in MIT's Technology Review (2008, cited in Pentland, 2009). Reality mining pulls together digital traces of our daily experience "using statistical analysis and machine learning methods." The technique provides extensive information on a multitude of aspects of our daily lives. Already in 2009, mobile data from the global positioning system (GPS) could be used in the analysis of mobile users' travel patterns and explore popular "hangouts" of various demographic subgroups. Since then the mobile technology has gone far ahead. Now, when smart mobile devices are packed with even more data-collecting sensors, mobile data mining seems to be inevitable.

Sustaining Brand Equity

While establishing and measuring brand equity requires a fair amount of mastery, sustaining and maximizing brand equity for even newly launched products and services is no less important.

Cross-Cultural Branding

Due to the growing speed of data exchange, expansion of a brand to global markets seems to be quite trivial in this globalized age of digital mobile technologies. However, despite vanishing physical distances, the cultural borders are still in place. As previously discussed in the section about brand loyalty of mobile consumers, Chinese mobile users appear to be the most mobile-driven group and differ significantly in their attitudes from their European counterparts. For example, Germans are significantly more privacy-concerned than Chinese. According to the Mobile People's study, while 89% of respondents from China liked sites to remember who they are, only 26% of German respondents did so.

Developing markets also showed a more dynamic growth of smartphone penetration as compared to established "big" smartphone markets (like USA, Japan, and UK) with an average growth rate of 32% against 17% for the established markets, according to the Kleiner Perkins Caufield Byers (KPCB) report on the Internet trends 2014 (Meeker, 2014). The most rapid expansion of smartphone penetration was marked in India with an annual growth rate around 50%. Yet the highest actual smartphone penetration was reported for Arabic countries: Saudi Arabia and UAE had 110% and 160% penetration respectively.

But how does this all translate into branding across borders? As mentioned above, India shows dramatic increases in smartphone penetration every year. With about 945 million mobile subscribers (Cerwall, 2015) and 74% mobile penetration, India is the second largest country in the world after China by the size of its mobile subscription base. With almost 70% of its population rural, with low literacy and media-reach rates among them, mobile phones have become one of the only media left to connect with consumers in isolated geographic and demographic segments. As a result, m-branding campaigns are successfully implemented by multinational corporations. Colgate was the first brand to use location-based voice communication in India in 2013 during the mass Hindu pilgrimage Maha Kumbh Mela that targeted millions of pilgrims from mostly rural areas. The campaign's goal was to attract visitors to the brand's stall and resulted in 300% increase in the number of attendees.

Cross-cultural mobile branding successfully works in reverse, too, such as when the United Nations Association of Germany (UNA-Germany) bridged the cultural problems of developing countries to the German publics. The UNA-Germany with the help of the agency Cheil Germany conducted an m-campaign to raise awareness among Western people about girls forced into arranged marriages. The

"Free the Forced" campaign organizers brilliantly used the context and powered it with a simple mobile technology. They selected one of the most romantic places in Cologne (the Hohenzollern Bridge) where happy newlyweds usually hang love padlocks as a symbol of their commitment and throw the keys into the canal, thus "locking" their love. The UNA-Germany placed 3,500 custom-made padlocks in bright blue forming in all caps the words "Free the Forced." Each lock had the same bold headline "Free the forced!" and a QR-Code that redirected passersby to a page where they could make a donation via PayPal. They then received a code with which they could open the lock and symbolically free one girl who had been forced into marriage. After only three days of the campaign almost all the padlocks were opened, generating enormous publicity for the cause. Over half a million unique visitors spread their word on the campaign via popular social media; the organization's Facebook page activity increased by an astounding 400%; and around 5.3 million people were reached through subsequent publicity in traditional media.

Not always, though, can the behavior of local mobile users be easily anticipated. Global brands entering new cultural markets might both undergo and bring in considerable transformations to the local societies and create unique subcultures. Japan's youth mobile culture turned out to set a social media precedent for one of McDonald's restaurants after they launched a regular McDonald's promotion serving unlimited amount of French fries for a fixed price of ¥150. An innocent tweet of four teens in Okayama read:

A bunch of us are on the second floor of the McDonald's in front of Okayama Station, and we're doing a challenge to eat sixty large orders of fries. Everyone come over right away, or retweet us! (Hansen, 2014).

Immediately the restaurant was packed with a huge number of local Japanese teens, which created a lot of inconvenience for other dining folks and McDonald's staff. The four teens were strongly criticized in the social media for "their inconsiderateness in appropriating the restaurant space [during] the lunch rush and busiest time of day at this McDonald's." However, what underlies this media buzz is the cultural amalgam of the changing Japanese mobile youth culture, empowered by the feeling of the virtual community on social media, and the brand's "glocalization" that has set specific cultural norms of consumption and behavior. These norms, according to Kelly Hansen from the Center for Asian and Pacific Studies in San Diego State University, were the reason why "the potato party may have been reasonably perceived by the teens as an acceptable activity to carry out within the public space of the restaurant" (Hansen, 2014).

Brand Portfolio Management

Since brand equity is an intangible asset, managing a brand portfolio follows similar principles of managing a portfolio of financial assets (Aaker, 2004). Formulating a brand portfolio strategy is an important strategic management activity and is related to firms' marketing and financial performance (Morgan & Rego, 2009). According to Filipsson (2008), the main concern of brand managers today is not only to create competitive brands, but also to strategically manage a combination of multiple brands in one portfolio. "The portfolio brands must work together to form a coherent whole," wrote Aaker (2004), referring to a *brand system*, which was later replaced by the concept of *brand portfolio strategy*.

Brands in such brand systems might play varying roles. Facebook's new in-house built app *Paper* is believed to be a testing ground for the company's new mobile environment (Smith, 2014; Edwards, 2014). "The idea was to revisit Facebook as an experience on mobile devices – effectively from the ground up," said Scott Goodson, an ex-Apple engineer working on the project, in an interview to WIRED (Metz,

2014). With a more functional focus, a media and entertainment branch of the Korean corporation CJ Group, a major producer of top Korean movies, TV shows and a K-Pop celebrities' incubator, integrated its mobile app CGV to support the branch's main business activities. With it, users can buy tickets and find movie-related information from the screens of their smartphones (Borison, 2014).

Brand portfolios, in which various brands play supporting roles, create elaborate *brand ecosystems*. Understanding of a brand as an ecosystem is crucial for the mobile environment that is gravitating around the concept of a seamless experience. From the economic perspective, such integrated ecosystems set higher consumer switching costs and lead to consumer lock-in effects. The more seamless is the experience and the more tied-in the services it incorporates, the more difficult it is for a user to leave the ecosystem. According to Keefe (2013), brand ecosystems include three main elements: content (what we consume); form (communication devices); and distribution (how we buy and pay).

One example of such an integrated mobile-oriented ecosystem is Apple. Its iPhone and iPad devices run on Apple's mobile operating system iOS, and incorporate such content distribution and payment systems as iTunes and Apple Pay. Amazon, which first entered the "Strategy & Global Information, Communications, and Technology 50 study" in 2013 (Acker et al., 2013), also has a sharp focus on building its own ecosystem. Apart from its online store, Amazon has built an expansive physical distribution system, owns the electronic Kindle device, and offers a video-on-demand service, Amazon Instant Video, to be a solid Neflix competitor (Kumparak, 2015) that is available on Amazon online stores and as a separate mobile app.

The brand ecosystem approach can also solve a problem for non-ICT-based companies in dealing with the mobile tide. The brand leverage, a form of brand extension, might be effective in the environments where an established brand has already gained its reputation and loyal customer base. However, new mobile reality shows that the brand leverage might also work in the other direction. As seen earlier, brand loyalty of new consumers strongly depends on their mobile experiences with a brand. Mobile environment, thus, rolls out an absolutely new battlefield for brands in which either no previous experience with a brand has been established or almost none of the previous experiences of consumers with physical products transfer to the mobile environment.

Building an experience-oriented brand ecosystem can help existing brands to connect with their mobile audiences. Nike's major move to launch an iPod-integrated Nike+iPod Sport Kit, and later its own fitness-tracking technology Nike+ FuelBand, grabbed hold of the ecosystem approach. Nike's VP of digital sports, Stefan Olander, claimed it to be a revolutionary turn in the company's strategy (Nudd, 2012):

It used to be that when you bought a product, that was the end of the relationship. It's classic marketing. "Great, you bought the product. See you in a year, when the next campaign comes along." That thinking has flipped on its head. Now, the purchase of any Nike product needs to be the beginning of the relationship we have with the consumer.

Today the co-branded Nike+ comprises a series of mobile apps, including Nike+ Running, Nike+ Training Club, and Nike+ FuelBand, that allow mobile users to collect Nike+ Fuel points and track their fitness activities. Studies, yet, show that despite its huge fitness community and opportunities for intercreativity (Spurgeon, 2009), the consumer resistance against Nike's initiative towards building deeper and more emotional relationships with its audiences still exists (Ringberg & Bjerregaard, 2012). This more skeptical attitude of consumers towards commercial actors should also be taken into consideration while establishing a brand ecosystem, since it can be interpreted as a marketing effort, rather than an attempt to bring enhanced consumer experiences through better products and services.

Experts predict a "battle of the ecosystems" to come to the fore, in which any companies that fail to build their own ecosystems are destined to lose (Acker et al., 2013):

We expect that the hardware and software sectors will continue to evolve into a 'battle of the ecosystems,' with players from both sides trying to build a suite of products and services grounded in an anchor platform they can control and monetize. This strategy is taking two forms: Companies like Apple, Google, and Samsung (with Microsoft playing catch-up) are combining mobile devices with cloud services focused on the end-user, while others, including Amazon, Microsoft, Oracle, Salesforce.com, and SAP, are building their own cloud services focused on solutions and infrastructure for the enterprise. This will likely leave the remaining pure-play hardware providers to pursue a commoditization strategy with little room for differentiation — success will depend on scale, operational excellence, and a relentless focus on costs.

The battle of the ecosystems can be well observed in the major M&A deals, as, besides building and nurturing their own brands in-house, companies can also acquire existing firms and expand their brand portfolio. The latter option is quite often considered to be more attractive (Chernatony & McDonald, 2003), and this seems to be true in the case of new mobile markets, as well. Even though the mobile landscape today is still fragmented, one can already see a gradual transition towards more concentration among such big players as Google and Microsoft, which routinely acquire new companies.

In its 2002 move, eBay acquired PayPal and grabbed the headlines as the most successful merger in the history of Silicon Valley. The acquisition of PayPal, as well as eBay's later purchase of Skype, was a part of the company's strategy to make eBay "more than just an auction business" (Mishkin & Waters, 2014). Yet eBay's attempt to build an ultimate shopping ecosystem was doomed to fail, since the company was unable to integrate its fast-growing acquired businesses with the slower auction and marketplace business. A recent decision of eBay to spin off its increasingly successful PayPal business was dubbed as a gradual "climb down" of the commerce giant (Kokalitcheva, 2015).

A recent Facebook acquisition of the text messaging mobile application WhatsApp got into media spotlight as "one of the largest M&A deals in history" (Knox, 2014). The new portfolio strategy of the technology giant is believed to be following the Procter & Gamble's (P&G) "house of brands." P&G is famous for managing competing brands with the goal to have a larger market share in total. Facebook's purchase of Instagram and WhatsApp might signal the same approach.

MAPPING THE FUTURE

As we showed in this chapter, the mobile revolution has impacted all stages of strategic brand management and should be considered at the highest strategic level. Hence, future research should address these changes in a comprehensive way. More studies are needed to capture the dynamic nature of mobile branding: with more longitudinal studies measuring brand performance and brand equity change over time, on the one hand; and the effects of fluid and co-creational brand identities within the mobile context, on the other hand.

Furthermore, as experience has become a key factor that drives loyalty, the importance of research on brand experience is growing, too. This relatively new research field is expanding rapidly with a multitude of mobile settings emerging as new consumer touch-points. According to Schmitt, Brakus, and Zarantonello (2015), who previously conceptualized the brand experience scale (Brakus, Schmitt, & Zarantonello, 2009), the future research in this field should study "the role of brands in consumption experiences" and "the need of brand experiences to reach positive psychological outcomes." But as new scales that capture brand equity and brand value emerge, there is a growing need for studies with large-scale validation of these scales. This is getting more viable with the growing amount of mobile-driven data. And perhaps even more focus can be placed on behavioral data extracted directly from consumers.

Since much data is now getting more personalized – driven from smartphones and tablets (and from smart wearables and smart homes in the nearest future), there is a possibility for companies to attend even to subtle changes and differences in individual consumer behavior. Hence, crude correlations and self-reported scales that have been used for decades by traditional brand management and marketing might not be as accurate as behavioral data from technology-based interactions with products and brands. The academic research of brand management should also (yet wisely) delve into the big data paradigm.

There is a growing number of tools for big data analytics offered for industry ad hoc research. However, this does not preclude the use of these tools by scholars. Analysis of huge amounts of data can provide insights into specific manifestations of more obscure psychological variables in consumer behaviors and choices. Numerous studies within computational scholarship suggest that contemporary computer algorithms are successful at predicting even sensitive personal information, including gender, age, personality, and even religious and political affiliations, of social media users based on their publicly accessible profiles, by analyzing interaction styles, network and semantic analysis, and analysis of Facebook likes (Youyou et al., 2014; Kosinski et al., 2012; Bachrach et al., 2012; Quercia, 2011; Schwartz et al., 2013). This is the evidence that rich data can and should be used in academic research. The big data analytic tools can be possibly applied to internal data of diverse companies for more precise cluster analyses and brand strategy identifications. As the result, interdisciplinary collaborations, especially with researchers from the fields like computer science and psychology, seem to be inevitable.

Of further consideration are the implications of mobile on brand portfolio management. As discussed previously, mobile subscription and smartphone penetration rates in such developing markets as China, India and Indonesia are growing rapidly. As a result, gaining a considerable mobile market share in these regions becomes a top strategic priority for many companies. One such case is Twitter. The company extended its portfolio and bought an Indian mobile marketing startup Zipdial, which allows people disconnected from the Internet to receive advertising on their cellphones (Frizell, 2015). It is difficult to evaluate the prospects of this acquisition, but it is clear that brand extensions across geographic borders and product categories should be studied more extensively. This should be done to understand the true value of mobile technologies in branding.

A comprehensive view in branding is crucial, and this view needs to embrace the new reality, too. The editors of the European Journal of Marketing – Melewar, Gotsi, and Andriopoulos (2012) in a special issue on corporate branding – highlighted the importance of addressing multiple levels of analysis, including the impact of individuals on branding, as well as of group and organizational dynamics. Thus, new comprehensive models are needed that could put various levels and/or dimensions of branding into a coherent system. Keller and Lehmann (2006), for instance, propose one such systems model that might be used as a framework to study technological and social transformations that mobile brings to organizations and consumers in terms of value chains and consumer behavior. But more research is needed to learn how these different levels of analysis interact with each other; how they are manifested in consumer behavior; and how they can be measured and validated. As an alternative to this, the brand ecosystem approach might have a potential to link multiple levels of branding, as also suggested by Bergvall (2006). But more academic research should be done to validate the industry insights and make more grounded conceptualizations. This is especially crucial within the context of emerging mobile platforms – new touchpoints with brands, like branded apps, mobile wallets and geo-fencing techniques – that generally remain under-researched with marginal contribution from mainly niche journals and industry.

CONCLUSION

This chapter has supported revolutionary new approaches and understandings of branding in the age of mobile. New paradigms are replacing traditional branding concepts, like the long-held funnel metaphor of branding. Instant connectivity and mobility of communication have drastically changed the landscape of brand touchpoints, making the funnel metaphor obsolete and replacing it with the "consumer decision journey." The new paradigm explains how that journey might pass through experience-oriented brand ecosystems. The new reality also pushes brand managers to replace old measures of brand equity with perhaps the new QRS approach. However, all these newly articulated research strands represent only a miniscule portion of the research opportunities that mobile opens up.

Changing behaviors, new industries, new technological advancements, and their impact on brand equity all raise new questions about branding and the nature of the mobile revolution. Yet branding-related questions are not fully addressed by a new conceptual framework. And even within that framework, there are many new or newly augmented strategies and tactics to employ. All of this raises questions about their most effective implementation – tactics and strategies, for example, related to augmented reality, gamification, geo-targeting, personalization, and co-creation. As consumers pass through the various newly emerging touchpoints, differences between their journeys should be studied, as well. And while industry is always anxious to gain insights from research to implement as best practices – which ones really are best when scrutinized scientifically? Finally, long-studied perspectives, such as cross-cultural branding, are now undergoing dramatic change in the global online community. New research opportunities abound under such circumstances.

REFERENCES

Aaker, D. A. (2004). *Brand portfolio strategy*. New York, NY: Free Press/Simon & Schuster.

Aaker, D. A., & Joachimsthaler, J. (2000). *Brand leadership*. London: Free Press.

Accenture. (2013). *Accenture 2013 Global Consumer Pulse Survey*. Retrieved 13 April 2015 from http://www.accenture.com/SiteCollectionDocuments/PDF/Accenture-Global-Consumer-Pulse-Research-Study-2013-Key-Findings.pdf

Acker, O., Geerdes, H., Gröne, F., & Schröder, G. (2013). Builders of the digital ecosystem: The 2013 Strategy & global ICT 50 study. *PricewaterhouseCoopers*. Retrieved from http://www.strategyand.pwc.com/media/file/Strategyand_Builders-of-the-Digital-Ecosystem.pdf

Adobe. (2013). *The marketer's guide to personalization: Improving customer engagement in a digital world*. Retrieved 29 March 2015 from http://offers.adobe.com/content/dam/offer-manager/en/na/marketing/Target/Guide%20to%20personalization.pdf

Adobe. (2015). *Quarterly Digital Intelligence Briefing: Digital Trends 2015*. Retrieved 20 Feb. 2015 from http://offers.adobe.com/content/dam/offer-manager/en/na/marketing/Target/Adobe%20Digital%20Trends%20Report%202015.pdf

Analytics, S. (2014, April 29). *Samsung & Apple slip as global smartphone shipments reach 285 million units in Q1 2014* [Press release]. Retrieved from http://blogs.strategyanalytics.com/WSS/post/2014/04/29/Strategy-Analytics-Global-Smartphone-Shipments-Reach-285-Million-Units-in-Q1-2014.aspx

Ayres, N. (2014). Adventures in Publishing: The new dynamics of advertising. *Brand Perfect*. Retrieved 14 April 2015 from http://brandperfect.org/brand-perfect-dynamics-of-advertising-report.pdf

Bachrach, Y., Kosinski, M., Graepel, T., Kohli, P., & Stillwell, D. (2012, June). Personality and patterns of Facebook usage. In *Proceedings of the 4th Annual ACM Web Science Conference* (pp. 24-32). ACM. doi:10.1145/2380718.2380722

Banfi, F., Begonha, D., Hazan, E., & Zouaoui, Y. (2014). Mobile must migrate: Digital as an imperative, not an option. *McKinsey publication of the Telecommunications, Media, and Technology Practice, 24*(7).

Basole, R. C., & Karla, J. (2012). Value transformation in the mobile service ecosystem: A study of app store emergence and growth. *Service Science, 4*(1), 24–41. doi:10.1287/serv.1120.0004

Bellman, S., Potter, R. F., Treleaven-Hassard, S., Robinson, J. A., & Varan, D. (2011). The effectiveness of branded mobile phone apps. *Journal of Interactive Marketing, 25*(4), 191–200. doi:10.1016/j.intmar.2011.06.001

Benjamin, B. (2006). The case study: Storytelling in industrial age and beyond. *On the Horizon, 14*(4), 159–164. doi:10.1108/10748120610708069

Bergvall, S. (2006). Brand ecosystems: Multilevel brand interaction. In J. E. Schroeder, M. Salzer-Mörling, & S. Askegaard (Eds.), *Brand culture* (pp. 166–175). New York, NY: Taylor & Francis.

Bertoni, S. (2014, February 21). How do you win the mobile wallet war? Be like Starbucks. *Forbes*. Retrieved from http://www.forbes.com/sites/stevenbertoni/2014/02/21/how-do-you-win-the-mobile-wallet-war-be-like-starbucks/

Borison, R. (2014, June 17). The 15 most successful app companies. *Business Insider*. Retrieved from http://www.businessinsider.com/15-most-successful-app-companies-2014-6?IR=T#7-cj-group-9

Brakus, J. J., Schmitt, B. H., & Zarantonello, L. (2009). Brand experience: What is it? How is it measured? Does it affect loyalty? *Journal of Marketing, 73*(3), 52–68. doi:10.1509/jmkg.73.3.52

Brown, B., Court, D., & McGuire, T. (2014, March). Views from the front lines of the data-analytics revolution. *The McKinsey Quarterly*.

Brown, R. L., & Harmon, R. R. (2014). Viral geofencing: An exploration of emerging big-data driven direct digital marketing services. *Proceedings of the International Conference on Management of Engineering & Technology (PICMET)*.

Brown, S., Kozinets, R. V., & Sherry, J. F. Jr. (2003). Teaching old brands new tricks: Retro branding and the revival of brand meaning. *Journal of Marketing, 67*(3), 19–33. doi:10.1509/jmkg.67.3.19.18657

Bulearca, M., & Tamarjan, D. (2010). Augmented reality: A sustainable marketing tool? *Global Business and Management Research: An International Journal, 2*(2), 237–252.

Carufel, R. (2013). Connection Disconnection: 80 Percent of Brands Don't Know Their Customers, New Yesmail and Gleanster Study of Executive-Level Marketers Reveals. *Bulldog Reporter's Daily Dog*. Retrieved on 10 April 2015 from https://www.bulldogreporter.com/dailydog/article/pr-biz-update/connection-disconnection-80-percent-of-brands-don-t-know-their-custom

Cerwall, P. (Ed.). (2015, February). *Ericsson mobility report: On the pulse of the networked society, Mobile World Congress Edition*. Retrieved from http://www.ericsson.com/res/docs/2015/ericsson-mobility-report-feb-2015-interim.pdf

Chang, C. (2009). 'Being hooked' by editorial content: The implications for processing narrative advertising. *Journal of Advertising, 38*(3), 51–65. doi:10.2753/JOA0091-3367380304

Cianci, A. M., & Kaplan, S. E. (2010). The effect of CEO reputation and explanations for poor performance on investors' judgments about the company's future performance and management. *Accounting, Organizations and Society, 35*(4), 478–495. doi:10.1016/j.aos.2009.12.002

Clifton, R., & Simmons, J. (Eds.). (2003). *Brands and branding*. London: Profile Books Ltd, The Economist Newspaper Ltd.

Consulting, F. (2013b). *Your Customers Are Demanding Omni-Channel Communications. What Are You Doing About It?* Retrieved 13 April 2015 from http://www.kana.com/customer-service/white-papers/forrester-your-customers-are-demanding-omni-channel-communications.pdf?_ga=1.38271843.13123 02063.1428939602

Corpuz, J. (2015, March 4). Ten best mobile racing games. *Tom's Guide*. Retrieved from http://www.tomsguide.com/us/best-mobile-racing-games,review-2351.html

Danova, T. (2014, January 5). The mobile video revolution: How Netflix, Vevo, and YouTube have thrived on smartphones and tablets. *Business Insider*. Retrieved from http://www.businessinsider.com/mobile-video-statistics-and-growth-2013-12#ixzz3XODJpOxH

Davis, S. M. (2002). *Brand asset management: Driving profitable growth through your brands*. San Francisco, CA: Josey Bass.

Duncan, E., Hazan, E., & Roche, K. (2014). Digital disruption: Evolving usage and the new value chain. *McKinsey publication of the Telecommunications, Media, and Technology Practice, 24*(1).

Econsultancy. (2015). *The Consumer Conversation: The experience void between brands and their customers*. Retrieved from https://www14.software.ibm.com/webapp/iwm/web/signup.do?source=swg-smartercom_medium&S_PKG=ov33876&dynform=18187&S_TACT=C348047W&cm_mmc=EconsultancySurvey-_-social-_-infographic-_-Apr2015

Edelman, D. (2010, December). Branding in the digital age: You're spending your money in all the wrong places. *Harvard Business Review*.

Edwards, J. (2014, February 3). Facebook's 'Paper' app is basically a massive redesign of Facebook – and it's excellent. *Business Insider*. Retrieved from http://www.businessinsider.com/facebook-paper-app-redesign-of-facebook-2014-2?IR=T

Edwards, L. H. (2012). Transmedia storytelling, corporate synergy, and audience expression. *Global Media Journal*, *12*(20), 1–12.

Elmieh, B., Austin, D. M., Collins, B. M., Oftedal, M. J., Pinkava, J. J., & Sweetland, D. P. (2015). *U.S. Patent No. 20,150,026,576*. Washington, DC: U.S. Patent and Trademark Office.

eMarketer. (2013, April 4). *Facebook to see three in 10 mobile display dollars this year*. Retrieved from http://www.emarketer.com/Article/Facebook-See-Three-10-Mobile-Display-Dollars-This-Year/1009782#sthash.3gy64laM.dpuf

eMarketer. (2015, March 26). *Facebook and Twitter will take 33% share of US digital display market by 2017*. Retrieved 20 Dec. 2015 from http://www.emarketer.com/Article/Facebook-Twitter-Will-Take-33-Share-of-US-Digital-Display-Market-by-2017/1012274#sthash.o1A1Vm4z.dpuf

Evans, C. (2014, March 17). 10 examples of augmented reality in retail. *Creative Guerilla Marketing*. Retrieved from http://www.creativeguerrillamarketing.com/augmented-reality/10-examples-augmented-reality-retail/

Evergage. (2014). *Real-time for the rest of us: Perceptions of real-time marketing and how it's achieved*. Retrieved 25 March 2015 from http://info.evergage.com/perceptions_of_realtime_marketing_survey_results

Feijoo, C., Gómez-Barroso, J. L., Aguado, J. M., & Ramos, S. (2012). Mobile gaming: Industry challenges and policy implications. *Telecommunications Policy*, *36*(3), 212–221. doi:10.1016/j.telpol.2011.12.004

Feldwick, P. (1996). Do we really need "brand equity? *The Journal of Brand Management*, *4*(1), 9–28. doi:10.1057/bm.1996.23

Fournier, S., & Avery, J. (2011). The uninvited brand. *Business Horizons*, *54*(3), 193–207. doi:10.1016/j.bushor.2011.01.001

Frizell, S. (2015, January 20). Twitter buys Indian mobile marketing startup. *Time*. Retrieved from http://time.com/3674300/twitter-zipdial-india/

Google. (August, 2012). The new multi-screen world: Understanding cross-platform consumer behavior. *Think with Google*. Retrieved from https://www.thinkwithgoogle.com/research-studies/the-new-multi-screen-world-study.html

Graffin, S., Pfarrer, M., & Hill, M. (2012). Untangling executive reputation and corporate reputation: Who made who. In M. L. Barnett & T. G. Pollock (Eds.), *The Oxford handbook of corporate reputation*. Oxford, UK: Oxford University Press. doi:10.1093/oxfordhb/9780199596706.013.0011

Granitz, N., & Forman, H. (2015). Building self-brand connections: Exploring brand stories through a transmedia perspective. *Journal of Brand Management*.

Guzman, F. (2005). *A brand building literature review*. Barcelona: Esade Business School.

Hansen, K. (2014). When French fries go viral: Mobile media and the transformation of public space in McDonald's Japan. *Electronic Journal of Contemporary Japanese Studies, 14*(2). Retrieved from http://www.japanesestudies.org.uk/ejcjs/vol14/iss2/hansen.html

Harvey, K. (2011). *Approaching PR Like a Journalist*. Retrieved 10 April 2015 from https://youtu.be/y10jCpR2akE

Herskovitz, S., & Crystal, M. (2010). The essential brand persona: Storytelling and branding. *The Journal of Business Strategy, 31*(3), 21–28. doi:10.1108/02756661011036673

Husson, T. (2015, February 9). The future of mobile wallets lies beyond payments. *Forrester Researcher.* Retrieved from https://s3.amazonaws.com/vibes-marketing/Website/Reports_$folder$/Forrester+-+The+Future+of+Mobile+Wallets+Report.pdf

IBM. (2015). *U.S. Online Holiday Benchmark Recap Report 2014*. Retrieved 10 April 2015 from http://www-01.ibm.com/common/ssi/cgi-bin/ssialias?subtype=WH&infotype=SA&appname=SWGE_ZZ_JV_USEN&htmlfid=ZZW03362USEN&attachment=ZZW03362USEN.PDF#loaded

Interbrand. (2014, October 9). *Apple and Google each worth more than USD $100 billon on Interbrand's 15th annual Best Global Brands Report* [Press release]. Retrieved from http://interbrand.com/en/news-room/15/interbrands-th-annual-best-global-brands-report

Interbrand. (2014). *The four ages of branding*. Retrieved from http://www.bestglobalbrands.com/2014/featured/the-four-ages-of-branding/

International Telecommunications Union. (2015). *Global ICT developments*. Retrieved 20 March 2015 from http://www.itu.int/en/ITU-D/Statistics/Pages/stat/default.aspx

Ivanov, C. (2005, November 18). Three new NBC Sports-branded mobile games in early 2006. *Softpedia*. Retrieved from http://archive.news.softpedia.com/news/Three-New-NBC-Sports-branded-Mobile-Games-In-Early-2006-12728.shtml

Jian, M., & Lee, K. W. (2011). Does CEO reputation matter for capital investments? *Journal of Corporate Finance, 17*(4), 929–946. doi:10.1016/j.jcorpfin.2011.04.004

Keefe, D. (2013). Branding in the new age of ecosystems. *Landor Associates*. Retrieved from http://landor.com/pdfs/DKeefe_Ecosystem_26June2013.pdf?utm_campaign=PDFDownloads&utm_medium=web&utm_source=web

Keller, K. L. (2013). *Strategic brand management: Building, measuring, and managing brand equity*. Harlow, UK: Pearson Education Ltd.

Keller, K. L., & Lehmann, D. R. (2006). Brands and branding: Research findings and future priorities. *Marketing Science, 25*(6), 740–759. doi:10.1287/mksc.1050.0153

Kim, Y. S., Han, M. J., Hong, C. J., Ko, S. B., & Kim, S. B. (2012). Mobile Digital Storytelling Development for Energy Drink's Promotion and Education. In *Advances in Automation and Robotics* (Vol. 1, pp. 557–566). Springer Berlin Heidelberg.

Knox, D. (2014, February 20). Is Facebook building a P&G-style house of brands? *Advertising Age*. Retrieved from http://adage.com/article/digitalnext/facebook-buys-whatsapp-facebook-house-brands/291798/

Kokalitcheva, K. (2015, April 9). After planned split, eBay and PayPal promise to stay friends. *Fortune*. Retrieved from http://fortune.com/2015/04/09/ebay-paypal-split/

Kosinski, M., Stillwell, D., & Graepel, T. (2013). Private traits and attributes are predictable from digital records of human behavior. *Proceedings of the National Academy of Sciences of the United States of America*, *110*(15), 5802–5805. doi:10.1073/pnas.1218772110 PMID:23479631

Kumparak, G. (2015, April 7). Amazon Instant Video finally comes to Android tablets. *TechCrunch*. Retrieved from http://techcrunch.com/2015/04/07/amazon-instant-video-finally-comes-to-android-tablets/

Lajoie, M., & Shearman, N. (Prod.). (2014). What is Alibaba? *Wall Street Journal*. Retrieved from http://projects.wsj.com/alibaba/

Lee, K. (2015). *The 5 Keys to Building a Social-Media Strategy for Your Personal Brand*. Retrieved 13 April 2015 from http://www.entrepreneur.com/article/243079

Liao, T. (2015). Augmented or admented reality? The influence of marketing on augmented reality technologies. *Information Communication and Society*, *18*(3), 310–326. doi:10.1080/1369118X.2014.989252

Lieb, R., Silva, C., & Tran, C. (2013). *Organizing for content: models to incorporate content strategy and content marketing in the enterprise*. Altimeter Group. Retrieved from http://www.ciccorporate.com/edm/2013/201305en/r2.pdf

Los Cannes Blog [loscannesblog]. (2014). *Nivea protection ad - Mobile Cannes 2014* [Video file]. Retrieved from https://www.youtube.com/watch?v=L9ZDyLlcdww

Low, G. S., & Fullerton, R. A. (1994). Brands, brand management, and the brand manager system: A critical-historical evaluation. *JMR, Journal of Marketing Research*, *31*(2), 173–190. doi:10.2307/3152192

Lundqvist, A., Liljander, V., Gummerus, J., & van Riel, A. (2012). The impact of storytelling on the consumer brand experience: The case of a firm-originated story. *Journal of Brand Management*, *20*(4), 283–297. doi:10.1057/bm.2012.15

Mackalski, R., & Belisle, J. F. (2015). Measuring the short-term spillover impact of a product recall on a brand ecosystem. *Journal of Brand Management*, *22*(4), 323–339. doi:10.1057/bm.2015.19

Marchesi, M., & Riccò, B. (2013, November). Augmented graphics for interactive storytelling on a mobile device. In *SIGGRAPH Asia 2013 Symposium on Mobile Graphics and Interactive Applications* (p. 59). ACM. doi:10.1145/2543651.2543683

Martin, K., & Todorov, I. (2010). How will digital platforms be harnessed in 2010, and how will they change the way people interact with brands? *Journal of Interactive Advertising*, *10*(2), 61–66. doi:10.1080/15252019.2010.10722170

McGonigal, J. (2011). *Reality is broken: Why games make us better and how they can change the world*. New York, NY: The Penguin Press.

McGrane, K. (2012). Content strategy for mobile. New York, NY: A Book Apart.

Meeker, M. (May 28, 2014). Internet trends 2014 – Code Conference [Presentation slides]. *Kleiner Perkins Caufield Byers*. Retrieved from http://www.kpcb.com/InternetTrends

Melewar, T. C., Gotsi, M., & Andriopoulos, C. (2012). Shaping the research agenda for corporate branding: Avenues for future research. *European Journal of Marketing, 46*(5), 600–608. doi:10.1108/03090561211235138

Men, L. R. (2012). CEO credibility, perceived organizational reputation, and employee engagement. *Public Relations Review, 38*(1), 171–173. doi:10.1016/j.pubrev.2011.12.011

Men, L. R., & Tsai, W. H. S. (2015). Infusing social media with humanity: Corporate character, public engagement, and relational outcomes. *Public Relations Review, 41*(3), 395–403. doi:10.1016/j.pubrev.2015.02.005

Mercator Advisory Group. (2013, August 1). *Mobile wallets: The business and the brand* [Press release]. Retrieved from https://www.mercatoradvisorygroup.com/Press_Releases/Mobile_Wallets___The_Business_and_the_Brand/

Metz, C. (2014, March 6). Facebook Paper has forever changed the way we build mobile apps. *Wired*. Retrieved from http://www.wired.com/2014/03/facebook-paper/

Mishkin, S., & Waters, R. (2014, September 30). Ebay offers ultimate 'Buy It Now' with plans to spin off PayPal. *Financial Times*. Retrieved from http://www.ft.com/intl/cms/s/0/a14eb3a6-4896-11e4-9d04-00144feab7de.html#axzz3WvBTc4EB

Mixpo. (2015). *The state of video advertising.* Retrieved 1 May 2015 from http://marketing.mixpo.com/acton/fs/blocks/showLandingPage/a/2062/p/p-00b4/t/page/fm/0

Mobile Marketing Association. (2013). Free the Forced. *MMA Case Study*. Retrieved from http://www.mmaglobal.com/case-study-hub/upload/pdfs/mma-2013-577.pdf

Mobile Marketing Association. (2013). How Colgate Turned the Maha Kumbh Mela into the Maha Tech Mela. *MMA Case Study*. Retrieved from http://www.digitaltrainingacademy.com/casestudies/mma-COLGATE.pdf

Morgan, N. A., & Rego, L. L. (2009). Brand portfolio strategy and firm performance. *Journal of Marketing, 73*(1), 59–74. doi:10.1509/jmkg.73.1.59

Muniz, K. M., Woodside, A. G., & Sood, S. (2015). Consumer storytelling of brand archetypal enactments. *International Journal of Tourism Anthropology, 4*(1), 67–88. doi:10.1504/IJTA.2015.067644

New, B. (January 4, 2011). *Vul het zelf maar in (Fill in the blank).* Retrieved from http://www.underconsideration.com/brandnew/archives/vul_het_zelf_maar_in_fill_in_the_blank.php#.VRl55vmsWSo

Nudd, T. (2012, June 20). How Nike+ made 'Just do it' obsolete: Stefan Olander on building a new ecosystem of engagement. *Adweek*. Retrieved from http://www.adweek.com/news/advertising-branding/how-nike-made-just-do-it-obsolete-141252

Ogilvy Action Germany [OgilvyActionGER]. (2013, June 19). *Back to vinyl – The office turntable* [Video file]. Retrieved from https://www.youtube.com/watch?v=Xg1pzEF3HTw

Parganas, P., Anagnostopoulos, C., & Chadwick, S. (2015). 'You'll never tweet alone': Managing sports brands through social media. *Journal of Brand Management, 22*(7), 551–568. doi:10.1057/bm.2015.32

Pearson, L. (2013). Fluid marks 2.0: Protecting a dynamic brand. *Managing Intellectual Property, 26*, 26–30.

Pearson, L. (2014). Fluid trademarks and dynamic brand identities. *The Trademark Reporter, 104*, 1411–1433.

Pentland, A. (2009). Reality mining of mobile communications: Toward a new deal on data. *The Global Information Technology Report 2008-2009: Mobility in a Networked World*, (pp. 75-80). World Economic Forum.

Pimentel, R. W., & Heckler, S. E. (2003). Changes in logo designs: Chasing the elusive butterfly curve. In L. M. Scott & R. Batra (Eds.), *Persuasive Imagery: A Consumer Response Perspective* (pp. 105–128). Mahwah, NJ: Lawrence Erlbaum Associates.

Pulizzi, J. (2012). The rise of storytelling as the new marketing. *Publishing Research Quarterly, 28*(2), 116–123. doi:10.1007/s12109-012-9264-5

Pulizzi, J. (2013). *Epic content marketing: How to tell a different story, break through the clutter, and win more customers by marketing less*. Boston, MA: McGraw-Hill.

Quercia, D., Kosinski, M., Stillwell, D., & Crowcroft, J. (2011, October). Our Twitter profiles, our selves: Predicting personality with Twitter. In *Privacy, Security, Risk and Trust (PASSAT) and 2011 IEEE Third Inernational Conference on Social Computing (SocialCom), 2011 IEEE Third International Conference on* (pp. 180-185). IEEE.

Radic, K. (2013, July 29). 80 variations of the iconic Lacoste logo. *Branding Magazine*. Retrieved from http://www.brandingmagazine.com/2013/07/29/lacoste-anniversary-polo-logo-peter-saville/

Ramsey, G. (2015). *The State of Mobile 2015*. eMarketer live webcast 25 March 2015.

Reporter, B. (2015). *The Great Customer Experience Divide: 4 Out Of 5 Consumers Declare Brands Don't Know Them as Individuals, IBM Study Finds*. Retrieved from https://www.bulldogreporter.com/dailydog/article/pr-biz-update/the-great-customer-experience-divide-4-out-of-5-consumers-declare-bra?utm_source=Bulldog+Reporter&utm_campaign=15bd0a556f-Daily_Dog_Monday_April_6&utm_medium=email&utm_term=0_200eecab38-15bd0a556f-154734149

Review, T. (2008). *Ten emerging technologies that will change the world*. Retrieved from http://www.technologyreview.com/article/13060/

Ringberg, T., & Bjerregaard, S. (2012). Brand Relationships 2.0: A Fundamental Paradox of Proactive Relational Branding and Critical Consumer Culture. In *The 41th EMAC Annual Conference 2012*.

Robideaux, D., & Robideaux, S. C. (2012). Convergent storytelling as promotion: A model for brand engagement and equity. *Innovative Marketing, 8*(1), 48–51.

Sääksjärvi, M., van den Hende, E., Mugge, R., & van Peursem, N. (2015). How exposure to logos and logo varieties fosters brand prominence and freshness. *Journal of Product and Brand Management, 24*(7), 736–744. doi:10.1108/JPBM-06-2014-0648

Sabatier, L. A. (2012). The new way to manage content across platforms. *Publishing Research Quarterly*, *28*(3), 197–203. doi:10.1007/s12109-012-9274-3

Salmeron, J. M. (February 8, 2013). If you love your brand, set it free. *Smashing Magazine*. Retrieved from http://www.smashingmagazine.com/2013/02/08/if-you-love-your-brand-set-it-free/

Schmitt, B. H., Brakus, J., & Zarantonello, L. (2015). The current state and future of brand experience. *Journal of Brand Management*, *21*(9), 727–733. doi:10.1057/bm.2014.34

Schwartz, H. A., Eichstaedt, J. C., Kern, M. L., Dziurzynski, L., Ramones, S. M., Agrawal, M., & Ungar, L. H. et al. (2013). Personality, gender, and age in the language of social media: The open-vocabulary approach. *PLoS ONE*, *8*(9), e73791. doi:10.1371/journal.pone.0073791 PMID:24086296

Seo, H. (2014). *Bowling online: smartphones, mobile messengers, and mobile social games for Korean teen girls*. (Unpublished master's thesis). The University of Texas at Austin, TX.

Shandwick, W. (2015). *The CEO Reputation Premium: Gaining Advantage in the Engagement Era*. Retrieved on 10 April 2015 from http://www.webershandwick.com/news/article/81-percent-of-global-executives-report-external-ceo-engagement-is-a-mandate

Sheng, M. L., & Teo, T. S. H. (2012). Product attributes and brand equity in the mobile domain: The mediating role of customer experience. *International Journal of Information Management*, *32*(1), 139–146. doi:10.1016/j.ijinfomgt.2011.11.017

Singh, S., & Sonnenburg, S. (2012). Brand performances in social media. *Journal of Interactive Marketing*, *26*(4), 189–197. doi:10.1016/j.intmar.2012.04.001

Sinton, J. (2014). *Why having a mobile site should be just the start*. TNS Global Connected Life 2014 (Special edition). Retrieved from http://www.tnsglobal.com/sites/default/files/tns-why-mobile-first-design-is-just-the-start_0.pdf

Smith, J. (2014, February 3). Facebook on Paper app: No plans for international release, iPad or Android app. *Pocket-lint*. Retrieved from http://www.pocket-lint.com/news/127001-facebook-on-paper-app-no-plans-for-international-release-ipad-or-android-app

Sprosen, S. (2014). *The influence of branded mobile applications on consumers' perceptions of a brand: the importance of brand experience and engagement*. (Unpublished master's thesis). Dublin Business School, Ireland.

Spurgeon, C. L. (2009). From mass communication to mass conversation: Why 1984 wasn't like 1984. *Australian Journal of Communication*, *36*(2), 143–158.

Statista. (2015). *Second screen usage among mobile internet users in selected countries as of January 2014*. Retrieved from http://www.statista.com/statistics/301187/second-screen-usage-worldwide/

SXSW. (2015). *Mecosystem 2020*. Retrieved from http://panelpicker.sxsw.com/vote/41042

Terlutter, R., & Capella, M. L. (2013). The gamification of advertising: Analysis and research directions of in-game advertising, advergames, and advertising in social network games. *Journal of Advertising*, *42*(2-3), 95–112. doi:10.1080/00913367.2013.774610

TNS Global. (2014). *Connected Life 2014*. Retrieved on March 29 from http://connectedlife.tnsglobal.com/

Valvi, A. C., & West, D. C. (2015). Mobile Applications (Apps) in Advertising: A Grounded Theory of Effective Uses and Practices. In Ideas in Marketing: Finding the New and Polishing the Old (pp. 349-352). Springer International Publishing.

Wagner, K. (2012). The most popular iPhone and iPad apps of all time. *Gizmodo*. Retrieved from http://gizmodo.com/5890602/the-most-popular-iphone-and-ipad-apps-of-all-time

Walsh, M. F., Winterich, K. P., & Mittal, V. (2010). Do logo redesigns help or hurt your brand. *Journal of Product and Brand Management*, *19*(2), 76–84. doi:10.1108/10610421011033421

Walsh, M. F., Winterich, K. P., & Mittal, V. (2011). How re-designing angular logos to be rounded shapes brand attitude: Consumer brand commitment and self-construal. *Journal of Consumer Marketing*, *28*(6), 438–447. doi:10.1108/07363761111165958

Wang, W.-H., & Li, H.-M. (2012). Factors influencing mobile services adoption: A brand-equity perspective. *Internet Research*, *22*(2), 142–179. doi:10.1108/10662241211214548

Wei, P. S., & Lu, H. P. (2014). Why do people play mobile social games? An examination of network externalities and of uses and gratifications. *Internet Research*, *24*(3), 313–331. doi:10.1108/IntR-04-2013-0082

Welsbeck, D., & Berney, P. (Eds.). (2014). People's web report II. *Netbiscuits*. Retrieved from http://www.netbiscuits.com/news/press/2014/10/28/brand-loyalty-dead-on-the-mobile-web-netbiscuits-peoples-web-report-ii/

Woodside, A. G., Sood, S., & Miller, K. E. (2008). When consumers and brand talk: Storytelling theory and research in psychology and marketing. *Psychology and Marketing*, *25*(2), 97–145. doi:10.1002/mar.20203

Yamakami, T. (2012, September). Transition Analysis of Mobile Social Games from Feature-Phone to Smart-Phone. In *15th International Conference on Network-Based Information Systems (NBiS)*, (pp. 226-230). IEEE. doi:10.1109/NBiS.2012.73

Yannopoulou, N., Moufahim, M., & Bian, X. (2013). User-generated brands and social media: Couchsurfing and AirBnb. *Contemporary Management Research*, *9*(1), 85–90. doi:10.7903/cmr.11116

Youyou, W., Kosinski, M., & Stillwell, D. (2015). Computer-based personality judgments are more accurate than those made by humans. *Proceedings of the National Academy of Sciences of the United States of America*, *112*(4), 1036–1040. doi:10.1073/pnas.1418680112 PMID:25583507

Chapter 13
Public Relations and Mobile:
Becoming Dialogic

Yulia An
The Ilmenau University of Technology, Germany

Kenneth E. Harvey
KIMEP University, Kazakhstan

ABSTRACT

The chapter discusses impact of mobile technologies on public relations practice and scholarship by tracking the historical development of public relations as a distinct field, thus mapping a structural framework for the further discussion of new trends and areas of academic and industry research. Moving from functional to co-creational perspective, public relations enters the conversation age in which the nature of mobile technologies forces practitioners to adopt the dialogic approach to build trust and nurture relationships. Existing theoretical frameworks are supplemented with industry examples within the field of crisis communication, corporate social responsibility and customer and employee relations. An overview of some of the latest trends in social media research, public segmentation and video marketing applied to co-creational perspective of public relations organizes new trends around a more fundamental paradigm shift. This structure places industry practices within broader academic research, providing both tactical and strategic views on public relations in the mobile age.

INTRODUCTION

With inception of advanced mobile connectivity technologies, mobile devices like smartphones and tablets, and wearables like a smartwatch, an array of new challenges and opportunities for public relations practitioners and scholars took off. Increasing multi-functionality of the instantly connected environment is gradually occupying media space and thus leads to a socio-technological lock-in.

New technologies become more and more pervasive and store a rapidly growing amount of personal information. Smartphones are now tied to our bank accounts, social media profiles, instant messaging applications, digital photo albums, and personalized mobile applications. As a result, a single mobile

DOI: 10.4018/978-1-5225-0469-6.ch013

phone now incorporates a growing number of value-added mobile and Internet services that are tied to a mobile device and reinforce the adoption effect (Tojib et al., 2014).

The reality is now digital and mobile – it changes rapidly and is always on hand in the screen of your smartphone. New publics switch between media platforms every other minute and rarely keep their smartphones more than an arm's length away (Time Inc., 2012). With transformation of communication per se, the essence of connecting with various publics also changes. Social media and instant messaging allow organizations to connect with their publics directly passing over traditional mass media that in the older days held a role of an information gatekeeper.

Researchers and practitioners of public relations are thus facing a major change in how and where communication takes place. Yet to embrace this technological change in its full swing, public relations scholars and practitioners need to change how they see new technology and expand their focus from exclusively tactical implementations of the public relations practice to wider social and technological issues (Kent & Saffer, 2014).

STATE OF THE ART

In the last few decades public relations have undergone a considerable transformation from a narrowly instrumental practice of organizational communication to a more strategically oriented co-creational perspective. The tremendous impact that technology has exerted on public relations was actively discussed in business and academic circles.

Already in the end of 1990s experts (Ross & Middleberg, 1999; Crawford, 1999; Holtz, 1999; Witmer, 2000, cited in Hurme, 2001) started making bold statements that those public relations practitioners who fall behind the Internet hype would have to leave the game.

As an array of new communication channels is burgeoning and the speed of communication rises rapidly, public relations practitioners (James, 2008) faced a sharp need for more technical skills, including search engine optimization (SEO), web analytics, web publishing, database management, and analytic software operation. New media forms are to be taken into consideration with video content proliferating and going mobile too. Forbes (Trautman, 2014), citing an eMarketer study, reported that in 2018 more than 70% of all online videos will be watched on a tablet.

However, apart from increased technological complexity and new opportunities for data visualization, integration, and measurement of communication effectiveness, a multitude of social changes have also come into play. One of the main changes is increased complexity of the environment and far less control over the communication process and messages disseminated that practitioners used to have in the early days of mass society. Planning campaigns is becoming more difficult with the high speed and uncertainty of communication (Argenti, 2006; James, 2008). As a result, the public relations practice turns from central planning and generally one-way communication to the dialogic two-way communication and even beyond (Botan & Taylor, 2004; Taylor & Botan, 2006; Botan & Hazleton, 2006). A co-creational perspective argues that a classic two-way symmetrical model is too procedural and still treats communication as instrumental (Edwards & Hodges, 2011, cited in Waddington, 2013; Phillips & Young, 2009, cited in Grunig, 2009), and thus needs to be replaced by a more strategic view of communication as creating shared meanings and interpretations rather than pursuing organizational goals.

Not only the older concepts are reconsidered but also new avenues of research emerge. One of the most rapidly growing fields is social media research. As reported by Gillin (2008), social media has now become a new industry standard with prominent bloggers, social media activists and other "new influencers" to have a considerable impact on what traditional media cover. A gross 70% of journalists who used blogs did so while looking for new stories and sources and conducting a journalistic research (Moyer, 2011). As a result, McCorkindale & DiStaso (2014) outlined three major streams in the social media research: ethics, public relations theories applied to the social media environment, and effective measurement of social media campaigns and activities.

Yet now we have entered a new socio-technological juncture with mobile Internet and smartphones leading the move. Pertti Hurme (2001:72) over a decade ago predicted that the "the amalgamation of the Internet and mobile communications to the mobile Internet will change the organisational media landscape profoundly," a change that is now taking a definite shape. Notwithstanding this insight, a recent mobile technological boom has been only marginally grasped by public relations scholars.

Studies analyzing the importance of smartphone users as an emerging new public only now slowly appear on the global research arena with a group of Israeli scholars (Avidar et al., 2013), among them, surveying young smartphone users and potential contribution of smartphones to the practice of public relations and the process of relationship building. Other studies took a more niche-oriented approach, focusing on specific applications and services that mobile media gave rise to, like location-based apps (e.g. Foursquare) and mobile social media (Humphreys, 2013; Frith, 2013). Call for more smartphone-oriented research was voiced in a "Smartphone psychology manifesto" by Miller (2012), who claimed that "smartphones could transform psychology even more profoundly than PCs and brain imaging did." Though addressing the audience of psychologists, this manifesto is yet an important insight for smartphone- and mobile-oriented research in public relations too.

SUMMARIZING THE PAST

With its roots in mass communication research, the initial focus of the public relations scholarship resided in the domain of persuasion and the use of formidable power of mass media (Grunig et al., 2006). This view, largely expressed in the one-way communication, was dominant up until the 1990s when James and Larissa Grunig and David Dozier first formulated their Excellence Theory that has become a classic theory of public relations. Thus, until recently the public relations have rested on four traditional models, as articulated by Grunig and Hunt (1984): press agentry, public information, two-way asymmetrical and two-way symmetrical communication models (Table 1).

Botan and Taylor (2004) refer to the early period of public relations as to a functional perspective. The functional perspective is characterized by one-way communication from an organization to its publics with an ultimate goal to deploy the publics in achieving its organizational goals. Public relations, thus, play an instrumental role by providing techniques of how to produce and disseminate pre-planned organizational messages. Referring back to the four models of public relations, this type of communication was especially emphasized in the press agentry and public information models. The two-way asymmetric communication steps only a bit forward – organizational communication is still instrumental but already contains a feedback loop for the further analysis and re-evaluation of the earlier developed messages.

As a result, key theories and methods of the functional perspective also reflect this instrumental and technical role of public relations. Traditional mass media models, like agenda-setting and the two-step flow of communication, probably are the most representative of this stream. Yet some theories within the field of economics and the public relations scholarship are also essential to understanding the functional

Table 1. An overview of four models of public relations

Characteristic	Model			
	Press Agentry / Publicity	**Public Information**	**Two-Way Asymmetric**	**Two-Way Symmetric**
Purpose	Propaganda	Dissemination of information	Scientific persuasion	Mutual understanding
Nature of communication	One-way; complete truth not essential	One-way; truth important	Two-way; imbalanced effects	Two-way; balanced effects
Communication model	Source → Recipient	Source → Recipient	Source → Recipient ← Feedback	Group ↔ Group
Nature of research	Little; "counting house"	Little; readability, readership	Formative; evaluative of attitudes	Formative; evaluative of understanding

(Adapted from Grunig & Hunt, 1984, p.22)

view. For example, a game-theoretical approach presents a framework for analyzing the decision-making schemes between a public relations practitioner and organization's one or more publics in reaching a common compromise (Murphy, 1989, 1991; Pompper, 2005). Further on, despite the controversy around the concept of *persuasion* as a function of public relations, it is believed to be an essential part of organizational communication that brings some degree of control into an otherwise uncertain and chaotic social environment, as it shapes the perceptions and cognitions of external publics (Miller, 1989; Pfau & Wan, 2006). The functional perspective, though being an offset of the past, still presents a practical framework for studying normative and functional aspects of public relations by providing formal methods of measuring effectiveness of organizational communication.

But however practical these theories might have been in the age of fixed organizational structures of the past, with the emergence of Internet and rapid expansion of mobile technologies, experts from both the industry and academia believe that these models won't hold water in the new complex digital environment. The functional perspective takes public relations as merely tactical and technical practice and totally disregards uncertainty and emerging complexity of the social landscape. Lee Edwards and Caroline Hodges in the introduction to *Public Relations, Society and Culture* (2011) claimed that the public relations scholarship is taking a turn towards the "socially constructed nature of practice, process and outcomes" that is supposed to more accurately capture the complex and intertwined social environment of the digital age than the dominant "Grunigian approach." What then follows is a major reconsideration of older theories and models in the light of new technologies.

MAPPING THE FUTURE

Whether the Excellence theory and the organization-centered "Grunigian approach" have really become obsolete in the new age or, in the opposite, have approached the point when they must finally be practiced in their truly symmetric nature, is still a question of the theoretical perspective. Grunig (2009), in response to the criticism, argued that it is not the public relations theory and those articulated "generic principles of public relations" that need to be changed in the face of the new media, but rather application of these principles by public relations practitioners.

The essence of this dispute is in any case unchanged – the old functional paradigm shows its age and by now has been almost completely replaced by the new one. The co-creational view on public relations is set to overcome limitations of the functional perspective and to engage with dynamically emerging publics of the new age. But even though the academics have largely agreed upon the new paradigm, practitioners are still slow to respond to the changes. As some studies show, public relations practitioners still work within the functional perspective. Instead of changing their understanding of public relations as a strategic management discipline and communication as a socio-cultural construction that is deeply rooted in its social context, they still devise older techniques but in the new context (Grunig, 2009; James, 2008).

Some key concepts of the new co-creational perspective include situated public relations research, co-orientation theory, rhetorical-organization theory, and organization-public relationships to name only few. Many of them have been in place since the late 1980s, but their profoundly qualitative and dialogic nature has only recently started taking shape and significance. Grunig's symmetrical excellence theory is also among them and still dominates the landscape (Botan & Taylor, 2004). Among others, complexity theory shows a huge potential as a research strand applied to public relations in the uncertain and dynamic environments of the modern age (Gilpin, 2010; Gilpin & Murphy, 2010; Murphy, 2000; Murphy, 2013; McKie, 2005).

But these are research strands and theories that have been in place for quite a while. The future as we see it today is mobile. So, how is the ongoing mobile revolution different from the dot-com revolution of the 1996-1999? A Morgan Stanley 2009 report (Meeker, 2009) on the state of technology argued that with tech cycles lasting on average around 10 years, we had already transitioned from the desktop Internet computing cycle to mobile Internet computing. Their prediction that users will be more likely to browse the Internet via mobile devices than desktop PCs in five years from then can be observed today in some countries of the world including India, Saudi Arabia and most of the African states (StatCounter Global Stats, 2014), where mobile browsing already overcame the desktop Internet use. In the USA the mobile Internet use increased by 73% in just one year.

But what do these changes have to do with communications and how can public relations be affected? Pew Research (2014) refers to mobile as to one of the three major technology revolutions that not only changes how people allocate their time and attention, but also "the way people think about how and when they can communicate and gather information by making just-in-time and real-time encounters possible."

James Fergusson from the Digital & Technology Practice at TNS reported that traditional voice calling is now receding into the background being replaced by a number of mobile value-added services. According to him, "with consumers using their phones for web browsing, social networking, mobile commerce, video and music, even controlling the home, it seems increasingly anachronistic that mobiles are referred to as phones at all" (PR Moment, 2012).

Key new research areas, thus, include focus on personalization, new audience's behaviors and attitudes, as well as their use of mobile apps. Some examples are studies of gamification and location-based mobile applications, mobile videos and even new publics. The research bridging mobile social media and crisis communication is also proliferating.

Relationship Management in the Age of Mobile

Relationship management is all about trust building. Yet, trust building is not new to the public relations and business research. The 80% of sales volume coming from the 20% of the sales force are to a great extent to the trust-based relationships that they built with their clients (Urban, Sultan, & Qualls, 2002). Trust was also one of six elements used in measuring organization's long-term relationships (Hon & Grunig, 1999). Similarly, it was one of three dimensions in the organization-public relationship (with openness and commitment to maintain a relationship) found to be "strongly related" to a consumer decision to stay with a current service provider as compared to switching to a competitor (Ledingham & Brunning, 1998) on relationship management in public relations. Trust and other constituents of quality organization-public relationships are integral to the concept of a dialogue (Pieczka, 2011). And as the concepts like symmetrical communication, relationship management, and corporate social responsibility are gaining weight, the concept of a dialogue is taking the stage again.

Yet the philosophical meaning of dialogue is not equal to the two-way symmetrical communication that is widely accepted in the public relations theory (Theunissen and Noordin, 2012). The nature of the dialogue is quite abstract, which makes it difficult to operationalize. To resolve this difficulty, Kent and Taylor (2002), conceptualized dialogue as an orientation system and came up with five tenets of the dialogic approach in public relations: mutuality, propinquity, empathy, risk, and commitment. The principle of *mutuality* recognizes both an organization and its publics as equal actors. *Propinquity* is a temporal feature that refers to immediacy and spontaneity of communication. *Empathy* characterizes truly dialogic organizations that are empathetic to their publics and act in the publics' best interests. With these three tenets in mind, it is yet very challenging to fully adopt the dialogic orientation, due to the *risk* of losing control over the communication outcomes. Risk is an important tenet to be aware of, since it is an indispensable part of communication between two equal sides. Thus, *commitment* to keep the dialogue developing is extremely important. This means to maintain the organization-public relationship on the basic principles of the dialogic theory and to adhere to those principles.

Social media theorist Clay Shirky's TED presentation (2009) about social media summarized the change that has occurred in professional communications.

We are increasingly in a landscape where media is global, social, ubiquitous and cheap. Now most organizations that are trying to send messages to the outside world, to the distributed collection of the audience, are now used to this change. The audience can talk back. And that's a little freaky. But you can get used to it after a while, as people do. But that's not the really crazy change that we're living in the middle of. The really crazy change is here: it's the fact that they are no longer disconnected from each other, the fact that former consumers are now producers, the fact that the audience can talk directly to one another; because there is a lot more amateurs than professionals, and because the size of the network, the complexity of the network is actually the square of the number of participants, meaning that the network, when it grows large, grows very, very large. As recently at last decade, most of the media that was available for public consumption was produced by professionals. Those days are over, never to return. ... Media, the media landscape that we knew, as familiar as it was, as easy conceptually as it was to deal with the idea that professionals broadcast messages to amateurs, is increasingly slipping away. In a world where media is global, social, ubiquitous and cheap, in a world of media where the former audience are now increasingly full participants, in that world, media is less and less often about crafting a single message to be consumed by individuals. It is more and more often a way of creating an environment for convening and supporting groups. And the choice we face, I mean anybody who has a message they want to have heard anywhere in the world, isn't whether or not that is the media environment we want to operate in. That's the media environment we've got. The question we all face now is, "How can we make best use of this media? Even though it means changing the way we've always done it."

Shirky foresaw in 2009, before the revolutionary emergence of mobile, that dialogue was to become the primary form of online communications. He noted that among the many imaginative uses by Obama campaigners in 2008 was the recognition of these vital characteristics of dialogue. After they established MyBarackObama.com (myBO.com), millions of supporters flocked in not just to support his campaign but to share their ideas – to dialogue. That summer he announced that he had changed his mind and would support the Foreign Intelligence Surveillance Act (FISA), even though a large portion of his supporters saw it as a system permitting warrantless spying on American citizens. Great debate and criticism occurred on Obama's own website, but eventually critics lauded the fact that his team never sought to hide, block or stifle the dialogue, as much as it might have been perceived to be detrimental to his election efforts. "They had understood that their role with myBO.com was to convene their supporters but not to control their supporters" said Shirky. "And that is the kind of discipline that it takes to make really mature use of this media." In other words, *commitment* is the key final ingredient to today's successful online communications (Figure 1) (Shirky, 2009).

The dialogic theory is at the heart of mobile communication. The study of the use of social media by the Red Cross shows strong compliance with the aforementioned dialogic principles (Briones et al., 2011). Mobile devices now allow all parties to overcome possible communication barriers, making it easier to publish content on the go and to communicate with publics instantly – thus committing to the *propinquity* principle. And with mobile communication permeating almost all aspects of our lives, incorporating the dialogue into mobile public relations is essential for effective relationship management.

Mobile and Corporate Social Responsibility (CSR)

Jones and Bartlett (2009) placed corporate social responsibility (CSR) within the framework of relationship management as a corporate activity that facilitates relationship building through the management of public's perceptions about an organization. Ubiquity and interactivity of mobile technology has also shaped this sector.

Figure 1. President Obama talks with first-graders at Tinker Elementary School on September 17, 2014 (Official White House)

With its mobile campaign "Breathe the Change," Halls raised consumer "breaths" to power a windmill in a small village in India and provided its inhabitants with electricity (MMA Global, 2013b). Launched on Earth Hour when all lights were off and lasting for 75 days, Halls collected 200,000 breaths from the company's consumers with 147,000 breaths coming from mobile contributions, or more than 70%. The company used technology of Interactive Voice Response (IVR) to gather and measure breaths of consumers via their mobile devices and a toll-free number. Callers could call the number and breathe into their mobile phones to leave a recorded message with their breaths.

An example of the "Text for Haiti" campaign showed that mobile communication reinforces more favorable perceptions about campaigning. A study by Weberling and Waters (2012) reported that those respondents who donated via text messaging possessed more positive attitudes than those who did not use mobile devices to donate. Similarly, the Swedish Committee for Afghanistan (SCA) that ran a text messaging campaign "Donate a Word" raised funds to support one year of schooling for over 1,250 girls in Afghanistan (MMA Global, 2013a). In cooperation with the Swedish writer Björn Ranelid, the organization launched a collaborative story that was written in 14 days using words donated as SMS. Donors could select words to use in a book and send them as a text message. A gradually unfolding story was available on a responsive website of the campaign, and all participants received a reply message with a link to the chapter where their selected word was used. With a stunning conversion rate of 28%, it transformed from a one-time campaign to an annual fund-raising event with new authors supporting the campaign each year.

Mobile and Crisis Communication

Coombs and Holladay (2001) integrated the relational management perspective with the symbolic approach to crisis communication, focusing on the history of an organization's relations with its publics and its influence on its publics' perceptions of a crisis and the organization itself. An effect termed as the Velcro effect suggested that historically formed negative relationships strongly affected the organizational reputation and perceptions of crisis responsibility.

Mobile technologies can have a mediating effect on crisis communication by involving publics in disseminating information about a crisis and in helping organizations to react to crisis situations. According to Veil, Buehner, & Palenchar (2011), "firsthand reporting by people on the scene, possessing nothing more than a cell phone, provides almost instantaneous news which then spreads rapidly among peoples' networks of contacts and friends." From sharing locations of spotted wildfires (Sutton et al., 2008) to responding to the earthquake in Haiti (Smith, 2010), user involvement in crisis communication is argued to be decisive in successfully fulfilling the functions of organizational public relations.

The best practices of organizational communication during crises involve a number of guidelines. Many of them touch upon the five tenets of the dialogic approach by Kent and Taylor discussed earlier. To illustrate this, let's compare the best practices summarized in Veil, Buehner, & Palenchar (2011, pp. 111-112) with the five tenets of the dialogic theory. As shown in Table 2, each of the articulated best practices corresponds to at least one of the tenets of the dialogic theory. For instance, mutuality was central to half of the practices, even prior to a crisis situation.

Table 2. Tenets of the dialogic approach compared to the best practices summarized by Veil, Buehner, & Palenchar

Tenets of the Dialogic Approach	Best Practices in Crisis Communication
Mutuality	*Collaborate and coordinate with credible sources. In addition to the development of relationships with the public, another best practice is to develop and maintain strong relationships with credible sources before a crisis. Gathering and disseminating accurate and consistent messages requires continuously validating credible sources, finding experts in proper areas, and creating relationships with them.*
Propinquity	*Meet the needs of the media and remain accessible. The public typically learns information about a crisis or risk from the media, and thus, remaining accessible to these outlets is crucial. The dissemination of accurate messages through crisis spokespersons identified before a crisis should be practiced continuously with media.*
Empathy	*Communicate with compassion, concern, and empathy. Spokespersons should demonstrate genuine concern for the situation and create a sense of legitimacy with the public. Humans respond well to other humans. Recognizing emotions as legitimate allows for an open exchange that humanizes the crisis response.*
Risk	*Accept uncertainty and ambiguity. Waiting until all information is known before responding to a crisis can put stakeholders and the organization in danger as other, potentially less credible, sources tell the story of the crisis. On the other hand, if a falsely reassuring statement is issued and later deemed false or exaggerated, the spokesperson risks losing trust and credibility.... Accepting uncertainty and avoiding overly confident statements allows the spokesperson to adjust messages as more information becomes available.*
Commitment	*Provide messages of self-efficacy. The final best practice outlined by the NCFPD involves allowing stakeholders in a crisis situation to gain a sense of control through meaningful actions that promote a sense of self-efficacy. ... Organizations should offer these recommended actions clearly and consistently and provide an explanation as to why the action is recommended.*
Mutuality Propinquity	*Partner with the public. During a crisis, the public has the right to know about the risks they face. Communicators should be deliberate in sharing available crisis information in a timely and accurate manner. Crises create a need for information; providing information may ease uncertainty. In addition, the public can provide essential information and assistance to mitigate the crisis.*
Mutuality Risk	*Listen to the public's concerns and understand the audience. Not only is it important for an organization to listen to the public; it is also imperative to act upon the concerns of risk and uncertainty and to establish dialogue before a crisis, regardless of whether the perceived risk is manifested.*
Mutuality Commitment Propinquity	*Communicate with honesty, candor, and openness. Sharing available information openly and honestly before and during a crisis is vital in minimizing additional threats as well as meeting the public's need for information so they do not turn to other sources. Once an organization is no longer considered a source of trustworthy information, management of the crisis is lost.*

Sources: Kent & Taylor (2002) and Veil, Buehner, & Palenchar (2011, pp. 111-112)

New Patterns of Building Relationships with Publics

Since the central idea of relationship management is building trust and connecting bridges between an organization and its publics, it would be worth mentioning significant changes that the latter underwent in recent years. Traditionally, when many people – even many organization executives – think of public relations, they think primarily of media and consumer relations. They think of using the media to favorably influence consumers. And while media and consumer relations have been the primary focus of this chapter, as well, it is important to also consider the impact of mobile technologies on other "external publics" and such "internal publics" as employees, investors, suppliers and distributors.

On the side of external publics, mobile technologies opened up new opportunities for community relations and connecting with action groups. Beyond mere social media activities, mobile technologies enable organizations to address community building in a number of novel ways. A telecommunications provider Vodafone launched mobile services that helped people in emerging markets without bank accounts to use basic banking services via the company's M-Pesa service (Vodafone, 2015). Among their other community-building initiatives were mobile health services to connect health workers with patients from distant locations, and a mobile learning platform "Learning with Vodafone" in India. Cisco (2014) in a similar fashion reported its support for the TaroWorks program, an initiative by the Grameen Foundation that utilizes "mobile technology to help social enterprise clients collect real-time data, manage field operations and customers, and measure impacts."

Instant connectivity and ubiquitous mobile technologies have also transformed such internal publics as shareholders and investors and forced investor relations towards more transparency. A study of a number of A-share companies listed on Chinese Shenzhen Stock Exchange by Bai et al. (2014) demonstrated empirical evidence that "a certain online/mobile information disclosure level may have different influence on the firm value in terms of different corporate performance levels." Argenti (2006) argues that borders between different publics are gradually blurring, and a message that was originally directed at employees can be spread to unexpected constituencies, like journalists, investors or financial analysts. An example of that was the case of Groupon in 2011. Groupon canceled its investor roadshow and put its IPO plans on hold one week before the scheduled date, supposedly after a leaked internal email of the company's CEO to its employees attracted attention of the industry regulator – the Securities and Exchange Commission (SEC) – and the news got into the press (Raice, 2011). In this case one email simultaneously affected at least four different internal and external publics – employees, investors, regulators, media, and even its customers, since the IPO was targeted at raising cash to better serve its customers.

Shifting Patterns of Internal Relations

Poor investor relations can leave an organization underfunded or its officials under attack from shareholders. Distributors, dealers and retailers are on the front line of sales for many B2C-oriented organizations and need constant attention. And there is no public more important to the success of any organization than its own employees. On one extreme, strong, supportive employees can lead an organization to unparalleled success, as seen in such recent examples as Facebook and Google. At the other extreme, organizations with labor union and other employee problems can literally be destroyed from within. Within the communications industry, major American newspapers have ceased operation despite many millions of dollars in monthly revenues and hundreds of thousand of subscribers. In a situation when newspaper are still losing money every month, seeing advertising revenues continuing to shrink as advertising moves to Internet-based media, the company must gain employee and labor union support in downsizing personnel since payroll is its single greatest expense. Poor, uncooperative relations between labor and management have killed many companies across the entire economy. But on a practical organizational level, enhanced internal communications through digital and, specifically, mobile technologies can improve cooperative efforts, innovation and productivity internally and external relations, as well.

A study by McKinsey Global Institute (Chui, Manyika, Bughin, Dobbs, et al, 2012) focused on how communications could be enhanced by social technologies, which are increasingly mobile in nature. The benefits for internal communications, they concluded, could be more valuable than new communication strategies with consumers. The potential, the report found, could raise the productivity of "interaction workers – high-skill knowledge workers, including managers and professionals — by 20 to 25%" (p. 1). Such employees, it noted, already spend an average of 28 hours a week in some form of collaborative communications that can be enhanced by up to 100% by social technologies. Of course it is impossible to precisely calculate the value of synergistic communications achieved "by streamlining communication, lowering barriers between functional silos, and even redrawing the boundaries of the enterprise to bring in additional knowledge and expertise in 'extended networked enterprises.'" And besides enhancing on-the-job communications, social technologies facilitate "cognitive surplus" when people contribute valuable ideas, knowledge and experience voluntarily during their leisure time (p. 2-4). In summary, the ways the McKinsey study found in which social technologies can create value internally include (p. 8) product co-creation and the use of social to forecast and monitor and to distribute business processes. Social technologies were also argued to have a potential in improving intra- and inter-organizational collaboration and communication and in matching talent to tasks.

Despite what seemed to be a fairly comprehensive study, the McKinsey report is already outdated and incomplete – in part because of the rapid spread of mobile technologies and development of new mobile apps since its publication. The Mobile & PR handbook produced by the Chartered Institute of Public Relations (Wilkinson, Miller, Burch & Halton, 2014) offered a more up-to-date perspective. Whole new dimensions of organizational communication can be created by replacing desktop telephones with smartphones. Creating an appropriate equipment agreement with employees is a challenge, but the benefits can be enormous. Should a company have a "bring your own device" (BYOD) policy or a "choose your own device" (CYOD) policy in which the company still owns the equipment? Those are details that can be worked out if the benefits are deemed adequate (p. 14).

Some of the benefits of making internal communications mobile are obvious for employees who themselves are relatively mobile, based on their organizational duties, such as many executives, external sales staff, engineers, factory operatives and retail employees. Other employees, especially of younger generations, would simply choose to be mobile technologically, even when working within a fairly constrained suite of offices. Even someone stuck at a desk most of the day can benefit from the power of a smartphone or tablet in collaborative and organizational terms, only some of which were considered in the McKinsey study. Such enterprise social network (ESN) apps as Jive, IBM Connections, Chatter and Yammer now facilitate the kind of collaborative communication in ways that McKinsey could only imagine. Such collaborations and internal communication can increasingly be enhanced by live video, as well as by recorded video created quickly with new apps that allow broadcast quality and quick editing from the latest mobile devices (Wilkinson, et al, p. 15-17).

The communication research organization Melcrum (2015) has developed dozens of case studies of organizations that are trail-blazing the new frontiers of internal relations. HP, for example, used to provide employees with interactive information from a wide variety of sources, at any time, from any device. This was fairly comprehensive but time-consuming to create and impossible to keep up to date. Slow systems of formalized communication did not work for a cutting-edge company in a rapidly changing industry. Using HTML5 for universal access, HP created HP News Now to integrate corporate news with employees' personal news feeds. They found that within three months, employee engagement with internal communications more than doubled. Melcrum (2014) suggests that the very nature of internal communications has dramatically shifted over the past 75 years. In the 1940s it focused on entertaining, in the 1950s on informing, in the 1960s on persuading, in the 1970s it shifted to engaging, in the 1980s it moved to managing change, in the 1990s to driving dialogue, but now to achieving greater productivity through collaboration, in many cases by using social technologies across all platforms. The fundamental premise of their study is that empowered employees can be a company's most valuable asset (p. 4).

Organizations succeeding in today's changing domain share two vital strategies in common: they're not merely engaging employees, they're actively empowering them. And they're connecting those empowered employees with each other in authentic, dynamic and creative ways so they can work together to bring about exceptional business results. These include increasing customer satisfaction by over 40 percent; improving profitability by nearly 30 percent; boosting overall performance by 36 percent and many other powerful measures.

The most important five internal communication innovations found in Melcrum's (2014, p. 6-10) international study included introduction of agile processes and cross-functional teams to improve planning and organizational flexibility, driving dialogue and collaboration to engage employees, and investment in internal social media platforms to increase performance and eliminate time waste on activities like email reading. The new technological age keeps redefining competencies within internal communications:

We're now as much organizational connectors as tactical experts, expected to encompass traditional functional responsibilities along with new competencies—and the change curve is getting steeper. ... Research indicates networks and collaboration, together with project and process management, are those internal communication capabilities most in-demand.

These studies primarily focused on employees, but it is easy to see that other internal publics such as investors, suppliers and distributors are stakeholders who would also have a vested interest in contributing to the success of an organization. That "cognitive surplus" available when people voluntarily contribute valuable ideas, knowledge and experience could come from these other internal sources. And with the use of social technologies, these other publics could feel more connected to the organization, more sympathetic to its challenges and more supportive in achieving solutions.

Across its various findings the Melcrum (2014) study identifies another distinct area in which mobile communications can be essential in achieving positive internal relations. Innovation had to begin in some cases with related training so that participating employees could contribute from a sound base of understanding. And many organizations are realizing that the core competencies of successful employees now are different than when they were hired. The need to provide ongoing, just-in-time training is even broader than suggested by Melcrum, however. A study by Forrester Research and the Business Marketing Association (Ramos, 2013) found that 97% of business-to-business marketing leaders around the world found themselves doing new tasks not previously expected. These new skills were deemed to be desperately needed going forward, and at this time of rapid technological change new unanticipated skills would likely be demanded in the future. Many of respondents felt like they were "navigating chaos" in trying to adapt to change. More than one-third felt "overwhelmed." More than 20% said the skills for which they were hired are now obsolete. About 60% were looking for younger employees they hoped would better understand the new technologies, but 47% said they could not fill those positions with adequately trained applicants. Meanwhile, without increased budgets, 85% said they were being asked to perform non-marketing tasks, and over 75% said their new duties included greater input on corporate strategy and greater voice in executive decision-making.

While employee training and development should be discussed in more detail elsewhere, it overlaps into employee relations and jeopardizes all the positive outcomes previously described. If other employees feel even a small portion of the frustration expressed by these marketing leaders, it means many employees, especially older ones, will feel frustrated and will resist positive change related to new technology. Frustration is incompatible with positive relations. Beer and Nohria (2000) noted that the four major obstacles to effective change are changing mindsets, complexity, resource constraint and know-how. Organizations may face several if not all of these obstacles as they navigate the chaos of disruptive innovation, but social technologies, in general, can help overcome resistive mindsets, and mobile technologies can provide the just-in-time training and support when employees face new tasks. No matter where employees are and what new challenges they encounter, mobile can provide support via searchable, detailed text and graphics, video demonstrations, or live support. In such a way employees can keep up with rapidly changing roles and tools, confident that the organization "has their back" and wants to help them be successful.

Public Segmentation to Enhance Relationship-Building

The study of new publics and their segmentation can help in better understanding new attitudinal and communication patterns to find grounds to connect with changing audiences. In this sense traditional methods of public segmentation based on simple demographics are not enough. A study on consumer readiness for mobile marketing via smartphones (Persaud & Azhar, 2012) showed that consumers' willingness to engage with mobile marketing campaigns were affected not only by age, gender and education, but rather by individual behavioral variables like trust, perceived value and consumer's shopping style.

An in-depth look into the customer segmentation brings up a multitude of behavioral attributes and media preferences that shape individuals' attitudes and spending patterns. Ernst&Young's study (2013) of how consumers use mobile data in the "mobile maze," a concept to describe a fast-paced market rolling out in the sophisticated technological environment, argued that customer segmentation brings opportunities to extract additional value from certain customer groups. For example, young urban consumers were found to be the most active smartphone users, yet, not necessarily conspicuous spenders. To find high-value segments, Ernst&Young argued for micro-segmentation that is localized to customer preferences in specific regions. In this case, such segmentation should be based not only on the age and location, but also on the type of services and purchasing patterns customers prefer. In Australia, 3G smartphone users opting for postpaid bill payment were, on average, higher spenders than prepaid users. However, taking a narrow age range of 18-24 showed that within this specific group prepaid users were, in the opposite, slightly outspending the postpaid group.

But even looking at broader customer groups characterized by generations can help better understand the new publics. Already by 2012 40% of the generation born into the digital age, the millennials, were using smartphones to access news, whereas only 12% of the boomers were doing so (Mutter, 2012a). While in 2014 older generations still preferred to watch video on PC and TV, half of the 18- to 34-year-old Netflix subscribers were already watching most of their videos on mobile devices (ComScore 2014). "For [millennials], technology is part of their everyday lives, not something in the workplace. … Their interactions increasingly tend to take place on smartphones and tablets," wrote Wilkinson, Miller, et al. (2014). They further cited the Project Tomorrow study of 3.4 million American schoolchildren, which found that by ages 8-11 more than half already used a smartphone or tablet, which number increased to 89% by age 14-18. By age 11 almost half of the children were using mobile social media apps such as Instagram, Snapchat and Vine, and one-third were specifically using Twitter.

Using Video to Connect with Mobile Publics

Facilitated by more powerful Internet devices, improved online server software, increasing bandwidth throughout the Internet, and rapidly spreading mobile broadband, in one year (June 2011 to June 2012) the number of videos viewed per month in America jumped 550%, from 6 billion to 33 billion (xStream 2013). The video revolution spread worldwide so that by August 2014 monthly video views surpassed 300 billion (Nguyen, 2014). Mobile broadband enhanced the video revolution because it rapidly increased the number of Internet devices capable of viewing video; because video was perhaps the easiest content to convert from PC to mobile, whereas most websites are still not fully mobile accessible; and because mobile users, especially the younger generations that are most mobile, have a strong preference for video content. The video trends are also driven by advertisers, 90-95% of whom believe video is the most powerful form of advertising (Aberdeen, 2014), and it is further supported by the fact that among the traditional media, TV has been the most favored content provider for decades and is still the only one still holding its own against the onslaught of online media (eMarketer, 2013). These intertwined trends will continue for the foreseeable future. Cisco (2014) projects Internet growth from 2013 until 2018 and estimates:

- *In 2013, only 33 percent of total IP traffic originated with non-PC [mostly mobile] devices, but by 2018 the non-PC share of total IP traffic will grow to 57 percent.*
- *By 2018, wired devices will account for 39 percent of IP traffic, while Wi-Fi and mobile devices will account for 61 percent of IP traffic. In 2013, wired devices accounted for the majority of IP traffic at 56 percent.*
- *It would take an individual over 5 million years to watch the amount of video that will cross global IP networks each month in 2018. Every second, nearly a million minutes of video content will cross the network by 2018.*
- *Internet video to TV doubled in 2013. Internet video to TV will continue to grow at a rapid pace, increasing fourfold by 2018.*
- *Consumer VoD traffic will double by 2018. The amount of VoD traffic by 2018 will be equivalent to 6 billion DVDs per month.*

So, what do these strong trends mean to public relations? Think video. An extensive analysis by the SocialBakers social marketing agency (Ross, 2015) showed that video posts now have by far the most reach. Companies still use photos (and graphics) most frequently to appeal to their fans, but the analysis showed they are actually the least effective type of posting in achieving reader response. The study looked at how many of 670,000 posts were liked or shared by those visiting 4,445 different brand pages. The average organic reach was 8.7% for native Facebook videos, 5.8% for status posts, 5.3% for links including non-native video links, and only 3.7% for photos, which means that video achieved 137% more promotion than photos. A separate analysis was done just of the fans of those respective brands, and the difference was even greater. Among fans, videos achieved 148% more reach. Brands during that four-month study ending Feb. 4, 2015, were responsible for posting 27% of all native videos on Facebook, and were reaping strong ROI for their efforts.

Aware that marketing budgets follow the audience and that video is the fastest-growing format on the Internet, social networks such as Facebook, Instagram and Twitter are making changes to help them challenge Google's YouTube and Vimeo for a bigger piece of the action (Mixpo, 2015). Facebook's changes in 2014 have allowed it to double its desktop video views and overtake YouTube, according to comScore statistics. In January 2014 Facebook's audience averaged 71% as many views as YouTube's audience, but by February 2015 the Facebook audience was viewing 55% more videos than YouTube's. However, Facebook says these statistics of desktop views pale in comparison with views from all Internet devices. In fact, they claim, the desktop views represent only 10% of Facebook's total video views. By far the most views now are coming on mobile. Mixpo surveyed 125 advertising agencies to see how they would respond. While the response by advertisers might be somewhat distinct from public relations, some of the findings seem to relate to both. For advertisers, Facebook will become the top medium in 2015, according to the survey. In 2014 about 78% of them ran video campaigns on YouTube while only 63% ran campaigns on Facebook. In 2015 that is expected to reverse, with 87% planning a campaign on Facebook versus 82% for YouTube, followed by Twitter, Instagram and LinkedIn.

The appeal of social networks Facebook and Twitter over YouTube relates to available data, according to the study. Being more "people based," they allow brands to collect individual user IDs from fans and followers for re-targeting that works across all devices. They also provide more behavioral data that allows more effective lookalike modeling. In an Adweek article, Jonathan Nelson, CEO of Omnicom Digital, summarized the advantages of such data, which could also be utilized by PR practitioners (Sloane, 2014):

When you have that much known information, tied to analytics and an ad server, you can start doing messaging in a way no one's ever done before. That's marketing nirvana. Facebook is different from every other media company because they actually know who you are. We will see this whole business go to another level because of this kind of knowledge that's about actual people. It's about real behavior.

Based on their numerous studies, Brightcove (2015) prepared a list of best practices in how practitioners can enhance their use of video in public relations and marketing:

1. Create a framework for how to use video to achieve brand goals.
2. Consider the emotions you want to evoke, the personas targeted and the desired action to be taken by viewers before scripting or producing a video.
3. Make sure the video includes quality visuals involving action, graphics and/or animations – not just talking heads.
4. Consider how the video will be showcased most effectively.

Once the video is produced, there is a list of opportunities for bringing it to publics. Brightcove suggests using various channels, including email and social platforms like Facebook, Twitter and YouTube, to increase reach and engagement. Videos are also effective instruments to drive traffic to an organizational website. Videos on landing pages contribute to an 86% increase in conversion rates and to 5300% higher chance to appear on the first page of search results on Google. In addition, creating teaser videos can help in driving traffic to corporate blogs.

Sarah Quinn of the Social Media Examiner (2015) has a separate set of recommendations for using video with Twitter. For example, videos can be creatively used to share real-time events on the go and to publish educational content. Apart from that, videos can even appear as responds to tweets, thus connecting with the audience on a personal level.

Another new twist to PR/marketing videos is the combination of personalization and video strategies. According to Oetting (2015), some organizations are achieving success by creating personalized videos including social media photos of each target recipient. Personalization and video have been shown to be two of the strongest PR/marketing strategies currently in use. Personalization, according to the report, increases clickthrough rates by 14%, conversion rates by 10%, and overall sales by 19%. On the other hand, video on a landing page increases conversions by 80%. So, combining video with personalization should multiply the benefits of both. One company, Switch Video, used a program to personalize videos at scale. Testing it on their own contacts, they saw a 1025% increase in email clickthrough rates. Using it for a client, they experienced a 272% increase in email open rates when the subject line mentioned the personalized video. It seems likely that programming will be further developed to facilitate personalized video, and available data and photos from social networks like Facebook will make large-scale application of this strategy even more effective. Perhaps a Photoshop strategy will be developed to automatically replace an actor's face with recipients'.

These industry best practices in many cases can and should lead to academic research that involves more strict research standards and less likelihood of researcher bias.

Customer Relations in the Age of Mobile

Mobile Internet has established a sense of truly instant connectivity and "just-in-time" communication. Cell phone and smartphone users increasingly rely on their mobile devices when coordinating personal meetings, solving unexpected problems and getting help overcoming crisis situations, or just deciding whether to visit a coffee shop around the corner (Pew Research, 2012).

The "just-in-time" proposition is thus an important concept for public relations practitioners to take into consideration. Focused on community building, a European telecommunications provider O2 launched its Priority Moments for Small Businesses campaign in London. The company's collaboration with over 17,000 local small businesses aimed at delivering unique location-based deals to O2's mobile users (MMA Global, 2013c). As an official partner to thousands of local businesses in the area, the company, on the one hand, positioned itself as "more than a telephone provider" and as a community builder. On the other hand, O2 was able to enhance its location-based services by providing a variety of unique deals to the customers it serves.

The O2's campaign is a good example of the "Four I's" framework suggested by Andreas Kaplan (2011). Kaplan's mobile social media proposition focuses on four essential activities that companies need to take into account while building relationships with their customers via mobile: *integrate*, *individualize*, *involve*, and *initiate*. This framework standing in the core of mobile communication is also practical to operationalize the "just-in-time" proposition.

First, integrating customer-relationship building activities into the actual contexts of mobile users helps public relations practitioners to avoid nuisance and the sense of intrusion into personal mobile space. This 'I' is important to take into account, since studies show that mobile devices are generally perceived as a means for private and personal communication (Häkkilä & Chatfield, 2005; Watson, McCarthy, & Rowley, 2013). Consumers are highly sensitive and resistant to mobile marketing (Watson, McCarthy, & Rowley, 2013) and meet commercially oriented messages with "irritation, intrusion and mistrust" (Grant & O'Donohoe, 2007). Consumer resistance to unsolicited promotional messages on mobile is shown to be mediated by privacy concerns and ad irritation. And this is why personalization of messages decreases the ad avoidance (Baek & Morimoto, 2012).

The latter point leads us to the second 'I' – individualization of messages and activities with respect to users' interests and preferences. According to Sweetser (2011), "tailoring the message to a specific target public to make the experience more personal is classic public relations." The current state of technology research enables a variety of message personalization options, and many of them involve personalization based on both user's location and preferences. For instance, a personalized mobile search engine (PMSE) proposed by Leung, Lee and Lee (2013) sorts out mobile users' preferences by first capturing their location and search queries and then tying them to users' clickthrough data. The algorithm then classifies the mobile search queries to location and content concepts and subsequently ranks the relevance of search results based on this personalized data. Personalization algorithms are also well presented in digital recommendation systems for travelers. Recommendations that are based on collaborative filtering use the feedback from users with similar preferences. In contrast, content-based recommendation leverages on user's own previous experience and preferences, and knowledge-based filtering matches user's individual preferences with search items' descriptions (Husain & Dih, 2012). Some other examples employ context-aware personalization (Skillen et al., 2012) and Bayesian Networks modeling to capture individual user preferences (Park, Hong, & Cho, 2007).

The third 'I' refers to the user *involvement* through an interactive conversation and goes in line with the co-creational approach discussed earlier. The dialogic nature of public relations is essential to the modern mobile age, as it complies with the recent shift toward a new "personal communication society" (Campbell & Park, 2008). A decade-old dialogic theory of Kent and Taylor (1998, 2002) for web-based communication has now taken a new turn. The difference between symmetric communication and a dialogue lies in the nature of the two concepts. Whereas symmetric communication has a more processual character describing the way organizations can listen to publics and how they can get their feedback, dialogue is rather a product of communication and to a greater extent focuses on building and nurturing relationships. Yet the field of public relations hasn't fully grasped the concept of a dialogue (Pieczka, 2011). Kent (2013) argues that, paradoxically, an explosion in the social media technology resulted in the regression of public relations practitioners' role from "organization–public relationship builders and counselors, to marketers, advertisers, and strategic communicators."

The last 'I' implies a central role of User-Generated Content (UGC) and calls to *initiate* its creation. From YouTube videos to Instagram pictures, UGC is a major driver of the new digital media economics. Even location-based check-ins have become a growing socializing tool and thus migrated into a distinct content category. For instance, Indonesian users were found to use Foursquare check-ins as a tool for self-expression to pin locations that they have never actually been to (Halegoua, Leavitt & Gray, 2012). Initiation of UGC, according to Kaplan, leads to word-of-mouth, but is not only limited to that. A study on social media brand communities (Goh, Heng, & Lin, 2013) shows that UGC has a stronger impact on consumer purchase expenditures than market-generated content (MGC).

The "just-in-time" concept has also impacted the way people prefer to get in touch with companies. Far over half of respondents (64%) in the survey by Harris Poll for OneReach (2014) said they would most prefer companies to text message them for customer service than "waiting on hold to speak with an agent." Customer relations used to be pretty simple – be friendly and helpful when people walked into the store or office, and be patient and supportive when they contacted customer service, most typically in person or by phone. But that was before the digital age when suddenly customers gained many new channels of communication and access to many more people with whom they could communicate. The importance of good customer relations has increased with those channels of access because those channels also amplify customers' voice. It is a PR coup to convert customers into advocates or evangelists for the organization, and it can be much more detrimental to have dissatisfied customers venting their frustrations on social media.

A study by Forester Research (2013, p. 3) found that significant numbers of consumers would like to be able to access organization support through any of 10 different channels (Figure 2):

While this was the overall response rate for all 7,440 adults surveyed, younger generations were more likely to use more options. For respondents 57 or older, no option was being used by over 30% of the respondents except telephone, FAQs and email. For respondents 24-32, all 10 options were being used by over 30% of the respondents. And for respondents 18-23, all 10 options were being used by over 40% of the respondents.

The Forester study (p. 2) concluded that 75% of the time, customers who fail to get the help they seek from their first channel of choice will move to another channel, but their patience does not endure long. One of the most common activities for which they seek help is in making a final decision on making a purchase. But if they don't get the help they seek quickly, 52% said they are likely to abandon the purchase. Indeed, IBM (2015) found that over 70% of all online shopping carts were abandoned during the busiest days of the 2014 Christmas shopping season. And Forester found that these kinds of experiences are common enough that survey respondents, on average, gave only 39% of all websites an "excellent" or "good" customer experience rating. Half of all business executives surveyed by Forester say there is impetus for enhancing customer relations through technology, and 50% of them are promoting this because they know that unhappy customers will switch to a competitor (p. 7).

In fact, brand loyalty can be won or lost in just 76 seconds – and an important key in the digital age is to maintain the human touch (LivePerson, 2013, pp. 2-4, 8). Of the 6,054 adults interviewed online, 77% want online service to be provided as quickly as possible. On average, if they don't get real-time help within 76 seconds, they won't be happy, and they will seek alternatives away from the organization's website. Most will turn to email (54%), which may make the situation worse, because over half (58%) have emailed a company in the past without any response. According to the study:

Figure 2. Consumer preferences for channels to access organization support
(Adapted from Forrester Research)

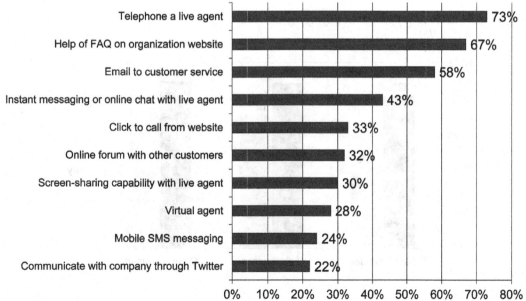

Every interaction with a brand is a defining moment – it can either drive customer loyalty, or on the contrary, lead to abandonment to a competitor. The repercussions of a negative digital experience have never been higher, and the result of a positive experience is becoming increasingly more valuable.

About three-quarters of the respondents from seven developed countries who make online purchases at least once a month said that they frequently choose one online vendor over another based on the ease of browsing and searching the site. Other key factors are previous positive experiences on the website (64%) and the availability of customer ratings or reviews (50%). When they seek help, it is most frequently to ask a question about a specific product or service (42%), to solve a problem in the purchase process itself (35%), or to request post-purchase assistance (35%). These are critical times in achieving customer loyalty and support, and if they have a positive online interaction with the company, they say, they are more likely to recommend the company to someone else (76%) or mention the company positively in social media (50%).

According to the respondents, their favorite ways to communicate with an organization online include having a live chat with an agent (87%), reconnecting with the same agent later (87%), having the option to transition to a live phone connection (81%), having the agent show an instructional video during the chat (79%), being able to share images and documents during the live chat (78%), sharing their own screen with the agent (78%), and being able to interact with the agent in a live video chat (p. 7). All of these technical capabilities are available 24/7 for as little as $45 a month (HotConference, 2015), but most organizations fail to implement them.

Besides access to live agents, simply being able to access a website in any acceptable fashion is a problem for mobile users. According to eMarketer co-founder Geoff Ramsey (2015), 86% of millennials complain that most websites are not mobile compatible, and 71% complain that most online merchants lack an app to use as a viable alternative. And yet almost as high a percentage of mobile users try to shop online as PC users – but many cannot complete their purchases using their favorite online device. Here are the statistics Ramsey offered for shopping and buying by device (Figure 3):

Figure 3. Mobile users' shopping and buying by devices
(Adapted from Ramsey, 2015)

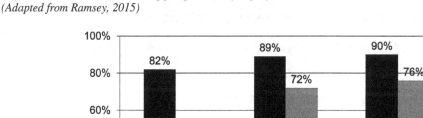

Of mobile users who are able to make purchases online, according to Ramsey, only 28% do so through websites; 72% through apps. HTML5 accommodates cross-platform adaptation of websites, but most companies have not yet implemented this programming language. A 2014 study of 100 leading consumer magazines found that even these communication companies were failing to communicate effectively (Ayres, 2014) – 93% of them failing to provide complete cross-platform access. Most chose apps over adaptable websites, and most of their apps also could not service all mobile devices.

In addition, audiences are transitioning to the person-to-person communication (James, 2008), and the rise of such instant messaging apps like WhatsApp, WeChat (in China) or Kakaotalk (in South Korea) is a compelling evidence for that. For public relations practitioners this is both a challenge and an opportunity to connect with their publics on a more personal level and engage them into active reciprocal communication.

Mobile devices are perceived to be highly personal and intimate means of communication and are usually used for one-to-one communication that necessarily incorporates the concept of interactivity. Studies show that higher degree of interactivity on the side of an organization can positively impact perceived quality of an organization-public relationship (Ballantine, 2005; Saffer, Sommerfeldt, Taylor, 2013; Voorveld, Van Noort, & Duijn, 2013). This effect is true for both functional and contingency interactivity. While the former type of interactivity focuses on a structural dimension of an organization's website as of a communication channel that allows for interactive communication, the latter one refers to a dialogic interactivity that presumes interchangeable roles of an organization and its publics that engage in a reciprocal conversation.

Studies explaining these consumer preferences and reasons why they remain with mobile messaging apps mostly focus on the concept of user satisfaction. Scholars argue that it is user satisfaction coupled with perceived switching costs that prevent smartphone users from changing mobile messengers (Kim, Kang, & Jo, 2014). The user satisfaction, in turn, is influenced by concepts like self-disclosure, flow and social presence (Park, Cho, & Lee, 2014; Park & Sundar, 2015), as well as self-esteem and thought suppression (Yang & Kim, 2014). These findings show importance of understanding the reasons that underlie the effects of interactivity and personal communication on forming quality relationships.

CONCLUSION

Future Research and Evolving Theories in the Mobile Age

Established theories within the co-creational perspective seem to have a good potential in dealing with the rising complexity of the digital (and mobile) media landscape. New theories can substantially enrich our understanding of publics and the ways relationships are built in the highly mobile environment, but the general direction has been already set. What we have yet to achieve are more exploratory empirical studies.

Much of the empirical research cited in this chapter came from industry, primarily because in such a rapidly changing field, academics have no chance of keeping up. Literally every month new technologies, new applications and new programmatic opportunities are occurring. Changes in online media immediately change the nature of the messages, and directly or indirectly create changes in senders and recipients relating to attitudes, activities, habits, culture, economics and more. The opportunity for academic research seems endless, and yet, like industry research, can be quickly outdated. It is a snapshot of a rapidly flowing stream of change. For example, the SocialBakers research involving 670,000 Facebook posts on 4,445 brand sites seems like a great research contribution, discovering that video posts achieve 137% more reach than the much more common photo post (Ross, 2015). But the researchers themselves admit that this snapshot may not be reliable for practitioner implementation very far into the future. Why? The surge in video posts is new, based on changes in Facebook programming and policies. Facebook made radical changes in order to successfully overtake YouTube in daily video views. Even if nothing changes within the competitive environment (which is unlikely), will the impact of video on Facebook users remain this powerful over the coming years? The researchers themselves think not. They believe the "newness" will wear off, and the contrast with current content will change. Whereas currently most postings by brands are photos and graphics, the current success of videos will cause more to be used and a relative reduction in photos. As the number of videos greatly increases, they may achieve greater impact overall, but the impact of each individual video may decrease. That is just one example of the complexity of new media research in this dynamic era.

The five most important issues (Verčič, Verhoeven, & Zerfass, 2014) articulated by public relations practitioners in Europe within the last seven years turned out to be closely connected to the new changing environment. Three of them link directly to the industry's response to the technological change by "coping with the digital evolution and social web," "dealing with the demand for more transparency and active audiences," and "dealing with the speed and volume of information flow." The remaining two reside within the strengthening co-creational perspective "linking business strategy and communication" and "building and maintaining trust".

Industry can be counted on to continue its ongoing research to track as many of these changes as possible because the changes have great financial consequence. But academics need to think bigger picture – which may come, in part, from lots of these little snapshots. Academics could replicate many of this industry research simply to apply stricter scientific standards and to eliminate bias, but, as in the Facebook video example, perhaps they are better served exploring the inherent advantages of video in communications, how different internal and external factors may mitigate the impact of video, and in furthering the development of related theory.

While considerable changes can be possibly brought into central research strands within public relations, these paradigm shifts should be explored too. For instance, an impact of the two issues – one of digital evolution and social web and the other of increasing speed and volume of the information flow – can significantly transform some basic theories within the issue management research field. A traditional concept of the life-cycle of an issue is now becoming obsolete, since many issues explode in the matter of minutes via social media. A Twitter study of network structures and intensity of individuals' activities (Yang and Leskovec, 2010) showed that even minority publics can bring their issues to wider exposure. Active users committed to their political agenda, even if they have only few connections, are able to generate trending messages passing over the involvement of big Twitter user hubs. These results, thus, highlight a phenomenon in which "a prevailing majority opinion in a population can be rapidly reversed by a small fraction of randomly distributed committed agents."

These issues are central to public relations both as a research field and as an industrial practice. Some of them can be already managed by the existing theories and practices that have evolved in the last decade. But the most cutting-edge practices are still to be explored.

REFERENCES

Argenti, P. A. (2006). How technology has influenced the field of corporate communication. *Journal of Business and Technical Communication, 20*(3), 357–370. doi:10.1177/1050651906287260

Avidar, R., Ariel, Y., Malka, V., & Levy, E. C. (2013). Smartphones and young publics: A new challenge for public relations practice and relationship building. *Public Relations Review, 39*(5), 603–605. doi:10.1016/j.pubrev.2013.09.010

Ayres, N. (2014). Adventures in Publishing: The new dynamics of advertising. *Brand Perfect*. Retrieved 14 April 2015 from http://brandperfect.org/brand-perfect-dynamics-of-advertising-report.pdf

Baek, T. H., & Morimoto, M. (2012). Stay away from me. *Journal of Advertising, 41*(1), 59–76. doi:10.2753/JOA0091-3367410105

Bai, Y., Yan, M., Yu, F., & Yang, J. (2014). Does market mechanism promote online/mobile information disclosure? Evidence from A–share companies on Shenzhen Exchange Market. *International Journal of Mobile Communications, 12*(4), 380–396. doi:10.1504/IJMC.2014.063654

Botan, C. H., & Hazleton, V. (Eds.). (2006). *Public relations theory II*. Mahwah, NJ: Lawrence Erlbaum Associaets, Inc., Publishers.

Botan, C. H., & Taylor, M. (2004). Public relations: State of the field. *Journal of Communication, 54*(4), 645–661. doi:10.1111/j.1460-2466.2004.tb02649.x

Brightcove. (2015). *The hero's guide to video marketing*. Retrieved 1 May 2015 from http://go.brightcove.com/brightcove-video-hero-guide?cid=70114000002QvKv&pid=70114000002Qu9G

Briones, R. L., Kuch, B., Liu, B. F., & Jin, Y. (2011). Keeping up with the digital age: How the American Red Cross uses social media to build relationships. *Public Relations Review, 37*(1), 37–43. doi:10.1016/j.pubrev.2010.12.006

Campbell, S. W., & Park, Y. J. (2008). Social implications of mobile telephony: The rise of personal communication society. *Sociology Compass, 2*(2), 371-387.

Cisco Systems Inc. (2014). *Cisco Visual Networking Index: Forecast and Methodology, 2013–2018*. Retrieved 1 May 2015 from http://www.cisco.com/c/en/us/solutions/collateral/service-provider/ip-ngn-ip-next-generation-network/white_paper_c11-481360.html

Cisco Systems Inc. (2014). *2014 Corporate Social Responsibility Report*. Retrieved from http://www.cisco.com/assets/csr/pdf/CSR_Report_2014.pdf

Coombs, W. T., & Holladay, S. J. (2001). An extended examination of the crisis situations: A fusion of the relational management and symbolic approaches. *Journal of Public Relations Research, 13*(4), 321–340. doi:10.1207/S1532754XJPRR1304_03

Crawford, A. P. (1999). When those nasty rumors start breeding on the Web, you're got to move fast. *Public Relations Quarterly*, 43-45.

Edwards, L., & Hodges, C. E. M. (Eds.). (2011). *Public relations, society and culture – Theoretical and empirical explorations*. London: Routledge.

eMarketer. (2013). *US Total Media Ad Spend Inches Up, Pushed by Digital*. Retrieved 20 Feb. 2015 from http://www.emarketer.com/Article/US-Total-Media-Ad-Spend-Inches-Up-Pushed-by-Digital/1010154

Forester Research. (2013). *Your Customers Are Demanding Omni-Channel Communications. What Are You Doing About It?* Retrieved 15 April 2015 from http://www.kana.com/customer-service/white-papers/forrester-your-customers-are-demanding-omni-channel-communications.pdf?_ga=1.38271843.131230 2063.1428939602

Frith, J. (2013). Turning life into a game: Foursquare, gamification, and personal mobility. *Mobile Media & Communication, 1*(2), 248–262. doi:10.1177/2050157912474811

Gillin, P. (2008). New media, new influencers and implications for public relations. *Society for New Communications Research*. Institute for Public Relations, Wieck Media.

Gilpin, D. R. (2010). Organizational image construction in a fragmented online media environment. *Journal of Public Relations Research, 22*(3), 265–287. doi:10.1080/10627261003614393

Gilpin, D. R., & Murphy, P. (2010). Implications of complexity theory for public relations: Beyond crisis. In R. L. Heath (Ed.), *The SAGE Handbook of Public Relations*. Los Angeles, CA: SAGE Publications.

Global, M. M. A. (2013a). *"Donate a Word via SMS." 2013, Gold Winner, EMEA, Social Impact/Not for Profit; 2013, Gold Winner, EMA, Messaging*. Swedish Committee for Afghanistan. Mobiento. Retrieved from http://www.mmaglobal.com/case-study-hub/upload/pdfs/mma-2013-501.pdf

Global, M. M. A. (2013b). *"Halls' Breathe the Change: 200,000 Breaths That Changed the Life of a Village." 2013 MMA Smarties Silver Winner Social Impact/Not For Profit*. Brand: Halls. Lead Agency: AD2C - Pinnacle. Retrieved from http://www.mmaglobal.com/case-study-hub/case_studies/view/27379

Global, M. M. A. (2013c). *"Priority Moments for Independent Business." 2013 MMA Smarties Silver Winner Location Based*. Brand: O2. Lead Agency: R/GA London. Retrieved from http://www.mmaglobal.com/case-study-hub/upload/pdfs/mma-2013-592.pdf

Goh, K. Y., Heng, C. S., & Lin, Z. (2013). Social media brand community and consumer behavior: Quantifying the relative impact of user-and marketer-generated content. *Information Systems Research, 24*(1), 88–107. doi:10.1287/isre.1120.0469

Grant, I., & O'Donohoe, S. (2007). Why young consumers are not open to mobile marketing communication. *International Journal of Advertising, 26*(2), 223–246.

Grunig, J. E. (2009). Paradigms of global public relations in an age of digitalization. *PRism, 6*(2).

Grunig, J. E., Grunig, L. A., & Dozier, D. M. (2006). The excellence theory. In C. Botan & V. Hazleton (Eds.), *Public Relations Theory II*. Mahwah, NJ: Lawrence Elbaum Associates.

Grunig, J. E., & Hunt, T. (1984). *Managing Public Relations*. Fort Worth, TX: Harcourt Brace.

Häkkilä, J., & Chatfield, C. (2005, September). 'It's like if you opened someone else's letter': user perceived privacy and social practices with SMS communication. In *Proceedings of the 7th international conference on Human computer interaction with mobile devices & services* (pp. 219-222). ACM.

Heath, R. L. (2006). A rhetorical theory approach to issues management. In C. H. Botan & V. Hazleton (Eds.), *Public relations theory II* (pp. 63–101). Mahwah, NJ: Lawrence Erlbaum Associaets, Inc., Publishers.

Holtz, S. (1999). *Public relations on the Net: Winning strategies to inform and influence the media, the investment community, the government, the public, and more!* New York, NY: AMACOM.

Hon, L. C., & Grunig, J. E. (1999). *Guidelines for measuring relationships in public relations.* Institute for Public Relations.

HotConference. (2015). *What can HotConference do for you?* Retrieved 16 April 2015 from http://www.hotconference.com/members/eduken/features.php

Humphreys, L. (2013). Mobile social media: Future challenges and opportunities. *Mobile Media & Communication, 1*(1), 20–25. doi:10.1177/2050157912459499

Hurme, P. (2001). Online PR: Emerging organizational practice. *Corporate Communications: An International Journal, 6*(2), 71–75. doi:10.1108/13563280110391016

Husain, W., & Dih, L. Y. (2012). A framework of a personalized location-based traveler recommendation system in mobile application. *International Journal of Multimedia and Ubiquitous Engineering, 7*(3), 11–18.

IBM. (2015). *U.S. Online Holiday Benchmark Recap Report 2014.* Retrieved 10 April 2015 from http://www-01.ibm.com/common/ssi/cgi-bin/ssialias?subtype=WH&infotype=SA&appname=SWGE_ZZ_JV_USEN&htmlfid=ZZW03362USEN&attachment=ZZW03362USEN.PDF#loaded

James, M. (2008). A review of the impact of new media on public relations: Challenges for terrain practice and education. *Asia Pacific Public Relations Journal, 8*, 137–148.

Jones, K., & Bartlett, J. L. (2009). The strategic value of corporate social responsibility: a relationship management framework for public relations practice. *PRism, 6*(1).

Kent, M. L. (2013). Using social media dialogically: Public relations role in reviving democracy. *Public Relations Review, 39*(4), 337–345. doi:10.1016/j.pubrev.2013.07.024

Kent, M. L., & Saffer, A. J. (2014). A Delphi study of the future of new technology research in public relations. *Public Relations Review, 40*(3), 568–576. doi:10.1016/j.pubrev.2014.02.008

Kim, B., Kang, M., & Jo, H. (2014). Determinants of Postadoption Behaviors of Mobile Communications Applications: A Dual-Model Perspective. *International Journal of Human-Computer Interaction, 30*(7), 547–559. doi:10.1080/10447318.2014.888501

Ledingham, J. A., & Bruning, S. D. (1998). Relationship management in public relations: Dimensions of an organization-public relationship. *Public Relations Review, 24*(1), 55–65. doi:10.1016/S0363-8111(98)80020-9

Leung, K. T., Lee, D. L., & Lee, W. C. (2013). PMSE: A personalized mobile search engine. *IEEE Transactions on Knowledge and Data Engineering, 25*(4), 820–834. doi:10.1109/TKDE.2012.23

LivePerson. (2013). *The Connecting with Customers Report A Global Study of the Drivers of a Successful Online Experience.* Retrieved 16 April 2015 from http://info.liveperson.com/rs/liveperson/images/Online_Engagement_Report_final.pdf

McCorkindale, T., & DiStaso, M. W. (2014). The state of social media research. *Research Journal of the Institute for Public Relations, 1*(1).

McKie, D. (2005). Chaos and complexity theory. In R. Heath (Ed.), *Encyclopedia of public relations* (pp. 121–123). Thousand Oaks, CA: SAGE Publications, Inc.; doi:10.4135/9781412952545.n58

Meeker, M. (2009). The mobile Internet report setup [Presentation slides]. *Morgan Stanley.* Retrieved from http://www.morganstanley.com/about-us-articles/4659e2f5-ea51-11de-aec2-33992aa82cc2.html

Miller, G. (2012). The smartphone psychology manifesto. *Perspectives on Psychological Science, 7*(3), 221–237. doi:10.1177/1745691612441215 PMID:26168460

Miller, G. R. (1989). Persuasion and public relations: Two "Ps" in a pod. In C. H. Botan & V. Hazleton (Eds.), *Public relations theory* (pp. 45–66). Hillsdale, NJ: Lawrence Erlbaum Associaets, Publishers.

Mixpo. (2015). *The state of video advertising.* Retrieved 1 May 2015 from http://marketing.mixpo.com/acton/fs/blocks/showLandingPage/a/2062/p/p-00b4/t/page/fm/0

Moment, P. R. (2012). *The global explosion in smartphone use has important implications for PR people.* Retrieved from http://www.prmoment.com/1025/the-global-explosion-in-smartphone-use-has-important-implications-for-PR-people.aspx

Moyer, J. (2011). Mapping the consequences of technology on public relations. *Institute for Public Relations.* Retrieved from http://www.instituteforpr.org/mapping-technology-consequences/

Murphy, P. (1989). Game theory as a paradigm for the public relations process. In C. H. Botan & V. Hazleton (Eds.), *Public relations theory* (pp. 173–192). Hillsdale, NJ: Lawrence Erlbaum Associaets, Publishers.

Murphy, P. (1991). The limits of symmetry: A game theory approach to symmetric and asymmetrical public relations. *Public Relations Review Annual, 3*(1-4), 115–131. doi:10.1207/s1532754xjprr0301-4_5

Murphy, P. (2000). Symmetry, contingency, complexity: Accommodating uncertainty in public relations theory. *Public Relations Review, 26*(4), 447–462. doi:10.1016/S0363-8111(00)00058-8

Murphy, P. (2013). *Coping with an uncertain world: Complexity theory's role in public relations theory-building.* Paper presented at the annual meeting of the International Communication Association, San Francisco, CA. Retrieved from http://citation.allacademic.com/meta/p174867_index.html

Mutter, A. (2012a). *The incredible shrinking newspaper audience.* Academic Press.

Nguyen, J. (2014). *Online Video Consumption in APAC and Total Video.* Retrieved from http://www.comscore.com/Insights/Presentations-and-Whitepapers/2014/Online-Video-Consumption-in-APAC-and-Total-Video

Oetting, J. (2015). The Next Phase in Personalization? Video Marketing. *HubSpot.* Retrieved 2 May 2015 from http://blog.hubspot.com/agency/personalization-video-marketing

OneReach. (2014, August). *The high demand for customer service via text message. 2014 U.S. Survey Report*. Retrieved from http://onereachcontactcenter.com/wp-content/uploads/2014/08/High-Demand-for-Customer-Service-via-Text-Message-2014-Report.pdf

Park, M. H., Hong, J. H., & Cho, S. B. (2007). Location-based recommendation system using bayesian user's preference model in mobile devices. In *Ubiquitous Intelligence and Computing* (pp. 1130–1139). Springer Berlin Heidelberg. doi:10.1007/978-3-540-73549-6_110

Park, S., Cho, K., & Lee, B. G. (2014). What Makes Smartphone Users Satisfied with the Mobile Instant Messenger?: Social Presence, Flow, and Self-disclosure. *International Journal of Multimedia & Ubiquitous Engineering*, *9*(11), 315–324. doi:10.14257/ijmue.2014.9.11.31

Persaud, A., & Azhar, I. (2012). Innovative mobile marketing via smartphones: Are consumers ready? *Marketing Intelligence & Planning*, *30*(4), 418–443. doi:10.1108/02634501211231883

Pew Research Center. (2012). Mobile technology factsheet. *Pew Internet Project*. Retrieved from http://www.pewinternet.org/fact-sheets/mobile-technology-fact-sheet/

Pew Research Center. (2014). *Three technology revolutions: Mobile*. Retrieved on March 18, 2015 from http://www.pewinternet.org/three-technology-revolutions/

Pfau, M., & Wan, H.-H. (2006). Persuasion: An intrinsic function of public relations. In C. H. Botan & V. Hazleton (Eds.), *Public relations theory II* (pp. 101–136). Mahwah, NJ: Lawrence Erlbaum Associaets, Inc., Publishers.

Phillips, D., & Young, P. (2009). *Online public relations: A practical guide to developing an online strategy in the world of social media*. London: Kogan Page.

Pieczka, M. (2011). Public relations as dialogic expertise? *Journal of Communication Management*, *15*(2), 108-124.

Pompper, D. (2005). Game theory. In R. L. Heath (Ed.), *Encyclopedia of public relations* (pp. 360–362). Thousand Oaks, CA: SAGE Publications, Inc.; doi:10.4135/9781412952545.n179

Quinn, S. (2015). *4 Ways to Use Twitter Video for Your Business*. Social Media Examiner.

Raice, S. (2011, September 7). IPO? Groupon plans in flux. *The Wall Street Journal*. Retrieved from http://www.wsj.com/articles/SB10001424053111904537404576554812230222934

Ramsey, G. (2015). *The State of Mobile 2015*. eMarketer live webcast 25 March 2015.

Reflections of a Newsosaur. (n.d.). Retrieved from http://newsosaur.blogspot.com/

Ross, C., & Middleberg, D. (1999). *Media in cyberspace study*. Fifth Annual National Survey 1998.

Ross, P. (2015). *Native Facebook Videos Get More Reach Than Any Other Type of Post*. Retrieved 1 May 2015 from http://www.socialbakers.com/blog/2367-native-facebook-videos-get-more-reach-than-any-other-type-of-post

Saffer, A. J., Sommerfeldt, E. J., & Taylor, M. (2013). The effects of organizational Twitter interactivity on organization–public relationships. *Public Relations Review*, *39*(3), 213–215. doi:10.1016/j.pubrev.2013.02.005

Sha, B.-L. (2007). Dimensions of public relations: Moving beyond traditional public relations models. In S. Duhé (Ed.), *New Media and Public Relations*. New York, NY: Peter Lang Publishing.

Shirky, C. (2009). *How social media can make history*. TED.com talk given in June 2009; video and script. Retrieved 28 April 2015 from http://www.ted.com/talks/clay_shirky_how_cellphones_twitter_facebook_can_make_history/transcript?language=en

Skillen, K. L., Chen, L., Nugent, C. D., Donnelly, M. P., Burns, W., & Solheim, I. (2012). Ontological user profile modeling for context-aware application personalization. In *Ubiquitous Computing and Ambient Intelligence* (pp. 261–268). Berlin: Springer Berlin Heidelberg. doi:10.1007/978-3-642-35377-2_36

Sloane, G. (2014). Facebook's New People-Based Ad Technology Is 'Marketing Nirvana'. Pepsi and Intel are early testers as social net unleashes data. *Adweek*. Retrieved 1 May 2015 from http://www.adweek.com/news/technology/facebooks-new-people-based-ad-technology-marketing-nirvana-160438

StatCounter Global Stats. (2014). *Mobile internet usage soars by 67%* [Press release]. Retrieved from http://gs.statcounter.com/press/mobile-internet-usage-soars-by-67-perc

Sutton, J., Palen, L., & Shklovski, I. (2008, May). Backchannels on the front lines: Emergent uses of social media in the 2007 southern California wildfires. In *Proceedings of the 5th International ISCRAM Conference* (pp. 624-632). Washington, DC: Academic Press.

Sweetser, K. (2011). Digital political public relations. In J. Strömbäck & S. Kiousis (Eds.), Political public relations: Principles and applications, (pp. 293-313). London: Taylor & Francis.

Taylor, M., & Botan, C. H. (2006). Global public relations: Application of a Cocreational approach. In M. Distasio (Ed.), *9th International Public Relations Research Conference Proceedings: Changing Roles and Functions of Public Relations*, (pp. 481-491). Miami, FL: University of Miami and Institute for Public Relations.

Theunissen, P., & Noordin, W. N. W. (2012). Revisiting the concept "dialogue" in public relations. *Public Relations Review*, *38*(1), 5–13. doi:10.1016/j.pubrev.2011.09.006

Time Inc. (2012). *Time Inc. Study Reveals That "Digital Natives" Switch Between Devices and Platforms Every Two Minutes, Use Media to Regulate Their Mood* [Press release]. Retrieved from http://www.businesswire.com/news/home/20120409005536/en/Time%C2%ADStudy%C2%ADReveals%C2%AD%E2%80%9CDigital%C2%ADNatives%E2%80%9D%C2%ADSwitch%C2%ADDevices#

Tojib, D., Tsarenko, Y., & Sembada A. Y. (2014). The facilitating role of smartphones in increasing use of value-added mobile services. *New Media and Society*, 1-21.

Trautman, E. (December 8, 2014). Five online video trends to look for in 2015. *Forbes Groupthink*. Retrieved from http://www.forbes.com/sites/groupthink/2014/12/08/5-online-video-trends-to-look-for-in-2015/

Veil, S. R., Buehner, T., & Palenchar, M. J. (2011). A Work-In-Process Literature Review: Incorporating Social Media in Risk and Crisis Communication. *Journal of Contingencies and Crisis Management*, *19*(2), 110–122. doi:10.1111/j.1468-5973.2011.00639.x

Verčič, D., Verhoeven, P., & Zerfass, A. (2014). Key issues of public relations of Europe: Findings from the European Communication Monitor 2007-2014. *Revista Internacional De Relaciones Públicas, 4*(8), 5–26.

Vodafone. (2015). *Driving economic and social development.* Retrieved on April 30, 2015 from http://www.vodafone.com/content/index/about/sustainability/our_vision.html

Voorveld, H. A., Van Noort, G., & Duijn, M. (2013). Building brands with interactivity: The role of prior brand usage in the relation between perceived website interactivity and brand responses. *Journal of Brand Management, 20*(7), 608–622. doi:10.1057/bm.2013.3

Waddington, S. (2013). A critical review of the Four Models of Public Relations and the Excellence Theory in an era of digital communication. *CIRP Chartered Practitioner Paper.* Retrieved from http://wadds.co.uk/2013/06/18/cipr-chartered-practitioner-paper-grunig-and-digital-communications/

Watson, C., McCarthy, J., & Rowley, J. (2013). Consumer attitudes towards mobile marketing in the smart phone era. *International Journal of Information Management, 33*(5), 840–849. doi:10.1016/j.ijinfomgt.2013.06.004

Weberling, B., & Waters, R. D. (2012). Gauging the public's preparedness for mobile public relations: The "Text for Haiti" campaign. *Public Relations Review, 38*(1), 51–55. doi:10.1016/j.pubrev.2011.11.005

Wilkinson, P., Miller, R., Burch, D., & Halton, J. (2014). *Mobile & public relations.* Retrieved 29 April 2015 from http://www.cipr.co.uk/sites/default/files/CIPRSM%20Guide%20-%20Mobile%20and%20PR.pdf

Witmer, D. (2000). *Spining the web: A handbook for public relations on the Internet.* New York, NY: Longman.

Xstream. (2013). *Online video trends & predictions for 2013.* Retrieved from http://www.xstream.dk/content/online-video-trends-predictions-2013

Yang, H.-C., & Kim, J.-L. (2013). The Influence of Perceived Characteristics of SNS, External Influence and Information Overload on SNS Satisfaction and Using Reluctant Intention: Mediating Effects of Self-esteem and Thought Suppression. *International Journal of Information Processing & Management, 4*(6), 19–30.

Yang, J., & Leskovec, J. (2010, December). Modeling information diffusion in implicit networks. In *2010 IEEE 10th International Conference on Data Mining (ICDM),* (pp. 599-608). IEEE. doi:10.1109/ICDM.2010.22

Chapter 14
Business and Mobile:
Rapid Restructure Required

Nygmet Ibadildin
KIMEP University, Kazakhstan

Kenneth E. Harvey
KIMEP University, Kazakhstan

ABSTRACT

This chapter will explore the peculiarities of business applications of mobile technologies, including a short history and a review of the current state of affairs, major trends likely to cause further change over the coming years, key theories and models to help understand and predict these changes, and future directions of research that may provide deeper scientific insight. M-commerce has many aspects from design and usability of the devices to monetization issues of mobile applications. M-enterprise is about drastic changes in internal and external communications and efficiency in the work of each business unit. M-industry reviews the impact of mobile technologies on traditional industries and the development of entirely new industries. M-style is how our everyday lives are changing in behavior, choices and preferences. After reading this chapter you will be able to differentiate m-business in many important areas: why is it important, where it is going, what is the value to consumers.

INTRODUCTION

Mobile technologies influence business in many different and sometimes unexpected ways. Mobile technologies influence public relations (including media, investor, employee, supplier, government and customer relations), marketing, accounting, banking, human resources, training, employee safety, supply chain, product development, and many other aspects of business and the entire environment in which the business operates (Ciaramitaro, 2011).

Several preconditions were necessary for mobile technologies to develop so fast and to become a norm in everyday life and business. Mobile technologies started to develop quickly when Internet connections became easy and massive volumes of information could move without difficulties. Then the

DOI: 10.4018/978-1-5225-0469-6.ch014

processors in mobile devices became powerful and affordable enough to work with these volumes of information for pictures, programs, video and audio contents. Software and applications followed the power of the servers and hardware, as providers of the devices created relatively low barriers for new applications to enter the market. So, mobile technologies emerged as a result of the combination of all these factors. Mobile technologies became fast, reliable, safe, cheap and convenient for customers. As infrastructure developed, mobile technologies created new opportunities, uncertainties and for some sectors they presented a clear threat, as do many innovations even in classic Schumpeterian terms (Corrocher & Zirullia, 2004; Perry, 2013).

Mobile technologies are relatively cheap for businesses or sometimes are even free in comparison with traditional tools and channels used to get products and services to market. It gives new advantages for small- and medium-size businesses with good new ideas for competing with giants in the marketplace (Boston Consulting Group, 2014).

MOBILE TECHNOLOGY ISSUES

Mobile technologies' rapid development brings forward certain issues regarding legal regulation, ethics, privacy and even personal safety. It has changed life styles and everyday communications (Parker, 2012; Zeichner, et al, 2014; Thayer, 2014). Mobile technologies are universal tools to reach customers and stakeholders because almost everyone has either a smartphone or opportunity to buy one, as prices are constantly declining.

The main issue for business has been how to monetize mobile technologies and deliver better value to customers. Are they tools to improve business or are they revolutionizing the whole concept of business? We would assume that mobile technologies for mature businesses create additional values as they are the continuation and substitute of e-business in a new mobile ever-accessible form, and they create a new layer of business organization and opportunities.

Fast changes in mobile technologies – when every year producers create new, powerful devices – are a challenge for all. Not everyone is able to follow the changes and features of each new model or device. The maturation of the products and services in the mobile world is so fast that investors in hardware and production lines are facing tough competition and cost challenges. Providers of applications and software have opportunities to reach the wide global audience but, on the other hand, they have to upgrade constantly and live in an environment of severe global competition with both other small and versatile companies and with huge digital giants.

Academic Research

The speed of change has created a situation where sometimes academic research cannot catch up with technical changes and market development, as it requires more profound work with data collection and testing of models and findings. The curve of maturity of a product from production to market, from achieving credibility and acceptance from the most enthusiastic customers and then from the more conservative ones, and then the obsolescing of the technology is so rapid that customers can feel frustrated and companies exhausted by the necessity of constant re-investment. These threats and shortcomings are definitely outweighed by the opportunities and profits for all participants of the mobile technologies market. Severe competition brings better value for the money to customers and helps the best technologies and innovations to achieve success.

New and Old Business Phenomena

Business models for mobile commerce can be very new, based on new technologies, or they can be improvements of existing and previously tested ones. Mobile technologies continue to improve on business models developed for e-business, such as freemium services, when customers have free access to certain valued services, and businesses monetize their volumes of data from the actions of the customers.

Mobile technologies with low barriers to enter and easy access to the global audience create unique opportunities for companies and entrepreneurs from all over the world, including developing countries. A good example of a productive strategy would be selling a fast-growing company to a major industry player, especially when the larger company is less than 10 years old and still maintains the spirit of a start-up.

A good business model and access to millions of customers provides an opportunity for a successful exit strategy for companies in developing countries as well as developed ones. As an example we can see the case when Twitter bought ZipDial (Economic Times, 2015). It is a new business phenomenon when giant Western companies are anxious to buy companies from developing nations in order to expand their global service, and it is one of the exit strategies for new companies and a chance for this type of company to succeed.

In this chapter we will explain and analyze the industry trends and academic research, future opportunities, and some risks and problems to be solved by mobile commerce. The topic is so wide and still seems to be exploding with new opportunities because mobile technologies and applications have not yet peaked; yet they are already dramatically changing our way of living and doing business worldwide.

INDUSTRY TRENDS AND ACADEMIC RESEARCH

Because of mobile technologies, people now have immediate access to information, and they demand that business and government provide this information. To be in business now means to be in mobile business.

Academic research and industry studies go hand in hand in relation to mobile technology. Some differences between academic research and industry white papers and marketing summaries is that academic research takes more time to conduct, and it should be more neutral in assessment of future trends and current impacts. M-commerce is not only about how companies use mobile to get to customers; it is also about how everyday operations and decision-making change, how industries change, how organizations appear or disappear from the competition, and how customer behavior and motivation change with the availability of mobile technology.

As mobile technologies change our lifestyles, they also have societal effects. Some academic research of the mobile technology revolution has been conducted over the past 15-20 years, although the rate of change has increased to such a degree as to make it impossible to keep up. We will briefly outline the main trends in academic research related to mobile technologies and their societal effects. The need for grouping is necessary, as mobile technologies have so vast and even unexpected impacts. Later we will consider evolving research in which people tackle the future challenges associated with mobile technologies.

Mobile technologies make communications and information exchange instantaneous. The rise of smart pads and smartphones, meant a new era for business and every aspect of society. For many people in developing nations a smartphone is their first computer with access to web services, browsing, entertainment, messaging, and all manner of business services.

There are also risks associated with rapid mobile development, such as the loss of distinct cultures, of written literary styles, of oral forms of communication as people begin to communicate with each other by short text or audio messages filled with an abbreviate vocabulary that's foreign to older generations. For theoretical assessment of the impact of mobile technologies in a broader context, we refer the reader to the works of Hungarian philosopher Kristóf Nyíri (2003a, 2003b, 2003c, 2005, 2006). He refers to the works of Wittgenstein in "images as pictorial languages" and possible spatiality of pictures (Nyíri, 2003a, p. 161). We enlist Nyíri's impressive works to show theoretical and philosophical aspects of mobile technologies. Nyíri edited volumes on mobile communications and cognition processes (2003a), mobile learning (2003b), mobile democracy (2003c), sense of place in mobile communications (2005), and the "epistemology of ubiquitous communications" (2006).

In *Mobile Communications. Essay on Cognition and Community* (2003a) the main idea is a very optimistic one that mobile phones return us to intense personal communications and help rebuild communities and reduce social "atomization." "Communicating synchronously in voice, writing, and graphics has the potential to create and maintain a higher level of human cohesion than could be achieved by any of these dimensions by themselves," writes Nyíri (p. 183-184). In this way immediate communications is a way to coming back to primordial times. It reflects Taleb's idea (2012) of the books' survival as an old innovation with little difference whether it is a plate, scroll or e-book. Taleb (p. 333) compares tablet PCs with ancient writing plates: "And the great use of the tablet computer (notably the iPad) is that it allows us to return to Babylonian and Phoenician roots of writing and take notes on a tablet (which is how it started*).*"

In *Mobile Learning. Essays on Philosophy, Psychology and Education,* Nyíri (2003b) analyzes important opportunities and problems associated with mobile technologies in the education process. One of the main ideas here is a phenomenon of collaborative learning for participants and the absence of codified space with hierarchies and power. Mobile learning makes up a major section of this chapter and furthers debate about advantages and disadvantages of this new way of learning. Nyíri suggests that mobile learning creates new educational opportunities, but many students still do not use them to full benefit.

In the 2007-2008 academic year Rave Mobile (now Rave Mobile Safety) initiated an experiment called Bronco Mobile, involving students from different colleges. Rave provided them with mobile services to facilitate the exchange of information with teachers and peers, to share locations for safety purposes, and to provide polling, emailing and other interactive services (Okunbor and Guy, 2008). The results of the experiment showed insignificant interest on the part of participants, as 70% never used these mobile applications for educational purposes (Guy, 2011, p.96). It was a surprising figure, but certain intervening factors might explain these statistics. The specific applications might not have been interesting for students; or students might have been satisfied with current study routines and not interested in expanding those routines. Some eight years have passed since the Bronco experiment, and more powerful iPhones and other smartphones may have changed this picture, and such powerful learning management systems as Moodle are now accessible by mobile. Still, the results of 2002-2008 experiments are interesting and deserve attention.

Other theoretical assessments of mobile communications relate to the concepts of place and time. The integration of mobile devices with geographic location opens new horizons. Space and time already became condensed with electronic devices, and now mobile communications is "anywhere and anytime." Nyíri's work on *Sense of Place in Mobile Communications. The Global and the Local in Mobile Communications* (2005) touches this issue from different angles. The main message of this work is similarly optimistic as in the previous ones, suggesting that mobile communications connect people both globally and locally, creating a new sense of community.

Another side of the mobile revolution is the exchange of information. Knowledge and technologies are becoming less and less a privilege of society's upper classes, and mobile has tremendous potential for promoting equality within countries and between countries. In human history, knowledge and production of knowledge have been connected with exercise of power. Digital communication has led to this information revolution, and now mobile technologies make information even more available for people all over the world. Dissemination of knowledge and new opportunities for education, healthcare and business have been facilitated as mobile communication has become part of everyday life. These themes are covered in Nyíri's book, *Epistemology of Ubiquitous Communications* (2006).

It can be said that the volumes edited by Nyíri are becoming outdated in terms of current mobile technology. Still they do underscore trends and possible impacts brought by mobile communications to our lives.

Big data analysis of such user preferences as online shopping, app use and website visits by mobile users gives business more targeted audiences to address while, on the other hand, raises issues of privacy and "big brother" commercialism. These debates relate not only to the legal gathering and use of data but also to the overall security of mobile devices and the sensitive information accumulated. Such data can be the target of criminals and foreign agents, as well. Vulnerable data includes stolen passwords, credit card and banking information, e-mail addresses, GPS coordinates where personal photos have been taken, and personal identification information, such as Social Security numbers. Attacks can come in the form of hacking, viruses, blackmail, publication of the private photos, and bullying. As Suarez-Tangil et al (2013) warn:

Malware in current smart devices – mostly smartphones and tablets – have rocketed in the last few years, in some cases supported by sophisticated techniques purposely designed to overcome security architectures currently in use by such devices.

These threats are increasing every day as people become more dependent on mobile devices and as criminals and foreign interests apply more powerful technology (Cavoukian and Prosch, 2010).

How do mobile devices influence children and young adults? This question has also been raised. Early access to the mobile world of games and entertainment, for example, raises concerns in child psychology. Children easily adopt interactive products and are often targeted by their producers. Children's overuse of such devices may reduce their creativity and physical fitness (Rowan, 2013).

Coming from academic research we will look at several major aspects of how mobile technologies affect business. These include the development of m-commerce, with a review of trends in m-advertising, m-shopping and m-sales; operational challenges for mobile enterprises, including adapting an organization's online storefront – its website – to mobile, developing new methods of mobile marketing, and changing everyday business operations because of mobile and related technologies; and a review of some of the maturing industries within the evolving mobile ecosystem.

Development of M-Commerce

The history of mobile technologies in commerce started back in 1997 in the Global Mobile Forum when major companies first discussed the main ideas and prospects for mobile commerce. Finland pioneered first mobile payments and banking transfers via mobile phones by SMS. After wide introduction of smartphones and improvement in broadband communication, m-commerce began developing explosively.

One of the first books to cover mobile commerce was *m-Profits: Making Money from 3G Services* by Tomi Ahonen (2002). He reviewed existing and possible future areas for m-commerce. While these principles have not changed, the industry changed dramatically with such devices as smartphones and tablet PCs, new mass markets, and growing sources of revenues. This book was a seminal piece with key concepts introduced, but the reality of mobile technology has been changing much faster than anyone anticipated.

Barnes (2002, p. 13) defined m-commerce as "Internet 'in your pocket' for which the consumer possibilities are endless, including banking, booking or buying tickets, shopping and real-time news." Now everything in this list is a reality, and much more can be added to the list, such as customer service, social media advertising, video-embedded advertising, real-time personalized messaging, etc.

Obsolescing is happening in this industry of mobile commerce so fast and new practices are becoming so widely accepted that the most common word to describe the situation is "ubiquitous."

The history of mobile technologies is happening right now. Eight years ago in 2007 Apple introduced the iPhone. Mobile technologies were developing before, but after the smartphone was introduced, m-commerce development became difficult to keep up with. The Economist (2014) provides these stunning figures: In every quarter of 2014 the number of new mobile subscriptions increased roughly by 120 million. At this speed, in 2015 mobile subscriptions was projected to exceed 7 billion, and 2.4 billion would be smartphones. Data volumes in 2015 were expected to increase by 70% over 2014.

Development of Effective M-Advertising

Early in 2011 eMarketer did a study comparing how much time people spent with different information sources, and then compared that to how much advertising was purchased from those media (Harvey, 2011, p. 2). What it found was that people already spent about as much time on their mobile devices as they did reading newspapers and magazines combined. However, advertisers were dedicating about 28% of their advertising budgets to newspapers and magazines and almost nothing to mobile. However, 2011 was the "welcome out party" for mobile. Before 2011, most of the mobile advertising was limited to SMS messaging. But with the rapid spread of smartphones and tablets, social networks suddenly became a viable advertising opportunity. In that one year, mobile use of social media increased 400%, and in 2012 it grew another 200%, followed by 93% in 2013. And in 2012 mobile video took off with a growth rate of 167%, followed by 175% in 2013 (eMarketer, 2014). Eventually advertising follows the audience, and for mobile, ad spending skyrocketed in the U.S. in 2013, growing by 120%, followed by 80% in 2014. Mobile ad expenditures in the U.S. are projected to surpass that of desktop ad expenditures during 2016, double desktop in 2017, as shown in Figure 1, and continue dominating digital advertising into the future (eMarketer, 2015b, p. 7). Demonstrating how quickly things have changed in mobile business, eMarketer in 2010 had predicted that between 2008 and 2013 mobile marketing would increase gradually by 37% (Okazaki, 2012). Even that cutting-edge company could not foresee how advertisers and online media would find a way to take advantage of all that time being spent on mobile media.

Figure 1. U.S. advertising expenditures by device, 2013-2017
(eMarketer, 2015b)

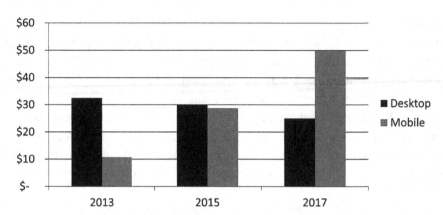

Mobile broadband connections to the Internet worldwide were already triple the fixed-wire connections (2.3 billion vs..7 billion) in 2014 (International Telecommunications Union, 2015), however that does not mean that all owners of mobile broadband use them frequently for the Internet. But the spread of smart mobile devices and the increased use of broadband applications are nearly certain to continue increasing at a rapid rate. Cisco (2015) predicts that between 2013 and 2018 the percentage of IP connection originations from non-PCs (mostly mobile) will grow from 33% to 57%. Most of these connections are by smartphones, but the number of tablet connections is also growing rapidly. In the USA, smartphones entered the market very slowly, starting around 2000, but began to surge with the introduction of iPhones in 2007. Smartphone penetration of the mobile phone market in 2006 was only 3%. By the end of 2007 it doubled to 6% and by the end of 2009 reached 17%. By the end of 2014 smartphones represented 75% of all mobile phones (comScore, 2015, p. 8). By June 2014 the number of smartphones in America reached 172 million. Meanwhile, tablets entered the market in 2010 and have skyrocketed to 93 million (comScore, 2014, p. 5). The spread of tablets will ensure that users will spend more time online per connection and be more involved commercially. In 2015 the number of tablets in use worldwide was expected to reach 1 billion, and by 2019 1.5 billion (eMarketer, 2015a), but there is no anticipation that tablets will ever surpass smartphones it total number.

Because mobile users so greatly outnumber desktop users around the world and the number of mobile shoppers is rapidly increasing, advertisers are rapidly turning their budgets to mobile (eMarketer, 2015b). In 2014 mobile advertising in America increased by about 80%. In 2015 it is expected to increase by another 50% and nearly catch up with desktop ad spending. By the end of 2017 mobile advertising will have zoomed ahead and doubled that of desktop advertising, eMarketer projects (p. 7).

Growth of Mobile Sales

The projections for mobile advertising, however, are based on projections for mobile sales. According to eMarketer (Ramsey, 2015) a higher percentage of tablet users shop and purchase than do smartphone users. About 89% of tablet users shop on their tablets, in contrast with the 82% of smartphone users. But the big difference is in actual purchasing – 72% of tablet users vs. 43% of smartphone users. Tablet statistics are just a little below those of desktop/laptop users, 90% of whom shop and 84% of whom make purchases. Nonetheless, people still spend much more time on their smartphones. In July 2014, Americans cumulatively spent 143 billion minutes on their tablets, 521 billion minutes on their smartphones, and 495 billion minutes on their desktop PCs (comScore, 2015, p. 6).

Despite mobile IP connections outnumbering PC connections more than 2:1, mobile advertising overtaking PC advertising, and time on mobile devices overtaking time on PCs, most online purchases are still made via PCs. Total e-commerce in the second quarter of 2014 reached $61.6 billion, but only $6.8 billion was made on mobile devices, the other $54.8 billion on PCs (comScore, 2014, p. 23). In Quarter 4 of 2014, comScore (2015, p. 19) reported that Americans spent 60% of their online time on mobile devices, as opposed to 40% on PCs, but 87% of their purchase dollars were spent on PCs and only 13% on mobile. So, is mobile advertising worth it? comScore research (comScore, 2014, p. 20) says it is, based on other metrics. Mobile advertising does a better job of building brand awareness (+4.1 vs. +1.8 for PCs). It is better in enhancing favorable attitudes (+2.2 vs. +1.3), increasing the likelihood that the audience will recommend the brand (+5.3 vs. +1.5), and achieving prospects' intent to purchase (+4.3 vs. +1.1). So, it seems that mobile advertising is much better at achieving brand awareness, favorable attitudes, customer evangelism, and intent to purchase, but mobile still does not accommodate purchasing as well as PCs, but the gap is closing. In the second quarter of 2014, the year-to-year growth of e-commerce overall was 10%, but the increase for mobile commerce was 43% (p. 26). And the increase of audience time spent on the various Internet devices over the four years between December 2010 and December 2014 was even more astounding: 37% for PCs, 394% for smartphones and 1,721% for tablets. And mobile is used far more by the younger generation 18-34, where only 5% don't yet use mobile Internet devices, and 21% are mobile only (comScore, 2015, p.4-5). In 2014, while older generations still preferred to watch video on PC and TV, half of the 18- to 34-year-old Netflix subscribers were already watching most of their videos on mobile devices (ComScore, 2014).

M-Enterprise: Operating in the Mobile Environment

A mobile-oriented enterprise faces many challenges and opportunities. Its "storefront" is on the Internet, and important adaptations need to be made to accommodate mobile users. There are also many new methods of marketing under development in order to reach prospective customers using any online device and no matter their channel preference. Big data is finally gaining the value long predicted, allowing m-enterprises to personalize messages to customers and prospects on their website, via email and through the variety of advertising tools.

Empowering Employees and Other Stakeholders

Mobile technologies and social media can be used to empower employees and make them up to 20-25% more effective, according to Chui et al. (2012). These technologies are particularly effective with "interaction workers" and "knowledge workers." Such employees, they noted, spend an average of 28 hours a week in collaborative communications, which can be enhanced up to 100% using social technologies internally, helping these workers to lower barriers and streamline communications and by extending internal networks to involve a broader range of expertise. The study concluded that social media, increasingly mobile in nature, could unlock $900 billion and $1.3 trillion in increased productivity, creativity and synergy.

The Melcrum organization (2015) has produced numerous case studies of how organizations are boosting performance by about 36% by employing these technologies to both engage and empower employees. An international study by Melcrum (2014, p. 6-10) agreed with Chui et al., in concluding that companies successfully utilizing these technologies to enhance internal communications are able to make internal processes more agile and improve employee engagement, organizational planning, and collaboration and synergy among cross-functional teams.

Other studies conclude that individual productivity is also enhanced by allowing employees to replace their wired office phone with a mobile smartphone capable not only of receiving live, recorded and text communication from any location but also providing quick access to the phone's social, organizational and informational applications (Wilkinson, Miller, Burch, & Halton, 2014).

Creating an M-Storefront

Why don't mobile users shop a lot but purchase relatively little, and what are the prospects for the future? There is increased pressure by consumers and by search engines for companies to make their websites mobile compatible. As of April 2015, Google has made mobile compatibility part of its search algorithm, which led many experts to call its new algorithm launch date of April 21 by such names as mobileged-don or mobocalypse (Search Engine Land, 2015). Google probably felt forced to add mobile access to its algorithm by such other search engines as Yahoo that had already done so. As of 31 December 2014, 50% of Yahoo's search traffic came from mobile users, compared to 43% of Google's and only 28% of Bing's (eMarketer, 2015c). Smaller search engines had given mobile even greater emphasis. If Google allowed other search engines to gain favored status with mobile users, it would almost certainly lose future market share overall.

Since Google is by far the most-used search engine, and 75% of Google searchers do not go past Page 1 of the search results and 60% don't go past the third organic listing (Hubspot, 2012, p. 33-34), not being mobile accessible after April 2015 could be catastrophic for many organizations. Competitors may leap ahead of them in search results, leading to loss of sales to those competitors. Nevertheless, research indicates that many online enterprises had not yet established websites that readily accommodate mobile users. Merkle/RKG (2015, p. 21) reported that among the biggest companies in the world, the Fortune 500, 46% of their websites would not meet Google's mobile-friendly criteria as that deadline approached. Among companies included in the Internet Retailer (IR) 500 – those that had made online sales their highest priority – still 29% did not yet qualify. Of all the companies in the world that would be expected to have mobile-friendly sites, these would be among them, but many did not. A 2014 study of 100 leading consumer magazines found that these major communication companies were also failing to communicate effectively with mobile users (Davies, 2013) – 93% of them failing to provide complete cross-platform access. Most chose apps over adaptable websites, and most of their apps also could not service all mobile devices.

Besides being mobile-accessible, the best digital storefronts have other high priorities, as well. An Adobe (2015, p. 5) survey of over 6,000 business executives found that the highest imperative for executives was to improve customers online experience. Of the 10 top priorities cited by executives, this was ranked No. 1 by the executives (p. 11). Components of achieving this enhanced customer experience were, in order of rankings by the executives (p. 17):

1.   Developing a cohesive strategy;
2.   Changing the culture within their organizations and creating greater cross-team cooperation;
3.   Enhancing related employee skills;
4.   Adopting new technologies that allow better use of data to achieve more compelling, personalized, real-time experiences for customers; and
5.   Developing enhanced data systems to make these strategies possible.

While these were the ranked priorities, still the most important overall, they said, was making the experience of customers and prospects "as personalized and relevant as possible" (p. 18). Of the executives, 38% indicated that before the end of 2015, they would achieve omnichannel personalization (p. 21).

Online shopping carts frequently have built-in capability of using data to make personalized product recommendations to shoppers, but not all organizations use them (Mallikarjunan, 2014, pp. 39-41). In a survey of business executives, 82% felt that personalizing website functions was at least somewhat important. Those same databases can typically be used for personalizing email solicitations, and even fewer organizations take advantage of that capability.

To accommodate modern marketing, some of the most important pages in a website are the "landing pages," where visitors arrive once they have clicked on a social media ad, email solicitation or other new media prompt. How those pages are constructed can make all the difference between success and failure. Industry researchers, for example, have found that content marketing videos can greatly enhance conversion rates. According to an Aberdeen Group study (2014), the conversion rate of companies using video is 4.8%, whereas the conversion rate for those not using video is only 2.9%.

Website search engine optimization (SEO) is vital to having a successful virtual storefront. However, websites that contain a lot of content marketing achieve far more search engine success. So-called inbound marketing or content marketing – the use of free content (usually educational, recreational or entertaining) attracts more traffic to the company website than only standard SEO techniques. Videos are one kind of content marketing, and their effectiveness in increasing the rate of conversion of visitors to customers, was demonstrated by the Aberdeen study (2014).

Companies that do content marketing should have much of that content on their website. They should never use a free blog site, for example, but rather have their company blogs on their own website, then promote and link them on social media. That will increase their search engine optimization and traffic. According to HubSpot (2012, p. 39), companies that blog have 97% more inbound links. That's important because another part of Google's search algorithm is based on the number and quality of inbound links. If other websites link to a company's website to share its content, Google considers that a vote of confidence. And if the linking website is itself very popular, such as the New York Times', its link counts more. More content on a website also leads to more pages being indexed by search engines. And the more website pages that are indexed, the more leads are generated by that company. A company that blogs at least once per week will achieve 100% more leads than companies that blog less than weekly (p. 95). So, the more content the better. Indeed, a website that has over 311 indexed pages will average more than 1000% more leads from their website than a website with fewer than 120 pages (p. 40), as shown in Figure 2.

Having a strong website is important, but having a strong, mobile-accessible website is more important. Of mobile users who are able to make purchases online, according to Ramsey (2015), only 28% do so through websites; 72% through apps. HTML5 accommodates cross-platform adaptation of websites, but many if not most companies, as noted, have not yet implemented this programming language

Figure 2. Number of indexed pages in relation to median monthly leads (Hubspot, 2010)

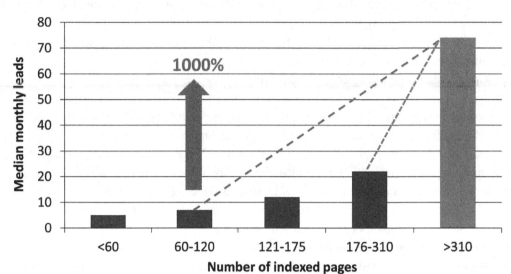

or other alternative. Another option is to create a separate website for mobile devices. Whatever code or format a website uses, these are the four primary rules used by Google to determine if a website is mobile-friendly (Google, 2014):

- "Avoids software that is not common on mobile devices, like Flash."
- "Uses text that is readable without zooming."
- "Sizes content to the screen so users don't have to scroll horizontally or zoom."
- "Places links far enough apart so that the correct one can be easily tapped."

New Methods of Mobile Marketing

With the new commercial environment have come many new methods of marketing, and major advertising hosts are quickly adapting to accommodate mobile, even if advertising buyers are not. Some of these marketing tools have been introduced previously in this chapter. We will expand on them somewhat in this section but leave greater detail for other chapters related specifically to marketing, advertising, branding and public relations:

- **Inbound Marketing:** Inbound marketing is typically free marketing, such as blogs, white papers, ebooks, webinars, videos and freemiums, sometimes promoted with other kinds of paid advertising. Because inbound marketing is free to the advertiser and the contents typically free information and education to the prospective customer, it could easily be considered an arm of public relations. Outbound marketing, in contrast, includes cold calling, trade shows, traditional advertising, and direct mail or email using purchased lists. Outbound is considered intrusive by many prospective customers, whereas inbound marketing is alluring and inviting. Most marketers agree there is some difference between inbound marketing and content marketing – that content marketing is a subset of inbound marketing – but most inbound marketing is content marketing, such as ebooks, webinars and videos (HubSpot, 2014). Inbound marketing such as standard search engine optimization techniques and freemiums, on the other hand, might not be considered content marketing, per se. Outbound marketing used to be the main approach used by most organizations, but a study by HotSpot (2014) indicated that 45% of company executives now cite inbound marketing as their primary source of leads, as opposed to only 22% from outbound marketing. The percentage identifying outbound marketing as their primary source dropped from 34% in 2013, whereas inbound marketing has stayed relatively steady (p. 29-30), as shown in Figure 3. Fewer executives cited online paid media as their primary source than either outbound or inbound. HubSpot considers online paid media as a different category from outbound advertising since much of it has also been adapted to be more inviting than intrusive. About 42% of B2B companies cited inbound as their primary source, as opposed to about 48% of B2C and 58% of non-profit organizations. The inbound marketing tool that most frequently correlates with a positive ROI is blogging – 13 times more likely to have increased ROI year after year than other inbound methods, according to HubSpot (p. 51). While companies do not have to pay anyone to make their content available to prospective customers, there are still costs, such as labor, that an organization has to cover. That cost, however, is typically much less than for outbound marketing. For companies of 1-25 employees, the average cost per inbound lead is $37, compared with $102 for outbound leads. For medium-size companies (51-200 employees), inbound leads average $70 each, compared with $220 for outbound leads. The largest companies with over 1,000 employees cite the lowest cost for both categories -- $27 for inbound leads and $45 for outbound leads.

- **Native Advertising:** Native advertising is a growing trend of advertising that mimics inbound marketing. In Facebook, for example, a regular posting by an organization from its own Facebook site, seen first by its own friends or fans and hopefully passed by them to their own friends, is inbound marketing, but Facebook now allows organizations to pay for advertising that looks like a regular post but with a small label of "Sponsored." Organizations pay for this advertising so it is posted on the sites of a selected set of Facebook users who are not yet following the sponsoring organization. Native advertising is now available on most social network sites and is even being offered by traditional media. Major newspapers like the Washington Post are allocating the time of their news journalists to write paid feature stories about advertisers. In most traditional media in America, these paid stories are placed and labeled in such a way that a discerning reading can detect that they are not news, but there is concern among some in journalism that many readers are not so discerning.

- **Video Advertising:** Video, according to marketing executives, is the most effective form of marketing communications, and 95% of the top websites use it. They are also more likely than other websites to feature videos created by their own staff (Aberdeen Group, 2014). A study by the SocialBakers social marketing agency (Ross, 2015) analyzed which of 670,000 Facebook posts by thousands of different brands achieved the most "reach" in terms of shares. The average reach was 8.7% for videos, 5.8% for status posts, 5.3% for links, and only 3.7% for photos, which means that video achieved 137% more promotion than more-often-used photos and graphics. Based on an analysis of several recent studies, Brightcove (2015) noted that emails with videos achieve 55% greater click-through rates; across such social channels as Facebook and Twitter, video achieves 1,200% more shares than links and text combined; web pages with video are 5,300% more likely to reach the first page of a Google search; and videos on a website landing page increases the conversion rate by 86%. Recognizing the growing power of video on the Web, Facebook changed policies and advertising opportunities to encourage the posting of more videos in 2014. In January 2014, 4.9 billion videos were viewed on Facebook by desktop users, but by February 2015 the monthly video views had nearly doubled to 9.6 billion (Mixpo, 2015, p. 3). During that same 13-month period, the number of videos viewed monthly by individual desktop users on Facebook soared from 58.7 to 107.4 – again, nearly double. In addition to these trends, projections are for TV and other video entertainment to begin making a very strong transition to the Internet, which will provide more video advertising opportunities. Everything is changing so quickly on the Internet that it is impossible sometimes to make good projections. The CEO of eMarketer admits that they look at other organizations' projections and try to find a middle ground that's not too conservative and not too optimistic (Ramsey, 2015). But, as discussed earlier, sometimes even the most optimistic projections fall short, as when eMarketer projected a 37% increase in mobile marketing revenues between 2008 and 2013 when in reality it turned out to be closer to 10 times that much. Their recent projection of the growth of video advertising could be just such an embarrassment for eMarketer. They predict that U.S. digital video ad revenues will increase about 34% in 2015 to about $7.8 billion and then nearly double to $14.4 billion by the end of 2019 (eMarketer, 2015d). That may seem optimistic enough, but doesn't take into account the possible "tipping point" that may soon hit television the same way it hit newspapers around 2007 (Downie, 2009). Even major news organizations like the Washington Post were caught by surprise at how quickly things changed. They expected that newspapers would gradually lose advertising to the Internet but not that they would lose half of their revenues in a space of about four years. There are indications that TV may be close to a similar tipping point. Between June 2011 and June 2012 the viewing of online videos jumped 550%, and the use of advertising embedded within videos increased 68% (Xstream, 2013). That could have been seen as the start of an online video revolution. In 2015 U.S. TV advertising is expected to approach $70 billion (eMarketer, 2013a). If the broadcasting tipping point hits in 2015, offline TV could lose half of that $70 billion to online TV by 2019, making a projection of $14.4 billion seem very silly. A survey by Belkin and Harris Interactive in early 2013 cited by eMarketer found that 30% of U.S. Internet users were already favorably considering replacing cable TV with an online streaming video subscription. eMarketer itself then projected that online TV viewers would cross the 50% tipping point in 2014 and surpass 145 million paid online viewers in America by 2017 (eMarketer, 2013a). Then ComScore (2014) found that half of the 18- to 34-year-old Netflix subscribers were already watching most of their videos on mobile devices. The latest data available confirms that the transition from TV/cable to online video services is speeding up. Blueshift Research's survey of 1,129 Americans in June 2015 showed that more now subscribe to online streaming video services than to pay-TV services. Those subscribing to at least one streaming video service is now 75% – up from 65% just six months ago (p. 4). And during that same time the percentage of Americans using pay-TV (cable) services dropped from 69% to 62% – and the rate of cancellations is increasing, with 1.3% canceling their pay-TV services in the previous month (p. 2). This revolution again is being led by the younger generations. Among those 18-29 years old, 47% have never signed up for pay-TV and 11.2% more have recently canceled such services (p. 3). This by itself indicates the eventual doom of cable/TV. According to eMarketer (2015d), the primary revenue even for most paid online TV services is still advertising, and a more recent eMarketer report (2015e) confirms this. Emarketer cites a survey by the Interactive Advertising Bureau that found that 78% of all respondents said they would prefer to watch free videos with advertising, while 15% preferred to pay for a video service without ads and 8% preferred to pay for individual videos without ads. Thus, more free TV is being offered online and competing with the growing number of online paid services. In its battle for video supremacy, YouTube, for example, is partnering with numerous organizations to bring more professional programming, such as TV reruns and brand-new programs, to their website. According to eMarketer (2015d), revenues passed on to YouTube's content partners have grown by over 50% each of the last three years. A study by Videology (2015) examined online video advertising in America during 2014. In Quarter 1, only 17% of video advertising targeted multiple online devices. The rest were aimed primarily at PCs. By Quarter 4, 51% of all video advertising campaigns were cross-device, and 41% of all video ad campaigns targeted Internet-connected TV in their cross-device mix, compared to only 7% in Quarter 1. An eMarketer report (2015e) also shows that smartphones are the preferred device for an increasing portion of the online video viewers worldwide. Between September 2013 and March 2015, the percentage of all online videos viewed by mobile devices increased from 15% to 41%, and the portion viewed on smartphones increased from 7% to 34% while the percentage of tablet video views stayed about the same. The portion viewed via mobile devices increased by 95% in just one year. All of this indicates that the move of video content and associated advertising from offline to online channels may be reaching its tipping point, and an increasing portion will be viewed on mobile devices.

- **Programmatic Advertising:** Because of the complexity of ordering, managing, tracking and developing content for online ads, many advertisers were previously buying primarily from only the largest online media, such as Facebook and Google. But technology is now allowing them to pursue low-cost advertising all over the Web. It is difficult to compare programmatic with video or native advertising because programmatic is not a specific kind of advertising but rather a way of locating, purchasing, managing and, in some cases, creating advertising in an automated way to save money and manpower. In 2014 mobile programmatic spending increased 291% and desktop programmatic spending 245% (PulsePoint, 2015, p. 4). Programmatic advertising is also able to better use Big Data to match the right advertising outlet for the right message for the right consumer with the use of decision-making algorithms. Many billions of impressions are now being purchased all around the world every day (p. 5).
- **Personalization:** The Adobe (2015) survey of 6,000 executives found that improving the customer experience was their most important goal over the next five years in differentiating their companies from their competitors (p. 15), but the key to improving the customer experience, they indicated, was to use data to make the experience as personalized and relevant as possible (p. 18). Programmatic and personalization strategies are distinct from one another but are developing in tandem. Programmatic would not be growing so rapidly if not tied to Big Data and the ability to personalize while at the same time automating the entire advertising process. Indeed, about 63% of the programmatic marketers are already using data signals to personalize their marketing messages (eMarketer, 2015f). Interviews by Evergage (2013, p. 2) of 114 digital marketers in 18 countries found that 88% believed that using real-time data in marketing was an important part of their 2014 plans. Facebook is growing rapidly largely because of its ability to share its pinpoint data with its advertisers. Tapping into data sources or developing a successful strategy for gathering data is essential to personalizing business communications. Free Wi-Fi is also a rapidly growing phenomenon, developing parallel to the rapid spread of tablets, 86% of which can only access the Internet via Wi-Fi (IEI-TV, 2015). But as financial incentives increase, the rate of Wi-Fi growth should also increase, and the greatest incentive may be that Wi-Fi providers can now collect so-

cial media data from Wi-Fi users and utilize that data for follow-up marketing. Many different kinds of stores, restaurants, entertainment and recreation providers, etc., may now offer Wi-Fi (1) because surveys show customers do making shopping decisions based on Wi-Fi availability, and (2) because the data is very valuable. This new programming asks Wi-Fi users to sign in through Facebook, vKontakte, Twitter or LinkedIn, which then gives the Wi-Fi provider access to the users name, email address, social media contact information, and other personal information provided in the user's profile. This technology could greatly expand organizations' data-collection capability over the coming years and help organizations to further personalize their communications with customers and prospective customers (IEI-TV, 2015).

- **Omnichannel Marketing:** The Adobe (2015, p. 29) survey of 6,000 business executives showed that 70% thought it was "very important" and 27% "somewhat important" to understand customers' journey across channels, while 66% thought it "very important" and 30% "somewhat important" to coordinate omnichannel marketing messages. A study by Lim, Ri, Egan and Biocca (2015) showed "participants exposed to repetitive ads on paired media of television, Internet, and mobile TV have greater perceived message credibility, ad credibility, and brand credibility than counterparts exposed to repetitive ads from a single medium. The multiple-media repetition also generated more positive cognitive responses, attitude toward the brand, and higher purchase intention than the single medium repetition. Finally, the cross-platform synergy effect remained robust for different levels of product involvement."

Impact on Entrepreneurial Organizations

Mobile technologies are also changing internal management and human resources within organizations. *Harvard Business Review* and SAP produced a series of reports (SAP, 2012) on mobile technologies in 2012 under the titles:

- "How Mobility is Changing the Enterprise" (*Harvard Business Review*, 2012a).
- "How Mobility is Transforming Industries" (*Harvard Business Review*, 2012b).
- "How Mobility is Changing the World" (*Harvard Business Review*, 2012c).

Figure 3. Leads from inbound vs. outbound
(Hubspot, 2014)

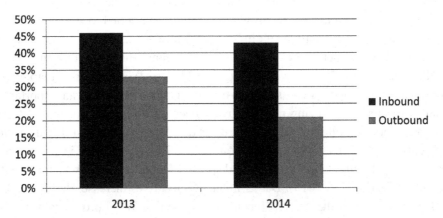

Mobile enterprise is a change of operations both in small and big transnational business because mobile technologies help to foster customer relations, increase overall efficiency and control costs. A poll conducted by AT&T and the Small Business & Entrepreneurship Council in the USA found that small businesses save more than $64 billion per year due to the use of mobile devices (AT&T, 2014). Small businesses use mobile technology in several ways. Based on U.S. polls in 2013, 97% of the small business owners used YouTube, Pinterest, Twitter, Facebook, and Instagram for business development (Nanji, 2013). Some 73% of them used mobile phones and tablet PCs for social media marketing. Thus, mobile technologies effect revolutionary change in the business landscape.

Internet marketing for small and medium businesses plays an important role in attracting customers. According to the SMB Internet Marketing Survey 2014, 32% of small and medium businesses say internet marketing is very effective. Fifty percent of SMBs spend less than $300 per month on it. In addition, 37% of SMBs plan to spend more on internet marketing in 2015, because it is an effective instrument to improve and expand their business (Anderson, 2015).

Mobile technologies are changing enterprises as part of "Mobile.2." Mobile.1 was when laptops surpassed desktop sales. Now tablet PCs might outsell laptops in 2015. According to IDG Research, there were three main drivers of PC tablets' demand starting in 2012: executives needing constant communication, the mobile work force, and customers wanting real-time support and action. HBR points to five areas of change in the work of enterprise in the era of mobile technologies: finance, sales, human resources, operations, and service (Harvard, 2012a).

Use of cloud technology for business is becoming a more and more usual thing in backing up data, CRM and EPR systems, and in leasing necessary software, as well. It enhances necessary frames and infrastructure for mobile technologies to be used in everyday operations.

One aspect of human resources changed by mobile technology is more voice to the employees and their increased feedback in the policy development of the company, its products, reactions of the market and deficiencies or errors in product development or company actions. Employees post their opinions on internal networks and provide better feedback for management (Kapko, 2015). It is a continuation of e-business, and the reaction is faster and easier to input.

M-enterprise is not some new fashion; it is a "must have" for businesses to stay competitive, to find new opportunities and to serve customers with better value. An impressive volume of papers in *Enterprise mobility: Applications, technologies and strategies* (Basole, 2008), published in 2008, demonstrated certain difficulties in introducing mobile technologies to enterprise.

The convergence of wireless, mobility, and the internet and its relevance to enterprises by Seybold (2008) described key areas of application for mobile in the everyday work of enterprise and the difficulties of convergence. There are a lot of issues, as businesses and employees struggle to stay abreast of new developments in mobile technology. Corporate culture can be a deterrent to m-enterprise (Hoang, Nickerson, Beckman, and Eng, 2008).

Development of 3G and 4G technologies makes a significant contribution in development of m-enterprise. Such technologies provide an opportunity to upload and download significant amounts of data in seconds.

M-enterprise solves a lot of issues, including speed of administrative decisions and meeting productivity, employee satisfaction and feedback, customers' satisfaction and feedback, sales cycles, and elimination of redundant and duplicated activities in everyday operations. More applications for m-enterprise are to be expected in the near future.

M-enterprise will support a demand for mobile devices with better capacities. It creates great demand for cheaper smartphones and makes the smartphone market in developing countries more attractive. If Apple in 2014 was generating more profits in China, then other giants could not ignore this market. New companies in China such as Xioami would follow the trend and try to enter the international markets with cheaper products with comparable power in terms of memory, speed and reliance (The Economist, 2014). Other Chinese companies would begin marketing on a global scale with their devices and applications. It is a story of how mobile creates opportunities for the developing world, as well.

New Challenges for M-Enterprises

With these new opportunities also come new challenges. Mobile technologies create new types of businesses, therefore businesses are doing new types of work and serving new kinds of customers. It means that new skills are required; otherwise moving forward will be impossible. Changing of responsibilities, duties and roles is not something new in the developing world, however, new technologies speed this process up (Carufel, 2013). Managers will face the problem of finding such talented people who are self-monitoring and flexible enough to adjust their work to meet new needs and challenges connected with adoption and implementation of new technologies.

Core competencies of many employees are quickly changing from when they were hired. A global study by Forrester Research and the Business Marketing Association (Ramos, 2013) found that 97% of B2B marketing executives are having to perform new tasks in order to be successful in the Digital Age (p. 2); 96% believed that the need to understand and implement new technologies will accelerate in the future (p. 3); 63% admit it is increasingly difficult to keep up with such demands (p. 3); and 34% are feeling "overwhelmed" (p. 4). And as they search for new employees with the right skills to help fill these needs, 44% said they cannot be found (p.4). If other employees feel even a small portion of the frustration expressed by these executives – and they might well feel more -- they will resist such changes related to new technology. Four major obstacles to effective change were noted by Beer and Nohria (2000): mindsets, complexity, constrained resources and adequate employee skills and knowledge. M-enterprises are almost certain to encounter such obstacles, however, mobile technologies can also help overcome these obstacles by providing just-in-time training and support for employees in the form of detailed instructional text and graphics, video-based training, or live support.

Maturation of Industries Within the New Commercial Ecosystem

Mobile technologies change not only the work of a particular enterprise, but entire industries are changing their operations. HBR underlines industries that are changing dramatically in Mobile.2. They include utilities, financial services, healthcare and retail.

Future use of mobile will expand to farming, sports, and education for testing, monitoring and using real-time data (Harvard Business Review, 2012b).

Mature businesses change with mobile and new businesses are initiated with mobile. New e-businesses from all over the world think globally right from the beginning and create unified service for all customers in the world. For e-business, mobile technology is a necessary tool to get to these global customers. New industries also appear, like m-learning and m-health, offering a combination of social and business impact.

M-Retailing

Mobile technologies demonstrate their power in retailing. M-retailing is a revolution in the industry, and "consumers have shaken retail to its foundations" (Kilcourse & Rowen, 2014). It is a clear example of an evolution from e-commerce to m-commerce. Industry is forced to change because the behavior of customers is changing during the buying process.

From the other side, m-retailing provides an opportunity for businesses to communicate with customers through more than one channel. It can keep the customer more engaged and enhance brand loyal by informing customers on their mobile devices about coupons, new products, sale updates, etc.

M-retailing is a clear sign that the line between virtual and real shopping is blurred. Delivery, variety and convenience enhance customer service in m-retailing. It is a set of activities engaged in direct selling of goods and services on several channels. Development of Internet and mobile technologies created a new channel called "online." It means that companies involved in multichannel retailing are able to sell merchandise online as well as in traditional retailing. Multichannel retailing helps businesses to rapidly attain great market share (Zhang, Irvin, et al, 2010). E-shoppers are becoming m-shoppers, and they are sensitive in their choices as to locations, time-task, social environment and control (Banerjee and Dholakia, 2013). M-retailing is another way for business to achieve consumer power. Such a way of running business is considered as more customer-focused. Development of mobile technologies has led to customers who are expecting retailers to move online and start retailing there as well as in the physical world. It is creating rapidly growing demand for m-retailing, however not all retailers are ready to implement such a strategy and are behind in technology applications (Carufel, 2015). M-retailing is more than a competitive advantage in the new world of mobile technologies; it is becoming a "must have" for the retailers. That is why m-retailing has been at the top of priorities for big retailing chains in 2015 (National Retail Federation, 2015). In 2014 mobile shopping grew 50% as a share of online shopping.

One of the challenges for m-retailing is the appearance and popularity of m-wallets, because customers may be prone to use them, and industry will not have a choice but to follow (Short, 2015). An m-wallet is a tool currently used mostly by such giants as Apple and Google to provide service to their huge army of loyal customers and to connect them even further to their devices and platforms. M-wallets decrease further the use of cash and credit cards in online transactions and make purchases safer, more convenient and perhaps more spontaneous and emotion-based. The limitations of bank credit cards and one-store m-wallets may be going away. Universality of m-wallets will be decisive for customers.

The biggest challenge, however, is the lack of mobile-friendly websites and a parallel of slow response times for those that are m-compliant (eMarketer, 2015g). According to *U.S. Mcommerce 2015: eMarketer's Forecast and Trends*, U.S. m-retail sales are expected to rise 32% to $77 billion in 2015, which will make it about 22% of all e-commerce. M-retailing sales are then predicted to double by 2019. In late 2013 less than 11% of the top 10,000 websites were mobile-friendly, according to eMarketer. One year later the percentage of mobile-friendly sites increased to almost 19%. Google's 2015 decision to give priority to mobile-accessible websites, however, had a majority of retailers in 2015 ranking this a major priority for the year. However, according to the report, mobile-friendly websites have a much slower response time, which was the No. 1 complaint of 42% of 1,000 m-shoppers surveyed. M-retailers hope improved technology and mobile broadband speed will solve this secondary problem over the next few years.

For those multichannel retailers who question whether online sales are adequate to justify increased digital marketing, a study by Deloitte Digital (2015) concludes that the influence of digital marketing needs to be measured in ways beyond m-sales. All online retail together led to $305 billion in sales in

2014 – a big number but only 6.5% of the $4 trillion in total sales. The other 93.5% of sales occurred at brick-and-mortar stores (p. 3). Deloitte's study was on the influence of online advertising on those brick-and-mortar sales. What they found was reflected by a quote from Mike Rodgers, executive vice president of J.C. Penney, who noted that traffic in their brick-and-mortar stores is down, but the conversion rate of shoppers into buyers is up by over 50%. In other words, shoppers have already done a lot of shopping online and go to the store for one final look and to buy (p. 4). So, while online sales, in general, and m-sales, specifically, may take a while before they approach the sales volume of brick-and-mortar stores, 64% of the sales will already be influenced by online marketing in 2015 (p. 7) and m-marketing will influence more than half of those sales (p. 11). The study did note, however, that the amount of influence varies, depending on the classification of goods purchased. The influence of digital and mobile marketing is twice as great for electronics, for example, than for food and beverages (p. 11). Overall, the very nature of shopping is being changed by the growth of retailers' presence online.

M-Utilities

Mobile technologies are making major changes in such a mature sector as utilities. Constant contact with employees in the field improves performance and increases productivity. Customers are more engaged with mobile applications and bill processing via mobile. Analytics from data helps to improve overall technical conditions and repair (Lewis, 2015).

M-Healthcare

The healthcare industry has been transformed by m-health, and it is a special and separate area deserving particular attention, which will be provided in more detail elsewhere in the book. In brief, healthcare as an industry was one of the earliest that took advantage of such mobile technologies as Palm Pilots, and even further with smartphones and tablets, as they are used in record-keeping, treatment, research, therapy and learning (Luanrattana, Win, & Fulcher, 2007). M-health has a lot of potential in terms of observation, research and overall prolongation of life expectancy and quality of life. M-health is projected to provide a big step for the health and pharmaceutical industries, as well as enhance related government programs. Consumers can be connected to constant monitoring of their health conditions, with the possibility for doctors to analyze data on a regular basis (Istepanian, Laxminarayan, & Pattichis, 2007). E-health changes to m-health policies and practices. Simple mobile reminders can decrease the risk of dangerous diseases like diabetes and coronary artery disease (Pop-Eleches et al., 2011). M-health invokes the development of new devices to regularly measure glucose and blood pressure, and can provide ever more sophisticated and detailed analysis of physical and mental conditions (Nemiroski et al., 2014). PA Consulting (2015) claims that mobile technologies cut costs in primary healthcare more than 7% and in acute care cases up to 10%.

M-Banking

Financial services are an industry that has pioneered the use of mobile technologies by SMS payment and access to bank accounts. The spread of smartphones changes how people use financial services on an everyday basis. A variety of services are available via mobile. Some of them are new, like peer-to-peer transfers without bank officer interference, and some of them, like regular check transactions, are an improvement over older ones. According to the Federal Reserve System (2015), 94% of mobile banking users in the USA checked their accounts and recent transfers in 2014.

Financial institutions can have access to potential customers with smartphones who have not used banking services -- so-called unbanked or underbanked clients. The Federal Reserve estimates the prevalence of smartphones possessed by these underbanked and unbanked residents in USA as a potential for banking service growth (Federal Reserve, 2015). The same tendency should be present for the rest of the world and especially for developing countries where unbanking or underbanking is even more common.

So, smartphones and broadband allow potential customers to be converted into real customers. Financial services by mobile can be various and different, including bank account management, money transfers, loan and mortgage payments, retail purchases, auctions and stock market operations.

M-Learning

M-learning, as an expansion of e-learning, is changing education and will impact business both as a commercial enterprise and as a way to enhance employee qualifications and company operations. The possibilities for monetization and commercialization of m-learning are growing, as well as its use as an additional tool in the general process of e-learning. It can make education more accessible for distant customers. In this regard m-learning is following the pattern of e-learning in its distance applications, and the main distant learning platforms, such as Moodle, have become mobile accessible. And one of the primary target markets of e-learning is now business itself.

According to Bersin (2014), HR departments are needing to retrain employees at a faster rate than ever before because of rapid changes occurring within business and industry. And training through traditional education channels does not provide the speed, convenience and on-demand access needed. Thus, online and mobile training is growing rapidly to accommodate business needs. Pappas (2015) notes that mobile education became a $107 billion industry by the end of 2015, and large companies represent 30% of all e-learning sales. Pappas adds that the greatest growth of m-learning was not in the United State, as one might expect, but rather in India, China, Malaysia, Romania and Poland, in that order.

M-learning has changed over time, as educators first looked to mobile devices as something disruptive and distracting (Sharples, 2003). Now educators look at how to involve mobile within the learning process (Jacobs, 2013; West, 2013). M-learning means for a student more interaction with the study material and a sense of ownership and control over the process of obtaining and mastering knowledge. In a practical way, m-learning can involve instructional games, personalized instruction, constant and instant availability of resources, modeling, and augmented reality. M-learning is cheap and inclusive, so future prospects for children from poor families and for developing countries look attractive. Education is becoming more available for people through m-learning. Constant and instant access to learning makes m-learning a more profound experience. The use of touch or voice commands in mobile devices, rather than a keyboard, makes the learning experience easy and even more similar to ancient use of clay tablets, as noted Taleb (2012).

A major survey of American university students in 2014 found that 99% own at least one mobile Internet device, and 92% own more than one, but relatively few instructors create assignments that take advantage of these tools. In fact, most instructors, the survey found, actually forbid the use of mobile devices in the classroom, although most students believe they can be extremely helpful academically. Most of the students want universities to use mobile-accessible online tools, and to provide "anytime, anywhere access to course materials" (Dahlstrom & Bichsel, 2014, pp. 14-27). This research suggests both how traditional universities can improve their education, and how for-profit competitors can make them wish they had.

M-Entertainment

M-learning frequently goes hand in hand with mobile entertainment. Many teachers and developers are trying to make the learning experience more like playing in what is called a gamification strategy. Still mobile entertainment is different and much bigger in terms of business. Mobile users, especially young ones, want first to play and then to study.

Mobile entertainment or m-entertainment can be divided in several areas: games of various kinds, PC consoles, TV and video streaming, chatting and social networks.

Entertainment was one of the first industries to benefit from mobile technologies. Games appeared with the first portable devices. At first mobile games were a continuation of regular electronic and computer games. Then mobile games started to be designed specifically for the portable devices. The mobile games market was estimated to reach a value of $25 billion in 2014. It is the fastest-growing segment of video and computer games, projected soon to become the largest segment (NewZoo Games Market Research, 2014). Microsoft (Mlot, 2012), Dimas Technologies (2014) and other companies are now rapidly developing 360-degree gaming and video equipment and products, facilitated by mobile technologies. The frequency of playing, gamer profiles, models for monetization, social network games and other issues are important for the mobile entertainment industry. An interesting profile observation is that the number of female gamers in mobile is higher than male ones, a new trend in the gaming industry (Information Solutions Group, 2013).

Mobile games are developing in several directions in terms of their device profile. We don't look to the classification of games into arcade, chasing, shooting or solving skills games. Our interest is integration of games with the mobile device, which is easily available for the gamer. One of the most important is continuation of computer games and their patterns to mobile devices. Another one is specifically mobile games designed specifically for these devices. Very promising is the integration of mobile devices with popular PC consoles. PC console entertainment is in itself a multibillion business with its own models, competitive devices and software, fans and monetization strategies. Mobile platforms cannot pass this opportunity.

Mobile TV and digital journalism are quickly integrating with mobile. Major TV providers and media outlets provide special digital conversion packages that take advantage of mobile's versatility and instant access, which benefit both producers and users. Digital conversion means reflection of the content in print and video for access on website and mobile.

Video streaming and personal channels are not only a hobby, they are a rapidly developing business. Online video is becoming more and more a personalized thing. Not only professional studios or news channels produce video. Teenagers creating videos in their home have attracted millions of followers and have become trendsetters and opinion-makers. And Facebook reports that by far most of their video views are now via mobile (Mixpo, 2015, p. 3). About 70% of all Internet users are now watching online TV-like entertainment (eMarketer, 2015h).

Transportation

Mobile technology, with its capacity to track devices globally, creates important changes in transportation and in the personal security of individuals, especially the elderly and children. Now with relatively cheap modifications, transport vehicles can be tracked and, in hazardous situations, stopped distantly by a command. Mobile technologies help to cut costs in transportation by using resources and time efficiently. They can cut accidents by controlling the speed of vehicles, decrease maintenance costs, and improve the overall comfort of passengers.

Transporters use mobile both in passenger and cargo transportation. Passengers are more engaged with new technologies, new services and more control. Traffic systems can avoid congestion with real-time data, and the level of satisfaction and efficiency is increased. It will reduce fuel use and carbon emissions, helping the environment (The Economist Intelligence Unit, 2014).

Taxi auctioning and Uber transport, when customers order taxis using their mobile devices and drivers follow with proposed prices, show the power of mobile technology and certain rigidity of current legal and market structures. Further developments are shared cars, and driverless or autonomous cars will require constant mobile access to "the Internet of all things."

GPS tracking and real-time data improve and recreate security services as an industry. Safety on the streets with cameras and safety of houses with a system of constant monitoring reduce crime, accidents, fire and apartment flooding, or at least make such incidents more difficult to occur. From the other side, criminals can actively use mobile technology, as well, and that is a challenge.

GPS technology as a component or supporter of mobile technology creates tremendous opportunities for people in many areas of life. GPS tracking along with mobile devices helps to track disabled people and children, it makes tourism different with references to hotels and restaurants, it changes personal car driving and hiking, it helps with weather forecasts and monitoring geo data and climate changes. Drones and their use in military or postal delivery would not be possible without GPS and mobile technologies. Tracking earthquakes, floods, other natural disasters or civil strife by web giants like Google and Facebook and regular mobile operators help to find people (Dewey, 2015; Alfred, 2015).

Other Mobile Applications and Opportunities

Mobile technology with GPS content is useful in agriculture for reacting to weather forecasts. Smallholder farmers would benefit most from that. Real-time price information is valuable, as well. Farmers are getting updates from conditions in the fields, or of their cattle, and it is possible because of real-time data collection (Baumüller, 2013).

One less noticeable application has been occurring with NFC (Near Field Communication) and RFID (Radio Frequency Identification). People, devices and items can be traced or scanned with these two different "tag" types. These tags are mobile, and use can be extensive in all mass production industries.

Mobile technologies provide both opportunities and threats related to mature businesses. They influence strategies and investment decisions. Physical limitations on devices and constraints on infrastructure and security are constantly evolving. Some people are even distrustful of mobile technologies as there are risks of privacy violation and personal security (Siau & Shen, 2004).

All researchers and industry leaders agree that mobile technologies are changing society. Using mobile devices regularly changes lifestyles, habits, communications, entertainment, dating, learning, playing, nurturing, leisure, training, shopping and many other areas of life. Some of these changes eliminate old business practices, and other ones create new opportunities. There are some activities in mobile that are difficult to monetize and will be free; others that are easier to monetize or require different models of commercialization.

EVOLVING THEORIES AND MODELS

Theories and models developed before the Internet Revolution are useful in analyzing the rapid changes now occurring. However, current circumstances also frequently expose their inadequacy. New experiences, such as online viral communications, were not possible before. Scholars studied how rapidly information spread through society, but the viral-like spread of news or information through society was typically associated with such shocking tragedies as the assassination of Martin Luther King Jr. and seldom with anything at all similar to an online viral video of a spinster's vocal audition for a TV talent show on the other side of the earth (Haroldsen & Harvey, 1979). Such experiences as viralization of online messages demand new research, theories and models.

How Past Theories Are Evolving in This New Business Environment

Mobile technologies have made formidable progress over the past 20 years. In 1995 the penetration rate globally of mobile devices was 1%; in 2014 it was 73% (Meeker, 2015, p. 5). Mobile data traffic grew 69% in 2014, 81% in 2013, and 70% in 2012 (Meeker, 2015, p. 13). All these statistics demonstrate active changes in the lives of the people by mobile technologies.

Still, the mobile revolution requires construction of certain theoretical language to describe the situation. Mobile technologies can be described as crucial innovation available to use, customize and improve virtually by anyone at any time. We can use the terminology of disruptive innovation by Christensen (2014) and Moore's *Crossing the Chasm* (2014) as a description and explanation of mobile technologies' impact on business and everyday life. These two theoretical frameworks provide language to describe the processes of change.

Regarding the specific areas of mobile technology applications, different theories are evolving as addition or variation of older theories like theory of mobile learning as a part of general theory of learning, based on constructivism (Sharples, Taylor and Vavoula, 2005; Shuler, 2009). Learning and healthcare as important aspects of social and individual life got a lot of attention in the mobile technologies research. Mobile reminders by SMS proved to be effective in preventing crises – whether they be medical or business crises (Dahlke et al., 2015). Reminders and other types of intervention are effective namely with mobile devices in comparison with desktop computers because frequency of intervention is much more effective with mobile devices. If reminders can effectuate positive health habits, they can also reinforce "habits of highly effective people" (Riley et al., 2011).

Health behavior models are built with the special attention to mobile input for preventive healthcare in general and even cancer survivorship (Riley et al., 2011, Dahlke et al., 2015; Nasi, Cucciniello, & Guerrazzi, 2015). In terms of impact to society we see sociological theory of mobile phone as a changing factor of society. Mobile Internet and its social impact also get theoretical attention as a new form of cyberculture (Herman, Hadlaw, & Swiss, 2015). Mobile technologies can bridge the digital divide between North and South and rich and poor in society as they are cheaper tools to get to new information, skills and knowledge (Akhras & de Resende, 2008). We see many remote areas are skipping an entire generation of technology and will never have fixed-wire broadband Internet but rather jump directly from dial-up technology to mobile broadband (Mutter, 2012).

Moore refers to Everett Rodgers to explain in *Crossing the Chasm* the way and stages of entering the market with high-tech products. Rogers (2003) developed a model of diffusive innovations in communications and human capital. He divided the life cycle of a product according to the expectations, choices and behavior of the customers. Now we see mobile technologies as a segmented market in terms of products or devices, applications for different groups, ages, professions and services. Each of the areas can be described in Rogers' terms. The first group in this life cycle are the innovators and technology enthusiasts; second are early adopters and visionaries. Both of these groups want to have high-performing technology. Then there is a chasm between these and the next group of prospective customers that might cause the product or service to disappear if producers lose confidence in the potential of their efforts to conquer the market. If this chasm can be crossed, then early majority pragmatists, late majority conservatives and, finally, laggards and skeptics will join the army of customers. After the product or service has crossed the chasm, then customers want solutions and convenience. In other words, latter groups of customers do not want sophistication or power; they want functionality and convenience in the use of the product (Rogers 2003, pp. 252-270). Rogers' model was widely accepted, then further developed by Moore (1995, 2014) and Norman (1998).

Norman (1998) further developed the ideas of Rogers and Moore. He suggested that good products and companies might not succeed if marketing efforts are directed too much to early adopters and not to late comers. It can happen because technological innovations require a lot of investment, but larger groups of customers value more the convenience than the technological power. All these things are applicable to mobile in terms of customer targeting and business strategies in quickly developing simple and convenient products and services for wider groups of customers. Marketing must be human-centered, not technology-centered.

Because of the speed of change in technology, some new devices or applications are becoming obsolete before some customer groups recognize their advantages and benefits. User experience and word-of-mouth promotion are vital to the success of high-tech marketing. Mobile devices, applications and services are among the best examples. Without the enthusiastic promotion by innovators and early adopters, a product faces a high risk of failure. Microsoft invented a tablet PC and could not sell it because the infrastructure in terms of Web access and applications was not ready. The product was not able to cross the chasm, but later tablet PCs became popular. This is a clear case of innovation diffusion. HP introduced its palm computer as a smartphone in its initial version. Still it was too early, too expensive and not valued by prospective customers. Now smartphone sales outperform other cell phones and are providing great value in a world with easy access to the Internet and related services. The technology adoption life cycle, with customers segmented into innovators, early adopters, early majority, late majority and laggards, relates to mobile technologies as they spread through the global market. The *Crossing the chasm* model describes and explains the market development of mobile technologies. As part of the high tech sector, they are disruptive and diffusive innovations within a rapidly reacting marketplace.

Segmentation of the market into such classifications as innovators, early majority or laggards is a kind of reductionism as any classification would be. Classification can be explained by the variety of factors, such as age, income, awareness, and regional or global placement. Why are some people more prone to these innovations? Is it psychological or do other factors influence, as well. These aspects have already attracted attention of researchers and need further research efforts. It is movement from grand theory to new theoretical development addressing specific details and features.

"Disruptive technologies" as a concept was introduced by Bower and Christensen (1995) and can be traced back to Josef Schumpeter with his idea of creative destruction, which in turn was adopted from philosophies of Marx and Hegel. Disruptive technologies conceptually are filling the gaps in the *Crossing the Chasm* model. Bower and Christensen paid attention to the loss of vision by the leading companies in the market when they disregard the technologies that are not yet popular among customers. Leading companies follow their customers too closely and ignore major technological breakthroughs because their mainstream customers' needs are not yet satisfied by these technologies. It is a paradox because customer needs are then formed in conjunction with the new technologies, starting perhaps with a small group of customers (innovators or early majority in the Moore model) who then change the market and preferences of mainstream customers. Christensen (2014) explained that it was for this reason that RCA ignored transistors, clung to its vacuum tube technology, and allowed Sony to control the lower end of the market with lower-quality, smaller and cheaper radios and TVs. But as the quality of transistors improved, Sony gained control of more and more of the marketplace until it was too late for RCA to maintain its position as the world's top manufacturer of audio-video equipment. Disruptive technologies continue to turn the world upside down. Consider who controlled the media industry just 10 years ago. It wasn't Google/YouTube, Facebook and Twitter. It was companies like Gannett, which is in the process of splitting off its publishing operations in order to protect its growing broadcasting and digital media. As explained in its announcement in USA Today: "The owner of USA Today, founded in 1906 as a small newspaper publisher in Upstate New York, joined a quickly growing list of media companies — including News Corp., Time Warner and Tribune media in about the past year — that have embraced the strategy to shield more profitable business lines from the decline in print advertising" (Yu, 2014). Other industries will see similar disruption as new technologies and new applications of technology open up new opportunities for upstarts and show other top-tier companies that success can no longer be maintained by continuing to do things the same way as in the past. And mobile technologies are at the cutting edge of these changes.

Another theoretical construct gaining importance relates to Gladwell's book, *The Tipping Point: How Little Things Can Make a Big Difference* (2000). Well ahead of the social media revolution, Gladwell described how "ideas and products and messages and behaviors spread like viruses do" (p. 7) until they effect such changes in society as to cause a situation to reach a tipping point, defined as "the moment of critical mass, the threshold, the boiling point" (p. 12). Due to the rapid yet subtle changes occurring in society because of disruptive technologies, we are seeing more frequent and important "tipping points" occurring. What subtle changes have to take place before suddenly an entire industry, such as the publishing industry, faces potentially catastrophic consequences. The Washington Post was hiring more news employees one year and laying them off the next (Downie, 2009). Everyone in the industry knew the Internet would probably destroy newspapers eventually – but how quickly the impact hit surprised everyone. On the more positive side, what subtle changes have to occur before a company like Facebook suddenly erupts to dominate the new media industry. In 2014 Facebook jumped from a distant second place in online video views past YouTube and beyond, and its advertising revenues reflect it. As noted earlier, between January 2014 and February 2015, Facebook's monthly desktop video views jumped 92%, and the number of videos viewed monthly via desktop per individual jumped 83%. But Facebook claimed that desktop had become a minor part of its audience (Mixpo, 2015, p. 3). Supporting this dramatic rise and Facebook's claim to leadership in mobile communications, AdWeek reported that Facebook's ad revenues jumped 46% in the first quarter of 2015 to $3.32 billion, and mobile advertising accounted for 73% of its ad sales, as opposed to 69% the previous quarter (Heine, 2015). Advertising follows the audience, so we can expect more advertising to move to Facebook over the subsequent months. Additional research on how online communications goes viral through society, and how this is causing major tipping points to occur more frequently is needed by the business community and important for academic researchers to pursue, as well.

Weaknesses of theoretical models are coming from their inability to explain new data and experiences. It makes sense to have plurality of theories with their own language to explain the complex and fast-changing reality related to mobile technologies. It brings forward new theories and models with different angles and perspectives.

One problem with theories related to disruptive technologies and innovations in high-tech industries is that they do not address such issues as the problem of copying. It can take so much money and research initially to make a product, but to copy it is much cheaper. Copyrights are difficult to protect in the environment of fast-paced and continuous introduction of new inventions and product improvements.

The pace of mobile technology development is a particular challenge for theoretical understanding and assessment because theories and research take time to develop and mobile developments are so fast. Research can take several years to conduct, but mobile devices are quickly becoming more powerful and accessible, as are mobile apps and business strategies to influence the behavior of mobile users, so that keeping up with academic research is impossible. For example, some healthcare research focuses on SMS reminding as a tool in preventive care, but new devices already operate with automatic healthcare applications with accompanying equipment. Thus, SMS as a reminding tool is becoming somewhat outdated and less relevant.

Mobile technologies in business are facing a continuous issue of monetization. Barriers are low for application developers and becoming lower for device producers, the market is competitive, growth is explosive, copying with small changes is easy, audiences are huge and mobile technology is so pervasive that to make money out of this combination of factors is possible but requires entrepreneurial ingenuity.

Mobile technologies change traditional or mature business sectors and they open new areas of creating value where people are willing to pay. Mature businesses use mobile as an additional, convenient and even necessary tool. However each application of mobile technologies might have a particular rather than universal business model and monetization scheme. Apple's business model is selling hardware or devices and charging fees on software and media downloads. Social networks work on focused advertisement and product promotion. Banking, auctions, learning and healthcare applications have different ways to achieve revenues for transactions or services.

Personal privacy and security are still important issues relating to ethics, commercial use, security monitoring by state agencies, the selling of individual data by commercial companies, and fraud. One of the most popular means of advertising, for example, is pay per click (PPC), but there is evidence that half of the clicks are machine rather than human generated. The short life cycle of the devices, the development of new programming, and the need for constant patching create additional technological security issues.

The advantages brought by Internet technologies and mobile technologies in particular are so huge that they outweigh the risks and make the issues solvable. Still these issues demand attention by researchers. And research without an immediate commercial value will probably not be pursued by anyone but academics.

MAPPING FUTURE RESEARCH

Because of the inclusive nature of business, almost every chapter of this book relates to business in some way. That's how broad business research needs to be. And digital business tools, applications and practices are seem to be changing by the minute.

- **Spreading Innovation:** Rodger's Crossing the Chasm and Christensen's disruptive technologies research discussed in the previous section is not only explaining how new innovations spread but how new businesses can overthrow well-established enterprises within the marketplace. This is ongoing research that demands further research in this rapidly changing business environment. Related to this is Gladwell's tipping point research. We have discussed the previous tipping point in the newspaper industry that has seen massive layoffs and loss of revenue, and what appears to be an imminent tipping point related to TV/cable. These represent massive disruption in the marketplace. The current state of innovation is unparalleled in human history, and right now mobile technologies are having some of the greatest impact worldwide. Christensen's innovation research (2014) seems to be some of the most useful currently because it does more than explain past changes; it actually suggests formulas for how entrepreneurs can successfully enter the marketplace and eventually overcome the top competitors. There is certainly room for additional research in this field.

- **Achieving Viralization:** Closely related to the tipping point concept is viralization of online messages. There are lots of opinions and useful best practices, but currently it is much more an art than a science. What does cause online videos or social media postings to go viral and spread to millions of people? More research can gradually add scientific clarity to this important topic.

- **Enhancing Marketing:** As reviewed in this chapter and expanded elsewhere in this book, new and/or rapidly growing concepts in advertising are developing before our eyes: inbound marketing, native advertising, programmatic advertising, online video, personalization, and omnichannel marketing. All of these demand more attention by researchers. While some of these have off-line parallels, the online environment and the growing move to mobile means research in these areas has had to begin almost from scratch. Research should address best and more effective practices, new related technologies and applications, ethical and legal considerations, regional differences within the global market, cultural/linguistic implications, and much more.

- **Expanding Distribution:** Long-tail strategies and now 3D printing services are important areas of research. Long-tail strategies refer to the advantages of online enterprises that offer digital products. A brick-and-mortar music store, for example, might stock the top 1,000 music albums while the online music store can digitally stock an almost infinite number. With digital storage costs so low, the sale of less popular music can be very profitable. If after the top 1,000 albums, the next 9,000 sell only one a month, that's still 9,000 album sales the brick-and-mortar store would not have. 3D printing is expanding the number of products that can now be digitized. Recent stories, for example, have shown disabled children being provided with plastic prosthetics they could not previously afford (Owen, 2015). And 3D printing is just catching hold. The major obstacle blocking 3D printing from achieving its tipping point is cost to consumers. But the recent survey by Blueshift Research (2015) found that while only about 5% of respondents had actually used a 3D printer in the last three months, 42% indicated that they are at least somewhat likely to buy a 3D printer once the price declines below $400, and about 10% are "extremely likely." What is its future of 3D printing, what obstacles need to be overcome to greatly increase the number of products available, and what strategies can be used to enhance this new method of digital distribution? This could have the greatest impact of all the current research.

- **Empowering Human Resources:** As discussed earlier in this chapter, mobile communications, internal social media, and related strategies can enhance the productivity of employees by 20-25% (Chui et al., 2012), and it is really impossible to put a value on the creativity and synergy that can be achieved by cross-functional teams these strategies enable (Melcrum, 2014). What is the most effective use of these technologies in achieving such results? There is much more research to be conducted relating to such an important topic. Melcrum's studies suggest that a key to making cross-functional teams more effective is providing necessary and timely training related to their assigned task. The use of mobile technologies in the field of employee training will be of increasing importance in this age of rapid technological change. As noted earlier in this chapter, 97% of the B2B marketing executives reported having to perform new tasks to keep up with these changes (Ramos, 2013), 96% believed the rate of change is increasing, and 34% are feeling overwhelmed. Similar impact is certainly not being felt by only marketing executives but by employees company-wide. While new technologies and applications are causing these challenges to occur, they can also help solve the accompanying problems. An important area of research, therefore, is how lifelong learning, just-in-time training and support for employees can be best implemented.

- **Spreading Application:** Overlapping with the need for training is the need to help employees and consumers understand and adopt effective new applications. This is becoming a field of its own, requiring expert assistance for lack of commonly used directories or support systems for those with less technical expertise or interest. Many people have smartphones and do not use them at full capacity. New applications for health and education, for example, can have great positive impact on society if employed – but most people don't know about them. How can existing technologies and applications be diffused more rapidly across society when the rate of innovation is so overwhelming? This field deserves more research and perhaps more innovation in monetizing such essential services.

- **Applying Big Data:** Big data is finally being used at least by big organizations to enhance automated services and personalized communications. Still the potential is greater than what is currently being achieved even by the largest organizations, much more for smaller organizations yet to employ the related technologies. This is a field that relates to both the potential to be achieved and the necessary limitations to be set. Research is needed in both areas. Business demands more effective marketing systems and strategies, but society demands less invasive and more ethical use of information previously considered private.

- **Humanizing The Digital World:** Researchers are discovering that in some cases it is more profitable to be less digital. Consumers want to feel that they are being treated like individuals. While Big Data can personalize messages to each, consumers can tell the difference between personalization and the true human touch. Artificial intelligence will gradually blur those lines, but for now there are technologies that allow customer service and other key departments of a business to cost-effectively provide consumers access to real people, as needed. LivePerson (2015), for example, is so sure that its technology will sell itself to businesses that it offers one log-in with unlimited engagements for $0. It and other organizations argue persuasively that their systems allow businesses to increase conversion rates by 25%, decrease bounce rates, and distinguish a brand from its competitors by offering a human personality. Some humanizing systems allow customer service to text chat, show prerecorded videos when helpful, talk live with audio only or with live video, and more. This book's chapter on mobile PR covers research related to this issue in more depth. But clearly this seemingly anti-technology strategy also deserves more study. What are the limitations of the digital approach, and when are real people more cost-effective than technology?

The field of business and the impact of mobile technologies is far too broad to touch on even a small portion of the important issues demanding increased research. These are just a few, and, as suggested earlier, most chapters in this entire book somehow relate back to business, since business is all about monetizing innovation.

REFERENCES

Aberdeen Group. (2014). *Analyzing the ROI of video advertising.* Retrieved February 16, 2015, from http://go.brightcove.com/bc-aberdeen-analyzing-roi-b

Adobe. (2015). *Quarterly digital intelligence briefing: Digital trends 2015.* Retrieved February 20. 2015, from http://offers.adobe.com/content/dam/offer-manager/en/na/marketing/Target/Adobe%20Digital%20 Trends%20Report%202015.pdf

Ahonen, T. (2002). m-Profits: Making money from 3G services. Hoboken, NJ: Wiley.

Akhras, F. N., & de Rezende, E. D. (2008). *Digital inclusion in social contexts: A perspective to the use of mobile technologies.* Retrieved June 10, 2015, from http://www.w3.org/2008/10/MW4D_WS/ papers/akhras.pdf

Alfred, C. (2015, April 27). How families abroad are tracking down missing loved ones after Nepal's earthquake. *The Huffington Post.* Retrieved May 7, 2015, from http://www.huffingtonpost.com/2015/04/27/ nepal-earthquake-missing_n_7147894.html

Anderson, M. (2015, January 7). *37% of SMBs plan to spend more on internet marketing in2015.* Retrieved from https://www.brightlocal.com/2015/01/07/37-smbs-plan-spend-internet-marketing-2015/

AT&T. (2014). *AT&T small business technology poll 2014.* Retrieved April 23, 2015, from http://about. att.com/mediakit/2014techpoll

Banerjee, S., & Dholakia, R. R. (2013). Situated or ubiquitous? A segmentation of mobile e-shoppers. *International Journal of Mobile Communications, 11*(5), 530–557. doi:10.1504/IJMC.2013.056959

Barnes, S. J. (2003). *Mbusiness: The strategic implications of mobile communications.* Oxford: Butterworth and Heinemann.

Basole, R. (Ed.). (2008). *Enterprise mobility: Applications, technologies and strategies.* Amsterdam: IOS Press.

Basole, R. C. (2008). Enterprise mobility: Researching a new paradigm. *Information-Knowledge-Systems Management, 7*(1-2), 1-7.

Baumüller, H. (2013). Mobile technology trends and their potential for agricultural development (November 2013). *ZEF Working Paper 123.* Retrieved April 23, 2015, from http://ssrn.com/abstract=2359465

Beer, M., & Nohria, N. (2000). Cracking the code of change. *Harvard Business Review.* Retrieved May 23, 2015, from http://webdb.ucs.ed.ac.uk/operations/honsqm/articles/change2.pdf

Bersin, J. (2014). The Red Hot Market for Learning Management Systems. *Forbes.* Retrieved July 31, 2015, from http://www.forbes.com/sites/joshbersin/2014/08/28/the-red-hot-market-for-learning-technology-platforms/1

Blueshift Research. (2015). *June 2015 trends tracker report.* Retrieved 2 July 2015 from https://smaudience.surveymonkey.com/download-trends-tracker-report-june-2015.html?utm_source=email&recent=email&program=Q2_15_June_Trends-Tracker_Wrap-Up&source=email&utm_campaign=trends-tracker&utm_content=trends-tracker-june-2015&mkt_tok=3RkMMJWWfF9wsRoiu6vJZKXonjHpfsX66e4vW6C1lMI%2F0ER3fOvrPUfGjI4FTcFiI%2BSLDwEYGJlv6SgFSrbFMaJy2LgJWBb0TD7slJfbfYRPf6Ba2Jwwrfg%3D

Boston Consulting Group. (2014). The mobile revolution: How mobile technologies drive a trillion-dollar impact. *Boston Consulting Group Report.* Retrieved May 27, 2015, from https://www.bcgperspectives.com/content/articles/telecommunications_technology_business_transformation_mobile_revolution/?chapter=4

Bower, J., & Christensen, C. (1995). Disruptive technologies: Catching the wave. *Harvard Business Review.* Retrieved June 12, 2015, from https://hbr.org/1995/01/disruptive-technologies-catching-the-wave

Brightcove. (2015). *The hero's guide to video marketing.* Retrieved May 1, 2015, from http://go.brightcove.com/brightcove-video-hero-guide?cid=70114000002QvKv&pid=70114000002Qu9G

Carufel, R. (2013, May 30). *Marketing's uncharted territory: 97% of B2B marketers are doing work they've never done before and expect the pace of change to accelerate as they "navigate chaos".* Retrieved April 23, 2015, from https://www.bulldogreporter.com/dailydog/article/pr-biz-update/marketings-uncharted-territory-97-of-b2b-marketers-are-doing-work-the

Carufel, R. (2015, March 31). *Technology PR standoff: Research finds consumers and retailers diverge on tech—buyers want more, retailers don't.* Retrieved April 23, 2015, from https://www.bulldogreporter.com/dailydog/article/pr-biz-update/technology-pr-standoff-research-finds-consumers-and-retailers-diverge

Castells, M., Fernández-Ardèvol, M., Qiu, J. L., & Sey, A. (2006). *Mobile communication and society: A global perspective (information revolution & global politics).* Cambridge, MA: MIT Press.

Cavoukian, A., & Prosch, M. (2010). *The Roadmap for Privacy by Design in Mobile Communications: A Practical Tool for Developers, Service Providers, and Users.* Retrieved December 16, 2015, https://www.privacybydesign.ca/content/uploads/2011/02/pbd-asu-mobile.pdf

Chernov, J. (2014). State of inbound 2014. *HubSpot.* Retrieved May 21, 2015 from http://www.stateofinbound.com/

Christensen, C. M. (2014). Disruptive innovation. In *The Encyclopedia of Human-Computer Interaction,* 2nd ed. Retrieved 26 June 2015 from https://www.interaction-design.org/encyclopedia/disruptive_innovation.html

Chui, M., Manyika, J., Bughin, J., Dobbs, R., Roxburgh, C., Sarrazin, H.,... Westergren, M. (2012). *The social economy: Unlocking value and productivity through social technologies*. Retrieved 29 April 2015 from http://www.mckinsey.com/insights/high_tech_telecoms_internet/the_social_economy

Ciaramitaro, B. L. (Ed.). (2011). *Mobile technology consumption: Opportunities and challenges*. Hershey, PA: IGI Global.

Cisco (2015). *Forecast projects nearly 10-fold global mobile data traffic growth over next five years*. Retrieved February 16, 2015, from http://newsroom.cisco.com/press-release-content?type=webcontent&articleId=1578507

comScore. (2014). *Half of millennial netflix viewers stream video on mobile*. Retrieved February 20, 2015, from http://www.comscore.com/Insights/Data-Mine/Half-of-Millennial-Netflix-Viewers-Stream-Video-on-Mobile

comScore. (2015). *U.S. digital future in focus 2015*. Retrieved May 19, 2015, from http://www.comscore.com/Insights/Presentations-and-Whitepapers/2015/2015-US-Digital-Future-in-Focus

Consulting, P. A. (2015). *Wearable Technologies Improves Health Outcomes and Reduces Costs*. Retrieved May 2, 2015, from http://www.paconsulting.com/introducing-pas-media-site/highlighting-pas-expertise-in-the-media/articles-quoting-pa-experts/wfmd-am-930-wearable-technology-improves-health-outcomes-and-reduces-cost-10-march-2015/

Corrocher, N., & Zulrilia, L. (2004). *Innovation and Schumpeterian competition in the mobile communications service industry*. Università Commerciale L. Bocconi, Milan. Retrieved May 1, 2015, from http://www2.druid.dk/conferences/viewpaper.php?id=2577&cf=17

Cristian, P. E., Thirumurthy, H., Habyarimana, J. P., Zivin, J. G., Goldstein, M. P., Damien, D. W., & Bangsberg, D. J. et al. (2011). Mobile phone technologies improve adherence to antiretroviral treatment in a resource-limited setting: A randomized controlled trial of text message reminders. *AIDS (London, England)*, 25(6), 825–834. doi:10.1097/QAD.0b013e32834380c1 PMID:21252632

Dahlke, D. V., Fair, K., Hong, Y. A., Beaudoin, C. E., Pulczinski, J., & Ory, M. G. (2015). Apps seeking theories: Results of a study on the use of health behavior change theories in cancer survivorship mobile apps. *JMIR mHealth and uHealth, 3*(1).

Dahlstrom, E., & Bichsel, J. (2014). *ECAR Study of Undergraduate Students and Information Technology, 2014*. Louisville, CO: ECAR, Oct. 2014, retrieved from http://www.educause.edu/library/resources/2014-student-and-faculty-technology-research-studies

Dahlstrom, E., Walker, J. D., & Dziuban, C. (2013). *ECAR Study of Undergraduate Students and Information Technology, 2013*. Louisville, CO: ECAR (EDUCAUSE Center for Analysis and Research), retrieved from http://www.educause.edu/library/resources/ecar-study-undergraduate-students-and-information-technology-2013

Darrell, M. W. (2013). *Mobile learning: Transforming education, engaging students, and improving outcomes*. Retrieved March 23, 2015, from http://www.brookings.edu/~/media/research/files/papers/2013/09/17-mobile-learning-education-engaging-students-west/brookingsmobilelearning_final.pdf

Davies, J. (2013, May 7). Consumer magazines are failing to capitalise on mobile ad opportunities, says Monotype report. *The Drum*. Retrieved from http://www.thedrum.com/news/2013/05/07/consumer-magazines-are-failing-capitalise-mobile-ad-opportunities-says-monotype

Dewey, C. (2015, April 27). How Google and Facebook are finding victims of the Nepal earthquake. *The Washington Post*. Retrieved May 7, 2015, from http://www.washingtonpost.com/news/the-intersect/wp/2015/04/27/how-google-and-facebook-are-finding-victims-of-the-nepal-earthquake/

Digital, D. (2015). *Navigating the new digital divide*. Retrieved May 25, 2015, http://www2.deloitte.com/content/dam/Deloitte/us/Documents/consumer-business/us-cb-navigating-the-new-digital-divide-v2-051315.pdf

Downie, L. (2009). *The future of journalism: Where we've been, where we're going*. Presentation delivered at Stanford University. Retrieved June 25, 2015, https://youtu.be/JYtOKnk5fiw?list=PLfSN69RCPDM-HID_MaJ7TxlfwMsnMAqPc

Duffy, M. (2011). Smartphones in the Arab Spring. *International Press Institute 2011 Report*. Retrieved May 2, 2015, from http://www.academia.edu/1911044/Smartphones_in_the_Arab_Spring

Economic Times. (2015). *Twitter buys Indian mobile marketing start-up ZipDial*. Retrieved May 1, 2015, from http://articles.economictimes.indiatimes.com/2015-01-21/news/58306142_1_zip-dial-amiya-pathak-sanjay-swamy

eMarketer. (2013a). *Digital TV, movie streaming reaches a tipping point*. Retrieved 20 Feb. 2015 from http://www.emarketer.com/Article.aspx?R=1009775

eMarketer. (2013b). *TV advertising keeps growing as mobile boosts digital video spend*. Retrieved 20 Feb. 2015 from http://www.emarketer.com/Article/TV-Advertising-Keeps-Growing-Mobile-Boosts-Digital-Video-Spend/1009780

eMarketer. (2014). *Despite time spent, mobile still lacks for ad spend in the US*. Retrieved 20 Feb. 2015 from http://www.emarketer.com/Article/Despite-Time-Spent-Mobile-Still-Lacks-Ad-Spend-US/1010788

eMarketer. (2015a). *Global tablet audience to total 1 billion this year*. Retrieved 19 May 2015 from http://www.emarketer.com/Article/Global-Tablet-Audience-Total-1-Billion-This-Year/1012451/1

eMarketer. (2015b). *Mobile content and activities roundup*. Retrieved 10 April 2015 from https://apsalar.com/blog/2015/04/emarketers-mobile-content-and-activities-roundup-now-available/

eMarketer. (2015c). *How much Yahoo traffic is mobile?* Retrieved 21 May 2015 from http://www.emarketer.com/Article/How-Much-Yahoo-Search-Traffic-Mobile/1012378/1

eMarketer. (2015d). *Most digital video monetization still comes from ads*. Retrieved 23 May 2015 from http://www.emarketer.com/Article/Most-Digital-Video-Monetization-Still-Comes-Ads/1012300/1

eMarketer. (2015e). *Mobile phones strengthen lead for mobile video viewing*. Retrieved 2 July 2015 from http://www.emarketer.com/Article/Mobile-Phones-Strengthen-Lead-Mobile-Video-Viewing/1012683?ecid=NL1001 eMarketer. (2015f). *Programmatic creative: Look to existing processes for guidance.* Retrieved 27 June 2015 from http://www.emarketer.com/Article/Programmatic-Creative-Look-Existing-Processes-Guidance/1012434#sthash.jZrmND1V.dpuf

eMarketer. (2015g). *Better site optimization lifts mobile conversion rates.* Retrieved 23 May 2015 from http://www.emarketer.com/Article/Better-Site-Optimization-Lifts-Mobile-Conversion-Rates/1012482/1

eMarketer. (2015h). *Seven in 10 US Internet Users Watch OTT Video.* Retrieved 15 Oct. 2015 from http://www.emarketer.com/Article/Seven-10-US-Internet-Users-Watch-OTT-Video/1013061?ecid=NL1001

Evergage. (2013). *Real-time for the rest of us: Perceptions of real-time marketing and how it's achieved.* Retrieved 18 May 2015 from http://info.evergage.com/perceptions_of_realtime_marketing_survey_results

Federal Reserve Board. (2015). *Consumers and Mobile Financial Services 2015.* Retrieved May 6, 2015, from http://www.federalreserve.gov/econresdata/consumers-and-mobile-financial-services-report-201503.pdf

Fulgoni, G. (2014). The State of Mobile Industry. *comScore.* Retrieved May 19, 2015, from http://www.comscore.com/Insights/Presentations-and-Whitepapers/2014/The-State-of-the-Mobile-Industry

Gafni, R., & Geri, N. (2013). Generation Y versus Generation X: Differences in Smartphone Adaptation. In *Proceedings of the Chais conference on Instructional Technologies Research 2013: Learning in the technological era.* Raanana: The Open University of Israel. Retrieved March 27, 2015, from http://www.openu.ac.il/innovation/chais2013/download/b1_4.pdf

Gladwell, M. (2000). *The tipping point: How little things can make a big difference.* New York, NY: Little, Brown and Company.

Global Pulse, U. N. (2015). *Using mobile phone data and airtime credit purchases to estimate food security 2015.* Retrieved May 30, 2015, from http://unglobalpulse.org/sites/default/files/Topups_Food%20Security_WFP_Final.pdf

Google. (2014). *Helping users find mobile-friendly pages.* Google Webmaster Central Blog. Retrieved May 21, 2015, from http://googlewebmastercentral.blogspot.com/2014/11/helping-users-find-mobile-friendly-pages.html

Guy, R. (2011). *Digitally Disinterested.* Santa Rosa, CA: Informing Science Press.

Hall, E., Charles S, Fottrell, S. W., & Byass, P. (2014). Assessing the impact of mHealth interventions in low- and middle-income countries – what has been shown to work? *Global Health Action, 7.*

Haroldsen, E., & Harvey, K. (1979). The diffusion of shocking good news. *The Journalism Quarterly, 56*(4), 771–775. doi:10.1177/107769907905600409

Harvard Business Review Analytic Services. (2012a). How mobility is changing the enterprise. *Harvard Business Review.* Retrieved from https://global.sap.com/campaigns/02_cross_mobile_havard_business_review_reports/HBR_HowMobilityIsChangingThe%20Enterprise.pdf

Harvard Business Review Analytic Services. (2012b). How mobility is transforming industries. *Harvard Business Review.* Retrieved from https://global.sap.com/campaigns/02_cross_mobile_havard_business_review_reports/HBR_HowMobilityIsTransformingIndustires.pdf

Harvard Business Review Analytic Services. (2012c). How mobility is changing the world. *Harvard Business Review*. Retrieved from https://global.sap.com/campaigns/02_cross_mobile_havard_business_review_reports/HBR_HowMobilityIsChangingTheWorld.pdf

Harvey, K. (2011). Using Internet tools to promote locally. *KIRC Proceedings 2011*. Retrieved 3 July 2015 from http://virtual-institute.us/Ken/KIRCharvey2011.pdf

Heine, C. (2015, April 22). *Facebook's Q1 ad revenue increased 46% to $3.3 billion*. AdWeek. Retrieved 30 June 2015 from http://www.adweek.com/news/technology/facebooks-q1-ad-revenue-increased-46-percent-33-billion-164234

Herman, A., Hadlaw, J., & Swiss, T. (Eds.). (2014). *Theories of the mobile Internet: Materialities and imaginaries*. New York, NY: Routledge.

Hoang, A. T., Nickerson, R. C., Beckman, P., & Eng, J. (2008). Telecommuting and corporate culture: Implications for the mobile enterprise. *Information, Knowledge, Systems Management*, 7(1), 77.

HubSpot. (2012). *120 awesome marketing stats, charts and graphs*. Retrieved May 20, 2015, from http://offers.hubspot.com/120-awesome-marketing-stats-charts-and-graphs

HubSpot. (2014). *State of Inbound 2014*. Retrieved 4 July 2014 from http://offers.hubspot.com/2014-state-of-inbound

IEI-TV. (2015). *What has your Wi-Fi done for you lately?* Retrieved 29 June 2015 from http://iei-tv.net/wifi.html

Information Solutions Group. (2013). *Pop Up Games Mobile Games Research, 2013*. Retrieved May 7, 2015, from http://www.infosolutionsgroup.com/popcapmobile2013.pdf

International Telecommunications Union. (2015). *Mobile-broadband penetration approaching 32 per cent, Three billion Internet users by end of this year* [Press release]. Retrieved from https://www.itu.int/net/pressoffice/press_releases/2014/23.aspx#.VZ_GcmCdIW0

Irwin, J. (2013). *Modernizing Education and Preparing Tomorrow's Workforce through Mobile Technology*. Paper presented at the i4j Summit. Retrieved June 27, 2015, from http://www.meducationalliance.org/sites/default/files/qualcomm_-_modernizing_education_and_the_workforce_through_mobile_technology.pdf

Istepanian, R., Laxminarayan, S., & Pattichis, C. S. (2007). *M-Health: Emerging mobile health systems*. New York, NY: Springer Science & Business Media.

Jacobs, I. M. (2013). *Modernizing education and preparing tomorrow's workforce through mobile technology. Innovation for Jobs Summit*. QUALCOMM.

Kapko, M. (2015). How mobile, social tech elevate enterprise collaboration. *The CIO Agenda*. Retrieved May 7, 2015, from http://www.cio.com/article/2873207/collaboration/how-mobile-social-tech-elevate-enterprise-collaboration.html

Kathy, G. (2015) *Mobile still tops retailers' priority lists, according to* Shop.org/Forrester *research report*. Retrieved April 23, 2015, from https://nrf.com/media/press-releases/mobile-still-tops-retailers-priority-lists-according-shoporgforrester-research

Kilcourse, B., & Rowen, S. (2014, February). *Mobile In retail: Reality sets in.* Retrieved April 23, 2015, from http://www.sap.com/bin/sapcom/en_us/downloadasset.2014-06-jun-10-18.mobile-in-retail-reality-sets-in--rsr-benchmark-report-2014-pdf.bypassReg.html

Lewis, J. (2015). *San Diego engages customers with new technologies.* PA Consulting. Retrieved May 2, 2015, from http://www.paconsulting.com/introducing-pas-media-site/highlighting-pas-expertise-in-the-media/opinion-pieces-by-pas-experts/intelligent-utility-san-diego-engages-customers-with-new-technologies-17-february-2015/

Lim, J. S., Ri, S. Y., Egan, B. D., & Biocca, F. (2015). The cross-platform synergies of digital video advertising: Implications for cross-media campaigns in television, Internet and mobile TV. *Computers in Human Behavior, 48,* 463–472. doi:10.1016/j.chb.2015.02.001

LivePerson. (2015). *LiveEngage transforms the connection between brands and consumers.* Retrieved 2 July 2015 from http://www.liveperson.com/

Luanrattana, R., Win, K. T., & Fulcher, J. (2007). Use of Personal Digital Assistants (PDAs) in Medical Education. *20th IEEE International Symposium on Computer-Based Medical Systems CBMS 2007.* Retrieved May 5, 2015, from http://ro.uow.edu.au/cgi/viewcontent.cgi?article=1657&context=infopapers

Mallikarjunan, S. (2014). The state of ecommerce marketing. *Hubspot.* Retrieved May 21, 2015, from http://offers.hubspot.com/state-of-ecommerce-marketing-2014

Meeker, M. (2015). Internet trends 2015 – Code Conference. *Kleiner Perkins Annual Report.* May 27, 2015. Retrieved May 28, 2015, from http://kpcbweb2.s3.amazonaws.com/files/90/Internet_Trends_2015.pdf?1432738078

Melcrum. (2014). *Inside Internal Communication: Groundbreaking innovations for a new future.* Executive summary retrieved 30 April 2015 from http://www.fruitfulconversations.co.uk/wp-content/uploads/2013/10/Melcrum-Inside-Internal-Comms-2.pdf

Melcrum. (2015). *Case studies.* Retrieved 30 April 2015 from https://www.melcrum.com/resources/case-studies

Merkle/RKG. (2015*). Digital marketing report Q1 2015.* Retrieved May 21, 2015, from http://www.rimmkaufman.com/content/quarterly/Merkle-RKG-Q1-2015-DMR.pdf

Mixpo. (2015). *The state of video advertising.* Retrieved May 1, 2015, from http://marketing.mixpo.com/acton/fs/blocks/showLandingPage/a/2062/p/p-00b4/t/page/fm/0

Mlot, S. (2012). Microsoft patent tips 360-degree gaming experience. *PC Magazine.* Retrieved June 26, 2015, from http://www.pcmag.com/article2/0,2817,2409621,00.asp

Mohamed, A. (Ed.). (2009). *Mobile learning: Transforming the delivery of education and training.* AU Press. Retrieved May 28, 2015, from http://www.zakelijk.net/media/boeken/Mobile%20Learning.pdf

Moore, G. (2014). *Crossing the chasm, marketing and selling disruptive products to mainstream customers.* Dunmore, PA: HarperCollins Publishers.

Moore, G. (2014). *Inside the Tornado: Marketing Strategies from Silicon Valley's Cutting Edge.* New York: HarperCollins Publishers.

Myles, A. (January 7, 2015). *SMB marketing survey 2014*. Retrieved April 23, 2015, from http://www. brightlocal.com/2015/01/07/37-smbs-plan-spend-internet-marketing-2015/

Nanji, A. (May 17, 2013) *How small-business owners are using mobile technology*. Retrieved April 23, 2015, from http://www.marketingprofs.com/charts/2013/10791/how-small-business-owners-are-using-mobile-technology

Nasi, G., Cucciniello, M., & Guerrazzi, C. (2015). The role of mobile technologies in health care processes: The case of cancer supportive care. *Journal of Medical Internet Research, 17*(2), e26. doi:10.2196/jmir.3757 PMID:25679446

National Retail Federation. (2015, February 6). *Mobile still tops retailers' priority lists, according to-Shop.org/Forresterresearch report* [Press release]. Retrieved from https://nrf.com/media/press-releases/mobile-still-tops-retailers-priority-lists-according-shoporgforrester-research

Nemiroski, A., Christodouleas, D. C., Hennek, J. W., Kumar, A. A., Maxwell, E. J., Fernández-Abedul, M. T., & Whitesides, G. M. (2014). Universal mobile electrochemical detector designed for use in resource-limited applications.*Proceedings of the National Academy of Sciences, 111(*33), 11984-11989. doi:10.1073/pnas.1405679111

NewZoo Games Market Research. (2014, October 29). *Global mobile games revenues to reach $25 billion in 2014*. Retrieved May 7, 2015, from http://www.newzoo.com/insights/global-mobile-games-revenues-top-25-billion-2014/

Norman, D. A. (1998). *The Invisible Computer: Why Good Products Can Fail, the Personal Computer Is So Complex and Information Appliances are the Solution*. Cambridge, MA: MIT Press.

Nyíri, K. (Ed.). (2003a). *Mobile Communication: Essays on Cognition and Community*. Vienna, Austria: Passagen Verlag.

Nyíri, K. (Ed.). (2003b). *Mobile Learning: Essays on Philosophy, Psychology and Education*. Vienna, Austria: Passagen Verlag.

Nyíri, K. (Ed.). (2003c). *Mobile Democracy: Essays on Society, Self and Politics*. Vienna, Austria: Passagen Verlag.

Nyíri, K. (Ed.). (2005). *A Sense of Place: The Global and the Local in Mobile Communication*. Vienna, Austria: Passagen Verlag.

Nyíri, K. (Ed.). (2006).*Mobile Understanding: The Epistemology of Ubiquitous Communication*. Vienna, Austria: Passagen Verlag.

Nyíri, K. (Ed.). (2007). *Mobile Studies. Paradigms and Perspectives*. Vienna, Austria: Passagen Verlag.

Okazaki, S. (2012). Teaching mobile advertising. In *Advertising Theory* (pp. 373–387). New York: Routledge.

Okunbor, D., & Guy, R. (2008). *Analysis Of A Mobile Learning Pilot Study*. Math and Computer Science working papers 01/2008

Owen, J. (2015). *E-NABLING hands across the world*. Retrieved 2 July 2015 from http://enablingthefuture.org/tag/3d-printed-prosthetics/

Pappas, C. (2015, January 25). The Top eLearning Statistics and Facts for 2015 You Need to Know. *eLearning Industry*. Retrieved August 5, 2015, from http://elearningindustry.com/elearning-statistics-and-facts-for-2015

Parker, M. (2012). Ethical considerations related to mobile technology use in medical research. *Journal of Mobile Technology in Medicine, 1*(3), 50–52. doi:10.7309/jmtm.23

Pearson, C., & Hussain, Z. (2015). Smartphone use, addiction, narcissism, and personality: A mixed methods investigation. *International Journal of Cyber Behavior, Psychology and Learning, 5*(1), 17–32. doi:10.4018/ijcbpl.2015010102

Perry, M. (2013). *Schumpeterian Gales of creative destruction in the personal computing market have destroyed the Windows' monopoly*. Retrieved May 1, 2015, from http://www.aei.org/publication/schumpeterian-gales-of-creative-destruction-in-the-personal-computing-market-have-destroyed-the-windows-monopoly/

Pop-Eleches, C., Thirumurthy, H., Habyarimana, J. P., Zivin, J. G., Goldstein, M. P., De Walque, D., & Bangsberg, D. R. et al. (2011). Mobile phone technologies improve adherence to antiretroviral treatment in a resource-limited setting: A randomized controlled trial of text message reminders. *AIDS (London, England), 25*(6), 825–834. doi:10.1097/QAD.0b013e32834380c1 PMID:21252632

PulsePoint. (2015). *Programmatic intelligence report 2014*. Retrieved May 6, 2015, from http://www.pulsepoint.com/resources/whitepapers/pulsepoint_intelligence_report.pdf

Ramos, L. (2013). *B2B CMOs must evolve or move on*. Retrieved May 23, 2015, from https://solutions.forrester.com/bma-survey-findings-ramos \

Ramsey, G. (2015). *The state of mobile 2015*. eMarketer live webcast 25 March 2015.

RazorFish. (2015). *Global digital marketing report 2015*. Retrieved May 7, 2015, from http://www.razorfish.com/binaries/content/assets/ideas/digitaldopaminereport2015.pdf

Riley, W. T., Rivera, D. E., Atienza, A. A., Nilsen, W., Allison, S. M., & Mermelstein, R. (2011). Health behavior models in the age of mobile interventions: Are our theories up to the task? *Translational Behavioral Medicine, 1*(1), 53–71. doi:10.1007/s13142-011-0021-7 PMID:21796270

Rogers, E. (2003). *Diffusion of innovations* (5th ed.). New York, NY: Free Press.

Ross, P. (2015). *Native Facebook videos get more reach than any other type of post*. Retrieved May 1, 2015, from http://www.socialbakers.com/blog/2367-native-facebook-videos-get-more-reach-than-any-other-type-of-post

Rowan, C. (2013). *The Impact of Technology on the Developing Child*. The Huffington Post.

SAP. (2012). *Mobile is transforming businesses, industries, and the world*. Retrieved April 23, 2015, from http://www.sap.com/netherlands/pc/tech/mobile/featured/offers/mobile-hbr-papers.html

Search Engine Land. (2015). *What is Mobilegeddon & the Google mobile friendly update?* Retrieved May 20, 2015 from http://searchengineland.com/library/google/google-mobile-friendly-update

Seybold, A. M. (2008). The convergence of wireless, mobility, and the Internet and its relevance to enterprises. *Information, Knowledge, Systems Management, 7*(1-2), 11–23.

Sharples, M. (2003). Disruptive devices: Mobile technology for conversational learning. *International Journal of Continuing Engineering Education and Lifelong Learning, 12*(5/6), 504–520. doi:10.1504/IJCEELL.2002.002148

Sharples, M., Taylor, J., & Vavoula, G. (2005). Towards a theory of mobile learning. *Proceedings of mLearn 2005, 1*(1), 1-9.

Short, K. (April 14, 2015). *Mobile wallets: A trend retailers can't ignore.* Retrieved April 23, 2015, from http://researchindustryvoices.com/2015/04/14/mobile-wallets-a-trend-retailers-cant-ignore//

Shuler, C. (2009). *Pockets of potential: Using mobile technologies to promote children's learning.* New York, NY: The Joan Ganz Cooney Center at Sesame Workshop.

Siau, K., & Shen, Z. (2004). Mobile communications and mobile services. *International Journal of Mobile Communications, 1*(1/2), 2003.

Suarez-Tangil, G., Tapiador, J., Peris-Lopez, P., & Ribagorda, A. (2013). Evolution, Detection and Analysis of Malware for Smart Devices. *IEEE Communications Surveya &Tutorials.* Retrieved December 16, 2015, from http://www.seg.inf.uc3m.es/~guillermo-suarez-tangil/papers/2013cst-ieee.pdf

Taleb, N. N. (2012). *Antifragile: Things That Gain from Disorder (Incerto).* New York: Random House.

Technologies, D. (2014). *Oculus Rift | 3d virtual gaming glass | VR glasses | 360 degree video | virtual reality glasses.* Retrieved June 26, 2015, http://www.360-degree-video.com/pages/posts/oculus-rift-3d-virtual-gaming-glass-vr-glasses-360-degree-video-virtual-reality-glasses-publication-date-october-2014-release-2014-242.php

Thayer, K. (2014, January 14). Mobile technology makes online dating the new normal. *The Forbes.* Retrieved May 27, 2015, from http://www.forbes.com/sites/katherynthayer/2014/01/14/mobile-technology-makes-online-dating-the-new-normal/

The Economist. (2014). *The world in numbers: Industries* Retrieved April 23, 2015, from http://www.economist.com/news/21632039-telecoms

The Economist Intelligence Unit. (2014). *How mobile is transforming passenger transportation. Clearing the way for more liveable cities.* Retrieved May 7, 2015, from http://www.sap.com/bin/sapcom/en_us/downloadasset.2014-11-nov-02-17.how-mobile-is-transforming-passenger-transportation-pdf.bypassReg.html

Turow, J. (2011). *The daily you: How the new advertising industry is defining your identity and your worth.* New Haven, CT: Yale University Press.

Venkatesh, S. A., Venkatesh, C. H., & Naik, P. (2010). Mobile marketing in the retailing environment: current insights and future research avenues. *Journal of Interactive Marketing.* Retrieved May 27, 2015, from http://merage.uci.edu/Resources/Documents/2010MobileJIM.pdf

Videology. (2015). *U.S. video market at a glance*. Retrieved May 23, 2015, from http://www.videolo-gygroup.com/files/m/research-data/US_Q4_2014_Video_Market_At-A-Glance.pdf

Watrtella, E., Rideout, V., Lauricella, A., & Connell, S. (2014). *Revised Parenting in the Age of Digital Technology A National Survey*. Retrieved December 16, 2015, from http://cmhd.northwestern.edu/wp-content/uploads/2015/06/ParentingAgeDigitalTechnology.REVISED.FINAL_.2014.pdf

West, D.M. (2013, September). *Mobile learning: Transforming education, engaging students, and improving outcomes*. Center for Technology Innovation at Brookings.

Wilkinson, P., Miller, R., Burch, D., & Halton, J. (2014). *Mobile & public relations*. Retrieved 29 April 2015 from http://www.cipr.co.uk/sites/default/files/CIPRSM%20Guide%20-%20Mobile%20and%20 PR.pdf

World Bank. (2012). *Information and communications for development 2012: Maximizing Mobile*. Washington, DC: World Bank. Retrieved May 2, 2015, from http://www. worldbank.org/ict/IC4D2012

Xstream. (2013). *Online video trends & predictions for 2013*. Retrieved May 31, 2013, http://www. xstream.dk/content/online-video-trends-predictions-2013

Yu, R. (2014, August 6). Gannett to spin off publishing business. *USA Today*. Retrieved 30 Jun 2015 from http://www.usatoday.com/story/money/business/2014/08/05/gannett-carscom-deal/13611915/

Zeichner, N., Perry, P., Sita, M., Barbera, L., & Nering, T. (2014, April). Exploring how mobile technologies impact pedestrian safety. *NYC Media Lab Research Brief*. Retrieved May 27, 2015, from http:// challenges.s3.amazonaws.com/connected-intersections/MobileTechnologies_PedestrianSafety.pdf

Zhang, J., Farris, P. W., Irvin, J. W., Kushwaha, T., Steenburgh, T. J., & Weitz, B. A. (2010). Crafting integrated multichannel retailing strategies. *Journal of Interactive Marketing*, 24(2), 168–180. doi:10.1016/j.intmar.2010.02.002

Chapter 15
Advertising and Mobile:
More than a Platform Shift

Kenneth E. Harvey
KIMEP University, Kazakhstan

Philip J. Auter
University of Louisiana – Lafayette, USA

ABSTRACT

There is no field that has experienced a more positive financial impact from mobile technology than advertising. This is evident by billions of dollars in traditional media fleeing to online media, and increasingly to mobile. Yet, it is difficult to distinguish mobile totally from other online advertising approaches. Mobile is certainly not diverging from the other platforms, but rather driving some of the strongest advertising trends. Because the trends of all online channels overlap with mobile, it will be difficult to address mobile without addressing all – then clarifying and exploring how mobile is driving and will continue to drive those trends.

INTRODUCTION: STATE OF THE ART

As traditional media lose audience, they are also losing advertising revenue. American newspapers now sell substantially less than half the advertising that they used to sell (Mutter, 2014). Those lost revenues are largely moving online, but not to the newspapers' websites. Mutter, a former media executive who now carefully tracks the rapidly changing media market in his weekly Newsosaur blog, noted that print advertising between 2005-2012 fell 25 times faster than the newspapers' digital advertising grew (Mutter, 2012). Between 2005 and 2013, American newspaper advertising, including digital, dropped from $49 billion to $21 billion. All digital ad sales combined for just $13 billion in 2005 but rose to $43 billion by 2013, of which only $3.4 billion (8%) went to the newspapers for their digital products. At that same time digital advertising, overall, caught up and surpassed broadcast TV advertising (Mutter, 2014). Instead of going to the long-established traditional media, paid advertising is moving rapidly to such new media sites as Facebook, Twitter and Google. Other advertising resources are being dedicated to "free"

DOI: 10.4018/978-1-5225-0469-6.ch015

inbound PR/marketing strategies, such as ebooks, white papers, webinars and viral videos, which still require organization resources to develop.

Broadcast, cable, and satellite TV are losing audiences at a much slower pace, but the trend could reach tipping point proportions in the near future. Due to numerous factors, including enhanced average bandwidth, improved server software, and the rapid overall growth of Internet devices, especially mobile broadband, the number of online videos viewed in June 2012 compared to June 2011 skyrocketed by 550%. Phenomenal growth has not stopped. In June 2011 the number of videos viewed online in America was 6 billion; one year later 33 billion (xStream, 2013a); and by August 2014 the number of videos view per month worldwide crossed 300 billion (Nguyen, 2014).

Video Advertising

In the same year that the viewing of online videos jumped 550%, the use of advertising embedded within videos increased 68%, and 64% of the online video advertisers said the online video spots were as effective as TV ads (Xstream, 2013b). The increase in video-embedded advertising, while still accompanying only about 23% of online videos, has more recently almost exactly mirrored the increase in online video views (comScore, 2013).

More recent data shows that more and more people are augmenting or replacing paid cable services with online services, such as Netflix and Hulu, and watching those on their digital TV. In this setting, these are called "over the top" (OTT) services, but, of course, most such services can also be viewed now on desktops, laptops, tablets and smartphones. Citing a Parks Associate study, Xstream (2015) notes that 57% of American homes with cable-type services and broadband subscribe to at least one of the online video services. Additionally, 7% (8.4 million) of American homes subscribe to online services with no traditional pay-TV service. U.S. revenues from such OTT services are expected to grow from $9 billion in 2014 to $19 billion in 2019. A separate analysis by Digital TV Research, also cited by Xstream, predicts global OTT revenues will grow from $26 billion in 2015 to $51 billion in 2020.

Another, more recent survey, suggests an even more rapid movement of TV advertising online. That a "tipping point" for TV advertising is approaching is indicated by Blueshift Research's survey in June 2015, which found that the percentage of Americans who subscribe to online streaming video services (75%) now exceeds those subscribing to pay-TV services (62%). And the rate of pay-TV cancellations is increasing, with an annualized rate in May 2015 of 15.6%, with 10.2% more saying they are likely to cancel their pay-TV subscription within six months – a 20.4% annualized rate (p. 2-4). Of course, trends in America tend to go global (Table 1).

Advertising follows audience – not immediately, but eventually. eMarketer's research in March 2011 was predictive of where advertising would gradually shift. They compared advertising revenues for each medium to the percentage of "media time" the American audience was spending with each medium. For example, they noted that TV was receiving about 43% of the ad revenues, and the audience, overall, was spending about 43% of their media time watching TV in 2011. Thus, it was predictable that TV's ad share would remain steady for the near future. However, newspapers were receiving about 17% of all ad revenues while Americans were spending only 5% of their media time reading newspapers. Thus, a steep drop in ad revenues for newspapers was predictable. Meanwhile, the Internet ad revenues had reached about 19% of the total, but the audience was spending 25% of its media time on the Internet, which made it clear that online ad revenues would continue to climb. And mobile in 2011 controlled about 8% of the audience's media time, but it was receiving only 0.5% – one half of 1% – of all ad revenues. So,

Table 1. Growth in average time spent per day with major media by U.S. adults, 2011-14 (Rounded to the nearest full percentage point)

CATEGORY	MEDIUM	2011	2012	2013	2014
DIGITAL		**19%**	**19%**	**15%**	**10%**
	Mobile (all non-voice)	100%	98%	46%	23%
	Video	N/A	167%	175%	50%
	Social networks	400%	200%	93%	21%
	Online (all)	8%	-4%	-5%	-5%
	Video	100%	75%	5%	0%
	Social networks	41%	5%	-1%	-5%
TV	**Broadcast & cable**	**4%**	**2%**	**3%**	**1%**
RADIO	**Broadcast & cable**	**-2%**	**-2%**	**-7%**	**-7%**
PRINT		**-12%**	**-14%**	**-16%**	**-19%**
	Magazines	-10%	-11%	-13%	-14%
	Newspapers	-13%	-15%	-18%	-22%
OTHER		**-18%**	**-24%**	**-29%**	**-30%**
TOTAL		**-5%**	**-5%**	**-2%**	**-2%**

Note: Ages 18+; time spent with each medium includes all time spent with that medium, regardless of multitasking. For example, 1 hour of multitasking online while watching TV is counted as 1 hour for TV and 1 hour for online. Source: eMarketer, April 2014.

again, the meteoric rise of mobile ad revenues was predictable. The shift in how advertising dollars are spent have proven eMarketer's premise that advertising will follow the audience – even though the shift sometimes is surprisingly slow, causes for which we will discuss later. Nonetheless, the ad expenditures are shifting. Digital advertising has already surpassed broadcast TV advertising. The figure below compares digital to the combination of broadcast and cable TV (Figure 1).

eMarketer (2013a) reported that digital advertising surpassed all print advertising (newspapers and magazines) in 2012 ($36.8 billion to $34.1 billion), and then only trailed TV (broadcast plus cable) in American advertising revenues ($68.5 billion for TV in 2014 to $47.6 billion for digital). eMarketer projected that digital advertising would continue to close the gap with TV by 2017 ($61.4 billion for digital to $75.3 billion for TV), and mobile advertising that year would represent more than half of all digital advertising ($31.1 billion). If the data revealed by Blueshift (2015) is accurate and scientifically valid, a major shift in video viewing is occurring, and the ad spending projections by eMarketer (2013a) illustrated above may be far too conservative. Table 2 below also suggests incentive for marketers to move more advertising online.

In a free economy, lower-cost opportunities tend to create a vacuum that sucks in new investment. The cost per "time spent" vacuum in the advertising market will almost certainly keep pulling more ad revenues into digital advertising and more specifically into mobile media, according to eMarketer's 2014 study. With the loss of additional audience share, the advertising cost per hard copy newspaper or magazine reader is increasing. In 2011 the cost per hour of a reader's time was 58 cents for magazine advertising and 55 cents for newspaper advertising. But that cost rose by 2014 to 84 cents per magazine reader and 82 cents per newspaper reader. TV advertising is far more competitive, at 17 cents per hour

Figure 1. U.S. TV vs. digital ad spending: 2011-2017 (in billions of dollars USD)

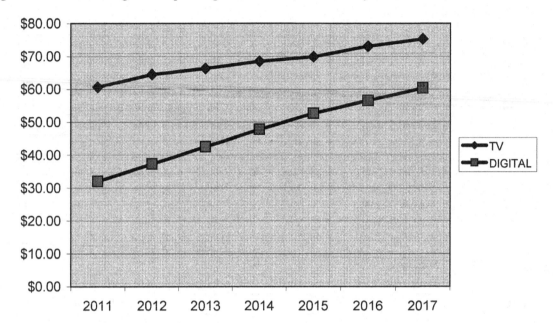

Table 2. U.S. major media ad cost per hour spent with each medium per adult, 2010-2014

CATEGORY	MEDIUM	2010	2011	2012	2013	2014
PRINT	All	$0.53	$0.56	$0.62	$0.70	$0.83
	Magazines	$0.53	$0.58	$0.62	$0.70	$0.83
	Newspapers	$0.53	$0.55	$0.59	$0.67	$0.82
TV	Broadcast only	$0.16	$0.15	$0.16	$0.17	$0.17
RADIO*	Broadcast only	$0.11	$0.11	$0.11	$0.12	$0.13
DIGITAL	All	$0.10	$0.10	$0.09	$0.09	$0.10
	Online**	$0.11	$0.12	$0.13	$0.13	$0.12
	Mobile	$0.02	$0.02	$0.03	$0.05	$0.07
TOTAL		**$0.17**	**$0.16**	**$0.16**	**$0.16**	**$0.16**

Note: Ages 18+; *Excludes off-air radio and digital; **Time spent online includes all non-mobile Internet activities. Source: eMarketer, April 2014

of viewer time, but rising. In comparison, the cost per hour of online audience time, however, is only 12 cents per hour. And the cost for mobile user time is only 7 cents per hour – arguably the best buy in advertising. Thus, eMarketer predicted continued advertising shift well into the future, with current projections to 2017 (eMarketer, 2014a).

However, "time spent" is a quantitative measure and not a qualitative one. A few years ago essentially the only mobile advertising was intrusive SMS messaging to which many users reacted negatively. With top social media transitioning effectively to mobile, and with the spread of mobile broadband, advertising opportunities have expanded greatly.

Programmatic Advertising

Expanding opportunities is also one of the biggest drivers behind programmatic advertising. Advertisers who were previously buying from only the largest online media, such as Facebook and Google, are now using technology to pursue low-cost advertising all over the Web and contract and fill those advertising opportunities through a technologically automated process. It is new enough that many advertising professionals have not yet grasped the concept, must less employed the strategy. In late 2012 it was estimated that fewer than 10% of all brand marketers had even heard of programmatic advertising (Gutman, 2012). By 2014, it was by far the fastest-growing trend in digital advertising, with mobile programmatic spending increasing 291% and desktop spending increasing 245% during the year (Pulsepoint, 2015). Programmatic advertising is also part of the answer to the oft-asked question of how advertisers can make best use of Big Data, which allows organizations to automate not just the purchasing and placement of ads, but even creation of advertising to match the right advertising outlet with the right message for the right consumer – all in the blink of an eye – considering any data an organization can code into its decision-making algorithm. PulsePoint, one of the companies that provides software and services for programmatic advertising, says every 24 hours it processes over 20 terabytes of data and five billion advertising impressions across over 200 countries (PulsePoint, 2015). Many predict that programmatic advertising will soon cross over and include traditional media outlets as well. So, programmatic advertising can involve everything – digital and traditional, desktop and mobile, and video, graphics and text. It's not what; it's how (Gutman, 2012).

Other growing trends in advertising include:

- **Social Media Marketing.** Social media represents what could be called the First Internet Revolution, which is best illustrated by the shocking use of social media by Barrack Obama in his first presidential campaign. Until then, most professional marketers saw social media as an interesting but not real effective sidelight to traditional PR/advertising. However, Obama used social media so effectively in rallying, educating, inspiring and organizing his followers, that it was arguably the most important factor in him winning the election. His online campaign donations surpassed $500 million – far more than Sen. John McCain's total donations of $358 million (Delaney, 2009). Altogether, Obama garnered $760 million – more than doubled McCain's total. Obama spent about $11 per vote and McCain $6 per vote (FEC, 2008). Just that difference in donations probably accounted for the margin of victory, since a 4% swing would have given McCain the popular majority. But every time Obama's team felt threatened, they would put out an appeal for more effort by volunteers and more money from donors, and social media provided an immediate response (Campaign Finance Institute, 2010). Obama's use of social media seemed to validate their use so that the next year, 2009, the number of Facebook users increased by about 350% and the number of Twitter users grew by 1500% (White, 2013). While social media networks have continued to grow, a 2015 survey of online marketers with 3,720 respondents suggests this trend has not yet peaked (Stelzner, 2015). Facebook is the most-used social medium, used by 93% of the marketers, followed by 79% who use Twitter, 71% who use LinkedIn, 56% who use Google+, and 55% who use YouTube (p. 23). As to which one is most important to marketers, 52% say Facebook, 21% LinkedIn, 13%, 13% Twitter and then it drops all the way to 4% favoring YouTube (p.28). Preferences differ greatly between B2C marketers, 65% of whom favor Facebook, and B2B marketers, who favor LinkedIn over Facebook 41% to 30% (p. 29). Of the respondents, 34% agreed and 58% strongly agreed that social media have become important to their business (p. 7), but only 42% agreed or strongly agreed that they can adequately measure their return on investment (ROI), while 35% were uncertain (p. 9). Facebook fared only slightly better, with 45% agreeing that it is effective and 35% uncertain (p. 11). So why do they do social media marketing? About 90% say it increases their exposure, 77% that it increases web traffic, 69% that it develops loyal fans, 68% that it provides marketplace insight, 65% that it generates leads, 58% that it improves search rankings, 55% that it grows business partnerships, 55% that it establishes thought leadership, 51% that it improves sales, and 50% that it reduces marketing expenses (p. 17). Regarding most of these benefits, the longer the marketers indicated that they had been using social media marketing, the more likely that they reported positive benefits. For example, only 37% of those using social media marketing less than one year reported increased sales, but for those using it more than five years, 70% reported increased sales (p. 18). With which media will they increase their efforts in the future? 66% say Twitter, YouTube and LinkedIn, and 62% Facebook (pp. 30-32). Another indication that this trend has not peaked is that most of the marketers recognize a need to learn more about how to use these media more effectively – 68% Facebook, 62% LinkedIn, 59% Google+, 58% Twitter and 56% YouTube (p. 38). Marketers can use social media as free PR services, or they can pay for advertising in a variety of ways, typically including pay per click (PPC) and paid "native ads" that mimic the free postings but can be placed beyond an organization's own fan/follower base. The vast majority of survey respondents are only routinely paying for Facebook ads (84%). The next highest was 41% for Google, 18% for LinkedIn ads, 17% for Twitter ads and 12% YouTube ads (p. 40). The most used content types used by marketers are visuals (71%), blogging (70%) and videos (57%), but the most important, they say, are blogs (45%), visuals (34%) and videos (19%). What kind of content do they plan to increase? Visuals (73%), videos (72%) and blogging (69%), but the one they say they need to learn more about is video (72%) (pp. 43-48). One reason why advertisers want more knowledge is that the social media are changing very rapidly and offering more and more advertising opportunities. Facebook, for example, saw great growth in ad sales as it began allowing native advertising mixed in with the news feed, but this is a relatively new offering not fully understood by many advertisers. Another is "carousel advertising," allowing an advertiser to rotate multiple images within an ad and achieving, according to Facebook, 30%-50% lower conversions costs and 20%-30% lower cost per click (Facebook, 2015). Other recent innovations in conjunction with Facebook include an enhanced Ads Manager app that allows advertisers to more easily track ad performance, edit existing ads, create new ads, and revise ad budgets and schedules; more precise targeting options, including the ability to exclude certain types of readers; the Qwaya tool that allows advertisers to easily conduct split A/B tests to determine which ad option is more effective; AdEspresso's Facebook Ads Compass conducts in-depth analysis of ad campaigns; and the Hootsuite ads program that allows small advertisers to use the power of programmatic advertising -- automatically generating native ads based on previous organic posts, and automatically selecting, targeting and bidding on ad space (Kim, 2015). New trends and opportunities in social media are difficult to keep up with, even for a full-time specialist. Facebook, in particular, had a breakout year in 2014 and continuing into 2015, largely because of its new emphasis on mobile. In the first quarter of 2015, Facebook's advertising revenues were up 46% over the same time in 2014, and mobile accounted for 73% of its total ad sales (Heine, 2015).

- **Inbound Marketing.** Inbound marketing is typically free marketing, which means it overlaps with social media. Social media postings are inbound marketing, but inbound marketing is not just social media postings. Inbound marketing tactics also include blogs, white papers, ebooks, webinars, videos and freemiums, some of which can be included in social media but most of which are not. When these elements are not part of social media, they are still promoted widely through social media. Inbound marketing draws prospective customers to the brand with free offerings considered of value to prospective customers. In contrast, outbound marketing includes cold calling, trade shows, traditional advertising, and direct mail or email, which is often considered intrusive. A 2014 HubSpot study of inbound marketing seemed to counter some of the findings of Stelzner (2015) and appears to have surveyed more top executives, including executives of major corporations. According to the HubSpot survey, 45% of company executives now cite inbound marketing as their primary source of leads. Outbound marketing and paid online media were primary lead sources for less than half as many as inbound marketing (p. 29-30). While there are company costs in employing inbound marketing tactics, small companies of 1-25 employees reported the average cost per inbound lead as $37, compared with $102 per outbound lead. At the other extreme, companies with over 1,000 employees cited $27 as the average cost per inbound lead and $45 per outbound lead.

- **Native Advertising:** Native advertising is also related to social media, as indicated previously. It is advertising that mimics organic postings but are indicated as "sponsored" postings Facebook and similarly by other social media. As such, advertisers can have them placed in the news feed of any potential customers meeting their criteria. According to projections by Business Insider (Hoelzer, 2015), native advertising in the age of social media is very effective and rapidly attracting a larger share of advertising dollars. On the back of new social media opportunities, online native advertising grew from $4.7 billion in 2013 to $7.9 billion in 2014, and it is expected to reach $21 billion by 2018. Native advertising is also being offered by more and more traditional offline media and can be purchased as display or sponsorship ads. But by far the fastest growth is in social media advertising where native ads embedded into a social medium's news feed is barely distinguished from organic posts by viewers. Part of Facebook's rapid growth in 2014 was due to its change in policy, allowing native video advertising. With that change, YouTube became the No. 1 medium for video viewership – nearly doubling its views and overtaking YouTube. Facebook went from 4.9 billion video views via desktop in January 2014 to 9.6 billion in February 2015, and videos viewed per user rose from 59 to 107. And advertisers were planning to take more of their video advertising to Facebook in 2015. (Mixpo, 2015, pp. 3-5).
- **Personalization:** A spring 2015 survey of marketing executives by Researchscape International (2015) had only 242 respondents from Evergage's email database and social media following but appears reflective of movement within the industry towards increased use of Big Data to personalize marketing messages to prospective customers (p. 2). It does not appear to contradict larger surveys by Adobe (2015), reported elsewhere in this book, but provides more in-depth insights. Nearly 80% of the participating companies had revenues exceeding $1 million a year, and 48% over $10 million. Three quarters were from the USA, and 25% were from nine other countries around the world (pp. 16-17). Some of the key findings of the Researchscape study include:
  - 91% of the marketers already used or intended to use personalization for online customer interaction in the subsequent 12 months (p. 2).
  - Of those using "real-time personalization" (accessing data within 1 second), 76% are using it on their standard website, but only 29% are using it on their mobile website (p. 2).

- ○ The primary uses or planned uses was in customizing the inline content of messages to prospects with text and images (57%), "call-outs" with short explanatory messages (43%), pop-up messages (41%) (p. 11).
- ○ The primary benefits they perceive from real-time personalization are increased visitor engagement (78%), improved customer experience (78%), increased lead generation/customer acquisition (60%), increased conversion rates from prospects to customers (58%), and improved brand perception (55%) (p. 10).
- ○ The biggest obstacles to making personalization a higher priority in their respective organizations are lack of knowledge (38%) and lack of budget (37%) (p. 14).
- ○ The type of data companies that are currently personalizing their content use to help personalize content for prospects and customers includes type of content they have viewed (48%), their geographic location (45%), how much time they spent on the site (36%), their navigation behavior (35%), their Internet device (33%), and demographic data (32%) (p. 6).

Evergage (2014), which provides personalization technology, says that most organizations are falling short in their use of personalization strategies. Besides the uses mentioned in the Researchscape survey, other opportunities include using low-cost pay-per-click (PPC) social media ads to take Web visitors directly to an appropriately customized landing page, which includes several options that help visitors differentiate themselves and receive appropriate content accordingly (p. 3); personalize promotional email, which most email marketers say can provide consistent ROI (p. 3); include a "social shop" on the website with special offers for people referred by friends on social media and who are advised what their friends have liked or purchased on the site (p. 4); and, of course, make sure that visitors using mobile devices receive the same personalized attention as those accessing through a PC. Evergage notes the success of some of their own customers using personalization. Publishers Clearing House (PCH.com) experienced a 36% lift in engagement with real-time messaging and message testing (p. 5). eMarketer (2015a) reviewed the Researchscape study but noted research by Econsultancy that found that only 18% of marketers felt that their ability to collect relevant website visitor data was strong. Another obstacle, according to an eMarketer (2015b) report, is the validating of data. While most IT decision-makers are very (28%), somewhat (57% or fairly confident (10%) that they have successfully transferred most customer data into digital channels, only 17% say they have fully automated systems in place to keep up with increased Big Data demands. Others admit that their systems require extensive manpower that may not keep up with future demands. When asked about important technology trends that would change how their businesses would operate over the next 10 years, 43% of executives cited "big data analytics" and 37% cited "automation" as the top responses, according to eMarketer.

- **Omnichannel Marketing:** Omnichannel marketing (aka cross-channel marketing) is closely related to personalized marketing because marketing messages are not fully personalized until they can follow the customer or prospective customer across different social, programming, advertising or technical channels. As noted elsewhere in this volume, an Adobe (2015, p. 29) survey of 6,000 business executives showed that 70% thought it was "very important" and 27% "somewhat important" to understand customers' journey across channels, while 66% thought it "very important" and 30% "somewhat important" to coordinate omnichannel marketing messages. Neustar (2015) argues that delivering customers and prospective customers the most "contextually relevant" messages is the key to omnichannel marketing, but tracking and communicating with a target audience across channels is not easy. Neustar's "5 Ways to Engage those Slippery Omni-channel Shoppers" ebook claims that the average online shopper now uses two or three devices and regulary navigates three or four social and programming channels where advertisers might expect to communicate with them. Neustar cites a study by Deloitte that claims that customer loyalty among omnichannel shoppers "is at an all-time low, averaging 58% and dropping. Other studies show that over 80% of shoppers research a product, brand and retailer online, peruse ratings and reviews, and check social media sites" (p. 2). Neustar says advertisers are missing the mark because of fragmented, inaccurate and static data, leading to lost sales and lost opportunities to up-sell or cross-sell, eroding customer loyalty, and lower-quality shopping experiences (p. 3). Keys to success suggested by Neustar are to:

- Maintain accurate cross-channel and cross-device identity data. They cite research that 83% of consumers now expect this, and that 46% will buy more if they receive targeted cross-channel offers (p. 5).

- Use rich insights to make a personal connection. Lewis Broadnax, executive director of Lenovo.com sales and marketing, is quoted by Neustar: "When we customized our home page offers with Neustar audience data, we saw over 40% lift in conversion rate." Neustar says it can coordinate a company's customer relations (CRM) data and point-of-sale (POS) data, along with verified third-party data to cover data "blind spots." Even big companies like Lenovo sometimes need to call in experts to accomplish this (pp. 5-6).

- Correlate CRM and other first-party data with third-party data to "save the sale." With a system that continually updates and applies such data, Robert O'Connel of Charming Shoppes believes his company can add $1 million in sales (p. 7). Keeping first-party data updated and mixing it with third-party data allows a company to target high-value or potentially high-value customers, customers who tend to abandon their shopping carts, others who tend to shop online but buy in-store, look-alike prospects who seem to mirror current high-quality customers, and even a competitor's customers and expose them to the right message at the right time. Neustar says the average company employing this strategy increases revenue per customer by 25% (p. 8).

- Attribute digital influence to the right channel. Neustar cites a study by Deloitte that 36% of in-store sales have been influenced by online messages and projected that the influence of digital marketing on in-store sales will increase to 50% by the end of 2014. Correlated omnichannel campaigns can increase online and in-store sales (p. 10).

- **Mobile Advertising:** All of these trends in advertising affect mobile advertising and are affected by mobile technologies. While we look at them separately, but we must understand the mobile element involved with each. The growth rate in mobile advertising in America is 50% per year and in 2015 represented half of all digital advertising, according to eMarketer (2015c). And the portion of the population using smartphones is rapidly growing. In 2015, 73% of U.S. Internet users and 59% of the overall population use a smartphone, and those numbers are expected to increase to 86% of Internet users and 71% of population by 2019. eMarketer also cited a Gallup poll that indicates that Americans are becoming more and more dependent on their smartphones, to the degree of being addicts. A growing portion of users feel compelled to check their mobile phone for messages every few minutes. More than half of all women under 49 said they would be very or somewhat anxious if they misplaced their smartphone for one day. For males of that same age group, the percentage of anxious respondents was about 40%. And yet many developing countries are even more dependent on smartphones than America. Another eMarketer report (2015d) noted that mobile retail sales in China are growing much faster than in the U.S. – 211% vs. 36% in 2014 and an estimated 85% vs. 32% in 2015. Mobile retails sales will reach 50% of all retail e-commerce in China in 2015, whereas it will represent only 22% in the U.S. By 2019 it is expected to reach 72% of e-commerce in China and 28% in America. This projection for America may be too low, however, because it was only in 2015 that Google began using mobile compatibility as part of its search engine algorithm to determine the sequence of listings. This will add impetus for all online businesses to upgrade their websites, whereas most shoppers in early 2015 were frustrated with the degree to which websites failed to accommodate mobile – especially smartphones. Consequently, while Americans spend more time on their smartphones than on desktop computers and far more time than on tablets, they will shop but most will not try to make purchases on their smartphone (Ramsey, 2015). However, the influence of mobile in America is much greater than just in mobile sales. A study by Deloitte (2015, p. 7) says 64% of all sales – online and in-store – are influenced now by online marketing, and half of those by mobile marketing (p. 11). Nevertheless, there is some confusion within the advertising industry as to how to even classify mobile. The idea of advertising channels and cross-channel advertising was developed before the Worldwide Web and included the different traditional media and the individual publications, programs or channels within those media. It was confusing then and more confusing now that we are adding different online devices and different websites and individual online publications, productions and programs. An ebook by Forrester (Ask, 2014) declares in its title, "Mobile is Not a Channel." Ask and her team of contributors believe that e-business professionals who treat mobile as just another channel will fail, and she says that currently includes about 62% of them. "Rather than leveraging mobile to do new things, they are still using mobile as a scaled-down version of the Web" (p. 1). What distinguishes mobile are the new opportunities mobile technologies provide, including the potential to (p. 2-3):

◦ Drive spontaneous purchases with flash sales directed at people based on their location. Geo-targeted marketing is enabled by mobile and can be sent, for example, just to those within one mile of a store.

◦ Influence sales in-store and online – not just on mobile. Mobile devices are more personal that any other, and people increasingly depend on them not only as convenient "Yellow Pages" but also for product and service descriptions, information, pricing, ratings, and reviews across all industries.

◦ Offer brand-new or greatly personalized services available for digital delivery, such as monitoring, coaching, just-in-time training, Uber-like services, etc.

◦ Expand potential customer base by offering cheaper pay-as-you-go services. In Africa this can include utilities like electricity. They can't afford 24/7 service but can order it as needed. Zipcars are an example in America, which allows customers to quickly arrange for rental of a car by the day or by the hour.

◦ Enhance pricing options to optimize profitability. As hotels and airlines do already, other industries could quickly adapt prices to availability, urgency, need, location, etc. Amazon has experimented with counter-geo-targeting. When it could detect that a consumer was shopping in a store, primed to buy, it sent special offers to steal those purchases.

◦ Accommodate loyal customers by the design of mobile apps especially for them. Starbucks has taken the lead in this approach. It now has 12 million active mobile app users who pay for purchases with their apps 6 million times per week, representing 15% of the company's revenue. Such customized apps could be offered by companies in many different industries.

∘ Enhance sales staff's ability to close a deal. Nike store employees, for example, can look up inventory and even ring up sales on their mobile devices. And Trane's iPad app gives HVAC salespeople the ability to create complicated proposals and demonstrate products and previous projects on the fly, nearly doubling their close rate.

Mobile devices allow companies to engage consumers in a much wider variety of "mobile moments," such as information at an impulse purchase moment, research during a purchase exploration moment, information during a customer's product setup moment, just-in-time self-service assistance in a loyalty moment, information and graphics at a social sharing moment. Forrester's research suggests that only 4% of enterprises are positioned to take advantage of the mobile mind shift. The most successful organizations create an inter-departmental mobile steering committee to create a new mobile vision, an "Agile" development team to build new apps, and a center of excellence with business and technical leaders to provide day-to-day guidance for mobile development. Lack of such resources tops the list of inhibitors for mobile development, according to Forrester's survey. Ask also co-authored a Forrester report (Husson and Ask, 2014) that predicted that most brands will greatly underinvest in mobile, leaving themselves susceptible to attack by competitors. According to Forrester research, consumers already expect to engage with brands seamlessly across platforms immediately and in context. However, at the time of the survey only 18% of U.S. companies had that same expectation, while 30% more were making an effort to achieve that mind shift and capability. With 42% of the world's population expected to own a smartphone by the end of 2015, Husson and Ask believe the gap between mobile-oriented enterprises and the myopic majority will increase. This will create a "mobile arms race" that could destroy those that leave themselves open to attack. One important area where late-starters may lose significant market share is in late development of mobile apps, which currently lead to the greatest volume of sales. The average American, note the authors, use 24 mobile apps per month but spend 80% of their time on just five apps. So-called "mobile app fatigue" may cause consumers to ignore many apps introduced in the future.

SUMMARIZING PAST RESEARCH

The expansion in digital advertising has led to a growth in the academic research surrounding digital marketing. Industry research indicates that individuals, especially college-aged students, can spend nearly 8 hours per day connected to technology; as this figure includes time spent on mobile devices, it would be ludicrous to ignore the potential that mobile advertising has for the business world (Godfrey and Duke, 2014). In 2010, eMarketer projected that spending on mobile marketing and advertisements would increase by 37% between 2008 and 2013 (Okazaki, 2012). That prediction looks silly in retrospect. Just in 2012 mobile advertising expenditures skyrocketed by about 120% to $8.8 billion, in 2013 that total doubled again (105%) to almost $18 billion, and in 2014 it was on pace to increase 75% and reach $31.5 billion (eMarketer, 2014b). An additional study found not only did the amount in 2013 increase at a nearly alarming rate, but also by 2017, that amount should exceed $62.8 billion, indicating a remarkable growth in the mobile advertising industry (Bart and Sarvary, 2014). These figures speak volumes on the pace and potential that mobile advertising has for the global market and therefore, must be closely studied.

Mobile advertising has also provided organizations a plethora of relationships that may have otherwise gone unnoticed. Partnerships between mobile providers and other companies can help to drive specific advertisements to specific users, based on their data usage, time spent on their devices, proximity to store locations, etc. (Bergen, 2014). No organization has a more direct line to the mobile user than the actual mobile provider, and gathering data via this method could be hugely rewarding for organizations seeking to increase their mobile advertising. Of course there is the need to respect user privacy, but if users would agree to let providers access to mobile data in order to improve the overall mobile experience -- which would include how, when, and how often the consumer was advertised to -- the relationship could prove beneficial for both parties.

Key Concepts and Themes

According to Harriet Swain (2007), more people owned a mobile device than a computer, so it was crucial for marketing professionals to stay ahead of the mobile trends, in order to compete for the attention of consumers. And when Swain wrote that in 2007, the mobile revolution was in its infancy. Apple was just announcing its iPhone. By 2011, according to the International Telecommunications Union (2015), mobile broadband connections to the Internet worldwide were already double those of fixed-wire connections (1.2 billion to.6 billion), and in 2014 mobile broadband tripled the fixed-wire connections (2.3 billion vs..7 billion). Fixed-wire broadband was increasing at a relative snail's pace compared to mobile. As more research on mobile mass communication is completed, studies have shown that user affinity for the smart phone continues to propel mobile marketing and advertising into the next phase of growth. An increase in "search-based Internet advertising, branded mobile entertainment such as computer games and user-generated content" (Shintaro and Barwise, 2011) demonstrates the need for continued monitoring of the mobile age's progress. It has been further observed that more mobile users are developing a sort of attachment to their mobile devices, as most mobile users can access everything from email to personal health documents via their mobile devices. Studies show that the more attached a user is to their device, the more likely they are to positively receive mobile advertisements (Kolsaker, 2009). Conversely, it is observed that users without an emotional attachment to their mobile device find mobile advertisements to be more annoying than helpful. But, as the number of mobile users continues to increase as well as the possibilities that mobile devices have for their users, it is reasonable to assume that most individuals will access their information via mobile devices within the next decade, essentially eliminating this issue. In essence, the mobile device itself is the key to accessing the consumer and, in turn, advertising to them.

Despite DeReyck and Degraeve's 2003 definition of mobile advertising as only involving text messages, of the 120% increase in mobile advertising in 2012, 105% increase in 2013 and 75% increase in 2014, very little came from text messaging (eMarketer, 2014b). It came mostly from social media advertising now adapted to mobile. In 2013 Google by itself accounted for almost 50% of all mobile ad revenues worldwide, according to eMarketer, while Facebook climbed from 5.4% to 17.5% in one year. Facebook was projected to finish 2014 with 21.7% of all mobile ad revenues, cutting into Google's market share. But Twitter and Yahoo also announced more advertising opportunities and tools in 2014 in an effort to become major players in the multi-billion-dollar competition. Part of the increase in mobile advertising is derived directly from organizations purchasing space within the timeline/news feed of social media websites. For many years organizations have attempted to send free viral messages through social media as part of a content marketing strategy. Now social media sites like Facebook, Instagram, and Twitter sell such "native advertising," labeled as "promoted" for full disclosure, that allows organizations to ensure that their messages reach a broader audience without depending solely on reader-dependent viralization of the message. This also allows advertisers more intimate access to the consumer. As companies have become more aware of how social media can work to their benefit, more time and funding has been put into social media development, as this is one of the primary places the modern day mobile user will interact with the mobile brand (McDermott, 2013).

Social media research, therefore, has become the most relevant at this time. There is still relatively little social media academic research, and most industry research is survey-based, good for identifying trends and correlations but not necessarily causation. A Hubspot's social media scientist Dan Zarrella, on the other hand, has used experiments and content analysis on a more scientific basis to better identify how the most successful social media marketers are succeeding. Since much of his research relates to inbound (content) marketing rather than paid advertising, we explore his research in more depth in the Mobile Marketing chapter. However, with the recent development of paid "native advertising," his findings have become more important than ever. And he found, even before native advertising was allowed, that social media viral content and social media advertising strongly supported each other. For example, in analyzing the organic search traffic to the HubSpot website itself, Zarrella (2013) found a positive correlation of higher organic search rates with both paid search tactics, such as paid-per-click advertising, and with such unpaid tactics as email campaigns, such as times when they are sending out information about new ebooks or webinars available to people in their database. So, the more active a company is in its PPC and emailing programs, the more likely visitors will also show up to the company website based on their own organic searches. Two possible causes of this are:

1. That people receiving non-organic prompts are also more likely to have questions of their own for which they go to a search engine for answers, and
2. That there may be a two-step opinion leader process in effect, as has been found with traditional media.

In other words, when opinion leaders (aka in social media as evangelists) are prompted by an organization, they are more likely to prompt their opinion followers online or offline.

Academic researchers Wu, Hu and Zhang (2013) conducted a simulation study in an effort to discover key elements to viral advertising in social networks, but the factors they explored were very limited because they had to be elements they could easily quantify and control. Viral information flow is not the same as general information diffusion, on which their research was based. Wu et al correctly noted the viral advertising is a new and relatively young research topic in the era of Web 2.0, but they still founded their research on old diffusion studies that fail to fully take into account some elements of rapid virus action seen on the Internet today. Perhaps the first hint of how viral actions are ignited was when Haroldsen and Harvey (1979) studied the first documented case of "shocking good news." Until then diffusion studies of "good news," such as the launch of Explorer I and the announcement of Alaskan statehood, never reached more than 49% of the studied population in the first day, and in most studies of "good news," the largest portion of respondents heard the news from the media. In contrast, the shocking news of John F. Kennedy's assassination reached 100% of the surveyed population in the first day, and over half heard the news from friends, family or even strangers. The shocking news ignited the interpersonal network very much as strong viral messages sweep through social media today. The 1979 study successfully demonstrated that "shocking good news" could do likewise even before the Worldwide Web was developed, but this is seen on a regular basis now that social media expedite interpersonal communication as never before.

Within the marketing industry today, professionals frequently use the word "shocking" when they describe the kinds of videos and other messages that organically achieve strong viral action in social media, but they, as Wu et al, find it very difficult to quantify viral factors in a way that creates any kind of reliable formula. While Zarrella studies such elements as specific word usage in viral headlines, time of day messages are initiated, etc., Wu et al were trying to determine how many people to select to optimally initiate a viral message at minimal cost. They were asking, essentially, what proportion of a group would need to receive the initial advertisement before they could expect it to spread by itself to most of the remaining portion of the group. They referred to that number as "the tipping point," as previously described by Gladwell (2000) and referred to in this chapter. They also discussed how to select people for this group but with little clarity and no specific recommendations useful in actual practice. Their suggestion that simultaneous mass media advertising enhances the viral action also needs more clarification and research. It seems difficult to employ their research without considering the message content and the degree to which the message content itself will initiate viral action, as discussed even in the 1979 "shocking good news" research.

Professionals frequently suggest that initiating a strong viral action is as much an art as a science, and, similar to a Hollywood movie producer, an advertiser might have to invest in numerous efforts before achieving blockbuster success in a large network like YouTube, even employing strategies acknowledged as "best practices" to achieve the shock, excitement, passion or other responses that would significantly increase the number of viewer shares. Nevertheless, further research into the artistic elements that help achieve viralization are certainly warranted.

The efforts of Wu et al, however, may have suggested another strategy that would facilitate real world success and be worthy of further research. In a large network like Facebook, how can a business create a specific pool of initial recipients in an effort to viralize a message? This also relates to new omnichannel (aka cross-channel or multichannel) strategies trending in the advertising industry. The survey of thousands of executives by Adobe (2015, p. 27) found that 59% of the respondents planned to have a cross-channel focus in 2015, and only 11% disagreed. This is part of companies' efforts to personalize their messages to customers and prospects, according to 38% of the respondents. But omnichannel efforts could go beyond sending the same message through every feasible online stream of information. Omnichanneling may also help companies to identify people who might serve as online opinion leaders on certain issues or products and help ignite viralization. How can the various channels interact in such a way? Which channels and strategies can best identify members of that original launch set, and how can they best be activated? Free webinars, for example, might tend to attract opinion leaders, and some viral action might be achieved by the webinar itself. Webinar organizers typically require participants to provide email addresses, which is perhaps the most important reason to hold webinars. Using webinar participants as a launch group for future messages might be effective, but even more so if an organization can get those individuals to connect with them via social media where viralization is easier to achieve. Perhaps promoting webinars more through social media messaging and paid advertising would achieve greater purpose. This is one way in which omnichanneling may achieve advertising synergy, where the total impact is greater than the sum of separate channels.

Key Theories and Methods

As sociologists and psychologists initiated scientific studies of mass communication in the 1930s, they began with the assumption that the media had a powerful direct and sometimes controlling influence on their audience. The mass society theory was a grand theory that original researchers thought could explain all aspects of the media phenomenon. They quickly found it could not. Citizens were not voting a certain way because a politician inundated the radio waves with his persuasive messages. They were influenced more by interpersonal relationships and by specific opinion leaders. Thus, the two-step or multi-step flow theory of communication was developed by Paul Lazarsfeld and Elihu Katz. That also was not comprehensive enough, so additional communication theories have been developed over the decades. Media sometimes changes attitudes rather than opinions and knowledge sets, so Carl Hovland led the way in developing the attitude change theory and related theories of dissonance, selective exposure, selective perception and selective retention. Later theory introduced the idea that media don't just use their audience, but the audience also uses the media. Thus, the uses and gratifications theory was developed. But psychological researchers countered with the social cognitive theory, relating to how the audience copy or model behaviors they see in the media through imitation and identification (Baran, 2006, pp. 420-431).

These general theories of communication have also been adapted more specifically to advertising, along with related models and strategies. Selective perception, for example, causes consumers to focus on some stimuli and ignore other stimuli. Why? Advertising researchers explore perceptual screens of both physiological nature (the five senses) and of psychological nature (such innate factors as personality and instinctive human needs and such learned factors as self-concept, attitudes, past experiences and lifestyle). If advertising messages can get past the perceptual screens, then cognition can occur in the form of learning and persuasion. Cognition theory considers the processes of memory, thinking and rational application of knowledge, while conditioning theory (aka, stimulus-response theory) explores how an advertising stimulus causes a need arousal, leading to a behavioral response (e.g., the purchase of a product), leading to either a satisfied or unsatisfied need and a subsequent response – a different response if unsatisfied and a repeat response if satisfied. These theories have then led to exploration of the different kinds of needs, such as Maslow's hierarchy of needs: physiological, safety, social, esteem and self-actualization, and the matching then of products and services with appropriate advertising messages to match the perceived needs. Also explored more thoroughly is the impact of opinion leaders within the family, society and personalized reference groups on consumers' decision-making. Models such as AIDA (Attention-Interest-Desire-Action) have been developed to help guide these concepts in actual application. And, finally, precise advertising strategies and tactics have been explored in advertising research (Arens, 2002, pp. 139-159).

All of these theories, models, strategies and tactics, however, need to be re-examined in light of the new marketing ecosystem. Never before have marketing messages had such an opportunity to viralize and spread to millions of people in a very short time – sometimes at no cost to the advertiser. The multi-step flow theory of communication serves as a foundation for new research into how to identify and use opinion leaders to viralize online messages, but many strands of new research are likely to develop in this effort. Online advertising also dictates a total rethinking of how ads are designed. A small PPC (pay per click) ad on Facebook has little in common with a full-page newspaper or magazine ad. Ads in traditional publications are most frequently portrait-shape in their dimensions, whereas most online display ads are small and frequently have a square, strong vertical or strong horizontal layout. A print ad typically has four primary parts – a visual to attract attention, a headline to help develop interest, body copy to develop a desire and to issue a call to action, and logo to achieve brand recognition/credibility and to provide essential contact information. Some of these traditional elements are missing from many online ads. Many online display ads lack space for body copy, but a visual and headline may attract enough attention and interest to get consumers to click a hyperlink for more information. The hyperlinked headline and graphic then take the consumer to the organization's website landing page, where all the traditional advertising elements are typically employed to achieve the desired action. The theories may stand up under this new scrutiny, but how they are applied is definitely changing.

More recent developments in mobile technology continue to yield better results for both the consumer and the advertiser, and increased access to mobile technology benefits both parties in the advertising process. The increased development and advancement of mobile technology has attracted the attention of more users and, therefore, more mobile users are engaging with mobile advertising. Mobile advertising has become more of a common practice, and is more widely accepted as a means of communicating with the consumer. Users no longer view mobile advertisements as simply spam. Users can trust that mobile advertisements are legitimate, and therefore, are more likely to have their purchase intention increased (Kim, 2014).

With this kind of trend line, retailers must now use mobile advertising as a means to increase consumer interest in their products as well as keep the customer engaged. The instant communication and, in turn, instant results of communication makes mobile use so appealing to the consumer. The ease of mobile advertising allows both the consumer and the sender to communicate in real time, creating an actual conversation between parties rather than an archaic message delivery system with no instantaneous response capabilities. In one study, 612 participants agreed to receive mobile ads on their wireless phones. The results indicated that "attitudes exert positive influences on intentions to receive advertising, especially among those who already have access to the Internet on their mobile phones" (Izquierdo-Yusta et al, 2015), meaning that these mobile advertisements are even more effective on users who are already linked into the mobile market. Additionally, consumers noted that they were more satisfied with advertisements that were location-based. This means that the advertisements sent to them were for products and services nearby, making the advertisements much more useful to the consumer. This suggests that merely spamming as many recipients as possible with mobile advertisements can work against the advertiser, instead of in their favor. Relevant and pertinent mobile advertisements will direct consumers to businesses that they can access (Reichhart, 2014). This can form a positive relationship between the consumer and the advertiser, and can ultimately increase the accuracy and efficiency of the mobile advertisements being delivered. Mobile advertising can also be widely effective in building brand communication; again, there is a distinct need for message relevance. This is largely due to the idea that the more relevant a message is to the consumer, the more pleased the consumer will be to receive the message. In turn, this happiness can inspire the consumer to interact more with the brand, thus building brand familiarity and loyalty (Varnali, 2014). Chen, Su and Yen (2014) also identified a link between how the advertisement is formatted, what product is involved, and brand credibility in relationship to location-based advertising. While the results of the study were unsurprising, they are important to consider:

Results show that animated location-based advertising, less personal information embedded advertisement message and foreign-brand products advertised through location-based advertising are significant contributors to a positive attitude towards location-based advertising. Moreover, there is a statistically significant relationship between attitude towards the location-based advertising and attitude towards brands (p. 291).

Clearly, more than just the color or the message medium can make a difference in the overall reception of the advertisement. Research has also shown that messages delivered in other media, aside from SMS messaging, have seen positive results. Banners and other types of mobile display forms of advertising (MDAs) have gained popularity in recent years. One study conducted an experiment involving MDAs that ran between 2007 and 2010. The results indicated that overall, the MDAs were able to increase "consumers' favorable attitudes and purchase intentions" as they are effective in their ability to cause "consumers to recall and process previously stored product information" (Bart, Stephen and Sarvary, 2014). Other studies show that while mobile advertising is indeed growing in popularity, it is important to not overwhelm the consumer with too many advertisements, or too many unimportant messages. Due to the increasing rate of mobile advertising, some mobile users have begun to regard SMS advertisements as spam or junk, and therefore fail to use them for their intended purpose, even though this is a commonplace form of legitimate advertising. It is important to diversify how messages are sent to consumers, as well as to monitor how often they are sent, depending what the sender hopes to gain from the message receiver, as well as what the advertisements intends to sell. Addressing this issue in their own study, Chou and Lien found that "results indicate that for SMS teaser ads featuring high-familiarity brands, a more likeable/familiar spokesperson reduces consumer curiosity. For ads pertaining to low-familiarity brands, spokesperson likeability/familiarity positively affects curiosity for consumers with more favorable SMS attitudes. Spokesperson variables, however, do not influence reactions from consumers with less favorable SMS attitudes" (2014). Additionally, it is important for the advertiser to keep track of how frequently their mobile advertisements are being sent to their consumer base. While it is true that repetition can make an advertisement more effective, sending more than three messages in a day will overwhelm the receiver, making the advertisement totally ineffective. It is also worth noting that putting any sort of time limit on the advertisement, such as "Act Now!" or "In the next five minutes" is likely to put unnecessary pressure on the customer, once again, making the advertisement ineffective (Rau, Zhou, Chen, and Lu, 2014).

Research also points to a need for the advertiser to understand how and how often the consumer wants to be communicated with. Because mobile technology allows the advertiser a unique and direct link to the consumer, it is important to not abuse this relationship. If consumers feel that they receive too many mobile advertisements or that the advertisements are worthless to them, they will reject the advertisement altogether. However, if consumers feel in control of the advertisements that they receive, they are more likely to have a positive attitude towards mobile advertising (Akpojivi and Bevan-Dye, 2015). Advertisers should also be aware of the type of data and mobile access that their advertisements require. SMS advertisements are sent as a text message, and therefore require almost no outside resources on the consumers' part. Messages with external links, videos, and/or images, however, may put a strain on the consumer and limit what parts of the message can be seen. Akpojivi and Bevan-Dye also found that the more in control the consumer felt about the data usage -- i.e. the mobile data required to view the message -- the more likely they would feel positively about the mobile advertisement, and in return, the more likely they would be to follow through on a purchase (2015).

Key Findings and Achievements

A study by Chinese researchers (Gao & Zang, 2014) compiles a series of hypotheses based on previous research and tests them as a group, employing questions used in previously published studies. They noted that while China has a billion mobile users, more than any other country, their smartphone users at the time of their survey-based study typically accessed the Internet through 3G and WiFi connections and seldom used their smartphones for more purposes than telephony and SMS. Thus, this study focused on SMS advertising. While their hypotheses (reworded for clarification sake) may seem obvious, compiled as a group with research citations is helpful in considering application.

1. Prior attitudes of consumers toward mobile advertising affects their response to mobile advertising (DeReyck and Degraeve, 2003; Elliot and Speck, 1998; Zanot, 1984; Ducoffe, 1996; Mohd et al, 2013; Chowdhury, 2010; Sandra et al, 2010; Drossos et al, 2007; Parissa et al, 2005; Tsang et al, 2004; Davis, 1989; Hsu, 2004; Brackett and Carr, 2001).
2. Irritation factors in mobile advertising have a negative effect on mobile advertising (DeReyck and Degraeve, 2003; Chowdhury, 2010; Chowdhury, 2010; Aaker and Bruzzone, 1985; Ducoffe, 1996; Altuna et al, 2009).
3. Entertainment qualities improve users' attitudes toward mobile advertising (DeReyck and Degraeve, 2003; Ducoffe, 1996; Schlosser et al, 1999; Chia-Ling et al, 2010; Chowdhury, 2010; Chowdhury, 2010; Shavitt et al, 1998).
4. Credible, useful information leads to more positive attitudes toward mobile advertising (DeReyck and Degraeve, 2003; Ducoffe, 1996; Schlosser et al, 1999; Chia-Ling et al, 2010; Chowdhury, 2010; Parissa et al, 2005; Chowdhury, 2010; Tsang et al, 2004; McKenzie and Lutz, 1989; Brackett and Carr, 2001; Merisavo et al, 2007).
5. Personalization of messages leads to more positive attitudes toward mobile advertising (Chellappa and Sin, 2005; Robin, 2003; Xu, 2006; Rao and Minakais,2003).
6. Providing incentives improves consumer attitudes toward mobile advertising (Drossos et al, 2007; Pietz and Storbacka, 2007; Hanley et al, 2006; Parissa et al, 2005).

Gao and Zang (2014) found additional support for all six of these hypotheses. While their study focused primarily on SMS advertising, it is reasonable to consider these factors with all digital and mobile advertising.

While SMS advertising is the oldest form of mobile advertising, these recommendations by Gao and Zang may be particularly useful to one of the newest. An Adobe survey (2015) of some 6,000 business executives found that 24% of the respondents will have started using geo-targeting technology by the end of 2015, and 25% more are actively exploring it. This technology takes advantage of the GPS function of smartphones to allow businesses to send messages only to those within a certain radius of the store. A study by Eddy (2014) found that consumers are more inclined to click on mobile ads when they relate to businesses in their proximity, even more so if the message includes a coupon or special offer. That further confirms Gao and Zang's study and the many studies on which theirs was based. The six elements they identified as essential to successful SMS advertising can likely be applied to geo-targeted ones.

Programmatic advertising very much relates to the search for viral opinion leaders, the desire to individualize messages to customers and prospective customers, and the goal to send coordinated advertising messages across all channels. As explained previously, programmatic advertising is not limited to any specific channel, Internet device, target audience or message content. Some charts confuse the issue by separating programmatic advertising from other types of advertising, such as video, but programmatic ads can be video, and video ads can be purchased through programmatic technology. And the real key to programmatic is that as many of the decisions as possible are made by computers in order to deal with the extremely fragmented online media market and the tons of consumer data being generated. With that in mind, here is some additional research that helps clarify the growth and use of programmatic advertising.

While programmatic spending increased 291% for mobile in 2014 and 245% for desktop, following triple-digit increases in 2013, as well (PulsePoint, 2015), eMarketer (2015e) estimates that the growth rate in America will continue strong but come down to a more earthly 50% in 2015. Despite all this growth, it is important to understand that programmatic is still a relatively small part of most brands' advertising budget and, given the right tools and opportunities, might continue its meteoric growth. Only 8% of American marketers put over 20% of their budget into programmatic. However, 62% of all brands surveyed planned to increase their programmatic budget in 2015. This to some degree is being driven by mobile, where respondents saw the greatest opportunity for programmatic advertising in 2015 – 33%, compared to just 20% for the next highest, video advertising – which, we previously noted, could also be nearing a growth tipping point and which can easily overlap with programmatic advertising. In 2014 mobile was the second most purchased type of programmatic advertising at 69%, only behind display (86%) and ahead of video (67%). Even with this increased growth, mobile is still predicted to trail display in total programmatic expenditures. In 2015, buyers were planning to allocate 41% of their programmatic budgets to display advertising, 30% to mobile and 29% to video advertising. What could become a problem, however, is that sellers only thought their inventory of advertising to sell through programmatic would include 28% display advertising -- well below buyer purchasing plans (eMarketer, 2015f). The growth of programmatic is a worldwide phenomenon. In fact, Asian advertisers increased programmatic significantly more than North American advertisers in 2014, increasing their programmatic spending by 329% (PulsePoint, 2015).

Developing and tracking programmatic advertising is currently very challenging. In a study by WBR Digital reviewed by eMarketer (2015g), retailers reported good return on investment (ROI), but multichannel analytics were found lacking. Only 8% reported having holistic analytics on a single platform with clear visibility. About 36% reported analytics that were spread across multiple platforms – inefficient but with solid visibility. Another 36% reported that they were still developing their analytics capabilities, and 20% reported analytics spread across several platforms with fractured visibility. Advertisers responding to the Advertiser Perceptions (2015) study suggested they would spend more money on programmatic advertising if a few problems were fixed. About 40% felt the need to better demonstrate ROI, and 30% cited a need for greater transparency in ad placement, as well as a need for assurance that their ads are placed in an environment compatible with the brand image. Because of the technical complexity of programmatic advertising, many companies have been contracting agencies to help them. Bringing that work in-house might also enhance their use of programmatic, which 59% of the advertisers say is their intention. But 61% of the agencies are skeptical the advertisers will accomplish that.

Nevertheless, the top three advantages to programmatic advertising, according to the WBR Digital (7 May 2015) survey, were:

1.   Improved media buying efficiency and targeting, according to 91% of respondents,
2.   Improved customer experience through relevant messaging, with 90%, and
3.   Improved media ROI or conversions, 87%.

As mentioned, before programmatic tools were developed, many advertising sellers were ignored by most advertisers as not cost-effective to find, contract and work with in their efforts to individualize advertising messages. Programmatic software immediately made it cost-effective, but there remains the challenge of creating so many different messages for so many individual using so many online media.

An eMarketer report (2015h), "Creating ads on the fly: Fostering creativity in the programmatic era," said to do this effectively will require a change in infrastructure and mindset. Brands that have been heavily involved in social media frequently find this easier than others, since social media engagement frequently requires responding to situations and personalizing messages "on the fly." Others who use customer service management systems or direct mail marketing are also used to personalizing messages. But programmatic takes personalization of advertising to a whole new level and requires use of data from many different data management platforms. To help them, about 63% of the programmatic marketers are using data signals with the personalization. Understanding how some individual agencies handle this work provides an inkling about others, as well. OneSpot is an agency that helps companies take their social media content and convert it into multiple ad units that can be applied in an automated process as data suggests. Matt Cohen, OneSpot founder and president, told eMarketer:

For each piece of content we'll create several different sizes of display ads, mobile ads, social ads and soon, native ads. We have a crawler that automatically looks at the client's content, pulls out the headlines, the video, thumbnail, the full video, the photo. It also does an analysis of the content to figure out what it's about, what are the keywords in it and so forth. Then it goes into an ad studio where the client can review it, and alternate variations can be created. ... [T]he chance that you're going to have something for anybody in your target audience is extremely high, and you're likely to have something that's actually really good for that specific person.

Gurbaksh Chahal, founder, chairman & CEO of another agency, RadiumOne, shared some of his company's methods in Forbes (Olenski, 2013):

Our platform powers today's real-time approach to advertising, reaching 26 billion impressions each day across 700 million users from its own proprietary first-party products. ... RadiumOne is different in how it weaves in social signals as well as campaign performance when amplifying and delivering advertising campaigns across the Web. It captures first-party social intent data from its mobile applications ... and then combines this information with our intelligence layer Sharegraph, to reveal not only the right audiences, but also the first and second degree relationships that are also most likely to respond to a particular ad. If a consumer gravitates to a Hyundai advertisement, it is likely that his friend is also interested in learning more about the same brand. RadiumOne is also the first-to-market for hashtag targeting, creating a way for brands to leverage the public hashtags people use on Twitter to target them with relevant display ads on mobile and desktop. ... Advertising is moving towards capitalizing on "now moments." We launched our innovative hashtag targeting technology back in Q4 of last year with great success. We found that people are already engaging in online brand conversations on social channels like Twitter, but to date, no one had found a way to monetize this form of social interaction. We are the first-to-market for this targeting technology to allow brands to purchase hashtags like keywords ... and then serve ads to the people who are explicitly using these terms. ... We create a social feedback loop so that we can pull from campaign data across the Internet, from external social channels and then back into Facebook. Hashtags can encompass much more than just a company or a brand name. They can convey an emotion, a category or theme. For example, during holidays like Valentines Day, an advertiser like flowers.com or Mars may wish to target users who tweet #love, #commitment or #gift. ... [I] t's about capturing the right elements from a "big data" set and making it work that shows ROI. They can connect their earned media to make their paid media that much more smarter.

As explained earlier, programmatic and video are not mutually exclusive. To the contrary, a large amount of programmatic is video. Indeed, there are indications that the greatest advertising potential on mobile and all things digital may soon be video, as suggested by trends previously discussed. But there are other factors, as well. A 2014 study by Aberdeen Group showed that online marketers almost unanimously (90-95%) proclaimed that video is the most powerful tool for digital marketing. Video helps "marketing and sales cut through the clutter with a differentiated voice, thereby increasing both information retention and the perception of the quality of the message it delivers." And those websites that use video to help convert Web visitors into clients/customers are 60% more successful in their rate of conversion (Aberdeen, 2014).

There is growing evidence that video-embedded advertising, somewhat mimicking commercial TV, represents the greatest growth opportunity for mobile advertising. A survey by Belkin and Harris Interactive in early 2013 found that 30% of U.S. Internet users were already favorably considering replacing cable TV with an online streaming video subscription. eMarketer's subsequent projection was that online TV viewers would cross the 50% tipping point in 2014 and reach a total of 145.3 million paid online viewers in America by 2017 (eMarketer, 2013b). ComScore (2014) reported that half of the 18- to 34-year-old Netflix subscribers were already watching most of their videos on mobile devices. In this setting the Millennials could be seen as the early adopters, with older viewers soon following suit. This trend could expedite the growth both of digital video viewing, in general, and specifically the growth of digital video advertising. IAB Europe (2010) reported that in the European Union people spend more of their free time on the Internet (18.2 hours per week) than on TV (13 hours per week), and 28% of those sitting in front of the TV are simultaneously using an Internet device – usually a mobile device (Cesar, Knoche and Bulterman, 2010).

While some online video services will offer video entertainment without advertising, the demand by advertisers would likely cause other providers to offer blended cable-like systems if not free commercial TV-like entertainment subsidized by advertisers. While digital advertising overall is catching up to TV advertising, digital video ad spending is still far behind. Total TV advertising was expected to reach $68.5 billion in 2014, and total digital advertising $47.8 billion. Of that, less than $6 billion was expected to be digital video advertising. However, eMarketer says since 2011 digital video advertising has been increasing about 40% per year. eMarketer predicts that growth rate will slow down (eMarketer, 2013c), but new video advertising opportunities could cause a major shift in advertising dollars. After all, traditional television still controls the biggest share of all advertising, but why would that advertising not shift online as viewers move to online TV? Even the shift of a relatively small portion of current TV advertising would make a big difference. In 2015, TV and cable advertising was still expected to reap 40% of all advertising expenditures in America while all digital video would attract less than 5% of ad revenues. That presumed that the percentage of media time spent on TV would be about 36% and time on digital video about 11% down (eMarketer, 2015i, p. 4). So, just based on "time spent," digital video appears to deserve much more video advertising than it receives.

In eMarketer's Cross-Platform Video Trends Roundup (2015j), executives already embedding advertising into online entertainment video expressed strong feelings about it for the following reasons:

- 77% because it has superior targeting and is more measurable.
- 61% because it provides higher ROI than TV.
- 54% because it is maturing and becoming a viable competitor to TV.
- 42% because online/mobile video ads are increasingly being viewed as equivalent to TV.

These seem to be very strong reasons for advertisers to make the move, so why aren't more executives doing so? Is their a tipping point coming where movement could occur very rapidly?

And not all marketing videos are embedded into entertainment videos. Website videos are also helpful in marketing. Of U.S. digital video viewers polled, 83% said product demonstrations are helpful, 77% said product overview videos are helpful, 59% said videos showing how a product is made are helpful, 56% said customer testimonials are helpful, and 49% said even "about the company" videos are helpful (eMarketer, 2015i, p. 7).

Marketo (2015) also reviewed the benefits of video marketing, based on a variety of research findings by it and other organizations (p. 3):

- Video achieves greater engagement. It cites research that 65% of viewers watch more than three-fourths of each video consumed – much higher than text-based content.
- Video achieves higher conversion rates, according to 70% of marketers, and 85% greater product purchase intent.
- Video boosts the effectiveness of other content that surrounds it.
- Adding video to email increases click-through rates by 100%-200%.
- Adding video to the front page of a website increases the chance of a first-page Google search by 5300%.
- Adding video to a landing page increases conversion by 80%.
- And adding video to social media mix enhances audience engagement by 1000%.

Lopez-Nores, Blanco-Fernandez and Pazos-Arias (2012) have suggested even more technological reasons for advertisers to shift. Video on computers, smartpads and smartphones can do a lot more with advertising than regular TV. Besides easier access to video entertainment on more platforms, such technologies as tactile screens, touchpads, geo-targeting, and cloud-based personalization engines will provide more opportunities for advertisers to provide non-invasive, personalized and socially interactive advertising. What one person sees on TV may be different than another person. One person may see a Mercedes-Benz driving down a road while another would see a Toyota – customized product placement. A consumer could touch such an item on the screen to find out more about it. Some overt advertising might be displayed to act as a prompt, but it might be shorter, more interactive and more integrated into the entertainment.

Key Weaknesses and Problems

While it is clear that mobile marketing appeals to both the consumer and the marketer, there is a lack of academic research to parallel industry practice (Okazaki, 2012). Perhaps this is due to the speed at which the mobile market changes and evolves, leaving academia in the wake of its progress. Indeed, 69% of the thousands of executives responding to an Adobe (2015) survey agreed that they "are going to be experimenting heavily with digital next year." They would not be "experimenting" if research could already provide the answers. It is reasonable to assert that unless further research is done to more clearly define the theories that drive mobile advertising, advertisers will be unable to harness its full power. And yet the research seems destined to struggle to keep apace of new technologies, applications and programs to become available to this rapidly changing field. Technology under development now was the stuff Star Trek producers imagined might occur hundreds of years from now – voice-operated computers that can understand and respond to human requests; 3D printing that can essentially transport items across great distances and have them reappear; small computing devices connecting to larger computers with seemingly infinite knowledge and enough artificial intelligence to correlate, analyze, synthesize and extrapolate educated opinions on questions with no certain answers; and mobile medical devices linked to the Center for Disease Control or to top hospitals that can quickly analyze blood and saliva samples to recommend treatment and block epidemics from getting started. We will soon have refrigerators passing digital messages to our mobile devices to alert us that we have run out of milk. As such startling new technologies are developed to help society as a whole, they will be adapted to help individual enterprises and consumers. And as Clay Shirky (2009) has pointed out, "These tools don't get socially interesting until they get technologically boring." In other words, most application of technologies in society always trails the development of those technologies. Research also indicates that the success of mobile advertising requires more than an affinity for mobile devices and ease of use. For instance, in 2007, Japan had a much better wireless infrastructure than places such as Europe of the United States, and therefore had a better grasp on mobile advertising and what it was capable of. An analysis of this infrastructure revealed that "informativeness and credibility of the advertising message have the greatest impact on consumers' attitude towards advertising on the mobile Internet" (Haghirian and Inoue, 2007). Of course, over the last decade, the mobile market has increased exponentially all across the globe, meaning that mobile infrastructure in most places has improved. Therefore, thanks to this infrastructure development, the use of mobile advertising can only continue to grow and improve.

Twenty-first century studies of mobile advertising have still focused primarily on SMS advertising messages. DeReyck and Degraeve (2003) defined mobile advertising as text messages that target potential customers for commercial purposes. Their definition failed to encompass other possibilities. The results of most past studies relating to mobile advertising are interesting and still applicable in many settings, but they are focusing on a kind of advertising that is likely to fade away as preferred means of communication requiring greater bandwidth take its place.

In order to fully understand how SMS advertising messages can be useful to both the consumer and the advertiser, it is important to consider the consumer's attitude about SMS advertising, and specifically, how this attitude can affect the receipt of the advertisement itself. While research does indicate that mobile advertising through SMS messages is received well by consumers overall, advertisers should "exercise caution around the factors that will determine consumer acceptance" (Carroll, Barnes, Scornavacca, and Fletcher, 2007). As introduced by Carroll et al, the four identified factors are:

1. The "permission," or whether or not the receiver has given the advertiser permission to send mobile advertisements via SMS.
2. The "content" factor, referring to the text, images, audio, etc., within the actual advertisement. What does the message contain? What is being advertised?

3. "The Wireless Service Provider Control" factor, which explores factors related to the consumers actual wireless provider, meaning does the provider allow for these sorts of messages, are there ways to opt out of receiving messages, etc.
4. Finally, the "delivery of message" factor, which pertains to how the advertisement was delivered, including both the medium of the message as well as the type of technology necessary to receive the message (2007).

Without respecting these four factors, the advertiser runs the risk of pushing the consumer away instead of drawing them closer. The discussion of SMS advertisements must include considerations for the content and the delivery time. In a 2011 survey of mobile users who received SMS advertisements, results indicated that the more relevant the advertisement was to the individual, the more pleased the individual was and, thus, the more likely the message receiver was to purchase the advertised product or service. Another factor, delivery time, also played directly into overall consumer happiness. Here, delivery time refers to the time and day of the week that the message was delivered, rather than the timeliness of delivery. Results of the study indicated that the best times for consumers to receive message advertisements were weekends and Mondays. Additionally, there was a higher message acceptance as well as a higher purchase intention in the afternoon/evening hours (Rau, Zhang, Shang and Zhou, 2011).

MAPPING THE FUTURE

There are many subjects to explore concerning mobile advertising, including:

*   **Keep Exploring Social Media Marketing:** Social media sparked the First Internet Revolution, as seen with Obama's first campaign, but years later 87% of the 3,720 online marketers who responded to a 2015 survey still felt they had lots of unanswered questions (Stelzner, 2015, p. 6). Part of the problem, the study suggests, is not the lack of research but the lack of adequate information being disseminated to those who need it. But another problem is certainly the speed with which social media marketing is changing. Most of the social media introduced new tools and tactics within the past year. New tools and tactics demand new research – and new training for marketers. The key questions raised by the online marketers included:

    ◦   92% wanted to know what social tactics are most effective.
    ◦   91% wanted to know the best ways to engage their audience with social media.
    ◦   88% wanted to know how to best measure ROI with social media marketing.
    ◦   87% wanted to know how to find their target audience with social media.
    ◦   87% wanted to know which are the best tools for managing social media marketing.

Part of the problem may be incomplete, inconsistent and unscientific research within industry with inadequate research and outreach by academia. But with such a high percentage of professionals feeling lost within the new world of online marketing, there are clearly questions to be explored and better outreach to be provided.

- **Explore Video:** With online video spearheading the Second Internet Revolution, in part, this, too, demands more research. Conditions that foster digital video advertising will continue to be enhanced – the technology, bandwidth, accessibility, strategies, opportunities and investments. The unexpected 550% explosion of online videos viewed in 2012 could be mirrored by continuing improvement of services. Average broadband speed will nearly quadruple between 2011 and 2016 from 9 megabits per second to 34 mbps, meaning the quality of online video will be much greater (Cisco 2012). This, too, is in part because of the spread of mobile broadband. And Internet connections, again largely because of the rapid spread of mobile devices, will nearly double worldwide by 2016, at which time there will be an estimated 2.5 connections for every man, woman, child and infant in the world (Cisco 2012). Integrated use of such technologies as tactile screens, touchpads, geo-targeting, and cloud-based personalization engines, as suggested by Lopez-Nores et al (2012), could also enhance the experience for both advertisers and consumers. There is a need to explore how video advertising can be used most effectively in the mobile environment when embedded as commercials within entertainment videos, used as standalone marketing tools, developed as inbound marketing content, or launched in social media as potential viral videos.

The same could be said about consumer viewing of digital video in general and mobile video specifically. The viewing of digital video in the U.S. (1:16 hours per day) is still dwarfed by TV (4:15 hours per day). While the time spent viewing digital video per American has increased about 550% between 2011-2015, why hasn't it been more effective in closing that gap with TV, and, again, could a tipping point be closer than we think (eMarketer, 2015i)? The tipping point for online advertising, in general, took experts by surprise. In one year many newspapers were still hiring people with confidence, and the next year they were laying people off. Could that kind of tipping point surprise occur for digital video viewership and advertising? Why has video advertising not migrated to the Internet as quickly as audience? What is required to create that tipping where broadcast TV advertising begins fleeing to the Internet? Still more bandwidth? More viewing options? Higher quality free video entertainment offered more like free broadcast TV?

- **Explore How Different Digital Media Affect the Message:** The eMarketer Video Trends Roundup (2015j, p.4) cited experimental research suggesting that the digital medium is the message in the sense of how much attention viewers pay digital video ads. High attention was achieved by 64% of the smartphone viewers, 54% of the tablet viewers and 52% of the PC viewers. What causes the difference in focus between advertising viewers using these different devices? How reliable is this experimentation, and what other psychological differences might be detected in how digital consumers respond to the different digital media?
- **Explore 'Time-Spent' Predictive Research:** While time spent with different media can undoubtedly suggest some future movement in advertising expenditures, as discussed previously, the degree of movement one might expect from such a large gap comparing newspapers and mobile devices, for example, has been years in the making. Can this quantitative tool be truly useful when so many elements within the advertising field are qualitative in nature? As suggested earlier, a few years ago essentially no quality means of advertising existed with mobile devices. Smartphones were just starting to become popular, and the only mobile advertising available on earlier mobile phones was intrusive SMS messaging. Many more mobile advertising opportunities have now become available, so will their low cost per audience minute now lead to rapid expansion of the

mobile market. Projections suggest it may. Nonetheless, it has taken a long time for advertising to follow the audience. The same thing is happening with video advertising. In 2013 U.S. adults spent 7.3% of their media time watching online video, but video ad spending was only 2.4%. In 2014, 9.2% time spent but only 3.5% of ad share. And in 2015 the projection is for 10.9% of the time spent on media but only 4.4% of the ad share (eMarketer, 2015i, p. 4). Can a formula be created to accurately predict how long it takes for advertising to catch up with audience shifts?

- **Explore Strategies and Benefits of Omnichanneling:** While the main goal of omnichanneling now seems to be to help personalize messages, there may be other opportunities, as well. There has been suspicion for some time, even in traditional media, that multi-channeling creates some synergy, but it is difficult to document because some of that synergy is suspected to be achieved long-term and by broadening the audience. Wu et al (2013) tried to create measurable synergy by cross-channeling viral messages with broadcast TV. They concluded that a combination of viral advertising and broadcast TV is better to diffuse advertisements than either method by itself, but they stopped short of claiming true synergy in so doing. As suggested earlier, omnichanneling does offer some opportunities to achieve measurable synergy not necessarily by the way they channel to separate audiences but by the way the can interact. One example is using webinars to collect a database of active social opinion leaders who can then maximize viralization of later messages.

- **Explore Mobile ROI:** Executives' rapid move to digital reflects in part their continued distrust in the ROI of traditional media in comparison to digital. More than half (51%) of Adobe's (2015) survey respondents gave only 1-2 points on a 5-point scale (5 being maximum trust) to their ROI confidence in offline ad expenditures and 45% for offline marketing programs. Only 27% of them had the same concerns for digital ad expenditures and 26% for digital marketing programs. ROI concern for mobile advertising and marketing fell in between. About 36% of the executives rated mobile ad expenditures as 1 or 2 in ROI confidence, and 35% for mobile marketing programs. Why do executives distrust mobile more than digital in general, and how can mobile media raise that trust level to mirror that of other digital media? The answer is measurability. Eight out of 10 U.S. marketers said they would invest more into mobile if they could track ROI better (eMarketer, 2015k). In another study cited by the eMarketer report, 88% of marketers felt confident tracking ROI with Web advertising, 76% with email, 66% with mobile Web, but only 37% with mobile apps. Most mobile purchases are now being made on mobile apps while organizations websites are mostly not yet accommodating easy use by mobile devices. Besides Google now making mobile accessibility part of its search algorithm, will this concern that app ROI is not adequately measurable lead to apps being replaced by mobile-friendly websites? Or can new means be created to measure mobile ROI?

- **Explore Programmatic Advertising:** Programmatic advertising is one of the newest and most powerful advertising trends, encompassing most types of digital advertising – social, video, display, mobile, etc. Since this process has only come to the attention of most marketers since 2012, relatively few academics have ventured into this field or know what research questions need to be answered. In general, key questions to consider might relate to the best practices and future better practices and the problems and possible solutions for how ads are purchased and managed, for how big data is used to personalize ads, and for how advertising content is automated. But programmatic advertising also highlights another problem related to this and other rapidly changing, technology-related practices in advertising – staying abreast of change. Starcom agency felt the impact of programmatic was so dramatic and the lack of knowledge so great among staff members that they developed a 20-month training program for all 1,200 employees (Pathak 2015). And among business-to-business marketing executives surveyed by Forrester Research, 47% said they have not been able to find employees trained adequately to address such new challenges (Ramos, 2013). Important academic research in the field of programmatic advertising may include how universities can keep apace with changes in the industry in order to adequately prepare their graduates.

CONCLUSION

This chapter is designed to shed light on the ever-evolving and quickly changing world of mobile advertising. While some groundwork and theory will continue to apply to this arm of advertising, it is clear that the rapid pace of mobile advertising will demand a network of theory that changes as quickly as the industry does. However, universal access to the Internet in conjunction with real-time data collection resources suggest that it is possible for academics and industry professionals alike to keep in step with the progress of mobile advertising.

REFERENCES

Aberdeen Group. (2014). *Analyzing the ROI of video advertising.* Retrieved 16 Feb. 2015 from http://go.brightcove.com/bc-aberdeen-analyzing-roi-b

Adobe. (2015). *Quarterly Digital Intelligence Briefing: Digital Trends 2015.* Retrieved 20 Feb. 2015 from http://offers.adobe.com/content/dam/offer-manager/en/na/marketing/Target/Adobe%20Digital%20Trends%20Report%202015.pdf

Advertiser Perceptions. (2015). *As the industry continues heavy shift to programmatic buying, disconnects grow between buyers and sellers, agencies and marketers.* Retrieved 6 May 2015 from http://www.advertiserperceptions.com/blog

Akpojivi, U., & Bevan-Dye, A. (2015). Mobile advertisements and information privacy perception amongst South African Generation Y students. *Telematics and Informatics, 32*(1), 1–10. doi:10.1016/j.tele.2014.08.001

Arens, W. F. (2002). *Contemporary Advertising.* New York City: McGraw-Hill Companies Inc.

Ask, J. A. (2014). *Mobile is Not a Channel.* Forrester Research Inc. Retrieved 1 August 2015 from http://urbanairship.com/lp/forrester-report-mobile-is-not-a-channel?ls=EmailMedia&ccn=150730eMarketer ForrChannel&mkt_tok=3RkMMJWWfF9wsRohvK3IZKXonjHpfsXw6egoUa%2BxlMI%2F0ER3fOv rPUfGjI4JRcRlI%2BSLDwEYGJlv6SgFQrbCMbNs3bgEWxA%3D

Baran, S. (2006). *Introduction to Mass Communication: Media Literacy and Culture.* New York City, NY: McGraw-Hill.

Bart, Y., Stephen, A., & Sarvary, M. (2014). Which Products Are Best Suited to Mobile Advertising? A Field Study of Mobile Display Advertising Effects on Consumer Attitudes and Intentions.[JMR]. *JMR, Journal of Marketing Research, 51*(3), 270–285. doi:10.1509/jmr.13.0503

Bergen, M. (2014). Verizon Wireless Dives into Mobile-Ad Business. *Advertising Age, 85*(13), 10.

Blueshift Research. (2015). *June 2015 Trends Tracker Report.* Retrieved 2 July 2015 from https://smaudience.surveymonkey.com/download-trends-tracker-report-june-2015.html?utm_source=email&recent=email&program=Q2_15_June_Trends-Tracker_Wrap-Up&source=email&utm_campaign=trends-tracker&utm_content=trends-tracker-june-2015&mkt_tok=3RkMMJWWfF9wsRoiu6vJZKXonjHpfsX66e4vW6C1lMI%2F0ER3fOvrPUfGjI4FTcFiI%2BSLDwEYGJlv6SgFSrbFMaJy2LgJWBb0TD7slJfbfYRPf6Ba2Jwwrfg%3D

Brackett, L. K., & Carr, B. N. Jr. (2001). Cyberspace advertising vs. other media: Consumer vs. mature student attitudes. *Journal of Advertising Research, 41*(5), 23–32. doi:10.2501/JAR-41-5-23-32

Brister, K. (2010). Making the Most of Mobile; Banks can capitalize on marketing strategies for handheld devices, even if mobile banking is still short of being mainstream. *US Banker,* 17-17.

Campaign Finance Institute. (2010). *All CFI Funding Statistics Revised and Updated for the 2008 Presidential Primary and General Election Candidates.* Retrieved 30 July 2015 from http://www.cfinst.org/press/releases_tags/10-01-08/Revised_and_Updated_2008_Presidential_Statistics.aspx

Carroll, A., Barnes, S. J., Scornavacca, E., & Fletcher, K. (2007). Consumer perceptions and attitudes towards SMS advertising: Recent evidence from New Zealand. *International Journal of Advertising, 26*(1), 79–98.

César, P., Knoche, H., & Bulterman, D. (2010). *Mobile TV: Customizing Content and Experiences.* Heidelberg: Springer.

Chen, J. V., Su, B., & Yen, D. C. (2014). Location-based advertising in an emerging market: A study of Mongolian mobile phone users. *International Journal of Mobile Communications, 12*(3), 291–310. doi:10.1504/IJMC.2014.061462

Choi, Y. K., Hwang, J., & McMillan, S. J. (2008). Gearing up for mobile advertising: A cross-cultural examination of key factors that drive mobile messages home to consumers. *Psychology and Marketing, 25*(8), 756–768. doi:10.1002/mar.20237

Chou, H., & Lien, N. (2014). Effects of SMS teaser ads on product curiosity. *International Journal of Mobile Communications, 12*(4), 328–345. doi:10.1504/IJMC.2014.063651

Chowdhury, H. K., Parvin, N., & Weitenberner, C. et al.. (2010). Consumer attitude toward mobile advertising in an emerging market: An empirical study. *Marketing, 12,* 206–216.

Cisco. (2012). *Cisco's VNI Forecast Projects the Internet Will Be Four Times as Large in Four Years.* Retrieved 16 Feb. 2015 from http://newsroom.cisco.com/press-release-content?type=webcontent&articleId=888280

Cisco. (2015). *Forecast projects nearly 10-fold global mobile data traffic growth over next five years.* Retrieved 16 Feb. 2015 from http://newsroom.cisco.com/press-release-content?type=webcontent&art icleId=1578507

comScore. (2013). *U.S. digital future in focus 2013.* Retrieved 16 Feb. 2015 from http://www.comscore. com/Insights/Presentations-and-Whitepapers/(offset)/10/?cs_edgescape_cc=KZ

comScore. (2014). *Half of Millennial Netflix Viewers Stream Video on Mobile.* Retrieved 20 Feb. 2015 from http://www.comscore.com/Insights/Data-Mine/Half-of-Millennial-Netflix-Viewers-Stream-Video-on-Mobile

Content Marketing Institute. (2013). *B2B Content Marketing 2014 Benchmarks, Budgets, and Trends – North America.* Retrieved 20 Feb. 2015 from http://contentmarketinginstitute.com/wp-content/uploads/2013/10/B2B_Research_2014_CMI.pdf/

Delany, C. (2009). *Learning from Obama's Financial Steamroller: How to Raise Money Online.* Retrieved 30 July 2015 from http://www.epolitics.com/2009/05/15/learning-from-obamas-financial-steamroller-how-to-raise-money-online/

Deloitte. (2015). *Navigating the new digital divide.* Retrieved 25 May 2015 from http://www2.deloitte. com/content/dam/Deloitte/us/Documents/consumer-business/us-cb-navigating-the-new-digital-divide-v2-051315.pdf

Drossos, D., Giaglis, G. M., Lekakos, G., Kokkinaki, F., & Stavraki, M. G. (2007). Determinants of effective SMS advertising: An experimental study. *Journal of Interactive Advertising, 7*(2), 16–27. doi:10.1080/15252019.2007.10722128

Ducoffe, R. H. (1996). Advertising value and advertising the web. *Journal of Advertising Research, 36,* 21–35.

Eddy, N. (2014). Mobile Advertising, Push Notifications Beneficial for Retailers. *EWeek,* 7-7.

Elliott, M. T., & Speck, P. S. (1998). Consumer perceptions of advertising clutter and its impact across various media. *Journal of Advertising Research, 38,* 29–42.

eMarketer. (2011). *Ad Dollars Still Not Following Online and Mobile Usage.* Retrieved 20 Feb. 2015 from http://www.emarketer.com/Article/Ad-Dollars-Still-Not-Following-Online-Mobile-Usage/1008311

eMarketer. (2013a). *US Total Media Ad Spend Inches Up, Pushed by Digital.* Retrieved 20 Feb. 2015 from http://www.emarketer.com/Article/US-Total-Media-Ad-Spend-Inches-Up-Pushed-by-Digital/1010154

eMarketer. (2013b). *Digital TV, Movie Streaming Reaches a Tipping Point.* Retrieved 20 Feb. 2015 from http://www.emarketer.com/Article.aspx?R=1009775

eMarketer. (2013c). *TV Advertising Keeps Growing as Mobile Boosts Digital Video Spend.* Retrieved 20 Feb. 2015 from http://www.emarketer.com/Article/TV-Advertising-Keeps-Growing-Mobile-Boosts-Digital-Video-Spend/1009780 eMarketer. (2013d). *Most Digital Ad Growth Now Goes to Mobile as Desktop Growth Falters.* Retrieved 20 Feb. 2015 from http://www.emarketer.com/Article/Most-Digital-Ad-Growth-Now-Goes-Mobile-Desktop-Growth-Falters/1010458

eMarketer. (2013e). *Tablets, Smartphones Drive Mobile Commerce to Record Heights.* Retrieved 20 Feb. 2015 from http://www.emarketer.com/newsroom/index.php/emarketer-tablets-smartphones-drive-mobile-commerce-record-heights/

eMarketer. (2014a). *Despite Time Spent, Mobile Still Lacks for Ad Spend in the US.* Retrieved 20 Feb. 2015 from http://www.emarketer.com/Article/Despite-Time-Spent-Mobile-Still-Lacks-Ad-Spend-US/1010788

eMarketer. (2014b). *Driven by Facebook and Google, Mobile Ad Market Soars 105% in 2013.* Retrieved 20 Feb. 2015 from http://www.emarketer.com/Article/Driven-by-Facebook-Google-Mobile-Ad-Market-Soars-10537-2013/1010690

eMarketer. (2015a). *Real-Time Personalization Affects the Bottom Line.* Retrieved 1 August 2015 from http://www.emarketer.com/Article/Real-Time-Personalization-Affects-Bottom-Line/1012689

eMarketer. (2015b). *Can Companies Validate All of Their Customer Data?* Retrieved 1 August 2015 from http://www.emarketer.com/Article/Companies-Validate-All-of-Their-Customer-Data/1012693

eMarketer. (2015c). *How Many Smartphone Users Are Officially Addicted?* Retrieved 1 August 2015 from http://www.emarketer.com/Article/How-Many-Smartphone-Users-Officially-Addicted/1012800?ecid=NL1001]

eMarketer. (2015d). *Mobile Accounts for Almost Half of China's Retail Ecommerce Sales.* Retrieved 1 August 2015 from http://www.emarketer.com/Article/Mobile-Accounts-Almost-Half-of-Chinas-Retail-Ecommerce-Sales/1012793?ecid=NL1001

eMarketer. (2015e). *Expect more programmatic ads to pop up on mobile.* Retrieved 6 May 2015 from http://www.emarketer.com/Article/Expect-More-Programmatic-Ads-Pop-Up-on-Mobile/1012291/1

eMarketer. (2015f). *Programmatic Sellers vs. Buyers: Who's More Hooked On Mobile?* Retrieved 6 August 2015 from http://www.emarketer.com/Article/Programmatic-Sellers-vs-Buyers-Whos-More-Hooked-On-Mobile/1012374

eMarketer. (2015g). *Why Retailers Are Buying In to Programmatic.* Retrieved 6 August 2015 from http://www.emarketer.com/Article/Why-Retailers-Buying-Programmatic/1012448

eMarketer. (2015h). *Programmatic Creative: Look to Existing Processes for Guidance.* Retrieved 6 August 2015 from http://www.emarketer.com/Article/Programmatic-Creative-Look-Existing-Processes-Guidance/1012434

eMarketer. (2015i). *Mobile Video Roundup.* Retrieved 30 July 2015 from http://www.emarketer.com/articles/roundups

eMarketer. (2015j). *Cross-platform video trends roundup.* Retrieved 20 Feb. 2015 from https://www.emarketer.com/public_media/docs/eMarketer_Cross_Platform_Video_Trends_Roundup.pdf eMarketer. (2015k). *Measuring mobile effectiveness still challenges marketers.* Retrieved 4 August 2015 from http://www.emarketer.com/Article/Measuring-Mobile-Effectiveness-Still-Challenges-Marketers/1012797?ecid=NL1001

Europe, I. A. B. (2010). *European Media Landscape Report Summary.* Retrieved 1 May 2015 from http://www.iabeurope.eu/research/about-mediascope.aspx

Evergage. (2014). *The ROI of Relevant Marketing.* Retrieved 1 August 2015 from http://info.evergage.com/the-roi-relevant-of-relevant-marketing

Facebook. (2015). *Improving Ad Performance with the Carousel Format.* Retrieved 30 July 2015 from https://www.facebook.com/business/news/carousel-ads

FEC.gov. (2008). *Financial Summary Report Search Results.* Retrieved December 22, 2008.

Gladwell, M. (2000). *The tipping point: how little things can make a big difference.* London: Little Brown.

Godfrey, R. R., & Duke, S. E. (2014). Leadership in Mobile Technology: An Opportunity for Family and Consumer Sciences Teacher Educators. *Journal of Family and Consumer Sciences, 106*(3), 36–40.

Gutman, B. (2012). Kellogg proves ROI of digital programmatic buying. *Forbes.* Retrieved 6 May from http://www.forbes.com/sites/marketshare/2012/11/09/kellogg-proves-roi-of-digital-programmatic-buying/

Haghirian, P., & Inoue, A. (2007). An advanced model of consumer attitudes toward advertising on the mobile internet. *International Journal of Mobile Communications, 5*(1), 48–67. doi:10.1504/IJMC.2007.011489

Haghirian, P., & Madlberger, M. (2005). Consumer attitude toward advertising via mobile devices – An empirical investigation among Austrian users. *Proceedings of 13th European Conference on Information Systems.* Retrieved 20 March 2015 from http://aisel.aisnet.org/ecis2005/44

Hanley, M., Becker, M., & Martinsen, J. (2006). Factors influencing mobile advertising acceptance: Will incentives motivate college students to accept mobile advertisements? *International Journal of Mobile Marketing, 1,* 50–58.

Haroldsen, E., & Harvey, K. (1979). The Diffusion of Shocking Good News. *The Journalism Quarterly, 56*(4), 771–775. doi:10.1177/107769907905600409

Heine, C. (2015). Facebook's Q1 Ad Revenue Increased 46 Percent to $3.3 Billion. *AdWeek.* Retrieved 30 July 2015 from http://www.adweek.com/news/technology/facebooks-q1-ad-revenue-increased-46-percent-33-billion-164234

Hershkowitz, S., & Lavrusik, V. (2013). 12 Best Practices for Media Companies Using Facebook Pages. *Facebook Media.* Published May 2, 2013. https://www.facebook.com/notes/facebook-media/12-pages-best-practices-for-media-companies/518053828230111

Hoelzel, M. (2015). Spending on native advertising is soaring as marketers and digital media publishers realize the benefits. *Business Insider.* Retrieved 30 July from http://www.businessinsider.com/spending-on-native-ads-will-soar-as-publishers-and-advertisers-take-notice-2014-11

HubSpot. (2014). *State of Inbound 2014*. Retrieved 4 July 2014 from http://offers.hubspot.com/2014-state-of-inbound

International Telecommunications Union. (2015). *Global ICT developments*. Retrieved 20 March 2015 from http://www.itu.int/en/ITU-D/Statistics/Pages/stat/default.aspx

Izquierdo-Yusta, A., Olarte-Pascual, C., & Reinares-Lara, E. (2015). Attitudes toward mobile advertising among users versus non-users of the mobile Internet. *Telematics and Informatics*, *32*(2), 355–366. doi:10.1016/j.tele.2014.10.001

Kim, K. J. (2014). Can smartphones be specialists? Effects of specialization in mobile advertising. *Telematics and Informatics*, *31*(4), 640–647. doi:10.1016/j.tele.2013.12.003

Kim, L. (2015). 5 Facebook Advertising Tools that Save Time and Improve your ROI. *Social Media Examiner*. Retrieved 30 July 2015 from http://www.socialmediaexaminer.com/5-facebook-advertising-tools/?awt_l=BKVE2&awt_m=3ixhuI4YULr.ILT&utm_source=Newsletter&utm_medium=NewsletterIssue&utm_campaign=New

King, C. (2015). 28 Social Media Marketing Predictions for 2015 from the Pros. *Social Media Examiner*. Retrieved 13 Feb. 2015 from http://www.socialmediaexaminer.com/social-media-marketing-predictions-for-2015/

Kolsaker, A., & Drakatos, N. (2009). Mobile advertising: The influence of emotional attachment to mobile devices on consumer receptiveness. *Journal of Marketing Communications*, *15*(4), 267–280. doi:10.1080/13527260802479664

Liu, C.-L. E., Sinkovics, R. R., Pezderka, N., & Haghirian, P. (2012). Determinants of consumer perceptions toward mobile advertising — a comparison between Japan and Austria. *Journal of Interactive Marketing*, *26*(1), 21–32. doi:10.1016/j.intmar.2011.07.002

López-Nores, M., Blanco-Fernández, Y., & Pazos-Arias, J. (2013). Cloud-Based Personalization of New Advertising and e-Commerce Models for Video Consumption. *The Computer Journal*, *56*(5), 573–592. doi:10.1093/comjnl/bxs103

MacKenzie, S. B., & Lutz, R. J. (1989). An empirical examination of the structural antecedents of attitude toward the ad in an advertising pretesting context. *Journal of Marketing*, *53*(2), 48–65. doi:10.2307/1251413

MarketingSherpa (2010). *2010 Social Media Marketing Benchmark Report*. SherpaStore.com.

Marketo. (2015). *How to Use Video Content and Marketing Automation*. Retrieved 30 July 2015 from http://www.marketo.com/ebooks/how-to-use-video-content-and-marketing-automation-to-better-engage-qualify-and-convert-your-buyers/

McDermott, J. (2013). How mobile ads are bought at agencies today. *Advertising Age*, *84*(29), S26.

Mixpo. (2015). *The state of video advertising*. Retrieved 1 May 2015 from http://marketing.mixpo.com/acton/fs/blocks/showLandingPage/a/2062/p/p-00b4/t/page/fm/0

Mohd, N., Mohd, N., Jayashree, S., & Hishamuddin, I. (2013). Malaysian consumers attitude towards mobile advertising, the role of permission and its impact on purchase intention: a structural equation modeling approach. *Asian Social Science*. Retrieved 20 March 2015 from http://ccsenet.org/journal/index.php/ass/article/view/26973/16462

Mutter, A. (2012). Print ads fell 25x faster than digital grew. *Reflections of a Newsosaur*. http://newsosaur.blogspot.com/2012_09_01_archive.html

Mutter, A. D. (2014). The plight of newspapers in a single chart. *Reflections of a Newsosaur*. Retrieved 20 Feb. 2015 from http://newsosaur.blogspot.com/2014/04/the-plight-of-newspapers-in-single-chart.html.http://newsosaur.blogspot.com/2014/04/the-plight-of-newspapers-in-single-chart.html

Neustar. (2015). *5 Ways to Engage those Slippery Omni-channel Shoppers*. Retrieved 3 August 2015 from https://hello.neustar.biz/omnichannel_wp_retail_mktg_lp.html?utm_medium=Advertising-Online&utm_source=Shop.org&utm_campaign=2015-eBook-Shop.org

Nguyen, J. (2014). *Online Video Consumption in APAC and Total Video*. Retrieved 1 May 2015 from http://www.comscore.com/Insights/Presentations-and-Whitepapers/(offset)/10/?cs_edgescape_cc=KZ retrieved 16 Feb. 2015.

Okazaki, S. (2012). *Teaching mobile Advertising. In Advertising Theory* (pp. 373–387). New York: Routledge.

Olenski, S. (2013). What is programmatic advertising, and is it the future? *Forbes*. Retrieved 6 May 2015 from http://www.forbes.com/sites/marketshare/2013/03/20/what-is-programmatic-advertising-and-is-it-the-future/

Pathak, S. (2015). How Starcom trained 1,200 employees to speak programmatic. *Digiday*. Retrieved 6 May 2015 from http://digiday.com/agencies/starcom-trained-talent-speak-programmatic/

Pepper, R. (2012). *The impact of mobile broadband on national economic growth*. Retrieved 16 Feb. 2015 http://newsroom.cisco.com/press-release-content?type=webcontent&articleId=1110575

Pew Research Center. (2014). *The revenue picture for American journalism, and how it is changing*. Retrieved 17 Feb. 2015 http://www.journalism.org/files/2014/03/Revnue-Picture-for-American-Journalism.pdf

Pietz, M., & Storbacka, L. (2007). *Driving Advertising into Mobile Mediums: Study of Consumer Attitudes Towards Mobile Advertising and of Factors Affecting on Them*. Umea University.

PulsePoint. (2015). *Programmatic intelligence report 2014*. Retrieved 6 May 2015 from http://www.pulsepoint.com/resources/whitepapers/pulsepoint_intelligence_report.pdf

Ramos, L. (2013). *The B2B CMO role is expanding: Evolve of move on*. Retrieved 30 April 2015 from http://solutions.forrester.com/Global/FileLib/Reports/B2B_CMOs_Must_Evolve_Or_Move_On.pdf

Ramsey, G. (2015). *The State of Mobile 2015*. eMarketer live webcast 25 March 2015.

Rau, P. L. P., Zhang, T., Shang, X., & Zhou, J. (2011). Content relevance and delivery time of SMS advertising. *International Journal of Mobile Communications*, 9(1), 19–38. doi:10.1504/IJMC.2011.037953

Rau, P. P., Zhou, J., Chen, D., & Lu, T. (2014). The influence of repetition and time pressure on effectiveness of mobile advertising messages. *Telematics and Informatics, 31*(3), 463–476. doi:10.1016/j.tele.2013.10.003

Reichhart, P. (2014). Identifying factors influencing the customers purchase behaviour due to location-based promotions. *International Journal of Mobile Communications, 12*(6), 642–660. doi:10.1504/IJMC.2014.064917

Researchscape International. (2015). *Trends & Priorities in Real-Time Personalization.* Retrieved 1 August 2015 from http://www.evergage.com/blog/survey-findings-trends-priorities-in-real-time-personalization/

Reyck, B. D., & Degraeve, Z. (2003). Broadcast scheduling for mobile advertising. *Operations Research, 51*(4), 509–517. doi:10.1287/opre.51.4.509.16104

Ross, P. (2015). *Native Facebook Videos Get More Reach Than Any Other Type of Post.* Retrieved 1 May 2015 from http://www.socialbakers.com/blog/2367-native-facebook-videos-get-more-reach-than-any-other-type-of-post

Schlosser, A. E., Shavitt, S., & Kanfer, A. (1999). Survey of Internet users' attitudes toward Internet advertising. *Journal of Interactive Marketing, 13*(3), 34–54. doi:10.1002/(SICI)1520-6653(199922)13:3<34::AID-DIR3>3.0.CO;2-R

Shintaro, O., & Barwise, P. (2011). Has the Time Finally Come for the Medium of the Future? *Journal of Advertising Research*, 5159–5171.

Shirky, C. (2009) *How social media can make history.* TED.com talk recorded in June 2009, retrieved 20 March 2015.

Soroa-Koury, S., & Yang, K. C. (2010). Factors affecting consumers' responses to mobile advertising from a social norm theoretical perspective. *Telematics and Informatics, 27*(1), 103–113. doi:10.1016/j.tele.2009.06.001

Stelzner, M. A. (2015). 2015 Social Media Marketing Industry Report. *Social Media Examiner*. Retrieved 30 July 2015 from http://www.socialmediaexaminer.com/SocialMediaMarketingIndustryReport2015.pdf

Stevenson, A., & Wang, S. (2014, December 20). A Jump to Mobile Ads in China. *The New York Times*, p. 2.

Swain, H. (2007, January26). Have You Got The Message Yet? *Times Higher Education Supplement*, 54–54.

Tsang, M. M., Hom, S.-C., & Liang, T.-P. (2004). Consumer attitudes toward mobile advertising: An empirical study. *International Journal of Electronic Commerce, 8*, 65–78.

Varnali, K. (2014). SMS advertising: How message relevance is linked to the attitude toward the brand? *Journal of Marketing Communications, 20*(5), 339–351. doi:10.1080/13527266.2012.699457

When will mobile advertising become the next big thing? (2008). *Communication Arts, 50*(2), 18.

White, D. S. (2013). *Social Media Growth 2006-2012.* Retrieved 16 Feb. 2015 from http://dstevenwhite.com/

Wu, J., Hu, B., & Zhang, Y. (2013). Maximizing the performance of advertisements diffusion: A simulation study of the dynamics of viral advertising in social networks. *Simulation: Transactions of the Society for Modeling and Simulation International*, *89*(8), 921–934. doi:10.1177/0037549713481683

Xstream. (2013a). *Online video trends & predictions for 2013* (Part 1). Retrieved 31 May 2015 from http://xstream.net/news-xstream

Xstream. (2013b). *Online video trends & predictions for 2013* (Part 2). Retrieved 31 May 2015 from http://xstream.net/content/online-video-trends-predictions-2013

Zanot, E. J. (1984). Public attitudes toward advertising: The American experience. *International Journal of Advertising*, *3*, 3–15.

Zarrella, D. (2011). *Zarrella's Heirarchy of Contagiousness: The Science, Design and Engineering of Contagious Ideas*. Dobbs Ferry, NY: Do You Zoom Inc.

Zarrella, D. (2013). *The Science of Marketing: When to Tweet, What to Post, How to Blog, and Other Proven Strategies*. Hoboken, NJ: John Wiley & Sons.

Chapter 16
Examining the Role of WeChat in Advertising

Qi Yao
University of Macau, China

Mei Wu
University of Macau, China

ABSTRACT

WeChat, the most popular social networking service mobile app in China, enables users to contact friends with text, audio, video contents as well as get to know new people within a range of a certain distance. Its latest version has gone beyond the pure social function and opened a window into marketing. This study applies activity theory as a framework to explore new features of WeChat as a new platform for advertising. Then the authors qualitatively analyze the implications that marketers adapted and appropriated WeChat to engage with their customers and in turn how these implications modified the ways of advertising. The significance of this study is that it applied activity theory as an attempt to complement the theoretical pillars of communication study. Activity theory focuses on the activity itself rather than the interaction between WeChat and the users. Second, activity theory reveals a 2-way process which emphasizes both on how advertisers adapt WeChat based on everyday practice; and how WeChat modifies the activities which the advertisers engage in.

INTRODUCTION

Integrated with multimedia and Internet functions, a smartphone allows people to access diverse social communication spaces, where the existing types of conventional communication (calling and texting) mix with types of internet communication (email and social networking service). Smartphone makes the online space available 24/7, and thus develops ubiquitous sociality (Lee, 2013). Moreover owing to the fast growth of information technology, more types of mobile social media have been created, such as WhatsApp, Line, and WeChat. These mobile social networking services have been growing their influence among people, especially youth, around the world. Their user-friendly interfaces conveniently

DOI: 10.4018/978-1-5225-0469-6.ch016

present recent information with text, photos, and videos (Pelet, 2014). All of these benefit marketing practitioners to wage their campaigns and engage the audiences.

The prevalence of smartphones in people's daily life has critically changed the consumer's decision-making process (Pelet, 2014). The consumers now could easily access professional opinions, others' reviews, price comparisons and other convenient functions, which help consumers to make full-scale evaluation on products and services. This suggests marketers to take account of new ways to influence consumers and to carefully implement social networking service as a marketing tool (Pelet, 2014).

WeChat is the most popular mobile application in relation to social networks in current China. It was developed by Tencent company in China and released in January 2011.WeChat provides users with multimedia communication function including speaker phone, photo and mood sharing, text message, corporation official account, location sharing, location based searching and service. The latest version of WeChat added more functions such as game center, online payment, emoticon shop, street view scanning and scanning for translation. It served different countries, languages, operating systems and network formats. It took 56 months for Facebook to reach 100 million users and 30 months for WhatsApp. But when it comes to WeChat, it took only 15 months to reach 100 million users (Wu, Jakubowicz & Cao, 2013). Nowadays in China, WeChat users reached around 500 million. Moreover, the overseas user amount has broken through 100 million by the end of 2014 (Tencent, 2015).

While academic studies have been focusing on usage, society, behavior and other particular issues in the mobility context, advertisers also started to pay attention to the advantage offered by mobile social media such as WeChat. Compared with conventional mobile advertising methods such as sending SMS and MMS, now advertising through apps became the trend. As discussed before, the advertising on WeChat is based on user's strong-tied social connectedness. The marketers cannot depend on mass messaging to WeChat users, but take advantage of the users' social connectivity for propagation.

For the companies and advertisers, the official accounts of WeChat enable them to engage with their customers with a new method. Through WeChat official account, the company is able to approach the customers privately since all the conversation between them cannot be seen or commented on by others. Since WeChat official account is smart phone-based, consumers could read the messages anytime as well as forward them to Weibo, Moments (Peng You Quan[1]), friends and save them in WeChat. Since its 5.0 version, one of the most significant changes is that WeChat has separated the official account into two different forms: service account[2] and subscription account[3]. Owing to these new features, the customers seized the initiative to receive the messages only if they want to. Meanwhile, the new situation reminds advertisers not to use WeChat as a broadcast tool, which has been widely applied in other social media such as Qzone and Weibo. Instead of message bombardment, companies are demanded to offer more valuable contents as well as a high level of interactive experience with its audience (Leung, 2013).

The academia and industry are always sensitive to any new-born advertising media. However WeChat, which has already attracted almost one tenth of the population on this planet, has not brought enough attention to the marketing communication area. This study applied activity theory (AT) as framework to explore new features of WeChat as a new platform for advertising. Then we qualitatively analyzed the implications that marketers adapted and appropriated WeChat to engage with their customers and in turn how these implications modified the ways of advertising.

MOBILE SOCIAL MEDIA AND WECHAT IN CHINA

Mobile social media is always applied by companies and advertisers in areas such as promotions, communication, relationship development and marketing research (Kaplan & Haenlein, 2011). Both Facebook and Twitter are world-famous media platform for target-oriented advertising and online business communication. Although these two biggest mobile social media are blocked by the government of China, there are several native mobile social media which are very similar to them--Renren and Weibo, which could be deemed as the Chinese version of Facebook and Twitter respectively. Especially Weibo, which has already owned 249 million users, is beloved by companies and advertisers in China (CNNIC, 2015).

On April 11, 2014, Tencent QQ, the first instant messaging and social media software in China, has reached a new concurrent user record of 200 million. Tencent took advantage of QQ's huge volume of users and allowed users to sign in WeChat using QQ ID. Until May, 2014, Tencent announced that WeChat climbed to 396 million monthly active users. Beside the quantity advantage, WeChat is able to achieve this business model as its parent company Tencent already owns a complete line of e-commerce and e-payment services including QQ Online Shop and Yixun.com. These facilities and expertise will provide strong backing for the future growth of WeChat. Moreover, with the rise of E-business such as Taobao, Jumei and Jingdong, the companies and advertisers could no longer ignore the power of mobile social media as marketing communication platform. Meanwhile, people also became aware of the different online marketing communication strategies applied for different kinds of social media.

WeChat is more conducive to information diffusion than former social media because of its unique functions and information flow mechanism, it combines mobile phone, social media and Internet in one platform (Wu, Jakubowicz & Cao, 2013). Since its intrinsic instant messaging nature WeChat's users keep relatively strong ties. The friends in one's WeChat are mostly their family members, schoolmates and colleagues who are familiar with them in real life. It is also why people will believe the information which was shared on WeChat moment rather than Weibo or blogs. Companies and advertisers have found that it is increasingly important to promote their company on WeChat not only for its strong word of mouth effect but also for its various supported service (Figure 1).

Compared with blogs and Weibo users, WeChat users are more equal, more interconnected and better able to establish relationships with those who are outside their original social networks. Beside the ac-

Figure 1. WeChat brings together three overlapping communication services in one display

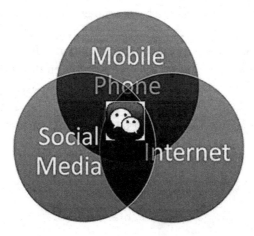

quaintance mentioned above, some companies also utilize WeChat to engage their customers: customers can get useful information through instant interaction. Under such circumstances, the information flow is not superincumbent but mutual. WeChat, which combines SNS application and instant messaging application, not only possesses the common features of former mobile social media, but also has its own new highlights. Its combination deserves deeper consideration and study.

MOBILE SOCIAL MEDIA STUDY

It is not easy to draw a clear definition of mobile social media because nowadays social media tools are getting more merged into diverse facets of mediated communication (Lüders, 2008). Therefore mobile social media can be roughly defined as a kind of software, application, or service which could be accessed via mobile devices such as PDA, tablet PC and smartphone. Mobile social media enables users to connect with others people as well as to generate, exchange and share various contents (Lee, 2013).

What is the changing characteristic of sociality through mobile social media? In the early stage, mobile social media was designed to emphasize the presumed mobility of users, usually via location-based service: through various wireless communication technologies, mobile device users are able to recognize others within certain vicinity (Wu & Yao, 2014). Dodgeball (Boyd & Ellison, 2007), Foursquare (Frith, 2012), Digu and Momo (Wu & Yao, 2014) are the examples of location-based mobile social media. Many cross-platform social media, such as Facebook and QQ, have specific locative functions like check-in and searching people nearby, however this cannot define its mobile presence. Actually according to a study of Facebook (Protalinski, 2012), among 901 million monthly active users, more than half of them (488 million) logged in Facebook by a mobile phone or tablet. Nowadays more and more people are accessing social media on mobile devices. Same phenomenon also happens in China. According to Nielsen, until June 2014 the number of mobile netizens in China has reached 530 million, which means 83.4% of the netizens chose to access the Internet through a mobile device. Among these people, 60.2% spend on average 1.5 hours every day using mobile social networking services. This report also revealed that 67% of the users consider mobile social networking services to be very important to them (Nielsen, 2014).

Users perform social interactions and develop social connections through adding or forwarding text, photo and other multimedia contents with peers on the mobile social media. These activities can be deemed as the extension of conventional sociality to online space. When accessing mobile social media, the users' workflow includes matching and integrating social functionalities, messaging, online contents and publishing tools within a complicated ecosystem of online social behavior. Therefore one should be cautious when studying the characterizations and usage of mobile social media, since mobile social media is defined by functions and uses instead of designs or forms (Brusse, Gardner, McAullay, & Dowden, 2014).

SOCIAL MEDIA ADVERTISING STUDY

Traditionally the process of advertising is often marketer-controlled. Based on one-way communication, the marketers decide on the way of communication and channels (face-to-face, survey, online panel and etc.) depending on demographic setting within a pre-determined space and time. However the social media altered the process, creating a new system where consumers are the core of content-generation process (Singh, Veron-Jackson, & Cullinane, 2008). The biggest difference between advertising on traditional media and social media is the development of technology has shifted from the passive consumers (in the context of traditional media) into positive and creative talent (in the context of social media). For marketers, the power has been altered from the company to the consumers (Pierre, 2012). Today's consumers complete the purchase activities with various stages, ranging from the first engagement with a brand, understanding a product, to hear about the product from social circles (Powers, Advincula, Austin, Graiko, & Snyder, 2012).

Speaking of social circles, consumers now possess more power than ever before: the accessibility of online social networks and digital devices like smartphone, PDA and tablet PC enables consumers to share with others discussing brands, services and products as well as interacting with brands conveniently and efficiently (Powers, Advincula, Austin, Graiko, & Snyder, 2012).

In order to make advertisement effective, the advertisers are aware of how to deliver the message to customers on the market. Academics and experts created several theories and models which demonstrate how customers react to the advertising message they are exposed to (Ciadvertising, 2007). AIDA model, one of the earliest as well as the most popular effect of advertising models, was designed to describe the different stages through which an advertiser should take an outlook. "A" refers to draw the *attention* of the customer. "I" means to raise customers' *interest* by concentrating and presenting advantages and benefits of a product. "D" is *desire*; it is the stage that the advertisement successfully persuades customers to desire the product or service which could satisfy their needs. The last "A" means *action*: the advertisement finally conducts customers making certain action or purchasing (Wijaya, 2012).

In the following part, activity theory will be introduced as the analytical framework to explore new features of WeChat as a new platform for advertising and how the AIDA model works within WeChat.

ACTIVITY THEORY

Since this study is among the first which introduces activity theory into communication study, it is critical to explain the reason why a psychology theory could be applied to smartphone app research. To

Table 1. The difference of traditional media, Weibo and WeChat

	Traditional Media	**Weibo**	**WeChat**
Communication Channel	One-way communication	Two-way communication	Two-way communication
	Opaque[4]	Transparent	Opaque
	Centralized[5]	Decentralized	Decentralized
Content	One-to-many	One-to-many	One-to-one
	Company-generated content	User/company -generated content	User/company -generated content
Influencers	Elites & Celebrities	Big Vs[6]	Users with large numbers of friends circle
Platform	Paid	Free	Free
	Pre-scheduled	Real-time	Real-time
Evaluation	Reach and frequency	Interaction	Interaction

elaborate the reasons, first this study will examine the most popular theories in communication study and their weakness. Then activity theory will be presented and explained for its complementary role in communication study.

To date, mobile services have been widely studied by the various theories. Among these, technology acceptance model (TAM), use & gratification theory are the most applied. TAM was developed to fulfill the need to understand users' satisfaction rate, and in turn to apply this rate to predict the success of a system (Davis, 1989). TAM mainly contains 2 attitudinal dimensions: Perceived Ease of Use and Perceived Usefulness, which are considered as direct factors related to the use of information system. Perceived Ease of Use refers to users' self-evaluations on the relationship between their cognitive work and the use of the system. Perceived Usefulness means how a particular system could improve users' job performance (Davis, 1989).

However despite its frequent application, TAM has been widely criticized. One part of the criticism is that TAM failed to consider the influence of social organization. Second is the insensitivity to diverse contexts of usage. TAM did not consider the probability that certain technology would be accepted in the beginning and then abandoned, or vice versa (Chuttur, 2009). The third and the most important reason is the studies that applied TAM are based on the fundamental premise that the scholars have the right to decide what shall be evaluated, even it is not certain whether the users indeed represent that type of use. In the real context, the relationship between the functionalities of the system and the tasks may be different among users and thus the ends would be completely different than scholars' anticipation (Salovaaraa & Tamminenb, 2008).

Use and gratification theory (U&G) was firstly developed in the studies of effectiveness of radio communication in the 1940s, when researchers were interested in different forms of media behaviors (Wimmer & Dominick, 1994). The primary goal of the U&G theory is to seek the psychological desires that shape users' choice for applying the mass media and the motivations which encourage them to take advantage of certain media for gratifications which satisfy people's inherent desires and needs (Elliott, 1974; Ruggiero, 2000; Zhu, 2004). However U&G faces various critics. One of them is that the data behind the theory is difficult to conclude and sometimes cannot be found (Littlejohn, 2002). Another problem is U&G is inclined to neglect the "dysfunctions of media in both culture and society" (Littlejohn, 1989). Lastly, U&G tends to consider media in positive ways and as able to provide audiences' needs. Less attention was paid on how the media cast negative cultural influences on society (Griffin, 2009). Such criticism makes U&G hard to analyze in the larger societal context (Severin & Tankard, 1997).

Therefore as mentioned above, communication study requires a theory that understands the users' real behaviors by investigating social-cultural contexts as users engage in media. Also attention should be drawn on how different parts of a community or organization could influence the implementation of the media. Moreover the communication study needs to understand that the media may be assimilated by the users and then in turn, the real practices could be modified when users are looking for adapting their activities with newly perceived possibilities of the media (Wu & Lin, 2011; Wu & Li, 2011).

Activity theory stems from a historical and cultural psychology originated by Vygotsky (1978) and Leont'ev (1978) from the former Soviet Union. Activity theory has been defined as a framework for understanding various forms of human activities and practices (Kuutti, 1996). Activity theory provides a broad range of conceptual frameworks that facilitate the research and design of Human Computer Interaction (HCI). Meanwhile it is also applied to determine how to provide users with the necessary tools to work with the interface to obtain the needed outcomes without participating in a long period of training (Gould, Verenikina, & Hasan, 2000).

An activity is comprised of subject, object and tools. According to Vygotsky (1978), human activity can be interpreted as a mediation process triggered by tools. The relations between subject and object are not direct but mediated. Tools play the part of mediating between them. Activity is a long-term formation, thus the outcome would not be transformed from object immediately, but through a procedure including several phases. Engeström (1987) improved this system and added 3 new relations: community, rules and division of labor. All of these components are mediated as shown below. Tool is applied by subject to complete an object, in order to fulfill a goal. Rules are then required between subject and other members within a community, meanwhile a division of labor is also indispensable between members of the community (Pang & Hung, 2001) (Figure 2).

- **Subject:** An actor (usually someone) who is involved in the activity and is chosen as the perspective in the analysis. In this study subject refers to the supervisor of social marketing from Company C, whom was interviewed.
- **Object**: Anything which could be manipulated or transformed by actors in an activity; an object could be material, rarely tangible (such as a plan), or totally intangible (such as a thought). The object for the supervisor as well as the company is to persuade more consumers to purchase its products.
- **Tool:** Anything utilized in the process of transformation. In Yamagata-Lynch's study (2001), the author pointed out that there are two kinds of tools available in the social environment: artifacts (also called "technical tools") and signs (or "psychological tools"). This study focuses on WeChat and its new features as a new tool for advertising.
- **Community**: A group of persons who pursue the same object. Here community means the social marketing team.
- **Rules**: Consist of laws, both implicit and explicit criterions, social relations and regulations within a community.
- **Division of Labor:** Involves both the vertical division of status and power, and the horizontal division of roles and tasks within a community.

Hence the activity system of this study can be shown as below (Figure 3).

Figure 2. Activity system

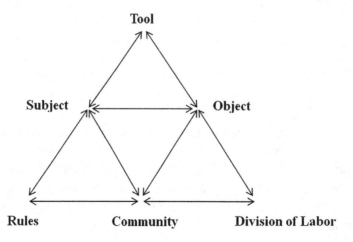

Figure 3. Activity system of the study

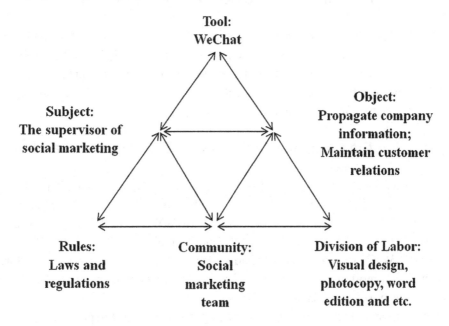

Another important concept of activity theory is contradiction, because it is the motive for the development of activity system. Activities are not inert but dynamically developing and changing, thus activities play as nodes within hierarchies and networks. The nodes are affected by other changes and activities in the work context and cause inequality between them (Kuutti, 1996). Therefore "contradiction" is introduced to understand these changes within and among components.

The primary contradiction found within a sole node of an activity. This contradiction appears from the tension between value of use and value of change. It exists in every corner of the triangle structure of activity and is the essential source of unsteadiness and development. The secondary contradiction emerges between the compositive nodes such as between the subject and the tool or between the tools and rules[7] (Engeström, 1993).

The reason this study applied activity theory to evaluate the advertising on WeChat is its focus on the real field usage. Rather than other new technology researches, activity theory does not emphasize on the history of adoption, but on how such technology is applied in practice and how it fits into peoples' everyday life.

Previous studies have pointed out that the strength of applying activity theory is because of the concept "tool meditation as the core to all human activities". Activity theory focuses on tools, such as WeChat in this study, as the meditators; and meanwhile instead of simply interaction, it emphasizes on the activity itself. In another word, the users here are deemed to be doing something rather than just using WeChat: WeChat is the tool through which users could fulfill or accomplish their goals. Therefore in this study, the focus is not only on the evaluation of WeChat as an advertising tool, but also on its usability and how well WeChat integrated into advertisers' activity to achieve their objective.

CASE STUDY: WECHAT OFFICIAL ACCOUNT OF COMPANY C

A case study of a WeChat official account has been conducted since winter 2014. The information provided in this part is according to a series of interviews with the supervisor of social marketing team of Company C[8]. The two interviewees were selected from the management level of Company C. The person who provided the majority of information was the Supervisor of Social Marketing Team, the other one was the manager of WeChat Operation Team.

The Background of the Company, Its WeChat Official Account and the Team

Located in a Southern province, Company C is one of the top smartphone manufacturers in China. The WeChat official account of Company C was established on Nov. 13, 2013. Its updating frequency is stochastic: usually 1-7 days per post. However, different from Weibo or Twitter, everyday WeChat official account could only send users one post, which may contain up to 8 columns of messages. Aside from traditional SNS, WeChat provides not only advertisement, but also includes other information such as store introduction, promotion, new arrival, etc. Moreover Company C irregularly launches events on WeChat such as bonus quizzes and games.

Social marketing Team, which is responsible for designing, updating and pushing content on WeChat, has 10 staff members. All of them graduated from electronic engineering major thus are well-acquainted with soft/hard ware knowledge. The operation of this community is very flexible. There is no particular division of labor, everyone who has time, or idea will be put in charge of the job.

Except WeChat, this team is also in charge of Weibo, QZone, Lofter[9] and web radio. Among all of these, Weibo is the most eye-catching because they invest most of the energy and resource in it. However they noticed that Weibo is declining because of a decrease of user numbers and degree of activity. More and more young people shift their attention to WeChat, which is more suitable for mobile devices. It seems Weibo is trying to find another business model because its business model is highly dependent on advertisements and is unable to get benefits from other channels. On the other hand, the advertisement bombardment severely ruins the user experience. Compared to Weibo, WeChat is a still in its developing stage, thus they dare not invest too much in WeChat due to its uncertainty.

The Object of WeChat Official Account

The WeChat official account provides 2 kinds of services. First is about the product, such as product introduction, A/S service and reservation. Second is to satisfy the consumers with living information, chatting service, games and etc. Therefore the purpose of WeChat and Weibo is the same: to solve customer's problems and form a positive image of the company. Compared to Weibo, WeChat is a combination of instant messaging and SNS. Its features enable the company to foster intimate relationships with customers as friends, because it is more private, more direct and more efficient in communication.

The supervisor mentioned "we hope after purchase consumers could still remember our company and we are happy if consumers are happy." In order to make consumers feel that they and the company have the same pursuit as well as to encourage consumers to spread the content, the official account sometimes rewards consumers with material objects like earphones, virtual bonus such as E-coupons and psychological consolation from the company.

The Process of Making Content for WeChat

First the team would have a meeting to set up the goal, exchange visual effects; after this the supervisor will arrange different jobs to the members, as mentioned above. An email will be sent to everyone to confirm the division of work, time, and resource distribution. If anything changed, they will use email to inform everyone. For example before the press conference, they will first, according to the official website, design a WeChat page with a link to the official website for reserving the product. The visual effect and style are consistent on the website, which contains more information; then they will extract the most essential and important things as the source material. The content including photos and texts are customized and optimized to fit in the screen of mobile devices. In another words, the work of the social marketing team is to integrate all the resources from website and other departments.

Referring to the rules, during this process except for professional ethics and laws, the team deems privacy protection the most. Since WeChat is a messaging app, it contains a lot of users' private information. The events they hold on WeChat always require user's authorization such as requiring gender, age and nick name. When reward or gift is provided, they also need users' address, phone number and the real name.

The Target Audience and the Content Pushing

According to the supervisor, generally they do not set the target audience specifically nor separate user groups when pushing contents, because according to their statistics (the official account server could tell the operating system of subscribers' smartphone) most people who subscribed to their official account are currently using their products. In another words the audience and the consumer group have been already prefixed both for the company and the official account. Although the marketing department has a total investigation data for all the consumers and they can use the data directly, they do not separate users in detail any further. The contents on WeChat are just for their product users.

When pushing the content, the time is very crucial. Usually the contents are pushed at 5-10 pm during weekends, and 3-4 pm and 7-9 pm on weekdays. "We must be careful to avoid message bombardment, which we usually do on Weibo. If the timing is wrong, the intimate relationship between customers and company will be ruined, since no one wants to hear the vibration or ringtone when they are busy with work or enjoy leisure." said the supervisor.

Viral Marketing Strategy

Mobile social media have shown as a powerful platform particularly in both helping new product launches (Kaplan & Haenlein, 2012) and developing viral marketing phenomena (Kaplan, 2011). Speaking of the offline strategy, the company pays a lot of attention to its "fan culture" and has a group of hard-core fans. Once the company pushes content on WeChat, these hard-core fans are the first wave that voluntarily spread out the message. In order to cultivate more fans, the company has dozens of "Home of C's Friend" (C stands for the name of the company) located in big cities such as Beijing, Shanghai, Chengdu and etc. These homes would hold offline events regularly, for example they would invite its loyal consumers to watch "Avengers" right after the movie was released. The aim of such offline activities is to encourage consumers to spread more online.

The online strategy to spread the virus is the game on WeChat (a flash game which could be operated directly on WeChat without downloading or redirection), which would be pushed to audience before the press conference. Even the game itself has nothing to do with the product or the conference, it attracts more than 100 thousand people to play, to win gift and, most important to forward the game to Moments.

Referring to the content design, the most popular content is interestingness. The team would combine the hot topic with its product. For example, when "Big Hero Six" was popular, the team designed a post with Baymax to promote its voice assistant as "your sincere life companion as Baymax". The usefulness is also the key to spread the virus, such as the information of the operating system upgrade, A/S service notice, smartphone insurance and etc.

Interaction and Evaluation

Different from Weibo and other social media, the interaction in WeChat is very limited. The users could comment on the content. However the comments would not be seen by the public until the company checks every comment (or comments) through the server and then gives permission to communicate them. Actually WeChat provides company a premium channel to build custom features, interact with customers and get the feedback, only if the company programs a set of API (Application Programming Interface). However such situation will be improved as the supervisor said they will develop a real-person and AI customer service within this year. The evaluation of the pushing content is also finite--the social circle of WeChat is relatively closed, thus the communication effect is not so prominent as Weibo, for instance the company does not know who reads or re-posts the content since WeChat only offers the number of reading hits without providing any details.

Problem and Challenge

The problems mainly come from WeChat. Since even the WeChat official does not know its course for the future, WeChat is inconsistent in its policy. For example Since July 2013, the subscription account could push one message per day and did not have a customized menu. The service account could push one message per month. But in the beginning of 2014, WeChat allowed the subscription accounts of government and charity organizations to have a customized menu. On April 15th, 2014, the service account was allowed to push four messages per month. Another example is since March 2014, some businesses started a marketing campaign: if the user called in his/her friends to "like" a message, when the "like" achieved a certain amount the user would receive reward such as a gift or even a tour. But on June 9th 2014, WeChat pronounced that such campaign was not permitted.

Therefore they have to both keep close contact with officials and predict the next step from even circumstantial sources. On the technical level, the functions of official account server have certain limits which impede them offering more diversified visual effects in photo and text. Moreover the life span of the content in WeChat is short, audiences tend to forget or ignore the messages quickly. Therefore the messages require a high level of time effectiveness.

FINDINGS

Technical advantages and the relentless march of technology offer unprecedented convenience for corporations to engage with their customers. In light of Activity Theory, this study investigated new features of WeChat as a new platform for advertising. Then we qualitatively analyzed the implications that marketers adapted and appropriated WeChat to engage with their customers and in turn how these implications modified the ways of advertising. In this part, concepts of activity system would be used to investigate WeChat as a new platform for advertising.

According to the interview, we can conclude that WeChat is still deemed an auxiliary method in the whole propagation system. It is mainly used to support the official website and Weibo. Under most situations, the advertising and propaganda in WeChat are based on the content of their official website. The process of making content is as follows (Figure 4).

The Subject

In the activity system, the activities are based on the unique combination of past experience and personal preferences. In this case, the supervisor and his team are building WeChat as an advertising and relationship managing platform. Since WeChat is a totally new app and has its own unique features, the

Figure 4. The process of making content

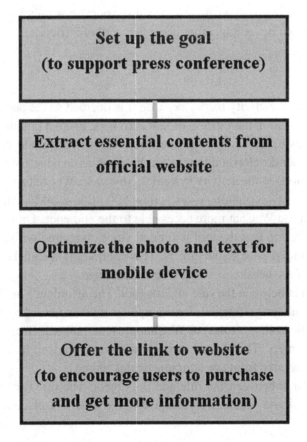

past experiences with other SNS media would not be very helpful in operating WeChat. Everything they have done with it is somehow conservative–"crossing the river while feeling the stones". However the supervisor still, based on his experience dealing with Weibo and Qzone, deems WeChat to be a powerful broadcasting tool, the most obvious example is the majority of messages in WeChat were about the products.

The Community and the Division of Labor

In terms of community, none of the members of the social marketing team was specially trained for operating WeChat. The divisions of labor are flexible, and the sources for WeChat come second-hand from other departments or the website, the highly homogenized information made WeChat sterile. Compared with Weibo, Company C invested too little in WeChat and developed WeChat as a broadcasting tool just as other social media.

The Object

According to activity theory, object motivates and directs different activities and the activities are coordinated around the object. In this case we can conclude that the object of the official account is, except for broadcasting, to form positive impressions, develop affinity, and reduce interpersonal uncertainty—in another words "to be good friend with the customers". Sharing is the key to extend the influence within WeChat, because different from Weibo or Renren, WeChat does not have a broadcasting function. The content, comment and "like" are only visible to the subscribers, thus only by sharing with subscribers' "Moment", could the content be seen by other users (subscribers' friends).

The Contradiction

Contradiction is important in Activity theory because it is deemed as the source of development–contradictions are always involved in the process of real activities. From a practical perspective, exploring the contradictions helps us to understand the characteristics of each node in an activity system, and indicates to us how to further develop and improve it. WeChat, as an advertising platform, both resolved and introduced contradictions in the activity system. In this case, WeChat resolved the contradiction of the tool: given the number of smartphone users is now 500 million (eMarketer, 2014), portability and mobility make the content on WeChat more accessible to the audience. On the other hand, however, it also introduced a contradiction that the tool is limited since compared with a website it contains less information, functions and has poor visual effects. This contradiction somehow explains why the marketer has to put the link of its website in every WeChat message.

Another contradiction is between the rule and the tool. The affordance of WeChat as an advertising platform is open to new marketing strategies, for example, to win the reward by accumulating "likes". On the other hand, the unstable policies (rules) of WeChat prevent marketers from innovating new methods to engage with their customers. Thus a new activity was introduced--to both keep close contact with officials and predict its next step from even circumstantial sources.

The third contradiction is between the rule and the object. The rule (privacy protection) here is a double-edged sword: the closed social circle made WeChat a perfect place to communicate privately with selected people. This private space enables one-to-one communication and is beneficial to creating a better, intimate relationship between customers and the company (the object). But such closed circle also prevents the company from reaching more audiences and evaluating their influence.

The Tool

1. **Subsidiary Role of WeChat:** Although more people access the Internet via mobile devices than computers, WeChat still serves as a subsidiary tool. The first reason is due to the limited size of the smartphone, the message has to be designed to fit into the screen. During this process, only the most important information would be selected to present on WeChat. The link to the official website indicates that the web is what the advertisers want to show the audiences. Moreover the rules—the bad usability of official account server also highly limit the expression of the content.

Second reason is the insufficient customer service. WeChat officials claimed that it should be used as a tool providing services and encouraging companies to offer value-added services for the customers. However instead of offering a service, the present official account serves merely as a shop window that exhibits its products.

2. **Broadcast Pushing as Existing Social Media:** In this case the use of "networkedness" has been found, the company intentionally takes advantage of social circles to expand the messages. However we noticed that such usage is very limited in its influence. WeChat was made as a two-way communication platform, but referring to the official account the communication is unidirectional. Scarcely could comments be seen in the posts, and all of them have already been inspected thoroughly by the "gatekeepers"—the social marketing team.

The official account did not divide its audiences demographically. It is more like a tool to spread the messages among already-fixed audiences. WeChat here is used as a tool to maintain the relationship with its fans, not the tool to expand its influence. It is like a gig where advocates are standing there and listening to the messages from their idol. Within this asymmetric communication, the outsiders could hardly step into the company's circle and become the potential customers due to the lack of impression formation. This problem is determined by its nature of closed social circle, but still the company innovated a new strategy to "bridge" outsiders to its social circle. This will be discussed later.

3. **Lack of Innovative Strategy:** Compared to Weibo, which broadcasts to the public, the dissemination circle of WeChat is confidential to everyone except among its friends. Therefore "water army", "zombie fans" and other traditional e-marketing strategies, which are widely used in other social media to boost advertising effect, were not found in this study. Even though some new methods could be applied to enhance its effect, Yao (2014) pointed out that positive emotions such as surprise, joy and other virusworthy elements like nationalism, livelihood and timely topic are useful to spread the virality, to entice audiences forwarding and sharing with others. But what we found in this was merely copy and paste from the official website with no innovative strategies.

4. In terms of AIDA model, usually on the traditional media, a brand starts drawing people's attention through highly popular celebrities in the commercials, so that the message could catch people's eyeballs and provide a distinct benefit (Thomas & Howard, 1990). But WeChat is a closed

social media; it is very hard to penetrate into the outsiders' social circle. Therefore the company sometimes holds collective events to "bridge" 2 previously separated social circles together. For example one of their events was associated with a bank. If the WeChat audiences of Company C open an account of the bank within the event period, they would gain a chance to win a gift such as a smartphone, earphones or coupons. In like manner, if the WeChat audiences of the bank purchase a smartphone of Company C, they would be rewarded with some gifts from the bank and the company too. Referring to "interest", we found that presenting advantages and features of the product along could not consistently keep audiences' interests. In contrast, if audiences are exposed too much to such information, they would probably unsubscribe the account. Thus as the "friend" of the audiences, the company holds various events irregularly. In the existing social media such as blog and cyber community, the administrator would give its users a "psychological reward[10]" to draw their interest. However, as a closed social community, any psychological reward would be useless to its users in WeChat since nobody else could see such a reward. Therefore in this case Company C has to offer material rewards.

CONCLUSION AND DISCUSSION

This study applied activity theory as a framework to investigate WeChat and analyzed the implications that marketers adapted and appropriated WeChat to engage with their customers and in turn how these implications modified the ways of advertising. We came up with three conclusions as follows:

Weibo is used as a broadcasting tool; its content is visible to everyone no matter if he/she is the follower of an account. Thus the dissemination structure of Weibo is extensive--the re-post, comment, feedback and "like" could be easily observed and measured. However the dissemination mechanism of WeChat is only available for the strong tie relationship and the content is closed to outsiders, which means the interactivity in WeChat is unobservable and hard to measure. The difference of advertising dissemination mechanism between Weibo and WeChat can be illustrated as follows (Figures 5 and 6).

The privacy protection has been treated more seriously in WeChat. As discussed above, the social circle of Weibo is accessible to everyone due to its openness. In another words, it is easy to know new friends, although most of them probably are only based on weak ties. But in WeChat in order to be the friend of the user, the account has to be very cautious with every user's information. Most users WeChat

Figure 5. The extensive structure of Weibo (the interactivity and dissemination are visible)

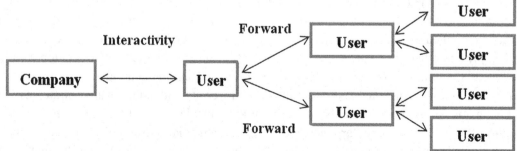

Figure 6. The restricted structure of WeChat

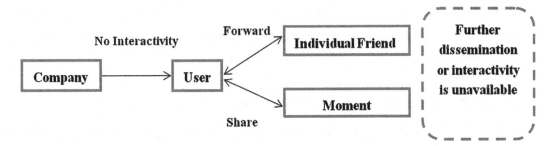

ID are just their QQ ID or cellphone number, which may contain a lot of personal information. Moreover the events held on WeChat also require users' real names, address and etc. Any misuse of such information would grievously jeopardize both user's privacy and company's reputation.

Low pushing frequency is another finding in this study. Compared to Weibo, which could send limitless messages every day, WeChat allows only one message per day. This called out a new challenge to the marketers–how to grab users' attention and interest with exquisitely designed content. However as WeChat is a smartphone-based app, the small screen, poor editing server and limited information largely obstruct the advertising dissemination. Therefore a new form of advertising is demanded to meet the needs. Unfortunately no new strategy has been found in this case.

The limitation of this study is first we cautiously claim the representativeness since only one official account was involved in this study. As an app WeChat evolves quickly, the official account in this study would be more sophisticated in the future since the supervisor claimed they will introduce customer service into it in this year. By then some of the conclusions in this study may not be applicable anymore. Second filed observation could be a complementary method to self-report methods. However since the company once encountered the issue of leakage of information, any form of observation was prohibited.

For the future study, more sophisticated studies are required to understand not only the company but how audiences interact with the official account and how its contents influence the perception of the advertisements and the decisions to forward or share. Comparative studies among different media platforms are also needed to investigate how activity system varies with different media and how the system could be applied not only in advertising but also in other aspects such as online-purchasing.

REFERENCES

Barker, M., Donald, N., & Krista. (2008). *Social Media Marketing: A Strategic Approach.* Cengage Learning.

Boyd, D., & Ellison, N. (2007). Social Network Sites: Definition, History, and Scholarship. *Journal of Computer-Mediated Communication, 13*(1), 210–230. doi:10.1111/j.1083-6101.2007.00393.x

Brusse, C., Gardner, K., McAullay, D., & Dowden, M. (2014). Social media and mobile apps for health promotion in Australian Indigenous populations: Scoping review. *Journal of Medical Internet Research, 16*(12), e280. doi:10.2196/jmir.3614 PMID:25498835

Chuttur. (2009). Overview of the Technology Acceptance Model: Origins, Developments and Future Directions. *Information Systems*. Indiana University.

Ciadvertising. (2007). *Hierarchy-of-effects models*. Retrieved from Ciadvertising: http://www.ciadvertising.org/studies/student/97_fall/theory/hierarchy/modern.html

CNNIC. (2015). *China Internet Network Information Center*. Retrieved from CNNIC: http://www.cnnic.cn/gywm/xwzx/rdxw/2015/201502/t20150203_51631.htm

Davis. (1989). Perceived usefulness, perceived ease of use, and user acceptance of information technology. *MIS Quarterly*, 319–340.

Elliott. (1974). Uses and gratifications research: A critique and a sociological alternative. In *The Uses of Mass Communications: Current Perspectives on Gratifications Research* (pp. 249-268). Beverly Hills, CA: Sage.

eMarketer. (2014). Retrieved from technews.cn: http://technews.cn/2014/12/25/china-will-top-500-million-smartphone-users-for-the-first-time-see-more-at-httpwww-emarketer-comarticle2-billion-consumers-worldwide-smartphones-by-20161011694sthash-1do3dlqq-dpuf/

Engeström. (1987). *Learning by expanding*. Cambridge University Press.

Engeström. (1993). Developmental studies of work as a test bench of activity theory: the case of primary care medical practice. In *Understanding Practice: Perspescives on Activity and Context*. Cambridge, UK: Cambridge University Press.

Frith. (2012). Location-based social networks and mobility patterns: An empirical examination of how Foursquare use affects where people go. *Mobilities Conference: Local & Mobile*.

Gould, Verenikina, & Hasan. (2000). *Activity Theory as a Basis for the Design of a Web Based System of Inquiry for World War 1 Data. Department of Information Systems*. Wollongong: Academic Press.

Griffin, E. M. (2009). *A first look at communication theory*. New York: McGraw-Hill Companies, Inc.

Huang, J. (2015). *Building Brand-Consumer Relationships on WeChat: the Case of IKEA*. Unpublished Master dissertation.

Ito, & Okabe. (2005). Technosocial situations emergent structuring of mobile e-mail use. In *Personal, portable, pedestrian: Mobile phones in Japanese life* (pp. 257–273). The MIT Press.

Kahle, & Valette-Florence. (2012). *Marketplace Lifestyles in an Age of Social Media: Theory and Method*. Academic Press.

Kaplan, A. (2012). If you love something, let it go mobile: Mobile marketing and mobile social media 4x4. *Business Horizons*, *55*(2), 129–139. doi:10.1016/j.bushor.2011.10.009

Kaplan, A. M., & Haenlein, M. (2011). The early bird catches the news: Nine things you should know about micro-blogging. *Business Horizons*, *54*(2), 105–113. doi:10.1016/j.bushor.2010.09.004

Kaplan, & Haenlein. (2011). Two hearts in three-quarter time: How to waltz the social media/viral marketing dance. *Business Horizons*, 253—263.

Kaplan, & Haenlein. (2012). The Britney Spears universe: Social media and viral marketing at its best. *Business Horizons*, 27-31.

Kuutti, K. (1996). Activity Theory as a Potential Framework for Human-computer Interaction Research. In Context and Consciousness: Activity Theory and Human-computer Interaction (pp. 17-44). Cambridge, MA: MIT Press.

Lee, D.-H. (2013). *Smartphones, mobile social space, and new sociality in Korea*. Mobile Media & Communication.

Leont'ev. (1978). *Activity, consciousness, and personality*. Englewood Cliffs, NJ: Prentice-Hall.

Leung, R. (2013). *WeChat Official Accounts and What This Means for Marketers*. Retrieved from ClickZ: http://www.clickz.com/clickz/column/2307141/wechat-official-accounts-and-what-this-means-for-marketers

Littlejohn. (1989). *Theories of Human Communication*. Belmount: Wadsworth.

Littlejohn. (2002). *Theories of human communication* (7th ed.). Belmont: Wadsworth.

Lüders. (2008). Conceptualizing personal media. *New Media & Society*, 683–702.

Nielsen. (2014). *Report of China Mobile SNS Users' Needs and Behaviors*. Retrieved from 199it.com: http://www.199it.com/archives/272681.html

Pang, P., & Hung, D. (2001). Activity Theory as a Framework for Analyzing CBT and E-Learning Environments. *Educational Technology*, 36–42.

Pelet, J. B. (2014). *Determinants of effective learning through social networks systems: An exploratory study*. IGI Global.

Pelet, J.-E., & Papadopoulou, P. (2014). Consumer behavior in the mobile environment: An exploratory study of M-commerce and social media. *International Journal of Technology and Human Interaction*, *10*(4), 36–48. doi:10.4018/ijthi.2014100103

Pierre, L. K. (2012). Marketing meets Web 2.0, social media, and creative consumers: Implications for international marketing strategy. *Business Horizons*, 261–271.

Powers, T., Advincula, D., Austin, M. S., Graiko, S., & Snyder, J. (2012). Digital and Social Media In the Purchase Decision Process. *Journal of Advertising Research*, *52*(4), 479–489. doi:10.2501/JAR-52-4-479-489

Protalinski. (2012). *Facebook has over 901 million users, over 488 mobile users*. Retrieved from ZDNet: http://www.zdnet.com/blog/facebook/facebook-has-over-901-million-users-over-488-million-mobile-users/12105

Ruggiero. (2000). Uses and gratifications theory in the 21st century. *Mass Communication & Society*, 3-37.

Salovaaraa, A., & Tamminenb, S. (2008). Acceptance or appropriation? A design-oriented critique on technology acceptance models. In *Future Interaction Design II* (pp. 157–173). London: Springer.

Severin, & Tankard. (1997). *Communication theories: Origins, methods, and uses in the mass media* (4th ed.). New York: Longman.

Shrum, L. (2004). *The Psychology of Entertainment Media: Blurring the Lines Between Entertainment and Persuasion*. Lawrence Erlbaum.

Singh, V.-J., Veron-Jackson, L., & Cullinane, J. (2008). Blogging: A New Play in Your Marketing Game Plan. *Business Horizons*, *51*(4), 281–292. doi:10.1016/j.bushor.2008.02.002

Tencent. (2015). *The New Milestone of WeChat*. Retrieved 12 20, 2015, from Techweb: http://www.techweb.com.cn/internet/2015-08-12/2188504.shtml

Thomas, B., & Howard, D. (1990). A Review and Critique of The Hierarchy of Effects in Advertising. *International Journal of Advertising*, 98–111.

Tuten, T., & Solomon, M. (2012). *Social Media Marketing*. Englewood Cliffs, NJ: Prentice Hall.

Vygotsky. (1978). *Mind in society: The development of higher psychological processes*. Cambridge, MA: Harvard University Press.

Walther, J. B. (2002). Cues filtered out, cues filtered in. In *Handbook of interpersonal communication*. Thousand Oaks, CA: Sage Publications.

Walther, J. B. (2005). Let Me Count the Ways: The Interchange of Verbal and Nonverbal Cues in Computer-Mediated and Face-to-Face Affinity. *Journal of Language and Social Psychology*, *24*(1), 36–65. doi:10.1177/0261927X04273036

Wijaya, B. S. (2012). *The Development of Hierarchy of Effects Model in Advertising*. International Research Journal of Business Studies.

Wimmer, & Dominick. (1994). *Mass media research: An introduction*. Belmont: Wadsworth.

Wu, M., Jakubowicz, P., & Cao, C. (2013). *Internet Mercenaries and Viral Marketing: The Case of Chinese Social Media*. IGI Global.

Wu, M., & Li, H. (2011). The Triumph of Shanzhai:No Name Brand: Mobile Phones and Youth Identity in China. In D. Y. Jin (Ed.), Global Media Convergence and Cultural Transformation: Emerging Social Patterns and Characteristics (pp. 213-232). IGI Global.

Wu, M., & Lin, H. (2011). Office on the move. In E. Coakes (Ed.), Knowledge Development and Social Change through Technology: Emerging Studies (pp. 232-246). IGI Global. doi:10.4018/978-1-60960-507-0.ch018

Wu, M., & Yao, Q. (2014). Location-Aware Mobile Media and Advertising: A Chinese Case. In X. Xu (Ed.), Interdisciplinary Mobile Media and Communications: Social, Political, and Economic Implications (pp. 228-244). IGI Global. doi:10.4018/978-1-4666-6166-0.ch013

Yamagata-Lynch, L. (2001). *Using activity theory for the sociocultural case analyses of a teacher professional development program involving technology integration*. (Doctoral dissertation). Indiana University.

Yao, Q. (2014). How Virusworthy Elements Affect Viral marketing Patterns in WeChat. 6th Chinese News and Communication Association.

Zhu. (2004). Competition between alternative sources and alternative priorities: A theory of weighted and calculated needs for new media. *China Media Report*, 16-24.

ENDNOTES

[1]   Moments or Peng You Quan, refers to "friends' circle", is one of the social functions on WeChat. Users could share text, photo, article and music to Moment. "Comment" or "like" about the sharing is only visible to user's friends.

[2]   Service account aims to offer various services such as online payment and customer service. The company could apply customized menu and connect to third-party platforms. Service account could only push four broadcast messages per month.

[3]   Subscription account only pushes the updated information about the company. Subscription account could push one broadcast message per day.

[4]   The dissemination route, which is unknown within traditional media, however is visible within social media such as Facebook and Weibo.

[5]   In the context of centralized traditional media, company is the only authority to offer the information or the services and to control users' data. However with the decentralized social media, the information is produced in several different places concurrently.

[6]   Big V refers to the popular microbloggers whose accounts have been verified and thus stamped with a "V" next to their names.

[7]   There are also tertiary and quaternary contradictions, but they are not involved in this study.

[8]   The company demanded to be anonymous for privacy protection. Therefore all names and brands used in this study are pseudo or acronyms names.

[9]   Lofter is the Chinese version of "Tumblr". Lofter is to provide a simple, easy-to-use microblogging platform which encourages users to post original content to a short-form blog.

[10]   For example in the cyber community, a new user would get the rank as "private". With his or her involvement in the community, the rank will be upgraded. According to the interview, such strategy is very effective to motivate users to participate.

Chapter 17
Reviewing Gratification Effects in Mobile Gaming

Yuchan Gao
The University of Nottingham, China

ABSTRACT

This chapter aims to critically review previous studies on the gratification effects mobile gaming provides in a variety of aspects including play experience, education and social networking. Previous studies reveal that mobile gaming gratifies a new play experience in casual, flexible and ubiquitous nature. Additionally, mobile gaming generates widely accepted gratifications in technology-based education, improving learning and teaching performance in education in general and specific courses. Meanwhile, mobile gaming demonstrates great gratification effect in quality, ability and physical training. Furthermore, scholars have widely researched on the gratification effect of mobile gaming to promote social networking with the applications of socially interactive technologies. Finally, this chapter delivers a forecast about the future development of mobile gaming into HR recruitment and management.

INTRODUCTION

Nowadays we have entered the era of mobile phone with billions of mobile phone owners use the device not only as an extension of line telephone to reach and contact each other but also as enhanced multimedia platform for social-networking, internet use, photo shoot, music, video and content-sharing along with other numerous functions. One of the striking differences between mobile phone and traditional line telephone installed at home or in office is that the former provides amazing experience of game play that could only be achieved with game consoles and personal computers in the past. This function, along with other functions that promote social networking, application usage as well as entertainment defines mobile phone's role as a portable integrated multi-functional device in the size of a hand that has brought new ideas and revolution to a variety of fields ranging from ultrathin laptop, portable tablet, cross-platform porting and cloud technology.

DOI: 10.4018/978-1-5225-0469-6.ch017

The past decade or so has witnessed a boom in the mobile gaming industry. Started from the 1997 *Snake* game on Nokia 6110 and revolutionized by the first generation of Apple phone—a new standard for smart phone for the time in 2007 (Tercek, 2007), mobile gaming now has developed into one of the most widely spread and commonly seen entertainments of our time with incredible industrial revenue. In an overall prosperity of all kinds of mobile applications, game applications accounted for 90% of the 2014 TOP 10 grossing applications on Applestore (Leonov, 2014). In 2014, mobile gaming revenue reached 21.1 billion dollars worldwide with extremely remarkable performance in North America and Asia (Superdata, retrieved January 22, 2015).

SNL Kagan's report in 2014 reveals that a new record of $5.23 billion was contributed by the mobile games segment in America that year. However, behind the recording-break performance of US mobile gaming industry there lies a shrink in the yearly growth of and percentage account of the segment in the overall mobile entertainment market. Compared to a surprising increase of 93% in 2013, the year 2014 only witnessed a rise of 27% in US mobile gaming, while the percentage account decreased from 63% in 2013 to 57% of the $9.14 billion profit obtained by the total mobile entertainment market revenue in America (Sinclair, 2015). In Asia, China, Japan and South Korea are the three countries whose market figures contributed the most to both regional growth of mobile gaming in Asia (a global share of 45%) and that in the world (82% of worldly growth) in 2014 (Newzoo, retrieved April 22, 2015). The total revenue achieved in the region's mobile gaming industry is $12.2 billion, but monthly spending on mobile games is the lowest of $4.17 per capita compared to $4.95 spent by Western European players and $6.21 by North American players (Statista, retrieved April 22, 2015).

As for the year of 2015, report from Newzoo considers 2015 the year that mobile game market--"the industry's most lucrative sector" will outperform the console game market. The total revenue is expected to surpass 25 billion dollars, which remarks a huge leap of 42% compared to the figure in 2013. Although some statistical organizations have expressed their concerns about the coming of saturation in mobile game growth in mature western markets such as US, just like what seems to be interpreted from the shrinking figure in US market aforementioned, Newzoo is simply not convinced by such pessimistic anticipations and comes up with their own neutral forecast. It is believed that mobile game market will continue surprising industry practitioners and investors with an estimated figure of 40.9 billion dollars in 2017 (Pearson, 2014). Tim Merel, the managing director of game investment bank Digi-Capital casts a way more optimistic eye on the prospect of mobile gaming development in the future. He anticipates that an annual increase rate of 23.6% is to be seen, reaching a new revenue record of $60 billion in mobile gaming segment and $100 billion in the whole game software industry by 2017. He also notes that Asia remains as "the biggest driver of economic value" when it comes to mobile and online games, "with the best companies' revenue growth and profit margins becoming the envy of foreign competitors". Therefore, he predicts that along with Europe Asia will account for over 80% of the worldwide revenue share in these segments (Takahashi, 2014). All these striking figures indicate a promising market condition for mobile game development, launch, retail and purchase at present and an even more prosperous condition likely to happen in the future. Regional leaps such as those in North America, Asia and Europe will lead to record-breaking sales and an overall boost in global mobile gaming industry as well as in game software industry as a whole.

Mobile gaming has indeed become one of the most profitable industries that change instantaneously with a blink of eyes. Along with such fast development there comes a flourish in mobile gaming studies as well. Mobile gaming is no longer, what Mayra (2008) claims, an upswept corner of academic research that had largely been ignored. It is in the heart of many scholars to uncover the secrets behind its gigantic success, examine its current mechanism, solve its unfathomed problems, forecast its future development, and more importantly, explore the boundless territory it is likely to expand to because of its limitless potentials. With every gratification combined mobile gaming will generate, as Bell (2006) believes, an integrated gaming activity that interweaves everyday mobile life.

This chapter aims to critically review previous studies concerning the gratifying effects of mobile gaming in a couple of well-applied areas and propose new research directions of mobile gaming application to other areas. The term "mobile gaming" used in this chapter specifically refers to gaming activities carried out with mobile games offered on mobile phones. Previous studies about both online and offline mobile gaming fall into the categories of review in this chapter. The first section is a comparative and critical review of existing literature on the gratifying effect of mobile gaming in three aspects—gaming experience, education and training as well as social-networking. This section introduces and examines the major scholastic concepts, theories and crucial findings about the gratifying effect of mobile gaming in the aforementioned aspects, as well as problems remaining to be solved in each aspect of mobile gaming application. It is hoped that this section can help the readers to form a general understanding of the present scholastic efforts on mobile gaming in game theories, interactive education studies and social networking researches. It is also an attempt to list out the difficulties and challenges faced by mobile gaming application to these areas. The second section is to map the direction of future studies on the possibility of mobile gaming application to human resource analytics, recruitment and management. With the help of this section, it is expected that it will encourage new researches on and development of mobile gaming to further promote and improve human social interaction with mobile phones.

MOBILE GAMING AS NEW GAMING EXPERIENCE

Main Concepts, Theories, and Findings

As a new member of the video game family, mobile gaming is notable for its convenient portability compared to traditional video game on PC or game consoles with gamepad. The displayer and the controller are integrated into one small device in the size of a hand which players have the largest convenience to carry around. Although such an easily portable nature is also seen in gaming on other handheld game consoles like iPad, Nintendo 3DS and PlayStation Vita, the dominant user coverage and constant portability of its carrier, mobile phones, speak the difference. Unlike other handheld game consoles whose central function is to play games, mobile phone is a multi-media platform that supports not only gaming but also numerous other equally, if not more, crucial functions such as communication, social-networking, internet use, photo shoot, music and video play. This comprehensive nature of mobile phone makes it possible that the users of any function mentioned above automatically have access to the other functions when necessary, and therefore makes them potential mobile game players.

The International Telecommunication Union estimated up to 6.9 billion mobile subscriptions around the world in 2014, a figure that is equivalent to 95.5% of the world's population (MobiThinking, retrieved May 2, 2015). This is a user coverage rate that would arouse envy in any handheld game console manufacturers. Although the primary function of mobile phones, as Martin (2014) points out, is not for gaming, this does not prevent users from playing casually or seriously the games available on their mobile phones. Additionally, for location-based hybrid-reality games such as *Mogi: Item Hunt*, the always-at-hand mobile phone also preponderates other portable game consoles because the virtual scenarios of such games are consistent with players' exploration of the real world. It is usual for mobile phone users to carry the devices along with them whenever and wherever in their daily routine (Mainwaring, Anderson & Chang, 2005), but it is less likely so for other handheld game console users to do so routinely because the device is normally heavier in weight, larger in size and exists for fewer functions than multi-functional mobile phones (Tassi, 2012). On the one hand, dominant user coverage rate of mobile phones creates an atmosphere that everyone around has an easy access to games and further paves a way for a more connected and interactive player social network within mobile games. On the other hand, constant portability makes casual and flexible gaming experience possible because no extra bulky device is required and game playing can be conducted, paused and resumed anytime anywhere in our daily life. All of these factors together generate a ubiquitous gaming experience which everyone can enjoy at the minimum cost whenever and wherever possible.

Therefore, scholars tend not to consider mobile games merely a low-quality extension of games from high definition console and computer to less technologically sophisticated mobile phone. They believe mobile gaming gratifies a brand-new way of casual, ubiquitous and flexible gaming. When it comes to such a new gaming experience, de Souza e Silva and Hjorth (2009) do not agree with some game researchers' specific restriction of the "play" activity within the magic circle, that is, "an activity separate from the ordinary aspects of life, with specific boundaries of time and place" (de Souza e Silva & Hjorth, p.605). Instead, they use a new definition of play as "casual play" to refer to games like mobile games played within/as/between the daily schedule of ordinary life—what Bell and his co-researchers call a gaming activity that "interweaves" everyday life (Bell, Chalmers, Barkhuus, Hall, Sherwood & Tennent, 2006). In this way, mobile games are no longer a separated part from ordinary routine in daily life but become an indispensable component of everyday activities. Because of the aforementioned new features of mobile phone as a gaming platform, playing mobile games seems to be integrated into other daily activities as ordinary as eating, sleeping and traveling. Here it is worth to clarify that the word "casual" does not refer to that mobile games are designed with easier levels or that mobile game players play games in a simpler way. As Juul (2010) argues, the casual nature of mobile games lies not in content or context but in providing "a meaningful experience within a short time frame" (Juul, p.8). In other words, mobile gaming does not gratify a way of playing with easier contents or simpler game tasks, instead it gratifies a way of playing that players receive a complete and gratifying gaming experience even just from a very limited time of playing.

In traditional perception mobile games are believed to help people pass the extra time before the next errand on schedule such as waiting for a friend, taking a public transportation or lunch break before the afternoon work. Yamakami (2013) believes that the essentially "asynchronous" nature of social interaction generates plenty of fragments time and "mobile games are ideal candidates for such time-killing purposes." (Yamakami, p.738) Also Lee, Goh and Chua (2010) argue that mobile games make players relaxed and effectively reduce the boredom during interstices of their daily schedule. Chan (2008), in his analysis of the pass-time function in Japanese mobile games, describes "these 'pick-up and play' games are geared for intermittent bouts of mobile play, ostensibly for 'killing time' or while traveling on Japan's ubiquitous public transport network" (Chan, p.16). He also believes "the ease of play" is a paramount factor in mobile games and an effective means for the industry to attract new casual gaming consumers instead of only targeting "hard-core" players who play more intensively. Richardson (2010) therefore concludes that mobile games becomes a way to cope with "impatience, aloneness and boredom" in modern life as well as to make full use of the interstices of fragmented pieces of time, but at the same time players can remain environmentally conscious and ready to resume the busy life anytime anywhere (Richardson, p.436).

Additionally, researchers such as Ito, Okabe and Matsuda (2005) and Chan (2008) who study on the locations of mobile gaming argue that mobile games are not only played shortly "on the go" or "in waiting" as perceived. They are also intensively played in a variety of flexible places including home and other stable environments when people are not at all waiting for the coming errands. Juul (2010) argues that casual games like mobile games do not prevent players from "engaging in longer sessions" (Juul, p.36). The difference between casual game and hardcore game, as he sees it, lies in that "a casual game is sufficiently flexible to be played with a hardcore time commitment, but a hardcore game is too inflexible to be played with a casual time commitment" (Juul, p.10). In other words, casual games permit a flexible way with regard to gaming time and location for both casual players who pick up the game for minutes and hardcore players who spend hours to thoroughly experience every element in the game. Thereby, instead of merely promoting casual gaming as a result of the portable nature of mobile platform, mobile games in fact reach a wider range of both casual and hardcore players by facilitating the flexibility of playing time and location and guaranteeing a kind of flexible gaming that interweaves anytime anywhere of everyday life. And that is what Juul calls flexible design of games that "fit into the lives of players" (Juul, p.2) rather than "flexible" players who have to fit into the rigid design of games.

All the aforementioned advantages of mobile gaming contribute together to its "ubiquity" (Zhou, 2013), that is, easily portable platform as well as casual and flexible design regarding playing time and location enable a gaming experience anytime anywhere in everyday life. Bissell (2007) claims that this gaming experience, as in one carried out in an everyday ubiquitous and casual context, successfully fills and saturates the interstices and fragmented pieces of time. Also, Richardson (2010) describes the corporeality of everyday ubiquitous casual mobile gaming as "a way to extend our understanding of both the hybrid experience of habitual mobile phone use in the everyday life world, and the human-technology relation or techno-somatic coupling it applies to mobile media practices more generally" (Richardson, p.435). Therefore, Feijoo, Go´mez-Barroso, Aguadoc and Ramos (2012) conclude that mobile gaming scenario should not be considered "a delayed or modest extension of console or PC games". Instead, it is "a distinct user experience with a number of unexplored avenue" (Feijoo et al, p.214). In mobile gaming, we see a platform with the largest user coverage and most convenient constant portability; we see a casual playing experience fulfilling our fragmented pieces of time; we see a flexible way of playing that places no restriction to time or location; more importantly, we see the opportunity to gratify a ubiquitous gaming activity that is blended into whenever wherever possible of everyday life.

Remaining Weaknesses and Problems

Concerns are also expressed about problematic issues with the attributes mobile gaming offers to satisfy a new way of gaming. Even with F2P (free to play) mobile games, players express "disillusionment" towards the lack of good service quality and regulations, and "indeed general lack of interest by regulators in the world of free services" (de Kervenoael, Palmer & Hallsworth, 2013, p.444). Also innovation in genres and types of game on the mobile platform is perceived as insufficient, as "many apps were free but very few worth the download time"(de Kervenoael et al., p.444). As pointed in de Kervenoael and her colleagues' research, many mobile game players felt uncomfortable with skill levels, upset with aim of the games or worried about the total cost such as for items and equipment generated from playing online games. Additionally, problems are also centralized with inaccessibility, the lack of in-advance knowledge, information and understanding of a targeted game as well as the absence of follow-up services and a standardized refunding system.

From industrial perspective, a successful mobile game or games in general, consists of the following characteristics (Park & Kim, 2013):

1. The quality of the product is high;
2. Innovative or novel elements can be traced from within;
3. It has unique features compared to other games;
4. Consumer needs are better fulfilled with this game than with the competitors' games; and
5. It is less costly.

Park and Kim also believe that above all the factors contributing to a good game design, good graphics, animation and sound are inclined to be valued more than other factors, because these factors "provide sensory proof of the reality of a game, supporting and enhancing the impact of the whole game fantasy experience" (Park & Kim, p.1355). They also argue that "the higher the level of reality or fantasy, the more entertaining a game becomes." (Park & Kim, p.1355) That is to say, when game developers design a game, they should indeed pay great attention to the representation of the game concept such as better scenario display or more detailed avatar portrayal in the released game to attract consumers.

But Liu and Li (2011) question this idea by asking "why would a user tolerate a mobile game on a 3-inch screen with relatively limited playability but not accept a similar game on a desktop computer?" (Liu & Li, p.891) When it comes to whether players want to play a particular mobile game, technological standards are not the only criterion to consider. In this respect, Liu and Li (2011) give further explanation–"factors influencing attitude and the behavioral intention to play a mobile game are different. Both perceived usefulness and perceived enjoyment are direct predictors of attitude, although not for behavioral intention. Their indirect impacts on intention are mediated via attitude." (Liu and Li, p.896) Although the visual image of a mobile game might indicate its useful promotion of life quality and possibility to be an enjoyable means of entertainment, they can only lead to a positive attitude and fail to directly improve or change players' intention to play. The intention for them to get involved in actual attempt, although indirectly influenced by the aforementioned indications, are also crucially determined by many intangible factors like good advertisement of the game concept, good publisher and/ or well-established reputation among the existing players.

The problems remaining within mobile gaming gratifying a new way of gaming lie in that game developers have to make "industry-wide implications" to reach "a broader audience" including hardcore players, casual players and flexible players who shift now and then from casual to serious gaming and backwards (Juul, p.7), while maintaining high standards in terms of product content, innovation, player gratification as well as service quality. At the same time, they should also make full use of the unique features of mobile gaming as mentioned above—dominant user coverage of platform, convenient and constant portability as well as flexible and ubiquitous gaming experience. Only in this way can they compensate the disadvantages of mobile gaming in technological sophistication, scenario and image representation compared to high definition console and computer games. The future of gaming will not be one with games designed with the best images and the most advanced engines on the latest mobile phone platforms; the future will be one with games adhering to the highest standard in innovation and player service, promoting the most flexible and ubiquitous gaming experience, fitting the best into the lives of players and interweaving everyday life.

MOBILE GAMING AS INTERACTIVE EDUCATION AND TRAINING

In addition to gratify a new gaming experience, mobile games are also widely introduced to current system of education to assist interactive teaching, learning and training. There are abundant researches in this area focusing on how to apply well-established learning theories to new models of mobile game-based learning (mGBL). Also, there are a wide range of evaluation studies on the conceptualization and operationalization of new technologies and elements in (mGBL), while many researchers specifically aim to examine the more prominent effects of mGBL in some particular courses. Attention is also paid to the effects and changes mGBL brings to ability and quality training outside of the curriculum category. Additionally, previous researchers also make great effort to explore the development and improvement of mobile games and mGBL models from the industrial designing perspective to realize better educational effects.

Theories and New Elements Applied to Mobile Game-Based Learning

In order to promote educational effects, learning theories need to be applied to the use of mobile gaming as an instructional tool. Zaibon and Shiratuddin (2010) apply learning theories of Behaviorism, Cognitivism and Constructivism to mGBL and reach important theoretical findings. Behaviorism related to mGBL functions to fragment learning objectives into approachable steps, guide and reinforce learners' performance to desired behaviors as well as provide positive feedback to the learners. Cognitivism is beneficial for organizing and connecting new information to the learners' established knowledge system, strengthening learners' memories as well as providing good, diverse and stimulating game design and resources. Constructivism helps to lead the knowledge construction process, present different game challenges related to learning and problem-solving as well as encourage group learning activities. With successful application of these theories in mGBL, mobile games can be contributive to promote "Linguistic", "logical-mathematical", "interpersonal", "intrapersonal", "spatial", "bodily-kinesthetic", "musical", "naturalist" and "existential" intelligence as well as to assist a series of educational events such as attention stimulation and learning interaction as well as better communication of learning materials and guidance (Zaibon & Shiratuddin, p.126).

Meanwhile, mobile gaming also possesses a series of brand-new elements that are not found in traditional education system. Visual instructor, for instance, a mobile heads-up display learners wear in playing games, can greatly add to the learning transparency and efficiency in mobile augmented reality learning (MARL) (Doswell & Harmeyer, 2007). The heads-up display provides hints for accomplishing learning objective through demonstrating graphical annotations placed on real-world objects and locations of other learners in one's team. Also, instruction and guidance concerning academic skills required to finish learning challenges can be abundantly given to learners through virtual instructor. In this way, Doswell and Harmeyer conclude "visual instructor may serve as a mission leader or guide for the player's real-world quest" (Doswell & Harmeyer, p.524). Additionally, location-based service is of great use in enhancing context-awareness in mGBL. Learners can get an individual-based learning material after location-based service embedded in mobile games analyze their current location, the surrounding environment, context and learners' personal profile. Lu, Chang, Kinshuk, Huang and Chen (2011) find that location-based service in context-aware mobile educational games accelerates the speed of learners to familiarize with the surrounding and increases their interaction with objects in learning or working environment.

Specifically, Lecture Quiz, a new concept of Wang's, Ofsdahl's and Morch-Storstein's (2008) is believed to be able to pave a new way for modern interactive teaching and learning. It is designed as a game run on the teacher's laptop, projected through multi-media devices and interacted with students on their mobile phones with internet connection. Interestingly unlike traditional game concept that the gaming experience happens on only one platform, Lecture Quiz separates the overall gaming experience into different parts—the running of games on the teacher's laptop, the display of the games on large screen in the lecture hall and the completion of game tasks in the students' mobile phones. A registration on the game client is necessary for the students before the games start. Then the students get to challenge the tasks in the games displayed by the teacher on the large screen (normally lecture questions with multiple alternatives). The students only have some seconds to answer these questions and the collective performance will be immediately shown on the large screen and students' individual performance on their game clients. Two game modes have been developed in Lecture Quiz Version 1.0—score distribution and last man standing. The former mode is to demonstrate the distribution of student answers to different question alternatives with a bar chart visualizing the result. In the latter game mode, students need to provide correct answers to play the next round. The game displays visual avatars with each participant's client nickname winning the round or losing the round and getting shot.

Interestingly, the two game modes can not only operate separately but also complement each other. In the first mode, students get to learn the correct answers to each lecture question and those who can remember all the correct answers get to win to the final round and become "the last man standing" in the second mode. After experiment and evaluation, the Lecture Quiz researchers find it is "most useful for testing and rehearsing theory" (Wang et al., p.198) because it is easy for the teacher to obtain a quantitative understanding of how much students have mastered the theory. As an unusual concept of mobile gaming application in education, Lecture Quiz does not show resemblance to other mGBL concepts in which learners are asked to play a specifically designed or chosen mobile game to facilitate learning. It demonstrates a way in which game display, participation and performance are separated parts on different ends but altogether constitute an integrated model of lecture interaction between the learners and the instructors. As such, this particular concept of mGBL helps to improve students' willingness to get involved in learning both more frequently and more actively. More importantly, Lecture Quiz helps the teachers to immediately get visually quantified feedback of the students' real-time condition in knowledge obtaining and memorizing. Also, it expands the "variation in how lectures are taught" (Wang et al., p.198) and thus the teachers' lecturing performance and efficiency can be greatly enhanced.

There are constructive outcomes obtained from the applications of the aforementioned new theories concept and elements embedded in mobile gaming for education. Kim and his co-researchers (2012) find even in low socioeconomic region children with little expose to technology display surprisingly high adaptation to mobile learning and good problem-solving ability to learning challenges, thus mobile gaming may provide a better way-out for educational improvement in developing countries. The same is true with secondary education which demonstrates equally high efficiency in learning engagement and learner motivation when mobile game learning is adopted. Huizenga, Admiraal, Akkerman and Dam (2009) report "situated and active learning with fun" (Huizenga et al, p.341) in 458 pupils who were asked to engage in learning of history in a mobile game about medieval Amsterdam. Gwee, Chee and Tan (2010a) reveal improved social practice in 15-year-old students receiving mGBL. In another paper, they research on the overall outcome of mGBL through analysis of the students' homework quality and find that students engaged in mGBL show higher score in the quality of their homework in terms of "personal voice", "perspective" and "relevance of content" (Gwee, Chee & Tan, 2010b, p.415).

Therefore, in term of applying existing learning theories and brand-new elements to mGBL, there are overall gratifying effects reported in both learning performance and teaching outcome. With the practices of such new concepts, theories, elements, it is expected that mGBL can greatly help to avoid learning boredom, alienation and disinterest, improve learning excitement, fun, motivation, performance and outcome as well as enhance a more interactive way of teaching in terms of effect, efficiency and variation.

Notable Effects of Mobile Game-Based Learning in Certain Specific Courses

Previous knowledge has revealed better performance of mGBL in certain courses than others. The most outstanding results come from the education of art and history, business, mathematics, citizenship as well as computer science and programming.

History study is so far the area that draws the most scholastic attention mainly because the ancient context of this course often cause "student disengagement and alienation" (Admiraal, Huizenga, Akkerman & Dam, 2011, p.1185) and result in bad learning performance or failing class. Ardito and his co-researchers specify this failure of education in student visits to historical site and state that "traditional visits to such (historical) parks tend to generate little interest in young students, especially when they are faced with the ruins of ancient settlements whose current appearance no longer reflects their initial purpose" (Ardito, Buono, Costabile, Lanzilotti, Pederson & Piccinno, 2008, p.76). Huizenga, Admiraal, Akkerman and Dam (2009) attribute such a phenomenon to the fact that the new generation of pupils who are brought up in an environment saturated with new information and communication technology is still "being educated with old paradigms and methods" (Huizenga et al., p.332).

On the contrary, new technology-oriented education such as mGBL can bring out empathy in the new generation of students and make gigantic differences in their learning performance and outcome. One of the effective strategies mGBL has been using in history education is to create narratives to "move beyond knowing fragmented facts of historical figures and events" and help students memorize abstract historical knowledge in a process of storification (Akkerman, Admiraal & Huizenga, 2009, p.449). With the help of a mobile game called Frequentie 1550 in which students got to walk in and experience the city and events of Amsterdam in medieval era, Akkerman et al. find that students' engagement in learning was highly evoked by 3 kinds of storification—receiving stories as spectators, constructing stories as directors and participating in stories as actors. Thus, students did not experience "failure to grasp the nature of historical context" nor "view the past through the lens of the past" (Akkerman et al, p.449) but were able to master historical knowledge that was hard to comprehend and memorize otherwise. Similarly, Winkler, Ide-Schoening and Herczeg (2008) use mobile co-operative game-based learning to evoke students' discovery of medieval artifacts in museum and find similar improvement in aesthetical perception and learning as well as comprehensive understanding of historical context.

Such improvement in learning performance, understanding, motivation and engagement are also revealed in the studies of other courses. In business education, location-based mobile game helps to simulate a visual environment in which students become consultants to provide advice to visual companies so that their knowledge of strategic management is exercised in visual reality context (Puja & Parsons, 2011). In mathematical education, students are involved in mGBL for practice of mathematics so that the fun of playing mobile games gets to compensate the learning of abstract and difficult mathematics (Kalloo, Kinshuk & Mohan, 2010). A mobile game called MobileMath successfully grasped students' attention by recommending personalized math tasks each student encountered based on their choice of play time, level of knowledge and previous performance in the task accomplishment. In citizenship education, mobile game is introduced to promote sense making and identity construction (Chee, Tan & Liu, 2010). In a mobile game called Statecraft X, students are able to enact the role of governors of different towns so that they obtain first-hand practice of what it takes them to govern a town with diverse races of fantasy NPCs all embracing different ideologies and demanding different sometimes contradictory things. This in fact "mirror(s) the political reality of diverse races and religions" of their country (Chee, et al, p.223) and thus helps them to acquire an in-depth understanding of real-world political survival. In computer science education, developing mobile games becomes a preliminary stage for students to understand basic knowledge of programming and obtain sense of achievement in learning before they are engaged in more sophisticated stages of game design and development (Kurkovsky, 2009; Hafizah, Hamid & Fung, 2007). All these practices of mGBL in the specific courses aforementioned have proved its potential to bring new blood to traditional education system which faces difficulties and deficiencies in such courses. These practices also prove the effects of mGBL in gratifying a more motivating, engaging, stimulating and interactive way of education with improved learning performance and outcome from students and ameliorated teaching efficiency and effect from teachers.

Mobile Game-Based Learning in Quality, Ability, and Physical Training

In terms of quality and ability training, mobile gaming also demonstrates its unique advantages in a variety of areas. Playing games has always been a process of solving problems and facing challenges encountered in games, especially in cooperation with other players, so it is presumable that one's ability of problem solving and collaboration can be formed and enhanced with mobile game-based training. Sánchez and Olivares (2011) employ mobile serious game (MSG), games to foster and improve problem-solving and collaboration abilities of students in secondary schools. They find that the experiment group got higher score in the evaluation of these abilities than the non-equivalent control group because the game offers opportunities to interact with other players, which becomes an incentive for collaboration. Also, the game assigns different roles for each player who plays as a complementary part to each other to achieve a communal goal. Each student showed high-level caution with execution of personal plan and its effect on the team. In this way, their "problem solving process" becomes "collaborative" (Sánchez & Olivares, p.1944). The same is true with training of visual perception skills, which are, to be specific, "identification, organization, and interpretation of sensory data received by the individual through the eye" (Balayan, M., Conoza, V., Tolentino, J., Solamo, R. & Feria, R., 2014, p.1). Balayan and his co-researchers reveal great potential in an educational mobile game called SkillVille to assist training of children's 3 visual perception skills for language and spelling, searching and sorting information as well as memorizing. In the game, the training of each skill is allocated to one of the three mini games designed with different difficulty levels provided. More importantly, leaning analytics can be carried out based on each player's performance in the current game and can be compared with previous performance to show the changes in the children's mastery of the three visual perception skills. Feedback can also be sent to their teachers or parents as updates of learning performance.

Outside of student education category, mobile gaming also demonstrates excellent performance in stress management and training of high pressure groups. The research of Smith, Woo, Parker, Youmans, LeGoullon, Weltman and de Visser (2013) examines Stress Resilience Training System (SRTS®), a game-based application to help players better cope with negative effects of stress. Although SRTS is more referred as a mobile application rather than a pure mobile game used for training purpose, but its game-based nature largely stimulates the trainees' interest and persistence in training as well as helps them to carry on stress management training with less stress generated from traditional cognitive training. The researchers reveal the application displays high effectiveness and ease of use in young military soldiers "with high subjective resilience" (Smith et al., p.2080). A heart rate monitor is attached to the application so that the trainers can keep an immediate track of the trained soldiers' biofeedback. SRTS is believed to pave a new way for skill and emotional training because "it utilizes both internal interactions, via videos and games, as well as external links to the world via the heart rate sensor" and should be able to set a model for future training applications to combine "the presentation of games and simulations, cognitive rules and exercises, and biofeedback-based personalization for individualized experiences within the program" (Smith et al., p.2076-2077). Therefore it is likely to lead to more iterative design and testing of systems similar to SRTS.

In addition to quality and skill training, mobile gaming is also believed to aid the build of physical fitness and healthier lifestyles. Buttussi and Chittaro (2010) examine a context-aware and user-adaptive fitness mobile game called Monster & Gold which monitors "users' hear rate, age, fitness level, and exercise phase to help them exercise at the optimal intensity to gain cardiovascular benefits" (Buttussi & Chittaro, p.51). Their findings indicate fitness mobile games help to strengthen and prolong regular physical training and exercise, which possesses great hope in lessening possibility of cardiovascular disease, diabetes and cancer resulted from obesity.

Thus, in terms of ability, quality and physical training, mGBL does a good job in transforming the painstaking training process into motivating, stimulating and pleasant gaming experience. Training objectives are unconsciously fulfilled either respectively or collaboratively, and can be persevered in the long run.

Better Development of Mobile Gaming Models for Education and Training

Scholastic attention is also paid in terms of how to develop adaptable mobile games that better facilitate education and training. Although most aforementioned researchers propose their own perception with regard to realizing such improvement, this issue is addressed specifically from the perspective of industrial design and development by Spikol and Milrad (2008), Lavin-Mera, Torrente, Moreno-Ger, Vallejo-Pinto and Fernández-Manjón (2009) and Zaibon and Shiratuddin (2009).

Professional game developers are first of all making their endeavors to make mobile educational games more gratifying to teaching and learning purpose. Zaibon and Shiratuddin (2009) propose a model for customized phases and steps to develop mGBL applications by breaking game development into 3 new steps different from traditional mobile game development. First of all, they suggest that pre-production phrase should be brainstorming of mGBL concept rather than a pure mobile game concept, which means the developers need to familiarize themselves with mGBL theories and preferably come up with their model of mGBL. Also they need to design features and challenges that help the players to meet the learning objectives. In the second production phrase, they need to prepare mGBL elements, develop learning contents as well as present these elements, features and contents with a suitable engine on a suitable platform with a learner-oriented principle. Finally, in post-production phrase they need to review and test whether the prototype game at hand caters to their learning model and contents and make relevant adjustment and improvement before they finally release the game. These phrases can be flexible and customized but the developers need to guarantee maximize learning benefits players can get from the gaming experience.

On the other hand, user-centric design and development are also a major proposal in this field. Spikol and Milrad (2008) argue it can be quite useful to involve the players in co-design and human-centric practices, especially in mobile outdoor game. Students can provide storification, game concept and refinement to the game. Their physical activities and emotions are integrated into mobile games when students explore the field in reality, "communicate with the game server that provides the logic and scoring for the game" (Spikol & Milrad, p.32), navigate the location-based map and solve the mystery embedded in the game. Lavin-Mera et al. also share similar notion that mobile educational games should not only developed by professional game developers but also by the users such as school instructors who have a thorough in-depth understanding of the teaching and learning objectives. Providing "an authoring environment" to develop mobile educational games by the instructors themselves greatly reduce the technological barrier and cost to involve both game developers and instructors (Lavin-Mera et al, p.22). Also it will erase the communication and understanding problems between teachers and game developers.

Hence, both practitioners of mobile gaming industry and beneficiaries of mobile education games have seen the effect of mobile gaming to promote technology-based interactive education and training as well as its boundless potential to bring more revolutionary changes to current education system. Both parties are making their own efforts and collaborating jointly to realize these potentials and changes not only in the post-production stage of usage and application in teaching and learning environment but also in the initial stage of designing and developing customized and productive learning-oriented mobile games for human-mobile interactive education.

MOBILE GAMING AS LOCATION-BASED INTERACTIVE SOCIAL NETWORKING

Social Networking

Long before the latest social networking-related technologies become as prevailingly embedded in mobile gaming as nowadays, some pioneer researchers had evaluated the alternative technologies in their time and prophesized the important effects mobile gaming could bring to social-networking and vice versa. Before the popular application of Global Positioning System (GPS) and location-based service as well as ubiquitous internet access, Bluetooth and short messaging service (SMS) bore great significance in promoting social-networking in mobile games. The research of Baber's and Westmancott's (2004) on a multiplayer mobile card game called BELKA reveal that while visibility of other players in the real world (in the case that players sit together and play) often leads to adoption of similar card strategies among different players, mobility supported by the use of Bluetooth (in the case that players move around the room and play with Bluetooth-enabled mobile phones) "changes the nature of play and alters the social aspects of gaming" (Baber & Westmancott, p.105). Players are more interactive with the Dealer of the card game and pay more attention to observe the timely change in trumps and other players' trade with the dealer. Therefore, they conclude mobility realized by the use of Bluetooth is "both a feature of and a consequence" of playing mobile social games because it assists gaming experience while changing the way of playing, interacting and socializing with other players. With the differences of social-networking patterns players demonstrate in two types of playing, they propose to industrial practitioners that the development of mobile social games is "not simply about placing conventional play onto a handheld platform, but requires consideration of the interplay between the technical, virtual and social aspects" (Baber & Westmancott, p.106). Crabtree and his co-researchers (2007) explore how SMS helps to create an interactive gaming experience when players co-build and explore an imaginary city based on the instruction written in the text message sent by other players. Their research findings reveal that the existence of cooperative work is a major source to provide players awareness and coordination, and that "gaming evidently relies on cooperative work and is an essential part of it" (Crabtree, Benford, Capra, Flintham, Drozd, Tandavanitj, Adams and Farr, p.196). Like Baber and Westmancott, Crabtree et al. also propose to future mobile social game developers that "the availability of the cooperative work of gaming is...of great value to the developers of new forms of interactive experience" (Crabtree, et al., 195-196).

The motivation of socializing in mobile games or games in general refers to the desire to establish new or maintain and further develop existing social relationships (House, Davis, Ames, Finn & Viswanathan, 2005). As de Souza e Silva (2006) argues, mobile phone technology has triggered "new types of sociability and new perceptions of physical spaces" (de Souza e Silva, p.19-20). de Souza e Silva and Hjorth (2009) claim that location-based mobile social games are "social experiences", and "the coordination with other players becomes critical for the creating of the play activity" (de Souza e Silva & Hjorth, p.618). In other words, mobile games with social-networking function encourage players to cooperate with each other so that they can acquire a connected gaming experience that is entirely different from traditional single-player game on consoles. "In the context of mobile gaming, which combines the material built or urban environment, the mobility of the body, and the virtual or online game space, alternative metaphors of porosity and networking are becoming more prevalent, aiming to capture how the corporeality of the everyday becomes quite literally merged with the embodiment of play" (Richardson, p.434).

Despite that social networking element is prevailing in mobile gaming nowadays, whether to experience this function, as to establish friendships and seek opportunities for collaboration (Bell et al, 2006), remains in the choice of players. In his research on Japanese mobile games, Chan (2008) finds that "the potentiality for making and unmaking these alignments is suggestive of how co-presence may be reflexively and provisionally ruptured at will" (Chan, p.22). Additionally, social networking function in mobile gaming is not necessarily carried out only among existing players. "Collaboration within teams naturally depended on how much time the participants spent in the company of team-mates" (Bell et al., p.422), but since existing teammate players are inevitably occupied with other errands that are time consuming, social-networking in mobile gaming could also refer to the phenomenon that one existing player invites other non-players in his real-world social network, such as family members, friends and colleagues, to play the game. Thus, Bell and his co-researchers argue that "successfully interweaving the game with everyday life also involved managing interactions with non-players including family, partners, colleagues and strangers" (Bell et al., p.423).

It is also worth noting that mobile social games are believed to be extremely beneficial to players of certain demographic groups by strengthening social networking within the group. Researchers like Mubin, Shahid and Mahmud (2008) as well as Oliveira, Cherubini and Oliver (2010) focus on how mobile games gratify certain aspects of the elderly group by improving their social network. In the study of Mubin et al.'s, a social interactive mobile game called Walk 2 Win is developed to promote social network amongst the elderly and increase their enjoyment accordingly. They find that elderly players are strongly inclined to form teams in the game not only with other elderly players but also with younger generations which they may have less access to communicate with in reality. In the study of Oliveira et al.'s, a mobile game called MoviPill is engaged to persuade elderly patients to stick to their medication with the help of social competition with other patients playing the game, and they find that the game greatly improves the patients' compliance to take pills as well as the accurate timing for pills due to these elderly patients' intention to win over other patients in their in-game social network.

Location-Based Service and Sharing

Many researchers believe that the present and future of mobile social games lie more in the use of new technological inventions like GPS and location-based service (LBS) to promote social networking and sharing in a play field of urban city. The reason for such a belief largely derives from the fact that mobile phone has become a platform that exceedingly emphasizes local elements in the use of location-based technologies. Hjorth and Khoo (2007) even consider mobile phone "a compelling phenomenon that demonstrates the importance of the local in shaping and adapting the technology" (Hjorth & Khoo, retrieved April 2, 2015). Because of the location-aware/location-based nature of mobile social gaming, the virtual game world that used to exist only behind the cold screen of our displayer get to fuse and interact with the local environment everyone lives in as reality. In such "co-presence" of both virtual and reality spaces, solitary gaming experience in the past revolutionizes into a locally connected and environmentally interactive game adventure in an enormous play field of urban city. Like Hjorth (2011) argues, "by using both online and offline spaces, pervasive games can offer new ways of experiencing place, play and identity" (Hjorth, p.360).

Apart from sharing the playfield of urban city with other players locally connected, there are more to be shared by and interacted with players in the social network promoted by mobile social games. Lee, Goh and Chua (2010) argue that "the portability of mobile devices adds a new dimension to social computing in which users can now co-create, seek, and share multimedia information anytime, anywhere and do these in new ways not possible with desktop applications " (Lee et al., p.1244). The content to share, according to findings of their research, could be either personal or social. Players are given the opportunity to share with the social network they establish in the games their personal feature (e.g. individual information and performance) or knowledge (e.g. understanding and tactics to complete game tasks and challenges) as well as team performance and accomplishment they co-create with other players. The work of Celino's, Cerizza's, Contessa's, Corubolo's, Dell'Aglio's, Valle's and Fumeo's (2012) even reveal players are willing to share through location-based service their current position in reality to their friends in the game.

Players of mobile games with content sharing function choose to share different contents based on different motivations driving them to play the game. Goh, Lee and Low. (2012) reveal six motivations to play mobile content-sharing gaming:

1. Awareness, which refers to the purposeless content gathered to be shared. Players "simply wanted to discover what was around them, without any immediate need for such information" (Goh et al., p.793);
2. Task performance, which refers to the information shared concerning "the completion of specific tasks or for decision-making" (Goh et al., p.793);
3. Competitive play–higher rankings players obtain in competition with other individual players or group players;
4. Exploratory play, showing players' efforts to "explore all the encounters simply to review what each game looks like" (Goh et al., p.795);
5. Reminder of experiences, which involves "creating content as a record and reminder of individual and collective experiences and may include key moments, everyday activities or even mundane content related to oneself, others within a social circle or even the public" (Goh et al., p.798) and

6. Self-presentation, referring to "creating content to promote an impression of themselves or to influence how other people view them" (Goh et al., p.798).

Through the use of location-based technologies and the function of sharing, certain social patterns are able to be established in the visual community in mobile games. Casey, Kirman and Rowland (2007) reveal reality-reflective social order is displayed in player behaviors of sharing information, content and initiating conversation with strangers and acquaintances in mobile social game. Licoppe and Inada (2006) find out players adapt to a "form of life" in which they "adjust their displacements to the 'augmented' ecologies that they encounter in relation to a game environment enhanced with relevant localized informational objects", while onscreen encounters in the case that players' positions are noticed by others through screen-mediated display, "testifies a player's commitment to the game" (Licoppe and Inada, p.57).

Weaknesses and Problems Remaining in Mobile Social Games

According to Juniper research (2011), by the end of 2014, revenue generated from mobile LBS are expected to top $ 12.7 billion. de Kervenoael, Palmer and Hallsworth (2013) therefore suggest that "this rapid growth implies that context-aware mobile services, argmented-reality and sensors require further regulation to allow a better development and understanding of mobile gaming" (de Kervenoael et al., p.440). Indeed, there are more efforts to be made to develop safer mobile game-based social network, and standardized regulations need to be executed to guarantee players' online and offline security. As aforementioned, the shape and promotion of social network by mobile social gaming is often accompanied by exposure of personal information and location in reality, therefore players are likely to bear potential risks of privacy invasion and safety threat. Licoppe and Inada (2006) express their worries about the risk of mobile social games to provide "resources for a player to stalk and prey on another", which is a real concern "at the local level of the management of interaction" and "at the public debate level" (Licoppe and Inada, p.57). The territories of players' personal life in reality become vulnerable to be invaded along with the use of LBS and the function of sharing. The situation is even deteriorated by the convenient portability of mobile phone. The fact that players carry the device around with them, a commonly seen phenomenon in modern life, makes them easily located targets and they can be involved in great danger such as property loss, physical injury or even threat to their family and friends.

Social networking sites (SNS) also become a major concern of privacy and safety issue in mobile social games. Social networking sites have become a prevailing platform to propagandize third-party mobile games as well as their own independently developed games. It is quite common that SNS users invite their real-world friends, family, SNS friends, followers and acquaintances to collaborate or compete in certain mobile games. But Wei, Lee, Lu, Tzou and Weng (2015) reveal that players who are repeatedly annoyed by "frequent notifications on SNS platforms" abandon the mobile social games offered to them in the fear that "SNSs could share their personal information and interactions with third parties in ways they might find objectionable" (Wei et al, p.16).

Therefore, weaknesses and problems of mobile social games remain in the concerns for privacy and safety. While promoting a gaming experience that makes full use of location-based and location-aware elements and enhances social networking through sharing and interaction, mobile game developers and operators need to take preventive measures to protect their players' benefits and deter safety threat and misuse of personal information for commercial or illegal purposes.

MAPPING THE FUTURE

In his book *People analytics: How social sensing technology will transform business and what it tells us about the future of work*, Waber (2013) describes a new way of human resource evaluation called "people analytics", that is, to use recording and assessing technology, especially sensors and software to trace, analyze and examine the way employees work and collaborate. With data and results obtained from such analysis and examination, employers are able to trace the daily pattern of task accomplishment, problem-solving and team cooperation in the workplace as well as discover the weakness in human resources and business management. Waber's research reveals extremely remarkable improvement in the working performance of service-related department such as call center and customer service with the help of people analytics. It also helps employers to discover the staff who are excellent in professionalism, work ethics and creativity, and therefore is beneficial to the establishment of efficient, effective and profitable organization as well as the overall improvement of business performance and employee satisfaction.

Peck (2013) mentions games as particularly easy and effective means to assess potential candidates for job positions among all the technologies currently embodied in people analytics. In fact, games have been widely employed in the labor market to evaluate the suitability of candidates to certain positions in the recruiting company. In these specifically designed games, job candidates encounter a variety of tasks and challenges that aim to test their abilities in multiple aspects such as decision-making and problem-solving, use of strategy, management of stress as well as cooperation and competition. Compared to traditional assessing software which embodies abstract and often vague questions in the form of questionnaire, people analytics games are able to provide vivid scenarios catering to specific evaluation objectives and provide candidates opportunity for real practice and execution of their strategies. Peck points out this brand-new way of assessment is becoming more popularly applied also because it saves the employers huge amount of time to review each candidate, which is normally done with time-consuming one-by-one interviews. Now the employers can have a large number of candidates to play the games at the same time and observe their capabilities and personal qualities through their performance in the game. More visually, these games offer explicit marks for each candidate based on their performance in the game so that the employers can easily shortlist the most promising candidates from a narrow-down group of applicants who get high marks. At the same time, employment bias against educational background can be largely prevented because so far no study or practice of people analytics shows that candidates with higher educational degrees definitely get high marks in their evaluation by the games. To Peck's mind, people analytics and the relevant games are the future of job employment. In the future, existing employees as well can be assessed of their working efficiency, performance and promotion possibility with the help of people analytics games which at the same time will offer customized function for employers to vary their requirements according to different job and promotion descriptions.

The author believes that a future development of mobile into people analytics deserves closer attention and more researches. Such development is likely to share the same pattern with the development of technology-based learning and training aforementioned. In the initial stage, learning and education were carried out on computer and other console-based hardware, and then gradually shifted to the mobile platform because of the prevalent use of mobile phones and tablets that enjoy outstanding convenience, portability and ubiquity. Currently, game-based people analytics is largely taken in the company venue or specific assessment organizations because the games are restrictively installed to computer or hardware there. In the future, if the platform is shifted to mobile phone, anyone who applies for the job can download the game to their mobile phones and access to the evaluation anytime anywhere. It will further accelerate the process of recruitment by saving both the employer and the employee time to get the evaluation results. Better still, accompanied by the location-based technologies offered in mobile phones, people analytics mobile games can also enable employers' access to location-based information of potential candidates. They can be offered with overall as well as individual information of possible candidates in the neighborhood or certain chosen regions with explicit scores or rankings quantifying their different abilities and qualities. On the side of job hunters, they can equally get access to and compare job descriptions and information in multiple recruiting companies. In the case of unsuccessful application, job hunters will be able to engage in the next application in the minimum time. Therefore, the convenience, portability, flexibility, ubiquity, interactivity and location-aware nature of mobile gaming are expected to refine job recruitment to a greater extent and gratify a better version of interactive people analytics.

CONCLUSION

The new century has witnessed the prosperity of mobile gaming industry and the development of mobile game-related technologies. While being surprised by record-breaking revenue generated from mobile gaming consumption, scholastic attention has been more and more drawn to studies of mobile gaming applied to a variety of fields. This chapter critically reviews previous studies on the gratifying effects of mobile gaming in promoting new and interactive ways of gaming, education and social networking as well as its likely potential in gratifying a new mechanism of interactive human resource recruitment and management in the future.

In terms of gaming experience, mobile gaming gratifies a new concept of casual, flexible and ubiquitous gaming different from traditional console or computer gaming. It promotes the changing nature of modern gaming not only as a means to acquire meaningful experience in a short time but also as a gaming experience open to the flexibility of player-centric gaming behavior. While fulfilling players' need for games available whenever and wherever possible, mobile gaming contributes to a ubiquitous gaming experience that fits into the lives of players and interweaves everyday life. But concerns are equally expressed with regard to flawed game content, insufficient game concept innovation as well as poor service quality. So it is high time for industrial practitioners to make amendment in the development and operation of mobile games according to these research findings while making full use of the casual, flexible and ubiquitous nature of mobile gaming to reach and gratify a broader range of players.

Additionally, mobile gaming gratifies a new way of technology-based interactive learning and teaching which enhances student engagement, motivation and performance as well as teaching efficiency and variation in a pleasant, stimulating and interactive education environment. Previous studies reveal great improvement in education system in general and also specific courses that often cause learning disengagement and alienation. Simultaneously, mobile gaming effectively assists training of various abilities, qualities and physical fitness. In terms of future amelioration, both scholars of social science and humanity studies as well as engineering researchers have expressed their efforts to engage teaching instructors and learners more into the co-design and co-development of mobile educational games.

Furthermore, mobile gaming serves to gratify a new way of interactive social networking with the application of location-based technologies such as GPS, location-based service, short messaging service and Bluetooth. These applications contribute respectively and jointly a socially interactive and location-aware gaming experience that facilitates content sharing and social order establishment in the games. Meanwhile, concerns are also expressed regarding the high risk embodied in mobile social games to privacy and personal safety. In the overall flourish of interactive mobile social games, industrial practitioners are expected to come up with more precautious measures to protect player information and personal well-being.

As for the future direction of mobile gaming, it is expected to become a fresh member strengthening people analytics—a new way to recruit and manage human resource with technology-based evaluation system. In the newly emerging field of people analytics, mobile gaming can be forecasted to possess better efficiency and effectiveness in portability, accessibility, time-saving and interactivity while displaying unique potentials in providing location-based information for both job applicants and recruiting companies. Therefore, it is hoped that industrial practitioners, academic scholars as well as recruiting employers can pay close attention to the development and application of such games.

The present of mobile gaming is about its gratifying effects in brand-new gaming experience, education and social-networking. And the future will be about its expanded usage in the field of human resource analytics, recruitment and management as well as many other unexplored fields demanding human social interaction with mobile phones.

REFERENCES

Admiraal, W., Huizenga, J., Akkerman, S., & Dam, G. T. (2011). The concept of flow in collaborative game-based learning. *Computers in Human Behavior*, *27*(3), 1185–1194. doi:10.1016/j.chb.2010.12.013

Akkerman, S., Admiraal, W., & Huizenga, J. (2009). Storification in history education: A mobile game in and about medieval Amsterdam. *Computers & Education*, *52*(2), 449–459. doi:10.1016/j.compedu.2008.09.014

Ardito, C., Buono, P., Costabile, M. F., Lanzilotti, R., Pederson, T., & Piccinno, A. (2008). Experiencing the past through the senses: An m-learning game at archaeological parks. IEEE Computer Society, 79-81.

Baber, C., & Westmancott, O. (2004). Social networks and mobile games: The use of bluetooth for a multiplayer card game (S. Brewster & M. Dunlop, Eds.). Academic Press.

Balayan, M. P. A., Conoza, V. V. B., Tolentino, J. M. M., Solamo, R. C., & Feria, R. P. (2014). On evaluating Skillville: An educational mobile game on visual perception skills. *IISA 2014, the 5th International Conference on Information, Intelligence, Systems and Applications*, (pp. 69 - 74).

Bell, M., Chalmers, M., Barkhuus, L., Hall, M., Sherwood, S., & Tennent, P. (2006). Interweaving mobile games with everyday life. *Proceedings of SIGCHI Conference on Human Factors in Computing Systems*, (pp. 417-426). New York: ACM Press. doi:10.1145/1124772.1124835

Buttussi, F., & Chittaro, L. (2010). Smarter phones for healthier lifestyles: An adaptive fitness game. *IEEE CS*, 51-57.

Casey, S., Kirman, B., & Rowland, D. (2007). The gopher game: A social, mobile, locative game with user generated content and peer review. *Proceedings of ACE '07 of the International Conference on Advances in Computer Entertainment Technology*, (pp. 9-16). New York: ACM. doi:10.1145/1255047.1255050

Celino, I., Cerizza, D., Contessa, S., Corubolo, M., DellAglio, D., Valle, E. D., & Fumeo, S. (2012). Urbanopoly—A social and location-based game with a purpose to crowdsource your urban data. *2012 ASE/IEEE International Conference on Social Computing and 2012 ASE/IEEE International Conference on Privacy, Security, Risk and Trust*, (pp. 910-913). doi:10.1109/SocialCom-PASSAT.2012.138

Chan, D. (2008). Convergence, connectivity, and the case of Japanese mobile gaming. *Games and Culture*, *3*(1), 13–25. doi:10.1177/1555412007309524

Chee, Y. S., Tan, E. M., & Liu, Q. (2010). Statecraft X: Enacting citizenship education using a mobile learning game played on Apple iPhones. *The 6th IEEE International Conference on Wireless, Mobile, and Ubiquitous Technologies in Education*, (pp. 222-224).

Crabtree, A., Benford, S., Capra, M., Flintham, M., Drozd, A., Tandavanitj, N., & Farr, J. R. et al. (2007). The cooperative work of gaming: Orchestrating a mobile SMS game. *Computer Supported Cooperative Work*, *16*(1-2), 167–198. doi:10.1007/s10606-007-9048-1

de Kervenoael, R., Palmer, M., & Hallsworth, A. (2013). From the outside in: Consumer anti-choice and policy implications in the mobile gaming market. *Telecommunications Policy*, *37*(6-7), 439–449. doi:10.1016/j.telpol.2012.06.008

de Souza e Silva, A. (2006). Interfaces of hybrid spaces. In A. Kavoori & N. Arceneaux (Eds.), The Cell Phone Reader: Essays in Social Transformation (pp.19–44). Pieterlen: Peter Lang International Academic Publishers.

de Souza e Silva, A., & Hjorth, L. (2009). Playful urban spaces: A historical approach to mobile games. *Simulation & Gaming*, *40*(5), 602–625. doi:10.1177/1046878109333723

Doswell, J., & Harmeyer, K. (2007). Extending the 'serious game' boundary: Virtual instructors in mobile mixed reality learning games. *Proceedings of DiGRA 2007 Conference*, (pp. 524-529).

Feijoo, C., Go'mez-Barroso, J., Aguadoc, J., & Ramos, S. (2012). Mobile gaming: Industry challenges and policy implications. *Telecommunications Policy*, *36*(3), 212–221. doi:10.1016/j.telpol.2011.12.004

Fields, T. (2012). *Social game design: Monetization methods and mechanics*. Waltham: Morgan Kaufmann Publishers.

Goh, D., Lee, C., & Low, G. (2012). I played games as there was nothing else to do: Understanding motivations for using mobile content-sharing games. *Online Information Review*, *36*(6), 784–806. doi:10.1108/14684521211287891

Gwee, S., Chee, Y., & Tan, E. (2010a). Spatializing social practices in mobile game-based learning. *2010 10th IEEE International Conference on Advanced Learning Technologies*, (pp. 555-557).

Gwee, S., Chee, Y., & Tan, E. (2010b). Assessment of student outcomes of mobile game-based learning. *Proceedings of the 18th International Conference on Computers in Education. Putrajaya, Malaysia: Asia-Pacific Society for Computers in Education.*

Hafizah, S., Hamid, A., & Fung, L. Y. (2007). Learn programming by using mobile edutainment game approach. *The First IEEE International Workshop on Digital Game and Intelligent Toy Enhanced Learning (DIGITEL'07).*

Hjorth, L. (2011). *Games and gaming: An introduction to new media*. Oxford, UK Berg.

Hjorth, L. (2011). Mobile @ game cultures: The place of urban mobile gaming. *Convergence (London)*, *17*(4), 357–371. doi:10.1177/1354856511414342

House, N. V., Davis, M., Ames, M., Finn, M., & Viswanathan, V. (2005), The uses of personal networked digital imaging: An empirical study of cameraphone photos and sharing.*Proceedings of CHI '05 Extended Abstracts on Human Factors in Computing Systems* (pp. 1853-6). ACM Press. doi:10.1145/1056808.1057039

Huizenga, J., Admiraal, W., Akkerman, S., & Dam, G. T. (2009). Mobile game-based learning in secondary education: Engagement, motivation and learning in a mobile city game. *Journal of Computer Assisted Learning*, *25*(4), 332–344. doi:10.1111/j.1365-2729.2009.00316.x

Ito, M., Okabe, D., & Matsuda, M. (Eds.). (2005). *Personal, portable, pedestrian: Mobile phones Japanese life*. Cambridge, MA: MIT Press.

Juniper Research. (2006). *Press release: casual gamers and female gamers to drive mobile games revenues over the $10 billion mark by 2009*. Retrieved March 27, 2015 from http://juniperresearch.com/viewpressrelease.php?pr=16S

Juul, J. (2010). *A casual game revolution: Reinventing video games and their players*. Cambridge, MA: MIT Press.

Kalloo, V., Kinshuk, & Mohan, P. (2010). Personalized game based mobile learning to assist high school students with mathematics. *2010 10th IEEE International Conference on Advanced Learning Technologies*, (pp. 485-487).

Kim, P., Buckner, E., Kim, H., Makany, T., Taleja, N., & Parikh, V. (2012). A comparative analysis of a game-based mobile learning model in low-socioeconomic communities of India. *International Journal of Educational Development*, *32*(2), 329–340. doi:10.1016/j.ijedudev.2011.05.008

Kurkovsky, S. (2009). Can mobile game development foster student interest in computer science? *ICEGIC 2009. International IEEE Consumer Electronics Society's Games Innovations Conference 2009*, (pp. 92-100).

Kurkovsky, S. (2009). Engaging students through mobile game development.*Proceedings of SIGCSE '09 of the 40th ACM Technical Symposium on Computer Science Education*, (pp. 44-48). New York: ACM Press. doi:10.1145/1508865.1508881

Lavin-Mera, P., Torrente, J., Moreno-Ger, P., Vallejo-Pinto, J., & Fernández-Manjón, B. (2009). Mobile game development for multiple devices in education. *International Journal of Emerging Technologies in Learning*, *4*, 19–26.

Lee, C., Goh, D., & Chua, A. (2010). Indagator: Investigating perceived gratifications of an application that blends mobile content sharing with gameplay. *Journal of the American Society for Information Science and Technology*, *61*(6), 1244–1257.

Leonov, I. (2014). *Trends and next steps for mobile games industry in 2015*. Gamasutra.

Licoppe, C., & Inada, Y. (2006). Emergent uses of a multiplayer location-aware mobile game: The interactional consequences of mediated encounters. *Mobilities, 1*(1), 39–61. doi:10.1080/17450100500489221

Low, G., Goh, D. & Lee, C. (2011). A study of motivations for using mobile content sharing games. *2011 IEEE*, (pp. 258-263).

Lu, C., & Chang, M., Kinshuk, Huang, E., & Chen, C. (2011). Usability of context-aware mobile educational game. *Knowledge Management & E-Learning: An International Journal, 3*(3), 448–477.

Mainwaring, S. D., Anderson, K., & Chang, M. F. (2005). *Living for the global city: Mobile kits, urban interfaces, and UbiComp.* Paper presented at UbiComp. New York, NY.

Martin, P. (2014). Socio-spatial relations in mobile gaming: Reconfiguration and contestation. In X. Xiaoge (Ed.), *Interdisciplinary mobile media and communications: Social, political and economic implications* (pp. 260–277). Hershey, PA: IGI Global.

Mayra, F. (2008). *An introduction to game studies: Games in culture.* London: Sage.

Mobile games market: Market insight beyond the rankings for the global mobile games market. (n.d.). Retrieved January 22, 2015 from http://www.superdataresearch.com/market-data/mobile-games-market/

MobileHCI. (2004). Berlin: Springer-Verlag.

MobiThinking. (2014). *Global mobile statistics 2014 part A: Mobile subscribers; handset market share; mobile operators.* Retrieved February 7, 2015 from http://mobiforge.com/research-analysis/global-mobile-statistics-2014-part-a-mobile-subscribers-handset-market-share-mobile-operators#subscribers

Mubin, O., Shahid, S., & Mahmud, A. A. (2008). Walk 2 Win: Towards designing a mobile game for elderly's social engagement. *Interaction, 2,* 11–14.

Newzoo. (2014). *Asia-Pacific contributes 82% of the $6Bn global games market growth in 2014.* Retrieved April 22, 2015 from http://www.newzoo.com/insights/asia-pacific-contributes-82-6bn-global-games-market-growth/

Okazaki, S. (2008). Exploring experiential value in online mobile gaming adoption. *Cyberpsychology & Behavior, 11*(5), 619–624. doi:10.1089/cpb.2007.0202 PMID:18785820

Oliveira, R. D., Cherubini, M., & Oliver, N. (2010). MoviPill: Improving medication compliance for elders using a mobile persuasive social game. *Proceedings of UbiComp '10 of the 12th ACM International Conference on Ubiquitous Computing,* (pp. 251-260). New York: ACM. doi:10.1145/1864349.1864371

Park, E., Baek, S., Ohm, J., & Chang, H. (2014). Determinants of player acceptance of mobile social network games: An application of extended technology acceptance model. *Telematics and Informatics, 31,* 3–15.

Park, H., & Kim, S. (2013). A Bayesian network approach to examining key success factors of mobile games. *Journal of Business Research, 66*(9), 1353–1359. doi:10.1016/j.jbusres.2012.02.036

Pearson, D. (2014). *Report: Mobile to become gaming's biggest market by 2015.* Newzoo.

Peck, D. (2013). They're watching you at work. *Atlantic (Boston, Mass.).*

Puja, J., & Parsons, D. (2011). A location-based mobile game for business education. *2011 11th IEEE International Conference on Advanced Learning Technologies*, (pp. 42-44).

Richardson, I. (2010). Ludic mobilities: The corporealities of mobile gaming. *Mobilities, 5*(4), 431–447. doi:10.1080/17450101.2010.510329

Sánchez, J., & Olivares, R. (2011). Problem solving and collaboration using mobile serious games. *Computers & Education, 57*(3), 1943–1952. doi:10.1016/j.compedu.2011.04.012

Schwabe, G., & Goth, C. (2005). Mobile learning with a mobile game: Design and motivational effects. *Journal of Computer Assisted Learning, 21*(3), 204–216. doi:10.1111/j.1365-2729.2005.00128.x

Sinclair, B. (2015). US mobile gaming market tops $5.2 billion. *Gamesindustry.biz*. Retrieved March 22, 2015 from http://www.gamesindustry.biz/articles/2015-01-13-us-mobile-gaming-market-tops-usd5-2-billion

Sintoris, C., Stoica, A., Papadimitriou, L., Yiannoutsou, N., Komis, V., & Avouris, N. (2010). Museum-Scrabble: Design of a mobile game for children's interaction with a digitally augmented cultural space. *International Journal of Mobile Human Computer Interaction, 2*(2), 19–38. doi:10.4018/jmhci.2010040104

Smith, M. A., Woo, H. J., Parker, J. P., Youmans, R. J., LeGoullon, M., Weltman, G., & de Visser, E. J. (2013). Using iterative design and testing towards the development of SRTS(R): A mobile, game-based stress resilience training system. *Proceedings of the Human Factors and Ergonomics Society Annual Meeting, 57*(1), 2076-2080. doi:10.1177/1541931213571463

Spikol, D., & Milrad, M. (2008). Combining physical activities and mobile games to promote novel learning practices. *Fifth IEEE International Conference on Wireless, Mobile, and Ubiquitous Technology in Education*, (pp. 31-38). doi:10.1109/WMUTE.2008.37

Statista. (n.d.). *Statistics and facts about gaming in Asia*. Retrieved March 22, 2015 from http://www.statista.com/topics/2196/video-game-industry-in-asia/

Takahashi, D. (2014). Mobile gaming could drive entire video game industry to $100B in revenue by 2017. *GamesBeat*. Retrieved March 22, 2015 from http://venturebeat.com/2014/01/14/mobile-gaming-could-drive-entire-game-indus try-to-100b-in-revenue-by-2017/

Tassi, P. (2012). Is the PlayStation Vita a dinosaur already? *Forbes*. Retrieved March 14, 2015 from http://www.forbes.com/sites/insertcoin/2012/02/17/is-the-playstation-vita-a-dinosaur-already/

Tercek, R. (2007). 1997-2007: The first decade of mobile games. *Game Developers Conference*, San Francisco.

Waber, B. (2013). *People analytics: How social sensing technology will transform business and what it tells us about the future of work*. FT Press.

Wang, A. I., Ofsdahl, T., & Morch-Storstein, O. K. (2008). An evaluation of a mobile game concept for lectures. *Software Engineering Education and Training, 2008. CSEET '08. IEEE 21st Conference*, (pp. 197-204). Charleston: IEEE.

Wei, P., Lee, S., Lu, H., Tzou, J., & Weng, C. (2015). Why do people abandon mobile social games? Using candy crush saga as an example. *International Journal of Social, Education, Economics and Management Engineering, 9*(1), 13–18.

Winkler, T., Ide-Schoening, M., & Herczeg, M. (2008). Mobile co-operative game-based learning with moles: Time travelers in Medieval. *Proceedings of SITE, 2008,* 3441–3449.

Wohn, D. Y., Lampe, C., Wash, R., Ellison, N., & Vitak, J. (2011). The "s" in social network games: Initiating, maintaining, enhancing relationships. In *Proceedings of the 44th Hawaii International Conference on System Sciences (HICSS '11),* (pp. 1 –10). doi:10.1109/HICSS.2011.400

Yamakami, T. (2013). Historical view of mobile social game evolution in Japan: Retrospective analysis of success factors. *ICACT2013,* (pp. 375-379).

Zaibon, S. B., & Shiratuddin, N. (2009). Towards developing mobile game-based learning engineering model.*2009 World Congress on Computer Science and Information Engineering,* (pp. 649-653). doi:10.1109/CSIE.2009.896

Zaibon, S. B., & Shiratuddin, N. (2010). Adapting learning theories in mobile game-based learning development.*2010 IEEE International Conference on Digital Game and Intelligent Toy Enhanced Learning,* (pp. 124-128). doi:10.1109/DIGITEL.2010.37

Zhou, T. (2013). Understanding the effect of flow on user adoption of mobile games. *Personal and Ubiquitous Computing, 17*(4), 741–748. doi:10.1007/s00779-012-0613-3

Chapter 18
Youth and Mobile:
An Investigation of Socialization

Zeinab Zaremohzzabieh
UPM, Malaysia

Bahaman Abu Samah
IPSAS, Malaysia

Seyedali Ahrari
IPSAS, Malaysia

Jamilah Bt. Othman
IPSAS, Malaysia

ABSTRACT

While the rapid growth in studies on the effects of mobile phones has deepened our understanding of the role mobile phones play in the socialization process of youth, further work is required in reviewing the growing influence of mobile phones for continuing socialization. The objective of this paper therefore is to assess literature from a range of selected studies and in doing so, highlight the role of mobile phones in contributing to youth socialization. This state-of-the-art review demonstrates that mobile phones are a powerful socializing tool that can lead to plentiful consequences. It will show that the influence of mobile phones can be beneficial. It explores the harmful effects of mobile phones. Finally, this chapter will incorporate previous advancements in research to inform forthcoming research and identify new concepts, themes and theories to support or improve the role of mobile phones in increasing the socialization skills of youth.

INTRODUCTION

The wide scale usage of mobile phones by youth has been a worldwide miracle, playing an essential role in youths' lives, especially in electronic communication. The mobile phone was invented a decade before youths began to use it for their daily lives. The reduction in prices for mobile phones and the creation of pre-paid phone cards in the last decade had led to widespread usage by youths (Ling, 2004). The new generations are very energetic users of these communication tools. During the year 2010, Pew's review reported that 75% of 12-17 year-olds now have cell phones, up from 45% in 2004 (Pew Internet & American Life Project, 2010). The percentage had also increased from 58% to 83% for youths between

DOI: 10.4018/978-1-5225-0469-6.ch018

age 12 and 17. In 2012 there were around 5,020,000 mobiles in New Zealand whereas the population was only 4,433,087 (TNS Global Market Survey, cited in New Zealand Herald, 2012). The demand for mobile phones are worth researched on as it has the capability to considerably alter social interactions. The usage of mobile phones has augmented a novel aspect of virtual mobility to a continuing trend for geographically extended, quicker, and more individualized social interaction. These usages of mobile phones have altered the forms of coordination in many parts of our public life, including friends, relatives, and work. Particularly between youngsters, the mobile phone has two functions: on the one hand it enables parents' interaction and care; however, on the other it facilitates massive social interactions with groups of friends outside of parental influence (Oksman & Rautiainen, 2002).

Furthermore, research has revealed the benefits of its usage, including the ability to maintain social relationships, intensify individual and collective identities, and enlarges social circles (wherein friends are vastly prominent (Smetana, Campione-Barr, & Metzger, 2006). Dresler-Hawke and Mansvelt (2008) explain that mobile phones play an important role in youths' lives, and are accepted as a required part of social communication through the upkeep of main social networks. Consequently, the mobile phone is found to be the starting place for youth socialization (Harper & Hamill, 2005). Previous research has shown that fast usage of mobile phones increases the socialization process with an individual's family and friends (Ling & Yttri, 2006).

In recent years, there has been an increasing amount of studies on young people and mobile phones carried out in different settings. Largely, social scientists consider the popularity of mobile phones in many areas in addition to the social function of the device in the lives of youths. Some researchers have paid attention to the social characteristics of mobile phone diffusion (e.g., Rheingold, 2002), while others (e.g., Ling, 2004) have investigated the influence of mobile phones in daily interactions (i.e. teenagers' social relationships and networks). However, limited researches have studied the function of mobile phones in youth socialization (e.g., Kalogeraki & Papadaki, 2010). The purpose of the current review is therefore to increase our understanding of where this tool fits in the association between youths, and to identify whether mobile phone use has an effect on youth socialization. Our aims are, thus, to prepare an inclusive evaluation of the writings published to date which has reported directly on mobile phones and their effect on primary sources of socialization such as peers and family. This review will attempt to integrate these past advances in research in order to inform future research and identify new findings to support or improve the role of mobile phones in increasing youths' socialization skills.

STATE OF THE ART

The role of this literature review is to consider major current studies in the field of mobile phone use and youth socialization. It will examine two divergent areas: how young people utilize a mobile phone's email and texting functions as tools to socialize with their family and peers, and the usage of mobile phones in the context of socialization. For the purposes of this review, we restricted our search to encompass studies that measure the association between mobile phone and socialization in youths. Electronic research was conducted using Google Scholar and ISI Web of Science, using the following key phrases: mobile devices and youth socialization (i.e., social networking, group/self-identity, social interaction with families/ peer groups, etc.). All journal articles identified were written in English and published between January 2004 and January 2015. The reason behind this was to enhance understanding of available research and avoid missing any relevant high quality research. Ancestral studies were conducted by studying the reference parts of each of the papers exposed and all obtainable review articles. To keep the review practicable, we decided to exclude unpublished papers (i.e., theses and conference papers). A total of 24 published articles were initially reviewed for inclusion in this chapter (see Table 1). For each article, we extracted information on the author(s), publication year, study aim(s), method(s), theory, participants, and major findings. All studies included youths, but the age range varies. Of the 24 studies, eleven of them directly used the qualitative method, twelve used the quantitative method, and one study used a mixed method. The studies were conducted in fifteen different countries: seven of the studies were conducted in five European and Nordic countries, while the rest were completed in Africa, Asia, Asia-Pacific, and America.

SUMMARIZING PREVIOUS STUDIES

This chapter addresses five common themes for youths' mobile phone usage. These themes are self-socialization, social networks, contact and social activity patterns, romantic relationships, and attachment to parents. These themes relate to the multiple facets of a mobile phone's email and texting functions, which participants in each study perceived as being beneficial or useful tools for their own socialization.

Self-Socialisation

The first theme is self-socialization. The course of self-socialization proposes that youths draw on their own sense of agency to choose the best social settings to support their growth, and this progression is both a product of and a contributor to young individuals' socialization. Because of the fast shifting nature of modern civilizations such as mobile technologies, novel types of communication, intensifying new socialization forms, diverse family forms, and expanding globalization, which open many potential paths for youth development, it is needed for youths to reflect on the self that they want or expect to become (Table 1). It has been found that regularity of mobile phone behavior in youths represents their identity and reinforces youth self-conception (Dittmar, Long, & Bond, 2007; Walsh, White, & Young, 2008).

The findings of current studies conducted in Australia showed an association between self-socialization and mobile phone behavior in young users (Walsh et al., 2010) and this links with processes of self-socialization that impacts behavior. According to Walsh et al., (2010), participants noticed that their peers use mobile phones when they felt a need to belong. In addition, it was found that participants with most common mobile phone use among a group of friends have stronger ties with each other. As Smetana et al.,(2006) argued, youths develop their identities in a bigger societal context and are affected by the behavior of other groups in society, and this view affects the frequency and approval of youths' behavior with regard to mobile phones (Igarashi et al., 2008).

Additionally, young people groups are not similar or motionless units; rather, the relationship between youth groups are very complex and influences such as behavior and connectedness with family, peers, and technology all come together to shape their identities (E. Green & Singleton, 2007). In a study in the northeast of Britain, Green and Singleton (2007) examined the connection between mobile phone usage and diverse youth identities, for instance British Asians and young mothers. Their results show that mobile phone use is interconnected through gender, race and social status, and mobile phones are the key to youth identity. For example, they believe that since young Pakistani-British Muslim females are bi- or multilingual, text and chat is also expressed through languages, signs and altered local cultural identities, where various customs, accounts and discussions are combined to make novel and varied identities within groups.

Table 1. A summary of the 24 relevant studies included in this review

Reference	Country of Study	Study Aim	Method	Theory/ Model	Participants	Study Findings
Walsh et al., (2008)	Australia	To discover the role of mobile phone use in youth's lives and whether addiction to mobile phone is signified in this cohort	A qualitative method (i.e. focus group discussions)	Theoretical framework of Brown's addiction criteria (1993, 1997)	32 participants, aged between 16–24 years	Cell phones have become a vital part of the social lifestyle of young Australians. They have become a tool for communication and ease of contact, with some young people showing signs of addictive behavior.
Walsh, White, Cox, & Young (2011)	Australia	To assess the role of psychosocial features in establishing youth's mobile phone behavior	A quantitative method (i.e. online surveys)	Psychosocial Theory	292 respondents, aged between 16–24 years	Age (juvenile) and self-identity meaningfully predicted the rate of mobile phone use. In contrast, age (younger), gender (female), self-identity and in-group norm predicted youth's mobile phone connection.
Hakoama & Hakoyama (2011)	The USA	To examine college students' cell phone use	Quantitative method (i.e. surveys)	Psychosocial Theory	501 college students from Midwestern University	• Financial responsibility for the cell phone cost depends on students 'gender, age, and work status. • Length of mobile phone possession, age at first mobile phone possession, length of mobile phone use per day and gender are vital for owning a mobile phone. • The students' social networks are expanded and increased through owning mobile phones • The mobile phone has become indispensable for the students, to the extent of pathological dependency.
Ali, Rizvi, & Qureshi (2014)	The South Punjab-Pakistan	To study overall utilization patterns of mobile phones among youth	Quantitative method (i.e. surveys)	—	317 young males and 310 young females belonging to several educational institutions	• Teenagers employ mobile phones to keep in touch with others. • The adolescents questioned expressed that other people use cell phones inappropriately. • Teenage girls were found to attend phone calls during lectures less often than boys. • The majority of the teenagers believe that mobile phones should not be given to youth before the age of 19.
Kamran (2010)	Pakistan	To investigate mobile phone texting and calling patterns among university students	Qualitative method (i.e. mobile phone communication diaries & in-depth interviews)	—	77 university students aged between 17–21 years	• The students were found to do most of their mobile phone communications with their peers, communicating in both positive and negative manners. • Most mobile phone communication among students was classified as challenging with reference to the setting, time and reason.
Lee, Meszaros, & Colvin (2009)	The USA	To examine the dynamics of mobile phone use among students and their relationship opinions to their parents	Quantitative method (i.e. surveys)	Attachment Theory	568 undergraduate college students	• The results failed to precisely gauge feelings towards mobile phone use in social interactions. • All three user groups and females scored their highest means on the factor of security/safety. • Students in the connected user group reported significantly higher linking to mothers toward mobile phones.

continued on following page

Table 1. Continued

Reference	Country of Study	Study Aim	Method	Theory/ Model	Participants	Study Findings
E. Green & Singleton (2007)	The Northeast of England	To study mobile phone usage by a young marginal racial group as a medium through which to explore diversity and technology use in everyday life	Qualitative method (i.e. focus group narratives and in-depth interviews)	Feminist theories of technology	Young Pakistani-British Muslim females aged between 15–25 years	• Appropriation and use of mobile phones is interconnected with gender, race and social status. • Mobile phones are key parts in the production of youth's identity projects.
Haverila (2011)	Finland	To study mobile phone use and broad mobile phone attribute favorites, particularly from the gender standpoint	Quantitative method (i.e. surveys)	The framework of Aoki & Downes, (2003)	Undergraduate students in Tampere aged between 19–25 years	• Significant differences between genders regarding the behavioral factors and specific behaviors were identified in three out of six behavioral factors and in 14 out of 30 specific behaviors. • Differences between genders in the conceptualization of cellphone behaviors were discovered. • Gender was analyzed in order to find separate and unique user groups on the basis of the differences in the most important behavioral variables (i.e. price efficiency, functionality, need in modern times, and security/ safety).
Thulin & Vilhelmson, (2007)	Sweden	To discover how youths' daily patterns of social communication are influenced by the amplified use of mobile phones	Qualitative method (i.e. diaries and in-depth interviews)	—	40 urban Swedish youth	• Youth total interactions with their social setting improve through using mobile phones (e.g. contact has become closer and more frequent). • The main purposes of youth's mobile communications are functional (or practical), sensitive and symbolic.
Boase & Kobayashi (2008)	Japan	To examine the extent to which Japanese adolescents use mobile phone–accessed e-mail to bond, bridge, and break with social ties	Qualitative method (i.e. surveys)	The Theory of Socio-emotional Selectivity	501 Tokyo high school students	The findings showed youths used mobile phone–accessed e-mail for both bonding and bridging, but this did not disturb social ties. These findings apply both to normal users and serious users.
Blackman (2010)	The USA	To examine mobile phone usage with parents, peers, and romantic partners in university freshmen, as well as gender differences in mobile phone use	Quantitative method (i.e. surveys)	—	105 college freshmen aged between 18–21 years	Participants employed voice calls more than text messaging in contact with parents, but researchers discovered no dissimilarities in rates of calls and text messages with friends or romantic associates.
Yi-Fan Chen (2006)	Taiwan	To examines the relationship between mobile phone use, mobile phone addiction, social capital, and academic performance in college students	A quantitative method (i.e., online surveys)	Media Dependency Theory	166 Taiwanese college students	A positive relationship was found to exist between heavy mobile phone usage and improving relationships with friends.

continued on following page

Table 1. Continued

Reference	Country of Study	Study Aim	Method	Theory/ Model	Participants	Study Findings
North, D. et al (2014)	South Africa	To investigate the uses and gratification of mobile phones (e.g. usage pattern, purchasing factors, reasons to use and behavioral issues) by university students	Quantitative method (i.e. surveys)	The Uses & Gratification Theory	362 respondents aged 20 and thereabouts	• For students, the most common motives for using mobile phones were socializing, safety and privacy, with female participants emphasizing more the benefits of safety and socializing. • The respondents showed signs of addictive behavior to their mobile phones.
Economides & Grousopoulou (2008)	Greece	To examine the comparative use of mobiles in various places	Quantitative method (i.e., surveys)	—	416 university students aged between 18–25 years	• Students use their mobiles mostly for phone calls and SMS (short message service). • They use their mobiles to communicate (via calls, SMS and email) mostly with their boy/girlfriend, then with their friends.
Oksman & Turtiainen (2004)	Finland	To analyze the meanings and context of use of mobile communication culture among youth	Media ethnographic method (i.e. interviews and collected material for observations)	Symbolic Interactionism	168 Finnish teenagers in urban and rural areas	• Mobile phones are an important social adhesive and engender the construction of a social identity among young people. • The use of chat and IRC are connected with the process of socialization and autonomy.
Abeele (2014)	Belgium	To identify and characterize potential mobile lifestyles within mobile youth culture	Quantitative method (i.e. surveys)	Apparatgeist Theory	605 adolescents in 18 different high schools (9th to 12th grade)	• It found support for the same gratifications-ranking structure, supporting the evidence for differentiated mobile lifestyles and underscoring the truism that there is no such thing as the typical adolescent mobile phone user. • Disinterested youth, who are more mature in their psychological development, seek more pragmatic gratifications from mobile communication, whereas those who are younger display a more dependent mobile lifestyle. For girls, mobile phones provide opportunities for emotional and social communication.
Mihailidis (2014)	The USA	To explore college students using mobile phones for communication and information needs in their everyday life	Media ethnographic method (i.e. gather detailed data on what students did with their mobile phones over a 24-hour period)	—	793 students from 8 colleges on 3 continents, collectively representing 52 countries	• The study unveiled a population tied to their mobile phones mainly through social networking apps • Their dependence on mobile phones for connecting to peers left them skeptical of the mobile phone's efficacy for useful connectivity, vivid communication, and diverse information use in everyday life.
Vacaru, Shepherd, & Sheridan (2014)	New Zealand	To examine the role of mobile phone use in young people's lives and to develop the current perspective on problematic mobile phone use	A qualitative method (i.e. focus group discussions)	—	45 participants aged 13 to 18	Young people believe one can become overly attached, or "addicted" to their mobile phone, or communication via their device, and negative consequences may be experienced as a result.

continued on following page

Table 1. Continued

Reference	Country of Study	Study Aim	Method	Theory/ Model	Participants	Study Findings
Lesitaokana (2015)	Botswana	To explore the resulting cultures from the interactive relationship between youth and their mobile phones	A qualitative method (i.e. focus group discussions, nonparticipant observation, and semi-structured interviews)	—	Local college students & workplace participants owning mobile phones	• Mobile phones promise the participants extensive connectedness with their family, relatives, and friends, and also access to public and social services. • It influences cultural practices among youth, such as constantly carrying a mobile handset, dependency on texting and social networks accessed through mobile phones, and the inappropriate use of mobile phones at funerals.
Miller-Ott, Kelly, & Duran (2014)	The USA	To examine mobile phone rules in emerging adult college students' relationships with their parents and determine whether these rules influence their satisfaction with the use of cell phones, satisfaction with the relationships, and closeness to their parents	Quantitative method (i.e. surveys)	—	207 undergraduate students aged 20 & thereabouts	The results suggested that mobile phones are used frequently between emerging adults and their parents, particularly their mothers, and emerging adults are highly satisfied with their use of mobile phones with their mothers and fathers.
Chen & Katz (2009)	The USA	To understand if there is a pattern in the use of mobile phones between college students and their family members at home, and to what degree mobile phones affect students' college life	A qualitative method (i.e. focus group interviews)	Media Dependency Theory	40 undergraduate students	• College students use mobile phones to have more frequent contact with their family • College students utilize mobile phones to share experiences and access emotional and physical support from their parents. • Mobile phones were found to be an umbilical cord between college students and their families, especially between students and their mothers.
Balakrishnan, Guan, & Raj (2011)	Malaysia	To explore mobile phone usage, extending work beyond teenage years to examine the role of mobile phones among urbanized youth, specifically university students	Mixed method (i.e. surveys and 24-hour diaries)	Uses and Gratification theory	417 participants aged 17 to 27 years	• Brand, trend and price tend to be the three most important purchasing factors among participants • Socializing and privacy emerged as the two most important reasons to use mobile phones. • Behavioral issues related to addiction and inappropriate use of mobile phones was also observed among the respondents. • Gender analysis revealed that females use their mobile phones more to socialize, gossip and as a safety device.
Kalogeraki & Papadaki (2010)	Greece	To explore the impact of mobile adoption on teenagers' social relationships	Quantitative method (i.e. surveys)	—	Students aged 12-18 years	The analysis highlights some important aspects of mobile communication to teenagers' social interactions and ties with peers and family.
Liu, Liu, & Wei (2014)	Tibet	To examine how teens use mobile phones to maintain their social networks and how their social network connections via the mobile phone in turn affect their psychological well-being	Quantitative method (i.e. surveys)	Uses and Gratification theory	Students aged 12–20 years	Teens actively use mobile phones as an important means of communication and a source of social support to help reduce loneliness.

In a study conducted in Botswana, Lesitaokana (2015) observed the cultures which result from interactive relationships between youths and their mobile phones and argued that some young people regard luxury smartphones as symbols of their identity. Several youth in Botswana are very keen on the visual value of a mobile phone. A number of youths use their mobile phones to express their personalities, while others use mobile phones to gratify their sense of identity among their friends. Parallel findings were stated by Balakrishnan et al., (2011) who surveyed a sample of 417 participants aged 17 to 27 years old in Malaysia. The results showed that certain characteristics such as mobile phone brand, trend, and price emerge as the most vital issues considered when buying a mobile phone. Owning a mobile phone functions as a way of shaping the youths' identity, and brands can serve as a personality factor for the owner. The findings of this study also show that females stress more on visual standards and usability of mobile phones compared to their male counterparts. These findings show that mobile phones enable youths to explore new ways of shaping their individual and collective identities through their phones.

Social Networks

The use of mobile social networks is a very effective tool and is frequently used as a way for socialization and social attachment by modern youths. Direct social communication is frequently substituted by communication through mobile phones which takes places in the cyber space and has an excessive reputation for experience and socialization of new generations. Thus, the second theme that emerged from the review of literature is mobile social networks, which are developed through social contact with parents and peers, trust, and bonding and bridging through social ties.

Social Interaction with Family and Friends

Researchers have considered the impact of mobile phones on social interaction (Aoki & Downes, 2003). Kamran (2010) who studied the forms of mobile phone texting and calling among college students in Pakistan found that improving social networks in youths depends on the convenience of social communication. Results revealed that students mainly communicate with members of their own age group, including peers, classmates and relatives. Similarly, another group study showed that Swedish youth social networks can be expanded through mobile phone usage (Thulin & Vilhelmson, 2007). Furthermore, this panel study indicated that short message services via mobile phones is the most important tool for social interaction, even at long distances (Thulin & Vilhelmson, 2007). Indeed, mobile phone functions such as email and text message seem to be ordinary technology that dynamically increases local interaction and the continuation of the trans-nationalization of daily social spaces.

Additionally, mobile phones totally change how people act within their social circles. In addition to an increased social life pace, mobile phones can ease personal interaction. Numerous scholars have found that a majority of people use mobile phones to connect to their kin and acquaintances. For example, the result of a study on 362 college students aged 20 and thereabouts in South Africa showed that a majority of participants socialize through their mobile phones to connect with parents and peers, with female respondents stressing more on the family aspect as opposed to their male acquaintances (North, Johnston, & Ophoff, 2014). Similarly, Lesitaokana (2015) observed that mobile phones are vital tools that allow the youth in Botswana chances to manage their lives and link and interact with their relatives and peers both locally and internationally. Mobile phones have also become the main device for social interaction between these young people. In another study conducted among urbanized Malaysian youths, social interaction with friends and family was discovered to be the key impetus for buying a mobile phone (Balakrishnan et al., 2011).

Trust

These widened networks of young mobile users rely on trust. For instance, the results of focus group interviews among female students in the USA report that through using mobile phones, their families are more likely to believe them and do not tend to call to control where they are (Chen & Katz, 2009). The respondents also agree that if there were no mobile phone linking, their fathers and mothers would not let them do many things, for example meet their friends in other towns or travel.

Bonding and Bridging through Social Ties

This subtheme refers to bonding through social ties, which describes the relationships among homogenous groups ranging from casual to close family bonds and friendships, and bridging through social ties, which describes the relations between distant friends, acquaintances, and social groups, as well as associations. These relations may in turn lead to others, and thus provide numerous links which lead to networks.

Hakoama and Hakoyama (2011) conducted a quantitative study on college students' cell phone use in the USA. Their study showed that students are inclined to utilize their cell phones to expand their social networks; for example, among the participants under 20 years old, almost two-thirds (61.5%) reported the reinforcement of current friendships, an increased family bond (20.7%) and an extended social network (17.8%). In this study, the students prefer to use their mobile phones to connect with their friends. Similarly, a study carried out by Boase and Kobayashi (2008) on Japanese students in Tokyo who profit from mobile phone email in forming social networks do not agree with the common opinion that mobile phones cause weaker ties and breaking of relationships; rather, they reported that this tool helps them to bridge to new social groups by allowing them to stay in close interaction with recently-met connections. The result was in contrast with Putnam's (2001) idea of declining levels of new relationship rates among youth due to new forms of media such as mobile phones. Haverila (2011) studied mobile phone use and features and discovered that respondents use cell phones for an array of reasons, like keeping in touch with parents and peers. In particular, the increasing profits of mobile phone companies are irrefutable and these findings emphasize the advantages of mobile phone use—from social networks, which foster strong social ties among peers and improved social connectedness to the strong bonds created in peer groups and particularly in close groups of acquaintances and friends.

Contact and Social Activity Patterns

Another theme to emerge was that of contact and social activity patterns. Young people use their mobiles as important tools for establishing their social actions. Oksman and Turtiainen (2004) consider mobile phone usage can help organize daily life and play a key part in communication between family and friends, thus creating a method of communication that is more socially grounded. By way of illustration, the results of a one panel study indicated that mobile phones affect the regular social practice of inner-city Swedish youths in frequent and interrelated ways (Thulin & Vilhelmson, 2007). The researchers found that spending time outside of the home can enhance social contacts and communication, and that respondents in this study often use their mobile phones for timetabling and organizing their social life and encounters within their circle of family and friends. The researchers also believe that by increasing mobile phone usage, other kinds of social contact and communication, including face-to-face interaction and Internet-based communication will be increased. In another study, Lesitaokana (2015) examined the use of mobile phones by young people in Botswana to access public and social services. The findings of this study revealed that mobile phones are properly used by the younger generation to manage aspects of their routines as part of their contemporary lifestyle and to re-form their social interactions with close relatives and friends as part of the Setswana custom that they have been raised in (Lesitaokana, 2015).

Furthermore, according to Mihailidis (2014), every function of mobile phone use—from communicating with inner social networks (family and friends) and interacting with local governments, to connecting with like-minded network societies—the mobile phone now has the capacity to enable widespread and dynamic collaboration, organization, and involvement. The results of a study from the USA, encompassing 793 students from eight universities on three continents, collectively representing 52 nationalities, demonstrate the role of the mobile phone as a tool for personal and public engagement. The participants in this study saw their mobile communication as contributing a distinct voice to active, engaged, and diverse narratives (Mihailidis, 2014). Certainly, the mobile phone seems to be an ordinary piece of technology that dynamically maintains local interaction and community, in addition to an unending trans-nationalization of regular social activities.

Romantic Relationships

Romantic involvements may have a role in the growth of a sense of self-socialization in two ways. Firstly, youths develop separate insights of themselves in the romantic age. They do not only have a concept of themselves with friends, but have diverse self-schemas of themselves with the overall peer group, with near friends, and in romantic relations (Connolly & Konarski, 1994). Thus, the fifth theme is the importance of mobile phones in romantic relationships. In general, mobile phones unite friends and encourage the potential for new associations. For instance, text messaging can establish the engagement necessary for starting a relationship: loving relationships among opposite sexes frequently begin through messaging. Lee et al., (2009) found that the majority of undergraduate college students use mobile phones to talk to each other as girl/boyfriends, thus supporting the finding that mobile phones are commonly used as a tool for building and maintaining new relationships. Similarly, another study showed that many young male and female Pakistani-British Muslims use mobile phones to form similar relationships (E. Green & Singleton, 2007). Blackman (2010) looked at mobile phone usage patterns by college freshmen with family, peers, and loving companions, and found support for the assumption that students largely used the call function to connect with their families and parents. However, no dissimilarity was found in the amount of calls and text messages with friends and romantic partners. In another study, participants employ mobile phones when forming new relationships; this is largely shown in young males trying to find a girlfriend, who begins by forming links with girls and sharing their thoughts with them through mobile phones, rather than via traditional ways of communication (Ali et al., 2014).

Furthermore, Economides and Grousopoulou (2008) examined the comparative use of mobiles at various places among university students. The findings of this study demonstrated that both genders spend most of their time on their mobile phones speaking with their boy/girlfriend. Specifically, men speak for about eleven minutes per day with their girlfriends, while females speak for fifteen minutes a day with their boyfriends. It can be observed that females tend to talk more on the phone than males at various places every day. In another study, which analyzed the results of a survey of 605 adolescents in different high schools (Abeele, 2014), it was revealed that the fifth most common use of a mobile phone was to express a participants' love to someone (M=3.45, *SD*=1.1). Obviously, this was found to be more important for adolescents who reported having a boy- or girlfriend (29.5%): for these people, expressing love was the second most important use of their mobile phones (M=4.07, *SD*=0.81). In another study, Vacaru et al. (2014) found that participants aged 13 to 18 use mobile phones as an instrument for the start of novel associations, both romantic and platonic: "Every so often, some persons just form relations over texts but they haven't got any contacts in person". On the other hand, the usefulness and viability of relationships which were initiated without past knowledge of the other person was brought into question on numerous events.

Attachment to Parents

The last theme to arise is that mobile phones are used for attachment to parents. Conventionally, entering college helps youths in their transition towards adulthood, as they become physically independent from their family and home to inhabit new settings with potentials for growth and development. However, mobile phones have changed communication between university students and families, allowing for instant connection and immediate response. Miller-Ott et al. (2014) found that young adult college students stress on their close relationships with their parents and frequent contact with them (mainly their mothers) via mobile phone. In similar findings, Blackman (2010) found that students contact their mothers more than their fathers via mobile phone. Lee d et al., (2009) investigated the dynamics of mobile phone use among 568 undergraduate university students and their feelings towards their parents in the USA, and addressed a similar concern. Results revealed that connected students show a more significant attachment to their mothers than the useful and efficient user groups which did not diverge in attachment scores for either fathers or mothers. Remarkably, female university students are meaningfully more connected to both parents/adult protectors than male students are.

In addition, the study conducted by Chen and. Katz (2009) in the USA showed that university students also use cell phones to share experiences with their parents and to keep in contact with their family (Chen & Katz, 2009). The results propose that college students using mobile phones have more regular contact with their family and are able to fulfil their family roles. University students also use mobile phones to receive expressive and physical support from their parents. The present study found that mobile phones have become like an "umbilical cord" (Geser, 2005; Spungin, 2006) between university students and family. Respondents in this study approve that the mobile phone is the most vital form of technology when connecting with parents. This is in line with Balakrishnan et al. (2011) who stated that female participants connect more with their families for expressive support.

Key Theories and Methods

Out of the 24 studies, six studies used communication theories, including feminist theories of technology, media dependency theory, and the Uses and Gratification theory. One study, which was carried out in the Northeast of England to explore diversity and technology use in the everyday lives of young Pakistani-British people had used the communication theory (E. Green & Singleton, 2007). This study synthesized a feminist theory of technology to enable an improved understanding of gender, identity and mobile phone use. Lohan (2001) claimed that the social influence of these links with previous work on communication technologies postulate the phone as a feminized and domesticated technology, undermining the old-style association of maleness with technologies, forming a valuable foundation for studies of gender, diversity and mobile phones. This reverberates with previous research which explored the manner in which different groups of young females and males embrace, negotiate, change and resist technologies and practices, and how these shape their personal and communal identities (E. Green & Singleton, 2007).

The Uses and Gratifications theory explains somewhat the way in which people use communication between other sources in their setting to satisfy their desires and attain their aims, and to do so by just asking them (E. Katz, Blumler, & Gurevitch, 1974, p. 21). This theory has been beneficial in reviewing users' mediated communication motivations (e.g., Dimmick, Sikand, & Patterson, 1994; Leung & Wei, 2000). According to the Uses and Gratifications theory, young people bend the media to their needs more readily than the media conquers them (E. Katz, Gurevitch, & Haas, 1973). It can also help to explain why people, especially youths choose certain technologies (e.g., mobile phones), how to employ it, and what are their feeling outcomes (Blumler, 1979). In four studies of phone usage in diverse cultural contexts, Liu et al. (2014) stated that the main gratifications of mobile phone use among Tibetan youths are similar—social networks—compared to youths from other cultures. The Uses and Gratification Theory has also been employed to investigate the incentives for use, buying factors, cell phone usage pattern and behavioral issues regarding mobile phone use by South African college students (North et al., 2014). The study found a general agreement between the students for the reasons of utilizing mobile phones, with the majority of them highlighting socializing. The parallel findings from urbanized Malaysian youths had resulted in socializing as the most common gratification for utilizing mobile phones (Balakrishnan & Raj, 2012). In another study, Abeele (2014) explored whether people can differentiate lifestyles within the mobile youth culture. In this study, they created a user typology of Flemish youth cell phone customers based on mobile phone gratifications. We again found support for the same gratifications-ranking structure, backing Apparatgeist's (J. E. Katz & Aakhus, 2002) claim that there is indeed a worldwide consideration of the (social) affordances of mobile communication tools.

Although media system dependency (MSD) theory is rooted in the tradition of the uses of the Uses and Gratification Theory, the theory is perhaps better aligned with the "media as system" research perspective championed by Ball-Rokeach and DeFleur (1986). The researchers conceived MSD theory as a by-product of the system's interaction with media, society, and individuals. In contrast with similar theories (e.g. Uses and Gratification Theory; Modeling Theory) that show the origins and impacts between audiences and media, it clarifies the association among mass media users and the community as a whole. It considers how media use and dependency influences behavioral intentions. This theory shows that people rely heavily on media to get information because the media serves them better than other sources. Another issue of dependency is social stability. When a society is facing a crisis, the media's impact on people becomes greater (Ball-Rokeach, 1998; Ball-Rokeach & DeFleur, 1976). In two other studies, the MD theory (Ball-Rokeach and DeFleur, 1976) provides the theoretical framework to explore whether American college students' mobile phone use furthers their dependency on their family while they are supposed to learn to be free in university, and to what degree this affects their college life (Chen & Katz, 2009). Similarly, media dependency theory was also used to demonstrate how Taiwanese college students depend on mobile phones to support from family and friends (Chen, 2006).

There have been a number of important theoretical contributions to the understanding of young people's usage of mobile phones in order to create a sense of belonging to their friends, thus influencing their insights of mobile phones as a socializing tool. One early contribution was made by Erikson (1968).

Erikson (1968), in his psychosocial theory, explains youth as a stage of seeking for identity. Erik Erikson highlighted the fact that the influence of a peer group increases adolescent identity development, while a previous study confirmed that young people are more willing to adopt new technological devices as they become readily available (Ling, 2001). In this case, the survey was conducted in a mid-size, Midwestern university, and applied Erikson's (1968) theory to assess how important it is for university students to own a mobile phone (Hakoama & Hakoyama, 2011).

Another previous study in this area used the traditional attachment theory as a theoretical framework. Lee et al. (2009) argued that this theory has inspired more study in the context of family in an attempt to gain greater understanding of the effect of modern tools such as mobile phones on family relationships and human growth. Numerous scholars in the area of adolescent and youth development deemed that the idea of a growth of identity shares similar features with conventional attachment theory, as defined by Bowlby (1980). In contrast to Freud's theory of development of personality, Bowlby's theory makes no formal attempt to define the procedures through which a child-parent connection might affect the appearance of pro-social behavior patterns. This theory offers a rich source for supposition about socialization and attachment. In a previous study, Lee et al. (2009) applied the attachment theory in investigating the dynamics of mobile phone use among 568 undergraduate college students and the sense of attachment to the participants' fathers and mothers. The findings of the study showed sophisticated research on mobile phone usage and parental elements by using the attachment theory to study the relationship between attachment to parents and mobile phone thoughts and doings.

The theory of socio-emotional selectivity proposes that mobile phones and e-mail might also be used to link to new ties and interrupt with old ties, sequentially allowing youths to access new collections of knowledge (Carstensen, Isaacowitzm, & Charles, 1999). This theory is coupled with the findings of a review study in Japan, which reveals that mobile phone e-mail is employed both to connect and associate, but not to break with bonds. We also discover that the strength with which Japanese youths use mobile phone e-mail is more often a result of bridging than bonding (Boase & Kobayashi, 2008). From the theoretical viewpoint delivered by symbolic interactionism, Oksman and Turtiainen (2004) analysed the meanings and contexts of mobile communication culture among 168 Finnish teenagers in urban and rural areas. The investigators detected features of the modern technology-mediated communiqué from the viewpoint of symbolic interactionism. The theoretical basis delivered by symbolic interactionism starts out from the supposition that culture and community are based on social interaction where people actively construct their everyday reality. The interaction is symbolic: it is rooted in signs and meanings, and social interaction produces shared meanings. Joint action needs the skill to think about ourselves as we do others. People fit lines of action jointly by first imagining how those with whom we are interacting might perceive us and then adjusting our behavior appropriately (Blumer, 1986). Thus, the symbolic interactionism relies on a main supposition; the persons' actions towards things are based on the senses that those things have for them (Blumer, cited in Denzin, 1992). Significantly, Blumer (1969) said that the task of media was to accurately reflect the empirical world. He also proclaimed a causal relationship between media and social behavior, but never developed a method of analysis that could pursue the relationship. In the aspect of this theory, the mobile communication of teenagers is symbolized by Goffman (1990) as a term of 'social stage', "what appearances of self-do youth produce in message and contact positions, and what type of a structure does communication via the mobile phone and other novel media establish?" (Oksman & Turtiainen, 2004, p. 323)

For research purposes, it is worth noting one methodological issue here. On the occasion of finding potentially problematic data, the traditional tendency of most research is to find weakness with the (methods of) observation, not with the theory itself. In addition, there is a multitude of methods that might be applied to accomplish such analytical work. These selected studies have highlighted an appropriate methodology comprising the qualitative, quantitative, as well as mixed methods. Of the twenty-four studies, eleven of them directly used the qualitative method, twelve used the quantitative method, and one used the mixed method (see Table 1), which can define traditional research methods (Hinton, 2013). Through employing a qualitative approach, the researchers can comprehend the experiences of the persons who are involved in the behavior of mobile usage (Walsh et al., 2008); by contrast, quantitative revisions provide useful broad-scale information regarding mobile user behavior.

Key Findings and Achievements

This chapter has yielded some noteworthy findings which reflect youth mobile users that are at a critical point in their socialization. Numerous understandings of the findings emphasize the role of mobile phones in making it much simpler to socialize, interact with parents, and see friends more frequently; it strengthens social contacts and bonding with friends and peers and allows youths to meet freely in everyday life. For instance, the key findings of earlier studies indicate that the main purpose of South African university students usage of mobile phones is for socializing (North et al., 2014). This is consistent with the study results of Balakrishnan and Raj (2012). The results of another study on 166 Taiwanese university students showed that heavy mobile phone users reported better relationships with their peers and family (Chen, 2006). Particularly, these findings have revealed that the mobile phone is proper for use by youths to manage aspects of their daily routines as part of their modern way of life and to reconstruct their social interactions with family and next of kin.

Additionally, some findings showed that identity development is the greatest predictor of both frequencies of use and mobile phone involvement. Similarly, some results adopted a critical approach which emphasizes the connections of features of identity. Green and Singleton (2007) argue that analysis of the difference is vital to achieving a broader, more diverse understanding of our mobile selves: that is how dissimilar groups of young females and males embrace, negotiate, adopt and oppose technologies and practices and how these shape their personal and shared identities. In the meantime, some results revealed that the ability to communicate freely with peers using mobile phones is an imperative part of identity development. Luxury mobile phones are also used as symbols of personal identity and modernity (Boase & Kobayashi, 2008; Lesitaokana, 2015).

Moreover, some of the findings help to further understand the influence of mobile phone communication on family relationships and paternal and maternal attachment (Haverila, 2011; Lee et al., 2009; Miller-Ott et al., 2014). This finding supports previous research which emphasizes the importance of parental contact via voice calls (Blackman, 2010). Another result proposes that university students use mobile phones to have more frequent interaction with their family and to fulfill family roles (Chen & Katz, 2009). Further, another study found the effects of mobile phones on Greek teenagers' social interactions and bonding with family and friends (Kalogeraki & Papadaki, 2010).

Moreover, the results of another study showed that youths' interactions with their social settings are boosted as mobile phones promote a flexible regime of direct discussion and endless updates (Thulin & Vilhelmson, 2007). The total conclusion of these findings is that young people's total interaction with their social environment may increase because of the usage of mobile phones. These results are valuable for raising sympathy about how youths use mobile phone technology to communicate with main contact groups.

Key Weaknesses and Problems

Many youths have mobile phones so they can keep in touch with friends through email, text messages and social media. Mobile phones have made it possible to connect with peers frequently, but it comes with disadvantages. Key findings from these studies challenge prior studies about the negative social effects of mobile phones. Unfortunately, mobile phone usage has developed into an addiction for many people. It totally obstructs an active and productive social life. People's lives become far less social and their interactions are far less genuine. This directs to lots of troubles in life regardless of age. Mobile phone usage has definitely become an addiction that impedes an active social life. This applies especially to youths; they are so addicted to their mobile phones and texting that they hardly interact face to face with each other. Youths spend all day and night texting instead of going out to the movies or the mall with friends. Nowadays mobile phones are perceived as significant in preserving social ties and conducting the more ordinary needs of daily life (Junco & Cole-Avent, 2008; Junco & Cotten, 2011).

A lot of young adults today cannot predict an existence without mobile phones. Chen (2006) found that Taiwanese college students have similar feelings when they do not have access to their mobile phones. This is consistent with the work of Walsh et al. (2010), in which respondents displayed feelings of frustration, anger, and worry when they were unable to employ their mobile phones, showing a sense of loss without their phones. The sign for addictive behavior can also be traced when most of the participants acknowledged to never turning off their mobile phones. Findings suggest that mobile phone use has become such an important part of student life that it is "invisible" and students do not necessarily realize their level of dependency on and/or addiction to their cell phones. Additionally, Vacaru et al. (2014) also stated several themes to exploring the problematic mobile phone use among New Zealand youths. For the purpose of their study, four key themes were analyzed; the practicability of using mobile phones, socialization through mobile phones, the damage produced by use of mobile phone technology, and the growing habit to having mobile phones. The researchers found that participants drew a difference between attachment and addiction, which has not been identified in previous studies. Respondents in this study viewed attachment to one's phone as a prevalent behavior, and one which is possibly helpful for the occurrence of challenges such as serious social vulnerabilities.

However, some studies reaffirmed the influence of mobile phone communication on social activity patterns of youths. Mobile phones can reduce the importance of social ties during the socialization process. Sociologists McPherson, Smith-Lovin and Brashears (2009) stated that the kind of social bonds reinforced by mobile phones are quite fragile and geographically dispersed, which is not the robust, always locally-based connections that tend to be a part of peoples' core discussion network. They showed the growth in mobile phone use as one of the chief trends that pulls people away from old-style social contexts, neighborhoods, volunteering, and social spheres that have been linked to big and varied core networks. In Botswana, the findings described the two visible inappropriate uses of mobile phones among youths comprising of obtrusive texting and social networking in household settings, and making and receiving voice calls while attending sacred gatherings such as funerals and family meetings (Lesitaokana, 2015). In addition, texting was also found to be a disturbing and addictive activity among young people in Botswana. This is because texting disrupts the propriety of founded settings, such as in the household and at special gatherings, when a user is in the company of other people, in many different ways.

As a final point, it is clear that a mobile phone is planned to keep social relations at all times. By calling or text messaging one's friends, one is able to never feel alone. As we all see, for youths, mobile phones are part and parcel of their newly-discovered private life, a space in which adults are not authorized without permission. This fact of modern life has greatly changed the family cell and shaken the rigid ideas that many parents may have had about parenting and parental concerns! Mobile phones have lessened the status of the family during the socialization process. Chen and Katz (2009) make an especially salient point about mobile phones and family interaction with the observation that "the mobile phone was employed by youth to negotiate the borders between their early years and later life with their families" (142). This study revealed that respondents who use mobile phones a great deal or display addictive use believed that mobile phones boost their social networks, but simultaneously weakens their relationships with their parents (Chen, 2006). In Pakistan, a research showed that youths use mobile phones to communicate within their age group in both constructive and destructive modes (Kamran, 2010). Moreover, Ling and Yttri (2006) argued that mobile phones have provided major levels of privacy and independence to young people by undermining the control forced by parents and other conventional means of socialization. Vacaru et al. (2014) found that participants in New Zealand aged 13 to 18 describe mobile phones as a source of personal socialization—by communicating through SMS, users have the skills to hold private conversations while in a public forum. Mobile phones also allow for quasi -emancipation by facilitating a way of communication which cannot be checked by families, either for content or for the timing of the communication. In addition, we can determine two significant effects of mobile phones on one's community relationship. While mobile phones improve relationships with peers and parents, it can also have a rather negative impact upon what was once called the family.

MAPPING THE FUTURE

New Concepts and Themes

This chapter broadly reveals several emerging concepts and themes related to the role of mobile phones in contributing to youth socialization. The pattern of evidence suggests that mobile phones have particular constructive influences on youth socializations, which include identity developments, family and peer social interactions, social activity patterns, and romantic relationships. In addition, other new themes common for youth are the issues of freedom and friendship. One can similarly observe the distinct needs of youth for autonomy play out through the ownership and use of mobile phones. Mobile phones also provide youths with convenient and easily identified signs of emancipation.

New Research Areas

In our interpretation, many research areas within this field hold the potential for productive ways for future studies. While a large quantity of findings on socializing impacts of the mobile phones have been conducted on young populations, a gap remains in understanding how other age groups such as adults use the mobile phone as a tool for maintaining socialization in relation to their friends and family. Generally, there is a requirement for more research on youths' desire to connect with their families via mobile phones. Further research is also needed on the influence of new social networking websites on youth socialization. Finally, much research is still required so as to achieve cross-natural assessment (Turel & Serenko, 2006).

New Theories and Methods

As we have seen, youths live in an increasingly complex new digital technology culture, and one of the key challenges for scholars is to select the suitable theoretical and methodological approaches. In this chapter, numerous study results help support communication and non-communication theories in this field. Communication theorists could take the mobile phone into contemplation when they search to determine how much mobile phone users depend on their mobile phones to get information as well as support to fulfil their needs in various contexts. Despite these theories, there is a need to discuss the role of mobile phones in the socialization process of youth to support the theory of primary socialization which can identify new technology such as mobile phones as secondary or primary factors in the socialization process.

Additionally, this chapter gives a literature review of methodologies used to study the impacts of mobile phone usage. Some of these methods comprise participant observation, mobile memoir, focus group interviews, assessment, and in-depth interviews. Though many methods have been planned particularly to study mobile phones, a multi-method approach can boost studies by offering verifying information to respondent self-reporting or to add contextual information to mobile phone data. Nevertheless, new growths in mobile and information technology deliver novel ways of leading research with youths. The universal use of mobile phones by youths lets scholars interact with participants who are neither actually present nor spatially fixed; so the data can be described as youths in their everyday lives (Hinton, 2013).

CONCLUSION

This is the comprehensive revision of past studies on mobile phones and its effect on youth socialization. This study identified 24 studies to define the extent to which mobile phones have entered youths' everyday lives, particularly the influence it has on socialization process. The literature review is presented in five parts:

Section 1: *Key themes* considers the multiple facets of mobile phone use, which participants in each study had noted as being useful for their own socialization (i.e., self-socialization, social networks, contact and social activity patterns, romantic relationships, and attachment to parents);

Section 2: *Key theories* and *methods* examines existing theories such as attachment theory, psychosocial theory, symbolic interactionism, and media dependency theory and methods such as qualitative, quantitative, and mixed methods — each providing a complementary basis for rationalizing the relationship between mobile phones and sociality as pertaining to youths;

Section 3: *Key findings* and *achievements* explores the literature relevant to considering the challenges to identify major concentrations of each study and to assess the impacts of mobile phone use, and from that, focuses on the interrelationships between mobile phones and young users;

Section 4: *Key weaknesses and problems* investigates emerging differences among young people in their judgments about the harmful social impact of mobile phones;

Section 5: We identify key questions relate to new themes, new research areas, and new theories and methods for future research.

In view of the above summary, it is apparent that mobile phones play a major role in youth social-ization. Results from these studies restate the want to use of mobile phones to build social identity and connection among youths. Even though the findings did not test directionality, significant empirical evidence illustrated that the mobile phone – just like mobile email and text messaging – looks to be an ordinary technology that dynamically reinforces social networks. These studies discovered that texting is increasingly becoming a key means of sustaining everyday social interactions and ties with peers and family. Some studies established that youths use mobile phone e-mail to bond with their strong ties. They also found that youths use mobile phone e-mail to bridge with ties, and that this bridging activity better accounts for the strength with which they use mobile phone email than does bonding action. Some studies found youths' texting patterns in their contact and social activities.

In addition, the study indicates that opposite to early predictions that mobile phones would direct to freedom of youth from their fathers and mothers, it looks to have had the contradictory outcome, namely it appears to enhance the connection and socialization of youth with their family unit. Some studies find that several respondents also realize the use of mobile phones to preserve the youth-parent relationship from isolated spaces and far distances. These studies reported that youths build their peer networks and develop their independence from parents' control via the mobile phone. These studies found that youths desire to connect with parents more than parents desire to supervise them. On the other hand, findings from some studies uncovered that mobile phones have developed to be an essential element of youths' everyday living, with some of them representing signs of addictive behavior, thus reducing the impor-tance of social ties and family during the socialization process.

Overall, this review significantly demonstrates that mobile phones are influential socializing agents that can lead to many constructive and destructive consequences. The conclusions drawn from this review add to the writings on mobile phones and youth socializations in different nations. This chapter establishes that in order for mobile phones to be able support positive outcomes on youth socialization, it is vital that they consider the role of youths mentioned in different approaches. Its effects can gener-ate helpful interventions that promote healthier mobile phone usage and decrease the negative effect of mobile phones on youth socialization. Findings concerning the socializing influences of mobile phones have implications for theory development, public policy decisions and interferences that can encourage healthier mobile phones usage among youths.

REFERENCES

Abeele, M. M. P. V. (2014). Mobile lifestyles: Conceptualizing heterogeneity in mobile youth culture. *New Media & Society*, 1–19.

Ali, S., Rizvi, S. A. A., & Qureshi, M. S. (2014). Cell phone mania and Pakistani youth: Exploring the cell phone usage patterns among teenagers of South Punjab. *FWU Journal of Social Sciences*, 8(2), 43.

Aoki, K., & Downes, E. J. (2003). An analysis of young people's use of and attitudes toward cell phones. *Telematics and Informatics*, 20(4), 349–364. doi:10.1016/S0736-5853(03)00018-2

Balakrishnan, V., Guan, S. F., & Raj, R. G. (2011). A one-mode-for-all predictor for text messaging. *Maejo International Journal of Science and Technology*, 5(2), 266–278.

Balakrishnan, V., & Raj, R. G. (2012). Exploring the relationship between urbanized Malaysian youth and their mobile phones: A quantitative approach. *Telematics and Informatics*, *29*(3), 263–272. doi:10.1016/j.tele.2011.11.001

Ball-Rokeach, S. J. (1998). A theory of media power and a theory of media use: Different stories, questions, and ways of thinking. *Mass Communication & Society*, *1*(1-2), 5–40. doi:10.1080/15205436.1998.9676398

Ball-Rokeach, S. J., & DeFleur, M. (1986). The interdependence of the media and other social systems. In G. Gumpert & R. Cathcart (Eds.), *Inter/media: Interpersonal communication in a media world* (pp. 81–96). Oxford, UK: Oxford University Press.

Ball-Rokeach, S. J., & DeFleur, M. L. (1976). A dependency model of mass-media effects. *Communication Research*, *3*(1), 3–21. doi:10.1177/009365027600300101

Blackman, S. L. (2010). *Cell phone usage patterns with friends, parents, and romantic partners in college freshmen*. Retrieved from http://trace.tennessee.edu/cgi/viewcontent.cgi?article=2380&context=utk_chanhonoproj

Blumer, H. (1969). *Symbolic interactionism: Perspective and method*. Berkeley, CA: University of California Press.

Blumer, H. (1986). *Symbolic interaction: Perspective and method*. Berkeley, CA: University of California Press.

Blumler, J. (1979). The role of theory in uses and gratifications studies. *Communication Research*, *6*(1), 9–36. doi:10.1177/009365027900600102

Boase, J., & Kobayashi, T. (2008). Kei-Tying teens: Using mobile phone e-mail to bond, bridge, and break with social ties – a study of Japanese adolescents. *International Journal of Human-Computer Studies*, *66*(12), 930–943. doi:10.1016/j.ijhcs.2008.07.004

Bowlby, J. (1980). Attachment and loss, volume 3: loss; sadness and depression. New York, NY: Basic Books.

Brown, R. I. F. (1993). Some contributions of the study of gambling to the study of other addictions. In W. R. Eadington & J. A. Cornelius (Eds.), Gambling behavior and problem gambling (pp. 241–272). Reno: Institute for the study of gambling and commercial gaming, University of Nevada.

Brown, R. I. F. (1997). A theoretical model of the behavioural addictions - Applied to offending. In J. E. Hodge, M. McMurran, & C. R. Hollin (Eds.), *Addicted to crime*. Chichester, UK: John Wiley.

Carstensen, L. L., Isaacowitzm, D. M., & Charles, S. T. (1999). Taking Time Seriously: A Theory of Socioemotional Selectivity. *The American Psychologist*, *54*(3), 165–181. doi:10.1037/0003-066X.54.3.165 PMID:10199217

Chen, Y.-F. (2006). Social Phenomena of Mobile Phone Use: An Exploratory Study in Taiwanese College Students. *Journal of Cyber Culture and Information Society*, *11*, 219–244.

Chen, Y.-F., & Katz, J. E. (2009). Extending family to school life: College students' use of the mobile phone. *International Journal of Human-Computer Studies*, *67*(2), 179–191. doi:10.1016/j.ijhcs.2008.09.002

Connolly, J. A., & Konarski, R. (1994). Peer self-concept in adolescence: Analysis of factor structure and of associations with peer experience. *Journal of Research on Adolescence, 4*(3), 385–403. doi:10.1207/s15327795jra0403_3

Denzin, N. (1992). Whose Cornerville is it, anyway. *Journal of Contemporary Ethnography, 21*(1), 120–132. doi:10.1177/0891241692021001007

Dimmick, J. W., Sikand, J., & Patterson, S. J. (1994). The gratifications of the household telephone sociability, instrumentality, and reassurance. *Communication Research, 21*(5), 643–663. doi:10.1177/009365094021005005

Dittmar, H., Long, K., & Bond, R. (2007). When a better self is only a button click away: Associations between materialistic values, emotional and identity-related buying motives, and compulsive buying tendency online. *Journal of Social and Clinical Psychology, 26*(3), 334–361. doi:10.1521/jscp.2007.26.3.334

Economides, A. A., & Grousopoulou, A. (2008). Use of mobile phones by male and female Greek students. *International Journal of Mobile Communications, 6*(6), 729–749. doi:10.1504/IJMC.2008.019822

Erikson, E. (1968). *Identity: Youth and Crisis.* New York, NY: Norton.

Fortunati, L., & Manganelli, A. M. (2002). Young people and the mobile telephone. *Revista de Estudios de Juventud, 52,* 59–78.

Goffman, E. (1990). Stigma: Notes on the Management of Spoiled Identity. Harmondsworth.

Green, E., & Singleton, C. (2007). Mobile Selves: Gender, ethnicity and mobile phones in the everyday lives of young Pakistani-British women and men. *Information Communication and Society, 10*(4), 506–526. doi:10.1080/13691180701560036

Green, N. (2003). Outwardly Mobile: Young People and Mobile Technologies. In J. E. Katz (Ed.), *Machines That Become Us: The Social Context of Personal Communication Technology* (pp. 201–218). New Brunswick, NJ: Transaction Publishers.

Haddon, L. (2004). *Information and communication technologies in everyday life: A concise introduction and research guide.* New York, NY: Berg.

Hakoama, M., & Hakoyama, S. (2011). The impact of cell phone use on social networking and development among college students. *The American Association of Behavioral and Social Sciences Journal, 15*(1), 20.

Harper, R., & Hamill, L. (2005). Kids will be kids: The role of mobiles in teenage life. In *Mobile World* (pp. 61–74). Berlin, Germany: Springer. doi:10.1007/1-84628-204-7_4

Haverila, M. (2011). Mobile phone feature preferences, customer satisfaction and repurchase intent among male users. *Australasian Marketing Journal, 19*(4), 238–246. doi:10.1016/j.ausmj.2011.05.009

Hinton, D. (2013). Private Conversations and Public Audiences: Exploring the Ethical Implications of Using Mobile Telephones to Research Young People's Lives. *Young, 21*(3), 237–251. doi:10.1177/1103308813488813

Igarashi, T., Motoyoshi, T., Takai, J., & Yoshida, T. (2008). No mobile, no life: Self-perception and text-message dependency among Japanese high school students. *Computers in Human Behavior*, *24*(5), 2311–2324. doi:10.1016/j.chb.2007.12.001

Junco, R., & Cole-Avent, G. A. (2008). An introduction to technologies commonly used by college students. *New Directions for Student Services*, *2008*(124), 3–17. doi:10.1002/ss.292

Junco, R., & Cotten, S. R. (2011). Perceived academic effects of instant messaging use. *Computers & Education*, *56*(2), 370–378. doi:10.1016/j.compedu.2010.08.020

Kalogeraki, S., & Papadaki, M. (2010). The impact of mobile use on teenagers' socialization. *International Journal of Interdisciplinary Social Sciences*, *5*(4), 121–134.

Kamran, S. (2010). Mobile phone: Calling and texting patterns of college students in Pakistan. *International Journal of Business and Management*, *5*(4), 26. doi:10.5539/ijbm.v5n4p26

Katz, E., Blumler, J. G., & Gurevitch, M. (1974). Utilization of mass communication by the individual. In J. G. Blumer & E. Katz (Eds.), The Uses of Mass Communication. Newbury Park, CA: Sage.

Katz, E., Gurevitch, M., & Haas, H. (1973). On the use of mass media for important things. *American Sociological Review*, *38*(2), 164–181. doi:10.2307/2094393

Katz, J. E., & Aakhus, M. (2002). *Perpetual contact: Mobile communication, private talk, public performance*. Cambridge, UK: Cambridge University Press. doi:10.1017/CBO9780511489471

Lee, S., Meszaros, P. S., & Colvin, J. (2009). Cutting the wireless cord: College student cell phone use and attachment to parents. *Marriage & Family Review*, *45*(6-8), 717–739. doi:10.1080/01494920903224277

Lesitaokana, W. (2015). Young People and Mobile Phone Technology in Botswana. In J. Wyn & H. Cahill (Eds.), *Handbook of Children and Youth Studies* (pp. 801–813). Springer Singapore. doi:10.1007/978-981-4451-15-4_30

Leung, L., & Wei, R. (2000). More Than Just Talk on the Move: A Use-and-Gratification Study of the Cellular Phone. *Journalism & Mass Communication Quarterly*, *77*(2), 308–320. doi:10.1177/107769900007700206

Ling, R. (2004). *The mobile connection: The cell phone's impact on society*. New York, NY: Morgan Kaufmann.

Ling, R., & Yttri, B. (2006). Control, emancipation, and status: The mobile telephone in teens' parental and peer relationship. In R. Kraut, M. Brynin, & S. Kiesler (Eds.), *Computer, phones, and the Internet: Domesticating information technology* (pp. 219–234). New York, NY: Oxford University Press.

Liu, X., Liu, X., & Wei, R. (2014). Maintaining social connectedness in a fast-changing world: Examining the effects of mobile phone uses on loneliness among teens in Tibet. *Mobile Media & Communication*, *2*(3), 318–334. doi:10.1177/2050157914535390

Lobet-Maris, C., & Henin, J. (2002). Talking without communicating or communicating without talking: From the GSM to the SMS. *Estudios de Juventud*, *57*, 101–114.

Lohan, M. (2001). Men, Masculinities and „Mundane" Technologies: The Domestic Telephone. In E. Green & A. Adam (Eds.), *Virtual Gender: Technology, Consumption and Identity* (pp. 189–205). London, Britain: Routledge.

Mathews, R. (2004). The psychosocial aspects of mobile phone use amongst adolescents. *InPsych*, *26*(6), 16–19.

McPherson, M., Smith-Lovin, L., & Brashears, M. E. (2009). Models and marginals: Using survey evidence to study social networks. *American Sociological Review*, *74*(4), 670–681. doi:10.1177/000312240907400409

Mihailidis, P. (2014). A tethered generation: Exploring the role of mobile phones in the daily life of young people. *Mobile Media & Communication*, *2*(1), 58–72. doi:10.1177/2050157913505558

Miller-Ott, A. E., Kelly, L., & Duran, R. L. (2014). Cell phone usage expectations, closeness, and relationship satisfaction between parents and their emerging adults in college. *Emerging Adulthood*, *2*(4), 313–323. doi:10.1177/2167696814550195

New Zealand Herald. (2012). *More mobile phones in NZ than people: Study*. Retrieved from http://www.nzherald.co.nz/nz/news/article.cfm?c_id=1&objectid=10801183

North, D., Johnston, K., & Ophoff, J. (2014). The Use of Mobile Phones by South African University Students. *Issues in Informing Science and Information Technology, 11*. Retrieved from http://iisit.org/Vol11/IISITv11p115-138North0469.pdf

Oksman, V., & Rautiainen, P. (2002). I've got my whole life in my hand. *Revista de Estudios de Juventud, 52*, 25–32.

Oksman, V., & Turtiainen, J. (2004). Mobile communication as a social stage meanings of mobile communication in everyday life among teenagers in Finland. *New Media & Society*, *6*(3), 319–339. doi:10.1177/1461444804042518

Palen, L., Salzman, M., & Youngs, E. (2000). Going wireless: behavior & practice of new mobile phone users. In *Proceedings of the 2000 ACM conference on Computer supported cooperative work* (pp. 201–210). Philadelphia, PA: ACM. doi:10.1145/358916.358991

Pew Internet & American Life Project. (2010). *Teens and Mobile Phones: Text messaging explodes as teens embrace it as the centerpiece of their communication strategies with friends* (A project of the Pew Research Center). Retrieved from http://pewinternet.org/Reports/2010/Teens-and-Mobile-Phones.aspx

Putnam, R. D. (2001). *Bowling alone: The collapse and revival of American community*. New York, NY: Simon & Schuster.

Rheingold, H. (2002). *Smart mobs: the next social revolution*. Cambridge, MA: Perseus Publishing.

Smetana, J. G., Campione-Barr, N., & Metzger, A. (2006). Adolescent development in interpersonal and societal contexts. *Annual Review of Psychology*, *57*(1), 255–284. doi:10.1146/annurev.psych.57.102904.190124 PMID:16318596

Taylor, A., & Harper, R. (2003). The gift of the gab?: A design oriented sociology of young people's use of mobiles. *Computer Supported Cooperative Work*, *12*(3), 267–296. doi:10.1023/A:1025091532662

Thulin, E., & Vilhelmson, B. (2007). Mobiles everywhere: Youth, the mobile phone, and changes in everyday practice. *Young*, *15*(3), 235–253. doi:10.1177/110330880701500302

Turel, O., & Serenko, A. (2006). Satisfaction with mobile services in Canada: An empirical investigation. *Telecommunications Policy, 30*(5), 314–331. doi:10.1016/j.telpol.2005.10.003

Vacaru, M. A., Shepherd, R. M., & Sheridan, J. (2014). New Zealand Youth and Their Relationships with Mobile Phone Technology. *International Journal of Mental Health and Addiction, 12*(5), 572–584. doi:10.1007/s11469-014-9488-z

Walsh, S. P., White, K. M., Cox, S., & Young, R. M. (2011). Keeping in constant touch: The predictors of young Australians' mobile phone involvement. *Computers in Human Behavior, 27*(1), 333–342. doi:10.1016/j.chb.2010.08.011

Walsh, S. P., White, K. M., Mc, D., & Young, R. (2010). Needing to connect: The effect of self and others on young people's involvement with their mobile phones. *Australian Journal of Psychology, 62*(4), 194–203. doi:10.1080/00049530903567229

Walsh, S. P., White, K. M., & Young, R. M. (2008). Over-connected? A qualitative exploration of the relationship between Australian youth and their mobile phones. *Journal of Adolescence, 31*(1), 77–92. doi:10.1016/j.adolescence.2007.04.004 PMID:17560644

Wei, R., & Lo, V.-H. (2006). Staying connected while on the move: Cell phone use and social connectedness. *New Media & Society, 8*(1), 53–72. doi:10.1177/1461444806059870

ADDITIONAL READING

Bianchi, A., & Phillips, J. P. (2005). Psychological predictors of problem mobile phone use. *Cyberpsychology & Behavior, 8*(1), 39–51. doi:10.1089/cpb.2005.8.39 PMID:15738692

Jarrett, R. L. (1995). Growing up poor: The family experiences of socially mobile youth in low-income African American neighborhoods. *Journal of Adolescent Research, 10*(1), 111–135. doi:10.1177/0743554895101007

Kreutzer, T. (2009). Generation mobile: online and digital media usage on mobile phones among low-income urban youth in South Africa. *Retrieved on March, 30*(1), 2009.

McEwen, R., & Fritz, M. (2011). EMF social policy and youth mobile phone practices in Canada. *Mobile Communication*, 133–156.

McLeod, J. M. (2000). Media and Civic Socialization of Youth. *The Journal of Adolescent Health, 27*(2), 45–51. doi:10.1016/S1054-139X(00)00131-2 PMID:10904206

Okada, T. (2005). Youth culture and the shaping of Japanese mobile media: Personalization and the keitai Internet as multimedia. *Personal, Portable, Pedestrian: Mobile Phones in Japanese Life*, 41–60.

Whittaker, R., Maddison, R., McRobbie, H., Bullen, C., Denny, S., Dorey, E., & Rodgers, A. et al. (2008). A multimedia mobile phone–based youth smoking cessation intervention: Findings from content development and piloting studies. *Journal of Medical Internet Research, 10*(5), e49. doi:10.2196/jmir.1007 PMID:19033148

Chapter 19
Left–Behind Children and Mobile:
A Critical Discourse Analysis

Janice Hua Xu
Holy Family University, USA

ABSTRACT

Through critical analysis of selected news stories from sina.com from 2010 to 2015 about "left-behind children" in China, the chapter examines media discourse on relationships between migrant families and communication technology. The author finds that the role of cell phones in their lives are portrayed in the following narratives: 1) Cell phones are highly valuable for connecting family members living apart; 2) Cell phones are used as a problem-solver in charity giving and rural development projects; 3) Cell phones can bring unexpected risks; 4) Cell phones could harbor or unleash evil—associated with increasing cases of crimes victimizing left-behind children and juvenile delinquency. The author discusses how institutional goals of social agencies, corporations, educators and law enforcement contribute to the polarity of cell-phone-related discourses, which reflect the societal anxieties over unsupervised access to technology by adolescents, as well as the cultural and political implications of empowering the "have-nots" of digital divide.

INTRODUCTION

Left-behind children refer to rural children under 18 who are left at home when both or one of their parents migrate to urban area for work. Across China, as a result of the rapid economic growth and gap between labor demands in urban and rural areas, more than 61 million children - nearly a quarter of children population in China - live in rural villages without the presence of their parents, who have migrated in search of work to provide a better life for their families. Recent findings showed that left-behind children were disadvantaged by developmental, emotional and social problems, while their parents, caught in what the media describes as "tide of migration," strive to empower the family financially (Su, Li, Lin, Xu, Zhu, 2013). Researchers found that due to a lack of family protection and educational op-

DOI: 10.4018/978-1-5225-0469-6.ch019

portunities, there have been growing signs of serious mental health problems and an increased criminal record among this vulnerable group (CCRCSR, 2014). Because migrant workers rarely get to spend time with their children, children often feel lonely and helpless, and sometimes have the fear of being abandoned. They are more prone to skipping class, fighting and even dropping out of school, as their caretakers are often unable or unwilling to monitor their study habits. It is also found that left-behind children had lower scores in health behavior and school engagement than rural children of non-migrant worker parents (Wen & Lin, 2012).

This chapter aims to situate the question of children and technology in the context of China's rural-to-urban labor migration, to address how popular discourses address the role of mobile phone technology in the lives of underprivileged children and youth in China. Using the theoretical framework of critical media studies, the paper examines how different narratives are weaved about their use of cell phones, and how the stories make sense of their unique family experiences with parents working far away, who might sometimes be reached through cell phones. With an intention to align the project with existing literature published in the field of communication about digital divide, migrant families, and mobile communication devices, the paper discusses how the popular media addresses the notions of class/inequality, family values and social network in China's trends of large-scale migration and urbanization through the reports on personal communication technologies. As Clifford Christians argues in his article titled "A theory of normative technology" (1989:124), technology is "the distinct cultural activity in which human beings form and transform neutral reality for practical ends with the aid of tools and procedures." The technological process—design, fabrication, and use—is usually value-laden, and is related to the definitions of problems and solutions by human beings in a specific cultural and social setting.

BACKGROUND

Left-Behind Children in Media

Although the concept of left-behind children in China already existed in the mid-1990s, it took almost a decade for the topic to receive wide attention in media and academia, as government agencies started to react to the special needs of this "new underprivileged group" (Jiang, 2011). For many years the discourses on migration in China's popular media had focused on the possible impacts of migrants' presence in the city communities, indicating wide anxieties by urban residents about the movement of mobile population from the countryside and its potential effect on social order, as reflected in media terms such as "tide of peasant workers," "floating population," and "outsider workers," which emphasized urban/rural difference. Media reports in China about left-behind children as a social issue started to appear in 2002, and increased in numbers in 2006, when a legislation was proposed at Chinese People Consultative Conference by 24 members to establish a mechanism to safeguard these children's healthy growth. In 2010 more news reports put the plights of these children in spotlight, as well as problem-solving measures that emerged in different local communities, when a few provinces passed laws to protect the rights of minors, with decrees referring to left-behind children (Zeng, 2013). However, media representations of these children in news reports were often stereotypes, usually as targets of charity or protective policy, or as "problematic children." Academic research papers by scholars in China on this topic proliferated since 2007 with occasional official grants, ranging from surveys of the sociological, educational, and psychological issues caused by absent parents, to studies of legal policies created to protect these children. Newspaper reports on the topic, while seeing a steady increase over the years, tend to revolve around the actions of government agencies or corporate sponsors and rarely depicting the children as center of action, and the stories were reported mostly from the large cities than from their rural locations (Sun, 2014).

As family communication is seen as craved by left-behind children, their access, or indeed lack of access, to cell phones has become an issue of concern, as seen in numerous news reports. With the rapid development of China's telecommunication industry, there has been a wide gap between computer and internet access of students in urban public schools and those in rural areas. The left-behind children are at a disadvantageous position as the digital "have-nots," compared to their urban peers and migrant students who follow their parents to the cities (Yang, et al., 2013).

There has been a rapid increase in the number of migrant workers owning mobile phones, which might cost three to four times their monthly income (Law & Peng, 2008). Mobile phones help workers to contact both their families in their home villages and kinsmen scattered in the city area. Among the left-behind children, the rate of mobile phone ownership is relatively low, but many children are close to peers or adults with mobile phones, such as teachers, neighbors, or village chiefs. There is very little statistics about the percentage of the children's phone ownership, although the rate is increasing as the overall cell phone ownership rate grows in rural China. A survey conducted by researchers in People's University of China (Deng, 2008) compared the communication technology habits of children of migrant workers living in cities with their parents and those in the countryside away from their parents, and found that cell phone ownership rate reached 38.2 percent among migrant children in the cities, and only 7.8 percent among left-behind children. The survey also found that the most commonly used phone feature for migrant children is texting, which is 77 percent. For left-behind children, the most commonly used phone feature is phone calls, 64.3%, while only 28.6% of them reported texting as the most commonly used phone feature. The researchers conclude that left-behind children are less familiar with the variety of phone functions as texting requires more skillful use of phones.

Mobile Phones, Public Discourse, and Youth

Globally, more and more rural families are beginning to be impacted by the increasing popularity of cell phone technology. According to International Telecommunication Union, an agency of the United Nations, mobile-cellular subscriptions reached almost 7 billion by the end of 2014, and 3.6 billion of these was estimated to be in the Asia-Pacific region (ITU 2014). The increase is primarily a result of growth in the developing world, where mobile-cellular subscriptions would account for 78 per cent of the global total. Africa, Asia and the Pacific, where penetration would reach 69 per cent and 89 per cent, respectively by the end 2014, are the regions with the strongest mobile-cellular growth. A World Bank report also estimates that as mobile devices are becoming cheaper and more powerful while networks are doubling in bandwidth roughly every 18 months and expanding into rural areas, around three-quarters of the world's inhabitants now have access to mobile phones (World Bank, 2012). In China, according to a report published by China's Ministry of Industry and Information Technology, there were 1.286 billion mobile phone users at the end of 2014 (Hwang, 2015). The report also indicates that 94.5% of China's population actually used a mobile phone at the end of 2014. In comparison, only 18.3% of China's population (249.43 million people) used a landline in that same period. The speed of growth in cell phone ownership is quite breathtaking, as many regions have skipped landline technology and moved straight to mobile. Meanwhile, A Pew Research Center report (2014) found that among the Chinese population, 37% own a smartphone, while in the US the percentage of smartphone users is 55%. Among 18- 29 year-olds in China, 69% own a smartphone, making it the predominant technological choice for the younger generation.

How mobile phone technology might affect the family and social communication patterns of left-behind children relates to the question of the role of mediated communication in young people's lives in general. Scholars have addressed the interconnected relationship of mediated and unmediated communication, which are both situated and context-dependent, as mediated practices are intricately embedded in the daily activities of users (Ling & Haddon, 2008). Some have observed that all information and communication technologies (ICTs) are accompanied by heightened popular reactions regarding their impact on the existing social order (Thurlow, 2006). Whether in terms of people's experiences of community life, their standards of morality, or the way they organize their personal relationships, public discourse about emerging technologies is typically polarized by judgments of their being either "all good" or "all bad." The mobile phone, as an increasingly popular communication tool, has affected family dynamics by changing parental surveillance patterns and young people's views of independence and negotiated freedoms. While family communication can benefit from increased coordination, mobile phones also brought new issues such as affordability, parent access, as well as over- reliance on the phone (Campbell, 2005; Devitt & Roker, 2009; Katz & Aakhus, 2002). For adolescents, cell phone ownership allows young people even more flexibility and spontaneity in their lives (Nawaz & Ahmad, 2012). Access to the internet through smartphones could provide many of the resources for explorations of identity, emotion and sexuality, for experimentation with self-disclosure and intimacy (Livingstone, 2009). It is becoming a highly sought after a status symbol among young people in the developing world, seen as an indication that one is socially connected, accessible and in demand (Ahmed & Qazi, 2011; Goggin, 2012). Concerns about the potential liberating or detrimental effects of mobile technologies is widely reported in popular media as well as academic publications, as many parents and educators worry about teen's social alienation and consequence on academic performance (Huang and Leung, 2009). In popular representations of young technology users, young people tend to be reduced to a homogenous group of users, often portrayed as innocent youth or youth at risk, vulnerable to the addictive power of cell phones, particularly the appeals of social networking and instant messaging.

Parent-Child Communication in Migrant Families

Addressing parent-child communication among migrant families, previous studies done in different countries found that while labor migration generates stress on familial connectedness, ICTs can ease this strain by allowing distant family members to participate in day-to-day life through technology. Use of ICTs to counter the families' physical fragmentation creates a virtual psychological effect that helps the "left behind" cope with the distance. Vertovec (2004) argues that low-cost calls serve as a kind of social glue for transnational El Salvadorian migrant families, facilitating parental role maintenance, joint decision making, and the continuance of a sense of family. Horst (2006:143) found that in rural Jamaica, the mobile phone is viewed as a blessing "transforming the role of transnational communication from an intermittent event to a part of daily life," which also brought unforeseen burdens and obligations. For instance, Jamaican children regard frequent phone calls from their parents abroad as surveillance. Other scholars are less positive about the role of mobile phones in migrant families that parent and children live apart. Madianou & Miller (2011) interviewed Filipina mothers working in UK and their children back home, and find that mobile phone communication allows the women to deal with the ambivalence that is deeply ingrained in their decision to migrate. It may even contribute to their decisions regarding the prolongation of migration abroad, which women can justify more easily to the extent that they believe phones allow them to become effective transnational parents. The researchers note that the children of these women are less confident about how the phone conversations function for child-parent communication, and the children do not associate frequency of calls with increased parental connectedness. In a study about migrant families in Jamaica, where the rate of migration and family separation is very high, Brown & Grinter (2014) find that as communication technologies become more affordable and ubiquitous, mobile phones are allowing migrant parents to maintain a caregiving network to remotely mediate household matters among the children, educators, and guardians. As there is also inevitable tension and inconsistency in the communication process involving shared responsibilities and authority, the mobile phones become tools to negotiate control. Also, among Filipina and Indonesian foreign domestic workers in Singapore, who often harbor a sense of anxiety about their children's well-being in their home countries, mobile phones are utilized actively as a tool to negotiate and redefine identities and relationships (Chib, 2014:73).

Most of the existing studies about cell phones and migrant families were conducted using interviews or survey methods, collecting data from parents, educators, and guardians, and sometimes the children themselves. Very little research is done about how the media portray the use of cell phones in the lives of the left-behind children and construct the role of mobile technology among this population. Meanwhile there have been a significant amount of news reports on this topic in China, as the usage of cell phones expands and the reports about left-behind children rise in number. As there is wide perception that the utility functions of cell phones might facilitate communication between family members that live apart from each other, the typical concerns by adults about children/teens and cell phones, usually in terms of addiction or detrimental effect on school performance or social activities, would likely have a different orientation when involving the left-behind children.

As Moscovici (1984) notes, social representations plays a vital role in constructing the knowledge systems on which we rely to interpret and react to events. Social representations are generated and maintained in the realm of public discourse, with "an implicit stock of images and ideas which are taken for granted and mutually accepted," constituting "a whole complex of ambiguities and conventions without which social life could not exist" (Moscovici, 1984:21). They enable communication to take place among members of a community by providing them with a code for naming and classifying individual and group history. Social representations, then, establish an order which can make the unfamiliar, familiar, enabling the new and the unknown to be included in a pre-established category; and they enable communication to take place, communication based on a shared code in a society. Such representations provide the building material for the framework of argument, opinion and explanation that are constructed by the news media around the events. Thus, how the Chinese news media presents news stories involving cell phones and left-behind children reveals important notions about the definition of the "left-behind" social problem, and the intended and unintended consequences of the use of communication technology by families with migrant parents. As these news stories are products of Chinese news agencies, which often serve both official and commercial purposes in China's transforming media landscape (Zhao, 2008), the paper tries to analyze the stories within the context in which they were produced. It explores the ways underprivileged children are represented when encountering the global force of mobile communication, which has affected their personal/family communication activities, their routine activities in their schools and communities, and potentially their social circles.

MAIN FOCUS OF THE CHAPTER

To analyze popular discourse on the relationship of left-behind children with mobile phone technology in China, this chapter examines reports in major newspapers through critical discourse analysis of selected samples from news stories online. In critical discourse analysis (Fairclough, 2003; van Dijk, 1993; Wodak & Chilton, 2005) news stories can be analyzed as texts, revealing the social, cultural and therefore national representations of an issue or group circulating in a society at a given time. Critical discourse analysis offers "not only a description and interpretation of discourses in social context but also offers an explanation of why and how discourses work" (Rogers 2004:2). It is a qualitative research method to demonstrate how text producers use language, wittingly or not, in a way that could be ideologically significant, which could influence a reader's view of the world. It could include analysis of spoken or written language texts, analysis of discourse practice, namely, processes of text production, distribution and consumption, and analysis of discursive events as instances of sociocultural practice (Fairclough, 2003). In China, reports about left-behind children have appeared in a broad range of media channels, from television, newspaper, magazines, to documentary film. They have instilled among the public certain images about this social group, generating pre-molded concepts about members of this category.

The stories for analysis are retrieved from sina.com, one of the largest internet portal in China, owned by China's leading media and Internet services company Sina Corp, through key word search of "left-behind children" and "cell phones." Sina.com's news site can be seen as the representative of China's enterprise-owned news databases, which include a broader range of stories than the official news portals Xinhuanet.com or People.com.cn. The search yielded 1867 news stories from various news organizations in mainland China, including national, provincial, and local newspapers, with a timeframe of January 1, 2010 to May 15, 2015. A small portion of these stories was not used for analysis when cell phones bear no direct relationship to the event reported in the story. The stories are coded by the different types of relationships reflected between cell phones and left-behind children or people related to them. Also analyzed is how the story headlines summarize and call attention to the stories. News headlines have characteristics which justify that particular attention be given to the stories, which includes the prominence they acquire, the role they play in orienting the interpretation of the reader, and the shared cultural context which they evoke (Develotte & Rechniewski, 2001). Headlines may influence how the stories will be understood and stored for later use in making sense of similar events and issues.

The stories are also analyzed by examining the ways the children are portrayed through their actions and emotions described, the news sources that provided the information to the news agencies, and the quotes used. The excerpts are translated from Chinese by the author, who is bilingual and has professional news translation experience.

RESEARCH FINDINGS

The author finds that the relationships of these children with mobile phones are portrayed in the following types of narratives:

1. Cell phones are highly valuable for connecting family members living apart, playing a crucial role for children and their migrant parents in faraway locations to keep in touch with each other to a limited extent, as used by the children, their families and relatives, or teachers and guardians;

2. Cell phones are used as a problem-solver for remote rural communities, highly popular in charity giving and rural development projects to benefit migrant families, with the promise of making them new phone users;

3. Cell phones sometimes bring unexpected risks–left-behind children's unfamiliarity, over-dependence or misuse of cell phones can lead to accidents or other risks, as often reported by educators or social service agencies handling their cases; and

4. Cell phones could harbor or unleash evil–as more and more cases of juvenile delinquency or crime stories are associated with cell phones, involving male and female offenders or victims.

It is evident that cell phones are seen in each scenario as playing either a positive or a negative role in the lives of the left-behind children.

In the following sections I am going to discuss each of these narratives, illustrating with examples and excerpts from selected news stories, and examine how each narrative focuses on one specific role that cell phones might potentially play in the lives of the left-behind children. I will then conclude by analyzing how each of these narratives relates to a strand of the flow of media discourse on cell phones and left-behind children, and coincides with the notion of communication technology as a source of problem or a source of solution for the left-behind population.

Cell Phones as Desired Facilitator of Family Communication

News stories frequently present a social reality that cell phone access is sparse and regarded as precious among rural migrant families in the inland provinces, focusing on the poor infrastructure and lack of communication facilities, and the desire for access. Landline or cell phones are rare resources, as the ownership and distribution of access often reflect the social stratification and power relations of these isolated communities. The problem definition focuses on the physical distance, highlighting the pain and anxiety of the children caused by separation from their parents. The "stock images" are the family members always yearning and anticipating the day of parent-child reunion.

As the reports indicate, many of the children only could talk to their parents through school office phones, or use cell phones owned by their teachers, relatives, or village chiefs to reach their parents. The children rarely have their own cell phones, and if they occasionally do, the news reporter notices the fact and mentions how the phones are used in the story. Many stories about schools with predominately left-behind children describe their eagerness to get access to communicate with their parents through phone calls, as letter writing seems less preferable.

A news story from *West China Metropolitan Daily* (2013-11-21) titled "Yibin city spreads family connection telephone network to console left-behind children" reports that in a boarding school in Jiangan county in Sichuan province, over 90% of students are left-behind children, many of them yearning to get a chance to talk with their parents over the phone. One 11-year-old boy who had not heard from his father in several months told the reporter "I wish dad could call me once a week, or just once a month." Another student also waits everyday to hear from her father:

Among the nearly 400 left-behind students in Yudinghua Elementary School, not many can remember their parents phone numbers, but Lin Liguo remembers her father's number firmly. To remind herself the number, she wrote it on her book and recites it everyday when she reads, but there have been few calls from this number. Each time she hears the school loudspeaker notifying a student to go to the office to answer the phone, she really hopes that one day her name would be called.

The story quotes Huang Chuanwei, a teacher who had worked 30 years at Jianshe village school at Jiangan county's town of Yile, saying that before the "family-connection phone" was installed, everyday after school some children would line up to use his cell phone to talk to their parents, occasionally more than 10 children in line. The parents usually would inform the teacher the time of the calls beforehand and Huang would tell the students to wait. "At that time Huang Chuanwei's salary was only around 1000 yuan a month, and the telephone calls would cost one tenth of his income." In other news stories profiling well-respected rural teachers in remote village schools, the teachers' cell phones are also routinely used by children to talk to their parents. The teachers might also contact the parents directly about issues of the children, or manage small allowances for the children, playing the role of a caretaker.

In another story about a village school, titled "10-year struggle of a school for left-behind children: 353 students have only 9 teachers," the school doorkeeper's cell phone is the designated communication device for the children when there is a need to reach their parents. The doorkeeper, who earns 600 yuan a month, gets a monthly phone cost compensation of a few dozen yuans from the school.

There is no sound of cell phone rings, as no children own cell phones. They have no pocket money, or occasionally less than10 yuan of it, usually handed to teachers to manage. Sometimes when they miss their parents, they would go to Liu Xianxi (the school doorkeeper) to borrow his cell phone to make a call, but this type of occasion is rare. Only when the children get sick, they would try to reach mom and dad faraway. (Qilu Evening News, 2014-7-22)

Occasionally news stories also indicate that village chiefs or other cell-phone owners lend their phones to left-behind children. A story from Chengdu Commerce Daily (2014-9-01) describes the lives of several left-behind children, one of them an outgoing 6-year-old boy who lives with his grandfather. Whenever he sees adults in the village walking by, he would run to them, grab their sleeves and beg for their cell phones to call his mother working in Jiangsu province:

Five years ago, when Linpeng was 1 year old, his parents went to work in Changshu, Jiangsu Province. Since then, little Linpeng could see his parents only once a year. His Grandpa Deng Guiquan said that since then whenever the boy sees a phone, he would try to get it: "Initially we thought he was just being naughty, but then when we looked at the phone, it showed he dialed his mother's number. We do not know where he got the number from."

As rural boarding schools for left-behind children become more common, some children staying in schools have less time with their relative/guardians in the hometown. One story indicates an old man whose monthly income is 160 yuan owns a cell phone, which he uses to contact his grandchild in a local boarding school. Obviously landlines are not available in his house.

Phone conversations are often the only channel of parent-child connection, and tend to be depicted in the stories as highly desired by them. However, some in-depth stories recently reveal that even when children can access their parents with phone calls, they could have difficulties in emotional bonding. A story in *China Youth Daily* (2015-3-2) reports that a survey by China Adolescence Research Center shows 60% of left-behind children talk with their parents over the phone at least once a week, but 81.9%

of the children believes their parents only care about their exam scores, while only 5.8% thinks their parents care about their mood. A high school student tells the reporter that his mother has been disappointed with his academic performance:

"Whenever my mother gives me a call, the first sentence would be, ' Where are you? ' The second sentence must be ' What score did you get in the exam? '" Said Xiaoguo.

In the story a boy living in Zhenxiong county of Yunnan province, where there are many left-behind children, says that whenever Spring Festival approaches the family's phone cost would be higher, with more conversations about whether or when the parents will come back:

Before the Spring Festival, migrant workers returned to the hometown one after another. When seeing people coming back, Li Xiaogang and his sisters would call their parents to ask about their return date. 'Although they told us the date of their train tickets, we were still worried, worrying that they were lying to us, and worrying that they would suddenly have something that stops them from coming back.'

Though most stories about left-behind children tend to describe family reunion as a cherished moment of happiness, the China Adolescence Research Center survey found that it is also a time of heightened conflict between family members. One boy indicated that the first day of his father's return would be like heaven for him, with new clothes and many exotic treats to eat, and the second day would be like hell, as the father would beat him up after seeing his transcripts. The expert quoted explains that as most migrant parents have only junior high school education or less, they work hard to make money doing physical work, and expect their children to get better education than themselves, but reality is often disappointing as their children could not keep good study habits with little adult supervision on a daily basis. In contrast, a story by CCTV's Finance Channel (2015-2-10) about Spring Festival travel rush (chun yun) describes a 10-year-old girl living with her grandparents in a mountain village in western Hubei province who studies extra hard and gets good grades, to ease the worries of her mother working in a movie theater in Zhejiang province. The girl begged her mother over the phone to come home for a visit this holiday season as she had not done so in four years, working hard to repay their family debt. Eventually the mother got an approval for 7 days off from her employer, and spent 4 days travelling on a bus and only 3 days with the family.

Stories about migrant families during Spring Festivals sometimes reveal that parents purchase cell phone for their children as a gift of compensation for not coming back for the Chinese New Year, or when they are shopping together in a happy reunion after being apart for long time. Overall, while the rate of cell phone ownership is low among the children, their access to cell phones is increasing, as the rate of ownership grows among adults in their lives. In some relatively affluent rural areas, cell phone ownership among left-behind teens is not uncommon. The phone communication plays important roles in the parent-child connection and in some instances between parents, teachers, and other guardians, although for some children, the phone calls can merely assure them of the existence of the relationship, and cannot provide the needed emotional support or activity supervision.

News stories reporting the establishment of "family-connection" phone lines in rural locations as a positive progress tend to emphasize the lack of existing communication facilities and the yearning of the children for conversations with their parents in other provinces; and stories around Spring Festival about migrant families tend to focus on the trips of reunion, which the families occasionally record in photos with cell phones as treasured moments. Stories with news sources from research institutes and social agencies usually have more quotes discussing the psychological needs of the children and the consequence of insufficient caretaking on their growth. Access to cell phones has become one of the recommended steps in solving the parent-child communication problem.

Cell Phones as Problem Solver for Isolated Communities

As the notion of "left-behind children" becomes widely recognized in recent years as a social problem associated with poverty and neglect, there are many news stories describing government measures and charity giving activities for children by different corporations, neighborhood committees, volunteer groups and nonprofit organizations. These stories reflect a social reality that China's economic development has enlarged the wealth gap between the urban regions and the rural areas where the migrant families originate from, In this narrative, mobile technology is seen not only given out as a device for family communication, but is often brought to the rural residents as a facilitator for them to connect with the "modern" outside world, and an educational tool supplied to the needy children along with gifts such as books, toys and stationeries. The "stock images" are grateful children happily receiving gifts from donors/administrators from the cities. Headlines tend to use slogan phrases such as "constructing dreams," ""delivering warmth," or "caring and love," to align with the government initiatives to build a "harmonious society" in which class conflict is minimized.

Many of these news stories emerge around the annual June 1st Children's Day, Spring Festival, and occasionally Mid-Autumn Festival. They often highlight the efforts to improve migrant families' access to communication technology, among other ways to help. As these projects roll out, some rural communities launched left-behind children activity centers equipped with communication facilities for parent-child phone connection, some with video call facilities. These reports often comes with news photos showing groups of children receiving gifts at donation ceremonies or utilizing the facilities provided to them with excitement. Cell phones for individual use are also reported as main donation items for left-behind children in news stories, as the donors intend to help the children connect with their parents faraway, offering phones and pre-paid cards.

A 2013-03-21 story from Red Network titled "Trip to fulfill dreams: online volunteers deliver positive energy with loving hearts" describes a group of 9 volunteers travelling to rural Anren county to visit left-behind children, bringing them "carefully-prepared gifts":

Among the loving-heart online volunteer dream-fulfillment team, there are businessmen, civil servants, corporation employees and other caring people from different professions. They come together for the purpose of spreading love. Left-behind child Tan Zhangqi lives alone. His father works outside all year round to make a living and rarely returns home throughout the year; his mother remarried. When he received the gift of cell phone from the volunteers, he could not suppress his feeling and came to tears. Tan Zhangqi said excitedly: 'Thank you so much, aunts and uncles (translation note: a respectful term for adults), with this phone, I can make calls to my mom and dad, and no longer need to go to my aunt's place to make calls. After I grow up I would help others just like you do.'

While depicting the difficult conditions of the rural schools or villages, these stories often appear with positive steps taken by the local agencies or volunteers to help the children. Terms such as "family-connection phone" (qinqing dianhua) or "warm booths" (wennuan xiaowu) are used to indicate organized effort with policy support from the government to alleviate local left-behind population problems. This also shows that there have been more grassroots activities among the urban middle-class to reach out to the rural regions to "do good" through self-organized travel or volunteerism, especially after tragic incidents of rural children's deaths due to accidents or suicide made headlines.

News stories about "inspiring" (lizhi) efforts sometimes feature young college students or graduate students sent to the rural schools as part of their education programs. Often familiar with personal communication technologies, the college students would help the children contact their parents on their personal cell phones, for instance, using QQ chat app, or in some cases manage to raise funds to acquire online chat facilities for the village, and helped the children access their parents regularly and learn information about the outside world.

The use of cell phones as problem solver for migrant families is also supposed to contribute to the safety of the children in some areas through high-tech positioning features of the phones, in addition to alleviating the consequences of lack of parent-child communication and supervision, and helping enrich the lives of the children with music and entertainment.

A story titled "20,000 left-behind children in Jiangsu province will receive free positioning wristwatch phone," by Yangzi Evening News (2012-05-29), focuses on the new GPS feature of the wristwatch phones that could offer migrant parents assurance of their children's whereabouts, as there have been more and more cases of missing or runaway children:

The reporters learned from the Provincial Women's Federation that in Jiangsu Province there are currently about 83 million children left behind in the province. To provide a more secure growth environment for these children whose parents are not around all year round, Jiangsu Children Safety Guard Program will equip satellite positioning wristwatch phones for children for free. In addition to free calls, SMS messages, alarm signals, and other functions, whenever a child presses the SOS button for help, the phone will send an SOS message of the location to the parent's cell phone.

These stories highlight children's eagerness to communicate with their parents by using the talk function, positioning them as recipients of charity and target of poverty-relief policies. Mobile phone, especially those with video chatting capacity or musical features, is seen as a magical gift that alleviates the problem of separation and poor communication, though occasionally recent news reports note the lack of depth of conversation and the inability to replace parental presence.

On the other hand, in charity initiatives collecting public donations to left-behind children, cell phone text messaging as the donation platform has been reported in many news stories in recent years. A China News Network story (2014-8-27) reports that Red Cross Society of China has launched a charity campaign called "Philanthropic Messenger", under which 10 million yuan ($1.63 million) will be used mainly to help underprivileged children left behind by their parents or suffering major diseases in Xinjiang, Tibet, Yunnan, Guizhou and Gansu. Through the project, 150 child-care centers will be set up within a year in designated areas, benefiting nearly 50,000 children. Sometimes, news agencies also engage in charity activities collecting donations from readers through the use of text messaging or SMS, as it is assumed that most of the readership of the media outlets is urban residents.

As these initiatives are reported as "achievements" by the different agencies, it is questionable whether they are effective in relieving poverty and improve the quality of life for the rural children. It is also questionable whether they feel empowered or enlightened after acquiring the mobile communication device.

Cell Phone Ownership Brings Unforeseen Risks and Hazards

For many rural children, cell phones enter their environment as a surprise and novelty, as they grew up with little access to toys or electronic devices that are familiar to urban children. Although this is not specifically emphasized, stories of accidents demonstrate that they encounter the new intruder as technology novice with limited past exposure to communication tools, thus their unfamiliarity with the functions of cell phones could lead to unexpected risks upon first encounter, or give them a false sense of security with a phone in hand. These stories usually become newsworthy to the media when local social service agencies or educators respond to cases endangering the safety of left-behind children "playing with" mobile phones or inadequately using them. The "stock images" are responsible professionals working hard to solve problem for children.

One news story in Red Network (2013-12-16) is titled "Mistaking cell phone rings for alarm clock, left-behind child goes to school at midnight, frightening family members." A middle-school student usually sets the alarm clock at 5:30 am and goes to school in darkness alone, as he lives far from school. As he was in bed one night, he heard his grandfather's phone ringing. Assuming it was the alarm clock, he got up and walked to school. Seeing the school gate locked, and shivering in the cold night, he went to a bus station nearby, climbed through a bus window and fell asleep inside. Meanwhile the worried grandfather gave a call to his teacher at 1 am, reporting the boy missing, leading to hours of search by two teachers before locating him.

Some news stories indicate that over-dependence on cell phones and a lack of basic communication/social skills could cause unanticipated problems for the children. There are rural children who use cell phones to communicate with their parents as a routine, but could not recall the phone number or their parents' address or workplace in the cities, which could frequently change. A story titled "More than 60 left-behind children were 'picked-up' in less than two months" by Wuhan Evening News (2013-8-18) reports cases of rural children arriving at the city of Wuhan during the summer holidays to visit their parents, getting lost on the streets and were sent to the city's Children Rescue and Protection Center. Many of the children ended up there because they could not meet their parents after losing their phones or wallets, and were lost. Similar centers have been established to shelter, identify, and counsel homeless children in large cities in China, different from traditional homeless centers for adults. The story describes the experience of a 16-year-old girl from Yunnan province who had never seen the outside world, planning to visit her father who did not return to the hometown at Spring Festival. Nervous about the trip, she asked her 17 year-old cousin to go together, but the two fell asleep on their train seats in the long ride and she found her cell phone held in hand gone. With money running out and no knowledge of her father's whereabouts, the two sought help from police and was sent to the protection center. The girl could not offer any clue to locate her father in Wuhan, as apparently the cell phone was the only way she communicated with him and she only knew he worked at a construction site but could not name it. The staff's attempt to contact her village did not yield any information either, as only an illiterate grandma and an 8-year-old brother were at home, which had no landline phone, and all of her uncles and aunts were working in Guangzhou province. Unable to help the girl meet her father, the staff eventually bought train tickets for the two teens to return to Yunnan.

As smartphones have social networking functions to bring the owner to new social circles, online friendship, fantasy, or love affairs could become new outlet for emotional needs of left-behind children. One story in Chongqing Morning News (2013-8-22) reports that Xiaojun, a 13-year-old boy, became addicted to QQ chat after receiving a smartphone from his parents, and fell in love with a girl online named "ice fish," who was also a left-behind child. After finding in surprise that they were both in the same first year junior high school class, they started dating, but this was found out by the girl's "sharp-eyed" grandmother. As the girl stopped dating Xiaojun out of family pressure, the distressed boy could not control his emotion. In a restaurant he smashed a glass bottle on the wall, wounding his hands, yelling about suicide, and was sent to hospital after the police arrived at the call of the frightened restaurant customers and contacted his parents.

Meanwhile, in schools, cell phone addiction as a problem has been bothering educators not only in urban schools, as some reports indicate, but also some rural boarding schools for migrant-children, where boredom and loneliness were widespread. A story titled "Childhood 'kidnapped' by cell phones" in Yanzhao Metropolitan News paper (2014-5-29) describes that cell phones have become the most desirable gift for Children's Day (June 1), replacing traditional gifts of toys, clothes, or travel. The story indicates that with more and more children addicted to smartphones, many parents and teachers find themselves failing to deal with the problem even when imposing punishment measures such as confiscation or cutting off internet access. The story reports that both urban and rural educators are worried about phones "invading classrooms," seeing students playing games in classes or sharing content of sex and violence, and the educators point out many adults are themselves addicted to cell phones.

Chongqing Xiushan county Songnong Middle School is a rural boarding school, with three school grades and four classes, totaling 192 students, among whom more than half are rural left-behind children.

Teacher Zhang of the school told the reporter that over ninety percent of the school's junior high third-year students are using internet- enabled smart phones. In order not to affect their learning, a school regulation states, 'when students return to school on Sunday they should hand over the phones to the teachers, then get the phones back when returning home on Friday.' But in reality, few students hand in their phones.

These stories mainly use educators and social service agencies as news sources, revealing the new challenges cell phones pose to their professional routines and their concerns. While the stories tend to highlight their dutiful performance in their caretaking roles, they inadvertently expose the contrast of the high-tech features of cell phones with the paucity of cultural life in the surroundings of these children. For these professionals, cell phones are seen as a negative factor, intruding into the familiar world of the educators/caretakers. As many children possess weak information literacy skills, and lack experience with devices such as calculators, landline phone, or computers, the sudden introduction of cell phones to their lives could bring unexpected consequences, leading to occasional mistakes, miscommunication, or overdependence. From time to time, it could lead to even bigger risks that could change their course of life.

Cell Phones Associated with Juvenile Delinquency and Crimes Victimizing Left-Behind Children

There have been a rising number of left-behind children involved in crimes in recent years, either as victims or as perpetrators, and the cases of juvenile delinquency among them have been worrisome to researchers. In contrast to news stories about communication technologies that often portray left-behind children as innocent and pure school students yearning for love and care, particularly chances to talk their parents, crime stories project a more greyish picture of their environment, with predators and dangerous temptations around, victimizing them or bringing out unanticipated motivations or reactions from some of the children. In this narrative cell phones often play a negative or even destructive role in their lives. These stories, often sensationally written, tend to emphasize the vulnerability of the adolescents without adult supervision, as well as the hazards of desires unleashed by mobile devices or other material goods, and point to the need of psychological counseling and guidance for left-behind children. In these narratives youth is considered as both at risk and a source of risk, while the police is portrayed as effective and dutiful in maintain law and order.

In a Xinhua News Agency report titled "Teen killers spur discussion on youth crime" (2013-09-23), Mei Zhigang, a sociology professor at Central China Normal University, said that an increasing number of juvenile delinquency cases was committed by members of China's "left-behind" migrant children. The professor is quoted as saying that lack of education and knowledge of the law is one of the main sources of the problem: "Many of them only regard killing others as a crime, thinking that theft, robbery and rape are only simple and forgivable mistakes." The news report states that according to a special report released by the All-China Women's Federation in May 2013, the problems facing left-behind and migrant children, including lack of family closeness, security, protection and educational opportunities, have been relieved but have not yet been resolved, and new problems keep emerging.

Juvenile delinquency among left-behind youths related to mobile phones are often caused by phone addiction, jealousy and greed related to ownership, or inappropriate social contacts. These stories are usually based on police or court reports and feature individual characters in a specific location identified in the report, often using aliases for minors. While an examination of the stories demonstrates that both male and female teens could be engaged in juvenile delinquency, a clear gendered pattern emerges. Among the boys, phone and online game addiction or desire for phones can sometimes lead to stealing, suicide, or even killing of family members who tried to block their access to mobile phones. Group fights or bullying could also occur among left-behind children or victimizing them. In some instances, an adolescent perpetrator could also use cell phone games to lure an unsuspecting peer, often a little younger, to an isolated location for robbery or even manslaughter or murder, according to news story descriptions.

Cell phones as object of theft or robbery appear to be more often associated with male adolescents than females. Qianjiang Evening News (2013-12-10) reports that in a peaceful village in Quzhou, where the residents usually don't lock their doors, a 12-year-old boy living with only his grandparents was found to have 18 records of theft in the township police department in less than one year, ranging from coins in a neighbor's drawer to cell phones owned by villagers. His explanation was he felt envious of classmates having "good things to eat and play with." The court indicated that as the total value of the stolen goods are assessed as more than 10,000 yuan, it fit the criteria of crime of theft, but the boy was not charged with a criminal offense as he was a minor. The news report uses the term "left-behind child" in the headline and explains his family background in the body of the story, bringing readers to the conclusion that his family situation led to his delinquent behaviors.

Crime stories referring to the term "left behind children" come not only from rural regions, but also in large cities, as some of the children whose parents were migrant workers in the 1980s or 1990s have grown up and left their hometowns. Some news stories of arrests of criminals attribute their behaviors to past experience as left-behind children, stating that the way they grew up with little adult supervision contributed to their potential tendency to violate laws without much knowledge of the consequences, leading to jail terms. Meanwhile the sporadic exposures to city life also make them discontent with staying in their barren rural hometowns living a life of farming. One feature story titled "The lost second generation migrant workers" by *Law and Life* (2010-5-17) starts with the following descriptions of this group:

They were 'left-behind children' before; they were the 'rootless children' who followed their 'first generation migrant' parents to search for dreams in the cities. As their 'first generation migrant' parents gradually withdraw from the stage of the city, they flow into large and small cities one after another-- seeking their dreams among the tall buildings that do not belong to them. This is a gigantic group floating at the margins of cities with hundreds of millions of people. Between their dreams and reality, there is a gap difficult to cross. Between the 'city impossible to stay' and the 'countryside impossible to return,' more and more second generation migrant workers have deviated from the navigation course of life, and got lost in the jungle of crimes.

Cell phones represented material temptations and a highly desirable status symbol unobtainable for the migrant youth. The story narrates the cases of three young people who grew up as left-behind children and decided to leave the countryside as teens to search for opportunities in the cities, where they later became criminals. The interviews reveal that the eye-opening wealth gap they witnessed between the city residents and the rural villagers led to their material desire, hatred against the rich, and wishful thinking to make a gain without being caught. The story features an interview with Lang Feng, a 19-year-old who came to the city of Ningbo from Xincai county, Henan province two years ago, and found a job as a waiter in a restaurant. There he was amazed at the wealth of some customers who "did not blink their eyes" at the huge bills of their banquets, and the presence of cell phones as a daily necessity among the urban residents:

It did not take long before the desire to own a cell phone pushed Liang Feng into the abyss. Liang Feng said, at that time he suddenly found that almost everyone had a cell phone; even the junk collector who often came to the restaurant regularly also had a cell phone. The market price of the cheapest cell phone was over a thousand yuan, but he only had less than 200 yuan in his pocket. The desire to get a cell phone was like a cat living in his heart, making his heart itchy with each scratch.

The story describes how he tried to snatch a woman pedestrian's cell phone and handbag one night after work, but was caught by security patrollers while running, and was sentenced to more than one year in a youth detention center, as he was under 18. Later he found a job at a jade factory after learning some work skills in the center. Another youth featured in the article committed a series of thefts of cash and cell phones in the suburb of Beijing, as he felt his job as a mover too physically demanding and low-paid. It is evident that smartphones have become highly tempting for youth with material ambitions, as they become more and more popular as a trendy consumer item.

Meanwhile, as news reports with statistical results indicate broad trends of vulnerability of left-behind girls to sexual assaults, there have been more and more stories in local newspapers reporting police arrests of sex offenders targeting these girls. Among the offenders there are teachers, neighbors, and other acquaintances. For adolescent girls, a cell phone as a gift or a promised gift is often used by criminals as bait, sometimes accompanied by cash, eventually turning them to victims of sexual abuse. A story in China News Network (2014-7-22) is titled "Man uses cash and cell phone to seduce and rape 13-year-old middle school student, making her pregnant." The following are excerpts:

Defendant Wang, nearly 50-year-old, is from the same village with victim Xiaohong, as the two families live less than one kilometer away. Xiaohong usually addressed politely Wang as 'uncle.' However, eyeing

Xiaohong's attractive figure, Wang's lust made him reckless, and stretched his evil claws to Xiaohong, his niece of the same clan.

Between November and December of 2013, while Xiaohong was going to and from school, or when her guardians were not at home, on six occasions Wang had sex with Xiaohong, by luring her with money and gifts, in the Plum Ditch hillside and tobacco toasting house and so on at Zhuotian town Shantian village. Each time Wang gave Xiaohong a little money, and he also promised to give her a cell phone as gift, but concerned that the family members would ask about the phone's origin, he did not give her the phone.

The story also reports that the court in Changting county of Fujian province sentenced the man to six years and six months with the crime of raping a minor under 14, after the family found out about her pregnancy in a Spring Festival reunion. Using legal terms such as "defendant" and reporting the court's reasoning for the sentence, without any quotes from an interviewee, the story clearly shows a reliance on court papers as news source.

As cautionary tales for both parents, educators, and youth, these stories emphasize that there were consequences to the choices young people made when faced with temptations, putting emphasis on the special vulnerability of left-behind children and youth who might make poor decisions or moral choices, getting them involved with the legal system. The social reality portrayed in these stories indicate that the social problem of left-behind children goes far beyond the issue of family separation, but is deep-rooted in poverty, social inequality, and gendered exploitations. It implicitly challenges the fantasy of upward social mobility for China's underprivileged population in the modernization process, as they make difficult decisions and strive to react to the vast changes in economy, even though these stories may make the events look like results of individual moral choices. These stories demonstrate that it would be a naïve wish to expect cell phones to play only a positive role in the left-behind children's lives by helping them improve communication with their parents. As earlier literature indicates, cell phone use can be a channel for youth to explore self-identity, sexuality, and independence (Livingstone, 2009). It is not surprising that among the millions of left-behind children, a fraction of them become involved in criminal activities with cell phones playing a role. The news discourse implies a recognized threat to social order posed by the migrant families, and often attributes it to "desires" of individuals and a lack of sense of law. By framing these stories in terms of personal choices, news stories produced by Chinese media organizations could safely downplay the responsibilities by policy makers and the educational system by highlighting the damaging power of cell phones, as this new technology enters social life with a crushing speed.

FUTURE RESEARCH DIRECTIONS

The experiences of children of migrant families with cell phone and other communication technologies are complex, and could vary among different regions, demographic groups, and personality types. There has been an increasing amount of media materials addressing the issue directly or touching upon the topic indirectly in national and local media outlets. As this research project only focuses on newspaper reports about left-behind children, future research designs could address different types of media representations of their use of technology, for instance, on television or entertainment programs. Researchers could also expand the questions to social presentations of their parents and educators, focusing on issues of technology in mother-child communication or teacher-student communication. To yield more pragmatic research results, future research projects can be also conducted among specific groups of children and youth to study their cell phone use habits, for example, among teenagers of a specific gender or age group in a region. Future research can be also conducted regarding the effect of specific content or softwares among youth and children in migrant families. A variety of methodologies could be applied, such as ethnography, interviews, and surveys, in addition to discourse analysis, which has the advantage of logistical convenience. Meanwhile, there could be restrictions researchers could face due to limitations of funding, time, and permission for access to conduct fieldwork. If possible, a collaborative approach by China-based researchers and Western researchers could also probably produce fruitful results.

CONCLUSION

A synthesis of news discourses on cell phone and migrant families brings together a picture of mobile technology as a double-edged sword for this social group. The analyses of news stories narratives reveal that there are polarized portrayals of mobile communication technology as either good and evil, as it could be seen as highly helpful to their dire situation, or highly dangerous to the vulnerable left-behind children. Cell phones in many cases are perceived as a powerful tool to solve the problem of serious lack of parental care, and in other occasions a source of temptation leading adolescents to take risks in a world of growing materialism and wealth gap in society.

Technology is value-laden and reflects the social/cultural settings of a given society. The glorious portrayal of mobile communication technology as a gift to the migrant population echoes the mainstream media's discourse of modernization, economic development, and progress, and supports the government initiatives of extensive infrastructure building. Meanwhile, the negative portrayals of cell phones in the lives of left-behind children showcase consequences of the vast social changes and a disturbing sense of moral crisis among the population, which have led to overloads in social agencies and the legal system. As social norms, communities and traditional family lives are threatened by uneven economic development and large-scale migration, the cautionary illustration of harmful consequences of adolescence using technology without supervision is aligned with the authorities' efforts of social control and moral education, which involved wide mobilization of youth leagues, neighborhood committees, and women's associations.

The contradictions of these stories demonstrate that there could be unanticipated cultural-economic and political implications in efforts to equip the "have-nots" of global and regional digital divide with mobile technology. A "Great Leap Forward" in telecommunication facilities may appear to be a significant achievement for the local administrators, but the consequence to the communities would be questionable, as many villages are only inhabited by children and the elderly. Earlier literature of migrant families indicates that increased phone communication is not equivalent to an enhanced sense of connectedness (Madianou & Miller, 2011). As more and more social groups attempt to address the yearning of the left-behind children for communication opportunities with their parents, the seemingly empowering capacities of cell phones could bring unanticipated hazards or risks, given the children's relative weakness in information literacy and their paucity of material or cultural life. Meanwhile the popular discourse also shows that there is a lack of coherence in addressing the societal anxieties over unsupervised access to mobile communication technology by underprivileged adolescents, with a discordance of intents to manage the risks and benefits of technology in the lives of millions of migrant families.

The social representations continuously supply images and ideas to public discourse about left-behind children, creating images of them as "have-nots" in digital divide eager for help or risky youth whose futures could be easily destroyed by the temptations of/from cell phones. The categorization of the individuals reproduces and reconfirms commonly held notions in society about left-behind children, highlighting their emotional and physical vulnerability, even though similar cases have occurred to children whose parents are around. The news editors might intend to call for attention and care for these children, or to attribute their involvement in accidents, addictions, or criminal behaviors to the fact that they were left-behind by their parents, as indicated by the news sources in these stories. As the news sources are often social agencies, corporations, educators, or law enforcement personnel in their professional roles, the stories about left-behind children could take on distinctly institutional perspectives, portraying them as children in poverty in need of charity assistance, or distracted students lacking guidance, or vulnerable population that potentially needs legal protection or monitoring. Each of these images corresponds to a line of popular discourse weaving the social representational of left-behind children in society, revealing contradictory notions about how cell phones could enter their lives and shift the balance. To a certain extent, the news media also serve as agents of various policy projects, creating visibility and publicity for their operations and organizational priorities, which need to turn to visible and coherent steps to prevent the widespread social problem from becoming a monumental crisis in the socio-economic transformations reshaping China.

REFERENCES

Ahmed, I., & Qazi, T. F. (2011). Mobile phone adoption & consumption patterns of university students in Pakistan. *International Journal of Business and Social Science, 2*(9), 205–213.

Brown, D., & Grinter, R. E. (2014). Aboard Abroad: Supporting Transnational Parent–School Communication in Migration-Separated Families. *Information Technologies & International Development, 10*(2), 49.

Campbell, M. A. (2005). *The impact of the mobile phone on young people's social life.* Social Change in the 21st Century Conference, QUT Carseldine, Brisbane.

CCR CSR News. (2014). *China's "Left-Over Children" Face Greater Chances of Psychological Problems, Juvenile Delinquency and Sexual Assault.* Retrieved from http://www.ccrcsr.com/news/china's-"left-over-children"-face-greater-chances-psychological-problems-juvenile-delinquency

Chib, A., Malik, S., Aricat, R. G., & Kadir, S. Z. (2014). Migrant mothering and mobile phones: Negotiations of transnational identity. *Mobile Media & Communication, 2*(1), 73–93. doi:10.1177/2050157913506007

Christians, C. (1989). A theory of normative technology. In *Technological Transformation* (pp. 123–139). Springer Netherlands. doi:10.1007/978-94-009-2597-7_9

Deng, Q. (2008). Did not imagine so many urban migrant children own cell phones. *China Youth Daily.* Retrieved from http://zqb.cyol.com/content/2008-04/14/content_2142436.htm

Develotte, C., & Rechniewski, E. (2001). Discourse analysis of newspaper headlines: a methodological framework for research into national representations. *The Web Journal of French Media Studies, 4*(1).

Devitt, K., & Roker, D. (2009). The role of mobile phones in family communication. *Children & Society*, *23*(3), 189–202. doi:10.1111/j.1099-0860.2008.00166.x

Fairclough, N. (2003). *Analysing Discourse: Text Analysis for Social Research*. London: Routledge.

Goggin, G. (2012). *Cell phone culture: Mobile technology in everyday life*. Routledge.

Horst, H. A. (2006). The blessings and burdens of communication: Cell phones in Jamaican transnational social fields. *Global Networks*, *6*(2), 143–159. doi:10.1111/j.1471-0374.2006.00138.x

Huang, H., & Leung, L. (2009). Instant messaging addiction among teenagers in China: Shyness, alienation, and academic performance decrement. *Cyberpsychology & Behavior*, *12*(6), 675–679. doi:10.1089/cpb.2009.0060 PMID:19788380

Hwang, A. (2015). China December mobile phone user base grows to 1.286 billion, says MIIT. *DIGITIMES*. Retrieved from http://www.digitimes.com/news/a20150123VL200.html

ITU. (2014). *ITU releases 2014 ICT figures: Mobile-broadband penetration approaching 32 per cent Three billion Internet users by end of this year*. Retrieved from https://www.itu.int/net/pressoffice/press_releases/2014/23.aspx#.VVD26Utu4dt

Jiang, L. (2011) Construction and research reflection of the issue of left-behind children. *Humanities Magazine (Renwen Zazhi)*. Retrieved from http://www.aisixiang.com/data/41509.html

Katz, J. E., & Aakhus, M. (Eds.). (2002). *Perpetual contact: Mobile communication, private talk, public performance*. Cambridge University Press. doi:10.1017/CBO9780511489471

Law, P. L., & Peng, Y. (2008). Mobile networks: migrant workers in southern China. Handbook of mobile communication studies, (pp. 55-64).

Ling, R., & Haddon, L. (2008). Children, youth and the mobile phone. In Drotner & Livingstone (Eds.), International Handbook of Children, Media and Culture. Sage Publications Ltd. doi:10.4135/9781848608436.n9

Livingstone, S. (2009). Children and the Internet. *Polity*.

Madianou, M., & Miller, D. (2011). Mobile phone parenting: Reconfiguring relationships between Filipina migrant mothers and their left-behind children. *New Media & Society*, *13*(3), 457–470. doi:10.1177/1461444810393903

Moscovici, S. (1984) The phenomenon of social representations. In Farr & Moscovici (Eds.), Social Representations. Cambridge, UK: Cambridge University Press.

Nawaz, S., & Ahmad, Z. (2012). Statistical study of impact of mobile on student's life. *IOSR Journal of Humanities and Social Science*, *2*(1), 43–49. doi:10.9790/0837-0214349

Pew Research Center. (2014). *Emerging nations catching up to U.S. on technology adoption, especially mobile and social media use*. Retrieved from http://www.pewresearch.org/fact-tank/2014/02/13/emerging-nations-catching-up-to-u-s-on-technology-adoption-especially-mobile-and-social-media-use/

Rogers, R. (2004). *An Introduction to Critical Discourse Analysis in Education*. Mahwah, NJ: Lawrence Erlbaum.

Su, S., Li, X., Lin, D., Xu, X., & Zhu, M. (2013). Psychological adjustment among left-behind children in rural China: The role of parental migration and parent–child communication. *Child: Care, Health and Development*, *39*(2), 162–170. doi:10.1111/j.1365-2214.2012.01400.x PMID:22708901

Sun, L. (2014). *Framing analysis of media reports on left-behind children in mainland China*. (Master's degree thesis). Shanghai Foreign Language Institute. Retrieved from http://cdmd.cnki.com.cn/Article/CDMD-10271-1014126282.htm

Thurlow, C. (2006). From statistical panic to moral panic: The metadiscursive construction and popular exaggeration of new media language in the print media. *Journal of Computer-Mediated Communication*, *11*(3), 667–701. doi:10.1111/j.1083-6101.2006.00031.x

van Dijk, T. (1993). Principles of critical discourse analysis. *Discourse & Society*, *4*(2), 249–283. doi:10.1177/0957926593004002006

Vertovec, S. (2004). Cheap Calls: The Social Glue of Migrant Transnationalism. *Global Networks*, *4*(2), 219–224. doi:10.1111/j.1471-0374.2004.00088.x

Wen, M., & Lin, D. (2012). Child development in rural China: Children left behind by their migrant parents and children of nonmigrant families. *Child Development*, *83*(1), 120–136. doi:10.1111/j.1467-8624.2011.01698.x PMID:22181046

Wodak, R., & Chilton, P. (2005). *A New Agenda in (Critical) Discourse Analysis: Theory, Methodology and Interdisciplinarity*. Amsterdam: John Benjamins. doi:10.1075/dapsac.13

World Bank. (2012). *Mobile Phone Access Reaches Three Quarters of Planet's Population*. Retrieved from http://www.worldbank.org/en/news/press-release/2012/07/17/mobile-phone-access-reaches-three-quarters-planets-population

Yang, Y., Hu, X., Qu, Q., Lai, F., Shi, Y., Boswell, M., & Rozelle, S. (2013). Roots of Tomorrow's Digital Divide: Documenting Computer Use and Internet Access in China's Elementary Schools Today. *China & World Economy*, *21*(3), 61–79. doi:10.1111/j.1749-124X.2013.12022.x

Zeng, X. (2013). *Analysis of media reports of left-behind children: Case study of China Youth Daily*. (Master's degree thesis). Southwest University. Retrieved from http://www.docin.com/p-805433612.html

Zhao, Y. (2008). *Communication in China: Political economy, power, and conflict*. Rowman & Littlefield Publishers.

Compilation of References

10 Advantages to Taking Online Classes. (2012, January 9). Retrieved from http://oedb.org/ilibrarian/10-advantages-to-taking-online-classes/

Aaker, D. A. (2004). *Brand portfolio strategy*. New York, NY: Free Press/Simon & Schuster.

Aaker, D. A., & Joachimsthaler, J. (2000). *Brand leadership*. London: Free Press.

Abdelghaffar, H., & Galal, L. (2012). Assessing Citizens Acceptance of Mobile Voting System in Developing Countries: The case of Egypt. *International Journal of E-Adoption*, *4*(2), 15–27. doi:10.4018/jea.2012040102

Abdelghaffar, H., & Magdy, Y. (2012). The adoption of mobile government services in developing countries: The case of Egypt. *International Journal of Information*, *2*(4), 333–341.

Abeele, M. M. P. V. (2014). Mobile lifestyles: Conceptualizing heterogeneity in mobile youth culture. *New Media & Society*, 1–19.

Aberdeen Group. (2014). *Analyzing the ROI of video advertising*. Retrieved 16 Feb. 2015 from http://go.brightcove.com/bc-aberdeen-analyzing-roi-b

Aberdeen Group. (2014). *Analyzing the ROI of video advertising*. Retrieved February 16, 2015, from http://go.brightcove.com/bc-aberdeen-analyzing-roi-b

Aberdeen Group. (2014). *Analyzing the ROI of video advertising*. Retrieved from http://go.brightcove.com/bc-aberdeen-analyzing-roi-b

Accenture. (2013). *Accenture 2013 Global Consumer Pulse Survey*. Retrieved 13 April 2015 from http://www.accenture.com/SiteCollectionDocuments/PDF/Accenture-Global-Consumer-Pulse-Research-Study-2013-Key-Findings.pdf

Accenture. (2015). *Seamless Retail Research Report 2015: Maximizing mobile to increase revenue*. Retrieved from https://www.accenture.com/_acnmedia/Accenture/Conversion-Assets/Microsites/Documents15/Accenture-Seamless-Retail-Research-2015-Maximizing-Mobile.pdf

ACE Electoral Knowledge Network. (2013). *The ACE Encyclopaedia: Civic and Voter Education*. ACE Electoral Knowledge Network. Retrieved from http://www.aceproject.org

ACE Electoral Knowledge Network. (2015, December 09). *Elections and Technology*. Retrieved from The Electoral Knowledge Network: https://aceproject.org/ace-en/topics/et/eth/eth02/eth02a

ACE Electoral Knowledge Network. (2015, December 09). *Voting From Abroad*. Retrieved from ACE Electoral Knowledge Network: http://aceproject.org/ace-en/topics/va/onePage

Acker, O., Geerdes, H., Gröne, F., & Schröder, G. (2013). Builders of the digital ecosystem: The 2013 Strategy & global ICT 50 study. *PricewaterhouseCoopers*. Retrieved from http://www.strategyand.pwc.com/media/file/Strategyand_Builders-of-the-Digital-Ecosystem.pdf

Adams, C., & Mouatt, S. (2010). Evolution of Electronic and Mobile Business and Services: Government Support for E/M-Payment Systems. *International Journal of E-Services and Mobile Applications*, 2(2), 58–0. doi:10.4018/jesma.2010040104

Admiraal, W., Huizenga, J., Akkerman, S., & Dam, G. T. (2011). The concept of flow in collaborative game-based learning. *Computers in Human Behavior*, 27(3), 1185–1194. doi:10.1016/j.chb.2010.12.013

Adobe. (2013). *The marketer's guide to personalization: Improving customer engagement in a digital world.* Retrieved 29 March 2015 from http://offers.adobe.com/content/dam/offer-manager/en/na/marketing/Target/Guide%20to%20personalization.pdf

Adobe. (2015). *Quarterly Digital Intelligence Briefing: Digital Trends 2015.* Retrieved 20 Feb. 2015 from http://offers.adobe.com/content/dam/offer-manager/en/na/marketing/Target/Adobe%20Digital%20Trends%20Report%202015.pdf

Adobe. (2015). *Quarterly digital intelligence briefing: Digital trends 2015.* Retrieved February 20. 2015, from http://offers.adobe.com/content/dam/offer-manager/en/na/marketing/Target/Adobe%20Digital%20Trends%20Report%202015.pdf

Advertiser Perceptions. (2015). *As the industry continues heavy shift to programmatic buying, disconnects grow between buyers and sellers, agencies and marketers.* Retrieved 6 May 2015 from http://www.advertiserperceptions.com/blog

Ahmed, I., & Qazi, T. F. (2011). Mobile phone adoption & consumption patterns of university students in Pakistan. *International Journal of Business and Social Science*, 2(9), 205–213.

Ahonen, T. (2002). m-Profits: Making money from 3G services. Hoboken, NJ: Wiley.

Aker, J. C., & Mbiti, I. M. (2010, June). Mobile Phones and Economic Development in Africa. *The Journal of Economic Perspectives*, 24(3), 207–232. doi:10.1257/jep.24.3.207

Akhras, F. N., & de Rezende, E. D. (2008). *Digital inclusion in social contexts: A perspective to the use of mobile technologies.* Retrieved June 10, 2015, from http://www.w3.org/2008/10/MW4D_WS/papers/akhras.pdf

Akkerman, S., Admiraal, W., & Huizenga, J. (2009). Storification in history education: A mobile game in and about medieval Amsterdam. *Computers & Education*, 52(2), 449–459. doi:10.1016/j.compedu.2008.09.014

Akpojivi, U., & Bevan-Dye, A. (2015). Mobile advertisements and information privacy perception amongst South African Generation Y students. *Telematics and Informatics*, 32(1), 1–10. doi:10.1016/j.tele.2014.08.001

Al Thunibat, A. A., Zin, N. A. M., & Ashaari, N. S. (2010, June). Mobile government services in Malaysia: Challenges and opportunities. In *Information Technology (ITSim), 2010 International Symposium in* (Vol. 3, pp. 1244-1249). IEEE.

Al Thunibat, A., Zin, N. A. M., & Sahari, N. (2011a). Identifying user requirements of mobile government services in Malaysia using focus group method. *Journal of e-Government Studies and Best Practices, 2011*, 1-14.

Al Thunibat, A., Zin, N. A. M., & Sahari, N. (2011b). Modelling the factors that influence mobile government services acceptance. *African Journal of Business Management*, 5(34), 13030.

Al-Fahad, F. N. (2008). Student Perspectives About Using Mobile Devices in theirStudies in the King Saud University, Kingdom of Saudi Arabia. *Malaysian Journal of Distance Education*, 10(1), 97–110.

Al-Fahad, F. N. (2009). Students' attitudes and perceptions towards the effectiveness of mobile learning in King Saud University, Saudi Arabia. *The Turkish Online Journal of Educational Technology*, 8(2), 111–119.

Alfred, C. (2015, April 27). How families abroad are tracking down missing loved ones after Nepal's earthquake. *The Huffington Post*. Retrieved May 7, 2015, from http://www.huffingtonpost.com/2015/04/27/nepal-earthquake-missing_n_7147894.html

Al-Hadidi, A., & Rezgui, Y. (2009). Critical success factors for the adoption and diffusion of m-government services: A literature review. In *9th European Conference on e-Government* (pp. 21-28). London, UK: Academic Publishing Limited.

Al-Hujran, O. (2012). Toward the utilization of m-Government services in developing countries: A qualitative investigation. *International Journal of Business and Social Science*, *3*(5), 155–160.

Ali, S., Rizvi, S. A. A., & Qureshi, M. S. (2014). Cell phone mania and Pakistani youth: Exploring the cell phone usage patterns among teenagers of South Punjab. *FWU Journal of Social Sciences*, *8*(2), 43.

Al-Khamayseh, S., Hujran, O., Aloudat, A., & Lawrence, E. (2006b). Intelligent m-government: application of personalisation and location awareness techniques. In *Second European Conference on Mobile Government*.

Al-Khamayseh, S., Lawrence, E., & Zmijewska, A. (2006a). Towards understanding success factors in interactive mobile government. In Proceedings of Euro mGov (pp. 3-5).

Al-Khamayseh, S., & Lawrence, E. (2006). Towards citizen centric mobile government services: A roadmap. *CollECTeR Europe*, *2006*, 129.

Ally, M. & Prieto-Blazquez, J. (2014). What is the future of mobile learning in education? *RUSC: Revista De Universidad Y Sociedad Del Conocimiento, 11*(1), 142-151.

Ally, M., & Tsinakos, A. (2014). *Increasing access through mobile learning*. Academic Press.

Ally, M., Schafer, S., Cheung, B., McGreal, R., & Tin, T. (2007). Use of mobile learning technology to train ESL adults. In A. Norman & J. Pearce (Eds.), *Making the Connections:Proceedings of the mLearn Melbourne 2007 Conference*. University of Melbourne.

Aloudat, A., & Michael, K. (2011). Toward the regulation of ubiquitous mobile government: A case study on location-based emergency services in Australia. *Electronic Commerce Research*, *11*(1), 31–74. doi:10.1007/s10660-010-9070-0

Aloudat, A., Michael, K., Chen, X., & Al-Debei, M. M. (2014). Social acceptance of location-based mobile government services for emergency management. *Telematics and Informatics*, *31*(1), 153–171. doi:10.1016/j.tele.2013.02.002

Altschull, J. H. (1996). A crisis of conscience: Is community journalism the answer? *Journal of Mass Media Ethics*, *11*(3), 166–172. doi:10.1207/s15327728jmme1103_5

Amailef, K., & Lu, J. (2008, November). m-Government: A framework of mobile-based emergency response systems. In *Intelligent System and Knowledge Engineering, 2008. ISKE 2008. 3rd International Conference on* (Vol. 1, pp. 1398-1403). IEEE.

Amailef, K., & Lu, J. (2011). A mobile-based emergency response system for intelligent m-government services. *Journal of Enterprise Information Management*, *24*(4), 338–359. doi:10.1108/17410391111148585

American Press Institute. (2006). *Newspaper Next: Blueprint for Transformation*. Retrieved 15 Jan. 2015 from http://www.americanpressinstitute.org/training-tools/newspaper-next-blueprint-transformation/

Analytics, S. (2014, April 29). *Samsung & Apple slip as global smartphone shipments reach 285 million units in Q1 2014* [Press release]. Retrieved from http://blogs.strategyanalytics.com/WSS/post/2014/04/29/Strategy-Analytics-Global-Smartphone-Shipments-Reach-285-Million-Units-in-Q1-2014.aspx

Anderson, L. (2014, August 6). *From #BringBackOurGirls to Syria, why do we forget about a crisis long before it's over?* Retrieved from Deseret News: http://national.deseretnews.com/article/2065/from-bringbackourgirls-to-syria-why-do-we-forget-about-a-crisis-long-before-its-over.html

Anderson, M. (2015, January 7). *37% of SMBs plan to spend more on internet marketing in2015.* Retrieved from https://www.brightlocal.com/2015/01/07/37-smbs-plan-spend-internet-marketing-2015/

Anderson, C. (2006). *The long tail: Why the future of business is selling less of more.* Hyperion.

Andoutsopoulos, J. (2006). Multilingualism, Diaspora, and the Internet: Codes and Identities on German-based Diaspora Websites. *Journal of Sociolinguistics, 10*(4), 429–450.

Antovski, L., & Gusev, M. (2005, July). M-government framework. In Euro mGov (Vol. 2005, pp. 36-44).

Aoki, K., & Downes, E. J. (2003). An analysis of young people's use of and attitudes toward cell phones. *Telematics and Informatics, 20*(4), 349–364. doi:10.1016/S0736-5853(03)00018-2

Appadurai, A. (1996). *Modernity at Large: Cultural Dimensions of Globalization.* Minneapolis: University of Minnesota Press.

Apriana, A. (2006). Mixing and Switching Languages in SMS Messages. *BAHASA DAN SENI, 34*(1), 36–57.

Arab, S. M. R. (2013). *Transforming education in the Arab world: Breaking barriers in the age of social learning.* Retrieved December 13, 2015, from http://www.arabsocialmediareport.com/home/index.aspx

Arab, S. M. R. (2014). *Citizen engagement and public services in the Arab world: The potential of social media.* Retrieved December 13, 2015, from http://www.arabsocialmediareport.com/home/index.aspx

Ardito, C., Buono, P., Costabile, M. F., Lanzilotti, R., Pederson, T., & Piccinno, A. (2008). Experiencing the past through the senses: An m-learning game at archaeological parks. IEEE Computer Society, 79-81.

Arens, W. F. (2002). *Contemporary Advertising.* New York City: McGraw-Hill Companies Inc.

Argenti, P. A. (2006). How technology has influenced the field of corporate communication. *Journal of Business and Technical Communication, 20*(3), 357–370. doi:10.1177/1050651906287260

Arit, J. (2014). What's Wrong With Our Well-Intentioned Boko Haram Coverage. The Wire –. *Atlantic (Boston, Mass.), 9*(May). Retrieved from http://www.thewire.com/politics/2014/05/whats-wrong-with-our-well-intentioned-boko-haram-coverage/361993/

Armatas, C., Holt, D., & Rice, M. (2005). Balancing the possibilities for mobile technologies in higher education. In *Proceedings of the 2005 ASCILITE conference.* Brisbane, Australia: ASCILITE.

Arulchelvan, S. (2014, July). New Media communication Strategies for Election Campaigns: Experiences of Indian Political Parties. *Online Journal of Communication and Media Technologies, 4*(3), 124–142.

Asabere, N. Y. (2013). Benefits and Challenges of Mobile Learning Implementation: Story of Developing Nations. *International Journal of Computers and Applications, 73*(1).

Ask, J. A. (2014). *Mobile is Not a Channel.* Forrester Research Inc. Retrieved 1 August 2015 from http://urbanairship.com/lp/forrester-report-mobile-is-not-a-channel?ls=EmailMedia&ccn=150730eMarketerForrChannel&mkt_tok=3Rk MMJWWfF9wsRohvK3IZKXonjHpfsXw6egoUa%2BxlMI%2F0ER3fOvrPUfGjI4JRcRlI%2BSLDwEYGJlv6SgFQr bCMbNs3bgEWxA%3D

AT&T. (2014). *AT&T small business technology poll 2014*. Retrieved April 23, 2015, from http://about.att.com/mediakit/2014techpoll

Atkinson, R., & Flint, J. (2001). Accessing hidden and hard-to-reach populations: snowball research strategies. *Social Research Update*, (33). Retrieved November 03, 2012, from http://sru.soc.surrey.ac.uk/SRU33.html

Avidar, R., Ariel, Y., Malka, V., & Levy, E. C. (2013). Smartphones and young publics: A new challenge for public relations practice and relationship building. *Public Relations Review*, *39*(5), 603–605. doi:10.1016/j.pubrev.2013.09.010

Ayres, N. (2014). Adventures in Publishing: The new dynamics of advertising. *Brand Perfect*. Retrieved 14 April 2015 from http://brandperfect.org/brand-perfect-dynamics-of-advertising-report.pdf

Baber, C., & Westmancott, O. (2004). Social networks and mobile games: The use of bluetooth for a multiplayer card game (S. Brewster & M. Dunlop, Eds.). Academic Press.

Bachrach, Y., Kosinski, M., Graepel, T., Kohli, P., & Stillwell, D. (2012, June). Personality and patterns of Facebook usage. In *Proceedings of the 4th Annual ACM Web Science Conference* (pp. 24-32). ACM. doi:10.1145/2380718.2380722

Baek, T. H., & Morimoto, M. (2012). Stay away from me. *Journal of Advertising*, *41*(1), 59–76. doi:10.2753/JOA0091-3367410105

Baetens Beardsmore, H. (1996). Reconciling Content Acquisition and Language Acquisition in Bilingual Classrooms. *Journal of Multilingual and Multicultural Development*, *17*(2-4), 114–127. doi:10.1080/01434639608666263

Bai, Y., Yan, M., Yu, F., & Yang, J. (2014). Does market mechanism promote online/mobile information disclosure? Evidence from A–share companies on Shenzhen Exchange Market. *International Journal of Mobile Communications*, *12*(4), 380–396. doi:10.1504/IJMC.2014.063654

Balakrishnan, V., Guan, S. F., & Raj, R. G. (2011). A one-mode-for-all predictor for text messaging. *Maejo International Journal of Science and Technology*, *5*(2), 266–278.

Balakrishnan, V., & Raj, R. G. (2012). Exploring the relationship between urbanized Malaysian youth and their mobile phones: A quantitative approach. *Telematics and Informatics*, *29*(3), 263–272. doi:10.1016/j.tele.2011.11.001

Balayan, M. P. A., Conoza, V. V. B., Tolentino, J. M. M., Solamo, R. C., & Feria, R. P. (2014). On evaluating Skillville: An educational mobile game on visual perception skills. *IISA 2014, the 5th International Conference on Information, Intelligence, Systems and Applications*, (pp. 69 - 74).

Ball-Rokeach, S. J. (1998). A theory of media power and a theory of media use: Different stories, questions, and ways of thinking. *Mass Communication & Society*, *1*(1-2), 5–40. doi:10.1080/15205436.1998.9676398

Ball-Rokeach, S. J., & DeFleur, M. (1986). The interdependence of the media and other social systems. In G. Gumpert & R. Cathcart (Eds.), *Inter/media: Interpersonal communication in a media world* (pp. 81–96). Oxford, UK: Oxford University Press.

Ball-Rokeach, S. J., & DeFleur, M. L. (1976). A dependency model of mass-media effects. *Communication Research*, *3*(1), 3–21. doi:10.1177/009365027600300101

Banerjee, S., & Dholakia, R. R. (2013). Situated or ubiquitous? A segmentation of mobile e-shoppers. *International Journal of Mobile Communications*, *11*(5), 530–557. doi:10.1504/IJMC.2013.056959

Banfi, F., Begonha, D., Hazan, E., & Zouaoui, Y. (2014). Mobile must migrate: Digital as an imperative, not an option. *McKinsey publication of the Telecommunications, Media, and Technology Practice*, *24*(7).

Baran, E. (2014). A Review of Research on Mobile Learning in Teacher Education. *Journal of Educational Technology & Society*, *17*(4), 17–32.

Baran, S. (2006). *Introduction to Mass Communication: Media Literacy and Culture* (4th ed.). New York City, NY: McGraw-Hill.

Barker, A., Krull, G., & Mallinson, B. (2005, October). A proposed theoretical model for m-learning adoption in developing countries. In Proceedings of mLearn.

Barker, M., Donald, N., & Krista. (2008). *Social Media Marketing: A Strategic Approach*. Cengage Learning.

Barnes, S. J. (2003). *Mbusiness: The strategic implications of mobile communications*. Oxford: Butterworth and Heinemann.

Barton, D. (1991). The social nature of writing. In D. Barton & R. Ivanic (Eds.), Writing in the community. Sage.

Barton, D. (1994). *Literacy: An introduction to the ecology of written language*. Oxford, UK: Basil Blackwell.

Barton, D. (2015). Tagging on Flicker as a Social Practice. In R. Jones, A. Chick, & C. Hafner (Eds.), *Discourse and Digital Practices: Doing discourse analysis in the digital age*. London: Routledge.

Bart, Y., Stephen, A., & Sarvary, M. (2014). Which Products Are Best Suited to Mobile Advertising? A Field Study of Mobile Display Advertising Effects on Consumer Attitudes and Intentions.[JMR]. *JMR, Journal of Marketing Research*, *51*(3), 270–285. doi:10.1509/jmr.13.0503

Basole, R. C. (2008). Enterprise mobility: Researching a new paradigm. *Information-Knowledge-Systems Management*, *7*(1-2), 1-7.

Basole, R. (Ed.). (2008). *Enterprise mobility: Applications, technologies and strategies*. Amsterdam: IOS Press.

Basole, R. C., & Karla, J. (2012). Value transformation in the mobile service ecosystem: A study of app store emergence and growth. *Service Science*, *4*(1), 24–41. doi:10.1287/serv.1120.0004

Baumüller, H. (2013). Mobile technology trends and their potential for agricultural development (November 2013). *ZEF Working Paper 123*. Retrieved April 23, 2015, from http://ssrn.com/abstract=2359465

Bauwens, M. (2005). The political economy of peer production. *CTheory, 1*. Retrieved from http://www.ctheory.net/articles.aspx?id=499

Baynham, M. (2004). Ethnographies of Literacy: Introduction. *Language and Education*, *18*(4), 285–290. doi:10.1080/09500780408666881

Baynham, M., & Prinsloo, M. (2009). *The Future of Literacy Studies*. Basingstoke, UK: Palgrave Macmillan. doi:10.1057/9780230245693

Beaudoin, C. (2009). Exploring the association between news use and social capital: Evidence of variance by ethnicity and medium. *Communication Research*, *36*(5), 611–636. doi:10.1177/0093650209338905

Beaudoin, C. (2011). News Effects on Bonding and Bridging Social Capital: An Empirical Study Relevant to Ethnicity in the United States. *Communication Research*, *38*(2), 155–178. doi:10.1177/0093650210381598

Beer, M., & Nohria, N. (2000). Cracking the code of change. *Harvard Business Review*. Retrieved May 23, 2015, from http://webdb.ucs.ed.ac.uk/operations/honsqm/articles/change2.pdf

Bell, M., Chalmers, M., Barkhuus, L., Hall, M., Sherwood, S., & Tennent, P. (2006). Interweaving mobile games with everyday life.*Proceedings of SIGCHI Conference on Human Factors in Computing Systems*, (pp. 417-426). New York: ACM Press. doi:10.1145/1124772.1124835

Bellman, S., Potter, R. F., Treleaven-Hassard, S., Robinson, J. A., & Varan, D. (2011). The effectiveness of branded mobile phone apps. *Journal of Interactive Marketing*, *25*(4), 191–200. doi:10.1016/j.intmar.2011.06.001

Benjamin, B. (2006). The case study: Storytelling in industrial age and beyond. *On the Horizon*, *14*(4), 159–164. doi:10.1108/10748120610708069

Bergen, M. (2014). Verizon Wireless Dives into Mobile-Ad Business. *Advertising Age*, *85*(13), 10.

Bergvall, S. (2006). Brand ecosystems: Multilevel brand interaction. In J. E. Schroeder, M. Salzer-Mörling, & S. Askegaard (Eds.), *Brand culture* (pp. 166–175). New York, NY: Taylor & Francis.

Berking, P., Haag, J., Archibald, T., & Birtwhistle, M. (2012). Mobile learning: Not just another delivery method. In *Proceedings of the 2012 Interservice/Industry Training, Simulation, and Education Conference*.

Berman, D. (2004, June 29). *Advocacy Journalism, The Least You Can Do, and The No Confidence Movement*. Retrieved from Independent Media Center: https://www.indymedia.org/en/2004/06/854953.shtml

Bernard, R. M., Abrami, P. C., Lou, Y., Wade, A., & Borokhovski, E. (2004). The effects of synchronous and asynchronous distance education: A meta-analytical assessment of Simonson's "equivalency theory". In M. Simonson & M. Crawford (Eds.), *2004 annual proceedings of selected research and development papers presented at the national convention of the Association for Educational Communications and Technology* (pp.102-109). Chicago, IL: AECT.

Bernard, R. M., & Amundsen, C. L. (1989). Antecedents to dropout in distance education: Does one model fit all? *Journal of Distance Education*, *4*(2), 25–46.

Bersin, J. (2014). The Red Hot Market for Learning Management Systems. *Forbes*. Retrieved July 31, 2015, from http://www.forbes.com/sites/joshbersin/2014/08/28/the-red-hot-market-for-learning-technology-platforms/1

Bersin, J. (2014a). The Red Hot Market for Learning Management Systems. *Forbes*. Retrieved July 31, 2015, from http://www.forbes.com/sites/joshbersin/2014/08/28/the-red-hot-market-for-learning-technology-platforms/1

Bersin, J. (2014b). *The Corporate Learning Factbook 2014: Benchmarks, trends, and analysis of the U.S. training market*. Retrieved Sept. 7, 2015, from http://www.bersin.com/uploadedFiles/012714WWBCLF.pdf

Bertlatsky, N. (2015, January 7). *Hashtag activism isn't a cop-out*. Retrieved from The Atlantic: http://www.theatlantic.com/politics/archive/2015/01/not-just-hashtag-activism-why-social-media-matters-to-protestors/384215/

Bertoni, S. (2014, February 21). How do you win the mobile wallet war? Be like Starbucks. *Forbes*. Retrieved from http://www.forbes.com/sites/stevenbertoni/2014/02/21/how-do-you-win-the-mobile-wallet-war-be-like-starbucks/

Bimber, B. (1990). *Karl Max and the Three faces of Technological Determinism*. Working Paper No.11. Massachusetts Institute of Technology.

Blackman, S. L. (2010). *Cell phone usage patterns with friends, parents, and romantic partners in college freshmen*. Retrieved from http://trace.tennessee.edu/cgi/viewcontent.cgi?article=2380&context=utk_chanhonoproj

Blanchard, A., & Horan, T. (1998). Virtual communities and social capital. *Social Science Computer Review*, *16*(3), 293–307. doi:10.1177/089443939801600306

Bloomberg. (2015). *USDKZT:CUR*. Retrieved 24 Dec. 2015 from http://www.bloomberg.com/quote/USDKZT:CUR

Blueshift Research. (2015). *June 2015 trends tracker report*. Retrieved 2 July 2015 from https://smaudience.surveymonkey. com/download-trends-tracker-report-june-2015.html?utm_source=email&recent=email&program=Q2_15_June_Trends-Tracker_Wrap-Up&source=email&utm_campaign=trends-tracker&utm_content=trends-tracker-june-2015&mkt_tok= 3RkMMJWWfF9wsRoiu6vJZKXonjHpfsX66e4vW6C1lMI%2F0ER3fOvrPUfGjI4FTcFiI%2BSLDwEYGJlv6SgFSrbF MaJy2LgJWBb0TD7slJfbfYRPf6Ba2Jwwrfg%3D

Blueshift Research. (2015). *June 2015 Trends Tracker Report*. Retrieved 2 July 2015 from https://smaudience.surveymonkey. com/download-trends-tracker-report-june-2015.html?utm_source=email&recent=email&program=Q2_15_June_Trends-Tracker_Wrap-Up&source=email&utm_campaign=trends-tracker&utm_content=trends-tracker-june-2015&mkt_tok= 3RkMMJWWfF9wsRoiu6vJZKXonjHpfsX66e4vW6C1lMI%2F0ER3fOvrPUfGjI4FTcFiI%2BSLDwEYGJlv6SgFSrbF MaJy2LgJWBb0TD7slJfbfYRPf6Ba2Jwwrfg%3D

Blumer, H. (1969). *Symbolic interactionism: Perspective and method*. Berkeley, CA: University of California Press.

Blumer, H. (1986). *Symbolic interaction: Perspective and method*. Berkeley, CA: University of California Press.

Blumler, J. (1979). The role of theory in uses and gratifications studies. *Communication Research, 6*(1), 9–36. doi:10.1177/009365027900600102

Boase, J., & Kobayashi, T. (2008). Kei-Tying teens: Using mobile phone e-mail to bond, bridge, and break with social ties – a study of Japanese adolescents. *International Journal of Human-Computer Studies, 66*(12), 930–943. doi:10.1016/j. ijhcs.2008.07.004

Bode, L., Vraga, E. K., Borah, P., & Shah, D. V. (2014). A new space for political behavior: Political social networking and its democratic consequences. *Journal of Computer-Mediated Communication, 19*(3), 414–429. doi:10.1111/jcc4.12048

Borison, R. (2014, June 17). The 15 most successful app companies. *Business Insider*. Retrieved from http://www. businessinsider.com/15-most-successful-app-companies-2014-6?IR=T#7-cj-group-9

Borucki, C., Arat, S., & Kushchu, I. (2005, July). Mobile government and organizational effectiveness. In *Proceedings of the First European Conference on Mobile Government, Brighton, UK:Mobile Government Consortium* (pp. 56-66).

Boston Consulting Group. (2014). The mobile revolution: How mobile technologies drive a trillion-dollar impact. *Boston Consulting Group Report*. Retrieved May 27, 2015, from https://www.bcgperspectives.com/content/articles/telecom-munications_technology_business_transformation_mobile_revolution/?chapter=4

Botan, C. H., & Hazleton, V. (Eds.). (2006). *Public relations theory II*. Mahwah, NJ: Lawrence Erlbaum Associaets, Inc., Publishers.

Botan, C. H., & Taylor, M. (2004). Public relations: State of the field. *Journal of Communication, 54*(4), 645–661. doi:10.1111/j.1460-2466.2004.tb02649.x

Bourdieu, P. (1991). *Language and Symbolic Power*. Cambridge, MA: Polity.

Bower, J., & Christensen, C. (1995). Disruptive technologies: Catching the wave. *Harvard Business Review*. Retrieved June 12, 2015, from https://hbr.org/1995/01/disruptive-technologies-catching-the-wave

Bowlby, J. (1980). Attachment and loss, volume 3: loss; sadness and depression. New York, NY: Basic Books.

Boyle, M., & Schmierbach, M. (2009). Media use and protest: The role of mainstream and alternative media use in predicting traditional and protest participation. *Communication Quarterly, 57*, 1-17.

Brackett, L. K., & Carr, B. N. Jr. (2001). Cyberspace advertising vs. other media: Consumer vs. mature student attitudes. *Journal of Advertising Research, 41*(5), 23–32. doi:10.2501/JAR-41-5-23-32

Brady, M. (1998). *Strategies for Effective Teaching: Using Interactive Video in the Distance Education Classroom.* Retrieved from http://www.designingforlearning.info/services/writing/interact.htm

Brakus, J. J., Schmitt, B. H., & Zarantonello, L. (2009). Brand experience: What is it? How is it measured? Does it affect loyalty? *Journal of Marketing, 73*(3), 52–68. doi:10.1509/jmkg.73.3.52

Brandão, C. A. L. (2008). A transdisciplinaridade. In *A transdisciplinaridade e os desafios contemporâneos* (pp. 17–39). Belo Horizonte, Minas Gerais: IEAT/UFMG.

Brandt, D., & And Clinton, K. (2002). Limits of the Local: Expanding Perspectives on Literacy as a Social Practice. *Journal of Literacy Research, 34*(3), 337–356. doi:10.1207/s15548430jlr3403_4

Bresser-Pereira, L. C. (1998). A Reforma do Estado dos anos 90: lógica e mecanismos de controle. In CLAD (Ed.), *Centro Latinoamericano de Administración para el Desarrollo.* Retrieved December 17, 2015, from http://old.clad. org/congresos/congresos-anteriores/ii-isla-de-margarita-1997/a-reforma-do-estado-dos-anos-90-logica-e-mecanismos-decontrole/?searchterm=A%20Reforma%20do%20Estado%20dos%20anos%2090

Brewer, D. (n.d.). *Are journalism and activism compatible?* Retrieved from Media Helping Media: http://www.media-helpingmedia.org/43-news-archive/global/341-are-journalism-and-activism-compatible

Briggs, M. (2008, Winter). The End of Journalism as Usual. *Nieman Reports*, 40-41.

Brightcove. (2015). *The hero's guide to video marketing.* Retrieved 1 May 2015 from http://go.brightcove.com/brightcove-video-hero-guide?cid=70114000002QvKv&pid=70114000002Qu9G

Brightcove. (2015). *The hero's guide to video marketing.* Retrieved May 1, 2015, from http://go.brightcove.com/brightcove-video-hero-guide?cid=70114000002QvKv&pid=70114000002Qu9G

Briones, R. L., Kuch, B., Liu, B. F., & Jin, Y. (2011). Keeping up with the digital age: How the American Red Cross uses social media to build relationships. *Public Relations Review, 37*(1), 37–43. doi:10.1016/j.pubrev.2010.12.006

Briones, R., Madden, S., & Janoske, M. (2013). Kony 2012: Invisible children and the challenges of social media campaigning and digital activism. *Journal Of Current Issues In Media & Telecommunications, 5*(3), 205–234.

Brister, K. (2010). Making the Most of Mobile; Banks can capitalize on marketing strategies for handheld devices, even if mobile banking is still short of being mainstream. *US Banker*, 17-17.

Brister, K. (2010, March 1). Making the Most of Mobile; Banks can capitalize on marketing strategies for handheld devices, even if mobile banking is still short of being mainstream. *US Banker*, 17-17.

Brown, D., & Grinter, R. E. (2014). Aboard Abroad: Supporting Transnational Parent–School Communication in Migration-Separated Families. *Information Technologies & International Development, 10*(2), 49.

Brown, R. I. F. (1993). Some contributions of the study of gambling to the study of other addictions. In W. R. Eadington & J. A. Cornelius (Eds.), Gambling behavior and problem gambling (pp. 241–272). Reno: Institute for the study of gambling and commercial gaming, University of Nevada.

Brown, T. (2003, June). The role of m-learning in the future of e-learning in Africa. In *21st ICDE World Conference.* Retrieved from http://www.tml.tkk.fi/Opinnot

Brown, B., Court, D., & McGuire, T. (2014, March). Views from the front lines of the data-analytics revolution. *The McKinsey Quarterly.*

Brown, R. I. F. (1997). A theoretical model of the behavioural addictions - Applied to offending. In J. E. Hodge, M. McMurran, & C. R. Hollin (Eds.), *Addicted to crime.* Chichester, UK: John Wiley.

Brown, R. L., & Harmon, R. R. (2014). Viral geofencing: An exploration of emerging big-data driven direct digital marketing services. *Proceedings of the International Conference on Management of Engineering & Technology (PICMET)*.

Brown, S., Kozinets, R. V., & Sherry, J. F. Jr. (2003). Teaching old brands new tricks: Retro branding and the revival of brand meaning. *Journal of Marketing, 67*(3), 19–33. doi:10.1509/jmkg.67.3.19.18657

Bruce Schneier, D. L. (2003, August). *Voting and Technology: Who Gets to Count Your Vote?* Retrieved from Schneier on Security: https://www.schneier.com/essays/archives/2003/08/voting_and_technolog.html

Brunton, C. (2009). *Final results: Voter and non-voter satisfaction survey 2008*. Colmar Brunton.

Brusse, C., Gardner, K., McAullay, D., & Dowden, M. (2014). Social media and mobile apps for health promotion in Australian Indigenous populations: Scoping review. *Journal of Medical Internet Research, 16*(12), e280. doi:10.2196/jmir.3614 PMID:25498835

Brynjolfsson, E., & McAfee, A. (2011). *Race against the machine*. Lexington, MA: Digital Frontier.

Brynjolfsson, E., & McAfee, A. (2014). *The second machine age: Work, progress, and prosperity in a time of brilliant technologies*. New York, NY: WW Norton & Company.

Budiman, R. (2015). Distance language learning: Students' views of challenges and solutions. *International Journal on New Trends in Education, 6*(3), 137–147.

Bulearca, M., & Tamarjan, D. (2010). Augmented reality: A sustainable marketing tool? *Global Business and Management Research: An International Journal, 2*(2), 237–252.

Burns, S. (2014, October 16). *'Advocacy' Is Not a Dirty Word in Journalism*. Retrieved from MediaShift: http://www.pbs.org/mediashift/2014/10/advocacy-is-not-a-dirty-word-in-journalism/

Bush, M. D. (2008). Computer-assisted language learning: From vision to reality? *CALICO Journal, 25*(3), 443–470.

Buttussi, F., & Chittaro, L. (2010). Smarter phones for healthier lifestyles: An adaptive fitness game. *IEEE CS*, 51-57.

Caballero, E. G. (2005). Pluralidad teórica, metodológica y técnica en el abordaje delas redes sociales: Hacia la "hibridación" disciplinaria. *Revista Hispana para el Análisis de Redes Sociales, 1*(9), 1–24.

Cakiroglu, U. (2014). Analyzing the effect of learning styles and study habits of distance learners on learning performances: A case of an introductory programming course. *International Review of Research in Open and Distance Learning, 15*(4).

Caldwell Community College and Technical Institute. (2015). Hudson, NC: Distance Learning Student Information Guide.

Cambridge Dictionaries Online. (2015, December 9). *Definition of interaction from the Cambridge Advanced Learner's Dictionary & Thesaurus*. Retrieved from Cambridge Dictionaries Online: http://dictionary.cambridge.org/dictionary/english/interaction

Campaign Finance Institute. (2010). *All CFI Funding Statistics Revised and Updated for the 2008 Presidential Primary and General Election Candidates*. Retrieved 30 July 2015 from http://www.cfinst.org/press/releases_tags/10-01-08/Revised_and_Updated_2008_Presidential_Statistics.aspx

Campbell, A. (2014, December 10). *#TBT: #BringBackOurGirls—Exploring The Potential (and Perils) of Hashtag Activism*. Retrieved from Medium: https://medium.com/@internetweek/tbt-bringbackourgirls-exploring-the-potential-and-perils-of-hashtag-activism-ab862425f2a0

Campbell, M. A. (2005). *The impact of the mobile phone on young people's social life*. Social Change in the 21st Century Conference, QUT Carseldine, Brisbane.

Campbell, S. W., & Park, Y. J. (2008). Social implications of mobile telephony: The rise of personal communication society. *Sociology Compass, 2*(2), 371-387.

Campbell, S. W. (2006). College Classrooms: Ringing, Cheating, and Classroom Policies. *Communication Education, 55*(3), 280–294. doi:10.1080/03634520600748573

Caronia, L. (2005). Mobile Culture: An Ethnography of Cellular Phone Uses in Teenagers' Everyday Life. *Convergence (London), 11*(96), 96–103. doi:10.1177/135485650501100307

Carr, D. (2012, March 25). *Hashtag Activism, and Its Limits*. Retrieved from New York Times: http://www.nytimes.com/2012/03/26/business/media/hashtag-activism-and-its-limits.html?_r=0

Carroll, J. (2005, July). Risky Business: Will Citizens Accept M-government in the Long Term?'. In Euro mGov (pp. 77-87).

Carroll, J. (2006). 'What's in It for Me?': Taking M-Government to the People. *BLED 2006 Proceedings*, 49.

Carroll, A., Barnes, S. J., Scornavacca, E., & Fletcher, K. (2007). Consumer perceptions and attitudes towards SMS advertising: Recent evidence from New Zealand. *International Journal of Advertising, 26*(1), 79–98.

Carstensen, L. L., Isaacowitzm, D. M., & Charles, S. T. (1999). Taking Time Seriously: A Theory of Socioemotional Selectivity. *The American Psychologist, 54*(3), 165–181. doi:10.1037/0003-066X.54.3.165 PMID:10199217

Carufel, R. (2013). Connection Disconnection: 80 Percent of Brands Don't Know Their Customers, New Yesmail and Gleanster Study of Executive-Level Marketers Reveals. *Bulldog Reporter's Daily Dog*. Retrieved on 10 April 2015 from https://www.bulldogreporter.com/dailydog/article/pr-biz-update/connection-disconnection-80-percent-of-brands-don-t-know-their-custom

Carufel, R. (2013, May 30). *Marketing's uncharted territory: 97% of B2B marketers are doing work they've never done before and expect the pace of change to accelerate as they "navigate chaos"*. Retrieved April 23, 2015, from https://www.bulldogreporter.com/dailydog/article/pr-biz-update/marketings-uncharted-territory-97-of-b2b-marketers-are-doing-work-the

Carufel, R. (2015, March 31). *Technology PR standoff: Research finds consumers and retailers diverge on tech—buyers want more, retailers don't*. Retrieved April 23, 2015, from https://www.bulldogreporter.com/dailydog/article/pr-biz-update/technology-pr-standoff-research-finds-consumers-and-retailers-diverge

Carufel, R. (Ed.). (2015). *Millennials Moving Away from TV: Boob Tube Content "Doesn't Cut It"—Digital Delivers More Relatable and Entertaining Programming*. Retrieved 15 Oct. 2015 from https://www.bulldogreporter.com/millennials-moving-away-from-tv-boob-tube-content-doesn-t-cut-it-digi/

Carvin, A. (2012). *Distant witness: Social media, the Arab Spring and a journalism revolution*. CUNY Journalism Press.

Casey, S., Kirman, B., & Rowland, D. (2007). The gopher game: A social, mobile, locative game with user generated content and peer review.*Proceedings of ACE '07 of the International Conference on Advances in Computer Entertainment Technology*, (pp. 9-16). New York: ACM. doi:10.1145/1255047.1255050

Castells, M., Fernández-Ardèvol, M., Qiu, J. L., & Sey, A. (2006). *Mobile communication and society: A global perspective (information revolution & global politics)*. Cambridge, MA: MIT Press.

Cavoukian, A., & Prosch, M. (2010). *The Roadmap for Privacy by Design in Mobile Communications: A Practical Tool for Developers, Service Providers, and Users*. Retrieved December 16, 2015, https://www.privacybydesign.ca/content/uploads/2011/02/pbd-asu-mobile.pdf

Cavus, N., & Ibrahim, D. (2009). m-Learning: An experiment in using SMS to support learning new English language words. *British Journal of Educational Technology, 40*(1), 78–91. doi:10.1111/j.1467-8535.2007.00801.x

CCR CSR News. (2014). *China's "Left-Over Children" Face Greater Chances of Psychological Problems, Juvenile Delinquency and Sexual Assault.* Retrieved from http://www.ccrcsr.com/news/china's-"left-over-children"-face-greater-chances-psychological-problems-juvenile-delinquency

Celino, I., Cerizza, D., Contessa, S., Corubolo, M., DellAglio, D., Valle, E. D., & Fumeo, S. (2012). Urbanopoly—A social and location-based game with a purpose to crowdsource your urban data. *2012 ASE/IEEE International Conference on Social Computing and 2012 ASE/IEEE International Conference on Privacy, Security, Risk and Trust,* (pp. 910-913). doi:10.1109/SocialCom-PASSAT.2012.138

Cerwall, P. (Ed.). (2015, February). *Ericsson mobility report: On the pulse of the networked society, Mobile World Congress Edition.* Retrieved from http://www.ericsson.com/res/docs/2015/ericsson-mobility-report-feb-2015-interim.pdf

César, P., Knoche, H., & Bulterman, D. (2010). *Mobile TV: Customizing Content and Experiences.* Heidelberg: Springer.

Chan, D. (2008). Convergence, connectivity, and the case of Japanese mobile gaming. *Games and Culture, 3*(1), 13–25. doi:10.1177/1555412007309524

Chang, A. M., & Kannan, P. K. (2002). Preparing for wireless and mobile technologies in government. *E-government,* 345-393.

Chang, C. (2009). 'Being hooked' by editorial content: The implications for processing narrative advertising. *Journal of Advertising, 38*(3), 51–65. doi:10.2753/JOA0091-3367380304

Chan, M. (2015). Mobile phones and the good life: Examining the relationships among mobile use, social capital and subjective well-being. *New Media & Society, 17*(1), 96–113. doi:10.1177/1461444813516836

Chao, P.-Y., & Chen, G.-D. (2009). Augmenting paper-based learning with mobile phones. *Interacting with Computers, 21*(3), 173–185. doi:10.1016/j.intcom.2009.01.001

Chee, Y. S., Tan, E. M., & Liu, Q. (2010). Statecraft X: Enacting citizenship education using a mobile learning game played on Apple iPhones. *The 6th IEEE International Conference on Wireless, Mobile, and Ubiquitous Technologies in Education,* (pp. 222-224).

Chen, W. P., Millard, D. E., & Wills, G. B. (2008). Mobile VLE vs. mobile PLE: How informal is mobile learning? In J. Traxler, B. Riordan & C. Dennett (Eds.), *Proceedings of the mLearn2008 Conference.* University of Wolverhampton.

Chen, J. V., Su, B., & Yen, D. C. (2014). Location-based advertising in an emerging market: A study of Mongolian mobile phone users. *International Journal of Mobile Communications, 12*(3), 291–310. doi:10.1504/IJMC.2014.061462

Chen, W., & Hirscheim, R. (2004). A paradigmatic and methodological examination of information systems research from 1991 to 2001. *Information Systems Journal, 14*(3), 197–35. doi:10.1111/j.1365-2575.2004.00173.x

Chen, Y.-F. (2006). Social Phenomena of Mobile Phone Use: An Exploratory Study in Taiwanese College Students. *Journal of Cyber Culture and Information Society, 11,* 219–244.

Chen, Y.-F., & Katz, J. E. (2009). Extending family to school life: College students' use of the mobile phone. *International Journal of Human-Computer Studies, 67*(2), 179–191. doi:10.1016/j.ijhcs.2008.09.002

Chernov, J. (2014). State of inbound 2014. *HubSpot.* Retrieved May 21, 2015 from http://www.stateofinbound.com/

Chernov, J. (2014). *State of Inbound 2014.* Cambridge, MA: Hubspot.

Cheung, S. (2004). Fun and games with mobile phones: SMS messaging in microeconomics experiments. In R. Atkinson, C. McBeath, D. Jonas-Dwyer & R. Phillips (Eds.), *Beyond the comfort zone:Proceedings of the 21st ASCILITE Conference*, (pp. 180-183). Perth, Australia: ASCILITE.

Chib, A., Malik, S., Aricat, R. G., & Kadir, S. Z. (2014). Migrant mothering and mobile phones: Negotiations of transnational identity. *Mobile Media & Communication*, *2*(1), 73–93. doi:10.1177/2050157913506007

Chiem, R., Arriola, J., Browers, D., Gross, J., Limman, E., Nguyen, P. V., & Seal, K. C. et al. (2010). The critical success factors for marketing with downloadable applications: Lessons learned from selected European countries. *International Journal of Mobile Marketing*, *5*(2), 43–56.

Choi, Y. K., Hwang, J., & McMillan, S. J. (2008). Gearing up for mobile advertising: A cross-cultural examination of key factors that drive mobile messages home to consumers. *Psychology and Marketing*, *25*(8), 756–768. doi:10.1002/mar.20237

Chou, H., & Lien, N. (2014). Effects of SMS teaser ads on product curiosity. *International Journal of Mobile Communications*, *12*(4), 328–345. doi:10.1504/IJMC.2014.063651

Chowdhury, H. K., Parvin, N., & Weitenberner, C. et al.. (2010). Consumer attitude toward mobile advertising in an emerging market: An empirical study. *Marketing*, *12*, 206–216.

Christensen, C. (2012). *Dr. Clayton Discusses Disruption in Higher Education*. [Video File]. Retrieved from https://www.youtube.com/watch?v=yUGn5ZdrDoU&feature=youtu.be

Christensen, C. M. (2014). Disruptive Innovation. In *The Encyclopedia of Human-Computer Interaction* (2nd ed.). Retrieved 26 June 2015 from https://www.interaction-design.org/encyclopedia/disruptive_innovation.html

Christensen, C. M. (2014). Disruptive innovation. In *The Encyclopedia of Human-Computer Interaction*, 2nd ed. Retrieved 26 June 2015 from https://www.interaction-design.org/encyclopedia/disruptive_innovation.html

Christians, C. (1989). A theory of normative technology. In *Technological Transformation* (pp. 123–139). Springer Netherlands. doi:10.1007/978-94-009-2597-7_9

Chuang, K.-W. (2009). Mobile Technologies Enhance The E-Learning Opportunity. *American Journal of Business Edueation*, *2*(9), 49–54.

Chui, M., Manyika, J., Bughin, J., Dobbs, R., Roxburgh, C., Sarrazin, H.,... Westergren, M. (2012). *The social economy: Unlocking value and productivity through social technologies*. Retrieved 29 April 2015 from http://www.mckinsey.com/insights/high_tech_telecoms_internet/the_social_economy

Chuttur. (2009). Overview of the Technology Acceptance Model: Origins, Developments and Future Directions. *Information Systems*. Indiana University.

Chyi, H. I. (2015). *Trial and Error: U.S. Newspapers' Digital Struggles toward Inferiority*. The University of Navarra. Retrieved 15 Oct. 2015 from http://irischyi.com/

Ciadvertising. (2007). *Hierarchy-of-effects models*. Retrieved from Ciadvertising: http://www.ciadvertising.org/studies/student/97_fall/theory/hierarchy/modern.html

Cianci, A. M., & Kaplan, S. E. (2010). The effect of CEO reputation and explanations for poor performance on investors' judgments about the company's future performance and management. *Accounting, Organizations and Society*, *35*(4), 478–495. doi:10.1016/j.aos.2009.12.002

Ciaramitaro, B. L. (Ed.). (2011). *Mobile technology consumption: Opportunities and challenges*. Hershey, PA: IGI Global.

Cisco (2015). *Forecast projects nearly 10-fold global mobile data traffic growth over next five years.* Retrieved February 16, 2015, from http://newsroom.cisco.com/press-release-content?type=webcontent&articleId=1578507

Cisco Systems Inc. (2014). *2014 Corporate Social Responsibility Report.* Retrieved from http://www.cisco.com/assets/csr/pdf/CSR_Report_2014.pdf

Cisco Systems Inc. (2014). *Cisco Visual Networking Index: Forecast and Methodology, 2013–2018.* Retrieved 1 May 2015 from http://www.cisco.com/c/en/us/solutions/collateral/service-provider/ip-ngn-ip-next-generation-network/white_paper_c11-481360.html

Cisco. (2012). *Cisco's VNI Forecast Projects the Internet Will Be Four Times as Large in Four Years.* Retrieved 15 May 2015 from http://newsroom.cisco.com/press-release-content?type=webcontent

Cisco. (2012). *Cisco's VNI Forecast Projects the Internet Will Be Four Times as Large in Four Years.* Retrieved 16 Feb. 2015 from http://newsroom.cisco.com/press-release-content?type=webcontent&articleId=888280

Cisco. (2012). *Cisco's VNI Forecast Projects the Internet Will Be Four Times as Large in Four Years.* Retrieved from http://newsroom.cisco.com/press-release-content?type=webcontent&articleId=888280

Cisco. (2014). *2014 Connected World Technology Final Report.* Retrieved 15 Oct. 2015 from http://www.cisco.com/c/dam/en/us/solutions/collateral/enterprise/connected-world-technology-report/cisco-2014-connected-world-technology-report.pdf

Cisco. (2015). *Forecast projects nearly 10-fold global mobile data traffic growth over next five years.* Retrieved 16 Feb. 2015 from http://newsroom.cisco.com/press-release-content?type=webcontent

Cisco. (2015). *Forecast projects nearly 10-fold global mobile data traffic growth over next five years.* Retrieved 16 Feb. 2015 from http://newsroom.cisco.com/press-release-content?type=webcontent&articleId=1578507

Cisco. (2015). *Forecast projects nearly 10-fold global mobile data traffic growth over next five years.* Retrieved from http://newsroom.cisco.com/press-release-content?type=webcontent&articleId=1578507

Clay, M. (1999). Development of Training and Support Programs for Distance Education Instructors. *Online Journal of Distance Learning Administration, 2*(3). Retrieved from http://www.westga.edu/~distance/clay23.html

Clifton, R., & Simmons, J. (Eds.). (2003). *Brands and branding.* London: Profile Books Ltd, The Economist Newspaper Ltd.

Clough, G. (2016). Mobile informal learning through geocatching. In J. Traxler & A. Kukulska-Hulme (Eds.), *Mobile learning: The next generation* (pp. 43–66). New York: Routledge.

Clough, G., Jones, A., McAndrew, P., & Scanlon, E. (2008). Informal learning with PDAs and smartphones. *Journal of Computer Assisted Learning, 24*(5), 359–371. doi:10.1111/j.1365-2729.2007.00268.x

CNNIC. (2015). *China Internet Network Information Center.* Retrieved from CNNIC: http://www.cnnic.cn/gywm/xwzx/rdxw/2015/201502/t20150203_51631.htm

Cochrane, T. (2005). Mobilising learning: A primer for utilising wireless palm devices to facilitate a collaborative learning environment. In *Proceedings of the 2005 ASCILITE conference.* Brisbane, Australia: ASCILITE.

Cohen, R. (2012, April 26). *Why did "Kony 2012" fizzle out?* Retrieved from NonProfit Quarterly: https://nonprofitquarterly.org/policysocial-context/20216-why-did-kony-2012-fizzle-out.html

Coiro, J., Knobel, M., Lankshear, C., & Leu, D. (Eds.). (2008). *Handbook of research on new literacies.* Mahwah, NJ: Erlbaum.

Coleman, R. (1997). The intellectual antecedents of public journalism. *The Journal of Communication Inquiry, 21*(1), 60–76. doi:10.1177/019685999702100103

Comitê Gestor de Internet do Brasil. (2014). Retrieved November 7, 2014, from http://www.cgi.br

comScore. (2013). *U.S. digital future in focus 2013.* Retrieved 16 Feb. 2015 from http://www.comscore.com/Insights/Presentations-and-Whitepapers/(offset)/10/?cs_edgescape_cc=KZ

comScore. (2014). *Half of Millennial Netflix Viewers Stream Video on Mobile.* Retrieved 20 Feb. 2015 from http://www.comscore.com/Insights/Data-Mine/Half-of-Millennial-Netflix-Viewers-Stream-Video-on-Mobile

comScore. (2014). *Half of millennial netflix viewers stream video on mobile.* Retrieved February 20, 2015, from http://www.comscore.com/Insights/Data-Mine/Half-of-Millennial-Netflix-Viewers-Stream-Video-on-Mobile

comScore. (2015). *U.S. digital future in focus 2015.* Retrieved May 19, 2015, from http://www.comscore.com/Insights/Presentations-and-Whitepapers/2015/2015-US-Digital-Future-in-Focus

Conejar, R. J., & Kim, H. K. (2014). The Effect of the Future Mobile Learning: Current State and Future Opportunities. *International Journal of Software Engineering & Its Applications, 8*(8).

CONIP SP. (2014). *Congresso de Inovação e Informática na Gestão Pública.* Retrieved November 02, 2014, from http://www.conipsp.com

Connolly, J. A., & Konarski, R. (1994). Peer self-concept in adolescence: Analysis of factor structure and of associations with peer experience. *Journal of Research on Adolescence, 4*(3), 385–403. doi:10.1207/s15327795jra0403_3

Constantinides, E. (2006). The marketing mix revisited: towards the 21st century marketing. *Journal of Marketing Management, 22*(3-4), 407-438.

Consulting, F. (2013b). *Your Customers Are Demanding Omni-Channel Communications. What Are You Doing About It?* Retrieved 13 April 2015 from http://www.kana.com/customer-service/white-papers/forrester-your-customers-are-demanding-omni-channel-communications.pdf?_ga=1.38271843.1312302063.1428939602

Consulting, P. A. (2015). *Wearable Technologies Improves Health Outcomes and Reduces Costs.* Retrieved May 2, 2015, from http://www.paconsulting.com/introducing-pas-media-site/highlighting-pas-expertise-in-the-media/articles-quoting-pa-experts/wfmd-am-930-wearable-technology-improves-health-outcomes-and-reduces-cost-10-march-2015/

Content Marketing Institute. (2013). *B2B Content Marketing 2014 Benchmarks, Budgets, and Trends – North America.* Retrieved 20 Feb. 2015 from http://contentmarketinginstitute.com/wp-content/uploads/2013/10/B2B_Research_2014_CMI.pdf/

Cook, F. L., Tyler, T. R., Goetz, E. G., Gordon, M. T., Protess, D., Leff, D. R., & Molotch, H. L. (1983). Media and agenda setting: Effects on the public, interest group leaders, policy makers, and policy. *Public Opinion Quarterly, 47*(1), 16–35. doi:10.1086/268764 PMID:10261275

Coombs, W. T., & Holladay, S. J. (2001). An extended examination of the crisis situations: A fusion of the relational management and symbolic approaches. *Journal of Public Relations Research, 13*(4), 321–340. doi:10.1207/S1532754XJPRR1304_03

Corpuz, J. (2015, March 4). Ten best mobile racing games. *Tom's Guide.* Retrieved from http://www.tomsguide.com/us/best-mobile-racing-games,review-2351.html

Corrocher, N., & Zulrilia, L. (2004). *Innovation and Schumpeterian competition in the mobile communications service industry*. Università Commerciale L. Bocconi, Milan. Retrieved May 1, 2015, from http://www2.druid.dk/conferences/viewpaper.php?id=2577&cf=17

Costa, A. de S. M. da, Barros, D. F., & Martins, P. E. M. (2010). Perspectiva histórica em administração: novos objetos, novos problemas, novas abordagens. *RAE – Revista Administração de Empresas, 50*(3), 288-299. doi: 10.1590/S0034-75902010000300005

Crabtree, A., Benford, S., Capra, M., Flintham, M., Drozd, A., Tandavanitj, N., & Farr, J. R. et al. (2007). The cooperative work of gaming: Orchestrating a mobile SMS game. *Computer Supported Cooperative Work, 16*(1-2), 167–198. doi:10.1007/s10606-007-9048-1

Crawford, A. P. (1999). When those nasty rumors start breeding on the Web, you're got to move fast. *Public Relations Quarterly*, 43-45.

Cristian, P. E., Thirumurthy, H., Habyarimana, J. P., Zivin, J. G., Goldstein, M. P., Damien, D. W., & Bangsberg, D. J. et al. (2011). Mobile phone technologies improve adherence to antiretroviral treatment in a resource-limited setting: A randomized controlled trial of text message reminders. *AIDS (London, England), 25*(6), 825–834. doi:10.1097/QAD.0b013e32834380c1 PMID:21252632

Cunha, M. A., Anneberg, R. M., & Agune, R. (2007). Prestação de serviços públicos eletrônicos ao cidadão. In P. T. Knight, C. C. C. Fernandes, & M. A. Cunha (Eds.), E-desenvolvimento no Brasil e no mundo: subsídios e Programa e-Brasil (pp. 559-584). São Caetano do Sul, São Paulo: Yendis.

Cunha, M. A., & Miranda, P. R. M. (2008). A pesquisa no uso e implicações sociais das tecnologias de informação e comunicação pelos governos no Brasil: uma proposta de agenda a partir da prática e da produção acadêmica nacional. In *Anais do XXXII ENANPAD*. Rio de Janeiro: ANPAD - Associação Nacional de Pós-Graduação e Pesquisa em Administração.

Dahlke, D. V., Fair, K., Hong, Y. A., Beaudoin, C. E., Pulczinski, J., & Ory, M. G. (2015). Apps seeking theories: Results of a study on the use of health behavior change theories in cancer survivorship mobile apps. *JMIR mHealth and uHealth, 3*(1).

Dahlstrom, E., & Bichsel, J. (2014). *ECAR Study of Undergraduate Students and Information Technology, 2014*. Louisville, CO: ECAR, Oct. 2014, retrieved from http://www.educause.edu/library/resources/2014-student-and-faculty-technology-research-studies

Dahlstrom, E., & Bichsel, J. (2014). *ECAR Study of Undergraduate Students and Information Technology, 2014*. Louisville, CO: ECAR. Retrieved from http://www.educause.edu/library/resources/2014-student-and-faculty-technology-research-studies

Dahlstrom, E., Walker, J. D., & Dziuban, C. (2013). *ECAR Study of Undergraduate Students and Information Technology, 2013*. Louisville, CO: ECAR (EDUCAUSE Center for Analysis and Research), retrieved from http://www.educause.edu/library/resources/ecar-study-undergraduate-students-and-information-technology-2013

Dahlstrom, E., Walker, J. D., & Dziuban, C. (2013). *ECAR Study of Undergraduate Students and Information Technology, 2013*. Louisville, CO: ECAR (EDUCAUSE Center for Analysis and Research). Retrieved from http://www.educause.edu/library/resources/ecar-study-undergraduate-students-and-information-technology-2013

Danova, T. (2014, January 5). The mobile video revolution: How Netflix, Vevo, and YouTube have thrived on smartphones and tablets. *Business Insider*. Retrieved from http://www.businessinsider.com/mobile-video-statistics-and-growth-2013-12#ixzz3XODJpOxH

Darrell, M. W. (2013). *Mobile learning: Transforming education, engaging students, and improving outcomes.* Retrieved March 23, 2015, from http://www.brookings.edu/~/media/research/files/papers/2013/09/17-mobile-learning-education-engaging-students-west/brookingsmobilelearning_final.pdf

Data-Pop Alliance. (n.d.). *Data "inflation" table.* Retrieved from http://www.datapopalliance.org/resources#data-inflation-table

Davies, J. (2013, May 7). Consumer magazines are failing to capitalise on mobile ad opportunities, says Monotype report. *The Drum.* Retrieved from http://www.thedrum.com/news/2013/05/07/consumer-magazines-are-failing-capitalise-mobile-ad-opportunities-says-monotype

Davis. (1989). Perceived usefulness, perceived ease of use, and user acceptance of information technology. *MIS Quarterly*, 319–340.

Davis, S. M. (2002). *Brand asset management: Driving profitable growth through your brands.* San Francisco, CA: Josey Bass.

de Kervenoael, R., Palmer, M., & Hallsworth, A. (2013). From the outside in: Consumer anti-choice and policy implications in the mobile gaming market. *Telecommunications Policy*, *37*(6-7), 439–449. doi:10.1016/j.telpol.2012.06.008

de Reuver, M., Stein, S., & Hampe, J. F. (2013). From eParticipation to mobile participation: Designing a service platform and business model for mobile participation. *Information Polity*, *18*(1), 57–73.

de Sa, M., & Carrico, L. (2010). Designing and evaluating mobile interaction: Challenges and trends. *Foundations and Trends in Human-Computer Interaction*, *4*(3), 175–243. doi:10.1561/1100000025

de Souza e Silva, A. (2006). Interfaces of hybrid spaces. In A. Kavoori & N. Arceneaux (Eds.), The Cell Phone Reader: Essays in Social Transformation (pp.19–44). Pieterlen: Peter Lang International Academic Publishers.

de Souza e Silva, A., & Hjorth, L. (2009). Playful urban spaces: A historical approach to mobile games. *Simulation & Gaming*, *40*(5), 602–625. doi:10.1177/1046878109333723

De Tocqueville, A., Mansfield, H. C., & Winthrop, D. (2002). *Democracy in America.* Folio Society.

de Zuniga, H. (2012). Social media use for news and individuals' social capital, civic engagement and political participation. *Journal of Computer-Mediated Communication*, *17*(3), 319–336. doi:10.1111/j.1083-6101.2012.01574.x

Definition of Screencast. (2010). *PC Magazine: Encyclopedia.* Ziff Davis.

Delany, C. (2009). *Learning from Obama's Financial Steamroller: How to Raise Money Online.* Retrieved 30 July 2015 from http://www.epolitics.com/2009/05/15/learning-from-obamas-financial-steamroller-how-to-raise-money-online/

Deloitte. (2015). *Navigating the new digital divide.* Retrieved 25 May 2015 from http://www2.deloitte.com/content/dam/Deloitte/us/Documents/consumer-business/us-cb-navigating-the-new-digital-divide-v2-051315.pdf

Deng, Q. (2008). Did not imagine so many urban migrant children own cell phones. *China Youth Daily.* Retrieved from http://zqb.cyol.com/content/2008-04/14/content_2142436.htm

Denzin, N. (1992). Whose Cornerville is it, anyway. *Journal of Contemporary Ethnography*, *21*(1), 120–132. doi:10.1177/0891241692021001007

Deuze, M. (2005). What is journalism? Professional identity and ideology of journalists reconsidered. *Journalism*, *6*(4), 442–464. doi:10.1177/1464884905056815

Develotte, C., & Rechniewski, E. (2001). Discourse analysis of newspaper headlines: a methodological framework for research into national representations. *The Web Journal of French Media Studies, 4*(1).

Devitt, K., & Roker, D. (2009). The role of mobile phones in family communication. *Children & Society, 23*(3), 189–202. doi:10.1111/j.1099-0860.2008.00166.x

Dewey, C. (2014, May 8). *#Bringbackourgirls, #Kony2012, and the complete, divisive history of 'hashtag activism'.* Retrieved from Washington Post: http://www.washingtonpost.com/news/the-intersect/wp/2014/05/08/bringbackourgirls-kony2012-and-the-complete-divisive-history-of-hashtag-activism/

Dewey, C. (2015, April 27). How Google and Facebook are finding victims of the Nepal earthquake. *The Washington Post.* Retrieved May 7, 2015, from http://www.washingtonpost.com/news/the-intersect/wp/2015/04/27/how-google-and-facebook-are-finding-victims-of-the-nepal-earthquake/

Digital, D. (2015). *Navigating the new digital divide.* Retrieved May 25, 2015, http://www2.deloitte.com/content/dam/Deloitte/us/Documents/consumer-business/us-cb-navigating-the-new-digital-divide-v2-051315.pdf

Dimmick, J. W., Sikand, J., & Patterson, S. J. (1994). The gratifications of the household telephone sociability, instrumentality, and reassurance. *Communication Research, 21*(5), 643–663. doi:10.1177/009365094021005005

Diniz, V., & Gregório, A. (2007). Do e-gov governo eletrônico para o M-gov Cidadania Móvel. In E-desenvolvimento no Brasil e no mundo: subsídios e Programa e-Brasil (pp. 688-702). São Caetano do Sul, São Paulo: Yendis.

Diniz, E. H., Petrini, M., Barbosa, A. F., Monaco, H., & Christopoulos, T. (2006). Abordagens epistemológicas em pesquisas qualitativas: além do positivismo nas pesquisas na área de sistemas de informação. In *Anais do XXX ENANPAD.* Salvador, Bahia: ANPAD - Associação Nacional de Pós-Graduação e Pesquisa em Administração.

Distance Learning Requirements & Competencies. (2015). Retrieved from http://www.blueridge.edu/academics/distance-learning/distance-learning-requirements-competencies

Distance Learning. (2015). Retrieved from http://www.sfmccon.edu/distance-learning/distance-learning.html

Dittmar, H., Long, K., & Bond, R. (2007). When a better self is only a button click away: Associations between materialistic values, emotional and identity-related buying motives, and compulsive buying tendency online. *Journal of Social and Clinical Psychology, 26*(3), 334–361. doi:10.1521/jscp.2007.26.3.334

DL Master of Engineering Graduate Program. (2015). Retrieved from https://engineering.tamu.edu/petroleum/academics/distance-learning/prospective-students/admission

Donnelly, J. (2012, February 15). *Freedom of the Press panel explores 'Arab Spring' aftermath.* Retrieved from The National Press Club: http://www.press.org/news-multimedia/news/freedom-press-panel-explores-arab-spring-aftermath

Doswell, J., & Harmeyer, K. (2007). Extending the 'serious game' boundary: Virtual instructors in mobile mixed reality learning games. *Proceedings of DiGRA 2007 Conference,* (pp. 524-529).

Downie, L. (2009). *The Future of Journalism: Where We've Been, Where We're Going.* Speech given at John S. Knight Fellowships Reunion 2009 given at Stanford University. Retrieved 15 Oct. 2015 from https://youtu.be/JYtOKnk5fiw?list=PLfSN69RCPDM-HID_MaJ7TxlfwMsnMAqPc

Downie, L. (2009). *The future of journalism: Where we've been, where we're going.* Presentation delivered at Stanford University. Retrieved June 25, 2015, https://youtu.be/JYtOKnk5fiw?list=PLfSN69RCPDM-HID_MaJ7TxlfwMsnMAqPc

Drossos, D., & Giaglis, G. M. (2010). Reviewing mobile marketing research to date: Towards ubiquitous marketing. In K. Pousttchi (Ed.), *Handbook of Research on Mobile Marketing Management* (pp. 10–35). Hershey, PA: IGI Global. doi:10.4018/978-1-60566-074-5.ch002

Drossos, D., Giaglis, G. M., Lekakos, G., Kokkinaki, F., & Stavraki, M. G. (2007). Determinants of effective SMS advertising: An experimental study. *Journal of Interactive Advertising, 7*(2), 16–27. doi:10.1080/15252019.2007.10722128

Ducoffe, R. H. (1996). Advertising value and advertising the web. *Journal of Advertising Research, 36*, 21–35.

Duffy, M. (2011). Smartphones in the Arab Spring. *International Press Institute 2011 Report.* Retrieved May 2, 2015, from http://www.academia.edu/1911044/Smartphones_in_the_Arab_Spring

Duncan, E., Hazan, E., & Roche, K. (2014). Digital disruption: Evolving usage and the new value chain. *McKinsey publication of the Telecommunications, Media, and Technology Practice, 24*(1).

Duncan, T., & Moriarty, S. E. (1998). A communication-based marketing model for managing relationships. *Journal of Marketing, 62*(2), 1–13. doi:10.2307/1252157

Dunne, F. P. (1968). *Observations.* Grosse Pointe, MI: Scholarly Press.

Dyck, J. J., & Gimpel, J. G. (2005, September). Distance, Turnout, and the Convenience of Voting. *Social Science Quarterly, 86*(3), 531–548. doi:10.1111/j.0038-4941.2005.00316.x

Eagle, L., & Kitchen, P. J. (2000). IMC, brand communications, and corporate cultures: Client/advertising agency co-ordination and cohesion. *European Journal of Marketing, 34*(5/6), 667–686. doi:10.1108/03090560010321983

Economic Times. (2015). *Twitter buys Indian mobile marketing start-up ZipDial.* Retrieved May 1, 2015, from http://articles.economictimes.indiatimes.com/2015-01-21/news/58306142_1_zip-dial-amiya-pathak-sanjay-swamy

Economides, A. A., & Grousopoulou, A. (2008). Use of mobile phones by male and female Greek students. *International Journal of Mobile Communications, 6*(6), 729–749. doi:10.1504/IJMC.2008.019822

Econsultancy. (2015). *The Consumer Conversation: The experience void between brands and their customers.* Retrieved from https://www14.software.ibm.com/webapp/iwm/web/signup.do?source=swg-smartercom_medium&S_PKG=ov33876&dynform=18187&S_TACT=C348047W&cm_mmc=EconsultancySurvey-_-social-_-infographic-_-Apr2015

Eddy, N. (2014). Mobile Advertising, Push Notifications Beneficial for Retailers. *EWeek,* 7-7.

Eddy, N. (2014, May 22). Mobile Advertising, Push Notifications Beneficial for Retailers. *EWeek,* 7-7.

Edelman, D. (2010, December). Branding in the digital age: You're spending your money in all the wrong places. *Harvard Business Review.*

Edelson, D., & O'Neill, D. K. (1994). *The CoVis Collaboratory Notebook: Supporting Collaborative Scientific Enquiry.* Academic Press.

Edwards, J. (2014, February 3). Facebook's 'Paper' app is basically a massive redesign of Facebook – and it's excellent. *Business Insider.* Retrieved from http://www.businessinsider.com/facebook-paper-app-redesign-of-facebook-2014-2?IR=T

Edwards, L. H. (2012). Transmedia storytelling, corporate synergy, and audience expression. *Global Media Journal, 12*(20), 1–12.

Edwards, L., & Hodges, C. E. M. (Eds.). (2011). *Public relations, society and culture – Theoretical and empirical explorations.* London: Routledge.

Eisinga, R. G., Te Grotenhuis, M., & Pelzer, B. (2012). Weather conditions and voter turnout in Dutch national parliament elections, 1971–2010. *International Journal of Biometeorology*, *56*(4), 783–786. doi:10.1007/s00484-011-0477-7 PMID:21792567

El Kiki, T., & Lawrence, E. (2006, April). Government as a mobile enterprise: real-time, ubiquitous government. In *Information Technology: New Generations, 2006. ITNG 2006. Third International Conference on* (pp. 320-327). IEEE.

El-Kiki, T., Lawrence, E., & Steele, R. (2005). A management framework for mobile government services. *Proceedings of CollECTeR, Sydney, Australia*, 2009-4.

Elliott, N. (2009). *The Easiest Way to a First-Page Ranking on Google*. Retrieved 20 August 2015 from http://blogs. forrester.com/interactive_marketing/2009/01/the-easiest-way.html

Elliott. (1974). Uses and gratifications research: A critique and a sociological alternative. In *The Uses of Mass Communications: Current Perspectives on Gratifications Research* (pp. 249-268). Beverly Hills, CA: Sage.

Elliott, M. T., & Speck, P. S. (1998). Consumer perceptions of advertising clutter and its impact across various media. *Journal of Advertising Research*, *38*, 29–42.

Ellison, N. B. (2007). Social network sites: Definition, history, and scholarship. *Journal of Computer-Mediated Communication*, *13*(1), 210–230. doi:10.1111/j.1083-6101.2007.00393.x

Ellison, N., Gray, R., Lampe, C., & Fiore, A. (2014). Social capital and resource requests on Facebook. *New Media & Society*, *16*(7), 1104–1121. doi:10.1177/1461444814543998

Ellison, N., Steinfield, C., & Lampe, C. (2007). The Benefits of Facebook "Friends:" Social Capital and College Students' Use of Online Social Network Sites. *Journal of Computer-Mediated Communication*, *12*(4), 1143–1168. doi:10.1111/j.1083-6101.2007.00367.x

Elmieh, B., Austin, D. M., Collins, B. M., Oftedal, M. J., Pinkava, J. J., & Sweetland, D. P. (2015). *U.S. Patent No. 20,150,026,576*. Washington, DC: U.S. Patent and Trademark Office.

eMarketer. (2011). *Ad Dollars Still Not Following Online and Mobile Usage*. Retrieved 20 Feb. 2015 from http://www. emarketer.com/Article/Ad-Dollars-Still-Not-Following-Online-Mobile-Usage/1008311

eMarketer. (2013). *Digital TV, Movie Streaming Reaches a Tipping Point*. Retrieved 20 Feb. 2015 from http://www. emarketer.com/Article.aspx?R=1009775

eMarketer. (2013). *US Total Media Ad Spend Inches Up, Pushed by Digital*. Retrieved 20 Feb. 2015 from http://www. emarketer.com/Article/US-Total-Media-Ad-Spend-Inches-Up-Pushed-by-Digital/1010154

eMarketer. (2013, April 4). *Facebook to see three in 10 mobile display dollars this year*. Retrieved from http://www. emarketer.com/Article/Facebook-See-Three-10-Mobile-Display-Dollars-This-Year/1009782#sthash.3gy64laM.dpuf

eMarketer. (2013a). *Digital TV, movie streaming reaches a tipping point*. Retrieved 20 Feb. 2015 from http://www. emarketer.com/Article.aspx?R=1009775

eMarketer. (2013a). *eMarketer: Tablets, Smartphones Drive Mobile Commerce to Record Heights*. Retrieved 20 Feb. 2015 from http://www.emarketer.com/newsroom/index.php/emarketer-tablets-smartphones-drive-mobile-commerce-record-heights/

eMarketer. (2013a). *US Total Media Ad Spend Inches Up, Pushed by Digital*. Retrieved 20 Feb. 2015 from http://www. emarketer.com/Article/US-Total-Media-Ad-Spend-Inches-Up-Pushed-by-Digital/1010154 eMarketer. (2013b). *Digital TV, Movie Streaming Reaches a Tipping Point*. Retrieved 20 Feb. 2015 from http://www. emarketer.com/Article.aspx?R=1009775

eMarketer. (2013b). *TV advertising keeps growing as mobile boosts digital video spend*. Retrieved 20 Feb. 2015 from http://www.emarketer.com/Article/TV-Advertising-Keeps-Growing-Mobile-Boosts-Digital-Video-Spend/1009780

eMarketer. (2013b). *US Total Media Ad Spend Inches Up, Pushed by Digital*. Retrieved 20 Feb. 2015 from http://www. emarketer.com/Article/US-Total-Media-Ad-Spend-Inches-Up-Pushed-by-Digital/1010154

eMarketer. (2013c). *TV Advertising Keeps Growing as Mobile Boosts Digital Video Spend*. Retrieved 20 Feb. 2015 from http://www.emarketer.com/Article/TV-Advertising-Keeps-Growing-Mobile-Boosts-Digital-Video-Spend/1009780

eMarketer. (2013d). *Most Digital Ad Growth Now Goes to Mobile as Desktop Growth Falters*. Retrieved 20 Feb. 2015 from http://www.emarketer.com/Article/Most-Digital-Ad-Growth-Now-Goes-Mobile-Desktop-Growth-Falters/1010458

eMarketer. (2013e). *Tablets, Smartphones Drive Mobile Commerce to Record Heights*. Retrieved 20 Feb. 2015 from http://www.emarketer.com/newsroom/index.php/emarketer-tablets-smartphones-drive-mobile-commerce-record-heights/

eMarketer. (2014). *Despite time spent, mobile still lacks for ad spend in the US*. Retrieved 20 Feb. 2015 from http://www. emarketer.com/Article/Despite-Time-Spent-Mobile-Still-Lacks-Ad-Spend-US/1010788

eMarketer. (2014). *Despite Time Spent, Mobile Still Lacks for Ad Spend in the US*. Retrieved 20 Feb. 2015 from http:// www.emarketer.com/Article/Despite-Time-Spent-Mobile-Still-Lacks-Ad-Spend-US/1010788

eMarketer. (2014). Retrieved from technews.cn: http://technews.cn/2014/12/25/china-will-top-500-million-smartphone-users-for-the-first-time-see-more-at-httpwww-emarketer-comarticle2-billion-consumers-worldwide-smartphones-by-20161011694sthash-1do3dlqq-dpuf/

eMarketer. (2014a). *Despite Time Spent, Mobile Still Lacks for Ad Spend in the US*. Retrieved 20 Feb. 2015 from http:// www.emarketer.com/Article/Despite-Time-Spent-Mobile-Still-Lacks-Ad-Spend-US/1010788

eMarketer. (2014b). *Driven by Facebook and Google, Mobile Ad Market Soars 105% in 2013*. Retrieved 20 Feb. 2015 from http://www.emarketer.com/Article/Driven-by-Facebook-Google-Mobile-Ad-Market-Soars-10537-2013/1010690

eMarketer. (2015). *Cross-platform video trends roundup*. Retrieved 20 Feb. 2015 from https://www.emarketer.com/public_media/docs/eMarketer_Cross_Platform_Video_Trends_Roundup.pdf

eMarketer. (2015, March 26). *Facebook and Twitter will take 33% share of US digital display market by 2017*. Retrieved 20 Dec. 2015 from http://www.emarketer.com/Article/Facebook-Twitter-Will-Take-33-Share-of-US-Digital-Display-Market-by-2017/1012274#sthash.o1A1Vm4z.dpuf

eMarketer. (2015a). *Global tablet audience to total 1 billion this year*. Retrieved 19 May 2015 from http://www.emarketer. com/Article/Global-Tablet-Audience-Total-1-Billion-This-Year/1012451/1

eMarketer. (2015a). *Real-Time Personalization Affects the Bottom Line*. Retrieved 1 August 2015 from http://www. emarketer.com/Article/Real-Time-Personalization-Affects-Bottom-Line/1012689

eMarketer. (2015a). *Worldwide ad spending*. Retrieved 30 May 2015 from http://www.emarketer.com/adspendtool

eMarketer. (2015b). *Can Companies Validate All of Their Customer Data?* Retrieved 1 August 2015 from http://www. emarketer.com/Article/Companies-Validate-All-of-Their-Customer-Data/1012693

eMarketer. (2015b). *Mobile Ad Spend to Top $100 Billion Worldwide in 2016, 51% of Digital Market*. Retrieved 30 May 2015 from http://www.emarketer.com/Article/Mobile-Ad-Spend-Top-100-Billion-Worldwide-2016-51-of-Digital-Market/1012299/1 eMarketer. (2015b). *Mobile content and activities roundup*. Retrieved 10 April 2015 from https://apsalar.com/blog/2015/04/emarketers-mobile-content-and-activities-roundup-now-available/

eMarketer. (2015c). *Cross-platform video trends roundup*. Retrieved 20 Feb. 2015 from https://www.emarketer.com/public_media/docs/eMarketer_Cross_Platform_Video_Trends_Roundup.pdf

eMarketer. (2015c). *How Many Smartphone Users Are Officially Addicted?* Retrieved 1 August 2015 from http://www.emarketer.com/Article/How-Many-Smartphone-Users-Officially-Addicted/1012800?ecid=NL1001]

eMarketer. (2015c). *How much Yahoo traffic is mobile?* Retrieved 21 May 2015 from http://www.emarketer.com/Article/How-Much-Yahoo-Search-Traffic-Mobile/1012378/1

eMarketer. (2015d). *Mobile Accounts for Almost Half of China's Retail Ecommerce Sales*. Retrieved 1 August 2015 from http://www.emarketer.com/Article/Mobile-Accounts-Almost-Half-of-Chinas-Retail-Ecommerce-Sales/1012793?ecid=NL1001

eMarketer. (2015d). *Most digital video monetization still comes from ads*. Retrieved 23 May 2015 from http://www.emarketer.com/Article/Most-Digital-Video-Monetization-Still-Comes-Ads/1012300/1

eMarketer. (2015d). *Seven in 10 US Internet Users Watch OTT Video*. Retrieved 15 Oct. 2015 from http://www.emarketer.com/Article/Seven-10-US-Internet-Users-Watch-OTT-Video/1013061?ecid=NL1001

eMarketer. (2015e). *Expect more programmatic ads to pop up on mobile*. Retrieved 6 May 2015 from http://www.emarketer.com/Article/Expect-More-Programmatic-Ads-Pop-Up-on-Mobile/1012291/1

eMarketer. (2015e). *Mobile phones strengthen lead for mobile video viewing*. Retrieved 2 July 2015 from http://www.emarketer.com/Article/Mobile-Phones-Strengthen-Lead-Mobile-Video-Viewing/1012683?ecid=NL1001

eMarketer. (2015f). *Programmatic creative: Look to existing processes for guidance*. Retrieved 27 June 2015 from http://www.emarketer.com/Article/Programmatic-Creative-Look-Existing-Processes-Guidance/1012434#sthash.jZrmND1V.dpuf

eMarketer. (2015f). *Programmatic Sellers vs. Buyers: Who's More Hooked On Mobile?* Retrieved 6 August 2015 from http://www.emarketer.com/Article/Programmatic-Sellers-vs-Buyers-Whos-More-Hooked-On-Mobile/1012374

eMarketer. (2015g). *Better site optimization lifts mobile conversion rates*. Retrieved 23 May 2015 from http://www.emarketer.com/Article/Better-Site-Optimization-Lifts-Mobile-Conversion-Rates/1012482/1

eMarketer. (2015g). *Why Retailers Are Buying In to Programmatic*. Retrieved 6 August 2015 from http://www.emarketer.com/Article/Why-Retailers-Buying-Programmatic/1012448

eMarketer. (2015h). *Programmatic Creative: Look to Existing Processes for Guidance*. Retrieved 6 August 2015 from http://www.emarketer.com/Article/Programmatic-Creative-Look-Existing-Processes-Guidance/1012434

eMarketer. (2015h). *Seven in 10 US Internet Users Watch OTT Video*. Retrieved 15 Oct. 2015 from http://www.emarketer.com/Article/Seven-10-US-Internet-Users-Watch-OTT-Video/1013061?ecid=NL1001

eMarketer. (2015i). *Mobile Video Roundup*. Retrieved 30 July 2015 from http://www.emarketer.com/articles/roundups

eMarketer. (2015j). *Cross-platform video trends roundup*. Retrieved 20 Feb. 2015 from https://www.emarketer.com/public_media/docs/eMarketer_Cross_Platform_Video_Trends_Roundup.pdf

eMarketer. (2015k). *Measuring mobile effectiveness still challenges marketers*. Retrieved 4 August 2015 from http://www.emarketer.com/Article/Measuring-Mobile-Effectiveness-Still-Challenges-Marketers/1012797?ecid=NL1001

Engeström. (1987). *Learning by expanding*. Cambridge University Press.

Engeström. (1993). Developmental studies of work as a test bench of activity theory: the case of primary care medical practice. In *Understanding Practice: Perspescives on Activity and Context*. Cambridge, UK: Cambridge University Press.

Erikson, E. (1968). *Identity: Youth and Crisis*. New York, NY: Norton.

EstonianE. (n.d.).

Ettema, J. S., & Glasser, T. L. (1984). *On the epistemology of investigative journalism*. Gainesville: Educational Resources Information Center.

Europe, I. A. B. (2010). *European Media Landscape Report Summary*. Retrieved 1 May 2015 from http://www.iabeurope. eu/research/about-mediascope.aspx

Evans, C. (2014, March 17). 10 examples of augmented reality in retail. *Creative Guerilla Marketing*. Retrieved from http://www.creativeguerrillamarketing.com/augmented-reality/10-examples-augmented-reality-retail/

Even, A. (2015, January 22). Big Data and mobile analytics: Ready to rule 2015. *Venture Beat*. Retrieved from http:// venturebeat.com/2015/01/22/big-data-and-mobile-analytics-ready-to-rule-2015/

Everett, S. P. (2007). *The Usability of Electronic Voting Machines and How Votes Can Be Changed Without Detection*. Rice University.

Evergage. (2013). *Real-time for the rest of us: Perceptions of real-time marketing and how it's achieved*. Retrieved 18 May 2015 from http://info.evergage.com/perceptions_of_realtime_marketing_survey_results

Evergage. (2014). *Real-time for the rest of us: Perceptions of real-time marketing and how it's achieved*. Retrieved 25 March 2015 from http://info.evergage.com/perceptions_of_realtime_marketing_survey_results

Evergage. (2014). *The ROI of Relevant Marketing*. Retrieved 1 August 2015 from http://info.evergage.com/the-roi-relevant-of-relevant-marketing

Facebook. (2015). *Improving Ad Performance with the Carousel Format*. Retrieved 30 July 2015 from https://www.facebook.com/business/news/carousel-ads

Fairclough, N. (2003). *Analysing Discourse: Text Analysis for Social Research*. London: Routledge.

Fallakhair, S., Pemberton, L., & Griffiths, R. (2007). Development of a cross-platform ubiquitous language learning service via mobile phone and interactive television. *Journal of Computer Assisted Learning*, *23*(4), 312–325. doi:10.1111/j.1365-2729.2007.00236.x

Fang, Z. (2002). E-Government in digital era: Concept, practice, and development. *International Journal of The Computer. The Internet and Management*, *10*(2), 1–22.

Farrell, L. (2009). Texting the Future: Work, Literacies and Economies. In M. Baynham & M. Prinsloo (Eds.), *The Future of Literacy Studies*. Basingstoke, UK: Palgrave Macmillan.

FEC.gov. (2008). *Financial Summary Report Search Results*. Retrieved December 22, 2008.

Federal Reserve Board. (2015). *Consumers and Mobile Financial Services 2015*. Retrieved May 6, 2015, from http:// www.federalreserve.gov/econresdata/consumers-and-mobile-financial-services-report-201503.pdf

Feijoo, C., Gómez-Barroso, J. L., Aguado, J. M., & Ramos, S. (2012). Mobile gaming: Industry challenges and policy implications. *Telecommunications Policy*, *36*(3), 212–221. doi:10.1016/j.telpol.2011.12.004

Feldwick, P. (1996). Do we really need "brand equity? *The Journal of Brand Management*, *4*(1), 9–28. doi:10.1057/ bm.1996.23

Ferdus, I. (2014). Photography as activism: The role of visual media in humanitarian crisis. *Harvard International Review*, (Summer), 22–25.

Fields, T. (2012). *Social game design: Monetization methods and mechanics*. Waltham: Morgan Kaufmann Publishers.

Figueiredo, A., Almeida, A., & Machado, P. (2004). Identifying and documenting test patterns from mobile agent design patterns. In A. Karmouch, L. Korba, & E. Madeira (Eds.), *Mobility Aware Technologies and Applications* (pp. 359–368). Berlin: Springer-Verlag. doi:10.1007/978-3-540-30178-3_35

Fiksu. (2015, June). *Fiksu's analysis: June 2015* [Press release]. Retrieved from https://www.fiksu.com/resources/fiksu-indexes#analysis

Firoozy-Najafabadi, H. R., & Pashazadeh, S. (2011, October). Mobile police service in mobile government. In *Application of Information and Communication Technologies (AICT), 2011 5th International Conference on* (pp. 1-5). IEEE. doi:10.1109/ICAICT.2011.6110902

Fisher, M., & Baird, D. (2006). Making mLearning work: Utilizing mobile technology for active exploration, collaboration, assessment, and reflection in higher education. *Journal of Educational Technology Systems*, *35*(1), 3–30. doi:10.2190/4T10-RX04-113N-8858

Fleming, K., & Thorson, E. (2008). Assessing the Role of Information-Processing Strategies in Learning From Local News Media About Sources of Social Capital. *Mass Communication & Society*, *11*(4), 398–419. doi:10.1080/15205430801950643

Forester Research. (2013). *Your Customers Are Demanding Omni-Channel Communications. What Are You Doing About It?* Retrieved 15 April 2015 from http://www.kana.com/customer-service/white-papers/forrester-your-customers-are-demanding-omni-channel-communications.pdf?_ga=1.38271843.1312302063.1428939602

Fortunati, L., & Manganelli, A. M. (2002). Young people and the mobile telephone. *Revista de Estudios de Juventud*, *52*, 59–78.

Fotouhi-Ghazvini, F., Earnshaw, R. A., Robison, D. J., & Excell, P. S. (2008). The MOBO City: A mobile game package for technical language learning. In J. Traxler, B. Riordan & C. Dennett (Eds.), *Proceedings of the mLearn2008 Conference*. University of Wolverhampton.

Fournier, S., & Avery, J. (2011). The uninvited brand. *Business Horizons*, *54*(3), 193–207. doi:10.1016/j.bushor.2011.01.001

Frangonikolopoulos, C. A., & Chapsos, I. (2012). xplaining the role and the impact of the social media in the Arab Spring. *Global Media Journal: Mediterranean Edition*, *7*(2), 10–20.

Frankola. (2000). *Why online learners drop out*. Workforce.com. Retrieved August, 2012, from: http://www.workforce.com/archive/feature/22/26/22/index.php

Frequently Asked Questions. (2015). Retrieved from http://bamabydistance.ua.edu/faqs/

Frith. (2012). Location-based social networks and mobility patterns: An empirical examination of how Foursquare use affects where people go. *Mobilities Conference: Local & Mobile*.

Frith, J. (2013). Turning life into a game: Foursquare, gamification, and personal mobility. *Mobile Media & Communication*, *1*(2), 248–262. doi:10.1177/2050157912474811

Frizell, S. (2015, January 20). Twitter buys Indian mobile marketing startup. *Time*. Retrieved from http://time.com/3674300/twitter-zipdial-india/

Fulgoni, G. (2014). The State of Mobile Industry. *comScore*. Retrieved May 19, 2015, from http://www.comscore.com/Insights/Presentations-and-Whitepapers/2014/The-State-of-the-Mobile-Industry

Gafni, R., & Geri, N. (2013). Generation Y versus Generation X: Differences in Smartphone Adaptation. In *Proceedings of the Chais conference on Instructional Technologies Research 2013: Learning in the technological era*. Raanana: The Open University of Israel. Retrieved March 27, 2015, from http://www.openu.ac.il/innovation/chais2013/download/b1_4.pdf

Galal, E., & Spielhaus, R. (2012). Covering the Arab Spring: Middle East in the Media – the Media in the Middle East. *The Editorial*. Retrieved from Academia.edu: http://www.academia.edu/2279607/Covering_the_Arab_Spring_Middle_East_in_the_Media_the_Media_in_the_Middle_East._The_Editorial

Ganguly, R. (2013). The 5 biggest mistakes in mobile app marketing. *Kissmetrics*. Retrieved from https://blog.kissmetrics.com/mistakes-in-app-marketing/

Gannes, L. (2009). *The Great Video SEO Frontier*. Retrieved 20 August 2015 from https://gigaom.com/2009/02/12/the-great-video-seo-frontier/

Gartner. (2013a, June 4). *Gartner says worldwide mobile payment transaction value to surpass $235 billion in 2013* [Press release]. Retrieved from http://www.gartner.com/newsroom/id/2504915

Gartner. (2013b, September 19). *Gartner says mobile app stores will see annual downloads reach 102 billion in 2013* [Press release]. Retrieved from http://www.gartner.com/newsroom/id/2504915

Georgiadis, C. K., & Stiakakis, E. (2010, June). Extending an e-Government Service Measurement Framework to m-Governement Services. In *Mobile Business and 2010 Ninth Global Mobility Roundtable (ICMB-GMR), 2010 Ninth International Conference on* (pp. 432-439). IEEE. doi:10.1109/ICMB-GMR.2010.31

Germanakos, P., Tsianos, N., Lekkas, Z., Belk, M., Mourlas, C., & Samaras, G. (2009). Human Factors as a Parameter for Improving Interface Usability and User Satisfaction. Academic Press.

Gheytanchi, E., & Moghadam, V. (2014). Women, social protests, and the new media activism in the Middle East and North Africa. *International Review of Modern Sociology, 40*(1), 1–26.

Ghyasi, F., & Kushchu, I. (2004). *Uses of mobile government in developing countries*. Retrieved from http://www.movlab.org

Gibbs, S. (2015, April 20). Google's 'mobilegeddon' will shake up search results. *The Guardian*.

Gillin, P. (2008). New media, new influencers and implications for public relations. *Society for New Communications Research*. Institute for Public Relations, Wieck Media.

Gillmor, D. (2006). *We the media: Grassroots journalism by the people, for the people*. O'Reilly Media, Inc.

Gilpin, D. R. (2010). Organizational image construction in a fragmented online media environment. *Journal of Public Relations Research, 22*(3), 265–287. doi:10.1080/10627261003614393

Gilpin, D. R., & Murphy, P. (2010). Implications of complexity theory for public relations: Beyond crisis. In R. L. Heath (Ed.), *The SAGE Handbook of Public Relations*. Los Angeles, CA: SAGE Publications.

Gladwell, M. (2000). *The tipping point: how little things can make a big difference*. London: Little Brown.

Gladwell, M. (2000). *The tipping point: How little things can make a big difference*. New York, NY: Little, Brown and Company.

Gladwell, M. (2000). *The Tipping Point: How Little Things Can Make a Big Difference*. New York: Little, Brown and Company.

Global Pulse, U. N. (2015). *Using mobile phone data and airtime credit purchases to estimate food security 2015*. Retrieved May 30, 2015, from http://unglobalpulse.org/sites/default/files/Topups_Food%20Security_WFP_Final.pdf

Global, M. M. A. (2013a). *"Donate a Word via SMS." 2013, Gold Winner, EMEA, Social Impact/Not for Profit; 2013, Gold Winner, EMA, Messaging.* Swedish Committee for Afghanistan. Mobiento. Retrieved from http://www.mmaglobal. com/case-study-hub/upload/pdfs/mma-2013-501.pdf

Global, M. M. A. (2013b). *"Halls' Breathe the Change: 200,000 Breaths That Changed the Life of a Village." 2013 MMA Smarties Silver Winner Social Impact/Not For Profit.* Brand: Halls. Lead Agency: AD2C - Pinnacle. Retrieved from http://www.mmaglobal.com/case-study-hub/case_studies/view/27379

Global, M. M. A. (2013c). *"Priority Moments for Independent Business." 2013 MMA Smarties Silver Winner Location Based.* Brand: O2. Lead Agency: R/GA London. Retrieved from http://www.mmaglobal.com/case-study-hub/upload/pdfs/mma-2013-592.pdf

Godfrey, R. R., & Duke, S. E. (2014). Leadership in Mobile Technology: An Opportunity for Family and Consumer Sciences Teacher Educators. *Journal of Family and Consumer Sciences, 106*(3), 36–40.

Goffman, E. (1990). Stigma: Notes on the Management of Spoiled Identity. Harmondsworth.

Goggin, G. (2012). *Cell phone culture: Mobile technology in everyday life.* Routledge.

Goh, D., Lee, C., & Low, G. (2012). I played games as there was nothing else to do: Understanding motivations for using mobile content-sharing games. *Online Information Review, 36*(6), 784–806. doi:10.1108/14684521211287891

Goh, K. Y., Heng, C. S., & Lin, Z. (2013). Social media brand community and consumer behavior: Quantifying the relative impact of user-and marketer-generated content. *Information Systems Research, 24*(1), 88–107. doi:10.1287/isre.1120.0469

Goh, T., & Kinshuk. (2006). Getting Ready for Mobile Learning—Adaptation Perspective. *Journal of Educational Multimedia and Hypermedia, 15*(2), 175–198.

Google. (2014). *Helping users find mobile-friendly pages.* Google Webmaster Central Blog. Retrieved May 21, 2015, from http://googlewebmastercentral.blogspot.com/2014/11/helping-users-find-mobile-friendly-pages.html

Google. (2015). *Public Data: Fertility rate.* Retrieved from https://www.google.kz/publicdata/explore?ds=d5bncppjof8f9_&met_y=sp_dyn_tfrt_in&idim=country:KAZ:UZB:KGZ&hl=en&dl=en

Google. (August, 2012). The new multi-screen world: Understanding cross-platform consumer behavior. *Think with Google.* Retrieved from https://www.thinkwithgoogle.com/research-studies/the-new-multi-screen-world-study.html

Google. (n.d.a). Micro-moments. *Think with Google.* Retrieved 1 August 2015 from https://www.thinkwithgoogle.com/micromoments/

Google. (n.d.b). Sephora turns smartphones into local store magnets. *Think with Google.* Retrieved 1 August, 2015 from https://www.thinkwithgoogle.com/interviews/sephora-turns-smartphones-into-local-store-magnets.html

Göth, C., & Schwabe, G. (2008). Designing tasks for engaging mobile learning, In J. Traxler, B. Riordan & C. Dennett (Eds.), *Proceedings of the mLearn2008 Conference.* University of Wolverhampton, UK.

Gould, Verenikina, & Hasan. (2000). *Activity Theory as a Basis for the Design of a Web Based System of Inquiry for World War 1 Data. Department of Information Systems.* Wollongong: Academic Press.

Government of Brunei Darussalam. (1972). *Report of the Brunei Education Commission.* Brunei: Government of Brunei.

Government of Brunei Darussalam. (1985). Education System of Negara Brunei Darussalam. Bandar Seri Begawan: Jabatan Perkembangan Kurikulum, Jabatan Pelajaran, Kementerian Pelajaran dan Kesihatan.

Graff, H. J. (1979). *The literacy myth: Literacy and social structure in the nineteenth century city.* Academic Press.

Graff, H. J. (1986). The history of literacy: Toward a third generation. *Interchange, 17*(2), 122–134. doi:10.1007/BF01807474

Graffin, S., Pfarrer, M., & Hill, M. (2012). Untangling executive reputation and corporate reputation: Who made who. In M. L. Barnett & T. G. Pollock (Eds.), *The Oxford handbook of corporate reputation.* Oxford, UK: Oxford University Press. doi:10.1093/oxfordhb/9780199596706.013.0011

Granitz, N., & Forman, H. (2015). Building self-brand connections: Exploring brand stories through a transmedia perspective. *Journal of Brand Management.*

Grant, I., & O'Donohoe, S. (2007). Why young consumers are not open to mobile marketing communication. *International Journal of Advertising, 26*(2), 223–246.

Green, E., & Singleton, C. (2007). Mobile Selves: Gender, ethnicity and mobile phones in the everyday lives of young Pakistani-British women and men. *Information Communication and Society, 10*(4), 506–526. doi:10.1080/13691180701560036

Greenemeier, L. (2015, December 09). *Ballot Secrecy Keeps Voting Technology at Bay.* Retrieved from Technology for Scientific American: http://www.scientificamerican.com/article/2012-presidential-election-electronic-voting/#

Green, N. (2003). Outwardly Mobile: Young People and Mobile Technologies. In J. E. Katz (Ed.), *Machines That Become Us: The Social Context of Personal Communication Technology* (pp. 201–218). New Brunswick, NJ: Transaction Publishers.

Griffin, E. M. (2009). *A first look at communication theory.* New York: McGraw-Hill Companies, Inc.

Grimus, M., Ebner, M., & Holzinger, A. (2012). Mobile Learning as a Chance to Enhance Education in Developing Countries-on the Example of Ghana. In mLearn (pp. 340-345).

Grönroos, C. (1997). From marketing mix to relationship marketing-towards a paradigm shift in marketing. *Management Decision, 35*(4), 322–339. doi:10.1108/00251749710169729

Gross, D. (2014). *Social Web tackles the #IceBucketChallenge.* CNN.com. Retrieved from http://www.cnn.com/2014/08/13/tech/ice-bucket-challenge/

Grunig, J. E. (2009). Paradigms of global public relations in an age of digitalization. *PRism, 6*(2).

Grunig, J. E., Grunig, L. A., & Dozier, D. M. (2006). The excellence theory. In C. Botan & V. Hazleton (Eds.), *Public Relations Theory II.* Mahwah, NJ: Lawrence Elbaum Associates.

Grunig, J. E., & Hunt, T. (1984). *Managing Public Relations.* Fort Worth, TX: Harcourt Brace.

Guo, B., Zhang, D., & Wang, Z. (2011, October). Living with internet of things: The emergence of embedded intelligence. In *Internet of Things (iThings/CPSCom), 2011 International Conference on and 4th International Conference on Cyber, Physical and Social Computing* (pp. 297-304). IEEE.

Gutman, B. (2012). Kellogg proves ROI of digital programmatic buying. *Forbes.* Retrieved 6 May from http://www.forbes.com/sites/marketshare/2012/11/09/kellogg-proves-roi-of-digital-programmatic-buying/

Guy, R. (2011). *Digitally Disinterested.* Santa Rosa, CA: Informing Science Press.

Guzman, F. (2005). *A brand building literature review.* Barcelona: Esade Business School.

Gwee, S., Chee, Y., & Tan, E. (2010a). Spatializing social practices in mobile game-based learning. *2010 10th IEEE International Conference on Advanced Learning Technologies,* (pp. 555-557).

Gwee, S., Chee, Y., & Tan, E. (2010b). Assessment of student outcomes of mobile game-based learning. *Proceedings of the 18th International Conference on Computers in Education. Putrajaya, Malaysia: Asia-Pacific Society for Computers in Education.*

Haddon, L. (2004). *Information and communication technologies in everyday life: A concise introduction and research guide*. New York, NY: Berg.

Hafizah, S., Hamid, A., & Fung, L. Y. (2007). Learn programming by using mobile edutainment game approach.*The First IEEE International Workshop on Digital Game and Intelligent Toy Enhanced Learning (DIGITEL'07).*

Haghirian, P., & Madlberger, M. (2005). Consumer attitude toward advertising via mobile devices – An empirical investigation among Austrian users. *Proceedings of 13th European Conference on Information Systems.* Retrieved 20 March 2015 from http://aisel.aisnet.org/ecis2005/44

Haghirian, P., & Inoue, A. (2007). An advanced model of consumer attitudes toward advertising on the mobile internet. *International Journal of Mobile Communications, 5*(1), 48–67. doi:10.1504/IJMC.2007.011489

Hagood, M. C., Leander, K. M., Luke, C., Mackey, M., & Nixon, H. (2003). Media and online literacy studies. *Reading Research Quarterly, 38*(3), 386–413. doi:10.1598/RRQ.38.3.4

Haji, H. A., Shaame, A. A., & Kombo, O. H. (2013, September). The opportunities and challenges in using mobile phones as learning tools for Higher Learning Students in the developing countries: Zanzibar context. In AFRICON, 2013 (pp. 1-5). IEEE.

Häkkilä, J., & Chatfield, C. (2005, September). 'It's like if you opened someone else's letter': user perceived privacy and social practices with SMS communication. In *Proceedings of the 7th international conference on Human computer interaction with mobile devices & services* (pp. 219-222). ACM.

Hakoama, M., & Hakoyama, S. (2011). The impact of cell phone use on social networking and development among college students. *The American Association of Behavioral and Social Sciences Journal, 15*(1), 20.

Ha, L., & James, E. L. (1998). Interactivity Reexamined: A Baseline Analysis of Early Business Web Sites. *Journal of Broadcasting & Electronic Media, 42*(4), 457–474. doi:10.1080/08838159809364462

Hall, E., Charles S, Fottrell, S. W., & Byass, P. (2014). Assessing the impact of mHealth interventions in low- and middle-income countries – what has been shown to work? *Global Health Action, 7.*

Hanjun, K. C.-H. (2005). Internet uses and gratifications: A Structural Equation Model of Interactive Advertising. *Journal of Advertising, 34*(2), 57–70. doi:10.1080/00913367.2005.10639191

Hanley, M., Becker, M., & Martinsen, J. (2006). Factors influencing mobile advertising acceptance: Will incentives motivate college students to accept mobile advertisements? *International Journal of Mobile Marketing, 1*, 50–58.

Hansen, K. (2014). When French fries go viral: Mobile media and the transformation of public space in McDonald's Japan. *Electronic Journal of Contemporary Japanese Studies, 14*(2). Retrieved from http://www.japanesestudies.org.uk/ejcjs/vol14/iss2/hansen.html

Hansen, A., Cottle, S., Negrine, R., & Newbold, C. (1998). *Mass Communication Research Methods.* New York: New York University Press. doi:10.1007/978-1-349-26485-8

Hare, K. (2013, October 28). *Keller, Greenwald debate whether journalists can be impartial.* Retrieved from Poynter: http://www.poynter.org/news/mediawire/227386/keller-greenwald-debate-whether-journalists-can-be-impartial/

Haroldsen, E., & Harvey, K. (1979). The Diffusion of Shocking Good News. *The Journalism Quarterly, 56*(4), 771–775. doi:10.1177/107769907905600409

Harper, R., & Hamill, L. (2005). Kids will be kids: The role of mobiles in teenage life. In *Mobile World* (pp. 61–74). Berlin, Germany: Springer. doi:10.1007/1-84628-204-7_4

Hartley, M. (1994). Generations of literacy among women in a bilingual community. In M. Hamilton, D. Barton, & R. Ivanic (Eds.), Worlds of literacies. Clevedon, UK: Multilingual Matters.

Harvard Business Review Analytic Services. (2012a). How mobility is changing the enterprise. *Harvard Business Review.* Retrieved from https://global.sap.com/campaigns/02_cross_mobile_havard_business_review_reports/HBR_HowMobilityIsChangingThe%20Enterprise.pdf

Harvard Business Review Analytic Services. (2012b). How mobility is transforming industries. *Harvard Business Review.* Retrieved from https://global.sap.com/campaigns/02_cross_mobile_havard_business_review_reports/HBR_HowMobilityIsTransformingIndustires.pdf

Harvard Business Review Analytic Services. (2012c). How mobility is changing the world. *Harvard Business Review.* Retrieved from https://global.sap.com/campaigns/02_cross_mobile_havard_business_review_reports/HBR_HowMobilityIsChangingTheWorld.pdf

Harvey, K. (2011). *Approaching PR Like a Journalist.* Retrieved 10 April 2015 from https://youtu.be/y10jCpR2akE

Harvey, K. (2011). Using Internet tools to promote locally. *KIRC Proceedings 2011.* Retrieved 3 July 2015 from http://virtual-institute.us/Ken/KIRCharvey2011.pdf

Harvey, K. (2015). *How to create your own informational, promotional & educational videos.* Retrieved 15 June 2015 from http://iei-tv.net/0MakeVideos.htm

Harvey, K., & Manweller, M. (2015). *PR, media executives support radical education reform.* Retrieved 24 December 2015 from http://virtual-institute.us/Ken/CurrentResearch.html

Hassan, M., Jaber, T., & Hamdan, Z. (2009, November). Adaptive mobile-government framework. In *Proceedings of the International Conference on Administrative Development: Towards Excellence in Public Sector Performance.*

Haverila, M. (2011). Mobile phone feature preferences, customer satisfaction and repurchase intent among male users. *Australasian Marketing Journal, 19*(4), 238–246. doi:10.1016/j.ausmj.2011.05.009

Heath, R. L. (2006). A rhetorical theory approach to issues management. In C. H. Botan & V. Hazleton (Eds.), *Public relations theory II* (pp. 63–101). Mahwah, NJ: Lawrence Erlbaum Associaets, Inc., Publishers.

Heath, S. B. (1983). *Ways with words: Language, life and work in communities and classrooms.* Cambridge, UK: CUP.

Heilbroner, R. (1994). *Technological Determinism Revisited.* EBSCOHOST.

Heine, C. (2015). Facebook's Q1 Ad Revenue Increased 46 Percent to $3.3 Billion. *AdWeek.* Retrieved 30 July 2015 from http://www.adweek.com/news/technology/facebooks-q1-ad-revenue-increased-46-percent-33-billion-164234

Heine, C. (2015, April 22). *Facebook's Q1 ad revenue increased 46% to $3.3 billion.* AdWeek. Retrieved 30 June 2015 from http://www.adweek.com/news/technology/facebooks-q1-ad-revenue-increased-46-percent-33-billion-164234

Hellström, J. (2012). Mobile Participation? Crowdsourcing during the 2011 Uganda General Elections. In Proc. M4D 2012 (pp. 411-424). New Delhi: Excel India Publishers.

Hellström, J. (2012). Mobile Governance: Applications, Challenges and Scaling-up. In M. Poblet (Ed.), *Mobile Technologies for Conflict Management: Online Dispute Resolution, Governance.* Participation.

Hellström, J. (n.d.). Mobile phones for good governance – challenges and way forward. *Stockholm University.*

Herman, J. (2014). Hashtags and human rights: Activism in the age of Twitter. *Carnegie Ethics Online,* 1-6.

Herman, A., Hadlaw, J., & Swiss, T. (Eds.). (2014). *Theories of the mobile Internet: Materialities and imaginaries.* New York, NY: Routledge.

Hermida, A. (2013). #Journalism. *Digital Journalism, 1*(3), 295–313. doi:10.1080/21670811.2013.808456

Hermida, A., Lewis, S. C., & Zamith, R. (2014). Sourcing the Arab Spring: A Case Study of Andy Carvin's Sources on Twitter During the Tunisian and Egyptian Revolutions. *Journal of Computer-Mediated Communication, 19*(3), 479–499. doi:10.1111/jcc4.12074

Herrington, A., Herrington, J., & Mantei, J. (2009). Design principles for mobile learning. In J. Herrington, A. Herrington, J. Mantei, I. W. Olney, & B. Ferry (Eds.), *New technologies, new pedagogies: Mobile learning in higher education* (pp. 129–138). Australia: University of Wollongong.

Hershkowitz, S., & Lavrusik, V. (2013). 12 Best Practices for Media Companies Using Facebook Pages. *Facebook Media.* Published May 2, 2013. https://www.facebook.com/notes/facebook-media/12-pages-best-practices-for-media-companies/518053828230111

Hershkowitz, S., & Lavrusik, V. (2013). 12 Best Practices for Media Companies Using Facebook Pages. *Facebook Media.* Retrieved from https://www.facebook.com/notes/facebook-media/12-pages-best-practices-for-media-companies/518053828230111

Hershkowitz, S., & Lavrusik, V. (2013). *12 Best Practices for Media Companies Using Facebook Pages.* Retrieved 15 Oct. 2015 from https://www.facebook.com/notes/facebook-media/12-pages-best-practices-for-media-companies/518053828230111

Herskovitz, S., & Crystal, M. (2010). The essential brand persona: Storytelling and branding. *The Journal of Business Strategy, 31*(3), 21–28. doi:10.1108/02756661011036673

Hinton, D. (2013). Private Conversations and Public Audiences: Exploring the Ethical Implications of Using Mobile Telephones to Research Young People's Lives. *Young, 21*(3), 237–251. doi:10.1177/1103308813488813

Hjorth, L. (2011). *Games and gaming: An introduction to new media.* Oxford, UK Berg.

Hjorth, L. (2011). Mobile @ game cultures: The place of urban mobile gaming. *Convergence (London), 17*(4), 357–371. doi:10.1177/1354856511414342

Hoang, A. T., Nickerson, R. C., Beckman, P., & Eng, J. (2008). Telecommuting and corporate culture: Implications for the mobile enterprise. *Information, Knowledge, Systems Management, 7*(1), 77.

Hoelzel, M. (2015). Spending on native advertising is soaring as marketers and digital media publishers realize the benefits. *Business Insider.* Retrieved 30 July from http://www.businessinsider.com/spending-on-native-ads-will-soar-as-publishers-and-advertisers-take-notice-2014-11

Holtz, S. (1999). *Public relations on the Net: Winning strategies to inform and influence the media, the investment community, the government, the public, and more!* New York, NY: AMACOM.

Hon, L. C., & Grunig, J. E. (1999). *Guidelines for measuring relationships in public relations.* Institute for Public Relations.

Horst, H. A. (2006). The blessings and burdens of communication: Cell phones in Jamaican transnational social fields. *Global Networks, 6*(2), 143–159. doi:10.1111/j.1471-0374.2006.00138.x

HotConference. (2015). *What can HotConference do for you?* Retrieved 16 April 2015 from http://www.hotconference.com/members/eduken/features.php

House, N. V., Davis, M., Ames, M., Finn, M., & Viswanathan, V. (2005), The uses of personal networked digital imaging: An empirical study of cameraphone photos and sharing. *Proceedings of CHI '05 Extended Abstracts on Human Factors in Computing Systems* (pp. 1853-6). ACM Press. doi:10.1145/1056808.1057039

Howard, P., & Hussain, M. (2011). The role of digital media: The upheavals in Egypt and Tunisia. *Journal of Democracy*, *22*(3), 35–48. doi:10.1353/jod.2011.0041

Hrastinski, S. (2008). Asynchronous and synchronous e-learning. *EDUCAUSE Quarterly*, *31*(4), 51–55.

Huang, J. (2015). *Building Brand-Consumer Relationships on WeChat: the Case of IKEA*. Unpublished Master dissertation.

Huang, H., & Leung, L. (2009). Instant messaging addiction among teenagers in China: Shyness, alienation, and academic performance decrement. *Cyberpsychology & Behavior*, *12*(6), 675–679. doi:10.1089/cpb.2009.0060 PMID:19788380

HubSpot. (2012). *120 awesome marketing stats, charts and graphs*. Retrieved May 20, 2015, from http://offers.hubspot.com/120-awesome-marketing-stats-charts-and-graphs

HubSpot. (2014). *State of Inbound 2014*. Retrieved 4 July 2014 from http://offers.hubspot.com/2014-state-of-inbound

HubSpot. (2014a). *Marketing benchmarks from 7000+ businesses*. Cambridge, MA: Hubspot.

HubSpot. (2014b). *State of Inbound 2014*. Retrieved 4 July 2014 from http://offers.hubspot.com/2014-state-of-inbound

Huizenga, J., Admiraal, W., Akkerman, S., & Dam, G. T. (2009). Mobile game-based learning in secondary education: Engagement, motivation and learning in a mobile city game. *Journal of Computer Assisted Learning*, *25*(4), 332–344. doi:10.1111/j.1365-2729.2009.00316.x

Hull, R., Facer, K., Stanton, D., Kirk, D., Reid, J., & Joiner, R. (2004). Savannah: Mobile gaming and learning? *Journal of Computer Assisted Learning*, (6): 399–409.

Humphreys, L. (2013). Mobile social media: Future challenges and opportunities. *Mobile Media & Communication*, *1*(1), 20–25. doi:10.1177/2050157912459499

Hung, S. Y., Chang, C. M., & Kuo, S. R. (2013). User acceptance of mobile e-government services: An empirical study. *Government Information Quarterly*, *30*(1), 33–44. doi:10.1016/j.giq.2012.07.008

Hurme, P. (2001). Online PR: Emerging organizational practice. *Corporate Communications: An International Journal*, *6*(2), 71–75. doi:10.1108/13563280110391016

Husain, W., & Dih, L. Y. (2012). A framework of a personalized location-based traveler recommendation system in mobile application. *International Journal of Multimedia and Ubiquitous Engineering*, *7*(3), 11–18.

Husbye, N., & Elsener, A. (2013). To Move Forward, We Must Be Mobile: Practical Uses of Mobile Technology in Literacy Education Courses. *Journal of Digital Learning in Teacher Education*, *30*(2), 46–51. doi:10.1080/21532974.2013.10784726

Husson, T. (2015, February 9). The future of mobile wallets lies beyond payments. *Forrester Researcher*. Retrieved from https://s3.amazonaws.com/vibes-marketing/Website/Reports_$folder$/Forrester+-+The+Future+of+Mobile+Wallets+Report.pdf

Hwang, A. (2015). China December mobile phone user base grows to 1.286 billion, says MIIT. *DIGITIMES*. Retrieved from http://www.digitimes.com/news/a20150123VL200.html

Hwang, G., & Chang, H. (2011). A formative assessment-based mobile learning approach to improving the learning attitudes and achievements of students. *Computers & Education*, *56*(4), 1023–1031. doi:10.1016/j.compedu.2010.12.002

IBM. (2015). *U.S. Online Holiday Benchmark Recap Report 2014.* Retrieved 10 April 2015 from http://www-01.ibm.com/common/ssi/cgi-bin/ssialias?subtype=WH&infotype=SA&appname=SWGE_ZZ_JV_USEN&htmlfid=ZZW033 62USEN&attachment=ZZW03362USEN.PDF#loaded

IEI-TV. (2015). *What has your Wi-Fi done for you lately?* Retrieved 29 June 2015 from http://iei-tv.net/wifi.html

Igarashi, T., Motoyoshi, T., Takai, J., & Yoshida, T. (2008). No mobile, no life: Self-perception and text-message dependency among Japanese high school students. *Computers in Human Behavior, 24*(5), 2311–2324. doi:10.1016/j.chb.2007.12.001

INDEPTH Network. (2007). South Africa Country Report. Report by the Sonke Gender Justice Network, Johannesburg/Cape Town.

Informações e Serviços de Telecomunicações Ltda. (2014). Retrieved November 1, 2014, from http://www.teleco.com.br

Information Solutions Group. (2013). *Pop Up Games Mobile Games Research, 2013.* Retrieved May 7, 2015, from http://www.infosolutionsgroup.com/popcapmobile2013.pdf

Ingram, M. (2014). NYT Reporter Shows the Power of Twitter in Journalism. *GigaOm.* Retrieved from https://gigaom.com/2011/05/27/nyt-reporter-shows-the-power-of-twitter-as-journalism/

Insook, H., & Seungyeon, H. (2014). Adoption of the Mobile Campus in a Cyber University. *International Review of Research in Open and Distance Learning, 15*(6), 237–256.

Interbrand. (2014). *The four ages of branding.* Retrieved from http://www.bestglobalbrands.com/2014/featured/the-four-ages-of-branding/

Interbrand. (2014, October 9). *Apple and Google each worth more than USD $100 billon on Interbrand's 15th annual Best Global Brands Report* [Press release]. Retrieved from http://interbrand.com/en/newsroom/15/interbrands-th-annual-best-global-brands-report

International Telecommunications Union. (2015). *Global ICT developments.* Retrieved 20 March 2015 from http://www.itu.int/en/ITU-D/Statistics/Pages/stat/default.aspx

International Telecommunications Union. (2015). *Mobile-broadband penetration approaching 32 per cent, Three billion Internet users by end of this year* [Press release]. Retrieved from https://www.itu.int/net/pressoffice/press_releases/2014/23.aspx#.VZ_GcmCdIW0

Irwin, J. (2013). *Modernizing Education and Preparing Tomorrow's Workforce through Mobile Technology.* Paper presented at the i4j Summit. Retrieved June 27, 2015, from http://www.meducationalliance.org/sites/default/files/qualcomm_-_modernizing_education_and_the_workforce_through_mobile_technology.pdf

Ishmatova, D., & Obi, T. (2009). m-government services: User needs and value. *I-WAYS-The Journal of E-Government Policy and Regulation, 32*(1), 39–46.

Islam, M. S., & Scupola, A. (2013). E-Service Research Trends in the Domain of E-Government: A Contemporary Study. In A. Scupola (Ed.), *Mobile Opportunities and Applications for E-Service Innovations* (pp. 152–169). Hershey, PA: Information Science Reference; doi:10.4018/978-1-4666-2654-6.ch009

Ismail, I., Azizan, S. N., & Azman, N. (2013). Mobile phone as pedagogical tools: Are teachers ready? *International Education Studies, 6*(3), 36–47. doi:10.5539/ies.v6n3p36

Ismail, M., & Razak, R. C. (2011). A Short Review on the Trend of Mobile Marketing Studies. *International Journal of Interactive Mobile Technologies, 5*(3), 38–42.

Istepanian, R., Laxminarayan, S., & Pattichis, C. S. (2007). *M-Health: Emerging mobile health systems*. New York, NY: Springer Science & Business Media.

Ito, & Okabe. (2005). Technosocial situations emergent structuring of mobile e-mail use. In *Personal, portable, pedestrian: Mobile phones in Japanese life* (pp. 257–273). The MIT Press.

Ito, M., Okabe, D., & Matsuda, M. (Eds.). (2005). *Personal, portable, pedestrian: Mobile phones Japanese life*. Cambridge, MA: MIT Press.

ITU. (2014). *ITU releases 2014 ICT figures: Mobile-broadband penetration approaching 32 per cent Three billion Internet users by end of this year*. Retrieved from https://www.itu.int/net/pressoffice/press_releases/2014/23.aspx#.VVD26Utu4dt

Ivanov, C. (2005, November 18). Three new NBC Sports-branded mobile games in early 2006. *Softpedia*. Retrieved from http://archive.news.softpedia.com/news/Three-New-NBC-Sports-branded-Mobile-Games-In-Early-2006-12728.shtml

Izquierdo-Yusta, A., Olarte-Pascual, C., & Reinares-Lara, E. (2015). Attitudes toward mobile advertising among users versus non-users of the mobile Internet. *Telematics and Informatics*, *32*(2), 355–366. doi:10.1016/j.tele.2014.10.001

Jacobs, I. M. (2013). *Modernizing education and preparing tomorrow's workforce through mobile technology*. *Innovation for Jobs Summit*. QUALCOMM.

Jahanshahi, A. A., Khaksar, S. M. S., Yaghoobi, N. M., & Nawaser, K. (2011). Comprehensive model of mobile government in Iran. *Indian Journal of Science and Technology*, *4*(9), 1188–1197.

James, J., & Versteeg, M. (2007). Mobile phones in Africa: How much do we really know? *Social Indicators Research*, *84*(1), 117–126. doi:10.1007/s11205-006-9079-x

James, M. (2008). A review of the impact of new media on public relations: Challenges for terrain practice and education. *Asia Pacific Public Relations Journal*, *8*, 137–148.

Jamnadas, H. K. (2015, October). Challenges & Solutions Of Adoption In Regards To Phone-Based Remote E-Voting. *Int J Scientific & Technology Research*, *4*(10).

Janssen, M. (2010). Electronic Intermediaries Managing and Orchestrating Organizational Networks Using E-Services. In Electronic Services: Concepts, Methodologies, Tools and Applications (pp. 1319-1333). Hershey, PA: Information Science Reference. doi:10.4018/978-1-61520-967-5.ch081

Jara, A. J., Bocchi, Y., & Genoud, D. (2014a, September). Social Internet of Things: The potential of the Internet of Things for defining human behaviours. In *2014 International Conference on Intelligent Networking and Collaborative Systems (INCoS)* (pp. 581-585). IEEE. doi:10.1109/INCoS.2014.113

Jara, A. J., Parra, M. C., & Skarmeta, A. F. (2014b). Participative marketing: Extending social media marketing through the identification and interaction capabilities from the Internet of things. *Personal and Ubiquitous Computing*, *18*(4), 997–1011. doi:10.1007/s00779-013-0714-7

Jarvis, J. (2009). *Newspapers in Decline?* Video presented on the CBC (Canadian) TV network. Retrieved 15 Oct. 2015 from https://youtu.be/YjUeJH4mdF4

Jayo, M. (2010). *Correspondentes bancários como canal de distribuição de serviços financeiros: taxonomia, histórico, limites e potencialidades dos modelos de gestão de redes*. São Paulo: Tese de Doutorado em Administração, Escola de Administração de Empresas de São Paulo –Fundação Getúlio Vargas.

Jensen, K. B. (2002). *A Handbook of Media and Communication Research: Qualitative and Quantitative Methodologies*. London: Routledge. doi:10.4324/9780203465103

Jeong, H., & Lee, M. (2013). The effect of online medía platforms on joining causes: The impression management perspective. *Journai of Broadcasting & Eiectronic Media, 57*(4), 439–455. doi:10.1080/08838151.2013.845824

Jiang, L. (2011) Construction and research reflection of the issue of left-behind children. *Humanities Magazine (Renwen Zazhi).* Retrieved from http://www.aisixiang.com/data/41509.html

Jian, M., & Lee, K. W. (2011). Does CEO reputation matter for capital investments? *Journal of Corporate Finance, 17*(4), 929–946. doi:10.1016/j.jcorpfin.2011.04.004

Jiugen, Y., Ruonan, X., & Jianmin, W. (2010, July). Applying research of mobile learning mode in teaching. In *Information Technology and Applications (IFITA), 2010 International Forum on, 3,* 417–420.

Jones, K., & Bartlett, J. L. (2009). The strategic value of corporate social responsibility: a relationship management framework for public relations practice. *PRism, 6*(1).

Jones, G. M. (1996). Bilingual Education and Syllabus Design: Towards a Workable Blueprint. *Journal of Multilingual and Multicultural Development, 17*(2-4), 280–293. doi:10.1080/01434639608666281

Joo, K. P., Andres, C., & Shearer, R. (2014). Promoting distance learners'; cognitive engagement and learning outcomes: Design-based research in the Costa Rican National University of Distance Education. *International Review of Research in Open and Distance Learning, 15*(8), 188–210.

Junco, R., & Cole-Avent, G. A. (2008). An introduction to technologies commonly used by college students. *New Directions for Student Services, 2008*(124), 3–17. doi:10.1002/ss.292

Junco, R., & Cotten, S. R. (2011). Perceived academic effects of instant messaging use. *Computers & Education, 56*(2), 370–378. doi:10.1016/j.compedu.2010.08.020

Juniper Research. (2006). *Press release: casual gamers and female gamers to drive mobile games revenues over the $10 billion mark by 2009.* Retrieved March 27, 2015 from http://juniperresearch.com/viewpressrelease.php?pr=16S

Juul, J. (2010). *A casual game revolution: Reinventing video games and their players.* Cambridge, MA: MIT Press.

Kahle, & Valette-Florence. (2012). *Marketplace Lifestyles in an Age of Social Media: Theory and Method.* Academic Press.

Kalloo, V., Kinshuk, & Mohan, P. (2010). Personalized game based mobile learning to assist high school students with mathematics. *2010 10th IEEE International Conference on Advanced Learning Technologies,* (pp. 485-487).

Kalman, Y. M., Raban, D. R., & Rafaeli, S. (2013). Netified: social cognition in crowds and clouds. In Y. Amichai-Hamburger (Ed.), *The Social Net: Human Behavior in Cyberspace* (2nd ed.). Oxford, UK: Oxford University Press. doi:10.1093/acprof:oso/9780199639540.003.0002

Kalogeraki, S., & Papadaki, M. (2010). The impact of mobile use on teenagers' socialization. *International Journal of Interdisciplinary Social Sciences, 5*(4), 121–134.

Kamdar, J. (2015). *Now on Twitter: group Direct Messages and mobile video camera.* Retrieved 15 Oct. 2015 from https://blog.twitter.com/2015/now-on-twitter-group-direct-messages-and-mobile-video-capture

Kamran, S. (2010). Mobile phone: Calling and texting patterns of college students in Pakistan. *International Journal of Business and Management, 5*(4), 26. doi:10.5539/ijbm.v5n4p26

Kang, H. D., Sung, K. W., Park, M., & Ahn, B. K. (2009). The Effect of Blended English Learning Program of CDI on Students' Achievement. *English Teaching, 64*(4), 181–201. doi:10.15858/engtea.64.4.200912.181

Kapko, M. (2015). How mobile, social tech elevate enterprise collaboration. *The CIO Agenda*. Retrieved May 7, 2015, from http://www.cio.com/article/2873207/collaboration/how-mobile-social-tech-elevate-enterprise-collaboration.html

Kaplan, & Haenlein. (2011). Two hearts in three-quarter time: How to waltz the social media/viral marketing dance. *Business Horizons*, 253—263.

Kaplan, & Haenlein. (2012). The Britney Spears universe: Social media and viral marketing at its best. *Business Horizons*, 27-31.

Kaplan, A. (2012). If you love something, let it go mobile: Mobile marketing and mobile social media 4x4. *Business Horizons*, *55*(2), 129–139. doi:10.1016/j.bushor.2011.10.009

Kaplan, A. M., & Haenlein, M. (2010). Users of the world, unite! The challenges and opportunities of Social Media. *Business Horizons*, *53*(1), 59–68. doi:10.1016/j.bushor.2009.09.003

Kaplan, A. M., & Haenlein, M. (2011). The early bird catches the news: Nine things you should know about microblogging. *Business Horizons*, *54*(2), 105–113. doi:10.1016/j.bushor.2010.09.004

Karp, J. A. (2000). Going postal: How all-mail elections influence turnout. *Political Behavior*, *22*(3), 223–239. doi:10.1023/A:1026662130163

Kathy, G. (2015) *Mobile still tops retailers' priority lists, according to*Shop.org/Forrester*research report*. Retrieved April 23, 2015, from https://nrf.com/media/press-releases/mobile-still-tops-retailers-priority-lists-according-shoporgforrester-research

Katz, E., Blumler, J. G., & Gurevitch, M. (1974). Utilization of mass communication by the individual. In J. G. Blumer & E. Katz (Eds.), The Uses of Mass Communication. Newbury Park, CA: Sage.

Katz, E., Gurevitch, M., & Haas, H. (1973). On the use of mass media for important things. *American Sociological Review*, *38*(2), 164–181. doi:10.2307/2094393

Katz, J. E., & Aakhus, M. (2002). *Perpetual contact: Mobile communication, private talk, public performance*. Cambridge, UK: Cambridge University Press. doi:10.1017/CBO9780511489471

Keating, J. (2014, May 20). *The less you know*. Retrieved from Slate: http://www.slate.com/blogs/the_world_/2014/05/20/the_depressing_reason_why_hashtag_campaigns_like_stopkony_and_bringbackourgirls.html

Keefe, D. (2013). Branding in the new age of ecosystems. *Landor Associates*. Retrieved from http://landor.com/pdfs/DKeefe_Ecosystem_26June2013.pdf?utm_campaign=PDFDownloads&utm_medium=web&utm_source=web

Keegan, D. (1996). *Foundations of distance education* (3rd ed.). London: Routledge.

Keengwe, J., & Bhargava, M. (2014). Mobile learning and integration of mobile technologies in education. *Education and Information Technologies*, *19*(4), 737–746. doi:10.1007/s10639-013-9250-3

Keller, K. L. (2013). *Strategic brand management: Building, measuring, and managing brand equity*. Harlow, UK: Pearson Education Ltd.

Keller, K. L., & Lehmann, D. R. (2006). Brands and branding: Research findings and future priorities. *Marketing Science*, *25*(6), 740–759. doi:10.1287/mksc.1050.0153

Kember, D. (1995). *Open Learning Courses for Adults: A Model of Student Progress*. Englewood Cliffs, NJ: Educational Technology Publications.

Kent, M. L. (2013). Using social media dialogically: Public relations role in reviving democracy. *Public Relations Review*, *39*(4), 337–345. doi:10.1016/j.pubrev.2013.07.024

Kent, M. L., & Saffer, A. J. (2014). A Delphi study of the future of new technology research in public relations. *Public Relations Review, 40*(3), 568–576. doi:10.1016/j.pubrev.2014.02.008

Khan, S. (2015). *Snapchat, the Arab world and global implications*. Retrieved December 13, 2015, from http://www.internationalpolicydigest.org/2015/07/27/snapchat-the-arab-world-and-global-implications/

Kiernan, P. J., & Aizawa, K. (2004). Cell phones in task based learning: Are cell phones useful language learning tools? *ReCALL, 16*(1), 71–84. doi:10.1017/S0958344004000618

Kilcourse, B., & Rowen, S. (2014, February). *Mobile In retail: Reality sets in.* Retrieved April 23, 2015, from http://www.sap.com/bin/sapcom/en_us/downloadasset.2014-06-jun-10-18.mobile-in-retail-reality-sets-in--rsr-benchmark-report-2014-pdf.bypassReg.html

Kim, L. (2015). 5 Facebook Advertising Tools that Save Time and Improve your ROI. *Social Media Examiner.* Retrieved 30 July 2015 from http://www.socialmediaexaminer.com/5-facebook-advertising-tools/?awt_l=BKVE2&awt_m=3ixhuI4YULr.ILT&utm_source=Newsletter&utm_medium=NewsletterIssue&utm_campaign=New

Kim, B., Kang, M., & Jo, H. (2014). Determinants of Postadoption Behaviors of Mobile Communications Applications: A Dual-Model Perspective. *International Journal of Human-Computer Interaction, 30*(7), 547–559. doi:10.1080/10447318.2014.888501

Kim, K. J. (2014). Can smartphones be specialists? Effects of specialization in mobile advertising. *Telematics and Informatics, 31*(4), 640–647. doi:10.1016/j.tele.2013.12.003

Kim, P., Buckner, E., Kim, H., Makany, T., Taleja, N., & Parikh, V. (2012). A comparative analysis of a game-based mobile learning model in low-socioeconomic communities of India. *International Journal of Educational Development, 32*(2), 329–340. doi:10.1016/j.ijedudev.2011.05.008

Kim, Y. S., Han, M. J., Hong, C. J., Ko, S. B., & Kim, S. B. (2012). Mobile Digital Storytelling Development for Energy Drink's Promotion and Education. In *Advances in Automation and Robotics* (Vol. 1, pp. 557–566). Springer Berlin Heidelberg.

Kim, Y., Yoon, J., Park, S., & Han, J. (2004). Architecture for implementing the mobile government services in Korea. In *Conceptual modeling for advanced application domains* (pp. 601–612). Springer Berlin Heidelberg.

King, C. (2015). 28 Social Media Marketing Predictions for 2015 from the Pros. *Social Media Examiner*. Published 1 Jan. 2015, downloaded 13 Feb. 2015 from http://www.socialmediaexaminer.com/social-media-marketing-predictions-for-2015/

King, C. (2015). 28 Social Media Marketing Predictions for 2015 from the Pros. *Social Media Examiner.* Retrieved 13 Feb. 2015 from http://www.socialmediaexaminer.com/social-media-marketing-predictions-for-2015/

King, C. (2015). *28 Social Media Marketing Predictions for 2015 from the Pros.* Social Media Examiner. Retrieved from http://www.socialmediaexaminer.com/social-media-marketing-predictions-for-2015/

Kitchen, P. J., Brignell, J., Li, T., & Jones, G. S. (2004). The emergence of IMC: A theoretical perspective. *Journal of Advertising Research, 44*(01), 19–30. doi:10.1017/S0021849904040048

Knox, D. (2014, February 20). Is Facebook building a P&G-style house of brands? *Advertising Age.* Retrieved from http://adage.com/article/digitalnext/facebook-buys-whatsapp-facebook-house-brands/291798/

Kogeda, O. K. (2013). Model for a Mobile Phone Voting System for South Africa.*Proc. 15th Annual Conference on World Wide Web Applications (ZAWWW 2013)*. Cape Town: Cape Peninsula University of Technology.

Kokalitcheva, K. (2015, April 9). After planned split, eBay and PayPal promise to stay friends. *Fortune*. Retrieved from http://fortune.com/2015/04/09/ebay-paypal-split/

Kolb, L. (2008). Toys to tools: Connecting student cell phones to education. Washington, DC: International Society for Technology in Education (ISTE).

Kolsaker, A., & Drakatos, N. (2009). Mobile advertising: The influence of emotional attachment to mobile devices on consumer receptiveness. *Journal of Marketing Communications*, *15*(4), 267–280. doi:10.1080/13527260802479664

Konrad, A. (2014, November 18). Teespring says it's minting new millionaires selling its T-shirts, raises $35 million of its own. *Forbes*. Retrieved 25 March 2015 from http://www.forbes.com/sites/alexkonrad/2014/11/18/teespring-says-it-making-millionaires-raises-millions/

KONY2012: What's the Real Story? (2014). *Reality Check: Uganda*. The Guardian.com. Retrieved from http://www.theguardian.com/politics/reality-check-with-polly-curtis/2012/mar/08/kony-2012-what-s-the-story

Koohy, H., Koohy, B., & Watson, M. (2014). A lesson from the ice bucket challenge: Using social networks to publicize science. *Frontiers in Genetics*, *5*, 1–3. doi:10.3389/fgene.2014.00430 PMID:25566317

Korucu, A. T., & Alkan, A. (2011). Differences between m-learning (mobile learning) and e-learning, basic terminology and usage of m-learning in education. *Procedia: Social and Behavioral Sciences*, *15*, 1925–1930. doi:10.1016/j.sbspro.2011.04.029

Kosinski, M., Stillwell, D., & Graepel, T. (2013). Private traits and attributes are predictable from digital records of human behavior. *Proceedings of the National Academy of Sciences of the United States of America*, *110*(15), 5802–5805. doi:10.1073/pnas.1218772110 PMID:23479631

Kovach, B., & Rosenstiel, T. (2001). Are watchdogs an endangered species? *Columbia Journalism Review*, (May/June), 50–53.

Kovach, B., & Rosenstiel, T. (2007). *The elements of journalism: What newspeople should know and the public should expect*. Three Rivers Press.

Krantz, M. (2015, July 23). Amazon just surpassed Walmart in market cap. *USA Today*. Retrieved 12 August 2015 from http://www.usatoday.com/story/money/markets/2015/07/23/amazon-worth-more-walmart/30588783/

Kristofferson, K., White, K., & Peloza, J. (2014). The nature of slacktivism: How the social observability of an initial act of token support affects subsequent prosocial action. *The Journal of Consumer Research*, *40*(6), 1149–1166. doi:10.1086/674137

Kukulska-Hulme, A. (2005). Introduction. In A. Kukulska-Hulme & J. Traxler (Eds.), *Mobile learning: A handbook for educators and trainers* (pp. 1–6). Wiltshire: The Cromwell Press.

Kukulska-Hulme, A., & Arcos, B. D. L. (2011). Researching emergent practice among mobile language learners. In *Proceedings of 10th World Conference on Mobile and Contextual Learning*.

Kumar, M. K. (n.d.). *Secure Mobile Based Voting System*. Academic Press.

Kumparak, G. (2015, April 7). Amazon Instant Video finally comes to Android tablets. *TechCrunch*. Retrieved from http://techcrunch.com/2015/04/07/amazon-instant-video-finally-comes-to-android-tablets/

Kurkovsky, S. (2009). Can mobile game development foster student interest in computer science? *ICE-GIC 2009. International IEEE Consumer Electronics Society's Games Innovations Conference 2009*, (pp. 92-100).

Kurkovsky, S. (2009). Engaging students through mobile game development.*Proceedings of SIGCSE '09 of the 40th ACM Technical Symposium on Computer Science Education*, (pp. 44-48). New York: ACM Press. doi:10.1145/1508865.1508881

Kuscu, H., Kushchu, I., & Yu, B. (2007) Introducing Mobile Government. In I. Kushchu (Ed.), *Mobile Government: An Emerging Direction in E-government* (pp.1-11). IGI Global.

Kushchu, I., & Kuscu, H. (2003, July). From E-government to M-government: Facing the Inevitable. In *the 3rd European Conference on e-Government* (pp. 253-260). MCIL Trinity College Dublin Ireland.

Kushchu, I., & Yu, B. (2004). *Evaluating mobility for citizens*. Niigata: mGovLab-Internacional University of Japan. Retrieved November 04, 2012, from http://www.mgovlab.org

Kuutti, K. (1996). Activity Theory as a Potential Framework for Human-computer Interaction Research. In Context and Consciousness: Activity Theory and Human-computer Interaction (pp. 17-44). Cambridge, MA: MIT Press.

Lai, C.-H., Yang, J., Chen, F., Ho, C., & Chan, T. (2007). Affordances of mobile technologies for experiential learning: The interplay of technology and pedagogical practices. *Journal of Computer Assisted Learning, 23*(4), 326–337. doi:10.1111/j.1365-2729.2007.00237.x

Lajoie, M., & Shearman, N. (Prod.). (2014). What is Alibaba? *Wall Street Journal*. Retrieved from http://projects.wsj.com/alibaba/

Lakatos, E. M., & Marconi, M. A. (2007). *Metodologia científica*. São Paulo, São Paulo: Atlas.

Lamarre, A., Galarneau, S., & Boeck, H. (2012). Mobile Marketing and Consumer Behaviour Current Research Trend. *International Journal of Latest Trends in Computing, 3*(1), 1–9.

Lam, W. S. E. (2000). Second Language Literacy and the Design of the Self: A Case Study of a Teenager Writing on the Internet. *TESOL Quarterly, 34*(3), 457–482. doi:10.2307/3587739

Lam, W. S. E. (2005). Second Language Socialization in a Bilingual Chat Room. *Language Learning & Technology, 8*(3), 44–65.

Lam, W. S. E. (2006). Re-envisioning Language, Literacy and the Immigrant Subject in New Mediascapes. *Pedagogies, 1*(3), 171–195. doi:10.1207/s15544818ped0103_2

Lam, W. S. E. (2009). Multiliteracies on Instant Messaging in Negotiating Local, Translocal, and Transnational Affiliations: A Case of an Adolescent Immigrant. *Reading Research Quarterly, 44*(4), 377–397. doi:10.1598/RRQ.44.4.5

Langley, A. (1999). Strategies for theorizing from process data. *Academy of Management Review, 24*(4), 691–710.

Lanza, B. B. B., & Cunha, M. A. (2012). Relations Among Governmental Project Actors: The case of Paraná mGov. *CONF-IRM 2012 Proceedings*. Retrieved from http://aisel.aisnet.org/confirm2012/64

Larivière, B., Joosten, H., Malthouse, E. C., van Birgelen, M., Aksoy, P., Kunz, W. H., & Huang, M. H. (2013). Value fusion: The blending of consumer and firm value in the distinct context of mobile technologies and social media. *Journal of Service Management, 24*(3), 268–293. doi:10.1108/09564231311326996

LaRose, R., & Eastin, M. S. (2004). A social cognitive theory of Internet uses and gratifications: Toward a new model of media attendance. *Journal of Broadcasting & Electronic Media, 48*(3), 358–377. doi:10.1207/s15506878jobem4803_2

Lassen, D. (2005, January). The Effect of Information on Voter Turnout: Evidence from a Natural Experiment. *American Journal of Political Science*, *49*(1), 103–118. doi:10.1111/j.0092-5853.2005.00113.x

Lauterborn, B. (1990). New Marketing Litany: Four Ps Passé: C-Words Take Over. *Advertising Age*, *61*(41), 26.

Lavin-Mera, P., Torrente, J., Moreno-Ger, P., Vallejo-Pinto, J., & Fernández-Manjón, B. (2009). Mobile game development for multiple devices in education. *International Journal of Emerging Technologies in Learning*, *4*, 19–26.

Law, P. L., & Peng, Y. (2008). Mobile networks: migrant workers in southern China. Handbook of mobile communication studies, (pp. 55-64).

Layne, K., & Lee, J. (2001). Developing fully functional E-government: A four stage model. *Government Information Quarterly*, *18*(2), 122–136. doi:10.1016/S0740-624X(01)00066-1

Ledingham, J. A., & Bruning, S. D. (1998). Relationship management in public relations: Dimensions of an organization-public relationship. *Public Relations Review*, *24*(1), 55–65. doi:10.1016/S0363-8111(98)80020-9

Lee, C. K. M. (2007) Linguistic Features of Email and ICQ Instant Messageing in Hong Kong. In B. Danet & S. Herring (Eds.), The Multilingual Internet: Language, Culture and Communication Online, (pp. 184-208). Oxford, UK: Oxford University Press.

Lee, K. (2015). *The 5 Keys to Building a Social-Media Strategy for Your Personal Brand*. Retrieved 13 April 2015 from http://www.entrepreneur.com/article/243079

Lee, C., Goh, D., & Chua, A. (2010). Indagator: Investigating perceived gratifications of an application that blends mobile content sharing with gameplay. *Journal of the American Society for Information Science and Technology*, *61*(6), 1244–1257.

Lee, D.-H. (2013). *Smartphones, mobile social space, and new sociality in Korea*. Mobile Media & Communication.

Lee, S. M., Tan, X., & Trimi, S. (2005). Current practices of leading e-government countries. *Communications of the ACM*, *48*(10), 99–104. doi:10.1145/1089107.1089112

Lee, S. M., Tan, X., & Trimi, S. (2006). M-government, from rhetoric to reality: Learning from leading countries. *Electronic Government. International Journal (Toronto, Ont.)*, *3*(2), 113–126.

Lee, S., Meszaros, P. S., & Colvin, J. (2009). Cutting the wireless cord: College student cell phone use and attachment to parents. *Marriage & Family Review*, *45*(6-8), 717–739. doi:10.1080/01494920903224277

Lennon, R. G. (2012, October). Bring your own device (BYOD) with cloud 4 education. In *Proceedings of the 3rd annual conference on Systems, programming, and applications: software for humanity* (pp. 171-180). ACM. doi:10.1145/2384716.2384771

Leonov, I. (2014). *Trends and next steps for mobile games industry in 2015*. Gamasutra.

Leont'ev. (1978). *Activity, consciousness, and personality*. Englewood Cliffs, NJ: Prentice-Hall.

Lesitaokana, W. (2015). Young People and Mobile Phone Technology in Botswana. In J. Wyn & H. Cahill (Eds.), *Handbook of Children and Youth Studies* (pp. 801–813). Springer Singapore. doi:10.1007/978-981-4451-15-4_30

Leung, R. (2013). *WeChat Official Accounts and What This Means for Marketers*. Retrieved from ClickZ: http://www.clickz.com/clickz/column/2307141/wechat-official-accounts-and-what-this-means-for-marketers

Leung, S. (2014, August 30). 5 ways the internet of things will make marketing smarter. *Forbes*. Retrieved from http://www.forbes.com/sites/salesforce/2014/08/30/5-ways-iot-marketing-smarter/

Leung, K. T., Lee, D. L., & Lee, W. C. (2013). PMSE: A personalized mobile search engine. *IEEE Transactions on Knowledge and Data Engineering, 25*(4), 820–834. doi:10.1109/TKDE.2012.23

Leung, L., & Wei, R. (2000). More Than Just Talk on the Move: A Use-and-Gratification Study of the Cellular Phone. *Journalism & Mass Communication Quarterly, 77*(2), 308–320. doi:10.1177/107769900007700206

Levitt, T. (1960). Marketing myopia. *Harvard Business Review, 38*(4), 24–47. PMID:15252891

Levy, M. (1990). Towards a theory of CALL. *CAELL Journal, 1*(4), 5–8.

Levy, M. (1997). *Computer-assisted language learning: context and conceptualization.* Oxford, UK: Clarendon Press.

Levy, M., & Stockwell, G. (2006). *CALL dimensions: options and issues in computer-assisted language learning.* Lawrence Erlbaum Associate, Inc.

Lewin, B. A., & Donner, Y. (2002). Communication in Internet Message Boards. *English Today, 18*(3), 29–37. doi:10.1017/S026607840200305X

Lewis, J. (2015). *San Diego engages customers with new technologies.* PA Consulting. Retrieved May 2, 2015, from http://www.paconsulting.com/introducing-pas-media-site/highlighting-pas-expertise-in-the-media/opinion-pieces-by-pas-experts/intelligent-utility-san-diego-engages-customers-with-new-technologies-17-february-2015/

Lewis, S. C., Holton, A. E., & Coddington, M. (2014). Reciprocal Journalism. *Journalism Practice, 8*(2), 229–241. doi:10.1080/17512786.2013.859840

Liao, T. (2015). Augmented or admented reality? The influence of marketing on augmented reality technologies. *Information Communication and Society, 18*(3), 310–326. doi:10.1080/1369118X.2014.989252

Licoppe, C., & Inada, Y. (2006). Emergent uses of a multiplayer location-aware mobile game: The interactional consequences of mediated encounters. *Mobilities, 1*(1), 39–61. doi:10.1080/17450100500489221

Lieb, R., Silva, C., & Tran, C. (2013). *Organizing for content: models to incorporate content strategy and content marketing in the enterprise.* Altimeter Group. Retrieved from http://www.ciccorporate.com/edm/2013/201305en/r2.pdf

Lim, J. S., Ri, S. Y., Egan, B. D., & Biocca, F. (2015). The cross-platform synergies of digital video advertising: Implications for cross-media campaigns in television, Internet and mobile TV. *Computers in Human Behavior, 48,* 463–472. doi:10.1016/j.chb.2015.02.001

Lim, J., & Golan, G. (2011). Social media activism in response to the influence of political parody videos on YouTube. *Communication Research, 38*(5), 710–727. doi:10.1177/0093650211405649

Lin, P., & Block, D. (2011). English as a "global language" in China: An investigation into learners' and teachers' language beliefs. *Science Direct, 39*(3), 392-402. Available from: http://www.sciencedirect.com/science/article/pii/S0346251X11000972

Lin, C., Huang, C., & Chen, C. (2014). Barriers to the adoption of ICT in teaching Chinese as a foreign language in US universities. *ReCALL, 26*(1), 100–116. doi:10.1017/S0958344013000268

Ling, R. (2005). The socio-linguistics of SMS: An analysis of SMS use by a random sample of Norwegians. In Mobile communications: Renegotiation of the social sphere (pp. 335–349). London: Springer.

Ling, R., & Haddon, L. (2008). Children, youth and the mobile phone. In Drotner & Livingstone (Eds.), International Handbook of Children, Media and Culture. Sage Publications Ltd. doi:10.4135/9781848608436.n9

Ling, R. (2004). *The mobile connection: The cell phone's impact on society.* New York, NY: Morgan Kaufmann.

Ling, R., & Yttri, B. (2006). Control, emancipation, and status: The mobile telephone in teens' parental and peer relationship. In R. Kraut, M. Brynin, & S. Kiesler (Eds.), *Computer, phones, and the Internet: Domesticating information technology* (pp. 219–234). New York, NY: Oxford University Press.

Littlejohn. (1989). *Theories of Human Communication.* Belmount: Wadsworth.

Littlejohn. (2002). *Theories of human communication* (7th ed.). Belmont: Wadsworth.

Liu, C. C., Tao, S. Y., & Nee, J. N. (2008). Bridging the gap between students and computers: Supporting activity awareness for network collaborative learning with GSM network. *Behaviour & Information Technology*, *27*(2), 127–137. doi:10.1080/01449290601054772

Liu, C.-L. E., Sinkovics, R. R., Pezderka, N., & Haghirian, P. (2012). Determinants of consumer perceptions toward mobile advertising — a comparison between Japan and Austria. *Journal of Interactive Marketing*, *26*(1), 21–32. doi:10.1016/j. intmar.2011.07.002

Liu, X., Liu, X., & Wei, R. (2014). Maintaining social connectedness in a fast-changing world: Examining the effects of mobile phone uses on loneliness among teens in Tibet. *Mobile Media & Communication*, *2*(3), 318–334. doi:10.1177/2050157914535390

LivePerson. (2013). *The Connecting with Customers Report A Global Study of the Drivers of a Successful Online Experience.* Retrieved 16 April 2015 from http://info.liveperson.com/rs/liveperson/images/Online_Engagement_Report_final.pdf

LivePerson. (2015). *LiveEngage transforms the connection between brands and consumers.* Retrieved 2 July 2015 from http://www.liveperson.com/

Livingstone, S. (2009). Children and the Internet. *Polity.*

Lobet-Maris, C., & Henin, J. (2002). Talking without communicating or communicating without talking: From the GSM to the SMS. *Estudios de Juventud*, *57*, 101–114.

Lohan, M. (2001). Men, Masculinities and „Mundane" Technologies: The Domestic Telephone. In E. Green & A. Adam (Eds.), *Virtual Gender: Technology, Consumption and Identity* (pp. 189–205). London, Britain: Routledge.

López-Nores, M., Blanco-Fernández, Y., & Pazos-Arias, J. (2013). Cloud-Based Personalization of New Advertising and e-Commerce Models for Video Consumption. *The Computer Journal*, *56*(5), 573–592. doi:10.1093/comjnl/bxs103

Los Cannes Blog [loscannesblog]. (2014). *Nivea protection ad - Mobile Cannes 2014* [Video file]. Retrieved from https://www.youtube.com/watch?v=L9ZDyLlcdww

Low, G., Goh, D. & Lee, C. (2011). A study of motivations for using mobile content sharing games. *2011 IEEE*, (pp. 258-263).

Low, G. S., & Fullerton, R. A. (1994). Brands, brand management, and the brand manager system: A critical-historical evaluation. *JMR, Journal of Marketing Research*, *31*(2), 173–190. doi:10.2307/3152192

Luanrattana, R., Win, K. T., & Fulcher, J. (2007). Use of Personal Digital Assistants (PDAs) in Medical Education. *20th IEEE International Symposium on Computer-Based Medical Systems CBMS 2007.* Retrieved May 5, 2015, from http://ro.uow.edu.au/cgi/viewcontent.cgi?article=1657&context=infopapers

Lu, C., & Chang, M., Kinshuk, Huang, E., & Chen, C. (2011). Usability of context-aware mobile educational game. *Knowledge Management & E-Learning: An International Journal*, *3*(3), 448–477.

Lüders. (2008). Conceptualizing personal media. *New Media & Society*, 683–702.

Lu, M. (2008). Effectiveness of vocabulary learning via mobile phone. *Journal of Computer Assisted Learning, 24*(6), 515–525. doi:10.1111/j.1365-2729.2008.00289.x

Lundqvist, A., Liljander, V., Gummerus, J., & van Riel, A. (2012). The impact of storytelling on the consumer brand experience: The case of a firm-originated story. *Journal of Brand Management, 20*(4), 283–297. doi:10.1057/bm.2012.15

Lynch, M. (2000). Against Reflexivity as Academic Virtue and Source of Privileged Knowledge. *Theory, Culture & Society, 17*(3), 25–54. doi:10.1177/02632760022051202

Mackalski, R., & Belisle, J. F. (2015). Measuring the short-term spillover impact of a product recall on a brand ecosystem. *Journal of Brand Management, 22*(4), 323–339. doi:10.1057/bm.2015.19

MacKenzie, S. B., & Lutz, R. J. (1989). An empirical examination of the structural antecedents of attitude toward the ad in an advertising pretesting context. *Journal of Marketing, 53*(2), 48–65. doi:10.2307/1251413

Madianou, M., & Miller, D. (2011). Mobile phone parenting: Reconfiguring relationships between Filipina migrant mothers and their left-behind children. *New Media & Society, 13*(3), 457–470. doi:10.1177/1461444810393903

Mainwaring, S. D., Anderson, K., & Chang, M. F. (2005). *Living for the global city: Mobile kits, urban interfaces, and UbiComp*. Paper presented at UbiComp. New York, NY.

Malinovski, T., Vasileva-Stojanovska, T., Jovevski, D., Vasileva, M., & Trajkovik, V. (2015). Adult Students' Perceptions in Distance Education Learning Environments Based on a Videoconferencing Platform -- QoE Analysis. *Journal of Information Technology Education, 14*, 2–19.

Mallat, N., Rossi, M., & Tuunainen, V. K. (2004). Mobile banking services. *Communications of the ACM, 47*(5), 42–46. doi:10.1145/986213.986236

Mallikarjunan, S. (2014). *The State of Ecommerce Marketing*. Cambridge, MA: Hubspot. Retrieved 15 August 2015 from http://offers.hubspot.com/state-of-ecommerce-marketing-2014

Mallikarjunan, S. (2014). The state of ecommerce marketing. *Hubspot*. Retrieved May 21, 2015, from http://offers.hubspot.com/state-of-ecommerce-marketing-2014

Mandhai, S. (2015). *Snapchat opens digital window on Mecca to millions*. Retrieved December 13, 2015, from http://www.aljazeera.com/news/2015/07/snapchat-opens-digital-window-mecca-millions-150714144609540.html

Marchesi, M., & Riccò, B. (2013, November). Augmented graphics for interactive storytelling on a mobile device. In *SIGGRAPH Asia 2013 Symposium on Mobile Graphics and Interactive Applications* (p. 59). ACM. doi:10.1145/2543651.2543683

MarketingSherpa (2010). *2010 Social Media Marketing Benchmark Report*. SherpaStore.com.

MarketingSherpa. (2010). *2010 Social Media Marketing Benchmark Report*. SherpaStore.com.

Marketo. (2015). *How to Use Video Content and Marketing Automation*. Retrieved 30 July 2015 from http://www.marketo.com/ebooks/how-to-use-video-content-and-marketing-automation-to-better-engage-qualify-and-convert-your-buyers/

Markoff, J. (2006, November 12). Entrepreneurs see a web guided by common sense. *The New York Times*. Retrieved from http://www.nytimes.com/2006/11/12/business/12web.html?ei=&_r=0

Martin, J. A. (2015). Mobile News Use and Participation in Elections: A bridge for the Democratic Divide? *Mobile Media @ Communication, 3*(2), 230-249. doi: 10.1177/2050157914550664

Martin, J. (2014). Mobile Media and Political Participation: Defining and Developing an emerging Field. *Mobile Media and Communication, 2*(2), 173–195. doi:10.1177/2050157914520847

Martin, K., & Todorov, I. (2010). How will digital platforms be harnessed in 2010, and how will they change the way people interact with brands? *Journal of Interactive Advertising, 10*(2), 61–66. doi:10.1080/15252019.2010.10722170

Martin, P. (2014). Socio-spatial relations in mobile gaming: Reconfiguration and contestation. In X. Xiaoge (Ed.), *Interdisciplinary mobile media and communications: Social, political and economic implications* (pp. 260–277). Hershey, PA: IGI Global.

Martin, P. W. (1999). Bilingual Unpacking of Monolingual Texts in Two Primary Classrooms in Brunei Darussalam. *Language and Education, 13*(1), 38–58. doi:10.1080/09500789908666758

Mason, J. (2002). *Qualitative Researching*. London: Sage Publications, Inc.

Mason, R. O., McKenney, J. L., & Copeland, D. G. (1997). An historical method for MIS research: Steps and assumptions. *Management Information Systems Quarterly, 31*(3), 307–320. doi:10.2307/249499

Mathews, R. (2004). The psychosocial aspects of mobile phone use amongst adolescents. *InPsych, 26*(6), 16–19.

Matthee, M., & Liebenberg, J. (2007). Mathematics on the move: Supporting mathematics learning through mobile technology in South Africa. In *Making the Connections:Proceedings of the mLearn Melbourne 2007 conference*.

Matyasik, M. (2014). Secure sustainable development: Impact of social media on political and social crisis. *Journal Of Security & Sustainability Issues, 4*(1), 5–16. doi:10.9770/jssi.2014.4.1(1)

Maumbe, B. M., Owei, V., & Taylor, W. (2007). Enabling M-Government in South Africa: An Emerging Direction for Africa. In I. Kushchu (Ed.), *Mobile Government: An Emerging Direction in E-government* (pp. 207–232). IGI Global. doi:10.4018/978-1-59140-884-0.ch011

Mayende, G., & Divitini, M. (2006, July). MOTUS goes to Africa: mobile technologies to increase sustainability of collaborative models for teacher education. In null (pp. 53-54). IEEE. doi:10.1109/TEDC.2006.22

Mayo, K., & Newcomb, P. (2008). How the Web was Won: An Oral History of the Internet. *Vanity Fair*. Retrieved from http://www.vanityfair.com/news/2008/07/internet200807

Mayra, F. (2008). *An introduction to game studies: Games in culture*. London: Sage.

Mazawi, A. (2007). 'Knowledge Society' or Work as 'Spectacle'? Education for Work and the Prospects of the Social Transformation in Arab Societies. In L. Farrell & T. Fenwick (Eds.), *Educating the Global Work Force: Knowledge, Knowledge Work and Knowledge Workers*. London: Routledge.

Mbati, L. S., & Brown, T. H. (2015). *Mobile Learning: Moving Past the Myths and Embracing the Opportunities*. Academic Press.

McAfee, A., & Brynjolfsson, E. (2012). Big Data: The management revolution. *Harvard Business Review, 90*(10), 60–68. PMID:23074865

McCarthy, E. J. (1964). *Basic Marketing. A managerial approach*. Homewood, IL: Richard D. Irwin.

McCorkindale, T., & DiStaso, M. W. (2014). The state of social media research. *Research Journal of the Institute for Public Relations, 1*(1).

McDermott, J. (2013). How mobile ads are bought at agencies today. *Advertising Age, 84*(29), S26.

McEvoy, C. (2015). Enrollment Declines Equal Bad Fiscal Q1 for Apollo Education APOL – Investors.com. *Investor's Business Daily*. Retrieved 20 July 2015 from http://news.investors.com/010815-733803-apollo-education-student-enrollment-declines.htm?ven=yahoocp&src=aurlled&ven=yahoo

McGonigal, J. (2011). *Reality is broken: Why games make us better and how they can change the world.* New York, NY: The Penguin Press.

McGrane, K. (2012). Content strategy for mobile. New York, NY: A Book Apart.

McKie, D. (2005). Chaos and complexity theory. In R. Heath (Ed.), *Encyclopedia of public relations* (pp. 121–123). Thousand Oaks, CA: SAGE Publications, Inc.; doi:10.4135/9781412952545.n58

McLellan, J. (2005). *Malay-English Language Alternation in Two Brunei Darussalam On-line Discussion Forum.* (Unpublished PhD Dissertation). Curtin University of Technology, Department of Language and Intercultural Education.

McPherson, M., Smith-Lovin, L., & Brashears, M. E. (2009). Models and marginals: Using survey evidence to study social networks. *American Sociological Review, 74*(4), 670–681. doi:10.1177/000312240907400409

Medeni, T. D., Medeni, I. T., & Balci, A. (2011). Proposing a Knowledge Amphora Model for Transition towards Mobile Government. *International Journal of E-Services and Mobile Applications, 3*(1), 17–38. doi:10.4018/jesma.2011010102

Meeker, M. (2009). The mobile Internet report setup [Presentation slides]. *Morgan Stanley.* Retrieved from http://www.morganstanley.com/about-us-articles/4659e2f5-ea51-11de-aec2-33992aa82cc2.html

Meeker, M. (2015). Internet trends 2015 – Code Conference. *Kleiner Perkins Annual Report.* May 27, 2015. Retrieved May 28, 2015, from http://kpcbweb2.s3.amazonaws.com/files/90/Internet_Trends_2015.pdf?1432738078

Meeker, M. (May 28, 2014). Internet trends 2014 – Code Conference [Presentation slides]. *Kleiner Perkins Caufield Byers.* Retrieved from http://www.kpcb.com/InternetTrends

Melcrum. (2014). *Inside Internal Communication: Groundbreaking innovations for a new future.* Executive summary retrieved 30 April 2015 from http://www.fruitfulconversations.co.uk/wp-content/uploads/2013/10/Melcrum-Inside-Internal-Comms-2.pdf

Melcrum. (2015). *Case studies.* Retrieved 30 April 2015 from https://www.melcrum.com/resources/case-studies

Melewar, T. C., Gotsi, M., & Andriopoulos, C. (2012). Shaping the research agenda for corporate branding: Avenues for future research. *European Journal of Marketing, 46*(5), 600–608. doi:10.1108/03090561211235138

Mellow, P. (2005). The media generation: Maximise learning by getting mobile. In *Proceedings of the 2005 ASCILITE conference.* Brisbane, Australia: ASCILITE.

Mengistu, D., Zo, H., & Rho, J. J. (2009, November). M-government: opportunities and challenges to deliver mobile government services in developing countries. In *Computer Sciences and Convergence Information Technology, 2009. ICCIT'09. Fourth International Conference on* (pp. 1445-1450). IEEE. doi:10.1109/ICCIT.2009.171

Men, L. R. (2012). CEO credibility, perceived organizational reputation, and employee engagement. *Public Relations Review, 38*(1), 171–173. doi:10.1016/j.pubrev.2011.12.011

Men, L. R., & Tsai, W. H. S. (2015). Infusing social media with humanity: Corporate character, public engagement, and relational outcomes. *Public Relations Review, 41*(3), 395–403. doi:10.1016/j.pubrev.2015.02.005

Mercator Advisory Group. (2013, August 1). *Mobile wallets: The business and the brand* [Press release]. Retrieved from https://www.mercatoradvisorygroup.com/Press_Releases/Mobile_Wallets___The_Business_and_the_Brand/

Mercuri, R. (2002, October). A better ballot box (S. Charry, Ed.). Academic Press.

Merkle/RKG. (2015). *Digital Marketing Report Q1 2015.* Retrieved 21 May 2015 from http://www.rimmkaufman.com/content/quarterly/Merkle-RKG-Q1-2015-DMR.pdf

Merkle/RKG. (2015*). Digital marketing report Q1 2015*. Retrieved May 21, 2015, from http://www.rimmkaufman.com/content/quarterly/Merkle-RKG-Q1-2015-DMR.pdf

Metz, C. (2014, March 6). Facebook Paper has forever changed the way we build mobile apps. *Wired*. Retrieved from http://www.wired.com/2014/03/facebook-paper/

Meyer, H. K., & Carey, M. C. (2014). In Moderation: Examining how journalists' attitudes toward online comments affect the creation of community. *Journalism Practice*, *8*(2), 213–228. doi:10.1080/17512786.2013.859838

Mihailidis, P. (2014). A tethered generation: Exploring the role of mobile phones in the daily life of young people. *Mobile Media & Communication*, *2*(1), 58–72. doi:10.1177/2050157913505558

Miller, G. (2012). The smartphone psychology manifesto. *Perspectives on Psychological Science*, *7*(3), 221–237. doi:10.1177/1745691612441215 PMID:26168460

Miller, G. R. (1989). Persuasion and public relations: Two "Ps" in a pod. In C. H. Botan & V. Hazleton (Eds.), *Public relations theory* (pp. 45–66). Hillsdale, NJ: Lawrence Erlbaum Associaets, Publishers.

Miller-Ott, A. E., Kelly, L., & Duran, R. L. (2014). Cell phone usage expectations, closeness, and relationship satisfaction between parents and their emerging adults in college. *Emerging Adulthood*, *2*(4), 313–323. doi:10.1177/2167696814550195

Minjuan, W., & Khan, M. (2014, April). Mobile Cloud Learning for Higher Education: A Case Study of Moodle in the Cloud. *International Review of Research in Open and Distance Learning*, *15*(2), 254–267.

Mishkin, S., & Waters, R. (2014, September 30). Ebay offers ultimate 'Buy It Now' with plans to spin off PayPal. *Financial Times*. Retrieved from http://www.ft.com/intl/cms/s/0/a14eb3a6-4896-11e4-9d04-00144feab7de.html#axzz3WvBTc4EB

Misuraca, G. C. (2009). e-Government 2015: Exploring m-government scenarios, between ICT-driven experiments and citizen-centric implications. *Technology Analysis and Strategic Management*, *21*(3), 407–424. doi:10.1080/09537320902750871

Mixpo. (2015). *The state of video advertising*. Retrieved 1 May 2015 from http://marketing.mixpo.com/acton/fs/blocks/showLandingPage/a/2062/p/p-00b4/t/page/fm/0

Mixpo. (2015). *The state of video advertising*. Retrieved May 1, 2015, from http://marketing.mixpo.com/acton/fs/blocks/showLandingPage/a/2062/p/p-00b4/t/page/fm/0

Mlot, S. (2012). Microsoft patent tips 360-degree gaming experience. *PC Magazine*. Retrieved June 26, 2015, from http://www.pcmag.com/article2/0,2817,2409621,00.asp

Mobile games market: Market insight beyond the rankings for the global mobile games market. (n.d.). Retrieved January 22, 2015 from http://www.superdataresearch.com/market-data/mobile-games-market/

Mobile Marketing Association. (2013). Free the Forced. *MMA Case Study*. Retrieved from http://www.mmaglobal.com/case-study-hub/upload/pdfs/mma-2013-577.pdf

Mobile Marketing Association. (2013). How Colgate Turned the Maha Kumbh Mela into the Maha Tech Mela. *MMA Case Study*. Retrieved from http://www.digitaltrainingacademy.com/casestudies/mma-COLGATE.pdf

MobileHCI. (2004). Berlin: Springer-Verlag.

MobiThinking. (2014). *Global mobile statistics 2014 part A: Mobile subscribers; handset market share; mobile operators*. Retrieved February 7, 2015 from http://mobiforge.com/research-analysis/global-mobile-statistics-2014-part-a-mobile-subscribers-handset-market-share-mobile-operators#subscribers

Mohamed, A. (Ed.). (2009). *Mobile learning: Transforming the delivery of education and training.* AU Press. Retrieved May 28, 2015, from http://www.zakelijk.net/media/boeken/Mobile%20Learning.pdf

Mohd, N., Mohd, N., Jayashree, S., & Hishamuddin, I. (2013). Malaysian consumers attitude towards mobile advertising, the role of permission and its impact on purchase intention: a structural equation modeling approach. *Asian Social Science.* Retrieved 20 March 2015 from http://ccsenet.org/journal/index.php/ass/article/view/26973/16462

Molina-Castillo, F., & Meroño-Cerdan, A. (2014). Drivers of Mobile Application Acceptance by Consumers: A Meta Analytical Review. *International Journal of E-Services and Mobile Applications, 6*(3), 34–47. doi:10.4018/ijesma.2014070103

Moment, P. R. (2012). *The global explosion in smartphone use has important implications for PR people.* Retrieved from http://www.prmoment.com/1025/the-global-explosion-in-smartphone-use-has-important-implications-for-PR-people.aspx

Moon, M. J. (2010). Shaping M-Government for emergency management: Issues and challenges. *Journal of e-Governance, 33*(2), 100-107.

Moon, M. J. (2002). The evolution of e-government among municipalities: Rhetoric or reality? *Public Administration Review, 62*(4), 424–433. doi:10.1111/0033-3352.00196

Moore, G. (2014). *Crossing the chasm, marketing and selling disruptive products to mainstream customers.* Dunmore, PA: HarperCollins Publishers.

Moore, G. (2014). *Inside the Tornado: Marketing Strategies from Silicon Valley's Cutting Edge.* New York: HarperCollins Publishers.

Morgan, A., & Kennewell, S. (2005). The Role of Play in the Pedagogy of ICT. *Education and Information Technologies, 10*(3), 177–188. doi:10.1007/s10639-005-2998-3

Morgan, N. A., & Rego, L. L. (2009). Brand portfolio strategy and firm performance. *Journal of Marketing, 73*(1), 59–74. doi:10.1509/jmkg.73.1.59

Moscovici, S. (1984) The phenomenon of social representations. In Farr & Moscovici (Eds.), Social Representations. Cambridge, UK: Cambridge University Press.

Moskowitz, G. (2013, November 15). Comedy nights and party supplies: How local video stores are scrambling to survive. *Time.* Retrieved from http://entertainment.time.com/2013/11/15/film-camp-and-party-supplies-how-local-video-stores-are-scrambling-to-survive/

Moyer, J. (2011). Mapping the consequences of technology on public relations. *Institute for Public Relations.* Retrieved from http://www.instituteforpr.org/mapping-technology-consequences/

Mpekoa, N. (2014). Designing, developing and testing a mobile phone voting system in the South African context. In J. V. Steyn (Ed.), *ICTs for inclusive communities in developing societies* (pp. 372-385). Port Elizabeth.

Mtebe, J. S., & Raisamo, R. (2014). Investigating students' behavioural intention to adopt and use mobile learning in higher education in East Africa. *International Journal of Education & Development Using Information & Communication Technology, 10*(3), 4–20.

Mubin, O., Shahid, S., & Mahmud, A. A. (2008). Walk 2 Win: Towards designing a mobile game for elderly's social engagement. *Interaction, 2*, 11–14.

Muneer, F. (2012, July 10). *The Kony 2012 controversy: A look at its coverage in American media vs. Ugandan media.* Retrieved from Huffington Post: http://www.huffingtonpost.com/fatima-muneer/the-kony-2012-controversy_b_1503990.html

Muniz, K. M., Woodside, A. G., & Sood, S. (2015). Consumer storytelling of brand archetypal enactments. *International Journal of Tourism Anthropology*, *4*(1), 67–88. doi:10.1504/IJTA.2015.067644

Murphy, P. (2013). *Coping with an uncertain world: Complexity theory's role in public relations theory-building*. Paper presented at the annual meeting of the International Communication Association, San Francisco, CA. Retrieved from http://citation.allacademic.com/meta/p174867_index.html

Murphy, P. (1989). Game theory as a paradigm for the public relations process. In C. H. Botan & V. Hazleton (Eds.), *Public relations theory* (pp. 173–192). Hillsdale, NJ: Lawrence Erlbaum Associaets, Publishers.

Murphy, P. (1991). The limits of symmetry: A game theory approach to symmetric and asymmetrical public relations. *Public Relations Review Annual*, *3*(1-4), 115–131. doi:10.1207/s1532754xjprr0301-4_5

Murphy, P. (2000). Symmetry, contingency, complexity: Accommodating uncertainty in public relations theory. *Public Relations Review*, *26*(4), 447–462. doi:10.1016/S0363-8111(00)00058-8

Murugesan, S. (2013). Mobile apps in Africa. *IT Professional*, *15*(5), 8–11. doi:10.1109/MITP.2013.83

Mutter, A. (2012). Print ads fell 25x faster than digital grew. *Reflections of a Newsosaur*. http://newsosaur.blogspot.com/2012_09_01_archive.html

Mutter, A. (2012a). *The incredible shrinking newspaper audience*. Academic Press.

Mutter, A. D. (10 Sept. 2012). Print ads fell 25x faster than digital grew. *Reflections of a Newsosaur*. Retrieved from http://newsosaur.blogspot.com/

Mutter, A. D. (2012). The incredible shrinking newspaper audience. *Reflections of a Newsosaur*.

Mutter, A. D. (2012a). *The incredible shrinking newspaper audience*. Academic Press.

Mutter, A. D. (2012b). Print ads fell 25x faster than digital grew. *Reflections of a Newsosaur*. Retrieved from http://newsosaur.blogspot.com/2012_09_01_archive.html

Mutter, A. D. (2014). The plight of newspapers in a single chart. *Reflections of a Newsosaur*. Retrieved 20 Feb. 2015 from http://newsosaur.blogspot.com/2014/04/the-plight-of-newspapers-in-single-chart.html

Mutter, A. D. (2014). The plight of newspapers in a single chart. *Reflections of a Newsosaur*. Retrieved 20 Feb. 2015 from http://newsosaur.blogspot.com/2014/04/the-plight-of-newspapers-in-single-chart.html.http://newsosaur.blogspot.com/2014/04/the-plight-of-newspapers-in-single-chart.html

Mutter, A. D. (2015). Should newspapers abandon digital? *Editor & Publisher*. Retrieved 15 April 2015 from http://newsosaur.blogspot.com/2015/10/should-newspapers-abandon-digital.html

Myers, M. D. (1997). Qualitative research in information systems. *Management Information Systems Quarterly*, *21*(2), 241–242. doi:10.2307/249422

Myles, A. (January 7, 2015). *SMB marketing survey 2014*. Retrieved April 23, 2015, from http://www.brightlocal.com/2015/01/07/37-smbs-plan-spend-internet-marketing-2015/

Nah, K. C., White, P., & Sussex, R. (2008). The potential of using a mobile phone to access the Internet for learning EFL listening skills within a Korean context. *ReCALL*, *20*(3), 331–347. doi:10.1017/S0958344008000633

Naismith, L., Lonsdale, P., Vavoula, G., & Sharples, M. (2004). *Literature review in mobile technologies and learning*. A Report for NESTA Futurelab.

Nalder, G., Kendall, E., & Menzies, V. (2007). Self-organising m-learning communities: A case study. In *Making the Connections:Proceedings of the mLearn Melbourne 2007 conference.*

Nanji, A. (May 17, 2013) *How small-business owners are using mobile technology.* Retrieved April 23, 2015, from http://www.marketingprofs.com/charts/2013/10791/how-small-business-owners-are-using-mobile-technology

Napoleon, A. E., & Bhuiyan, M. S. H. (2010). Contemporary research on mobile government. In *Scandinavian workshop on e-Government (SVEG)* (p. 61).

Nasi, G., Cucciniello, M., & Guerrazzi, C. (2015). The role of mobile technologies in health care processes: The case of cancer supportive care. *Journal of Medical Internet Research, 17*(2), e26. doi:10.2196/jmir.3757 PMID:25679446

Nasr, R. (2015). Munster: Facebook is Quickly Catching Up to Facebook. *CNBC*. Retrieved 15 Oct. 2015 from http://www.cnbc.com/2015/07/27/munster-facebook-is-quickly-catching-up-to-google.html

National Retail Federation. (2015, February 6). *Mobile still tops retailers' priority lists, according to*Shop.org/Forrester-*research report* [Press release]. Retrieved from https://nrf.com/media/press-releases/mobile-still-tops-retailers-priority-lists-according-shoporgforrester-research

Nawaz, S., & Ahmad, Z. (2012). Statistical study of impact of mobile on student's life. *IOSR Journal of Humanities and Social Science, 2*(1), 43–49. doi:10.9790/0837-0214349

NCSL. (2015, July 7). *Voting Equipment: Paper Ballots and Direct-Recording Electronic Voting Machines.* Retrieved from National Conference of State Legislators: http://www.ncsl.org/research/elections-and-campaigns/voting-equipment.aspx OECD/International

Nemiroski, A., Christodouleas, D. C., Hennek, J. W., Kumar, A. A., Maxwell, E. J., Fernández-Abedul, M. T., & Whitesides, G. M. (2014). Universal mobile electrochemical detector designed for use in resource-limited applications. *Proceedings of the National Academy of Sciences, 111*(33), 11984-11989. doi:10.1073/pnas.1405679111

Neustad, R. E., & May, E. R. (1996). *Thinking in time: the uses of history for decision makers.* New York: The Free Press.

Neustar. (2015). *5 Ways to Engage those Slippery Omni-channel Shoppers.* Retrieved 3 August 2015 from https://hello.neustar.biz/omnichannel_wp_retail_mktg_lp.html?utm_medium=Advertising-Online&utm_source=Shop.org&utm_campaign=2015-eBook-Shop.org

New Zealand Herald. (2012). *More mobile phones in NZ than people: Study.* Retrieved from http://www.nzherald.co.nz/nz/news/article.cfm?c_id=1&objectid=10801183

New, B. (January 4, 2011). *Vul het zelf maar in (Fill in the blank).* Retrieved from http://www.underconsideration.com/brandnew/archives/vul_het_zelf_maar_in_fill_in_the_blank.php#.VRl55vmsWSo

NewZoo Games Market Research. (2014, October 29). *Global mobile games revenues to reach $25 billion in 2014.* Retrieved May 7, 2015, from http://www.newzoo.com/insights/global-mobile-games-revenues-top-25-billion-2014/

Newzoo. (2014). *Asia-Pacific contributes 82% of the $6Bn global games market growth in 2014.* Retrieved April 22, 2015 from http://www.newzoo.com/insights/asia-pacific-contributes-82-6bn-global-games-market-growth/

Nguyen, J. (2014). *Online Video Consumption in APAC and Total Video.* http://www.comscore.com/Insights/Presentations-and-Whitepapers/(offset)/10/?cs_edgescape_cc=KZ retrieved 16 Feb. 2015.

Nguyen, J. (2014). *Online Video Consumption in APAC and Total Video.* Retrieved 1 May 2015 from http://www.comscore.com/Insights/Presentations-and-Whitepapers/(offset)/10/?cs_edgescape_cc=KZ retrieved 16 Feb. 2015.

Nguyen, J. (2014). *Online Video Consumption in APAC and Total Video*. Retrieved 16 Feb. 2015 from http://www.comscore.com/Insights/Presentations-and-Whitepapers/(offset)/10/?cs_edgescape_cc=KZ

Nguyen, J. (2014). *Online Video Consumption in APAC and Total Video*. Retrieved from http://www.comscore.com/Insights/Presentations-and-Whitepapers/2014/Online-Video-Consumption-in-APAC-and-Total-Video

Nielsen. (2014). *Report of China Mobile SNS Users' Needs and Behaviors*. Retrieved from 199it.com: http://www.199it.com/archives/272681.html

Nisbet, M. (2007). The future of public engagement. *Scientist (Philadelphia, Pa.)*, *21*(10), 38–44.

Norman, D. A. (1998). *The Invisible Computer: Why Good Products Can Fail, the Personal Computer Is So Complex and Information Appliances are the Solution*. Cambridge, MA: MIT Press.

North, D., Johnston, K., & Ophoff, J. (2014). The Use of Mobile Phones by South African University Students. *Issues in Informing Science and Information Technology, 11*. Retrieved from http://iisit.org/Vol11/IISITv11p115-138North0469.pdf

Ntaliani, M., Costopoulou, C., & Karetsos, S. (2008). Mobile government: A challenge for agriculture. *Government Information Quarterly*, *25*(4), 699–716. doi:10.1016/j.giq.2007.04.010

Nudd, T. (2012, June 20). How Nike+ made 'Just do it' obsolete: Stefan Olander on building a new ecosystem of engagement. *Adweek*. Retrieved from http://www.adweek.com/news/advertising-branding/how-nike-made-just-do-it-obsolete-141252

Nyíri, K. (Ed.). (2003a). *Mobile Communication: Essays on Cognition and Community*. Vienna, Austria: Passagen Verlag.

Nyíri, K. (Ed.). (2003b). *Mobile Learning: Essays on Philosophy, Psychology and Education*. Vienna, Austria: Passagen Verlag.

Nyíri, K. (Ed.). (2003c). *Mobile Democracy: Essays on Society, Self and Politics*. Vienna, Austria: Passagen Verlag.

Nyíri, K. (Ed.). (2005). *A Sense of Place: The Global and the Local in Mobile Communication*. Vienna, Austria: Passagen Verlag.

Nyíri, K. (Ed.). (2006). *Mobile Understanding: The Epistemology of Ubiquitous Communication*. Vienna, Austria: Passagen Verlag.

Nyíri, K. (Ed.). (2007). *Mobile Studies. Paradigms and Perspectives*. Vienna, Austria: Passagen Verlag.

O'Malley, C., Vavoula, G., Glew, J. P., Taylor, J., Sharples, M., & Lefrere, P. (2003). *Guidelines for learning/teaching/tutoring in a mobile environment*. Academic Press.

O'Brien, J. A. (2004). *Sistemas de Informação e as decisões gerenciais na era da internet* (2nd ed.). São Paulo, São Paulo: Saraiva.

OECD/ITU. (2011). *M-Government: Mobile Technologies for Responsive Governments and Connected Societies*. Paris: OECD Publishing; doi:10.1787/9789264118706-en

Oetting, J. (2015). The Next Phase in Personalization? Video Marketing. *HubSpot*. Retrieved 2 May 2015 from http://blog.hubspot.com/agency/personalization-video-marketing

Ogilvy Action Germany [OgilvyActionGER]. (2013, June 19). *Back to vinyl – The office turntable* [Video file]. Retrieved from https://www.youtube.com/watch?v=Xg1pzEF3HTw

Ojigho, O. (2014, November 16). *Hashtag activism makes the invisible visible*. Retrieved from The Mantle: http://www.mantlethought.org/other/hashtag-activism-makes-invisible-visible

Ojo, A., Janowski, T., & Awotwi, J. (2013). Enabling development through governance and mobile technology. *Government Information Quarterly*, *30*, S32–S45. doi:10.1016/j.giq.2012.10.004

Okazaki, S. (2008). Exploring experiential value in online mobile gaming adoption. *Cyberpsychology & Behavior*, *11*(5), 619–624. doi:10.1089/cpb.2007.0202 PMID:18785820

Okazaki, S. (2012). Teaching mobile advertising. In *Advertising Theory* (pp. 373–387). New York: Routledge.

Okazaki, S. (2012). Teaching mobile Advertising. In *Advertising Theory* (pp. 373–387). New York: Routledge.

Oksman, V., & Rautiainen, P. (2002). I've got my whole life in my hand. *Revista de Estudios de Juventud*, *52*, 25–32.

Oksman, V., & Turtiainen, J. (2004). Mobile communication as a social stage meanings of mobile communication in everyday life among teenagers in Finland. *New Media & Society*, *6*(3), 319–339. doi:10.1177/1461444804042518

Okunbor, D., & Guy, R. (2008). *Analysis Of A Mobile Learning Pilot Study*. Math and Computer Science working papers 01/2008

Olanrewaju, T., Smaje, K., & Willmott, P. (2014, May). The seven traits of effective digital enterprises. *McKinsey Insights*. Retrieved from http://www.mckinsey.com/insights/organization/the_seven_traits_of_effective_digital_enterprises

Olejnik, L., Acar, G., Castelluccia, C., & Diaz, C. (n.d.). *The leaking battery A privacy analysis of the HTML5 Battery Status API*. Retrieved from http://eprint.iacr.org/2015/616.pdf

Olenski, S. (2013). What is programmatic advertising, and is it the future? *Forbes*. Retrieved 6 May 2015 from http://www.forbes.com/sites/marketshare/2013/03/20/what-is-programmatic-advertising-and-is-it-the-future/

Olesen, T. (2008). Activist journalism? The Danish Cheminova debates, 1997 and 2006. *Journalism Practice*, *2*(2), 245–263. doi:10.1080/17512780801999394

Oliveira, R. D., Cherubini, M., & Oliver, N. (2010). MoviPill: Improving medication compliance for elders using a mobile persuasive social game.*Proceedings of UbiComp '10 of the 12th ACM International Conference on Ubiquitous Computing*, (pp. 251-260). New York: ACM. doi:10.1145/1864349.1864371

Oller, R. (2012). *The Future of Mobile Learning (Research Bulletin)*. Retrieved from https://net.educause.edu/ir/library/pdf/ERB1204.pdf

OneReach. (2014, August). *The high demand for customer service via text message. 2014 U.S. Survey Report*. Retrieved from http://onereachcontactcenter.com/wp-content/uploads/2014/08/High-Demand-for-Customer-Service-via-Text-Message-2014-Report.pdf

Orlando, J. (2010, August 18). *Save Time and Teach Better with Screencasting - Faculty Focus*. Retrieved July 25, 2015, from http://www.facultyfocus.com/articles/effective-teaching-strategies/save-time-and-teach-better-with-screencasting/

Orlando, J. (2014, March 7). *Screencasting feedback with Screencast-O-Matic*. [Video File]. Retrieved July 25, 2015, from https://www.youtube.com/watch?v=CDcfX2Qj6k0&feature=youtu.be

Orlikowski, W., & Baroudi, J. J. (1991). Studying information technology in organizations: Research approaches and assumptions. *Information Systems Research*, *2*(1), 1–28. doi:10.1287/isre.2.1.1

Osang, F. B., Ngole, J., & Tsuma, C. (2013). Prospects and Challenges of Mobile Learning Implementation in Nigeria. Case Study National Open University of Nigeria NOUN. In *International Conference on ICT for Africa 2013* (pp. 20-23).

Otte, E. & Rousseau, R. (2002). Social network analysis: a powerful strategy, also for the information sciences. *Journal of Information Science, 28*, 441–453. doi: 10.1177/016555150202800601

Owen, J. (2015). *E-NABLING hands across the world.* Retrieved 2 July 2015 from http://enablingthefuture.org/tag/3d-printed-prosthetics/

Ozbek, E. A. (2015). A classification of student skills and compentencies in open and distance learning. *International Journal on New Trends in Education*, *6*(3), 174–185.

Palen, L., Salzman, M., & Youngs, E. (2000). Going wireless: behavior & practice of new mobile phone users. In *Proceedings of the 2000 ACM conference on Computer supported cooperative work* (pp. 201–210). Philadelphia, PA: ACM. doi:10.1145/358916.358991

Palfrey, J., & Gasser, U. (2008). *Born digital: Understanding the first generation of digital natives.* Philadelphia: Basic Books.

Pang, P., & Hung, D. (2001). Activity Theory as a Framework for Analyzing CBT and E-Learning Environments. *Educational Technology*, 36–42.

Pappas, C. (2014). *The 20 Best Learning Management Systems.* Retrieved 20 July 2015 from http://elearningindustry.com/the-20-best-learning-management-systems

Pappas, C. (2015, January 25). The Top eLearning Statistics and Facts for 2015 You Need to Know. *eLearning Industry*. Retrieved August 5, 2015, from http://elearningindustry.com/elearning-statistics-and-facts-for-2015

Parganas, P., Anagnostopoulos, C., & Chadwick, S. (2015). 'You'll never tweet alone': Managing sports brands through social media. *Journal of Brand Management*, *22*(7), 551–568. doi:10.1057/bm.2015.32

Park, E., Baek, S., Ohm, J., & Chang, H. (2014). Determinants of player acceptance of mobile social network games: An application of extended technology acceptance model. *Telematics and Informatics*, *31*, 3–15.

Parker, M. (2012). Ethical considerations related to mobile technology use in medical research. *Journal of Mobile Technology in Medicine*, *1*(3), 50–52. doi:10.7309/jmtm.23

Park, H., & Kim, S. (2013). A Bayesian network approach to examining key success factors of mobile games. *Journal of Business Research*, *66*(9), 1353–1359. doi:10.1016/j.jbusres.2012.02.036

Park, M. H., Hong, J. H., & Cho, S. B. (2007). Location-based recommendation system using bayesian user's preference model in mobile devices. In *Ubiquitous Intelligence and Computing* (pp. 1130–1139). Springer Berlin Heidelberg. doi:10.1007/978-3-540-73549-6_110

Park, S., Cho, K., & Lee, B. G. (2014). What Makes Smartphone Users Satisfied with the Mobile Instant Messenger?: Social Presence, Flow, and Self-disclosure. *International Journal of Multimedia & Ubiquitous Engineering*, *9*(11), 315–324. doi:10.14257/ijmue.2014.9.11.31

Parsons, D. (2014). The future of mobile learning and implications for education and training. *Increasing Access*, 217.

Patel, M., & Veira, J. (2014, December). Making connections: An industry perspective on the Internet of Things. *McKinsey Insights*. Retrieved from http://www.mckinsey.com/insights/high_tech_telecoms_internet/making_connections_an_industry_perspective_on_the_internet_of_things

Pathak, S. (2015). How Starcom trained 1,200 employees to speak programmatic. *Digiday*. Retrieved 6 May 2015 from http://digiday.com/agencies/starcom-trained-talent-speak-programmatic/

Payne, M. (2014, August 20). *Tell Mel:Throwing cold water on Ice Bucket Challenge.* Retrieved from News Press: http://www.news-press.com/story/news/investigations/melanie-payne/2014/08/19/throwing-cold-water-ice-bucket-challenge/14309687/

Pearson, C., & Hussain, Z. (2015). Smartphone use, addiction, narcissism, and personality: A mixed methods investigation. *International Journal of Cyber Behavior, Psychology and Learning*, *5*(1), 17–32. doi:10.4018/ijcbpl.2015010102

Pearson, D. (2014). *Report: Mobile to become gaming's biggest market by 2015*. Newzoo.

Pearson, L. (2013). Fluid marks 2.0: Protecting a dynamic brand. *Managing Intellectual Property*, *26*, 26–30.

Pearson, L. (2014). Fluid trademarks and dynamic brand identities. *The Trademark Reporter*, *104*, 1411–1433.

Peck, D. (2013). They're watching you at work. *Atlantic (Boston, Mass.)*.

Pelet, J. B. (2014). *Determinants of effective learning through social networks systems: An exploratory study*. IGI Global.

Pelet, J.-E., & Papadopoulou, P. (2014). Consumer behavior in the mobile environment: An exploratory study of M-commerce and social media. *International Journal of Technology and Human Interaction*, *10*(4), 36–48. doi:10.4018/ijthi.2014100103

Penberthy, J. (2015). *Weird YouTube Method Pulls in $816,481.53*. Live webinar presented 30 July 2015.

Peng, H., Su, Y.-J., Chou, C., & Tsai, C.-C. (2009). Ubiquitous knowledge construction: Mobile learning re-defined and a conceptual framework. *Innovations in Education and Teaching International*, *46*(2), 171–183. doi:10.1080/14703290902843828

Pentland, A. (2009). Reality mining of mobile communications: Toward a new deal on data. *The Global Information Technology Report 2008-2009: Mobility in a Networked World*, (pp. 75-80). World Economic Forum.

Pepper, R. (2012). *The impact of mobile broadband on national economic growth*. Retrieved 16 Feb. 2015 from http://newsroom.cisco.com/press-release-content?type=webcontent

Pepper, R. (2012). *The impact of mobile broadband on national economic growth*. Retrieved 16 Feb. 2015 http://newsroom.cisco.com/press-release-content?type=webcontent&articleId=1110575

Pepper, R. (2012). *The impact of mobile broadband on national economic growth*. Retrieved from http://newsroom.cisco.com/press-release-content?type=webcontent&articleId=1110575

Pereira, D. C., & Meirelles, M. R. G. (2009). Uma abordagem transdisciplinar do método Análise de Redes Sociais. *Informação & Informação*, *14*(2), 84–99.

Perry, M. (2013). *Schumpeterian Gales of creative destruction in the personal computing market have destroyed the Windows' monopoly*. Retrieved May 1, 2015, from http://www.aei.org/publication/schumpeterian-gales-of-creative-destruction-in-the-personal-computing-market-have-destroyed-the-windows-monopoly/

Persaud, A., & Azhar, I. (2012). Innovative mobile marketing via smartphones: Are consumers ready? *Marketing Intelligence & Planning*, *30*(4), 418–443. doi:10.1108/02634501211231883

Pertierra, R. (2005, April). Mobile phones, identity and discursive intimacy. *An Interdisciplinary Journal on Humans in ICT Environments*, *1*(1), 23–44. doi:10.17011/ht/urn.2005124

Peterson, T. (2014, August 20). *Vice News keeps the lens open during Ferguson protests*. Retrieved from Ad Age: http://adage.com/article/media/vice-news-lens-open-ferguson-protests/294641/

Pew Internet & American Life Project. (2010). *Teens and Mobile Phones: Text messaging explodes as teens embrace it as the centerpiece of their communication strategies with friends* (A project of the Pew Research Center). Retrieved from http://pewinternet.org/Reports/2010/Teens-and-Mobile-Phones.aspx

Pew Research Center. (2012). Mobile technology factsheet. *Pew Internet Project*. Retrieved from http://www.pewinternet.org/fact-sheets/mobile-technology-fact-sheet/

Pew Research Center. (2014). *Emerging nations catching up to U.S. on technology adoption, especially mobile and social media use.* Retrieved from http://www.pewresearch.org/fact-tank/2014/02/13/emerging-nations-catching-up-to-u-s-on-technology-adoption-especially-mobile-and-social-media-use/

Pew Research Center. (2014). *State of the News Media 2014.* Retrieved 30 May 2015 from http://www.journalism.org/packages/state-of-the-news-media-2014/

Pew Research Center. (2014). *The revenue picture for American journalism, and how it is changing.* Retrieved 17 Feb. 2015 http://www.journalism.org/files/2014/03/Revnue-Picture-for-American-Journalism.pdf

Pew Research Center. (2014). *The revenue picture for American journalism, and how it is changing.* Retrieved from http://www.journalism.org/files/2014/03/Revnue-Picture-for-American-Journalism.pdf

Pew Research Center. (2014). *Three technology revolutions: Mobile.* Retrieved on March 18, 2015 from http://www.pewinternet.org/three-technology-revolutions/

Pew Research Center. (2015). *State of the News Media 2015.* Retrieved 30 May 2015 from http://www.journalism.org/files/2015/04/FINAL-STATE-OF-THE-NEWS-MEDIA1.pdf

Pfau, M., & Wan, H.-H. (2006). Persuasion: An intrinsic function of public relations. In C. H. Botan & V. Hazleton (Eds.), *Public relations theory II* (pp. 101–136). Mahwah, NJ: Lawrence Erlbaum Associaets, Inc., Publishers.

Phillips, D., & Young, P. (2009). *Online public relations: A practical guide to developing an online strategy in the world of social media.* London: Kogan Page.

Pieczka, M. (2011). Public relations as dialogic expertise? *Journal of Communication Management, 15*(2), 108-124.

Pierre, L. K. (2012). Marketing meets Web 2.0, social media, and creative consumers: Implications for international marketing strategy. *Business Horizons*, 261–271.

Pietz, M., & Storbacka, L. (2007). *Driving Advertising into Mobile Mediums: Study of Consumer Attitudes Towards Mobile Advertising and of Factors Affecting on Them.* Umea University.

Pimentel, R. W., & Heckler, S. E. (2003). Changes in logo designs: Chasing the elusive butterfly curve. In L. M. Scott & R. Batra (Eds.), *Persuasive Imagery: A Consumer Response Perspective* (pp. 105–128). Mahwah, NJ: Lawrence Erlbaum Associates.

Pintor, R. L. (n.d.). *Voter Turnout Rates from a Comparative Perspective. Academic Press.*

Poell, T., & Borra, E. (2011). Twitter, YouTube, and Flickr as platforms of alternative journalism: The social media account of the 2010 Toronto G20 protests. *Journalism, 13*(6), 695–713. doi:10.1177/1464884911431533

Pompper, D. (2005). Game theory. In R. L. Heath (Ed.), *Encyclopedia of public relations* (pp. 360–362). Thousand Oaks, CA: SAGE Publications, Inc.; doi:10.4135/9781412952545.n179

Pope, M., Pantages, R., Enachescu, N., Dinshaw, R., Joshlin, C., Stone, R., & Seal, K. (2011). Mobile payments: The reality on the ground in selected Asian countries and the United States. *International Journal of Mobile Marketing, 6*(2), 88–104.

Porter, M. E., & Heppelmann, J. E. (2014). How smart, connected products are transforming competition. *Harvard Business Review, 92*(11), 11–64.

Powell, W. W. (1990). Neither market nor hierarchy: network forms of organization. In B. M. Staw & L. Cummings (Eds.), *Research in organizational behavior.* Greenwich, UK: JAI Press.

Powers, T., Advincula, D., Austin, M. S., Graiko, S., & Snyder, J. (2012). Digital and Social Media In the Purchase Decision Process. *Journal of Advertising Research, 52*(4), 479–489. doi:10.2501/JAR-52-4-479-489

Prasertsilp, P. (2013). Mobile Learning: Designing a Socio-Technical Model to Empower Learning in Higher Education. *LUX: A Journal of Transdisciplinary Writing and Research from Claremont Graduate University, 2*(1), 23.

Protalinski. (2012). *Facebook has over 901 million users, over 488 mobile users.* Retrieved from ZDNet: http://www.zdnet.com/blog/facebook/facebook-has-over-901-million-users-over-488-million-mobile-users/12105

Protess, D. L. (1992). *The journalism of outrage: Investigative reporting and agenda building in America.* Guilford Press.

Protess, D. L., Cook, F. L., Curtin, T. R., Gordon, M. T., Leff, D. R., McCombs, M. E., & Miller, P. (1987). The impact of investigative reporting on public opinion and policymaking targeting toxic waste. *Public Opinion Quarterly, 51*(2), 166–185. doi:10.1086/269027

Puja, J., & Parsons, D. (2011). A location-based mobile game for business education. *2011 11th IEEE International Conference on Advanced Learning Technologies,* (pp. 42-44).

Pulizzi, J. (2012). The rise of storytelling as the new marketing. *Publishing Research Quarterly, 28*(2), 116–123. doi:10.1007/s12109-012-9264-5

Pulizzi, J. (2013). *Epic content marketing: How to tell a different story, break through the clutter, and win more customers by marketing less.* Boston, MA: McGraw-Hill.

Pulman, A. J. (2008). *The Nintendo DS as an assistive technology tool for health and social care students.* Paper presented at the mLearn 2008 Conference, Shropshire, UK.

PulsePoint. (2015). *Programmatic intelligence report 2014.* Retrieved 6 May 2015 from http://www.pulsepoint.com/resources/whitepapers/pulsepoint_intelligence_report.pdf

PulsePoint. (2015). *Programmatic intelligence report 2014.* Retrieved May 6, 2015, from http://www.pulsepoint.com/resources/whitepapers/pulsepoint_intelligence_report.pdf

Putnam, R. D. (2001). *Bowling alone: The collapse and revival of American community.* Simon and Schuster.

Putnam, R. D., Feldstein, L., & Cohen, D. J. (2004). *Better together: Restoring the American community.* Simon and Schuster.

Quercia, D., Kosinski, M., Stillwell, D., & Crowcroft, J. (2011, October). Our Twitter profiles, our selves: Predicting personality with Twitter. In *Privacy, Security, Risk and Trust (PASSAT) and 2011 IEEE Third Inernational Conference on Social Computing (SocialCom), 2011 IEEE Third International Conference on* (pp. 180-185). IEEE.

Quinn, C. N. (2012). *The mobile academy: mLearning for higher education.* Jossey-Bass.

Quinn, S. (2015). *4 Ways to Use Twitter Video for Your Business.* Social Media Examiner.

Radic, K. (2013, July 29). 80 variations of the iconic Lacoste logo. *Branding Magazine.* Retrieved from http://www.brandingmagazine.com/2013/07/29/lacoste-anniversary-polo-logo-peter-saville/

Rafaeli, S. (1986). The electronic bulletin board: A computer-driven mass medium. *Computers and the Social Sciences, 2*(3), 123–137. doi:10.1177/089443938600200302

Rafaeli, S., & Sudweeks, F. (1998). Interactivity on the Nets. In *Network Netplay: Virtual Groups on the Internet* (pp. 173–189). Menlo Park, CA: AAAI Press/MIT Press.

Raice, S. (2011, September 7). IPO? Groupon plans in flux. *The Wall Street Journal*. Retrieved from http://www.wsj.com/articles/SB10001424053111904537404576554812230222934

Rainie, L., Hitlin, P., Jurkowitz, M., Dimock, M., & Neidorf, S. (2012, March 15). *The viral Kony 2012 video*. Retrieved from Poynter: http://www.pewinternet.org/2012/03/15/the-viral-kony-2012-video/

Ramos, L. (2013). *B2B CMOs must evolve or move on*. Retrieved May 23, 2015, from https://solutions.forrester.com/bma-survey-findings-ramos \

Ramos, L. (2013). *The B2B CMO role is expanding: Evolve of move on*. Retrieved 30 April 2015 from http://solutions.forrester.com/Global/FileLib/Reports/B2B_CMOs_Must_Evolve_Or_Move_On.pdf

Ramsey, G. (2015). *The state of mobile 2015*. eMarketer live webcast 25 March 2015.

Ramsey, G. (2015). *The State of Mobile 2015*. eMarketer live webcast 25 March 2015.

Rau, P. L. P., Zhang, T., Shang, X., & Zhou, J. (2011). Content relevance and delivery time of SMS advertising. *International Journal of Mobile Communications*, *9*(1), 19–38. doi:10.1504/IJMC.2011.037953

Rau, P. P., Zhou, J., Chen, D., & Lu, T. (2014). The influence of repetition and time pressure on effectiveness of mobile advertising messages. *Telematics and Informatics*, *31*(3), 463–476. doi:10.1016/j.tele.2013.10.003

RazorFish. (2015). *Global digital marketing report 2015*. Retrieved May 7, 2015, from http://www.razorfish.com/binaries/content/assets/ideas/digitaldopaminereport2015.pdf

Reflections of a Newsosaur. (2012). Retrieved 15 April 2015 from http://newsosaur.blogspot.com/2012_10_01_archive.html

Reflections of a Newsosaur. (n.d.). Retrieved from http://newsosaur.blogspot.com/

Reichhart, P. (2014). Identifying factors influencing the customers purchase behaviour due to location-based promotions. *International Journal of Mobile Communications*, *12*(6), 642–660. doi:10.1504/IJMC.2014.064917

Reporter, B. (2015). *The Great Customer Experience Divide: 4 Out Of 5 Consumers Declare Brands Don't Know Them as Individuals, IBM Study Finds*. Retrieved from https://www.bulldogreporter.com/dailydog/article/pr-biz-update/the-great-customer-experience-divide-4-out-of-5-consumers-declare-bra?utm_source=Bulldog+Reporter&utm_campaign=15bd0a556f-Daily_Dog_Monday_April_6&utm_medium=email&utm_term=0_200eecab38-15bd0a556f-154734149

Requirements. (2015). Retrieved from http://distance.msstate.edu/geosciences/bmp/requirements

Researchscape International. (2015). *Trends & Priorities in Real-Time Personalization*. Retrieved 1 August 2015 from http://www.evergage.com/blog/survey-findings-trends-priorities-in-real-time-personalization/

Review, T. (2008). *Ten emerging technologies that will change the world*. Retrieved from http://www.technologyreview.com/article/13060/

Reyck, B. D., & Degraeve, Z. (2003). Broadcast scheduling for mobile advertising. *Operations Research*, *51*(4), 509–517. doi:10.1287/opre.51.4.509.16104

Rheingold, H. (2002). *Smart mobs: the next social revolution*. Cambridge, MA: Perseus Publishing.

Richards, I. (2012). Beyond city limits: Regional journalism and social capital. *Journalism*, *14*(5), 627–642. doi:10.1177/1464884912453280

Richardson, I. (2010). Ludic mobilities: The corporealities of mobile gaming. *Mobilities*, *5*(4), 431–447. doi:10.1080/17450101.2010.510329

Richardson, R. J. (1989). *Pesquisa social: métodos e técnicas*. São Paulo, São Paulo: Atlas.

Ricker, W. H. (1968, March). A Theory of the Calculus of Voting. *The American Political Science Review*, *62*(01), 25–42. doi:10.2307/1953324

Riecken, R. & Lanza, B. B. B. (2007). E-Paraná: a rede de informações e serviços eletrônicos do governo do Estado do Paraná. *Informação & Informação*, *12*(2).

Riley, W. T., Rivera, D. E., Atienza, A. A., Nilsen, W., Allison, S. M., & Mermelstein, R. (2011). Health behavior models in the age of mobile interventions: Are our theories up to the task? *Translational Behavioral Medicine*, *1*(1), 53–71. doi:10.1007/s13142-011-0021-7 PMID:21796270

Ringberg, T., & Bjerregaard, S. (2012). Brand Relationships 2.0: A Fundamental Paradox of Proactive Relational Branding and Critical Consumer Culture. In *The 41th EMAC Annual Conference 2012*.

Rius, A., Masip, D. & Clariso, R. (2014). Student projects empowering mobile learning in higher education. *RUSC: Revista De Universidad Y Sociedad Del Conocimiento*, *11*(1), 192-207.

Rivest, R. L. (n.d.). *Electronic Voting*. Massachusetts Institute of Technology.

Roberts, G., & Klibanoff, H. (2006). *The Race Beat*. New York: Vintage Books.

Robideaux, D., & Robideaux, S. C. (2012). Convergent storytelling as promotion: A model for brand engagement and equity. *Innovative Marketing*, *8*(1), 48–51.

Robin, J. (2012, February). Mobile-izing: Democracy, Organization and India's First "Mass Mobile Phone" Elections. *The Journal of Asian Studies*, *71*(1), 63–80. doi:10.1017/S0021911811003007

Rogers, Y., Price, S., Harris, E., Phelps, T., Underwood, M., Wilde, D.,... Neale, H. (2002). *Learning through digitally-augmented physical experiences: Reflections on the Ambient Wood project*. Equator IRC. Available http://www.equator.ac.uk/papers/Ps//2002-rogers-1.pdf

Rogers, E. (2003). *Diffusion of innovations* (5th ed.). New York, NY: Free Press.

Rogers, E. M. (1983). Elements of Diffussion. In E. Rogers (Ed.), *Diffusion Of Innovations* (pp. 1–37). London: Collier Macmillian Publishers.

Rogerson-Revelle, P. (2015). Constructively aligning technologies with learning and assessment in a distance education master's programme. *Distance Education*, *36*(1), 129–147. doi:10.1080/01587919.2015.1019972

Rogers, R. (2004). *An Introduction to Critical Discourse Analysis in Education*. Mahwah, NJ: Lawrence Erlbaum.

Roggenkamp, K. (2004). Development modules to unleash the potential of Mobile Government. In *European Conference on E-government*.

Rosenstone, J. (1982, February). Economic Adversity and Voter Turnout. *American Journal of Political Science*, *26*(1), 25–46. doi:10.2307/2110837

Ross, C., & Middleberg, D. (1999). *Media in cyberspace study*. Fifth Annual National Survey 1998.

Ross, P. (2015). *Native Facebook Videos Get More Reach Than Any Other Type of Post*. Retrieved 1 May 2015 from http://www.socialbakers.com/blog/2367-native-facebook-videos-get-more-reach-than-any-other-type-of-post

Ross, P. (2015). *Native Facebook videos get more reach than any other type of post*. Retrieved May 1, 2015, from http://www.socialbakers.com/blog/2367-native-facebook-videos-get-more-reach-than-any-other-type-of-post

Rossel, P., Finger, M., & Misuraca, G. (2006). Mobile" e-government options: Between technology-driven and usercentric. *The electronic. Journal of E-Government, 4*(2), 79–86.

Rowan, C. (2013). *The Impact of Technology on the Developing Child*. The Huffington Post.

Ruggiero. (2000). Uses and gratifications theory in the 21st century. *Mass Communication & Society*, 3-37.

Ruigrok, N. (2010). From Journalism of Activism to Journalism of Accountability. *The International Communication Gazette, 72*(1), 85–90. doi:10.1177/1748048509350340

Rukzio, E., Broll, G., Leichtenstern, K., & Schmidt, A. (2007). Mobile interaction with the real world: An evaluation and comparison of physical mobile interaction techniques. In B. Schiele, A. K. Dey, H. Gellersen, B. de Ruyter, M. Tscheligi, R. Wichert, E. Aarts, & A. Buchmann (Eds.), *European Conference Proceedings, AmI 2007*.

Russell, T. L. (2001). *The No Significant Difference Phenomenon* (5th ed.). IDECC.

Sääksjärvi, M., van den Hende, E., Mugge, R., & van Peursem, N. (2015). How exposure to logos and logo varieties fosters brand prominence and freshness. *Journal of Product and Brand Management, 24*(7), 736–744. doi:10.1108/JPBM-06-2014-0648

Sabatier, L. A. (2012). The new way to manage content across platforms. *Publishing Research Quarterly, 28*(3), 197–203. doi:10.1007/s12109-012-9274-3

Saffer, A. J., Sommerfeldt, E. J., & Taylor, M. (2013). The effects of organizational Twitter interactivity on organization–public relationships. *Public Relations Review, 39*(3), 213–215. doi:10.1016/j.pubrev.2013.02.005

Salim, A., & Wangusi, N. (2014). Mobile phone technology: an effective tool to fight corruption in Kenya. In *Proceedings of the 15th Annual International Conference on Digital Government Research* (pp.300-305). ACM. doi:10.1145/2612733.2612772

Salmeron, J. M. (February 8, 2013). If you love your brand, set it free. *Smashing Magazine*. Retrieved from http://www.smashingmagazine.com/2013/02/08/if-you-love-your-brand-set-it-free/

Salovaaraa, A., & Tamminenb, S. (2008). Acceptance or appropriation? A design-oriented critique on technology acceptance models. In *Future Interaction Design II* (pp. 157–173). London: Springer.

Sánchez, J., & Olivares, R. (2011). Problem solving and collaboration using mobile serious games. *Computers & Education, 57*(3), 1943–1952. doi:10.1016/j.compedu.2011.04.012

Sanders, E. (1980, August). On the Costs, Utilities and Simple Joys of Voting. *The Journal of Politics, 42*(3), 854–863. doi:10.2307/2130557

Santana, S., & Morwitz, V.G. (2013). *We're in this together: How sellers, social values, and relationship norms influence consumer payments in pay-what-you-want contexts*. Manuscript submitted for publication.

SAP. (2012). *Mobile is transforming businesses, industries, and the world*. Retrieved April 23, 2015, from http://www.sap.com/netherlands/pc/tech/mobile/featured/offers/mobile-hbr-papers.html

Sareen, M., Punia, D. K., & Chanana, L. (2013). Exploring factors affecting use of mobile government services in India. *Problems and Perspectives in Management, 11*(4), 86–93.

Saxena, M. (1994). Literacies among Panjabis in Southall. In M. Hamilton, D. Barton, & R. Ivanic (Eds.), Worlds of literacies. Clevedon: Multilingual Matters.

Saxena, M. (2007). Multilingual and multicultural identities in Brunei Darussalam. In A. B. M. Tsui & J. Tollefson (Eds.), *Language policy, culture and identity in Asian contexts*. Mahwah, NJ: Lawrence Erlbaum.

Saxena, M. (2009). Negotiating Conflicting Ideologies and Linguistic Otherness: Code-switching in English Classrooms. *English Teaching, 8*(2), 167–187.

Saxena, M. (2014). Critical Diglossia and Lifestyle Diglossia: National Development, English, Linguistic Diversity and Language Change. *International Journal of the Sociology of Language*. doi:10.1515/ijsl-2013-0067

Saxena, M., & Sercombe, P. (2002). Patterns and Variations in Language Choices and Language Attitudes among Bruneians. In D. W. C. So & G. M. Jones (Eds.), *Education and Society in Plurilingual Contexts* (pp. 248–265). Brussels: VUB Press.

Schlosser, A. E., Shavitt, S., & Kanfer, A. (1999). Survey of Internet users' attitudes toward Internet advertising. *Journal of Interactive Marketing, 13*(3), 34–54. doi:10.1002/(SICI)1520-6653(199922)13:3<34::AID-DIR3>3.0.CO;2-R

Schmitt, B. H., Brakus, J., & Zarantonello, L. (2015). The current state and future of brand experience. *Journal of Brand Management, 21*(9), 727–733. doi:10.1057/bm.2014.34

Schultz, C. (2011, March 4). *Why Connie Schultz Won't Give up on the Fight for Good Journalism*. Retrieved from Poynter: http://www.poynter.org/news/mediawire/106009/why-connie-schultz-wont-give-up-on-the-fight-for-good-journalism/

Schultz, D. E., & Patti, C. H. (2009). The evolution of IMC: IMC in a customer-driven marketplace. *Journal of Marketing Communications, 15*(2-3), 75–84. doi:10.1080/13527260902757480

Schwabe, G., & Goth, C. (2005). Mobile learning with a mobile game: Design and motivational effects. *Journal of Computer Assisted Learning, 21*(3), 204–216. doi:10.1111/j.1365-2729.2005.00128.x

Schwartz, H. A., Eichstaedt, J. C., Kern, M. L., Dziurzynski, L., Ramones, S. M., Agrawal, M., & Ungar, L. H. et al. (2013). Personality, gender, and age in the language of social media: The open-vocabulary approach. *PLoS ONE, 8*(9), e73791. doi:10.1371/journal.pone.0073791 PMID:24086296

Search Engine Land. (2015). *What is Mobilegeddon & the Google mobile friendly update?* Retrieved May 20, 2015 from http://searchengineland.com/library/google/google-mobile-friendly-update

Sebba, M. (2009). Unregulated spaces. Plenary lecture at the conference on "Language Policy and Language Learning: New Paradigms and New Challenges". University of Limerick, UK.

Sebba, M. (2007). Identity and Language Construction in an Online Community: The Case of 'Ali G. In P. Auer (Ed.), *Style and Social Identities: Alternative Approaches to Linguistic Heterogeneity*. Berlin: Mouton/de Gruyter.

Sefton-Green, J. (2006). Youth, technology, and media cultures. *Review of Research in Education, 30*(1), 279–306. doi:10.3102/0091732X030001279

Seo, H. (2014). *Bowling online: smartphones, mobile messengers, and mobile social games for Korean teen girls*. (Unpublished master's thesis). The University of Texas at Austin, TX.

Seong-Jae, M. (2010, February03). From the Digital Divide to the Democratic Divide: Internet Skills, Political Interest, and the Second- Level Digital Divide in Political Internet Use. *Journal of Information Technology & Politics, 7*(1), 22–35. doi:10.1080/19331680903109402

Severin, & Tankard. (1997). *Communication theories: Origins, methods, and uses in the mass media* (4th ed.). New York: Longman.

Seybold, A. M. (2008). The convergence of wireless, mobility, and the Internet and its relevance to enterprises. *Information, Knowledge, Systems Management, 7*(1-2), 11–23.

Sha, B.-L. (2007). Dimensions of public relations: Moving beyond traditional public relations models. In S. Duhé (Ed.), *New Media and Public Relations*. New York, NY: Peter Lang Publishing.

Shah, D. V., Kwak, N., & Holbert, R. L. (2001). Connecting' and 'disconnecting' with civic life: Patterns of Internet use and the production of social capital. *Political Communication, 18*, 141–162. doi:10.1080/105846001750322952

Shah, D., Friedland, L., Wells, C., Kim, Y., & Rojas, H. (2012). Communication, consumers, and citizens: Revisiting the politics of consumption. *The Annals of the American Academy of Political and Social Science, 644*(1), 6–18. doi:10.1177/0002716212456349

Shandwick, W. (2015). *The CEO Reputation Premium: Gaining Advantage in the Engagement Era*. Retrieved on 10 April 2015 from http://www.webershandwick.com/news/article/81-percent-of-global-executives-report-external-ceo-engagement-is-a-mandate

Shareef, M. A., Archer, N., & Dwivedi, Y. K. (2012). Examining adoption behavior of mobile government. *Journal of Computer Information Systems, 53*(2), 39.

Sharf, S. (2015, April 15). 62 million: The only number Netflix shareholders care about. *Forbes*. Retrieved from http://www.forbes.com/sites/samanthasharf/2015/04/15/netflix-subscriber-count-crosses-62-million-sending-stock-above-500/

Sharples, M., Taylor, J., & Vavoula, G. (2005). Towards a theory of mobile learning. In *Proceeding of the mLearn 2005 Conference*.

Sharples, M., Taylor, J., & Vavoula, G. (2005). Towards a theory of mobile learning. *Proceedings of mLearn 2005, 1*(1), 1-9.

Sharples, M. (2003). Disruptive devices: Mobile technology for conversational learning. *International Journal of Continuing Engineering Education and Lifelong Learning, 12*(5/6), 504–520. doi:10.1504/IJCEELL.2002.002148

Sheng, M. L., & Teo, T. S. H. (2012). Product attributes and brand equity in the mobile domain: The mediating role of customer experience. *International Journal of Information Management, 32*(1), 139–146. doi:10.1016/j.ijinfomgt.2011.11.017

Shintaro, O., & Barwise, P. (2011). Has the Time Finally Come for the Medium of the Future? *Journal of Advertising Research*, 5159–5171.

Shirky, C. (2009) *How social media can make history*. TED.com talk recorded in June 2009, retrieved 20 March 2015.

Shirky, C. (2009). *How social media can make history*. TED.com presentation. Retrieved from http://www.ted.com/talks/clay_shirky_how_cellphones_twitter_facebook_can_make_history

Shirky, C. (2009). *How social media can make history*. TED.com talk given in June 2009; video and script. Retrieved 28 April 2015 from http://www.ted.com/talks/clay_shirky_how_cellphones_twitter_facebook_can_make_history/transcript?language=en

Shirky, C. (2009). *How social media can make history*. TED.com. Retrieved 25 March 2015 from http://www.ted.com/talks/clay_shirky_how_cellphones_twitter_facebook_can_make_history

Short, K. (April 14, 2015). *Mobile wallets: A trend retailers can't ignore*. Retrieved April 23, 2015, from http://researchindustryvoices.com/2015/04/14/mobile-wallets-a-trend-retailers-cant-ignore//

Shrum, L. (2004). *The Psychology of Entertainment Media: Blurring the Lines Between Entertainment and Persuasion*. Lawrence Erlbaum.

Shuler, C. (2009). *Pockets of potential: Using mobile technologies to promote children's learning*. New York: The Joan Ganz Cooney Center at Sesame Workshop.

Siau, K., & Shen, Z. (2004). Mobile communications and mobile services. *International Journal of Mobile Communications*, *1*(1/2), 2003.

Simon, J. (2014, December 11). *What's the Difference Between Activism and Journalism?* Retrieved from Nieman Reports: http://niemanreports.org/articles/whats-the-difference-between-activism-and-journalism/

Sims, J., Vidgen, R., & Powell, P. (2008). E-learning and the digital divide: Perpetuating cultural and socio-economic letisim in higher education. *Communications of the Association for Information Systems*, *22*(1), 429–442.

Sinclair, B. (2015). US mobile gaming market tops $5.2 billion. *Gamesindustry.biz*. Retrieved March 22, 2015 from http://www.gamesindustry.biz/articles/2015-01-13-us-mobile-gaming-market-tops-usd5-2-billion

Singh, S., & Sonnenburg, S. (2012). Brand performances in social media. *Journal of Interactive Marketing*, *26*(4), 189–197. doi:10.1016/j.intmar.2012.04.001

Singh, V.-J., Veron-Jackson, L., & Cullinane, J. (2008). Blogging: A New Play in Your Marketing Game Plan. *Business Horizons*, *51*(4), 281–292. doi:10.1016/j.bushor.2008.02.002

Sinton, J. (2014). *Why having a mobile site should be just the start*. TNS Global Connected Life 2014 (Special edition). Retrieved from http://www.tnsglobal.com/sites/default/files/tns-why-mobile-first-design-is-just-the-start_0.pdf

Sintoris, C., Stoica, A., Papadimitriou, L., Yiannoutsou, N., Komis, V., & Avouris, N. (2010). MuseumScrabble: Design of a mobile game for children's interaction with a digitally augmented cultural space. *International Journal of Mobile Human Computer Interaction*, *2*(2), 19–38. doi:10.4018/jmhci.2010040104

Skillen, K. L., Chen, L., Nugent, C. D., Donnelly, M. P., Burns, W., & Solheim, I. (2012). Ontological user profile modeling for context-aware application personalization. In *Ubiquitous Computing and Ambient Intelligence* (pp. 261–268). Berlin: Springer Berlin Heidelberg. doi:10.1007/978-3-642-35377-2_36

Sloane, G. (2014). Facebook's New People-Based Ad Technology Is 'Marketing Nirvana'. Pepsi and Intel are early testers as social net unleashes data. *Adweek*. Retrieved 1 May 2015 from http://www.adweek.com/news/technology/facebooks-new-people-based-ad-technology-marketing-nirvana-160438

Smala, S., & Al-Shehri, S. (2013). Privacy and identity management in social media: Driving factors for identity hiding. In J. Keengwe (Ed.), *Research perspectives and best practices in educational technology integration* (pp. 304–320). Hershey, PA: IGI Global.

Smetana, J. G., Campione-Barr, N., & Metzger, A. (2006). Adolescent development in interpersonal and societal contexts. *Annual Review of Psychology*, *57*(1), 255–284. doi:10.1146/annurev.psych.57.102904.190124 PMID:16318596

Smith, J. (2014, February 3). Facebook on Paper app: No plans for international release, iPad or Android app. *Pocket-lint*. Retrieved from http://www.pocket-lint.com/news/127001-facebook-on-paper-app-no-plans-for-international-release-ipad-or-android-app

Smith, M. A., Woo, H. J., Parker, J. P., Youmans, R. J., LeGoullon, M., Weltman, G., & de Visser, E. J. (2013). Using iterative design and testing towards the development of SRTS(R): A mobile, game-based stress resilience training system. *Proceedings of the Human Factors and Ergonomics Society Annual Meeting*, *57*(1), 2076-2080. doi:10.1177/1541931213571463

Smutkupt, P., Krairit, D., & Esichaikul, V. (2010). Mobile marketing: Implications for marketing strategies. *International Journal of Mobile Marketing*, *5*(2), 126–139.

Society of Professional Journalists. (2014, September 6). *SPJ Code of Ethics*. Retrieved from Society of Professional Journalists: http://www.spj.org/ethicscode.asp

Solima, L., Della Peruta, M. R., & Del Giudice, M. (2015). Object-Generated Content and Knowledge Sharing: The Forthcoming Impact of the Internet of Things. *Journal of the Knowledge Economy*, 1-15.

Song, G., & Cornford, T. (2006, October). Mobile government: Towards a service paradigm. In *Proceedings of the 2nd International Conference on e-Government* (pp. 208-218). University of Pittsburgh, USA.

Song, G. (2005, July). Transcending e-government: A case of mobile government in Beijing. In *The First European Conference on Mobile Government*.

Soroa-Koury, S., & Yang, K. C. (2010). Factors affecting consumers' responses to mobile advertising from a social norm theoretical perspective. *Telematics and Informatics*, *27*(1), 103–113. doi:10.1016/j.tele.2009.06.001

So, S. (2008). A study on the acceptance of mobile phones for teaching and learning with a group of pre-service teachers in Hong Kong. *Journal of Educational Technology Development and Exchange*, *1*(1), 81–92.

Spikol, D., & Milrad, M. (2008). Combining physical activities and mobile games to promote novel learning practices. *Fifth IEEE International Conference on Wireless, Mobile, and Ubiquitous Technology in Education*, (pp. 31-38). doi:10.1109/WMUTE.2008.37

Sprosen, S. (2014). *The influence of branded mobile applications on consumers' perceptions of a brand: the importance of brand experience and engagement*. (Unpublished master's thesis). Dublin Business School, Ireland.

Spurgeon, C. L. (2009). From mass communication to mass conversation: Why 1984 wasn't like 1984. *Australian Journal of Communication*, *36*(2), 143–158.

Stanisic Stojic, S. M., Dobrijevic, G., Stanisic, N., & Stanic, N. (2014). Characteristics and activities of teachers on distance learning programs that affect their ratings. *International Review of Research in Open and Distance Learning*, *15*(4), 248–262.

StatCounter Global Stats. (2014). *Mobile internet usage soars by 67%* [Press release]. Retrieved from http://gs.statcounter.com/press/mobile-internet-usage-soars-by-67-perc

Statista. (2015). *Second screen usage among mobile internet users in selected countries as of January 2014*. Retrieved from http://www.statista.com/statistics/301187/second-screen-usage-worldwide/

Statista. (n.d.). *Statistics and facts about gaming in Asia*. Retrieved March 22, 2015 from http://www.statista.com/topics/2196/video-game-industry-in-asia/

Steel, E. (2014, August 21). *'Ice Bucket Challenge' donations for A.L.S. research top $41 million*. Retrieved from New York Times: http://www.nytimes.com/2014/08/22/business/media/ice-bucket-challenge-donations-for-als-top-41-million.html?_r=0

Stelzner, M. A. (2015). 2015 Social Media Marketing Industry Report. *Social Media Examiner*. Retrieved 30 July 2015 from http://www.socialmediaexaminer.com/SocialMediaMarketingIndustryReport2015.pdf

Stevenson, A., & Wang, S. (2014, December 20). A Jump to Mobile Ads in China. *The New York Times*, p. 2.

Stewart, C. A. III. (2008). *2008 Survey of the Performance of American Elections*. The Massachusetts Institute of Technology.

Stewart, C. III. (2014). *2014 Survey of the Performance of American Elections*. The Massachusetts Institute of Technology.

Stockwell, G. (2008). Investigating learner preparedness for and usage patterns of mobile learning. *ReCALL, 20*(3), 253–270. doi:10.1017/S0958344008000232

Storer, T. L. (2006, June). *ResearchGate*. Retrieved from ResearchGate: http://www.researchgate.net/profile/Ishbel_Duncan/publication/250889738_An_Exploratory_Study_of_Voter_attitudes_towards_a_Pollsterless_Remote_Voting_System/links/5447aac90cf22b3c14e0f845.pdf

Straumsheim, C. (2013). Tech as a Service. *Inside Higher Ed*. Retrieved 20 July 2015 from https://www.insidehighered.com/news/2013/10/17/survey-shows-it-service-dominates-top-priorities-among-university-it-officials

Straumsheim, C. (2014). Moodle for the Masses. *Inside Higher Ed*. Retrieved 20 July 2015 from https://www.insidehighered.com/news/2014/02/13/moodle-tops-blackboard-among-small-colleges-analysis-sayshttps://www.insidehighered.com/news/2014/02/13/moodle-tops-blackboard-among-small-colleges-analysis-says

Street, B. V. (1993). Cross-cultural perspectives on literacy. In J. Maybin (Ed.), *Language and literacy in social practice* (pp. 139–150). Clevedon: Multilingual Matters.

Suarez-Tangil, G., Tapiador, J., Peris-Lopez, P., & Ribagorda, A. (2013). Evolution, Detection and Analysis of Malware for Smart Devices. *IEEE Communications Surveya &Tutorials*. Retrieved December 16, 2015, from http://www.seg.inf.uc3m.es/~guillermo-suarez-tangil/papers/2013cst-ieee.pdf

Suki, N. M., & Suki, N. M. (2007). Mobile phone usage for m-learning: Comparing heavy and light mobile phone users. *Campus-Wide Information Systems, 24*(5), 355–365. doi:10.1108/10650740710835779

Sullivan, M. (2013, October 26). *As Media Change, Fairness Stays Same*. Retrieved from New York Times: http://www.nytimes.com/2013/10/27/public-editor/as-media-change-fairness-stays-same.html?_r=0

Sun, L. (2014). *Framing analysis of media reports on left-behind children in mainland China*. (Master's degree thesis). Shanghai Foreign Language Institute. Retrieved from http://cdmd.cnki.com.cn/Article/CDMD-10271-1014126282.htm

Sundsøy, P., Bjelland, J., Iqbal, A. M., & de Montjoye, Y. A. (2014). Big Data-Driven Marketing: How machine learning outperforms marketers' gut-feeling. In Social Computing, Behavioral-Cultural Modeling and Prediction (pp. 367-374). Springer International Publishing.

Su, S., Li, X., Lin, D., Xu, X., & Zhu, M. (2013). Psychological adjustment among left-behind children in rural China: The role of parental migration and parent–child communication. *Child: Care, Health and Development, 39*(2), 162–170. doi:10.1111/j.1365-2214.2012.01400.x PMID:22708901

Sutton, J., Palen, L., & Shklovski, I. (2008, May). Backchannels on the front lines: Emergent uses of social media in the 2007 southern California wildfires. In *Proceedings of the 5th International ISCRAM Conference* (pp. 624-632). Washington, DC: Academic Press.

Swain, M. (1983). Bilingualism without Tears. In M. Clarke & J. Handscombe (Eds.), On TESOL '82: Pacific Perspectives on Language Learning and Teaching. Washington, DC: Teachers of English to Speakers of Other Languages (TESOL).

Swain, H. (2007, January26). Have You Got The Message Yet? *Times Higher Education Supplement*, 54–54.

Sweetser, K. (2011). Digital political public relations. In J. Strömbäck & S. Kiousis (Eds.), Political public relations: Principles and applications, (pp. 293-313). London: Taylor & Francis.

Sweller, J. (1988). Cognitive load during problem solving: Effects on learning. *Cognitive Science, 12*(2), 257–285. doi:10.1207/s15516709cog1202_4

SXSW. (2015). *Mecosystem 2020*. Retrieved from http://panelpicker.sxsw.com/vote/41042

Taewoo, N. (2010). Whither Digital Equality?: An Empirical Study of the Democratic Divide. *Proc. 43rd Hawaii International Conference on System Sciences* (pp. 1-10). IEEE Computer Society.

Takahashi, D. (2014). Mobile gaming could drive entire video game industry to $100B in revenue by 2017. *GamesBeat*. Retrieved March 22, 2015 from http://venturebeat.com/2014/01/14/mobile-gaming-could-drive-entire-game-industry-to-100b-in-revenue-by-2017/

Taleb, N. N. (2012). *Antifragile: Things That Gain from Disorder (Incerto)*. New York: Random House.

Tangney, B., Weber, S., Knowles, D., Munnelly, J., Watson, R., Salkham, A. A., & Jennings, K. (2010). *MobiMaths: an approach to utilising smartphones in teaching mathematics*. Academic Press.

Tassi, P. (2012). Is the PlayStation Vita a dinosaur already? *Forbes*. Retrieved March 14, 2015 from http://www.forbes.com/sites/insertcoin/2012/02/17/is-the-playstation-vita-a-dinosaur-already/

Taylor, A., & Harper, R. (2003). The gift of the gab?: A design oriented sociology of young people's use of mobiles. *Computer Supported Cooperative Work*, *12*(3), 267–296. doi:10.1023/A:1025091532662

Taylor, M., & Botan, C. H. (2006). Global public relations: Application of a Cocreational approach. In M. Distasio (Ed.), *9th International Public Relations Research Conference Proceedings: Changing Roles and Functions of Public Relations*, (pp. 481-491). Miami, FL: University of Miami and Institute for Public Relations.

Technologies, D. (2014). *Oculus Rift | 3d virtual gaming glass | VR glasses | 360 degree video | virtual reality glasses*. Retrieved June 26, 2015, http://www.360-degree-video.com/pages/posts/oculus-rift-3d-virtual-gaming-glass-vr-glasses-360-degree-video-virtual-reality-glasses-publication-date-october-2014-release-2014-242.php

Tencent. (2015). *The New Milestone of WeChat*. Retrieved 12 20, 2015, from Techweb: http://www.techweb.com.cn/internet/2015-08-12/2188504.shtml

Tercek, R. (2007). 1997-2007: The first decade of mobile games. *Game Developers Conference*, San Francisco.

Terlutter, R., & Capella, M. L. (2013). The gamification of advertising: Analysis and research directions of in-game advertising, advergames, and advertising in social network games. *Journal of Advertising*, *42*(2-3), 95–112. doi:10.1080/00913367.2013.774610

Thayer, K. (2014, January 14). Mobile technology makes online dating the new normal. *The Forbes*. Retrieved May 27, 2015, from http://www.forbes.com/sites/katherynthayer/2014/01/14/mobile-technology-makes-online-dating-the-new-normal/

The Advantages of Distance Learning. (2015). Retrieved from http://www.usjournal.com/en/students/help/distancelearning.html

The Economist Intelligence Unit. (2014). *How mobile is transforming passenger transportation. Clearing the way for more liveable cities*. Retrieved May 7, 2015, from http://www.sap.com/bin/sapcom/en_us/downloadasset.2014-11-nov-02-17.how-mobile-is-transforming-passenger-transportation-pdf.bypassReg.html

The Economist. (2014). *The world in numbers: Industries* Retrieved April 23, 2015, from http://www.economist.com/news/21632039-telecoms

The Economist. (2014). *The world in numbers: Industries*. Available at http://www.economist.com/news/21632039-telecoms

Theunissen, P., & Noordin, W. N. W. (2012). Revisiting the concept "dialogue" in public relations. *Public Relations Review*, *38*(1), 5–13. doi:10.1016/j.pubrev.2011.09.006

Thomas, B., & Howard, D. (1990). A Review and Critique of The Hierarchy of Effects in Advertising. *International Journal of Advertising*, 98–111.

Thulin, E., & Vilhelmson, B. (2007). Mobiles everywhere: Youth, the mobile phone, and changes in everyday practice. *Young*, *15*(3), 235–253. doi:10.1177/110330880701500302

Thurlow, C. (2006). From statistical panic to moral panic: The metadiscursive construction and popular exaggeration of new media language in the print media. *Journal of Computer-Mediated Communication*, *11*(3), 667–701. doi:10.1111/j.1083-6101.2006.00031.x

Thurman, N. (2011). Making 'The Daily Me': Technology, economics and habit in the mainstream assimilation of personalized news. *Journalism*, *12*(4), 395–415. doi:10.1177/1464884910388228

Time Inc. (2012). *Time Inc. Study Reveals That "Digital Natives" Switch Between Devices and Platforms Every Two Minutes, Use Media to Regulate Their Mood* [Press release]. Retrieved from http://www.businesswire.com/news/home/20120409005536/en/Time%C2%ADStudy%C2%ADReveals%C2%AD%E2%80%9CDigital%C2%ADNatives%E2%80%9D%C2%ADSwitch%C2%ADDevices#

TNS Global. (2014). *Connected Life 2014*. Retrieved on March 29 from http://connectedlife.tnsglobal.com/

TNS. (2014). *Report into the 2014 General Election*. TNS New Zealand.

Tojib, D., Tsarenko, Y., & Sembada A. Y. (2014). The facilitating role of smartphones in increasing use of value-added mobile services. *New Media and Society*, 1-21.

Torres Diaz, J., Moro, A. & Torres Carrion, P. (2015). Mobile learning: perspectives. *RUSC: Revista De Universidad Y Sociedad Del Conocimiento*, *12*(1), 38-49.

Townsend, L. (2014). How much has the ice bucket challenge achieved? *BBC News Magazine*. Retrieved from http://www.bbc.com/news/magazine-29013707

Trading Economics. (2015). *Kazakhstan GDP Annual Growth Rate*. Retrieved 20 July 2015 from http://www.trading-economics.com/kazakhstan/gdp-growth-annual

Tran, M. (2014). *Revisiting disruption: 8 good questions with Clayton Christensen*. Retrieved 15 August 2015 from http://www.americanpressinstitute.org/publications/good-questions/revisiting-disruption-8-good-questions-clayton-christensen/

Trautman, E. (December 8, 2014). Five online video trends to look for in 2015. *Forbes Groupthink*. Retrieved from http://www.forbes.com/sites/groupthink/2014/12/08/5-online-video-trends-to-look-for-in-2015/

Traxler, J., & Leach, J. (2006, November). Innovative and sustainable mobile learning in Africa. In *Wireless, Mobile and Ubiquitous Technology in Education, 2006. WMUTE'06. Fourth IEEE International Workshop on* (pp. 98-102). IEEE. doi:10.1109/WMTE.2006.261354

Traxler, J. (2007). Defining, Discussing and Evaluating Mobile Learning: The moving finger writes and having writ..... *The International Review of Research in Open and Distributed Learning*, *8*(2).

Traxler, J., & Kukulska-Hulme, A. (2016). Introduction to the next generation of mobile learning. In J. Traxler & A. Kukulska-Hulme (Eds.), *Mobile learning: The next generation* (pp. 1–10). New York: Routledge.

Trimi, S., & Sheng, H. (2008). Emerging Trends in M-Government. *Communications of the ACM*, *51*(5), 53–58. doi:10.1145/1342327.1342338

Troussas, C., Virvou, M., & Alepis, E. (2014). Collaborative learning: Group interaction in an intelligent mobile-assisted multiple language learning system. *Informatics in Education*, *13*(2), 279–292. doi:10.15388/infedu.2014.08

Tsang, M. M., Hom, S.-C., & Liang, T.-P. (2004). Consumer attitudes toward mobile advertising: An empirical study. *International Journal of Electronic Commerce, 8*, 65–78.

Tsentas, T. (2011, November 12). *Challenging the information landscape: WikiLeaks' effect on the media, activism and politics.* Retrieved from Media @ McGill: http://media.mcgill.ca/en/content/challenging-information-landscape-wikileaks%E2%80%99-effect-media-activism-and-politics

Turel, O., & Serenko, A. (2006). Satisfaction with mobile services in Canada: An empirical investigation. *Telecommunications Policy, 30*(5), 314–331. doi:10.1016/j.telpol.2005.10.003

Turow, J. (2011). *The daily you: How the new advertising industry is defining your identity and your worth.* New Haven, CT: Yale University Press.

Tuten, T., & Solomon, M. (2012). *Social Media Marketing.* Englewood Cliffs, NJ: Prentice Hall.

Udell, J. (2004). Jon Udell: Name that genre: screencast. *InfoWorld.*

Uhm, J. H. (2015). *The goals and purposes of Seoul Cyber University.* Live presentation at KIMEP University.

USAspending.gov. (2008). *Top 100 Recipients of Federal Assistance for FY 2008.* Retrieved 20 July 2015 from usaspending.gov.

Using the Ushahidi Platform to Monitor the Nigeria Elections. (2011). Retrieved November 29, 2015, from: https://www.ushahidi.com/blog/2011/03/30/using-the-ushahidi-platform-to-monitor-the-nigeria-elections-2011

Vacaru, M. A., Shepherd, R. M., & Sheridan, J. (2014). New Zealand Youth and Their Relationships with Mobile Phone Technology. *International Journal of Mental Health and Addiction, 12*(5), 572–584. doi:10.1007/s11469-014-9488-z

Valentine, D. (2002). Distance Learning: Promises, Problems, and Possibilities. *Online Journal of Distance Learning Administration, 5*(3). Retrieved from http://www.westga.edu/~distance/ojdla/fall53/valentine53.html

Valentine, L. (2013). Payments Landscape: Still crazy after all these years: Few clear winners stand out, leaving banks mostly still watching. Herewith an informal "new payments scorecard.". *ABA Banking Journal, 105*(7), 24.

Valenzuela, S., Park, N., & Kee, K. (2012). Is There Social Capital in a Social Network Site? Facebook Use and College Students' Life Satisfaction, Trust, and Participation. *Journal of Computer-Mediated Communication, 14*(2009), 875–901.

Valvi, A. C., & West, D. C. (2015). Mobile Applications (Apps) in Advertising: A Grounded Theory of Effective Uses and Practices. In Ideas in Marketing: Finding the New and Polishing the Old (pp. 349-352). Springer International Publishing.

van Dijk, T. (1993). Principles of critical discourse analysis. *Discourse & Society, 4*(2), 249–283. doi:10.1177/0957926593004002006

Varnali, K. (2014). SMS advertising: How message relevance is linked to the attitude toward the brand? *Journal of Marketing Communications, 20*(5), 339–351. doi:10.1080/13527266.2012.699457

Vasquez-Cano, E. (2014). Mobile Distance Learning with Smartphones and Apps in Higher Education. *Educational Sciences: Theory and Practice, 14*(4), 1505–1520.

Veil, S. R., Buehner, T., & Palenchar, M. J. (2011). A Work-In-Process Literature Review: Incorporating Social Media in Risk and Crisis Communication. *Journal of Contingencies and Crisis Management, 19*(2), 110–122. doi:10.1111/j.1468-5973.2011.00639.x

Venkatesh, S. A., Venkatesh, C. H., & Naik, P. (2010). Mobile marketing in the retailing environment: current insights and future research avenues. *Journal of Interactive Marketing*. Retrieved May 27, 2015, from http://merage.uci.edu/Resources/Documents/2010MobileJIM.pdf

Verbergt, M., & Schechner, S. (2015, June 25). Taxi drivers block Paris roads in Uber protest. *The Wall Street Journal*. Retrieved from http://www.wsj.com/articles/taxi-drivers-block-paris-roads-in-uber-protest-1435225659

Verčič, D., Verhoeven, P., & Zerfass, A. (2014). Key issues of public relations of Europe: Findings from the European Communication Monitor 2007-2014. *Revista Internacional De Relaciones Públicas*, *4*(8), 5–26.

Vergeer, M. (2015). Peers and Sources as Social Capital in the Production of News: Online Social Networks as Communities of Journalists. *Social Science Computer Review*, *33*(3), 277–297. doi:10.1177/0894439314539128

Vergeer, M., & Pelzer, B. (2009). Consequences of media and Internet use for offline and online network capital and well-being. A causal model approach. *Journal of Computer-Mediated Communication*, *15*(1), 189–210. doi:10.1111/j.1083-6101.2009.01499.x

Vertovec, S. (2004). Cheap Calls: The Social Glue of Migrant Transnationalism. *Global Networks*, *4*(2), 219–224. doi:10.1111/j.1471-0374.2004.00088.x

Videology. (2015). *U.S. video market at a glance*. Retrieved May 23, 2015, from http://www.videologygroup.com/files/m/research-data/US_Q4_2014_Video_Market_At-A-Glance.pdf

Vis, F. (2013). Twitter as a Reporting Tool for Breaking News. *Digital Journalism*, *1*(1), 27–47. doi:10.1080/2167081 1.2012.741316

Vizeu, F. (2010). Potencialidades da análise histórica nos estudos organizacionais brasileiros. *RAE - Revista de administração de Empresas*, *50*(1), 37-47.

Vodafone. (2015). *Driving economic and social development*. Retrieved on April 30, 2015 from http://www.vodafone.com/content/index/about/sustainability/our_vision.html

Voorveld, H. A., Van Noort, G., & Duijn, M. (2013). Building brands with interactivity: The role of prior brand usage in the relation between perceived website interactivity and brand responses. *Journal of Brand Management*, *20*(7), 608–622. doi:10.1057/bm.2013.3

Vygotsky. (1978). *Mind in society: The development of higher psychological processes*. Cambridge, MA: Harvard University Press.

Waber, B. (2013). *People analytics: How social sensing technology will transform business and what it tells us about the future of work*. FT Press.

Waddington, S. (2013). A critical review of the Four Models of Public Relations and the Excellence Theory in an era of digital communication. *CIRP Chartered Practitioner Paper*. Retrieved from http://wadds.co.uk/2013/06/18/cipr-chartered-practitioner-paper-grunig-and-digital-communications/

Wagner, K. (2012). The most popular iPhone and iPad apps of all time. *Gizmodo*. Retrieved from http://gizmodo.com/5890602/the-most-popular-iphone-and-ipad-apps-of-all-time

Wallsten, K. (2015). Non-elite Twitter sources rarely cited in coverage. *Newspaper Research Journal*, *36*(1), 24–41.

Walsham, G. (1995). The emergence of interpretivism in IS research. *Information Systems Research*, *6*(4), 376–394. doi:10.1287/isre.6.4.376

Walsh, M. F., Winterich, K. P., & Mittal, V. (2010). Do logo redesigns help or hurt your brand. *Journal of Product and Brand Management, 19*(2), 76–84. doi:10.1108/10610421011033421

Walsh, M. F., Winterich, K. P., & Mittal, V. (2011). How re-designing angular logos to be rounded shapes brand attitude: Consumer brand commitment and self-construal. *Journal of Consumer Marketing, 28*(6), 438–447. doi:10.1108/07363761111165958

Walsh, S. P., White, K. M., Cox, S., & Young, R. M. (2011). Keeping in constant touch: The predictors of young Australians' mobile phone involvement. *Computers in Human Behavior, 27*(1), 333–342. doi:10.1016/j.chb.2010.08.011

Walsh, S. P., White, K. M., Mc, D., & Young, R. (2010). Needing to connect: The effect of self and others on young people's involvement with their mobile phones. *Australian Journal of Psychology, 62*(4), 194–203. doi:10.1080/00049530903567229

Walsh, S. P., White, K. M., & Young, R. M. (2008). Over-connected? A qualitative exploration of the relationship between Australian youth and their mobile phones. *Journal of Adolescence, 31*(1), 77–92. doi:10.1016/j.adolescence.2007.04.004 PMID:17560644

Walther, J. B. (2002). Cues filtered out, cues filtered in. In *Handbook of interpersonal communication*. Thousand Oaks, CA: Sage Publications.

Walther, J. B. (2005). Let Me Count the Ways: The Interchange of Verbal and Nonverbal Cues in Computer-Mediated and Face-to-Face Affinity. *Journal of Language and Social Psychology, 24*(1), 36–65. doi:10.1177/0261927X04273036

Wang, A. I., Ofsdahl, T., & Morch-Storstein, O. K. (2008). An evaluation of a mobile game concept for lectures. *Software Engineering Education and Training, 2008. CSEET '08. IEEE 21st Conference*, (pp. 197-204). Charleston: IEEE.

Wang, C. (2014). Antecedents and consequences of perceived value in Mobile Government continuance use: An empirical research in China. *Computers in Human Behavior, 34*, 140–147. doi:10.1016/j.chb.2014.01.034

Wang, C., Feng, Y., Fang, R., & Lu, Z. (2012). Model for Value Creation in Mobile Government: An Integrated Theory Perspective. *International Journal of Advancements in Computing Technology, 4*(2).

Wang, W.-H., & Li, H.-M. (2012). Factors influencing mobile services adoption: A brand-equity perspective. *Internet Research, 22*(2), 142–179. doi:10.1108/10662241211214548

Wang, Y.-S., Wu, M.-C., & Wang, H.-Y. (2009). Investigating the determinants and age and gender differences in the acceptance of mobile learning. *British Journal of Educational Technology, 40*(1), 92–118. doi:10.1111/j.1467-8535.2007.00809.x

Warschauer, M. (2009). Digital Literacy Studies: Progress and Prospects. In M. Baynham & M. Prinsloo (Eds.), *The Future of Literacy Studies*. Basingstoke, UK: Palgrave Macmillan.

Warschauer, M., El Said, G. R., & Zohry, A. (2002). Language Choice Online: Globalization and Identity in Egypt. *Journal of Computer-Mediated Communication, 7*(4). Retrieved from http://jcmc.indiana.edu/vol7/issue4/warschauer.html

Watrtella, E., Rideout, V., Lauricella, A., & Connell, S. (2014). *Revised Parenting in the Age of Digital Technology A National Survey*. Retrieved December 16, 2015, from http://cmhd.northwestern.edu/wp-content/uploads/2015/06/ParentingAgeDigitalTechnology.REVISED.FINAL_.2014.pdf

Watson, C., McCarthy, J., & Rowley, J. (2013). Consumer attitudes towards mobile marketing in the smart phone era. *International Journal of Information Management, 33*(5), 840–849. doi:10.1016/j.ijinfomgt.2013.06.004

Weberling, B., & Waters, R. D. (2012). Gauging the public's preparedness for mobile public relations: The "Text for Haiti" campaign. *Public Relations Review, 38*(1), 51–55. doi:10.1016/j.pubrev.2011.11.005

Weghorn, H. E. (2007, April 23). Mobile Ticket Control System with RFID Cards for Administering Annual Secret Elections of University Committees. *Informatica*, 161-166.

Wei, P. S., & Lu, H. P. (2014). Why do people play mobile social games? An examination of network externalities and of uses and gratifications. *Internet Research*, *24*(3), 313–331. doi:10.1108/IntR-04-2013-0082

Wei, P., Lee, S., Lu, H., Tzou, J., & Weng, C. (2015). Why do people abandon mobile social games? Using candy crush saga as an example. *International Journal of Social, Education, Economics and Management Engineering*, *9*(1), 13–18.

Wei, R., & Lo, V.-H. (2006). Staying connected while on the move: Cell phone use and social connectedness. *New Media & Society*, *8*(1), 53–72. doi:10.1177/1461444806059870

Welles, B. (2014, June 17). *3Qs: A closer look at hashtag activism*. Retrieved from News @ Northeastern: http://www.northeastern.edu/news/2014/06/3qs-hashtag-activism/

Wellman, B., Haase, A., Witte, J., & Hampton, K. (2001). Does the Internet increase, decrease, or supplement social capital? Social networks, participation, and community commitment. *The American Behavioral Scientist*, *35*(3), 436–455. doi:10.1177/00027640121957286

Welsbeck, D., & Berney, P. (Eds.). (2014). People's web report II. *Netbiscuits*. Retrieved from http://www.netbiscuits.com/news/press/2014/10/28/brand-loyalty-dead-on-the-mobile-web-netbiscuits-peoples-web-report-ii/

Wen, M., & Lin, D. (2012). Child development in rural China: Children left behind by their migrant parents and children of nonmigrant families. *Child Development*, *83*(1), 120–136. doi:10.1111/j.1467-8624.2011.01698.x PMID:22181046

West, D.M. (2013, September). *Mobile learning: Transforming education, engaging students, and improving outcomes*. Center for Technology Innovation at Brookings.

When will mobile advertising become the next big thing? (2008). *Communication Arts*, *50*(2), 18.

White, D. S. (2013). *Social Media Growth 2006-2012*. Retrieved 16 Feb. 2015 from http://dstevenwhite.com/

White, D. S. (2013). *Social Media Growth 2006-2012*. Retrieved from http://dstevenwhite.com/

Wihbey, J. (2013, August 1). *Digital activism and organizing: Research review and reading list*. Retrieved from Journalist's Resource: http://journalistsresource.org/studies/society/internet/digital-activism-organizing-theory-research-review-reading-list#

Wijaya, B. S. (2012). *The Development of Hierarchy of Effects Model in Advertising*. International Research Journal of Business Studies.

Wilkinson, M. B., & Halton. (2014). *Mobile and Public Relations*. Retrieved 25 May 2015 from http://newsroom.cipr.co.uk/ciprsm-launch-new-guide-to-mobile-and-public-relations

Wilkinson, P., Miller, R., Burch, D., & Halton, J. (2014). *Mobile & public relations*. Retrieved 29 April 2015 from http://www.cipr.co.uk/sites/default/files/CIPRSM%20Guide%20-%20Mobile%20and%20PR.pdf

Wimmer, & Dominick. (1994). *Mass media research: An introduction*. Belmont: Wadsworth.

Winkler, T., Ide-Schoening, M., & Herczeg, M. (2008). Mobile co-operative game-based learning with moles: Time travelers in Medieval. *Proceedings of SITE*, *2008*, 3441–3449.

Winter, M., & Pemberton, L. (2011). Unearthing invisible buildings: Device focus and device sharing in a collaborative mobile learning activity. *International Journal of Mobile and Blended Learning*, *3*(4), 1–18. doi:10.4018/jmbl.2011100101

Winters, N., Sharples, M., Shuler, C., Vosloo, S., & West, M. (2013). *UNESCO/Nokia The Future of Mobile Learning Report: Implications for Policymakers and Planners.* UNESCO.

Witmer, D. (2000). *Spining the web: A handbook for public relations on the Internet.* New York, NY: Longman.

Witschge, T. (2013). Transforming journalistic practice: A profession caught between change and tradition. In C. Peters & M. Broersma (Eds.), *Rethinking journalism: Trust and participation in a transformed news landscape* (pp. 160–172). New York: Routledge.

Wodak, R., & Chilton, P. (2005). *A New Agenda in (Critical) Discourse Analysis: Theory, Methodology and Interdisciplinarity.* Amsterdam: John Benjamins. doi:10.1075/dapsac.13

Wohn, D. Y., Lampe, C., Wash, R., Ellison, N., & Vitak, J. (2011). The "s" in social network games: Initiating, maintaining, enhancing relationships. In *Proceedings of the 44th Hawaii International Conference on System Sciences (HICSS '11)*, (pp. 1–10). doi:10.1109/HICSS.2011.400

Woodside, A. G., Sood, S., & Miller, K. E. (2008). When consumers and brand talk: Storytelling theory and research in psychology and marketing. *Psychology and Marketing, 25*(2), 97–145. doi:10.1002/mar.20203

World Bank. (2012). *Information and communications for development 2012: Maximizing Mobile.* Washington, DC: World Bank. Retrieved May 2, 2015, from http://www. worldbank.org/ict/IC4D2012

World Bank. (2012). *Mobile Phone Access Reaches Three Quarters of Planet's Population.* Retrieved from http://www.worldbank.org/en/news/press-release/2012/07/17/mobile-phone-access-reaches-three-quarters-planets-population

Wren, A. (2014, May 25). *Why #BringBackOurGirls is NOT another slacktivism campaign.* Retrieved from Asher Wren: http://asherwren.svbtle.com/why-is-bringbackourgirls-not-another-slacktivism-campaign

Wright, C. R., Lopes, V., Montgomerie, T. C., Reju, S. A., & Schmoller, S. (2014). Selecting a Learning Management System: Advice from an Academic Perspective. *EDUCAUSE Review Online.* Retrieved 20 July 2015 from http://www.educause.edu/ero/article/selecting-learning-management-system-advice-academic-perspective

WSJ. (2011, October 19). Apollo Group 4Q Net Soars on Fewer Charges: Enrollment Falls. *The Wall Street Journal.*

Wu, M., & Li, H. (2011). The Triumph of Shanzhai:No Name Brand: Mobile Phones and Youth Identity in China. In D. Y. Jin (Ed.), Global Media Convergence and Cultural Transformation: Emerging Social Patterns and Characteristics (pp. 213-232). IGI Global.

Wu, M., & Lin, H. (2011). Office on the move. In E. Coakes (Ed.), Knowledge Development and Social Change through Technology: Emerging Studies (pp. 232-246). IGI Global. doi:10.4018/978-1-60960-507-0.ch018

Wu, M., & Yao, Q. (2014). Location-Aware Mobile Media and Advertising: A Chinese Case. In X. Xu (Ed.), Interdisciplinary Mobile Media and Communications: Social, Political, and Economic Implications (pp. 228-244). IGI Global. doi:10.4018/978-1-4666-6166-0.ch013

Wu, J., Hu, B., & Zhang, Y. (2013). Maximizing the performance of advertisements diffusion: A simulation study of the dynamics of viral advertising in social networks. *Simulation: Transactions of the Society for Modeling and Simulation International, 89*(8), 921–934. doi:10.1177/0037549713481683

Wu, M., Jakubowicz, P., & Cao, C. (2013). *Internet Mercenaries and Viral Marketing: The Case of Chinese Social Media.* IGI Global.

Xstream. (2013). *Online video trends & predictions for 2013.* Retrieved 31 May 2013 from http://www.xstream.dk/content/online-video-trends-predictions-2013

Xstream. (2013). *Online video trends & predictions for 2013.* Retrieved from http://www.xstream.dk/content/online-video-trends-predictions-2013

Xstream. (2013). *Online video trends & predictions for 2013.* Retrieved May 31, 2013, http://www.xstream.dk/content/online-video-trends-predictions-2013

Xstream. (2013a). *Online video trends & predictions for 2013* (Part 1). Retrieved 31 May 2015 from http://xstream.net/news-xstream

Xstream. (2013b). *Online video trends & predictions for 2013* (Part 2). Retrieved 31 May 2015 from http://xstream.net/content/online-video-trends-predictions-2013

Yamagata-Lynch, L. (2001). *Using activity theory for the sociocultural case analyses of a teacher professional development program involving technology integration.* (Doctoral dissertation). Indiana University.

Yamakami, T. (2013). Historical view of mobile social game evolution in Japan: Retrospective analysis of success factors. *ICACT2013,* (pp. 375-379).

Yamakami, T. (2012, September). Transition Analysis of Mobile Social Games from Feature-Phone to Smart-Phone. In *15th International Conference on Network-Based Information Systems (NBiS),* (pp. 226-230). IEEE. doi:10.1109/NBiS.2012.73

Yang, J., & Leskovec, J. (2010, December). Modeling information diffusion in implicit networks. In *2010 IEEE 10th International Conference on Data Mining (ICDM),* (pp. 599-608). IEEE. doi:10.1109/ICDM.2010.22

Yang, H.-C., & Kim, J.-L. (2013). The Influence of Perceived Characteristics of SNS, External Influence and Information Overload on SNS Satisfaction and Using Reluctant Intention: Mediating Effects of Self-esteem and Thought Suppression. *International Journal of Information Processing & Management, 4*(6), 19–30.

Yang, Y., Hu, X., Qu, Q., Lai, F., Shi, Y., Boswell, M., & Rozelle, S. (2013). Roots of Tomorrow's Digital Divide: Documenting Computer Use and Internet Access in China's Elementary Schools Today. *China & World Economy, 21*(3), 61–79. doi:10.1111/j.1749-124X.2013.12022.x

Yannopoulou, N., Moufahim, M., & Bian, X. (2013). User-generated brands and social media: Couchsurfing and AirBnb. *Contemporary Management Research, 9*(1), 85–90. doi:10.7903/cmr.11116

Yao, Q. (2014). How Virusworthy Elements Affect Viral marketing Patterns in WeChat. 6th Chinese News and Communication Association.

Yin, R. K. (2005). Estudo de Caso: Planejamento e Métodos. 3a ed. Porto Alegre, Rio Grande do Sul. *The Bookman.*

Youngblood, S. (2014, May 12). *#BringBackOurGirls fills void left by sluggish press.* Retrieved from Peace & Collaborative Development Network: http://www.internationalpeaceandconflict.org/profiles/blogs/bringbackourgirls-fills-void-left-by-sluggish-press?xg_source=activity#.VUu4s5TF9rg

Youyou, W., Kosinski, M., & Stillwell, D. (2015). Computer-based personality judgments are more accurate than those made by humans. *Proceedings of the National Academy of Sciences of the United States of America, 112*(4), 1036–1040. doi:10.1073/pnas.1418680112 PMID:25583507

Yu, R. (2014, August 6). Gannett to spin off publishing business. *USA Today.* Retrieved 30 Jun 2015 from http://www.usatoday.com/story/money/business/2014/08/05/gannett-carscom-deal/13611915/

Zaibon, S. B., & Shiratuddin, N. (2009). Towards developing mobile game-based learning engineering model.*2009 World Congress on Computer Science and Information Engineering,* (pp. 649-653). doi:10.1109/CSIE.2009.896

Zaibon, S. B., & Shiratuddin, N. (2010). Adapting learning theories in mobile game-based learning development. *2010 IEEE International Conference on Digital Game and Intelligent Toy Enhanced Learning*, (pp. 124-128). doi:10.1109/DIGITEL.2010.37

Zanot, E. J. (1984). Public attitudes toward advertising: The American experience. *International Journal of Advertising*, *3*, 3–15.

Zarrella, D. (2011). *Zarrella's Heirarchy of Contagiousness: The Science, Design and Engineering of Contagious Ideas*. Dobbs Ferry, NY: Do You Zoom Inc.

Zarrella, D. (2013). *The Science of Marketing: When to Tweet, What to Post, How to Blog, and Other Proven Strategies*. Hoboken, NJ: John Wiley & Sons.

Zeichner, N., Perry, P., Sita, M., Barbera, L., & Nering, T. (2014, April). Exploring how mobile technologies impact pedestrian safety. *NYC Media Lab Research Brief*. Retrieved May 27, 2015, from http://challenges.s3.amazonaws.com/connected-intersections/MobileTechnologies_PedestrianSafety.pdf

Zeng, X. (2013). *Analysis of media reports of left-behind children: Case study of China Youth Daily*. (Master's degree thesis). Southwest University. Retrieved from http://www.docin.com/p-805433612.html

Zhang, J., Farris, P. W., Irvin, J. W., Kushwaha, T., Steenburgh, T. J., & Weitz, B. A. (2010). Crafting integrated multi-channel retailing strategies. *Journal of Interactive Marketing*, *24*(2), 168–180. doi:10.1016/j.intmar.2010.02.002

Zhao, Y. (2008). *Communication in China: Political economy, power, and conflict*. Rowman & Littlefield Publishers.

Zhou, T. (2013). Understanding the effect of flow on user adoption of mobile games. *Personal and Ubiquitous Computing*, *17*(4), 741–748. doi:10.1007/s00779-012-0613-3

Zhu. (2004). Competition between alternative sources and alternative priorities: A theory of weighted and calculated needs for new media. *China Media Report*, 16-24.

Zimmermann, A., Henze, N., Righetti, X., & Rukzio, E. (2009). Workshop on mobile interaction with the real world. In A. Zimmermann, N. Henze, X. Righetti, & E. Rukzio (Eds.), *Mobile Interaction with the Real World* (pp. 9-14).

About the Contributors

Xiaoge Xu, Ph.D, is Professor of Mobile Studies at Botswana International University of Science and Technology, Botswana, Founder of Mobile Studies International and MOBILE518, Singapore.

* * *

Seyedali Ahrari is currently is a PhD student in Universiti Putra Malaysia in the youth studies. He is working on youth civic development. He published some studies related to youth civic engagement and their commitment to participation. He graduated from major of Sociology from Islamic Azad University in Tehran, Iran.

Saleh Al-Shehri is an Assistance Professor of mobile learning and TESOL at College of Education, King Khalid University, Saudi Arabia. He received his PhD in mobile and contextual learning from the University of Queensland, Australia in 2012. Al-Shehri has published several articles and book chapters, and presented at different international conferences. His research interests include mobile learning, mobile language learning, mobile social media, behavior of mobile learners, design-based research, and connectivism. Saleh is currently focusing on prospects of mobile learning in Saudi Arabia and the Arab world, and the analysis of mobile social media interaction.

Yulia An, BA, is a graduate student in the Institute of Media and Communication Science at the Ilmenau University of Technology (TU Ilmenau) in Germany.

Philip Auter, Ph.D., is a professor and graduate program coordinator for the Department of Communication at the University of Louisiana. He's a member of the UL Distance Learning Leadership Council and the Executive Director of the American Communication Association. He has published 28 journal articles and presented at conferences worldwide.

Maria Alexandra Cunha holds a doctorate in business administration from Universidade de São Paulo (USP) and she is full professor at Escola de Administração de Empresas de São Paulo (Sao Paulo Business School) of Fundação Getulio Vargas (FGV), where her line of research is management of information technology.

Yuchan Gao, PhD candidate, University of Nottingham Ningbo, China

Keorapetse Gosekwang is a postgraduate student at Botswana International University of Science and Technology, studying MSc Information System and Data Management. He is a graduate from the University of Botswana with a Diploma in Archives and Records Management and Bachelor's Degree in Information Systems. His research interests are in the areas of data modelling as he is currently developing a theoretical framework of database models relevant to the management of electronic records.

Kenneth E. Harvey is an Associate Professor of Media & Communications and Distance Learning Coordinator at KIMEP University in Almaty, Kazakhstan. He is also president of The International Education Institute, which is developing an online public service TV network at http://IEI-TV.net.

Nygmet Ibadildin's research interests are political economy of natural resources, institutional development, and discourse analysis. Currently he is teaching at KIMEP University, Almaty, Kazakhstan.

Tebogo Kebonang is currently pursuing MSc Information Systems at Botswana International University of Science and Technology. He is a graduate of BSc (Hon) Software Engineering and Multimedia with Limkokwing University of Creative Technology. His area of interest is looking into job matching algorithm to enhance their accuracy.

Kushatha Kelebeng is a postgraduate student at Botswana International University of Science and Technology, studying MSc Computer Science. Her research interest includes mobile learning, big data analysis and Business Intelligence. She is currently developing a collaborative mobile learning application for senior secondary students in Botswana.

Beatriz Lanza, holds a doctorate in Business Administration at Federal University of the State of Paraná (UFPR) and Center for Technology in Government at State University of New York (CTG SUNY). She is Senior Analyst at IT Company of the State Government (CELEPAR). Currently, she is researching for designing a conceptual Business Model for SMS-based government services that can be used by governments around the world to reach more citizens and at a lower cost.

Wendy Li is a research assistant of MOBILE518 and Mobile Studies International. Currently, she is working on her thesis at the University of Melbourne, Australia.

Tebogo Mangwa is a graduate of Limkokwing University with a major in BSc (Hon) in Information Technology. She is currently pursuing her masters in Information and data management. Her research interests are in data management, networking and system administration.

Hans K. Meyer is an associate professor of journalism at the E.W. Scripps School of Journalism at Ohio University. His research explores how journalists can empower communities online and how they can adapt to new technologies.

Rebaone Mlalazi is a teaching and research assistant pursuing a master's degree in Computer Science with Botswana International University of Science and Technology in Botswana, Palapye. He has a degree in Computer Science from the University of Sunderland in United Kingdom. His research interests include big data analysis, e-learning, m-learning and artificial intelligence based expert systems. For the past 2 years he has been developing mobile applications for teaching and learning.

Thotobolo Morapedi is a postgraduate student at Botswana International University of Science and Technology, currently pursuing MSc. Information Systems and Data management. He graduated from Staffordshire University, UK with BSc. (Honors) Business Computing with specialism in Management. His area of Interests is in education expert system, Mobile Computing, Mobile Learning and Mobile Commerce.

Jamilah Othman is a lecturer at the Department of Professional Development and Continuing Education, Faculty of Educational Studies, University Putra Malaysia. Currently, she is appointed as Head of Laboratory for Social Change, Economy, and Peace Studies at Institute for Social Studies (IPSAS).

Fernando de la Cruz Paragas is a faculty member of the Department of Communication Research at the College of Mass Communication of the University of the Philippines Diliman (UPD). His research focuses on the intersections among mediated communication, international migration, and message design.

Yao Qi, PhD candidate, Department of Communication, Faculty of Social Sciences, University of Macau.

Mukul Saxena holds a Ph.D. in Sociolinguistics from the University of York, UK. Dr. Saxena is Director of Research and CRALC, Associate Professor, School of English, Faculty of Arts and Education, The University of Nottingham Ningbo, China.

Burton Speakman is a Ph.D. candidate at the E.W. Scripps School of Journalism at Ohio University.

Samantha Stevens is a Mass Communications graduate student at the University of Louisiana at Lafayette, specializing in Advertising and Communications Coordination. She received her undergraduate degree in English/Literature from Louisiana State University.

Mei Wu, PhD, Association Professor, Department of Communication, Faculty of Social Sciences, University of Macau, China.

Janice Hua Xu (Ph.D. in communication, University of Illinois at Urbana-Champaign) is assistant professor of Communication at Holy Family University, Pennsylvania. Prior to college teaching in the U.S., she worked as lecturer of international communication in Peking University, news assistant at New York Times Beijing Bureau, and radio broadcaster at Voice of America, Washington DC.

Zeinab Zaremohzzabieh is a PhD student in Universiti Putra Malaysia in youth studies. She obtained her Master in Social research from Islamic Azad University in Tehran, Iran in 2009. She is currently working on human development in relation with ICT. She published a number of studies in relation with youth and ICT.

4 Cs 220
4 Ps 220-221

A

activism 200-201, 203-206, 208, 211-212
Activity Theory 26, 30, 102, 386-387, 390-391, 393, 397-398, 400
AIDA model, 390
Arab Spring 32, 200-201, 208-209
attachment to parents 431, 439, 441

B

big data 194, 238-241, 253, 256, 266, 268, 273, 316, 319, 355, 358, 371
blended learning 73
bonding and bridging 436-437
brand ecosystem 271-273
brand equity 248-251, 254, 265, 268-270, 272-274

C

C2C 225
cable 53, 74, 172, 174, 178-179, 185, 187-188, 255, 352-353, 371
co-creation 253, 261, 268, 274, 294
co-creational perspective 284-285, 288, 303-304
communication channels 22, 32-33, 258, 261, 285
communication in migrant families 455
communication spaces 386
contact and social activity 431, 437
critical discourse analysis 452, 457
cross-cultural branding 269, 274
cross-platform branding 248, 257
cultural divide 38
cultural norms 32, 42-43, 270
customer experience 230-231, 235, 249-251, 255-256, 259, 265-266, 301, 320

D

democratic revolution 150-152, 162, 168
developing regions 134, 136-137, 145
Dialogic Theory 289-291, 300
digital divide 142, 165, 167, 334, 452-453, 468-469
distance education 62, 70, 105-106
distance learning 61-63, 65, 67, 70, 72, 82-83, 88, 91, 105

E

electronic government 111-113, 120-122, 127-128, 133-136, 138-142, 145-146
emancipation 444
emergency management 134, 136-137, 142
Excellence Theory 286-288

F

fluid brand 253
focus interviews 1

G

Game-Based Learning (GBL) 30, 412, 414-415
gaming industry 332, 407, 417, 422
geotargeting 228, 235
grammar skills 40, 43
gratification effect 406

H

human resource 408, 421-423
hybrid language 53-55

I

ice bucket challenge 201, 210, 212

IMC 220-222
in-app advertising 225
inbound marketing 220-221, 225-227, 229, 234, 321
industrial practitioners 417, 422-423
information literacy 464, 468
instant messaging 50, 284-285, 303, 388-389, 394, 455
international telecommunication union 150-151, 154-155, 159, 162-164, 166-167, 408, 454
internet of things 220, 237, 251

J

just-in-time 232, 288, 295-296, 299-300, 328, 361
juvenile delinquency 452, 464-465

K

knowledge network 151-152, 164-165

L

language learning 4-5, 22-23, 27, 29, 31-32, 37, 40-41, 43-45, 103
Learning Management System (LMS) 62, 81-82
learning purposes 24, 31-32, 36, 40, 45
learning tools 5, 23, 25-27, 30, 32, 38, 40, 42, 45, 100
left-behind children 452-469
lifestyle diglossia 49, 51, 56-57
Location-Based Technologies (LBT) 419-420, 422
long tail 223-224

M

m-advertising 316-317
Mason's historical method 111, 114, 125
maximum variation sampling 1, 7
m-business 312
m-commerce 137, 145, 228, 231, 233, 265, 312, 314, 316-317, 329
media communication 193
media players 23-24
media tools 23, 31, 39, 42, 389
m-enterprise 312, 319, 327-328
m-entertainment 332
m-health 328, 330
micro-moments 232
m-industry 312
m-learning 2-8, 11, 17-19, 96-100, 104-108, 328, 331-332
mobile apps 175, 224-225, 230, 257, 266-267, 271, 288, 294, 337, 361

mobile game 30, 407-411, 413-416, 418, 420
mobile gaming 267, 406-413, 415-418, 420, 422-423
mobile government 111-113, 118-123, 127-128, 133, 135-139, 141-146
mobile language learning 22-23, 31-32
mobile phones 1-7, 10-19, 23-27, 30, 34, 42, 53, 96-97, 101-103, 107, 136, 155-156, 159, 164, 167, 240, 251, 264, 267, 269, 291, 315, 317-318, 327, 366, 408-409, 413, 417, 421-423, 429-431, 436-446, 454-457, 463, 465
mobile search 300
mobile video 317, 375
mobile-first 248-249, 251
m-retailing 329
m-sales 316, 329-330
m-shopping 316
m-storefront 320
m-style 312
multilingual literacy 50, 56
multi-media platform 406, 408
m-utilities 330

N

national language 50-52, 57
neighborhood committees 461, 468
new media 53, 173-175, 180-181, 189-190, 194-195, 203, 209, 285, 287, 304, 321, 336, 351
next generation 23, 45, 168, 235

O

OECD 112, 150-151, 154-155, 159, 162-164, 166-167
organizational communication 285-287, 291, 294

P

Paraná government 117, 121-122, 124-125, 128
people analytics 421-423
personal branding 262
phone communication 150, 442-443, 455, 460, 468
privacy protection 395, 398, 400
public relations 69, 220, 225, 228, 284-292, 294, 297-300, 303-304, 312, 322
public sectors 103, 133, 143, 145
public service 151, 168, 207

R

real world 41, 69, 196, 200-204, 208, 211-212, 364, 409, 417

Volume I - pp. 1-797 · Volume II - pp. 798-1599 · Volume III - pp. 1600-2370 · Volume IV - pp. 2371-3135 · Volume V - pp. 3136-3911
Volume VI - pp. 3912-4689 · Volume VII - pp. 4690-5429 · Volume VIII - pp. 5430-6185 · Volume IX - pp. 6186-6955 · Volume X - pp. 6956-7709

547

recruiting companies 422-423
relationship management 288-290, 292
romantic relationships 431, 438, 444
rural boarding schools 459, 464

S

screencasting 63
self-developed skills 16, 19
self-socialization 431, 438
service delivery 134-135, 138, 141-142, 150-151
Short Message Service (SMS) 119, 136
simulcast 62, 81-82
sina.com 452, 457
slacktivism 203, 209, 211
social capital 200-204, 207-208, 211-212
social communication 37-38, 386, 430, 436, 455
social interactions 154, 168, 389, 430, 438, 442, 444, 446
social media 22-24, 31-32, 34-45, 96, 101-102, 113, 181-184, 192-196, 200-212, 220-221, 228, 234, 236, 238, 251, 255, 261, 270, 273, 284-286, 288-290, 292, 295-296, 298-300, 302, 304, 317, 319, 321, 327, 336, 354, 358, 362-364, 370, 386-390, 395-396, 398, 443
social media proposition 299
Social Network Analysis (SNA) 111-112, 114, 116, 125, 240
social networking 4, 11, 18, 106, 204, 252, 265-266, 288, 386-387, 389, 406, 408, 417-420, 422, 430, 443-444, 455, 463

Social Networking Service (SNS) 386-387
social technologies 112, 293-296, 319
socialization 429-431, 436, 441-446
symbolic capital 52, 57
synchronous 62, 64, 67, 71, 73, 89

T

teacher-student interaction 29, 32, 42
teaching experience 23, 33, 42
technological platform 2-4
technology-based learning 421
tertiary education 44
traditional media 171, 173-175, 177, 180-181, 183-184, 187-190, 194, 196, 202-203, 205, 270, 286, 297, 351, 355, 390
two-way communication 114, 134, 141, 193, 228, 285

U

unregulated spaces 49-50, 52-53, 56-57

V

videoconferencing 71

W

wealth gap 461, 466, 468
web-conferencing 62, 64, 67, 72, 74, 77, 80, 83-84
WeChat 303, 386-390, 393-401

Volume I - pp. 1-797 · Volume II - pp. 798-1599 · Volume III - pp. 1600-2370 · Volume IV - pp. 2371-3135 · Volume V - pp. 3136-3911
Volume VI - pp. 3912-4689 · Volume VII - pp. 4690-5429 · Volume VIII - pp. 5430-6185 · Volume IX - pp. 6186-6955 · Volume X - pp. 6956-7709

548

Printed in the United States
By Bookmasters